To Susan and Lou

The three dimensions of *John Logie Baird*

Dr Douglas Brown, MPhil, PhD, GM8FFH

Radio Society of Great Britain

Published by the Radio Society of Great Britain, 3 Abbey Court, Fraser Road, Priory Business Park, Bedford. MK44 3WH. Tel 01234 832700. www.rsgb.org

First published 2012

© Radio Society of Great Britain 2012. All rights reserved. No part of this publication may be reproduced, stored in a retrieval system, or transmitted in any form, or by any means, electronic, mechanical, photocopying, recording or otherwise, without the prior written agreement of the Radio Society of Great Britain.

ISBN 9781 9050 8679 5

Publisher's note

The opinions expressed in this book are those of the author and not necessarily those of the RSGB. While the information presented is believed to be correct, the author, publisher and their agents cannot accept responsibility for consequences arising from any inaccuracies or omissions.

Illustrations and sources

Illustrations included in the text relating to British, American and European patents, including selected abstracts from patents, have been sourced and reproduced with permission from the European Patent Office. http://gb.espacenet.com.

A special "thank you" is given to Dr Peter Waddell and Prof Malcolm Baird for their continued support, and to Lomond School in Helensburgh, Scotland, for permission to include the portrait of John Logie Baird by Emilio Coia as the frontispiece, which is also shown in colour on page C.16. While every effort had been made to trace copyright holders and obtain permission for material included in this publication this has not always been possible. The author and publisher would be grateful for further information relating to unidentified sources to ensure acknowledgement in future editions.

Cover design: Kevin Williams
Editing, typography and design: Mike Dennison, G3XDV of Emdee Publishing
Production: Mark Allgar, M1MPA

Printed in Great Britain by Latimer Trend & Co Ltd of Plymouth

Contents

	List of illustrations iv
	Introduction 1
1:	Television pre-history 3
2:	Baird the young entrepreneur 11
3:	Television begins 21
4:	The realisation of television 39
5:	Colour, light and human perception 69
6:	The colourful John Logie Baird 79
7:	Colour from cathode rays 107
8:	The third dimension and holographic television . . 125
9:	The final chapter 149
A1:	Appendix: Phonovision project 163
A2:	Appendix: Colour from a monochrome TV! 167
A3:	Bibliography 173
	Index 181

List of Illustrations

Figure	Caption	Page
	Portrait of JL Baird by kind permission of Lomond School (1990)	viii
1.1	The scene to be scanned	4
1.2	Receiving the Picture	4
1.3	Mechanical scanner	5
1.4	The scanning process and the signal from light cell	5
1.5	Paul Nipkow's German patent describing an 'Electric telescope'	8
2.1	Self Portrait of Baird (1900) enabled by his remotely triggered camera	12
2.2	Gas engine used by Baird to power his electricity generator (1901)	13
2.3	Baird at Rutherglen Power Station	14
2.4	Packaging and leaflet from a surviving pair of Baird Undersocks [courtesy of D P Waddell]	16
3.1	Advertisement placed in *The Times*	22
3.2	Abstract from Baird's premier television patent *GB222604*	24
3.3	Abstract from Baird's television patent *GB232619*	25
3.4	The first ever photograph of a television picture	28
3.5	Baird's image enhancing patent *GB270222*	30
3.6	Block diagram of the modern aperture circuit	30
3.7	Waveforms relating to the modern crispening circuit	31
3.8	Waveforms describing Baird's image correcting patent *GB270222*	31
3.9	Floodlight television transmitter	32
3.10	Floodlight television receiver	33
3.11	Flying-spot method of scanning	34
3.12	Baird's British patent *GB269658*	35
4.1	Abstract from Baird's patent *GB348211* for an active panel used for showing the end of the Derby in June 1931	47
4.2	Illustration from Baird's patent GB289104 describing a sound and vision videodisc	48
4.3	Abstract from Phonovisor patent *GB324049* (January 1930)	50
4.4	(a) 30-line image of Charlie Chaplin recovered from 78rpm videodisc; (b) 30-line image of man holding a pen, scanned by the author and recovered from an audio cassette tape; (c) 30-line image of cat, scanned by the author and recovered from an audio cassette tape	50
4.5	Abstract from Baird's patent *GB285738*	51
4.6	Abstract from Baird's Noctovisor patent *G288882*	51
4.7	The camera section of Baird's Noctovisor	52
4.8	Abstracts from John Baird's patents *GB292185* and *US1699270*	54
4.9	Baird televisor screen calibrated for ranging a reflected signal	55, C.1
4.10	Baird's intermediate film system	57
4.11	Example of an all-electronic Baird television receiver	58
4.12	The Baird Televisor	59
4.13	Baird Telecine (1939)	61

4.14	Baird electron camera	.61
4.15	Big screen 405-line image in the Tatler News Theatre London, using a Baird television projector	.63
4.16	Example of Baird Television products at the outbreak of WW2	.65
5.1	Planck energy distribution of the Sun	.69, C.1
5.2	Splitting white light into its spectral components	.70, C.1
5.3	Recombining the spectral colours to produce white light	.70, C.1
5.4	Additive mixing of light	.71, C.2
5.5	Subtractive mixing of pigments	.71, C.2
5.6	Wavelength and frequency of the visible spectrum	.71
5.7	CIE (1936) colour space diagram	.72, C.2
5.8	Obtaining and reproducing a colour image from black and white film	.73, C.3
5.9	Baird selected (red/orange) and (green/blue) to achieve a wide range of colours	.74, C.4
5.10	Colour interpretation relies on human perception	.75, C.4
5.11	Spectral response of the eye against wavelength (Vos, J J Walvarem, P L 1971)	.75
6.1	Emulation of a low-definition colour television image	.79, C.5
6.2	Abstracts from *GB266564*, a preliminary specification for the production of colour television	.80
6.3	Abstract from Baird's patent *GB267378*	.81
6.4	Abstract from Baird's patent *GB312560*	.82
6.5	Abstract from Baird's British colour television patent *GB314591*	.82, C.5
6.6	Baird's colour television patent *GB321389*	.83, C.5
6.7	Abstract from Baird's patent *GB322776*	.84
6.8	Abstract from Baird's patent *GB319307* showing the switching commutator	.85, C.6
6.9	Abstract from Baird's video wall patent *GB476195* (1935)	.86
6.10	Abstract from Baird's optical switch patent *GB473303* for scanning or recombining colour images	.87
6.11	The Dominion Theatre, London where Baird demonstrated large-screen television	.88
6.12	Spiral/slot scanning arrangement described in *GB418527*	.89
6.13	The improvement in blue response obtained by the caesium-antimony photosurface	.90
6.14	Abstract from Baird's patent *GB532259* describing the use of a Caesium-antimony photocell produced by Sommer (1938)	.91, C.6
6.15	Panchromatic (Bi-Ag-O-Cs) photosurface characteristic	.93
6.16	The three monochrome camera tube approach to colour separation	.93, C.7
6.17	The three monochrome CRT colour television receiver	.93, C.7
6.18	Colour filter discs for Baird's field-sequential colour television system	.94, C.7
6.19	Interlacing described in patent *GB508039*	.95
6.20	Abstract from Baird's patent *GB545078*, describing a method of interlacing	.95
6.21	Abstract from Baird's colour television patent *GB473323*	.95

6.22	Baird's two-colour field-sequential transmitter and receiver	96, C.8
6.23	Abstracts from J L Baird's patents *GB545462* and *GB545491* describing the replacement of the rotating filter disc with an electro-optical device	97
6.24	Cylindrical drum filter disc described in patent *GB545603*	97
6.25	Junction between filters is modified to perfectly match the storage matrix of an electronic camera device	98
6.26	Abstract from Baird's patent *GB555167* for a tri-band RGB cathode-ray tube	99
6.27	The receiver process described in *GB555167*	99, C.9
6.28	Abstract from Baird's patent *GB557992*	100
6.29	Abstract from Baird's Telechrome patent *GB562168*	101, C.9
6.30	John Logie Baird working on the Telechrome colour television receiver	101
6.31	Abstract from Baird's patent *GB562334*	102, C.10
6.32	Sleepers's tri-band colour tube built by Rauland (1947)	103
7.1	Abstract from Geer's patent *US2480848*	107
7.2	Geer's patent US2480848 (1944) based on Baird's (1942) Telechrome	108
7.3	Geer's alternative pyramid screen enables the cathodes to be located at the rear	108
7.4	Goldsmith's three-cathode colour television tube *US2481839*	108
7.5	DuMont's three-colour cathode colour television tube *US2544690* (1946)	109
7.6	Arthur Bronwell's Chromoscope *US2461515* (1945)	109
7.7	Werner Fleschsig's colour picture tube (1938)	111
7.8	Kaplan's parallax colour tube (1947)	112, C.10
7.9	Shadowmask principle	113
7.10	Extract from Baird's diary	114
7.11	Lawrence Chromatron colour tube	115, C.11
7.12	The Sony Trinitron	116, C.11
7.13	Comparison between Samuel Kaplan's Chromacolor tube and the standard Shadowmask	118
7.14(a)	Baird's influence on the evolution of frame sequential colour television	120, C.12
7.14(b)	Baird's influence on the evolution of the Shadowmask cathode-ray tube	121, C.13
7.14(c)	Baird's influence on the evolution of the Trinitron cathode-ray tube	122, C.14
8.1	Example of a parallel viewed stereo-pair	126, C.15
8.2	Anaglyph viewed in 3D with red colour filter over the left eye and a blue colour filter over the right eye	C.15
8.3	Abstract from Baird's 3DTV patent *GB292365*	128
8.4	Section of lenticular panel	129
8.5	Abstract from Baird's 3DTV patent *GB321441*	130
8.6	Abstract from the first volumetric imaging patent *GB373196*	131
8.7	An illustration from the first volumetric imaging patent *GB373196*	132
8.8	Baird manufactured matrix camera tube of the Zworykin type undergoing visual inspection (left) and (right) large quantities of Baird Iconoscopes, image dissectors and other cathode-ray tubes ready for testing in 1939	134

8.9(a)	Abstracts from Baird's patent *GB552582*, including Fig 1 from the drawing from Baird's auto-stereoscopic active shutter 3DTV (1941)	135
8.9(b)	Abstracts from Baird's patent *GB552582*, including Figs 2, 3 and 4 from the drawing for Baird's auto-stereoscopic active shutter 3DTV (1941)	136
8.9(c)	Abstracts from Baird's patent *GB552582*, including Figs 5, 6, 7 and 8 showing alternative multiple viewer passive polarised spectacles (1941)	137
8.10	Abstract from Baird's patent GB573008	138
8.11	Baird's description of a means to improve depth of focus in his provisional specification No *13887*	139
8.12	Plan view of scanned image	140
8.13	Persistence of vision enables a 3D image when the screen is vertically displaced at the frame rate of the scanner	141
8.14	Part of Baird's provisional specification No *19825/43*	143
8.15	Abstract from the completed specification in Baird's patent *GB573008*	143
8.16	Abstract from the completed specification in Baird's patent *GB573008*	143
8.17	Figures 1 to 11 from the completed specification GB573008.	144
8.18	The dynamic holographic television concept described in the completed specification *GB573008*	146
9.1	The impressive Grosvenor receiver	156
A1.1	Block diagram of the scheme used by the author to recover phonovision images	163
A1.2	30-line image reproduced from audio cassette on an oscilloscope	164
A1.3	30-line image drifting horizontally	164
A1.4	Stabilised image of Charlie Chaplin	165
A2.1	SCART socket, numbered from the front, and its pin connections	168
A2.2	Outputs required from the universal SCART lead	168
A2.3	Sequential timing is organised by the decimal counter from the field syncs provided by the sync separator	169
A2.4	Field sequential waveform constructed by sampling signals in correct RGB order and combining with sync pulse	170
A2.5	Tri colour disc and 5-inch monitor. Note the coupling method	170
A2.6	The disc may have straight or curved spokes, but they must be strong to support the circumference of the disc	171
A2.7	White disc with a black timing mark in conjunction with a reflective opto-electronic reflector generates a tachometer pulse once per cycle	171
A2.8	Complete operational block diagram	172
A2.9	Colour bars photographed from the author's field-sequential colour television display	172, C.15

vii

THE THREE DIMENSIONS OF JOHN LOGIE BAIRD

**Portrait of John Logie Baird, by kind permission of Lomond School
(see page C.15 for a colorised verion of this picture)**

Introduction

Did John Logie Baird contribute anything to modern television or was he simply a well-meaning technological extrovert whose only contribution to information science was the development of an outdated mechanical form of television?

THE AUTHOR HAS EXTENSIVELY researched the life and work of John Logie Baird (1888-1946) for over 25 years and analysed the technological expectations, successes and failures of this exceptional man of vision. The book commences with a journey describing the relevant technology of the nineteenth and early twentieth century, setting the scene for Baird's line of enquiry.

There is a biographical sketch of Baird's earliest days in Scotland and entrepreneurial exploits, followed by the developments in Hastings that led to the realisation of real television in London on 2 October 1925.

Baird's contribution to the evolution of the television industry and the inauguration of the first television company and television service is supported with a review of relevant patents. The main focus is to reveal Baird's lesser-known contributions including colour and 3DTV before and during the Second World War. It will be shown that many of Baird's innovations were advanced, including a futuristic demonstration in 1941 of 500-line 3DTV colour system viewed without the need for prismatic or colour filter spectacles.

The Three Dimensions of John Logie Baird celebrates the contributions made by this man of vision over two decades from 1926 to 1946, covering; the birth of monochrome television; the first working colour television from 30-lines to 600-lines of definition and three-dimensional television (3DTV) covering; stereoscopic 30-lines to auto-stereoscopic 500-line; parallax barrier, colour anaglyph 3D, and his lesser-known work on holographic television.

Baird's was the first to send recognisable moving images with a monochromatic tonal grey-scale across a distance of space, and in the nineteen thirties maintained the lead in television development despite a major financial world depression and serious competition from the American radio industry.

It will be shown that in the nineteen forties, the work of John Logie Baird was the catalyst for the development of colour television cathode-ray tube concepts by the American industry, and demonstrated many of the techniques now employed in stereoscopic television.

In 1969 an electro-mechanical video camera carried by *Apollo 11* used methods first demonstrated by Baird to send colour television pictures to the Earth from the surface of the Moon.

Based on research from this author, Baird is recognised as the father of a holographic television system, still under development and known as volumetric or cubic imaging.

In 1946, with the war at an end, Baird was recovering from a stroke and seemed to be making good progress. He had recently given evidence to a Government television committee chaired by Lord Hankey, that had been set up in 1943 to make recommendations on a post war television service. A great deal of Baird's contribution was accepted as the way forward, including television with a minimum of 600-lines in colour, to rise to 1000-lines with a future option of 3DTV.

There was much to do, a new company John Logie Baird Ltd had been set up with co-director Jack Buchanan, but after years of dogmatically fighting chronic ill-health Baird's inventive genius was brought to a premature close at the age of 57 on 14 June 1946.

Douglas Brown

1

Television Pre-history

Television is a unique word owing its origins to two languages, 'tele' from the Greek 'far' and 'visio' from the Latin 'sight'. The earliest known use of the term 'television' dates back to 1900, when Constantine Perskyi gave a talk at the International Congress of Electricity on the possibilities of electrically seeing moving images over long distances by wire. However, the concept of sending 'still' facsimile images by electricity had already been a fact for some time.

ALEXANDER BAIN, A SCOTTISH CLOCKMAKER described his idea for a photo-telegraphy system in an 1843 British patent, and by 1846 installed an image telegraph operating between Edinburgh and Glasgow. The system, able to reproduce outlines scanned by a metal stylus on a partly insulated block, used a clock escapement to synchronise a scanning stylus attached to a pair of pendulums.

There were two main disadvantages, the first was unreliable synchronisation between the sending and receiving station and the second was a complete lack of spontaneity. Bain's scanner could not send a message until it had been prepared as an etching on a copper plate.

At the transmitter the stylus systematically scanned the plate, line-by-line, producing pulses of electrical current when it made contact with the raised metal of the etching. At the receiver a similar arrangement of synchronised pendulums and stylus scanned a sheet of electrochemically treated paper to render the original message visible.

Bain's machine was superseded by Frederick Bakewell's 'Image Telegraph' demonstrated in 1851 at the World Fair in London. The most notable design improvement was that the messages could now be written directly on a piece of metal foil using non-conducting ink. While this device employed a simplified method of image scanning it failed to solve the problem of unreliable synchronisation that had troubled Bain's apparatus.

In 1862, Abbe Giovanna Caselli an Italian priest demonstrated the first reliably synchronised facsimile machine. The 'Pantelegraph' also scanned an image drawn on metal foil to be electrically reproduced on chemically sensitive paper. However, despite the technical success, the development of the facsimile was badly timed in history and doomed to commercial failure by the introduction of the teleprinter.

While the breakthrough in facsimile communication of text and line drawings was a step in the correct direction regarding the development of television, a great deal of technological development continued to stand in the way

Electromechanical Scanning

Imagine a picture, which in its simplest form comprises a number of small points of black, grey and white light. In order to transmit each of these points as a corresponding level of electricity, it must be systematically explored. At the receiver, the levels of electricity must be systematically converted back into levels of light, which when viewed through an identical scanning device operating in synchronisation with the transmitter reproduces a visual image of the original picture.

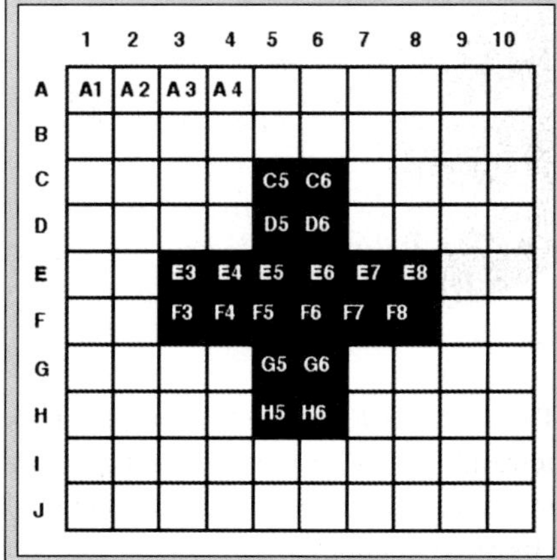

A simpler way of thinking about this process is to imagine a completed jigsaw of a black cross on a white background (**Fig 1.1**). There are one hundred square-shaped pieces in our jigsaw puzzle made up of ten pieces across by ten pieces down. Anyone familiar with the children's game of 'Battleships' may recognise the process to be described. This experiment may be tried over a telephone, but it is important to set the rules: the receiving person must first draw a 10 by 10 square and name each square using numbers '1 to 10' across the top and letters 'A to J' down the side.

Fig 1.1: The scene to be scanned

The sender using an identical grid will scan the picture shown in Fig 1.1 with a pointer starting with A1 at the top left corner and moving across to A10, indicating to the receiving person to shade only the appropriate boxes. In this example the first two rows have no black squares. At the end of each row the sender moves down to the next line of data, again moving from left to right until all the rows have been scanned and J10 is reached. The receiving person carefully follows the instructions by shading each named square as shown in **Fig 1.2**. At the end of the process, which will take a few minutes to complete, a copy of the sender's picture will emerge despite there being no actual visual contact.

Several devices were suggested to achieve this. The first electro-mechanical systems of television sent a picture remotely in the form of a signal to a distant location by wire, via a telephone network or by radio. To make the picture possible to send, the black and white levels were turned by means of a photocell into pulses of electricity representing picture brightness. Originally the scanning process was slow enough to enable the resulting video signal to be processed like sound.

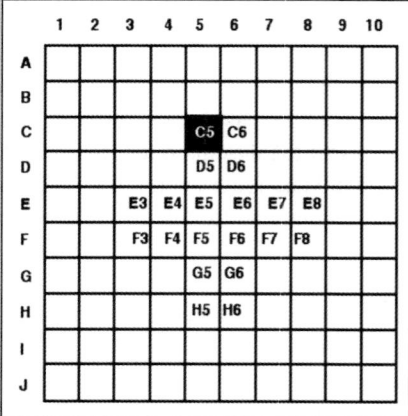

Fig 1.2: Receiving the picture

Fig 1.3: Mechanical scanner

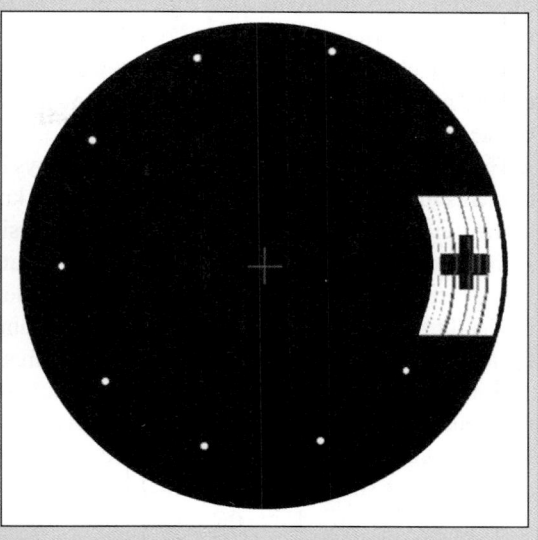

The general idea was to use some type of a revolving scanner shown in **Fig 1.3** with a spiral of holes, to pick out optically the series of points of light or picture elements, which could then be systematically converted to corresponding levels of electricity. **Fig 1.4** illustrates that by scanning the original scene of a black cross in Fig 1.1, the image is effectively viewed by the photocell as a continuous string of serial electrical information related to time. Note that the vertical lines are included in the drawing to mark the start of each row. The resulting waveform, which has been idealised in Fig 1.4, shows the corresponding levels of electricity representing the pattern of the black cross to be transmitted by wire or radio.

Although the example shown is a simple image in contrasting black and white, a detailed grey-scale picture of a real scene, would be scanned using precisely the same process although varying heights of pulses would emerge depending on the brightness of the particular point scanned. The early schemes scanned a grid similar in shape to Fig 1.1, but with each column slightly curved by the arc defined by the spirally positioned apertures on the rotating disc, shown in Fig 1.3. The solid disc demonstrates vertical scanning and has a spiral of ten holes, each aperture being the size of a picture element with the distance between apertures being equal to the height of the grid to be scanned. When the disc is rotated, a single aperture reveals the light level from one tiny picture element at a time and sends the light level to the light cell. When the first scanning hole has completed scanning the entire column, the next aperture in the disc scans the next column. This continues for one full turn of the disc by which time all ten apertures have in turn scanned the image and the first frame of video signal has been produced as a signal from the light cell. The cycle is repeated at speed to recover a moving image over a number of frames. The first Baird television service used 30-line images from a disc rotating at 750rpm and producing pictures at the rate of 12.5 frames per second (fps).

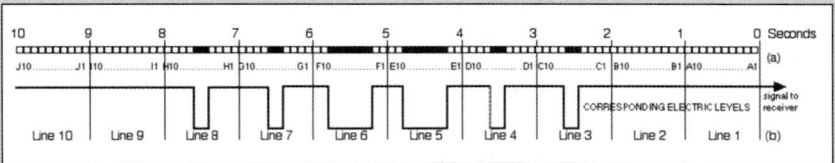

Fig 1.4: The scanning process and the signal from the light cell

of sending and receiving live motion pictures of real objects with light and shade by electricity.

To more clearly define the roots of television it is important to address the optical and psychological discoveries relating to human vision that began in the early eighteen thirties.

The fundamental concept of television employs knowledge gained from developments in cinematography and relies on a well known physical phenomenon of human eye-sight and perception known as 'persistence of vision'. This is the inability of the eye to follow optical changes that occur at rates faster than the storage time taken by the retinal chemical memory and the visual cortex. Basically this means that the human eye and brain combination stores an image for a short time before it may be updated or refreshed.

Storage time, which varies with brightness, was measured by Joseph Plateau who studied the inertia of the eye and discovered that it was about one sixth of a second. Plateau demonstrated 'persistence of vision' in 1832, with his invention of the 'Phenakistiscope'.

This was followed in the same year by Simon Ritter von Stampfer's 'Stroboscope,' which comprised a series of drawings around the edge of a disc with a series of slits cut through the perimeter. As a series of drawings are viewed through the slits of the revolving drum the impression of continuous movement is perceived.

In 1834, William George Horner made further improvements to apparatus showing the illusion of a moving image with the 'Zoetrope'.

Pioneers of moving photography moved one step further by replacing the series of drawings with sensitive photographic plates. By means of exposing the plates in fast succession, within an otherwise light-tight box using a shutter mechanism, pioneers were able to record the photographic impression of visual movement.

Many moving picture camera inventions intervened, notably the work of William Friese-Green, but it is generally conceded that Thomas Alva Edison made the motion picture a practical reality when he demonstrated the 'Kinetoscope' on 15 April 1894.

Cinema exploits persistence of vision by recording a series of still images filmed frame by frame at rates of 16 per second (amateur) to 24 frames per second (professional). To reduce the impression of perceived flicker on the cinema screen, some projectors are fitted with double shutters. When exposed to this series of ever-changing momentary still photographs, the retinal and visual cortex memory is regularly refreshed with new positional information, arriving a fraction of a second before the last perceived image disperses. The impression of natural movement viewed on a cinema screen is therefore an optical illusion, registered by the eye and translated by the brain.

While cinematography presents a succession of photographic slides to the eye in such a way that real movement is perceived, monochrome television deceives the eye and brain to a greater extent by constructing each frame from a rapidly moving small point of light in a linear pattern known as a raster.[1]

The process of obtaining a monochrome television signal whether scanned by mechanical disc or by an electronic video camera relies on the following process. Every element of an image to be televised is systematically scanned by a light sensor device and converted to a continuous series of electrical impulses known as video. Horizontal and vertical pulses combine with the video to enable synchronisation and the entire signal is known as composite video. This signal, therefore, contains all the necessary information to produce a dynamic facsimile of the grey scale of an original moving image.

When the composite video signal is received at some remote location the pulses are decoded and used to lock both the horizontal and vertical timing of the raster scan on the screen. The video part of the signal modulates the brightness of the scanning spot while persistence of vision enables the viewer to perceive a moving monochrome television image.

A non synchronised photograph of a cinema screen is likely to return a whole picture, two superimposed frames or a blank screen, while a photograph taken of the image on a television screen, exposed using the same unsynchronised shutter speed will most likely show a full or a partially constructed picture on a scanning raster. With only a single point of light tracing out the image, monochrome television scanners convey visual information at enormously fast rates to enable our eyes and brain to resolve the illusion of a single frame.

Knowledge of electricity and radio broadcasting were important prerequisites for the television pioneer together with an understanding of the optics and technology of the cinema. Fortunately radio technology had been developing in parallel.

Guglielmo Marconi exceeded the expectations of F B Morse's Electric Telegraph of 1837 by sending Morse code by wireless between America and Britain. However, television could not seriously be considered until some device could be found to convert levels of light intensity into corresponding levels of electricity.

Coincidentally there was such a material known to the electrical industry although its value as a light sensitive cell was yet to be discovered. When the largest ship of the time Isambard Kingdom Brunel's ship, *The Great Eastern* rolled out the first transatlantic telegraphy cable between Ireland and Newfoundland in 1866 [2] a discovery was made that would encourage a great many schemes for television.

Willoughby Smith, chief electrician for the Telegraph Construction and Maintenance Company, obtained bars of selenium, a material discovered by Berzelius in 1817 and known to have a very high electrical resistance. Willoughby Smith obtained several specimens of selenium and instructed Joseph May, the chief assistant at the Greenwich works, to establish a method of measuring cable continuity at the on-shore terminal. May discovered that a stick of crystalline selenium offered considerably less resistance to a battery current when exposed to light than when kept in the dark. This phenomenon became known as the photoconductive effect.

However, the resulting report to the Society of Telegraph Engineers in 1873 by Willoughby Smith on the sensitivity of selenium to light had little immediate effect on the scientific community.

It was not until the invention of the telephone by Alexander Graham Bell in 1875, Paul Nipkow's concept for a visual telegraph patented in 1884 and the radio experiments of Heinrich Hertz in 1887, that selenium was seriously applied in numerous schemes. Nipkow described one of the simplest means by which an

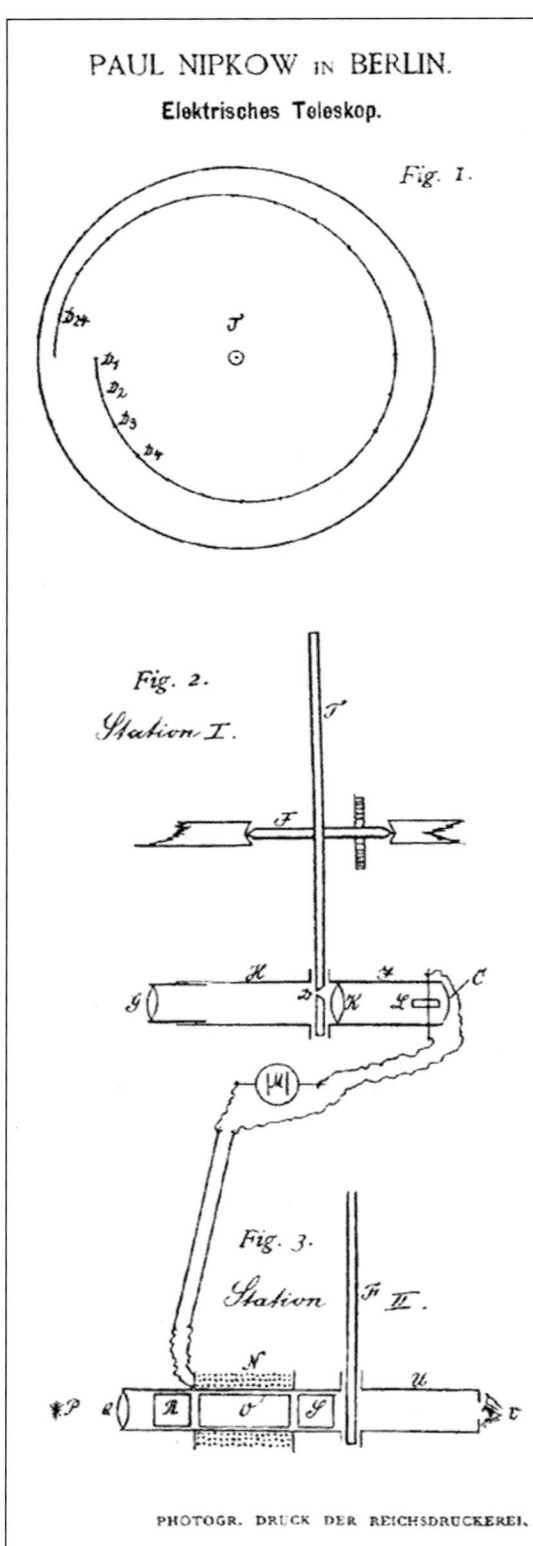

Fig 1.5: Paul Nipkow's German patent describing an 'Electric Telescope'

image could be systematically scanned for electrically seeing at a distance (**Fig 1.5**), but was unable to find the technology to build a working model. This was the basis from which many pioneers, including John Logie Baird, began their work.

Initially several devices were conceived to dissect an image optically into smaller picture elements to be converted from light directly into electricity by means of a selenium light cell. This was the age of mechanical devices and a great deal of thought went into devising a suitable light scanner. On paper there was a plethora of ingenious concepts involving rotating discs, lenses, mirror drums and prisms. Mechanical inertia, lack of optical sensitivity, low resolution, slow response time, loss of balance and synchronisation were amongst the many challenges to be faced by the early pioneers before a working system of television could be realised. While many ideas existed, the prospect of constructing and demonstrating a working system was beyond the reach of the supporting technology of the nineteenth century.

Contemporary radio technology was in its infancy, limiting any success in the field of television to that of a 'closed circuit' or a 'land-line' system. Marconi's 'Morse code' technique for wireless in 1896 was unable to cope with sending anything more complex than the interruptions of a radio wave. This was a situation that would dramatically change by the start of the twentieth century.

In 1900, Reginald Aubrey Fessenden was first to modulate a radio wave with speech and music by wiring a microphone in series with the aerial circuit of a radio transmitter. The microphone was heavily insulated from high voltage and high temperature due to its location in the transmitter. This method of achieving modulation was inefficient, crude and perhaps more than a little risky, but Fessenden's futuristic vision changed wireless forever, paving the way to a new and exciting age of radiotelephony and media entertainment. It must have come as quite a shock to telegraphy operators to hear distorted music instead of the expected dots and dashes that usually emanated from headphones. Fessenden continued to refine his modulation system and succeeded with the first two-way transatlantic radiotelephony transmission and the first broadcast of entertainment and music in 1906.

In the same year, it was the breakthrough in the development of the 'Audion' valve by Lee de Forest, contested by J Ambrose Fleming [3] with his 'triode' valve that promised much for the future of radio and television. Both devices were identical in function and capable of magnifying small electrical currents based on a modification to Fleming's earlier 'diode' vacuum tube. But, even with those technologies in place, the complexity of solving the problems associated with television would be delayed for another two decades.

During this time, two inherent characteristics of selenium, relating to light and electricity, were found to present serious obstacles. First, selenium proved so insensitive to the small levels of light provided from the apertures of a Nipkow disc that a valve amplifier was rendered useless. The other problem with selenium is electrical inertia, causing it to have a slow response time compared to the relatively high speed of signals produced by a mechanical image dissector. Despite these and many other problems, it is remarkable that John Logie Baird successfully translated television theory into practice over the comparatively short period from 1923 to 1925.

References

[1]   Describes analogue monochrome television; analogue colour television requires three points of light for each picture element.
[2]   The telegraph cable was laid between Valentia Island, Ireland and Hearts Content, Newfoundland.
[3]   Although history records Fleming as the inventor of the triode, de Forrest's claim is substantial.

THE THREE DIMENSIONS OF JOHN LOGIE BAIRD

2

Baird the Young Entrepreneur

The combination of dogmatism, ill health, entrepreneurial spirit, and an insatiable appetite for science fact and fiction, provided the young Baird with the necessary ingredients for the development of the world's first working system of television.

BORN ON 13 AUGUST 1888 IN HELENSBURGH, a small Scottish seaside town on the north bank of the Firth of Clyde, John Logie Baird was the youngest of the two sons and two daughters of the Reverend John and Jessie Baird. He was a healthy child until the age of two when he contracted a serious illness that would leave him frail and vulnerable to illness for the remainder of his life.

"I was ill for several months, and remained for a time a delicate weakling." [1]

Ailing health would influence Baird's formative years and shape important future career decisions. In 1894 Baird was enrolled in his first school, Ardenlee, a private school run by a teacher who spread terror among the pupils with vigorous and indiscriminate use of a cane.

"I was a day boy and also too young to go under Mr Porteous and therefore had a happy, carefree time learning to read and write without any particular effort. Mr Porteous went bankrupt (and did not even pay his church seat rent) so I was removed from his school and sent to Miss Johnson's Preparatory School." [2]

Miss Johnson turned out to be a fearsome middle-aged spinster who ruled the pupils with a rod of iron. Baird claims to have been terrorised, and considered the years spent at her school to have been the unhappiest of his life [3]. In 1899 Baird attended Larchfield [4] a private school run by three men fresh out of Oxford / Cambridge. He considered it a really dreadful place and rated it as a poor imitation of a public school. Sport was paramount on the curriculum enabling Baird to enjoy rugby in the winter and cricket in the summer, but he did not enjoy the mandatory icy cold showers and the effect they had on his delicate health.

Despite the enforced 'classical' education, Baird had an aptitude for mechanical and scientific subjects and was a good artist, draughtsman and photographer. As well as being a follower of the popular science fiction works of H G Wells, he was an avid reader keeping himself informed of any new technological advances. He was particularly interested in learning about the work of Alexander Graham Bell and read all he could about the function of the telephone. In 1900 the twelve year-old Baird decided to put his technical knowledge to the test. He constructed a telephone network with a small switchboard, which he soon had operating from the bedroom of his home at the 'Lodge' in Helensburgh. He stretched wires high across streets, anchored them to trees and routed them through bedroom windows to reach the telephone handsets he had constructed for his school friends. On a particularly stormy night one of Baird's telephone wires blew down from where it had once spanned high across a main road. Amidst the inclement weather the offending wire had fallen low enough to raise a cabman from the elevated seat of his horse-drawn carriage and deposit him unceremoniously in the mud below:

"*Under the natural impression that the newly formed National Telephone Company was to blame, the driver demanded damages.* [5]"

To put the situation into perspective, in 1900 there were around 18,000 telephones in Britain owned mainly by the wealthy. There were no private telephones in Scotland until March 1901 when the first municipal telephone exchange was inaugurated in Glasgow. With this background, the telephone company investigated, and on discovering the source of the wires, reprimanded Baird for setting up a private system and demanded that it be immediately dismantled. It seems that Baird's preoccupation with the basic components for carrying out future television research: imagery, electricity and mechanisms were clearly in place by 1900 when he decided to move on to his next project.

Not one to be easily dissuaded from his interests, he placed the disappointment of losing his telephone system behind him and turned his attention to another of his hobbies, photography. He constructed a remote trigger for the shutter of his camera to enable self-portraits and an example of this work is shown in **Fig 2.1**. Baird also attempted, with some mischievous assistance from his friends, to

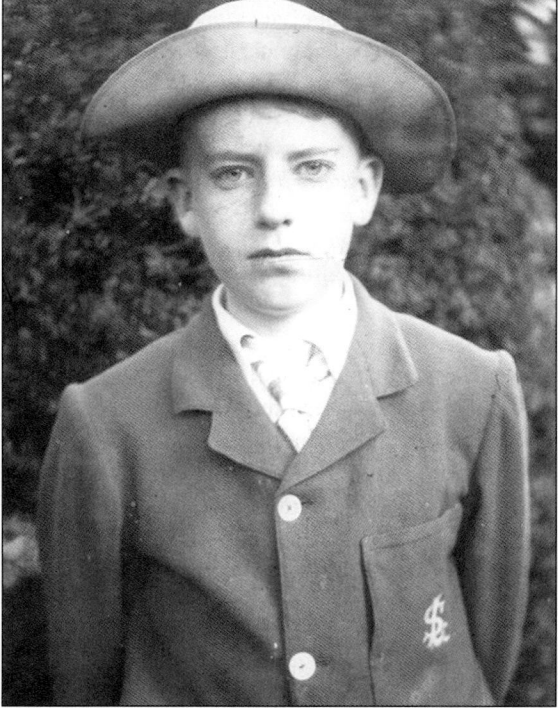

Fig 2.1: Self portrait of Baird (1900) enabled by his remote controlled camera [courtesy of Prof Malcolm Baird]

Fig 2.2: Gas engine used by Baird to power his electricity generator (1901)

fly from the roof of his home seated in a large home made glider kite. Emerging shaken, but otherwise unscathed from the remains of the glider he vowed never to venture into an aircraft again.

These events were the product of a developing and inventive mind, but it was a spark of genius that enabled him to generate domestic electricity at the age of thirteen. He purchased a second hand gas engine (**Fig 2.2**), which he coupled to the spindle of an electric motor to enable it to function as an electric dynamo. The output wires from the dynamo were connected across a battery of home made jam-jar accumulators to store the electricity for the house lights. The local newspaper reported:

> "The Lodge was enjoying the amenities of electric lights, thanks to the ingenuity of a youthful member of the household." [6]

It may have been around 1901 that the idea of experimenting with television first occurred to Baird. He was aware that much had already been written on the subject of television and researched the subject from Ernst Rhumer's book, *Das Sehen und Seinen Bedeuten in Electral Technique* (Seeing and its Meaning in Electrical Techniques) here he would have called upon his father the Reverend John Baird to assist him by translating the German text. He would later discover that the foundation for the talking picture was based on Rhumer's work on optically recording sound on film. He soon learned that only a few spots or tiny elements of a picture had been successfully transmitted. With this knowledge the twelve-year old Baird acquired selenium from a local chemist and set about the task of producing a homemade light cell in the kitchen.

> "I had many trials resulting chiefly in bad smells and burnt fingers. But I did learn one thing, which was that the current from selenium was infinitesimally small." [7]

On leaving school in 1906 Baird enrolled in an electrical engineering diploma course at the Glasgow and West of Scotland Technical College [8], but suffering from persistent ill health he was unable to qualify until 1914. His study was interspersed with work experience at Halley's Industrial Motors in Yoker, Clydebank, (1909-1911) and The Argyll Motor Company in Alexandria, Scotland (1911-1912). There is also some evidence to suggest that he carried out further experiments on television during 1912 [9]. By researching the subject energetically, Baird learned about the use of multiple shutters with selenium cells to transmit letters and numbers, and of similar experiments made by Rignoux and Fournier in 1906. Probably the most important factor gleaned from his research was that a breakthrough in television did not appear to be on the horizon.

During 1912 until 1913 Baird was an apprentice draughtsman with Springfield Electrical Works in Glasgow where he gained experience in drawing sophisticated switchgear. After receiving a Diploma and an Associateship from the Royal College (formerly the Glasgow and West of Scotland Technical College), Baird progressed to Glasgow University, where his qualifications enabled him to complete a degree in electrical engineering after a period of six-months study. The BSc should have been attained as a matter of course, but was interrupted by the hostilities of the First World War. Early in 1915 most young men of Baird's age were signing up for the armed services and, feeling that he

Fig 2.3: Baird at Rutherglen power station [picture courtesy of Prof Malcolm Baird]

CHAPTER 2: BAIRD THE YOUNG ENTREPRENEUR

should also do his duty, he presented himself for enlistment, but his fragile health was recognised immediately by the doctor who carried out a thorough physical examination declaring him unfit for any service.

Baird took the position of Assistant Mains Engineer for the Clyde Valley Power Station in Rutherglen, Glasgow (**Fig 2.3**). The job was slow and uninspiring and mostly involved sitting around for hours waiting for emergency call outs. With all this spare time he soon became bored and his inventive mind began to wander.

There followed a period of misadventure after he came up with the notion that it may be possible to manufacture synthetic diamonds by applying enormous forces to a piece of carbon. When he looked around the workplace he could see that the components required for such an experiment were easily at hand. First he attached a couple of stout pieces of wire to the ends of a carbon rod and set it hard in a pot of cement. Next he redirected the energy from the sub-station across the protruding wires leading to the potted carbon rod. In the resulting explosion the main circuit breaker for the station opened plunging a large area of Glasgow into darkness. After some awkward explaining and a reprimand, Baird was regarded as a rather hazardous character, and even worse, there was no diamond to be found!

Inevitably, to stave off the everyday boredom, his imagination wandered again but this time with much more useful entrepreneurial results. He had heard that many of the soldiers on the allied front were suffering from 'trench foot' a condition brought about by standing for long periods in pools of water. He came up with the idea of an undersock to provide double insulation for the feet, which together with medication would keep the soldiers feet warm and dry in the winter and cool in the summer. From personal experience, Baird had learned that the problem of constantly cold feet could best be resolved by lining the socks with newspaper, but that would have been impossible to market:

> "From Hinckley I got six dozen specially made unbleached half-hose. Then I sprinkled them with borax and put them in large envelopes printed with 'The Baird Undersock' containing a pamphlet describing their advantages and containing testimonials." [10]

Baird rented an office at 196 St Vincent Street in Glasgow and inserted an advertisement in a popular magazine, *The People's Friend*:

> "The Baird Undersock, Medicated, Soft, Absorbent. Keeps the feet warm in winter and cool in summer. 9d per pair, post free." [11]

Unfortunately the 30/- advertisement attracted only one customer. But, not one to be easily disillusioned he used the time when he was off duty to take his product around local chemists and stores. This was a great success achieving the sale of two-dozen pairs and orders for another six-dozen. After considering his rather precarious position at Clyde Valley Power Station he thought it wise to find others to do the selling and placed an advert in *The Glasgow Herald*:

> "Traveller wanted, visiting chemists and drapery stores, to carry sideline" [12]

Soon the Baird Undersock was being carried throughout Britain reaching as far south as London. The business was such a success that he could afford to invest some of his profits in a campaign of advertising and looked for people to carry sandwich boards. As most of the able men had been called up for active duty he hired Glasgow women and inadvertently received his first taste of editorial publicity. Baird was surprised to discover that sandwich board women made big news, appearing in some of the illustrated papers with photographs and captions:

> "'First sandwich women in Glasgow', 'New occupation for the ladies,' and such like headings. The name 'Baird Undersock' appeared prominently on the placards. Some of the newspapers published this without comment, but in two cases I had to pay a small fee to have the name reproduced in the paper. It was first class publicity." [13]

The undersock business was making two hundred pounds a week of which he was able to bank one hundred and ninety-five pounds in comparison to the wage he received from the Clyde Valley Sub Station of thirty shillings. When news of the Baird Undersock finally reached his employers Baird quickly tendered his resignation and in 1918 officially formed the Baird Undersock Company, registration certificate, Number 15348 [14].

> "Encouraged by his success Baird brought out a special shoe cleaner, the "Osmo Boot Polish", and within a year this young Scotsman had built up a prosperous little business and began to make money at a fairly rapid rate – a success well deserved." [15]

Surprisingly, a pair of Baird undersocks is still in existence with the packaging and pamphlet shown in **Fig 2.4**. The packaging is over-stamped with: "Improved pattern, February 1918" displaying a price of one shilling and six pence (7.5p) which is double the price of the original 9d/pair. In 1919 when poor

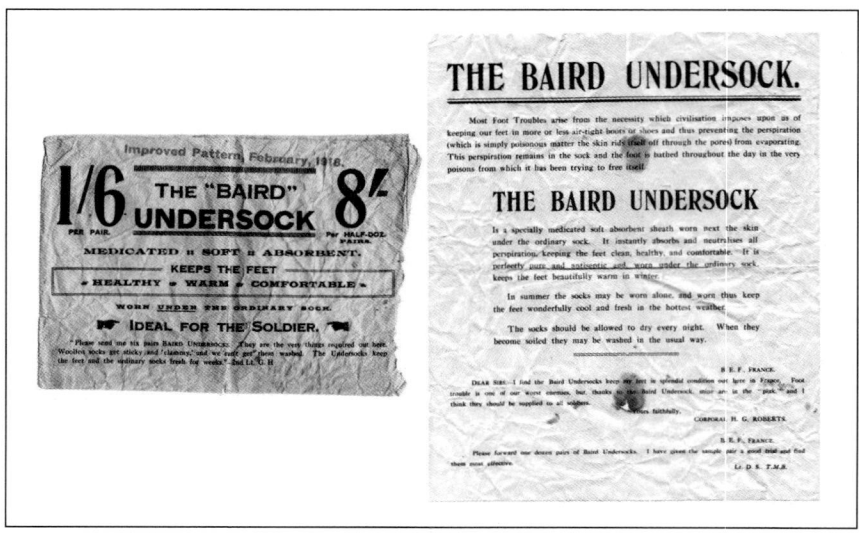

Fig 2.4: Packaging and leaflet from a surviving pair of Baird Undersocks [courtesy of Dr Peter Waddell]

CHAPTER 2: BAIRD THE YOUNG ENTREPRENEUR

health forced him to close the business he had banked one thousand, six hundred pounds, or the equivalent of twelve years earnings as an engineer.

With newly established wealth and in search for an improvement in his health, Baird sailed to the warmer climate of Trinidad, but ironically, soon after disembarking from the cargo ship, contracted dysentery. On recovering, he noticed an abundance of fruit growing freely on the island and turned his attention to jam making. With assistance from newly acquired friends Ram Roop and 'Tony', he built a small outdoor factory in Bourg Mulatrice to produce mango chutney, guava jelly, tamarind syrup and marmalade. In the process of heating the first batch of preserve, problems associated with outdoor cooking took them by surprise. Swarms of ants and mosquitoes, attracted by the compelling aroma invaded in thousands to happily sacrifice themselves to the boiling fluid. When this problem was resolved large quantities of non contaminated jam and chutney were packaged in jars, but there was a major flaw in the business plan; with such an abundance of fruit on the island there was simply no market for the preserve in the Port of Spain. Baird filled a large cask and other smaller containers in the prospect of selling the goods on his imminent return to London.

"With this cargo he set out for London, more dead than alive, and with three quarters of his capital gone. [16]"

On his return to London, he stayed for a short time with his elder sister Annie a district nurse at Deptford. According to Margaret Baird, after the warm climate of the West Indies, Baird found the English winter as intolerable as ever and on one occasion pulled the carpet up to add to the bedclothes.

In order to financially re-establish himself he took a small office at 166 Lupus Street in London where he set about the onerous task of selling the preserve, the quality of which he soon discovered was well below standard. Conceding to a financial loss, he sold the remains of the preserve for the trivial sum of fifteen pounds to a sausage maker as bulk additives for his products.

Recent events reminded Baird that the fragile condition of his health would undermine the likelihood of become a successful employee of some large well paying concern. With lessons learned and funds running low, Baird decided to continue along the path of entrepreneurialism with the immediate priority of improving his financial situation:

"To Annie he put the choice whether he should try television again or whether he should invent a razor that would not get blunt. She thought that there would be more money in the razor. John actually produced a razor blade of glass, but he cut himself badly with it and gave up the idea," [17]

He moved into an inexpensive room in the attic of a boarding house in Bloomsbury and began looking for suitable business opportunities by searching through advertisements in *The Times*:

"Not one single reasonable proposition resulted. The majority of the business opportunities were one form or another of patent medicines." [18]

Then as a saving grace, Baird learned that a considerable quantity of Australian honey was lying at the docks, which the owners were anxious to clear at almost any price. Down to two hundred pounds and completely undeterred by recent events, he located and purchased two tons of the honey at a very low price, which he immediately advertised in *The Times* and *The Morning Post*:

"28lb tins finest Australian honey, post free to your address 30/-." [19]

The response was good producing enough profit to enable the purchase of a horticultural business that included a small shop and the acquisition of a business partner. This time he took out an insurance policy to cover loss of earnings in the event of ill health. As the business prospered he also purchased a share in a company with a corner in the coir fibre market.

Unfortunately for Baird the shop was located under a railway bridge with walls that were often streaming with water. The damp environment proved too much and he contracted his usual cold resulting in several weeks in bed. He was forced to sell his share of the business to his partner for the sum of one hundred pounds and two hundred shares in what transpired to be a worthless oil company. On the advice of his doctor, Baird left London for the better air of Buxton to take a complete rest. With an income of six pounds a week from his insurer he was able to recuperate in the luxury of Buxton Hydro [20]. He was ill for nearly six months, but with characteristic determination returned again to risk the smog of London.

Despite a rapidly dwindling capital of one hundred pounds it was not long before Baird established another business. He located a source of twin tablet double wrapped soap and with the help of travelling salesmen there was soon a healthy demand for 'Baird's Speedy Cleaner.' [21].

The soap business quickly prospered beyond expectations with large quantities of soap being regularly imported from France and Belgium. As the director of a small limited company worth two thousand pounds Baird appeared to be on his way to becoming a soap magnate [22].

Two co-directors had been appointed and Baird was in the process of arranging a merger with Captain Hutchinson the owner of a rival soap company 'Hutchinson's Rapid Washer' when his health broke down once again. Unable to continue with his share in the business he was forced to sell out to the fellow directors. The proposed 'soap merger' was cancelled and Baird took to his bed.

Captain O. G. Hutchinson will return later in this account to become Baird's co-director in the world's first television company:

"Hutchinson appeared with a bottle of Eau de Cologne and was thoroughly alarmed at my state. The doctor was called in and I got steadily worse. He became concerned and told me that I must get out of London at once or he would not answer for my recovery." [23]

Tiltman's account illustrates the severity of his condition:

"Baird had the most severe set-back of all. He had a complete physical and nervous breakdown of a serious character, and after a thorough examination

his doctor informed him that he must abandon for ever all thoughts of a business career in London." [24]

Once again Baird had lost a promising business through chronic ill health. A close analysis of Baird's entrepreneurial career with the exception of the jam fiasco indicates that in reasonable health his success had little to do with the merchandise.

It was his imagination and determination to succeed that enabled him to promote and successfully market almost any product. These were the key characteristics when linked to engineering skill that would later enable Baird to develop and market the idea of television to the world.

References

[1] *Sermons Soap and Television*, John Logie Baird, Royal Television Society. 1988, p.1.
[2] *Sermons Soap and Television*, p.5.
[3] *Sermons Soap and Television,* p.5-6.
[4] Larchfield is now part of Lomond School in Helensburgh.
[5] *Baird of Television*, Ronald, F. Tiltman, London, Seeley Service & Co, 1933, p.29.
[6] *Vision Warrior*, Tom McArthur and Peter Waddell, Orkney: Scottish Falcon Books 1990. p.41.
[7] *Sermons Soap and Television*, John Logie Baird, Royal Television Society. 1988 p.22.
[8] Now the University of Strathclyde, Glasgow and West of Scotland Technical College formed in 1887 receiving a Royal charter from King George V to become The Royal Technical College in 1912.
[9] *Vision Warrior*, Tom McArthur and Peter Waddell, Orkney: Scottish Falcon Books 1990, p.76.
[10] *Sermons Soap and Television*, John Logie Baird, Royal Television Society. 1988 p.30-31.
[11] *Sermons Soap and Television*, p.31.
[12] *Sermons Soap and Television*, John Logie Baird, Royal Television Society. 1988, p.31
[13] *Sermons Soap and Television*, p.31-32.
[14] *Vision Warrior*, Tom McArthur and Peter Waddell, Orkney: Scottish Falcon Books 1990, p.70.
[15] *Baird of Television*, Ronald, F, Tiltman. London, Seeley Service & Co, 1933, p.40.
[16] *Television Baird*, Margaret Baird, Cape Town, Haum 1973. P.34-35.
[17] *Television Baird*. P.36.
[18] *Sermons Soap and Television*, John Logie Baird, Royal Television Society, 1988, p.37.
[19] *Sermons Soap and Television*, p.38.
[20] *Sermons Soap and Television*, John Logie Baird, Royal Television Society, 1988, p.38.
[21] *Baird of Television*, Ronald, F, Tiltman, London, Seeley Service & Co, 1933, p.45

[22] *Television Baird*, Margaret Baird, Cape Town, Haum 1973, p.40.
[23] *Sermons Soap and Television*, Royal Television Society, 1988, p.41.
[24] *Baird of Television*, London, Seeley Service & Co, 1933, p.46.

3

Television Begins

Baird moved to the coastal resort of Hastings to convalesce. This was a diversion that would give him time to consider the experiments that he, and others, had carried out on the problems of television.

DESPITE A STATEMENT IN HIS biographical notes [1], suggesting he had not experimented with television since those early days in the kitchen of his home in Helensburgh, Waddell and McArthur [2] seem convinced that by 1919, Baird had already developed a form of unrefined television.

"The young man, pale-faced and almost crippled with colds, who sailed out of Britain for the New World on a cargo boat in the winter of 1919 had almost certainly solved many of the problems that puzzled Bidwell." [3]

In 1908 Shelford Bidwell the English Physicist and inventor, best known for his work on telephotography had been pessimistic about the likelihood of anyone developing a practical working system of television:

"The process, could probably be done for a cost of around one and three quarter million pounds." [4]

His time in Hastings was critical for Baird, he dreamed of solving the problems of television, but it seemed that his health, which precluded any hope of full-time employment was now threatening to prevent him from reaching the self-sufficiency he desperately required for television research. His morale must have been low, but as before, Baird would find the strength and determination to succeed.

In the short term he would focus on developing television with whatever limited resources he could find. It was, however, a difficult situation; without financial support he could not advance his ideas and without a breakthrough in television, a backer would not easily be found.

In the winter of 1922, John Logie Baird moved into lodgings at Linton Crescent, Hastings. According to the traditional history, Baird spent the spring of 1923 sitting in the pale sunshine at the sea front and browsing in the public library. As he began to recover, his inventive mind went into action.

Margaret Baird wrote that his first idea:

". . . though ingenious contained the usual element of farce." [5]

The idea involved wearing a pair of oversized boots inside which he placed two partially inflated balloons. The idea was that people wearing these devices might be able to obtain the same advantages that a motor car gains from the use of pneumatic tyres:

"I walked a hundred yard in a succession of drunken and uncontrollable lurches, followed by a few delighted urchins. Then the demonstration was brought to a conclusion by one of my tyres bursting. More thought was needed." [6]

Although this is an amusing episode Baird's concept was being considered more seriously by others: On 21 August 1922, Charles Cooney applied for a US patent [7] for a boot and shoe having an inflated air cushion inserted in the sole and heel. The patent was approved and issued on 2 September 1924. This was followed by another patent on 18 November 1924 by Alfonso Miceli [8] who successfully patented his pneumatic sole in the United States of America as a shock absorbing, removable shoe attachment for medical or general usage. The description of Clooney's patent describes fairly accurately the concept of Baird's crude experiment. Today, pneumatic soles are found in many sport and casual shoe products.

With little financial support and no prospect of work on the horizon, Baird decided that his only option was to invest his remaining funds exclusively on television research and development. An article entitled 'Television' would later appear in *The Motor News* on 26 June 1926 indicating that John Logie Baird had commenced television experiments prior to 1916, suggesting that television equipment was either already at his disposal when he arrived at Hastings, or was reconstructed from earlier plans. The experimental apparatus consisted of an old tea chest, an old hat box, some darning needles, a bull's-eye lens from a local cycle shop, and a plentiful supply of sealing wax and glue.

The device grew until it filled his bedroom. Electric motors, batteries, wireless thermionic valves and transformers were added. Baird used this basic apparatus to transmit the crude silhouette of a cardboard Maltese cross on 27 June 1923, and placed an advertisement in the personal column of *The Times* (**Fig 3.1**), being careful not to ask for money at this stage. He was aware that to demonstrate television properly he would need to resolve the skin tones and facial characteristics of a moving person, but to achieve this, a number of unsolved problems would have to be addressed.

Fig 3.1: Advertisement placed in *The Times*

> SEEING by WIRELESS.—Inventor of apparatus wishes to hear from someone who will assist (not financially) in making working model.—Write Box S.686; The Times, E.C.4.

Dr Fournier D'Albe together with a great number of scientists insisted that television could not be accomplished on the principle Baird was applying.

"I was trying to get television by dividing the picture into a series of little sections of light and darkness and sending these in very rapid succession. That is the principle in general use today." [9]

CHAPTER 3: TELEVISION BEGINS

Almost all of these scientists proposed that a means must be discovered to send an entire picture simultaneously. Baird indicated in his autobiographical notes that D'Albe correctly recognised that the other barrier to television by his method was the problem of recovering enough light from the subject. The transmission of silhouettes, such as in the case of the Maltese cross, was fairly simple to achieve, especially when most of the direct light fell straight onto the photocell. To transmit a human face Baird's system relied totally on reflected or diffused light, but he soon discovered that the amount of light that could be gathered from a human face, illuminated by an electric arc, was so small it could hardly be measured.

A *Daily News* report of Baird's demonstration of the Maltese cross, attracted a gift of fifty pounds from a friend of his father. This enabled Baird to move the apparatus from his bedroom, and install it in a small attic at No 8 Queens Arcade rented from a Mr Tree [10] for five shillings per week. The advertisement in *The Times* attracted Mr Odhams, the chairman of Odhams Press, who was not convinced about the future of the equipment and put Baird in touch with the chief research engineer of the BBC, Captain A G D West (later to be at the helm of Baird Television), and Mr F H Robinson, the editor of *Broadcasting*. Both were impressed, but Mr Odhams would not invest in a television project that was yet unable to send the recognisable face of a living person over a distance by wire.

Robinson was however, inspired by what he had seen and published an article in the *Kinematograph Weekly*. The article was read by Wilfred E Day the owner of a prosperous wireless and cinema business. Day immediately recognised the potential of the device and became Baird's partner in television for the sum of two hundred pounds. It is interesting that Day is named as the co-author on John Logie Baird's premier television patent *GB222604* describing a television system using a Nipkow revolving disc for the camera, with a static active display screen for the receiver,

This patent (**Fig 3.2**) has an application date of 26 July 1923 although curiously their partnership began after this date on 17 April 1924. It must be assumed that Day insisted that his name be retrospectively added on the 21 May 1924, when the specification was completed.

In television patent *GB230576*, application date 29th December 1923, Baird described a similar Nipkow disc for collecting image data, but at the receiver he replaced the static matrix of electric light bulbs shown in *GB222604*, with a rotating plate fitted with a spiral of small lamps. A photograph appeared in *Baird of Television* [11] showing Baird giving a demonstration to William Le Queux, the spy thriller novelist and wireless enthusiast who wrote accounts of Baird's work for the *Radio Times*. When the spiral of lamps was set in motion they would represent the fixed matrix by persistence of vision so this was economic in reducing the number of lamps to the number of scanning lines.

The advantages of simplicity and economy were probably outweighed by the disadvantages of requiring two rotating discs instead of one at the transmitter only. This seems to be a passing experiment to prove the matrix concept and was soon superseded by the less complicated concept of using a transmitter disc fitted with a spiral of lenses, married up with a receiving disc with a spiral of small-holes. Baird returned to the static display using a live matrix of lamps for televising the Derby in 1930 and 1931.

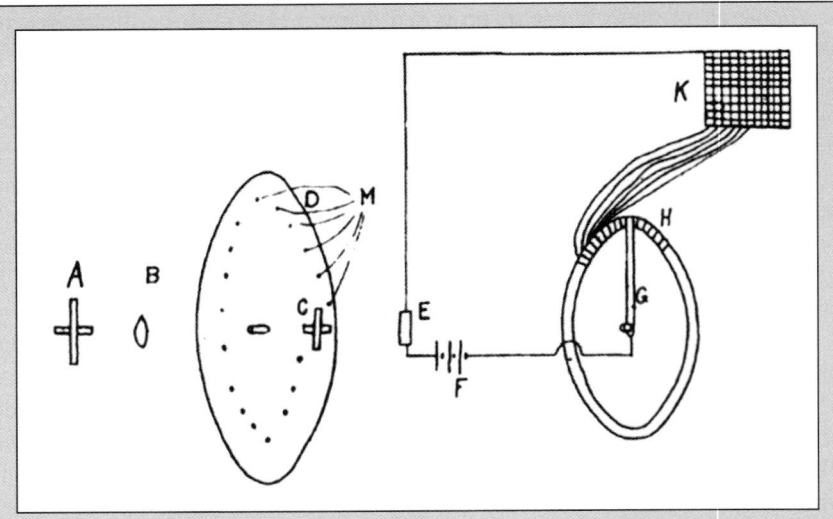

The scene or object (A) to be transmitted is focused by means of lens (B) to form an image (C) on a revolving plate (D) perforated with a series of small holes (M) arranged in a spiral. So that when the disc is revolved every part of the image transmitted passes through a hole. Behind the disc there is a light sensitive cell (E) on which the varying light falling on the perforations (M) causes the current flowing from battery (F) to vary. This signal is amplified by a thermionic valve and transmitted by wire or wireless to the receiving station. At the receiving station a brush fitted on the end of an arm (G) revolving in perfect synchronism with the disc (D) at the transmitting station, passes over a series of contacts (H). Each contact is connected to a small electric lamp and the lamps are arranged in rows to form screen (K). Each hole sweeps out a strip of the picture to be reproduced by a row of lamps on the receiving screen so that each hole has its corresponding row of lamps.

Fig 3.2: Abstract from Baird's premier television patent *GB222604*

After many experiments, Baird became aware that the low sensitivity and poor response time of selenium was restricting the scanning speed and seriously hampering progress. The only images resolved at this time were poorly defined silhouettes requiring a strong backlight on the subject. He discovered that the problem of the time lag of selenium became less important if a light chopper was introduced in the path of the light cell to stop the cell from charging up to its full potential. This initially took the form of a pair of counter-rotating slotted discs (**Fig 3.3**), interrupting the scanned image at a radio frequency, resulting in patent application *GB235619* dated 12 March 1924. This had two main effects:

- Light reaching the selenium cell, in the form of high frequency pulses, were small enough and fast enough to make the time-lag characteristic negligible.
- A smaller than normal video signal was produced in the form of a radio frequency modulated signal carrier, which may have been in the region of, as Baird's patent suggests, up to 500kHz.

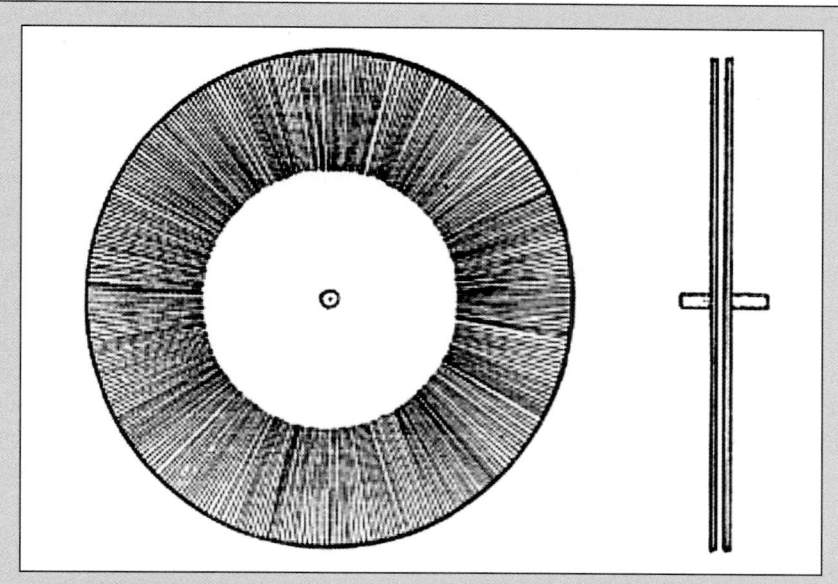

For the purpose of minimising inertia effects in the selenium or other light sensitive cells used, the light is interrupted at high frequency by means of two serrated discs revolving concentrically, close together, and in opposite directions. These discs may be serrated around their circumferences, or, alternatively, the discs may be made of glass, celluloid, or other transparent material and opaque lines drawn in the place of the teeth cut in the solid disc. By making the teeth very fine and numerous, frequencies of light interruptions of from ten thousand to over half a million per second may be obtained, and by the expression "interrupting at high frequency" in this specification I refer to frequencies of this order."

Fig 3.3: Abstract from Baird's television patent *GB232619*

This modification presented Baird with a completely different type of video signal. The image data of 10kHz bandwidth was now effectively a modulated radio frequency carrier. Here we have to surmise that he used the advantages gained from a resonant tuned radio frequency circuit before filtering and amplifying the signal in the normal way. This would have provided a significant increase in sensitivity and may have brought him closer to reproducing a grey-scale image. Experiments continued with the serrated disc and on 17 March 1924 he took out another British patent for an updated type of light chopper *GB236978*. This simplified the earlier proposal by enabling the spindle that carried the Nipkow disc to also carry the chopper disc.

While Baird was experimenting with the basic concept of television in Hastings, there is evidence that he was also experimenting with another device that used electromagnetic radio waves to detect distant objects. At first glance this would seem to be a form of radiolocation, a fact that may be substantiated by some of his later patents for a radio imaging system that alludes to a future ranging system. Norman Loxdale [12], a Hastings schoolboy and Baird follower, stated that Baird always seemed to refine his ideas before taking out patents. Loxdale recalled a strange event that took place early in Hastings in 1923. Baird asked Norman and some of his Boy Scout friends to haul a piece of equipment up the West Hill. He later described the apparatus as consisting of a parabola dish reflector with a lot of associated equipment including a power convertor that operated from a car battery. He recalled that Baird turned the reflector dish, which had a kind of spike in the middle, and pointed it towards the East Hill. At the same time he saw that Baird was looking into an opaque glass plate set into a case. He was looking under the sort of black cloth used with the portrait cameras of the time. Baird moved the reflector towards the sea and then up towards the sky and asked Loxdale to take a look. He saw a dull brownish image of something that looked like a moving wave on a screen. It was not a bright image. Loxdale recalled that the screen seemed to have been fabricated using small particles, like iron filings. Later as an adult, when he thought back on that scene, it seemed likely that Baird was getting radio reflections from the sea, the East Hill and possibly from the Appleton layer of the ionosphere.

The funds raised from the Will Day partnership enabled Baird to invest in several hundred torch batteries that he eagerly connected in series to realise his dream of a two-thousand-volt power supply. This was a painstaking task involving hours of work.

"He had finished the joining-up and was connecting the supply to some part of the wiring when his attention wandered and he received the full force of the two thousand volts through his hands. It was amply sufficient to cause death, but he was lucky; for a few seconds he was twisted into a knot in helpless agony, and then, fortunately fell backwards, breaking the circuit and saving his life." [13]

A small crowd gathered and the next day the newspapers reported: *"Inventor pinned to the ground by short-circuit"* and *"Serious Explosion in Hastings Laboratory"* [14].

The publicity surrounding this episode led to Baird's landlord sending him an ultimatum; Cease forthwith your experiments or vacate the premises. Baird was reluctant to do either and continued to work, but was eventually forced to depart from Hastings when he received a letter from Mr Tree's solicitor. Baird was urged to return to London by Wilfred Day who found him a small attic located at 22 Frith Street, Soho.

In March 1925 Gordon Selfridge, the owner of the famous Oxford Street store, heard about television and called on John Logie Baird at Frith Street. Selfridge asked Baird if he would be interested in demonstrating television at his store on the basis of three shows per week. This was agreeable, and Selfridge arranged to pay twenty pounds a week for a three-week period.

The public demonstrations at Selfridges involved the transmission of outlines and shapes over a few yards by means of a crude wireless transmitter. This provided a reasonable income and helped raise public awareness, but after the event Baird was ill for several weeks. Once again short of funds, Baird advertised for a company promoter and a Mr Brooks responded. He suggested that they should send a circular to all the general practitioners on the medical register, explaining the wonderful possibilities of the invention. It can only be assumed that to reach professional members of the community with their appeal it was a simple matter to obtain published lists of doctor's medical practices:

"Out of the 3,000, six replies were received. Mr. Brooks called on these six and collected £75. The expenses he claimed amounted to well over this figure and he wished to retain the whole amount. However, we finally settled for £25 and two double whiskies." [15]

John Logie Baird registered Television Limited with Wilfred Day on 11 June 1925, with a nominal capital of three thousand pounds. This was divided into two thousand-nine hundred 'Ordinary Shares of one pound' and two thousand 'Founders Shares of one shilling each.' Roger Brooks (most likely the aforementioned 'company promoter'), who identified himself on the document simply as an inventor, witnessed the transaction.

The registered office of Television Limited was 22 Frith Street, London, W1. The company was founded on assets comprising of two completed patents *GB222604* and *GB230576*, in the joint names of John Logie Baird and Wilfred Ernest Lytton Day, and five provisional specifications [16]. It should be noted that Day was a sleeping partner in the arrangement and that the relationship between the two men although always polite and cordial was a battle of wits often on the verge of collapse.

On 2 October 1925 John Logie Baird made the breakthrough that allowed him to successfully register a television grey scale. His first subject was the head of a ventriloquist's dummy 'Stookie Bill,' and later on the same day he transmitted the image of William Taynton a neighbouring office boy.

In the closed company file of Television Ltd, it was recorded that on 16 December 1925 Wilfred Day resigned, selling his share to Captain Oliver George Hutchinson and a close friend of Hutchinson's, Captain Broadrip. It is not known why Baird claims to have first met Hutchinson in 1922, when Baird's association with Hutchinson began long before the proposed merger of 'Baird's Speedy Cleaner' and 'Hutchinson Rapid Washer.' *The Motor News* of 26 June 1926 recorded:

"It is interesting to relate that Messrs Baird and Hutchinson were apprentices together at the Argyll Motor Works in Glasgow over 20 years ago." [17]

While this indicates that Hutchinson was no stranger to Baird, the statement they met over 20 years ago at the Argyll Works should be further corrected to fourteen years:

"Then Captain Hutchinson who had met Baird at the Argyll Works in 1912 came to his assistance and found capital when the inventor was in financial straights." [18]

It appears that Hutchinson had been close to Baird and his work for a very long time but for some reason they did not wish this to be known:

"Captain Oliver George Hutchinson had served in the army in the First World War. It is likely that he was one of the few who over the years had known just what Baird was up to. As his work became more involved the need for secrecy increased. Baird apparently found it necessary to pretend that Hutchinson and he had just met by chance." [19]

Margaret Baird [20] wrote that Hutchinson put most of his own capital into Television Ltd, while Baird's cousins the Inglis's bought some more shares and stated that Wilfred Day made a good profit in selling. Hutchinson became business director and Ben Clapp (callsign 2KZ) was hired as the first technical assistant.

On 26 January 1926, John Logie Baird demonstrated television at Frith Street, to nearly fifty members of the Royal Institute of Great Britain. This was the first public demonstration of true television. Dr Alexander Russell FRS and past president of the IEE and the Physical Society, wrote in *Nature* on 3 July 1926:

"The image cannot be compared with that produced from a good kinematograph film. The likeness, however, was unmistakable, and all the motions are reproduced with absolute fidelity. This was the first time we have seen real television. Mr. Baird is the first to have accomplished this marvellous feat." [21]

The first ever television photograph was taken of Hutchinson from the image as it formed on the Baird Televisor screen during the 1926 demonstration (**Fig 3.4**). The number of slightly curved scanning lines indicated that that it was produced from a scanning disc using thirty lenses or holes, while the earlier television apparatus now in the London Science Museum, has no photocell and is fitted with a disc of only sixteen scanning lenses.

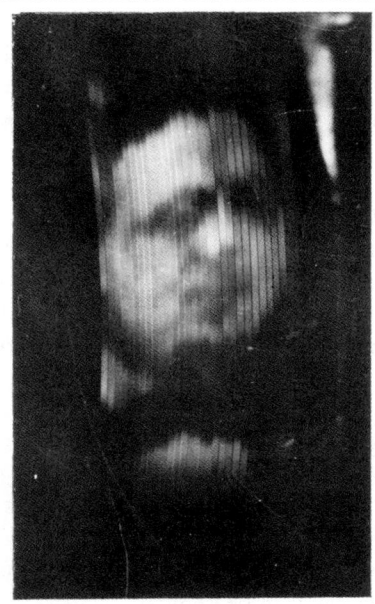

Fig 3.4: The first ever photograph of a television picture [picture courtesy of the National Media Museum]

This was a typical Baird tactic. He was careful not to reveal the secret of his success and was particularly anxious to withhold useful information from his American competitors. He stated that his breakthrough came when he was trying to solve the problem of the slow response time of selenium [22] while scanning the ventriloquist's dummy's head 'Stookie-Bill'. The inertia of the light cell was causing the subject to appear as an out-of-focus streaky blob. He learned that when light fell on the cell, the current, instead of jumping to its full value, rose slowly and continued to rise as long as light fell on it. When the light was removed the current did not stop instantly, but stopped increasing and began to fall slowly.

To overcome this he first tried to pulse the light reaching the cell through the slotted chopper disc previously described. The resulting fast pulses partly resolved the problem of electrical inertia in the selenium but produced radio frequency interference that obscured the image. After concluding that the chopper disc solution was impractical he returned to solve the problem by applying not mechanical, but electrical techniques. Baird theorised that by introducing a transformer into the light cell circuit and by superimposing a curve representing the time rate of change of the current, upon the curve of current with time, he may be able to dispense with the light-chopper and filter circuits by electronically compensating for the poor response time of selenium:

> *"The funds were going down - the situation was becoming desperate and we were down to our last £30 when at last, one Friday in the first week of October 1925, everything functioned properly. The image of the dummy's head formed itself on the screen and with what appeared to me almost unbelievable clarity. I had got it! I could scarcely believe my eyes, and felt myself shaking with excitement."* [23]

While Baird's new circuit configuration may have compensated for the poor response time of selenium, it is unlikely that it alone would have significantly improved the poor sensitivity governed by the floodlight principle. Baird's competitors became confused by his success and a great deal of doubt has existed to this day about the true nature of his 'breakthrough.' An indication of the level of industrial secrecy that Baird maintained can be seen from the apparatus that Baird claims was responsible for the first true television picture with a tonal grey scale.

It is interesting to note that the first photograph produced from this television system successfully produced an image of co-director Captain Hutchinson and was constructed from 30 vertical scanning lines (Fig 3.4).

In contrast, the television equipment gifted by Baird to the South Kensington Science Museum, London was an obvious decoy. There was no light cell and it included a confusion of components that appear to be no longer required, such as the chopper disc. The equipment was most likely cobbled together from a compilation of experimental parts including a scanning disc, which with only sixteen apertures could not have produced the 30-line photograph of Hutchinson.

There can be no doubt that Baird had certainly made a breakthrough in realising television in advance of countless other scientists and engineers. Research by the author has revealed that the secret of his success was the combination of two individual improvements.

The first was the addition of an electronic circuit to compensate for the poor response time of selenium and the second brought about a massive improvement in the sensitivity of the pick up device. Instead of floodlighting the subject, a technique which resulted in miniscule levels of reflected light passing through the apertures on the disc that were too dim for the selenium light cell to resolve, Baird reversed the process and scanned the object using a small, intensely bright, focused spot of light.

In this situation the small apertures or lenses on the disc were used to actively scan the object to be televised with an intense beam of light. The reflected

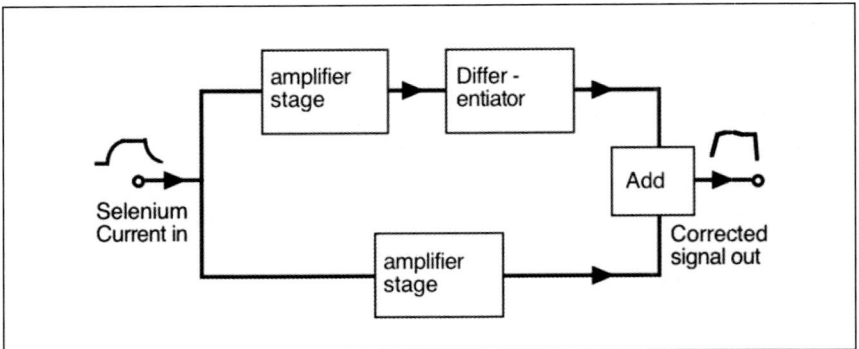

Fig 3.5: Baird's image enhancing patent *GB270222*

light, instead of passing through the apertures of the disc where it would become severely diminished, was bright enough to be easily collected using a number of external photocells placed appropriately close to the subject. This brought about a massive increase in the sensitivity of Baird's mechanical scanner. This process was later named the 'flying-spot scanner' and is still the basis of the preferred method used for scanning cinema films for television. Baird would not reveal this, or his circuit to overcome the sluggish response time of selenium, to his competitors until the 1930s.

Some modern reviewers of Baird's work suggested that a special circuit such as the introduction of a transformer was unlikely to produce the detail and grey scale that were apparently miraculously achieved. It has also been suggested that if such a circuit existed achieving the desired result, then why has it disappeared into obscurity, never to be heard of again? It has been concluded by most reviewers to date, that the discovery of the flying-spot scanner was Baird's only breakthrough. This is nonsense! Whilst adoption of the flying spot would certainly increase the sensitivity by a large margin it does not explain why the first images were in focus despite the inherent time lag of selenium. The answer may now be revealed.

The basic technique of pulse correction first appeared in Baird's British patent *GB270222* [24] (**Fig 3.5**), and today this is widely known as 'crispening' or 'aperture control.' The concept, which Baird undeniably invented in 1927 to

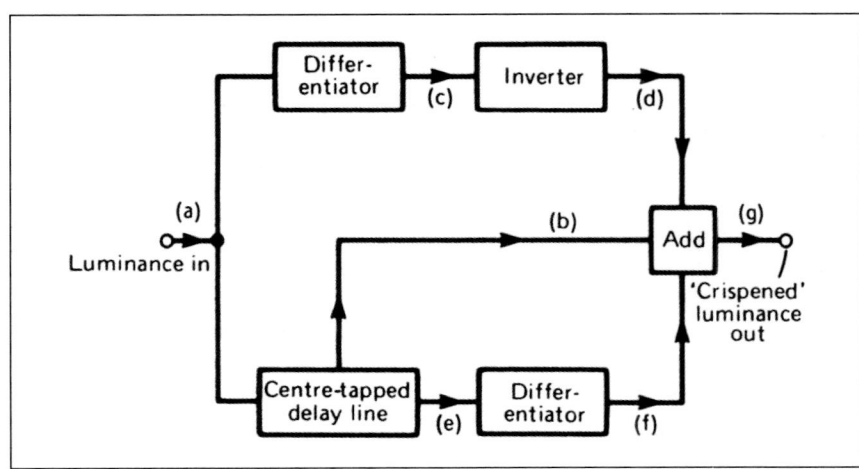

Fig 3.6: Block diagram of the modern aperture circuit

CHAPTER 3: TELEVISION BEGINS

solve the problem of the poor high frequency response of selenium, has since then found its way, in a modified form, into every VHS video recorder (**Fig 3.6**).

Baird's patent expired on 20 October 1945 to enter the 'public domain,' well in advance of the popularity of the technique in the modern television industry. Many adaptations and configurations of the technique have since been applied to CRT, LCD and plasma displays, television data projectors, television cameras, flying-spot scanners, camcorders, videocassette recorders and VHS video players, but it can now be revealed that the master patent belonged to Baird. Without this 'exotic' circuit in the replay path of VHS cassette recorders, circa 1970s to 1990s, the replayed pictures would have lacked sharpness, appearing unacceptably soft and poorly defined.

The modern application of the process, although not directly referenced to Baird, is described by Trundle:

> *"The crispening circuit goes some way to compensate for HF roll-off by artificially 'sharpening up' vertical edges in the picture. The technique has been used in broadcast and CCTV for many years to compensate for the finite scanning spot size in cameras and FSS (Flying Spot Systems). In such applications it is known as aperture correction because it has the effect of limiting the scanning spot size to that of a smaller 'aperture' inserted into the tube, and enhancing definition."* [25]

In the VHS circuit (**Fig 3.7**), the differentiated waveform from the luminance signal is summed with itself to produce artificial pre-shoot and over-shoot. These little spikes are used to exaggerate the soft edges of the image to bring about an artificially sharpened picture.

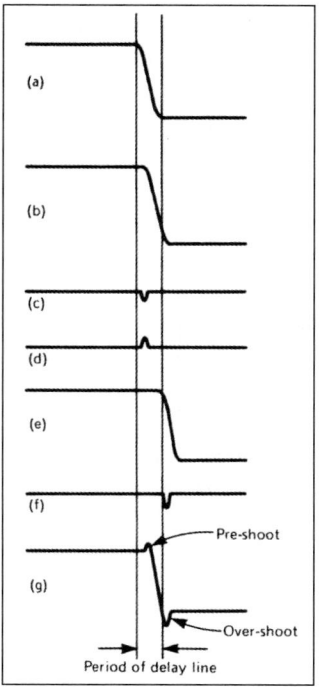

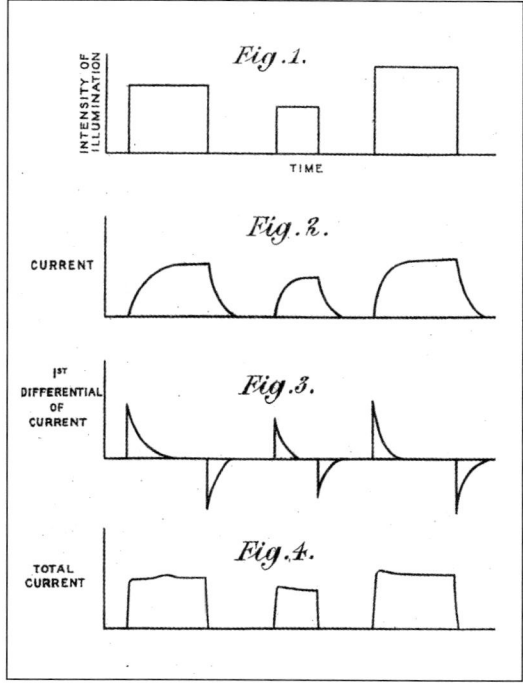

(left) Fig 3.7: Waveforms relating to the modern crispening circuit

(right) Fig 3.8: Waveforms describing Baird's image correcting patent *GB270222*

Fig 3.9: Floodlight television transmitter

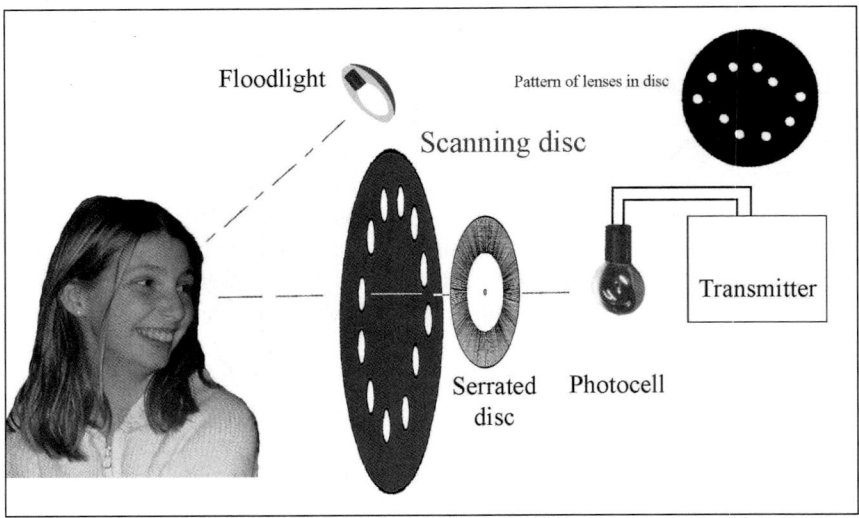

In Baird's patent (**Fig 3.8**), the differentiated waveform is directly added to compensate for the poor rise and fall times of the signal characterised by selenium. From Fig 3.5, it can be seen that the modern circuitry has the addition of a delay line to emphasise the overshoot of a pulse.

In each case the transitions of the signals are exaggerated to artificially sharpen the image. By using such a signal 'crispening' circuit Baird could noticeably improve the sensitivity of his system by removing the light chopper from the optical path and dispensing with the RF detection circuit. Despite these improvements, when using the floodlight method, the quantity of reflected light reaching the selenium from a safely directly illuminated subject remained below the signal to noise requirements of a valve amplifier. While it is unlikely, at this stage, that Baird would be successful in resolving a human face by the floodlight method, the inclusion of the crispening circuit would provide a dramatic improvement in the televising of backlit negatives, transparencies and moving film.

A closer look at the floodlight system, **Fig 3.9**, indicates that Baird's television apparatus employed two different arrangements of scanning. In the version illustrated a double spiral of lenses was utilised. On rotation of the disc, the spirals would appear to merge to give the impression of transparent scanning lines positioned in decreasing radii from the edge of the disc. This technique, known as interlacing, often credited to Alan Blumlein of EMI, was employed in analogue television to reduce the effective signal bandwidth without increasing perceived flicker.

In another embodiment of early apparatus, before settling on the 30-line standard, Baird used approximately sixteen lenses to form a single spiral for sequential scanning. It is interesting that although these items may have become obsolete in conjunction with Baird's crispening circuit, they were attached to the apparatus at the Science Museum in London.

If sufficient reflected light was focused on to the light sensitive cell, then its relative brightness level represented a corresponding change in the resistance of selenium. At each sampled point of an image an equivalent current was produced.

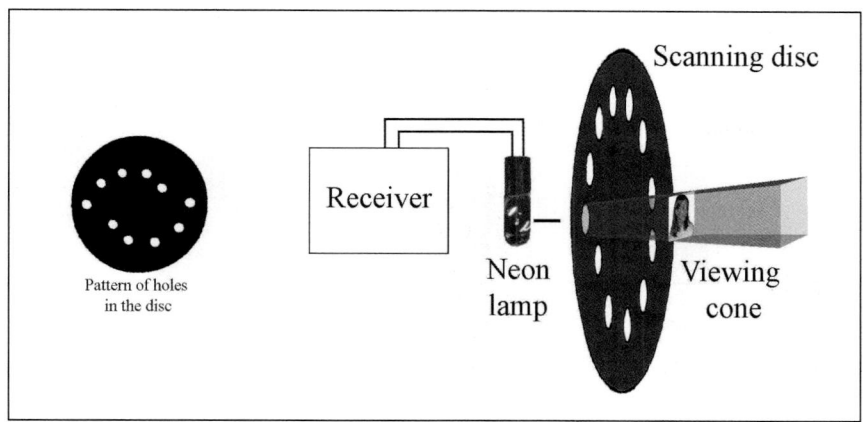

Fig 3.10: Floodlight television receiver

This dynamic process produced a complex analogue alternating current waveform, similar to that produced by the voice on a telephone. The 'Televisor' signal was then amplified and directed to a radio transmitter. A synchronising signal radiated via a second transmitter was obtained by coupling an AC generator to the shaft of the scanner's DC motor.

At the receiver, **Fig 3.10**, the vision signal was detected and amplified to modulate the neon lamp. The alternating current signal was also detected and amplified to drive an AC motor to synchronise the speed of the scanning disc. Persistence of vision enabled the viewer to see a reconstruction of the transmitted image, which appeared by viewing the brightness variations of the neon lamp through the holes in the synchronised receiving disc.

The quantity of light required by the floodlight system was extreme. Lieutenant Colonel Lefroy, from the British Air Ministry had been making abstracts of published accounts of John Logie Baird's work in connection with television since April 1924:

> *"The object to be seen is illuminated by using a large Osram 15,000 candle power lamp, of 1/2 watt type, on 200 volt mains. This lamp is placed 18in. from the object, and gives out such intense radiant heat that flesh cannot be kept near it more than a few seconds. His transmitter and receiver are synchronised by alternating current. He has invented, and uses, a special light sensitive cell of colloidal fluid. The pictures on his receiving screen are said to flicker."* [26]

This letter, dated 13 May 1926, does not offer any explanation regarding the televising of William Taynton in the previous year. If the heat from the floodlights had been so intense that flesh would have easily burned, then Baird must have used an alternative system. On 20 January 1926, Baird applied for British patent *GB269658* protecting the flying-spot method of television scanning [27].

Abramson speculated:

> *"It is not known when Baird first got this idea. But like most inventors, he was probably familiar with the inventions of the past. At this time, Baird had gone about as far as he could with his crude, insensitive equipment. By*

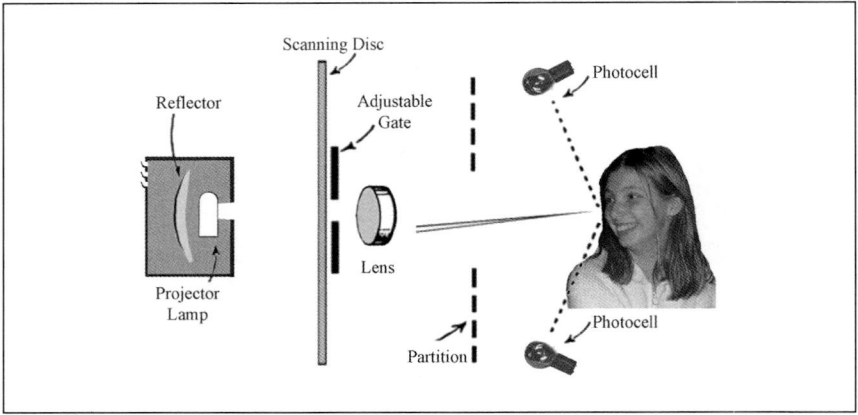

Fig 3.11: Flying-spot method of scanning

reversing the scanning process, as did Dr Gray of the Bell Telephone Labs, he obtained the results he had been seeking.
"All the evidence suggests that he first actually used this method for scanning an actual object ("Bill," his ventriloquist's dummy) and a young office boy (William Taynton) on October 2 1925." [28]

The flying-spot principle is explained with reference to **Fig 3.11**. A number of photocells may be placed close to the subject and a source of bright light (usually an arc lamp) is place behind the scanner. When the scanner rotates, a pinpoint of intensely bright light is projected through the lenses on the disc to scan the subject. The quantity of light collected by the photocells is now enormous compared with that obtained using the floodlight method. This process also made it practical to replace selenium cells with a more responsive but less sensitive photocell. The application date of the flying-spot patent indicates that Baird was aware of the advantages to be gained by using this method prior to January 1926! Further evidence has been found, accurately determining the year:

"Victor Mills assisted John Baird with his original experiments at Hastings. During a chat with the Editor he recalled the appearance at his front door one winter evening in 1923 of a figure clad in an old raincoat. It was JLB seeking advice in connection with excessive background noise from his selenium cell. Victor Mills suggested using a smaller cell and scanning the object thereby enhancing the signal to noise ratio. This was the beginning of a pleasant association which lasted until Baird moved to Frith Street." [29]

This reveals that although Baird may have given the impression of developing the floodlight system up to 1925, he may have known of the flying-spot principle from as early as the autumn of 1923. The three diagrams used by Baird in this 1926 patent, **Fig 3.12**, are carefully devised so that in each case the photocell is shown to be on the other side of the disc from the object to be televised. If Baird had been using this process since 1923, then it is likely that his desire for a high voltage supply at Hastings was to power an arc lamp for a flying-spot scanner. A search of Baird patents before 1926 assists the theory that Baird may have used the flying-spot scanner prior to October 1925.

CHAPTER 3: TELEVISION BEGINS

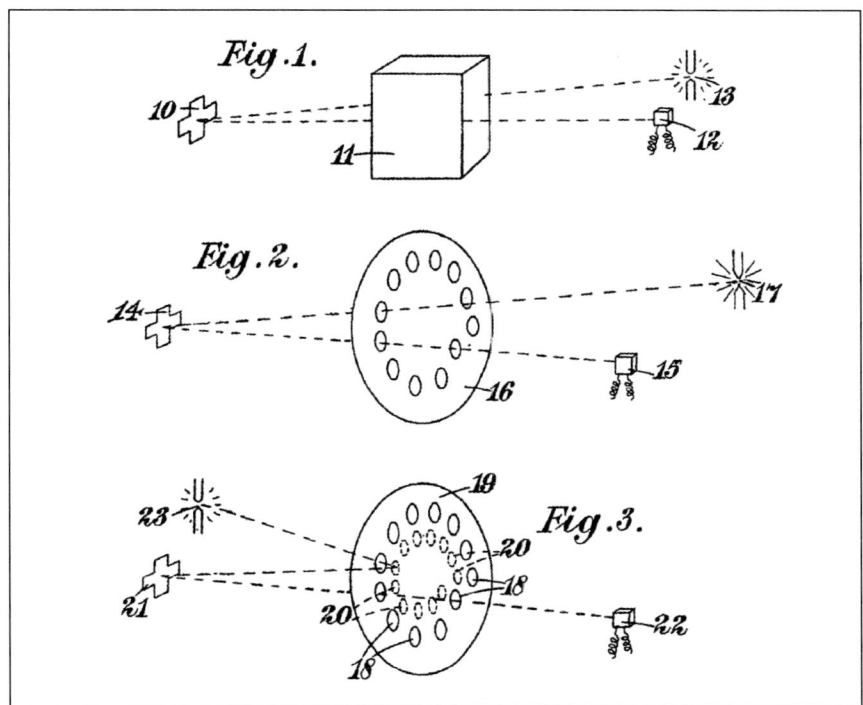

Fig 3.12: Baird's British patent GB269658

Although Baird's application of a flying-spot scanner was novel for use in television, it was not the first flying-spot scanner patent. The concept of a travelling light source was first suggested for use in motion pictures by Georges Rignoux, who applied for a French patent on 20 May 1908 [30] which was issued on 5 October 1909.

On 2 January 1910, Ekstrom applied for a Swedish patent [31] using the flying-spot scanner exclusively for scanning transparencies and film by the transmission method, and it was issued on 3 February 1912. On 15 August 1923, Hammond Jr applied for a US patent [32] in which he describes the scanning of live objects. This patent took six years to materialise due to his deep involvement with the US military; hence Baird could not have become aware of its existence until it was released on 20 August 1929.

On 25 November 1924, J E Gardner and H D Hineline applied for a British patent [33] for an all-electronic transparency flying-spot scanner. Despite its advanced technology it was quite unsuitable for scanning live objects. This patent has a convention date in the United States of November 1923, but no such patent has been found and was most likely rejected due to counter interference claims from Hammond Jr. Baird filed for a United States patent for the flying-spot scanner on 31 December 1926, which was eventually granted on 25 June 1935 [34].

It is therefore concluded that Baird's breakthrough on 2 October 1925 was achieved by a combination of the inclusion of the crispening circuit, the removal of the light chopper and the application of the flying-spot scanner. This would indicate that the demonstration on 26 January 1926 was also achieved using the flying-spot technique.

New source material that supports this argument also reveals that the first public demonstration of 'real television' that reproduced half tones occurred before 1926. In November 1925, John Logie Baird televised the face of Robert Shaw in the Temperance Cafe in Falkirk, Scotland. Mr. Shaw gave an account of his experience:

"It was autumn 1925. My night school teacher said to me, 'Would you go down to the Temperance Cafe, where I understand there is going to be a demonstration of seeing by wireless?' No one was in the room when I went in. Then two men came in, one was Logie Baird and the other was his great friend John Hart. They proceeded to couple-up to the mains electricity. Well after it was coupled up, Baird gave a talk, which I think was about half an hour. He went into it very comprehensively. After the talk was over he passed right by the chair I was sitting on and he said, 'Would you follow me?' Which I did! I followed him into the 'Quick Lunch Bar.' He said, 'Would you sit at the projector?' I never saw myself but the audience saw it, and they said it was very like me." [35]

The most interesting part of this evidence is that Shaw spoke of sitting in front of a projector, which strongly suggests he was being scanned by a flying-spot television scanner. If we are to accept that Baird was secretly using this form of scanning as early as 1925, there remains one final puzzle to solve. How did Baird conceal that he was employing the flying-spot scanner from the newspaper press, competitors and fifty members of the Royal Institute on 26 January 1926?

We will never know for sure, but I offer this theory: It would be important to install an inefficiently positioned floodlight to act as a decoy, carefully designed to illuminate the subject while having little or no effect on the hidden photocells. Light from the nearby floodlight would be positioned to provide sufficient glare and obscure the subject from the functional light, projected from the flying-spot scanner. To be successful the floodlight should be powered from a DC source to ensure that mains ripple did not reach the photocells and obscure the picture. Fortunately DC consumer mains supplies were common during the mid 1920s, but Baird would need to carry an AC to DC power supply when demonstrating in AC mains locations. If a quantity of floodlight inadvertently reached the photocells then this would only result in a negligible steady DC current level being superimposed on the returned AC television signal.

To recover a television image signal, an arc lamp would be positioned inside the apparatus at a position behind the scanning disc. Awareness of the travelling flying-spot would be reduced by the glare from the apparent floodlight and the carefully positioned angle of the beam. Reflected light from the subject would contain two components, a small steady brightness from the decoy floodlight and a more powerful alternating brightness from the flying-spot scanner constituting the image data. The photocells would convert this to a corresponding electrical current to be coupled to the first stage of an amplifier where a DC blocking capacitor would ensure that only the television component would be amplified.

If this clever method of concealment was used, it was sufficient to convince scientific and learned members of the Royal Institution that the first system of television, capable of televising a moving image in light and shade had been achieved by the floodlight method. Baird's decision not to reveal the secret of his success gave him a fourteen-month lead during which time he held a monopoly on the development of television.

Despite the intervening politics and problems related to the formation of the first television service, Baird found time to work on a plethora of television related projects as will be seen in the next chapter.

References

[1] *Sermons, Soap and Television,* John Logie Baird, Royal Television Society, 1988, p42.
[2] *Vision Warrior*, Tom McArthur and Peter Waddell, Scottish Falcon Books, Orkney Press, 1990, p84.
[3] *Vision Warrior*, Tom McArthur and Peter Waddell, Scottish Falcon Books, Orkney Press, 1990, p84.
[4] Shelford Bidwell, *Nature,* 4 June 1908.
[5] *Television Baird*, Margaret Baird, Haum, Cape Town, 1973, p41.
[6] *Sermons, Soap and Television,* John Logie Baird, Royal Television Society, 1988, p42.
[7] Charles Cooney. Boot and shoe having inflated air cushion inserted in the sole and heel thereof, *US1506975*, 21 August 1922, issued 2 September 1924.
[8] Alfonso Miceli. Shoe Attachment, *US1516395,* applied 14 November 1923, patent accepted 18 November 1923.
[9] *Sermons, Soap and Television,* John Logie Baird, Royal Television Society, 1988, p54.
[10] Mr Tree, nicknamed 'Mr Twigg' by Baird in his autobiographical notes is corrected in *John Logie Baird: a life,* Antony Kamm and Malcolm Baird, NMS Publishing Ltd, Edinburgh, 2002 p54.
[11] *Baird of Television*, Ronald Tiltman, Seeley Services, London, opp p64.
[12] *Vision Warrior*, Tom McArthur and Peter Waddell, Scottish Falcon Books, Orkney Press, 1990, p126.
[13] *John Baird: The romance and tragedy of the pioneer of television*, Sydney A Moseley, Odhams Press, 1952, p68.
[14] *John Baird: The romance and tragedy of the pioneer of television*, p69.
[15] *Sermons, Soap and Television,* John Logie Baird, Royal Television Society, 1988, p52.
[16] Television Limited: Public Records Office. London, call number 206588.
[17] *Sermons, Soap and Television,* John Logie Baird, Royal Television Society, 1988. p40.
[18] *Daily Express,* 8 January 1926.
[19] *Vision Warrior*, Tom McArthur and Peter Waddell, Scottish Falcon Books, Orkney Press, 1990, pp102-103.
[20] *Television Baird*, Margaret Baird, Haum. Cape Town, 1973, p60.
[21] Dr A Russell, *Nature*, 3 July 1926, pp18-19.
[22] *Sermons, Soap and Television,* John Logie Baird, Royal Television Society, 1988, p55.
[23] *Sermons Soap and Television*, p57.

[24] John Logie Baird. British Patent Office, London, *GB270222,* applied 21 October 1925, issued 21 April 1927.
[25] *Beginner's Guide to Videocassette Recorders*, E Trundle, Butterworth, London, 1983, p72-75.
[26] Lieut Col Lefroy, Air Ministry File No 43A, Public Records Office in London, call no. AIR2 269/S/24132.
[27] John Logie Baird, British Patent Office, London, *GB269658,* applied 20 January, 1926, issued 20 April, 1927.
[28] *The History of Television 1880 to1941*, Albert Abramson, McFarland, 1987, pp83-84.
[29] Ray Herbert, *Baird Newsletter,* No. 2, November 1983, p3.
[30] GPE Rignoux, French Patent Office, Paris, *FR390435,* applied 20 May, 1908, issued 5 October, 1909.
[31] A Ekstrom, Swedish Patent Office, Number 32,220, applied 24 January, 1910, issued 3 February, 1912.
[32] JH Hammond, US Patent Office, Washington DC, *US1725710,* applied 15 August 1923, issued 20 August 1929.
[33] JE Gardner and HD Hineline, British Patent Office. *GB225553,* applied 28 November 1923, issued 28 May 1925.
[34] John Logie Baird, US Patent Office, Washington DC, *US2006124,* filed 31 December 1926, granted 25 June 1935.
[35] *TV is King,* Jan Leman Productions. BBC, Scotland, April 1994.

4

The Realisation of Television

On 4 January 1926, the Baird Company made an application to the Post Office for a broadcasting licence with a view to creating the first independent television service and to encourage a market for Baird television receivers. Not being covered by the radio broadcasting legislation of the time, television presented an interesting dilemma for the Post Office. It was eventually ruled that: although a licence was not deemed necessary to broadcast a picture alone; it would be necessary if sound accompanied the picture. On that basis Television Ltd was granted two licences with the callsigns 2TV for their headquarters at Motograph House and 2TW for a station at Middle Road, Harrow, Middlesex.

PRIOR TO 1922, THE ACOUSTIC PHONOGRAPH and the telephone had been considered as the hi-tech domestic appliances of the day. By 1926 wireless was developing into a new mass media news and entertainment system. This set the scene for the Post Office issuing Baird with the first television broadcasting licenses on 5 August, which allowed him to operate with a maximum transmitter power of 250 watts on the 200 metre radio band. Both television stations began transmission on 10 August, and from October 2TV experimentally operated twice daily during weekdays.

To place this in context, when John Logie Baird registered the first independent British television company 'Television Limited' on 11 June 1925, less than three years had passed since the formation of the British Broadcasting *Company*.

Unlike the current British Broadcasting *Corporation*, the first 'BBC' was a commercial concern established on 18 October 1922 by a consortium of radio manufacturers to encourage sales of radio receivers.

The main radio manufacturers represented had both British and American interests: Metropolitan Vickers, originally British Westinghouse, a branch of the American Westinghouse Company, and subsequently owned by General Electric USA; British Thomson-Houston, also bought by GE of America; British GEC, 60 percent of whose shares were owned by GE USA; The Radio Communication Company, specialising in naval radio, which had information and patent swaps with Metro-Vickers; Marconi in Britain whose American firm was forced in 1919 by US Government order to join with GE to form RCA; Western Electric, also American-owned, part of the American Telephone and Telegraph Company who were linked by patent exchanges and trade agreements with GE USA in the 1930s.

Following the appointment of the General Manager, John Reith and the Chief Engineer, Captain Peter Eckersley, the BBC was formally registered as a limited company on 15 December 1922. The first BBC was financed partly from the sales of radio receivers and also from a Post Office licence fee of ten shillings (50p) collected annually from the owners of radio sets. BBC radio station 2LO (London) went on the air on the 14 November 1922, followed by 5IT (Birmingham) and 2ZY (Manchester).

Wireless listening figures steadily increased to 18,000 by October 1922 and during 1925 it was recorded that 55% of the population owned a receiver [1]. On 20 December 1926, only eleven months after Baird's first public demonstration of television, the British Broadcasting Company was dissolved to make way for the British Broadcasting Corporation established as a non-commercial public service under a Royal Charter.

As Baird struggled to achieve a domestic television service the radio industry paid little or no attention to the future potential of the new technology. The tiny red/orange 'living' pictures consisted of two-thousand one-hundred picture elements and despite the small number of only thirty scanning lines and some flicker, portraits were televised by wire with remarkable clarity.

Signals produced by this method consisted of frequencies which conveniently satisfied the requirements of a good audio amplifier, but broadcasting images by radio was a different proposition. Although only a few homemade television receivers existed at that time, wireless listeners were able to tune into the strange warbling tones from Baird's preliminary test transmissions. The spectrum of signals generated by 30-line television pictures ranged from 13.5Hz to approximately 13.5kHz (kilocycles per second) producing strange rhythmical sounds from a radio loudspeaker.

While this suggests that television signals are similar to the sound normally broadcast using a medium waveband radio transmitter, the resolved images lacked their original fidelity. Sound reproduced by wireless broadcasting on the medium waveband is of much lower quality than sound amplified using a good audio amplifier.

Received images were noticeably degraded by the signal bandwidth limits imposed on the medium waveband by international regulation. Despite there being a maximum channel allocation of 9kHz, the amplitude modulation process generates a double-sideband limiting the highest frequency that may be broadcast to 4.5kHz.

Although being of minor importance to the intelligibility of speech and while rendering music tolerable, the filtering out of Baird's television components above 4.5kHz removed three-quarters of the original content and all of the finer detail of an image. To compensate for this loss of definition, performers wore thick lines of facial make-up to exaggerate their features. Although the early television performers appeared rather bizarre in the studio they were surprisingly recognisable on the screen.

Barton Chapple: *"With an amplifier which does not cut off signals below 20 kilocycles, the wealth of resulting detail is astounding. We are, nevertheless, at once up against the 9 kilocycles limit now allowed to broadcasting stations. Even with this cut off artificially imposed as a result of deliberations based entirely upon the considerations of aural broadcasting, the detail which comes over is particularly good."* [2]

CHAPTER 4: THE REALISATION OF TELEVISION

In his autobiographical notes Baird indicated that the first television service through the BBC was more of a philanthropic cause than a profit making concern [3]. There was a surge of publicity followed by criticism and disdain, some of which eventually gave way to the acknowledgment that the successful development of television may have genuine potential for the generation of wealth. Ironically the positive publicity that Baird's television generated became the catalyst for competition that would eventually break his monopoly.

Support is not always a prerequisite of achievement, and while John Logie Baird's television schemes may have been revered in some quarters they were heavily criticised in others. Initially undermined when the quality was unfairly compared to the higher definition available from the cinema screen, it was argued that television although still under development, had advantages over the 'big screen' such as accompanying sound and spontaneity.

The silent movies then available were indeed visually superior, but it is important to keep the facts in perspective. Baird's demonstration of television on 26 January 1926 was a remarkable achievement considering that the first commercially successful feature film with sound and vision *The Jazz Singer* was not released until 1927.

The founding company Television Ltd established the Baird Development Company in 1927, Baird International in 1928 and Baird Television Ltd in 1930. Baird International's purpose was to establish television companies in France, Germany, Canada and the USA.

The electronics and wireless giants in the USA, Bell Laboratories and RCA with enormous resources at their disposal presented the greatest competition. Alternative systems for mechanical cameras and receivers emerged, image quality gradually improved and the first generation of higher definition [4] television was on the horizon. Leaders in electronic imaging, Farnsworth, Zworykin of RCA, Marconi-EMI and others would eventually demonstrate that the entire process of television from camera to screen could be made inertia-free, consisting of no moving parts.

As the industrial campaign for television raged in the USA, Baird and Hutchinson struggled to receive recognition and support from the British scientific community and the radio industry. There can be no doubt that between 1924 and 1930 Baird would have been better employed concentrating on the development of electronic television, but was more inclined to explore other avenues. This would result in his company lagging several years behind the competition in the search for a useful electronic television camera [5].

In 1926, the public relations for Television Limited became the responsibility of partner and business director, Captain Oliver George Hutchinson. Burns indicated [6] that Hutchinson's appointment enabled Baird to concentrate on his inventive genius while Hutchinson was able to attend to business matters. Burns is critical of the methods applied to promote television:

"Unfortunately, Hutchinson's business and publicity methods were justifiably regarded with some suspicion and concern in certain quarters, notably the BBC, and these caused Baird to suffer some criticism. There is little doubt that Hutchinson engaged in gross exaggeration to advance Baird Television." [7]

Hutchinson wrote to the Secretary of the Post Office on 11 January 1926, indicating that the television company had made 500 receiving sets. Although difficult to dispute at the time, commercial television receivers for public consumption were not available until February or March 1930.

Exwood [8] revealed that some members of the Royal Institution asserted that Baird and his associates used the Institution's name to promote Baird Television. This was in relation to demonstrations given before an audience of members of the Royal Institution. The first of these took place on 26 January 1926 at 22 Frith Street and the second on 30 December 1926 at Motograph House, when 'Television in the Dark' using infra red light was first demonstrated. The situation came to a head in June 1928 when Baird was asked to explain a quote he made for an article in *Television*:

> *"I gave a demonstration of Noctovision to the Royal Institution."* [9]

Exwood [10] explained that the reference of a 'demonstration being given to the Royal Institution' infuriated A A Campbell-Swinton, member and former manager of the Royal Institution, who on 12 June 1908 published in *Nature* advanced ideas for television suggesting an electronic pick-up with a cathode-ray display. Swinton followed this on 7 November 1911 by delivering a presentation entitled 'Distant Electric Vision' to the Roentgen Society.

Swinton never made any claim to building or testing his suggested system and made it clear that it was an idea alone requiring a great deal of experimentation and modification. It would not have pleased Swinton that Baird achieved television by a method he had personally written-off and published as impractical in 1908 [11].

While no such electronic system was then capable of producing comparable results to Baird's television pictures of 1926, similar concepts to Swinton's had already been demonstrated with some results in the USA. In 1925 Zworykin of Westinghouse demonstrated a crude electronic television system to his employers, but they were unimpressed by the results and told him to concentrate his efforts on something more useful. Zworykin would move to the Radio Corporation of America (RCA) where in the 1930s he would refine his ideas with more success.

Two factions emerged, those supporting Swinton's ideas and those impressed by Baird's achievements. This continues to be the basis of much debate. Swinton, who was potentially one of Baird's competitors, wrote to Sir William Bragg, Fullerian Professor of Chemistry and member of the Royal Institution. He was concerned that statements published about Baird's achievements, in relation to the Royal Institution's name, may be being used on the Stock Exchange to fleece the public. In response, Sir William Bragg wrote tactfully to Baird:

> *"It is quite true that you were kind enough to invite members of the Royal Institution, but no such demonstration was given at, or to, the Royal Institution."* [12]

Baird immediately agreed and had a correction printed in the July 1928 issue of *Television* magazine.

CHAPTER 4: THE REALISATION OF TELEVISION

There can be little doubt that these events played a part in the legacy of Baird's achievements being disregarded by some members of the professional community. Damage to a reputation is not easily repaired as demonstrated by an anti-Baird lobby that continues to exist.

It must be said that fault also lies on the side of Baird who seems to have been rather too eager to leave the politics to Hutchinson. Baird's compelling interest lay in the workshop of television and he considered board meetings to be analogous to going to church, functions to be slept through [13].

Baird had been absorbed in working on improvements and new applications for television. On 26 January 1926, *The Morning Post* published an article indicating that Baird had developed a stereoscopic, three-dimensional television system. It was revealed on 10 and 28 August 1926, that Baird had been experimenting with colour and 3D television from as early as 1925 [14].

Television Limited's monopoly was broken on 7 April 1927 when it was revealed that Bell Telephone Laboratories had discovered the advantages of Baird's flying-spot scanner.

"The Bell Telephone Laboratories of the American Telephone and Telegraph gave a demonstration of their television system over both wire and radio circuits. The demonstration was under the direction of Dr Herbert E Ives and Dr Frank Gray." [15]

Bell Laboratories had demonstrated closed circuit and broadcast television in light and shade. Pictures and sound were sent by wire from Washington DC to New York City and by wireless from Whippany, New Jersey to New York City. The 50-line pictures sent by wire over a distance of two hundred miles, required three separate lines, one for sound, one for synchronisation and the other to carry the television signal. The wireless broadcasting station 3XN sent images on 191 metres, with synchronisation on 1600m and voice on 207m over a distance of 22 miles.

When reports of the American achievement reached Baird's board of directors they were devastated and threatened to withdraw support unless Baird could redress the balance. Concerned by the challenge, Baird took immediate steps to demonstrate that the British were still ahead of the race and restored confidence on 24 May 1927 by transmitting television images by telephone from his laboratories in London to the Central Hotel in Glasgow. 438 miles became the new record for broadcasting television. In response to the report that Bell Labs had used three telephones lines and one thousand men to service their demonstrations, Baird proudly indicated to the board that he used two operators and two telephone lines.

With renewed confidence the Baird directors decided to protect their investments. The board was concerned about the vulnerable nature of their new business with regard to the frailty of their single most valuable asset, 'John Logie Baird' and insured his life for the sum of one hundred and fifty thousand pounds. Baird wrote:

"Two doctors prodded me about, they whispered together, did more prodding and listening, whispered again, obviously they did not like the proposition,

also obviously they were reluctant to turn down a magnificent bit of business, £150,000 policies are not dropped without very excellent and unanswerable reasons. Finally the insurance company decided to take the risk for twelve months at a whacking premium, £2,000 I think it was and the situation was saved." [16]

John Logie Baird would have preferred that television broadcasting be developed independently from the radio monopoly of the British Broadcasting Corporation, but circumstances made this difficult. In 1927, signals from the Baird television transmitting aerials at Long Acre caused interference with the Admiralty's radio communications at neighbouring Whitehall. The Admiralty, who also used the medium waveband, complained to the Post Office and notice was given to Baird to close down transmissions.

To resolve the problem two actions were taken. First it was decided that tentative arrangements should be made to transfer the television transmitters and aerials to Hendon and secondly, Television Limited opened up discussions to broadcast Baird's television programmes through the BBC.

Exwood: *"It appears that in June or early July 1927 Baird succeeded in persuading H L Kirke (Head of the BBC Research) to put television transmissions out over 2LO and this was done on three occasions before Kirke was stopped. I knew H L Kirke as an independent spirit and it seems quite in character that he arranged the experiments without consulting anyone above him."* [17]

Moseley: *"Let there be no mistake about the fact that the successful development of Baird's system of television was dependent utterly upon the goodwill of those who controlled the broadcasting monopoly in Great Britain. Baird had made many successful transmissions and had given notable demonstrations of his invention, and these had enabled him to raise very large sums of money from the public. But the hard fact remained that he could not make a commercial success of his invention unless the BBC allowed it to 'go on the air' or, most unlikely, television was granted an independent entity - which Baird preferred."* [18]

The major stumbling block between Baird and the BBC appeared to be Sir John Reith, the Director General of the BBC. Baird recalled that twenty or so years had elapsed since he first met Reith as a fellow student at the Royal Technical College where he found him to be bullish and self centred:

"Reith did not distinguish himself in his examinations; he was worse than I was without the excuse of ill health.
The examiners awarded no marks for impressive appearances, no marks for oracular booming voices, no marks for influential relatives." [19]

In his autobiographical notes, Baird indicated that during his first meeting with the BBC, Sir John Reith was affable, cordial and offering support. They parted company after the meeting in a friendly fashion, but this was an attitude that would not continue. Baird was suspicious there were too many agendas belonging to Hutchinson, Moseley and several members of the BBC. It has been suggested that there was a fear among some members of the BBC board of directors who maintained their commercial connections, that television could jeopardise the wireless industry. These were only suspicions, but it was an argument that also worked in reverse; the Baird's television display system the 'Televisor' was designed to work in conjunction with the wireless and could therefore benefit the market.

Perhaps television would have been perceived in a different light if Baird had been part of the radio industry and not an independent entity. With a Government order demanding that television transmissions from Long Acre should remain suspended, the BBC responded from a more comfortable position and refused their support. But the BBC had seriously underestimated Baird's determination. International medium-wave broadcasting is easily possible in the evenings, which set the scene in July 1929 for the first pirate television broadcast by Baird from the German Broadcasting House in Berlin. The successful reception of television programs in Britain originating from Baird in Germany brought embarrassment and a sudden change in the attitude of the BBC.

"Whatever the cause, the BBC certainly reacted swiftly." [20]

As a result, the BBC immediately adopted Baird's television with the first broadcast being sent out on 30 September 1929, albeit on the restricted scale of five hours a week and outside normal radio broadcasting times. It had taken three difficult years of campaigning by Baird, Hutchinson and Moseley, but this was not the full service that Baird had expected.

Despite Baird's first ever demonstration of simultaneous sight and sound during test transmissions to the BBC on 5 March 1929, and a showing of 'Tele Talkies' on 19 August 1929, the first 'official' television broadcast on 30 September 1929 from the BBC's 2LO transmitter could not offer the facility of accompanying sound.

When a single wavelength is utilised to convey both sound and picture they have to be sent alternately first with the picture on the screen followed by the sound component at the loudspeaker. Unfortunately when viewing the picture the unpleasant noise of the vision signal is impressed on the loudspeaker and when the sound is heard the screen flashes with interference from the sound signal. This retrograde step did little to further the public perception of the art.

Simultaneous transmissions of sound and vision were delayed until 31 March 1931 when the BBC station at Brookmans Park was put into operation and used to serve as the sound channel [21]. These delays did little to assist the credibility of the now struggling Baird Company and only added to the frustration of Baird. Ian Anderson, a partner in Vowler & Company and a major Baird investor wrote:

"On or about 1 March 1930, a scheme of arrangement and amalgamation between the original company Baird Television Development Co Ltd, and Baird International Television Co Ltd, came into force and the capital of the International Company was reorganised so as to be £825,000."

Anderson continued:

"Baird and Hutchinson were allotted some three or four hundred thousand one shilling deferred shares in the original international company, and these, under the scheme of amalgamation before referred to, were converted into some other similar stock. Baird and Hutchinson were appointed joint managing directors and were given contracts for a period of at least five years. Their actual remuneration I am afraid I do not know, but it must have been quite substantial." [22]

Anderson was authorised on behalf of an important company (which remains unknown to the author) to offer Baird and Hutchinson one hundred and twenty five thousand pounds each for their shares. Captain Hutchinson regarded one hundred and twenty five thousand pounds as completely inadequate and refused to sell, while Baird replied:

"Oh, Mr. Anderson, £125,000 - I just don't know what I would do with that amount of money and I would just hate to have it - I would not be able to sleep at night."

In his autobiographical notes Baird later wrote with some regret at having turned down a small fortune:

"If an inventor reads this book, let him by this be admonished to do what Graham Bell (the inventor of the telephone) did, and sell at once for cash. Inventors are no match for financiers where stocks and shares are concerned, and will if they hold on, find that the financiers have the cash and they have the paper." [23]

This is a revealing insight into the contrasting, motives, financial interests and expectations of Hutchinson and Baird.

On 14 February 1930, Baird took out a futuristic patent for an active television screen to be used for televising the Derby in June 1931. The illustration in **Fig 4.1** shows that the screen is made up from a matrix of small cubicles controlled by shutters each representing a picture element. At a first glance you would be forgiven for confusung it with the illustration of a modern flat panel colour television display. There are also similarities in its operation to a plasma panel and a shuttered LCD display. Further to this the shutters could be made to vibrate at audio frequencies to enable the entire screen to act as a large loudspeaker and an optical device may be used to magnify the resulting picture on a larger screen.

The intensity of publicity surrounding the first method of television resulted in John Logie Baird becoming synonymous with mechanical schemes of low-definition television. It is therefore, not surprising that a myth evolved depicting John Logie Baird as a string, sealing wax, and cardboard inventor, whose only useful contribution to the industry was through an ability to raise public awareness.

There can be no doubt that his disregard for personal appearance, his tousled and unruly hair, captured so accurately by artists such as Edmund J Sullivan,

CHAPTER 4: THE REALISATION OF TELEVISION

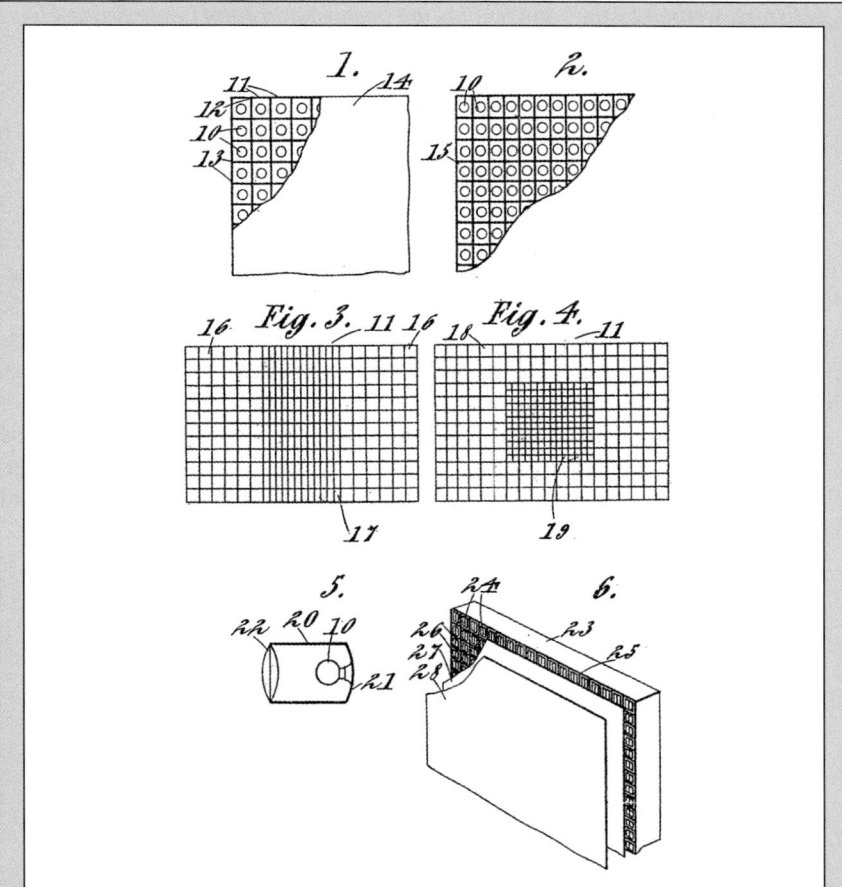

Fig 4.1: Abstract from Baird's patent *GB348211* for an active panel display used for showing the end of the Derby in June 1931

In a television system in which the signals are applied to a receiving system screen comprising a bank of lamps or a bank of shutters operating in conjunction with a constant light source the lamps and/or the shutters are arranged in cells to prevent the light emitted from any lamp reaching a part of the image field assigned to another lamp. The lamps may be silvered at the back, blackened at the sides and frosted on the front surface to enable a whole image to appear on a diffusion screen.

Anatas Botzaritch Sava and Emilio Coia (see page viii) perpetuate the colourful and eccentric image of Baird.

Publicity, encompassing the all-electronic and vacuum tube work of the Baird Company, carried out by S A Moseley in the magazine *Television* may have attempted to update the popular view, but his audience was limited to readers of the scientific and pseudo scientific press.

Baird became a legend in his own lifetime and was very quickly written off by 1937 when the BBC selected Marconi-EMI's electronic system of television in preference to Baird's high definition system. In 1986, William Phillips summed up the legend of Baird:

THE THREE DIMENSIONS OF JOHN LOGIE BAIRD

"History, written by the winners, has not been kind to John Logie Baird. His work is customarily dismissed as a tragic-comic dead end. The self-taught inventor, poor, sick and slightly cracked, blunders about in attics with bicycle batteries and hat box lids. He cobbles together a crude mechanical TV system, doomed to be swept aside by a giant industrial joint venture's research laboratories. RIP, Baird; all hail, Blumlein and Browne." [24]

Blumlein and Browne worked for McGee who led the development of the EMI Emiscope electronic television camera, based on RCA's Iconoscope. Phillips may have known of Eckersley's condescending opinion of John Logie Baird. Captain Eckersley, the chief engineer of the BBC and initially against television, wrote articles casting doubts with implications that there was something fishy about the whole business.

His stance as principal critic of the Baird system apparently changed after he witnessed an experimental broadcast of television through the BBC transmitter, 2LO in 1927. But despite this and an ensuing friendship with Baird the mould had been firmly cast.

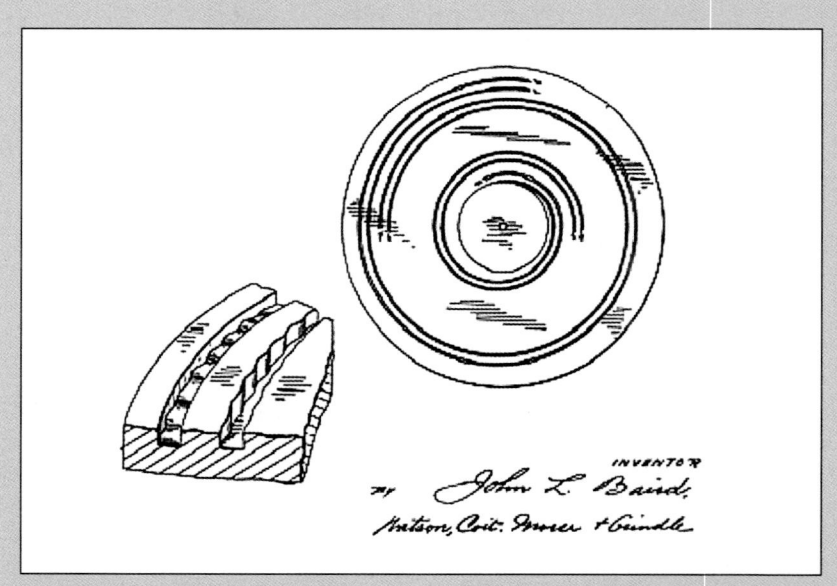

In a television system records may be made of the electrical fluctuations caused by the image traversing the cell. These records may be made by phonograph or by other like means and these records can be used to actuate the receiving television machine and reproduce the image . . . .

It also lies within the scope of this invention to combine the two records in a single grove, using the side walls of the groove for say, the sound record and the bottom of the groove for the sight record.

Fig 4.2: Abstract from Baird's patent *GB289104* describing a sound and vision videodisc

CHAPTER 4: THE REALISATION OF TELEVISION

In a letter to Sydney Moseley after Baird's death, Eckersley wrote:

"Neither Baird nor Marconi were pre-eminently inventors or physicists; they had, however, that flair for picking about on the scrap heap of unrelated discoveries and assembling the bits and pieces to make something work and so revealing possibilities if not finality." [25]

Baird continued to explore the wider applications of television and on 15 October 1926, applied for a British patent describing a videodisc recording system called 'Phonovision.'

By sending the signal from a televised image to a loudspeaker he discovered that each scanned object produced a unique audible sound. He was fascinated by this and claimed to be able to recognise some items by their sound alone. From this Baird conceived the idea of recording the audio signal emanating from the television images directly onto the surface of a 78rpm gramophone record disc. These recordings would effectively constitute the world's first videodisc system.

The concept involved replaying the videodisc to reconstruct the original moving images and accompanying soundtrack on a 'Baird Televisor' receiver. In his British and US patents, Baird described a method by which sound and vision were recorded simultaneously.

The illustration (**Fig 4.2**) from the original patent shows that the recording technique involved a double spiral with 'hill and dale' recording in one groove and 'sidewall' recording on the other.

Bairds patent *GB289104* [26] issued on 16 April 1928 is the first known published description of a practical dual track recording system (see Fig 4.2), anticipating the stereophonic recording technique later attributed to Alan Blumlein of EMI [27].

Although phonovosion did not reach a commercial stage, a number of amateurs and enthusiasts recorded the Baird television signal at home using stylus disc cutting equipment. Baird had a further surge of interest in 1930 when he applied for a second phonovision patent *GB324049* describing a portable Phonovisor player as shown in **Fig 4.3**.

The main concept of *GB324049* was the built-in monitor to reproduce the replayed programmes of sound and vision directly through a magnified viewing window without the requirement of a further screen.

In 1986 a working exhibit, that became part of an exhibition that toured Britain for two years, was developed by the author to celebrate Baird's Phonovision. For simplicity and reliability the gramophone disc was replaced with a cassette recorder and the mechanical components replaced with an oscilloscope. This project used virtually the same analogue techniques as Baird's original device by recording two tracks of data on magnetic tape in preference to the 78rpm disc.

Fig 4.4a is a photograph taken of an image recovered from an old phonovision disc showing the 30-line cartoon image of silent movie comedian Charlie Chaplin, while **Figs 4.4b and 4.4c** show 30-line images originated by the author and played back from cassette tape. Appendix 1 describes an analogue method for decoding images from existing phonovision discs.

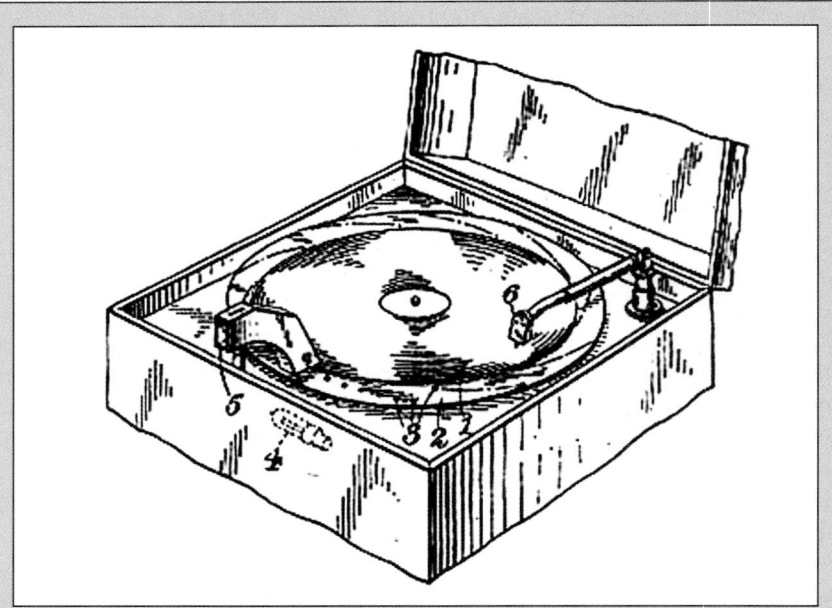

Fig 4.3: Abstract from Phonovisor patent *GB324049* (January 1930)

In an apparatus wherein vision is recorded upon a gramophone-like record, to aid in the reproduction of the visual image there is attached, in accordance with this invention, to the turning table of a gramophone, a disc containing a number of spirally-arranged sets of holes disposed consecutively round the circumference of the disc. Each spiral corresponds to one complete traversal of the image. Thus if, say, five spirals are arranged round the circumference, five explorations are accomplished at each revolution, thus obviating the need of any gearing in the recording or reproducing mechanism. Alternatively, the spirally arranged sets of holes may be formed in the margin of the record disc itself.

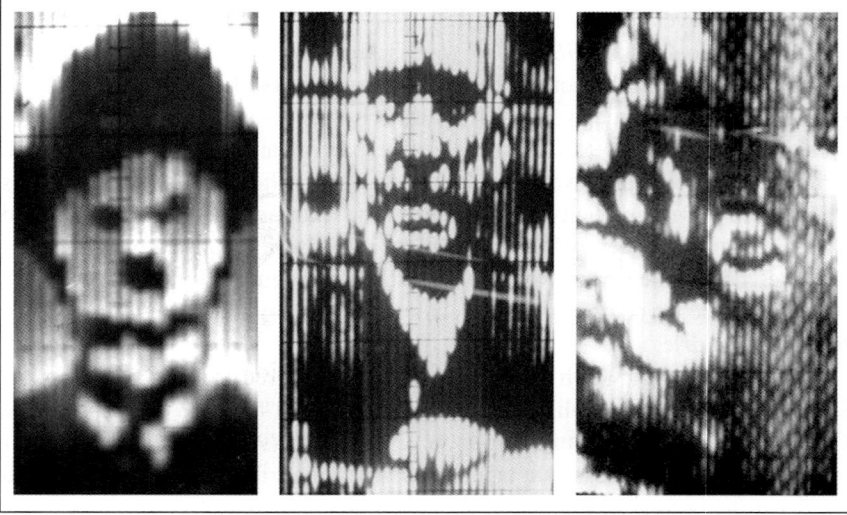

Fig 4.4: (a) 30-line image of Charlie Chaplin recovered from a 78rpm videodisc; (b) 30-line image of a man holding a pen, scanned by the author and recovered from an audio cassette tape; (c) 30-line image of a cat, scanned by the author and recovered from an audio cassette tape

CHAPTER 4: THE REALISATION OF TELEVISION

> *To produce an image without the use of a lens, a screen may be made out of a honeycomb device made for example by a large number of tubes. These tubes might be 1/10th inch diameter and 2 inches long and the screen might contain a bank of say 10,000 of them placed side by side longitudinally to form a square of 1000 a side. If a ground glass screen is placed at the back of such a screen an image appears on it of any object placed in front.*

Fig 4.5: Abstract from Baird's patent *GB285738*

For further information on the Baird videodisc system, Donald McLean (G2TV) has published details of extracting, enhancing and reproducing historic phonovision images from the dawn of television history, using digital technology [28].

On 15 October 1926, Baird, with typical foresight, applied for British patent *GB285738* describing a concept that would have major future implications for the medical and communications industry. This invention described a bundle of small metal tubes or quartz rods similar to a modern endoscope and capable of resolving an image in place of a lens (**Fig 4.5**).

In describing the advantages of total internal reflection Baird indicated that the thin flexible rods or tubes of glass, quartz or other transparent material could be bent or curved. The fibre-optic bundle was to be placed between the scanning disc and the photocell in an attempt to reduce critical light loss due to the inverse square law. An image presented at one end of the bundle was viewed correctly at the other end of the bundle in the same manner as a flexible endoscope.

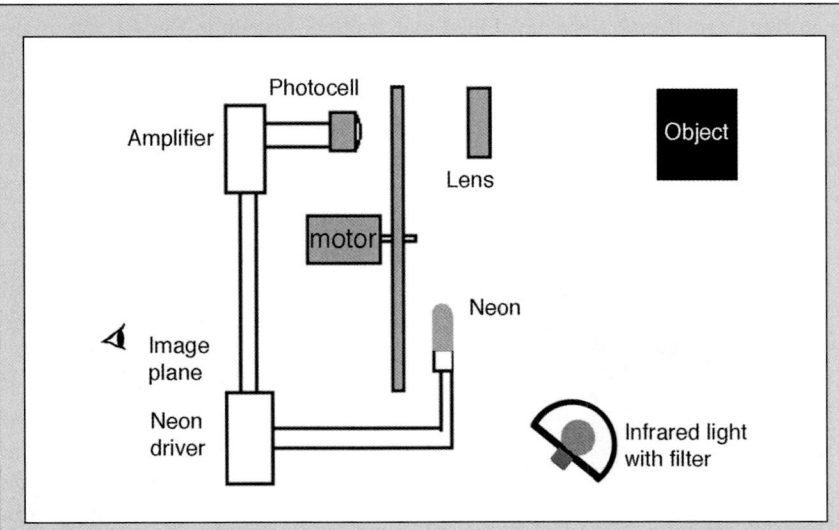

To see an object in darkness the infra-red or ultra-violet rays may be directed on the object or scene, which may then be received by a television apparatus using a cell sensitive to such rays. Thin sheet ebonite may be used as a filter for the infra-red rays.

Fig 4.6: Abstract from Baird's Noctovisor patent *GB288882*

Although it had been known that light could travel along a jet of water since the 1840s, Baird's use of the concept of total internal reflection is thought to be the earliest practical application of a coherent fibre-optics bundle.

The third patent to be applied for by Baird on 15 October 1926 was for a night vision viewing scope called a 'Noctovisor'. The completed specification was left on 15 August 1927 and accepted on 16 April 1928 as *GB288882*. This was a method of seeing in the dark by television where the object to be viewed was floodlit or scanned by a flying-spot using a lamp rich with infra red light as shown in **Fig 4.6**.

To achieve this, Baird operated the apparatus in complete darkness and covered the infra red light source with a pitch or ebonite filter to remove any visible wavelengths of light, rendering the scene effectively unlit with respect to human vision. In uncharacteristic fashion, Baird took the unusual step of demonstrating the concept prior to its final acceptance and publication on 16 April 1928.

A demonstration was given to Alexander Russell and W R Crookes for *Nature* on 23 November 1926, only one month after the date of the provisional patent application. Details were delayed until 5 February 1927, after a further demonstration of night vision at the British Association Meeting in Leeds in 1927 created quite a stir in the scientific world. Russell's letter to *Nature* stated [29]:

> *"The application of these rays to television enables us to see what is going on in a room which is in complete darkness. So far as I know, this achievement has never been done before."*

The Noctovisor of 1926, which produced the ghostly appearance of a subject was greatly improved by 1929 when a new generation of photocells with higher infrared sensitivity were developed and applied. Shown in Fig 4.6, the night vision device used a single rotating disc, with two sets of spiral holes enclosed in an otherwise light-sealed camera fitted with a lens. The first set of apertures scanned the infrared illuminated image while the other set of scanning holes on the disc were used for the display, ensuring perfect mechanical synchronisation.

Fig 4.7: The camera section of Baird's Noctovisor

The lens focused the infrared light being reflected from the subject onto the appropriate portion of the disc, where it was converted to an electrical signal by a sensitive photocell to be amplified.

The signal from the photocell amplifier was connected to the lamp driver circuit, which modulated the neon lamp positioned on the front face of the disc. The invisible image was rendered visible by viewing the modulated neon lamp through the second set of scanning apertures. This was achieved by means of a window located at the rear of the unit opposite the lower part of the disc.

CHAPTER 4: THE REALISATION OF TELEVISION

Fig 4.7 shows the camera section of the Baird night-vision scope, which was normally mounted on an adjustable base for accurate vertical, horizontal and rotational sighting.

While night vision was not a particularly useful addition to television for entertainment purposes it was recognised that it would be useful in the event of a war by secretly viewing an enemy in the dark. In summing up the article in *Nature*, Alexander Russell stated:

"The direct application of Mr Baird's invention in warfare to locating objects apparently in the dark seems highly probable, but I would hope that useful peace applications will soon be found for it."

Anderson who worked with Baird during WW2 wrote:

"John Baird also wanted to use a radio scanning beam and replace photocells with radio receivers, we now know this as Scanning Radar. This was in the early thirties." [30]

Night vision was later developed by the British and American military with no further reference to Baird.

On the 9 February 1928, Baird successfully spanned the Atlantic with television.

"Ben Clapp joined the Company in November 1926 and became the only technical assistant. In 1925 he had obtained special permission from the Post Office to use the unusually high power of 1kW for transoceanic tests from his amateur radio station (2KZ) at Coulsdon. There is little doubt regarding the role which John Baird envisaged for this privately owned wireless equipment." [31]

Baird sent Ben Clapp and Captain Hutchinson to the receiving station of Robert Hart (W2CVJ) in Hartsdale, New York. In the UK, Len Luger from the Marconi wireless station at Croydon Aerodrome, operated Clapp's 2kW transmitting station, 2KZ located at Coulsdon:

"The persons seated in London were clearly seen in New York. At midnight, London time, the image of a ventriloquist's doll was set before the transmitter at Long Acre, sent by land line to Couldson and thence transmitted across the Atlantic on a wavelength of 45metres. The signals were picked up by an amateur operating a station at Hartsdale, a few miles from New York." [32]

The image was resolved clearly and witnessed by a reporter from Reuter's Press. Baird was then asked to sit in front of the camera to become the first person to have his image successfully televised across the Atlantic.

Another experiment took place shortly after the transatlantic television test. This time, Baird's signal was received and viewed on-board a ship in the middle of the Atlantic. A television receiver located on the Cunard liner *Berengaria* enabled passengers and officers to see pictures from the Baird Laboratory in

In 2003 the Antique Wireless Assn Museum curator Ed Gables in New York, together with Narrow Band Television Association enthusiasts, worked together to celebrate the 75th anniversary by recreating the transatlantic television by wireless event on 15 metres. On 6 February the interference gave way to 30 seconds of a good television picture and history was repeated.

53

Fig 4.8: Abstracts from John Baird's patents *GB292185* and *US1699270*

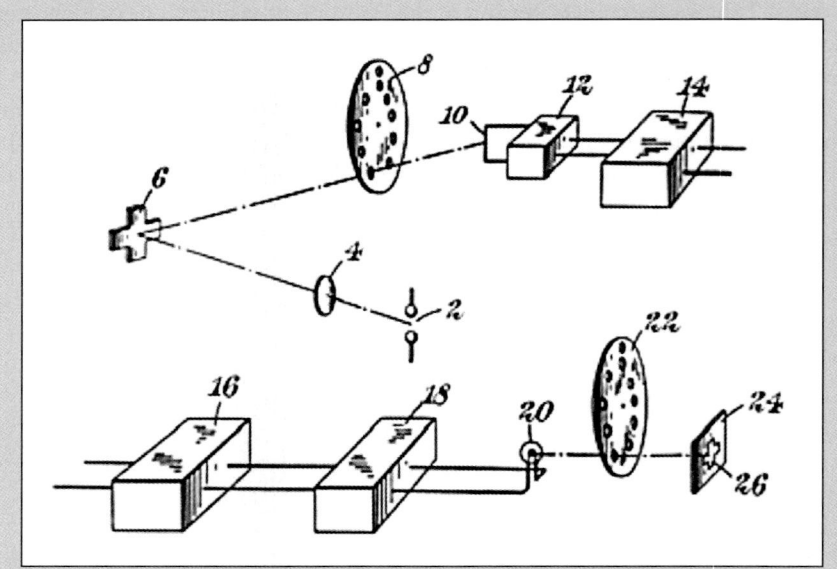

GB292185: It is well known that wireless waves can be refracted and reflected in a similar manner to visible light waves, and we therefore use these waves in place of light waves and replace the light-sensitive cell or cells by a wireless receiver or receivers. Any of the devices known to the art may be used, replacing the light source at the transmitter by a generator of wireless waves of suitable wavelength, and the light-sensitive cell by a wireless receiver tuned to the appropriate wavelength.

US1699270: The invention may thus be used for transmitting the waves, without the risk of detection by the object on which they are projected, and the method is thus extremely valuable in case the invention is used during a war, for instance, where it is desired to view the enemy's position without detection.

London. The chief wireless operator recognised the image of his fiancée, who had been invited by Baird to sit before the transmitter.

Baird was prolific throughout 1928 and on 4 May he described a method of radio imaging, **Fig 4.8**, in his master British patent *GB292185* and US patent: *US1699270*.

In the US patent Baird described how a distant region, or an area of terrain, could be scanned by sending out a narrow beam of electromagnetic short wireless waves and then be imaged from the reflected radio signal.

This was an invention that Baird considered to be of significant enough importance to patent in America. It is interesting that the American patent includes a statement, which suggests the future use of radar in war, but this is not present in the British patent.

It is almost certain that this patent would have been scrutinised by the US and British military and although it was almost certainly a corner stone for the

CHAPTER 4: THE REALISATION OF TELEVISION

development of radiolocation, the questions arise, what exactly did Baird mean by "to view the enemy's position". Does this mean that Baird's radio imaging television system was also capable of ranging? To answer this question I decided to take a closer look at the patent together with the timing of the Baird Televisor screen.

With reference to Fig 4.8, the Baird Televisor may be used as a ranging device by transmitting a 30-line, line sync pulse and locking a local receiver to the pulse. Only the sync locking mechanism is used at the local receiver causing a white line to appear across the top of an otherwise blank televisor screen. This pulse is shown as the transmitted pulse on **Fig 4.9** [also reproduced as a colour picture on page C.1]. A television line will take approximately 2.7mS, which becomes the effective time represented as the height of the screen.

The second requirement is that the receiving televisor remains locked to the local transmitting synchronising signal while being attached to an aerial and awaiting a distant signal. The transmitted electromagnetic radiation of the line sync pulse travels into the distance at the speed of light until it meets with a radio reflective medium, at which point the radio signal carrying the white line of information is reflected back to the source as a radio echo. Provided that the return time falls within the 2.7mS scan time of the line, an image of the reflected white horizontal line will appear some distance down the screen.

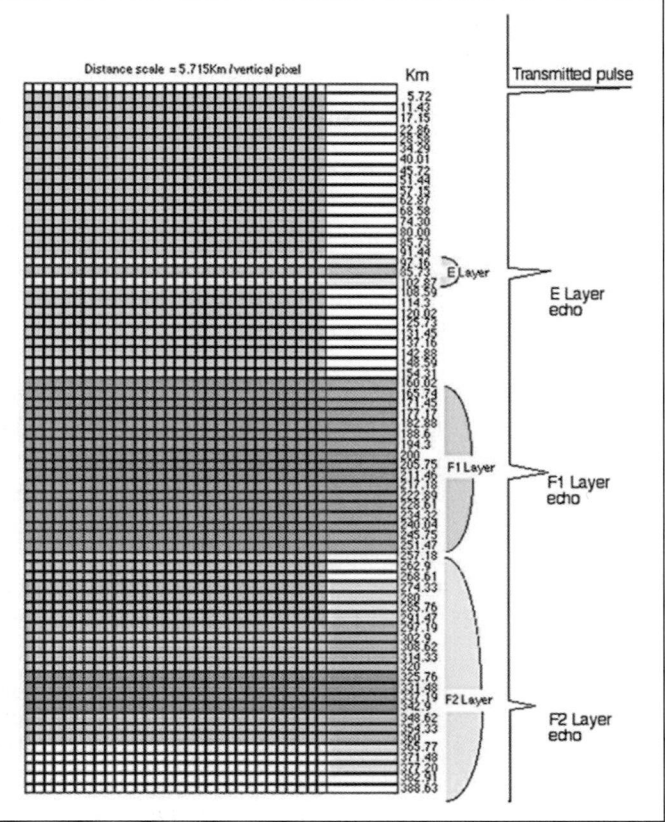

Fig 4.9: Baird televisor screen calibrated for ranging a reflected signal [see page C.1 for a colour version]

To find the range of the object it is a simple matter of measuring the distance from the top of the screen to the received image of the reflected line pulse. This may be done by means of placing a suitable graticule calibrated in kilometres in front of the screen and reading from the scale, where distance equals: C x T, where: C = 300,000,000 metres/second and T = the return path time in mS (one-thousand of a second).

As the total time taken for a reflected pulse to return will be twice the distance from the transmitter the range of the object is calculated by using the formula: C x T / 2.

Fig 4.9 shows a calibrated graticule for a 30-line televisor screen, based on a scale of 30 horizontal pixels by 70 vertical pixels accurately representing the screen. Each horizontal row of pixels relates to an object distance of 5.72km. The scale on the right of the illustration shows the initial pulse time (or zero time) relating to a white bar at the top of the screen. The relative positions that relate to reflections received from the Earth's ionisation layers E, F1 and F2 are shown on the scale. The maximum theoretical range of an object measured using the 30-line standard is 388.63km.

In 1931, Baird Television Ltd was running into financial difficulties due to the high cost of operating the 30-line television service through the BBC.

> *"It was indeed a sorry position, The Baird Company had introduced television to the BBC, but instead of the Corporation's paying us, as our German confreres did, we were actually made to pay the BBC for the use of their transmitter."* [33]

On the verge of collapse, the old parent company Television Ltd, which owned a block of a million deferred shares in the Baird Company went into voluntary liquidation in 1931. These were acquired and held for a short time by Moseley at a cost of £16,500. This was made possible by a cash injection of £5,800 from Isidore Ostrer, President of the Gaumont British Film Company, and a separate financial loan. In January 1932 Moseley sold the shares to Isidore Ostrer, transferring control of the Baird Company to Gaumont British.

John Logie Baird continued with television related projects and was particularly interested in the development of colour and stereoscopic television, which led him to set up a private laboratory adjacent to his home in London. This subject will be dealt with in the next chapter.

One of the first actions of Gaumont British was to hire Captain A G D West as Technical Director, John Logie Baird as Managing Director, Sydney Moseley as Director and T M C Lance as General Manager. The main concern of the company was to enter the electronic age of television.

It was from this point in time that the company began experimenting with 240-line and higher definition television. This was despite an earlier statement in 1931 when Baird said he believed the neon tube would remain the lamp of the home receiver. Later, he regretted having said this, particularly in connection with those wishing to discredit or demoralise. To give Baird some credit, at the time of the statement he was accurately reflecting on the problems that many of his competitors had faced.

In the late nineteen-twenties cathode-ray tube technology was in its infancy, involving short-lived devices requiring continuous evacuation by mechanical pumps to maintain their operation. However, Baird should have been aware that by the early nineteen-thirties the sealed-off cathode-ray tube had advanced sufficiently to implement the eventual demise of mechanical scanning.

It was not long before the Radio Corporation of America (RCA) appeared on the scene to exploit their new electronic television camera the 'iconoscope'. This device developed by Vladimir Zworykin incorporated a major breakthrough in camera technology known as image storage.

A storage camera gains sensitivity by exposing the whole of the image to a matrix of tiny light cells for the duration of the frame-scanning period. By this process the feeble energy from tiny light levels are allowed to accumulate until they reach useful values. The facility of storage resulted in vastly increased camera sensitivity in comparison to a 'real time' scanning disc, which is only

CHAPTER 4: THE REALISATION OF TELEVISION

Fig 4.10: Baird's intermediate film system

capable of exposing tiny fragments of a feeble image to a photocell for a fraction of a second. The Baird Company developed and built their own storage cameras similar to the Iconoscope but was unable to use them commercially due to patent legislation.

Baird's 240-line 'Intermediate film system' (**Fig 4.10**) used a standard 35mm camera, but it was cumbersome and bolted to the floor. The system produced a transmission delay time of 64 seconds between filming and transmission. To ensure synchronisation, the audio sound track was simultaneously recorded onto the film with the image [34]. The film could be used as an endless loop or as a fixed reel length for recording purposes. The problem that faced Baird was in developing a comparable non-storage electronic camera in response to RCA's challenge. While there was no question of obtaining a licence from RCA, the Baird Company employed almost every other conceivable method of television.

Over the years, television cameras have improved and changed in style, but Zworykin's invention of 'storage' has never been superseded.

Storage is a critical concept enabling a television camera to operate successfully under the widest range of lighting conditions. The Intermediate film system used the advantages gained from the high photosensitivity of photographic film, but it also had disadvantages including a total lack of portability. After being 'cine recorded' the film reel was quickly developed, washed, fixed and washed again, but there was a likelihood of unpleasant developer or fixing fluid leaking from the tanks. The film ran at 6,000rpm and was immediately scanned while wet by a 240-hole precision disc enclosed within a vacuum. The light source was derived from a 120 amp arc lamp and the resulting optical signal from the disc was converted by a photomultiplier to produce a clean television signal. This was a process that anticipated by many years the quick film processing methods developed in the 1950s, called 'Hot-kine' used by news services where speed was essential.

In 1934 a cross-licensing agreement with Farnsworth, the American electronic television pioneer enabled Baird Television to manufacture their own version of Farnsworth's electronic camera. To support this, Baird Television hired cathode-ray tube specialist Dr Szegho to establish a high vacuum facility in their new premises at the Crystal Palace in London.

RCA had a cross-licence 'storage' agreement with Marconi-EMI in Britain giving them a substantial advantage over Baird Television. Szegho designed the first Baird cathode ray tubes (cathovisors) and development began on the production of the first quality 240-line all-electronic television receivers, **Fig 4.11**. All of the television parts, both electrical and mechanical, were manufactured in-house with the exception of the amplifying valves. Dr Samson developed the Image Dissector camera tube, which became known as the Baird Electron Camera and Dr Sommer joined him in 1935, to concentrate on the development of photocells and specialised photomultiplier tubes.

The British Government became aware of the developments in higher definition television and asked that an advisory committee be set up to investigate and advise on the standardisation for the future of television broadcasting in Britain. The appointment of the Television Advisory Committee was announced in the House of Commons on 14 May 1934, with the following terms of reference:

Fig 4.11: Example of an all-electronic Baird television receiver

CHAPTER 4: THE REALISATION OF TELEVISION

"To consider the development of Television and to advise the Postmaster General on the relative merits of the several systems and on the conditions under which any public service of Television should be provided."

With higher definition television on the horizon, the future of the BBC low definition service was now becoming uncertain. The *Report of the Television Committee* led by the Right Honourable Lord Selsdon, (previously Sir William Mitchell-Thomson, the Postmaster General from 1924 to 1929), was presented by the Postmaster General to Parliament on January 1935. The first casualty was the existing 30-line television service. In summary it recommended:

a. No low definition system of Television should be adopted for a regular public service.

b. High definition Television has reached such a standard of development as to justify the first steps being taken towards the early establishment of a public television service of this type.

The recommendation for the BBC to discontinue the experimental low definition transmissions after only six years came as a serious blow to John Logie Baird and the Company. Many Baird enthusiasts and supporters of low definition television were to discover that their Baird Televisor receivers, manufactured by Ferranti for Baird Television (**Fig 4.12**) were now useless.

Fig 4.12: The Baird Televisor

The first public television service fell into history having failed to reach any level of profitability and most likely made a serious loss. Revenue obtained from the sale of Baird Televisors, Televisor kits and components had been continually offset by the high charges incurred by the BBC for the use of the transmitters.

Baird wrote:

"We were indeed acting more like a philanthropic institution for the benefit of the television-minded public, than a business concern, for there was a very big television public, even in those days. This was proved not by the number of television sets which we sold, although we had sold nearly 1,000, but the immense number of amateurs who had built their own sets; the sets in these days were comparatively simple and well within the range of the amateurs. The immense public interest is however proved by the fact that when we started a magazine, Television, *150,000 copies of the first issue were sold."* [35]

At close down there were about seven thousand low definition receivers in use [36]. The Baird 30-line televising apparatus returned by the BBC when the transmissions were ceased in 1935 is now in the London Science Museum at South Kensington [37].

"But Baird having lent his name to mechanical television systems inevitably had to face total obsolescence of every piece of equipment when these system were superseded. So his name becomes associated with a system which was abandoned."

The Selsdon Committee gave the following recommendations for the future high definition public service:

"It seems probable that the London area can be covered by one transmitting station and that two systems of Television can be operated from that station. On this assumption we suggest that a start be made in such a manner as to provide an extended trial of two systems, under strictly comparable conditions, by installing them side by side at a station in London where they should be used alternately - and not simultaneously - for a public service. There are two systems of high definition Television – owned by Baird Television Limited and Marconi-EMI Television Company Limited respectively - which are in a relatively advanced stage of development, and have indeed been operated experimentally over wireless channels for some time past with satisfactory results. We recommend that the Baird Company be given an opportunity to supply the necessary apparatus for the operation of its system at the London station, that the Marconi-EMI Company be given a similar opportunity in respect of apparatus for the operation of its system also at that station."

This report was detrimental to John Logie Baird, whose systems and reputation as the 'inventor of television' were for the first time to be challenged by a rival British company. Critics seized upon the opportunity of writing-off Baird's

CHAPTER 4: THE REALISATION OF TELEVISION

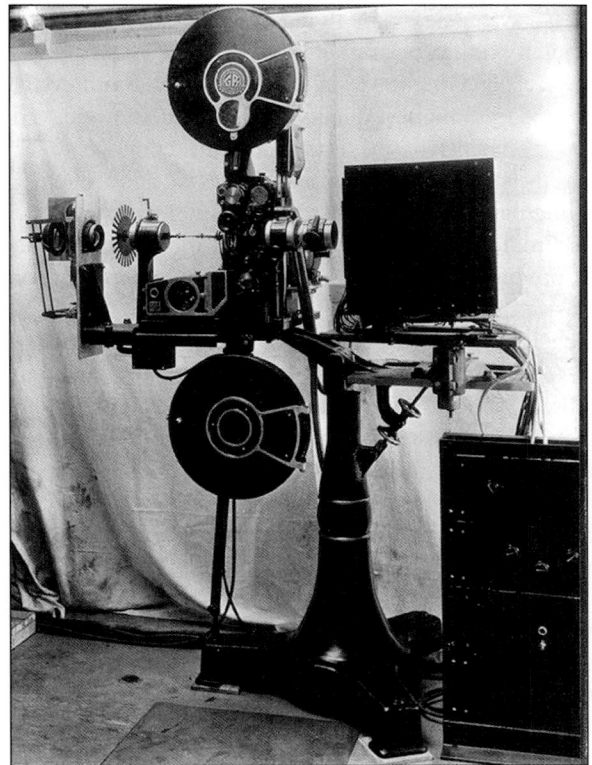 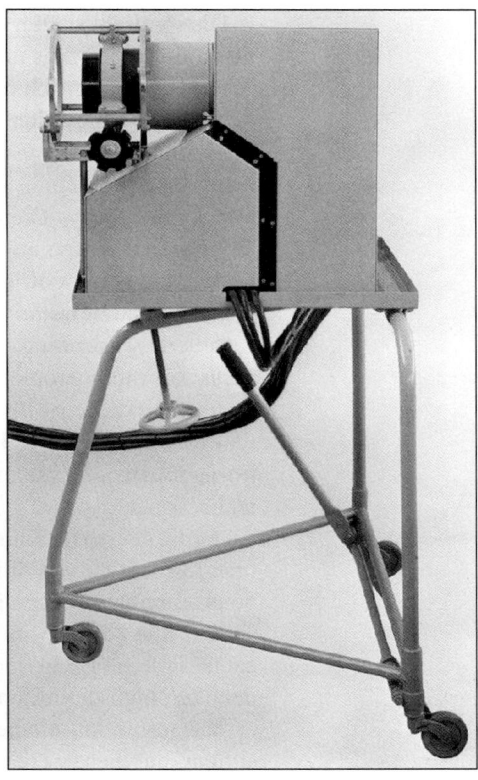

(left) Fig 4.13: Baird Telecine (1939)

(right) Fig 4.14: Baird electron camera

work as a failure at a time when it was not widely known or publicised that Baird Television was deeply involved in the development of high definition television. Baird would discover that despite his advanced research the general public would have difficulty in associating his name with this new higher definition television.

Baird Television developed a range of experimental electronic cameras and receivers. Many Baird Television patents were taken out covering alternative systems, which included the novel idea of storage cathode ray tubes, which would later find their place in the electronics industry.

The scientists and electronic engineers who contributed most to these designs were of the highest calibre [38]. Their work culminated in a range of advanced photomultiplier tubes, cathode ray tubes, image intensifier tubes and an electronic camera known as the 'Baird electron camera.' The Baird electron camera was part of the Baird, Farnsworth licence agreement and employed electron multiplication to compensate for storage. It was capable of producing good quality daylight images of up to seven hundred lines on a closed circuit, but to realise its full potential it required a broadcasting wavelength three-times shorter than was currently available.

Baird's high definition television methods comprised a 240-line sequential system at 25 frames per second utilising the 'spotlight' studio, two 'telecine' machines shown in **Fig 4.13** and to a smaller extent the Baird electron camera which failed to achieve the same sensitivity as a storage camera is shown in **Fig 4.14**.

The spotlight studio, which used an electronic flying-spot scanner in the darkness of a studio, was unpopular with the performers.

The Telecine machines were precision-made and the best that Baird could offer. Baird Television produced the highest quality of recorded and replayed television programmes. Developments of the Baird Telecine continue to be recognised by the industry as the standard for recording, televising and archiving television. Today, Cintel International, (formerly Rank Cintel and 'Cinema Television, Incorporating Baird Television Ltd) remain as world leaders in the design and supply of 'post production' television film scanning equipment for the film and television industry.

Generally speaking, when John Logie Baird's name is brought into a conversation most people relate him to the early mechanical 30-line television and are unaware of his more advanced developments. So potent is the myth that it is often believed that Baird competed against EMI-Marconi's all-electronic television system at Alexandra Palace in 1937 using a 30-line mechanical scanner.

During the trials Baird used an impressive 240-lines (220 effective picture lines) of vertical definition by 600 lines of horizontal definition, with synchronous scanning producing a high-definition image comprising of 132,000 pixels. This was in comparison to EMI's 405-line (385 effective picture lines) of vertical definition by 334 lines of horizontal definition, with interlaced scanning producing a high-definition image comprising 128,590 pixels [39]. To the observer, the resolution of the picture observed from both standards appeared very similar, but there were problems with both systems.

Baird's frame rate of 25 per second caused a visible 25Hz flicker on bright areas of the screen while EMI was having difficulty accurately interlacing two fields at 50 frames per second, resulting in 25Hz interlace flicker. Baird's Telecine system was ideal for film reproduction while EMI excelled in showing live outside broadcasts.

But the contest was not staged on a level playing field, nor was it wholly British. Baird included the technology of the American Philo T Farnsworth, and EMI, a company closely related to the Radio Corporation of America (RCA), was given Zworykin's Iconoscope technology.

This was effectively American technology competing for system supremacy on British soil. In the red corner stood EMI with their 'so called British' Emitron camera and in the blue corner stood Baird, but in reality the British television trial of 1937 was a complete fiasco. It was the Radio Corporation of America's all-electronic technology (presented by EMI) competing against the all-electronic Farnsworth technology (presented by Baird). Let there be no mistake, American television technology was fought and won on British soil.

To be fair, Baird Television used a couple of other systems of scanning that were inferior, but on a more positive side, Baird's film scanner would form the basis for all future analogue and digital video telerecording and film scanning.

It was decided that the 405-line system using the Emitron camera was indeed the superior system, resulting in the Baird system being dropped by the BBC. The loss of the Baird Laboratories in the Crystal Palace fire on the evening of 30th November 1936, during the television trials, had been yet another major setback.

CHAPTER 4: THE REALISATION OF TELEVISION

On the 4 February 1937 it was officially announced that the EMI-Marconi interlaced, 405-line system using the Emitron, was the technology adopted as the first British television standard for transmission through the BBC. This came as no surprise to most Baird Company engineers [40].

The Baird Company having previously manufactured both transmission and receiver equipment was now reduced to the production of high quality 405-line television receivers.

It seemed to Baird that:

"Being out of the BBC, the company should concentrate on television for the cinema, and should work hand in glove with Gaumont British, installing screens in their cinemas and working towards the establishment of a broadcasting company independent of the BBC for the supply of television programmes to cinemas." [41]

Isidore Ostrer the President of Gaumont British and the principal shareholder holding control of Baird Television Ltd, was thoroughly upset with the BBC's decision to go with the EMI system and hinted at withdrawing his support. Although disappointed in the loss of the BBC contract, John Logie Baird was as resilient as ever and could see another way forward without the BBC. He arranged a meeting with Ostrer to talk over the possibility of commercially exploiting in Gaumont Cinemas, Baird Television's 405- line monochrome projection cathode ray tubes, **Fig 4.15**, and his own ideas for large screen colour.

Ostrer, who had previously consented to Logie Baird installing and operating television projection equipment in the Dominion Theatre in London, agreed that

Fig 4.15: Big screen 405-line image in the Tatler News Theatre, in London, using a Baird television projector

63

large screen television had possibilities outside the BBC monopoly. To enable this, a new company was formed to install and operate large screen theatre television projectors in Ostrer's Gaumont chain of film theatres throughout Britain and the USA. This new Gaumont British company would virtually control Baird Television Ltd and was appropriately named 'Cinema Television Ltd'. This, however, was not a separate entity but was a business identity to be staffed and run by existing Baird Television employees.

Baird wrote:

"There was to be a reorganisation at Baird Television Ltd Board. I was to become President (at £4,000 per annum), Greer to remain Chairman and Clayton, Managing Director. I had some experience of the way arrangements of this sort fell through if not pressed, and I insisted on getting my contract at once." [42]

On 22 February 1938 Baird and his wife Margaret set sail from Marseilles on the *Strathaird* to give an address at a world convention of radio engineers in Sydney, Australia, held between 5 and 14 April. He was given a reception never previously accorded to any visitor to Australia, according to Les Bean who represented the Baird Company's Australian interests.

On 24 February 1939, Cinema Television screened the Boon-Danahar boxing contest (broadcast by the BBC) to a paying audience in three London cinemas. They used a special large screen Baird Television 405-line cathode ray tube projector, This 'live' television event was well received by the press. Rival television company, Scophony Ltd, also screened the event using a semi-mechanical electronic technique in a competing cinema in London, but Baird Television's all-electronic television projection system was the first of its type in the world.

The 405-line monochrome cinema projector apparatus was based on an outstanding new pipe-shaped cathode ray tube, named from its similarity in appearance to a tobacco pipe. This tube also exploited by Fernseh, Baird's sister company in Germany, was developed in-house by Dr Szegho and Gilbert Tomes quite separately from J L Baird's colour theatre apparatus, (to be described in the next chapter). The screening of the Boon-Danahar boxing match, which was the first time that public television had been made commercially available to paying audiences in a film theatre, although highly successful, received the disapproval of the BBC who complained accordingly.

While this technical success may have boosted the morale of the Baird Company the celebrations would be short lived as war clouds swiftly gathered. For a short period before the Second World War, Baird continued in his role as President of Cinema Television Ltd. The outbreak of war left Baird Television Ltd in serious financial circumstances. When television ceased broadcasting most of their assets were held in a large stock of, now redundant, television receivers typical of those shown in **Fig 4.16**.

Gaumont British held some three hundred thousand pounds in bonds in the Baird Television Company, which had a total capital of one million and eighty thousand pounds. Shortly after war was declared, bond holders, under the terms of their bond, put the company into liquidation, acquiring the company's assets in payment of their bonds.

CHAPTER 4: THE REALISATION OF TELEVISION

Fig 4.16: Example of Baird Television products at the outbreak of WW2

These assets were taken over by Cinema Television Ltd and under a Board of Trade agreement it was sanctioned that Baird Television Ltd would be allowed to continue to trade, as a means of goodwill to its many clients and customers from within Cinema Television Ltd.

John Logie Baird's contract was terminated on 17 December 1940 and 'Cinema Television Ltd' absorbed Baird Television Ltd, to become Cinema Television Ltd (Incorporating Baird Television Ltd). While many of the staff were dispersed into 'active duty' the company continued to function with those remaining Baird Television scientists, engineers and other staff members working under the technical direction of Captain West, with T M C Lance as Head of Research and Ben Clapp the General Manager.

From 3 February 1941, Cinema Television, which continued to pursue television research, survived mainly from working on secret wartime contracts run independently from two separate locations by Captain West and Captain Moon [43].

Dr Szegho, had been relocated to Baird Incorporated, New York, to set up a facility for the manufacture of theatre projection tubes. His task was to produce a steady supply of pipe-shaped tubes in support of Ostrer's on-going demonstrations of large screen television for his American chain of Gaumont Theatres. To succeed in this he had to win over the film magnates of Hollywood who were initially very impressed, but ultimately the film unions saw this as a threat to their livelihood. The project failed leaving Baird Incorporated on the verge of collapse and Szegho apparently stranded in USA.

It was revealed to Cinema Television, which was still actually Baird Television, on 15 January 1941 that their American branch was in financial difficulty. This was indicated in a telegram sent by Ostrer from the New York offices of Baird Television stating that the now failing Baird Incorporated in New York was prepared to take on any work, currently being undertaken by Cinema Television, to support the British War effort. While this initially appeared to be a practical solution to the American Baird Company it proved impossible, as the transfer of funds to the USA from Britain had been cancelled by the British Government, for the duration of the Second World War.

Although Szegho had stopped manufacturing tubes he remained at the company in a voluntary capacity until a solution could be found. Fortunately Szegho was a perpetual optimist. He considered this to be a situation that would eventually work to his advantage and spent his time preparing an inventory of the assets of the New York branch.

Ostrer sold the entire Gaumont British Corporation to J Arthur Rank in 1942 and the American Baird Company to the Rauland Corporation, a radio manufacturing company interested in television. Baird Incorporated was a timely acquisition for Rauland, giving them access to a specialist in the field. Szegho was hired as head of cathode ray tube research in time for America to enter the Second World War. Szegho's contribution to the war effort won Rauland multi-million dollar war contracts for the design, development and manufacture of specialised radar tubes.

With Baird's contract with Cinema Television Ltd severed he became a free agent able to concentrate on his ambition of leading the world in colour and stereoscopic television. John Logie Baird would be hired as Technical Consultant to Cable and Wireless and would developed colour and 3D television during the war years at his private laboratories leading up to his untimely death in 1946.

References

[1] www.localhistory.scit.wlv.ac.uk/Museum/Engineering/Electronics/history/ValveEra.htm.
[2] *Television Today and Tomorrow,* S A Moseley and H J & Barton Chapple, 2nd ed, Isaac Pitman & Sons, London, 1931, p45.
[3] Originally published in 1988 by the Royal Television Society as *Sermons Soap and Television* and revised by Malcolm Baird after new source material became available, *Television and Me* published my Mercat Press, 2004.
[4] Higher definition in the mid to late 1920s was considered to be any television system with 200 or more lines of definition.
[5] For detailed Information, refer to sister publication by the same author *Images Across Space*, RSGB Publications, ISBN 978-1-874289-21-0.
[6] 'J L Baird: Success and Failure', R W Burns, *Proceedings IEE,* Vol 126, No 9, Sept 1979, p922.
[7] *J L Baird: Success and Failure,* p923.
[8] 'John Logie Baird: 50 Years of Television', M Exwood, *IERE History of Technology Monograph*, IERE, 1976, pp4-5.
[9] 'Best Letters of the month', J L Baird, *Television,* July 1928, p39.
[10] 'John Logie Baird: 50 Years of Television', M Exwood. *IERE History of Technology Monograph*, IERE, 1976, pp4-5.
[11] *The History of Television 1880 to1941,* A Abramson, McFarland, North Carolina, 1987, p28.
[12] *John Logie Baird: 50 Years of Television*, p4.
[13] *Sermons Soap and Television, 2nd ed*, J L Baird, Royal Television Society, London, 1990, p75.

[14] John Logie Baird, British Patent Office, *GB266564*, applied 1 September 1925, issued 1 March 1927.
[15] *The History of Television: 1880 to 1941*, Albert Abramson, McFarland, 1987, p99.
[16] *Sermons Soap and Television, 2nd ed*, J L Baird, Royal Television Society, London, 1990, pp75-76.
[17] 'John Logie Baird: 50 Years of Television', M Exwood, *IERE History of Technology Monograph*, IERE, 1976, p10-11.
[18] *The Romance and Tragedy of the Pioneer of Television*, S A Moseley and H J Barton-Chapple, Odhams Press Ltd, London, 1952, p109.
[19] *Sermons Soap and Television*, John Logie Baird, Royal Television Society, 1988, p24.
[20] *Vision Warrior*, Tom McArthur and Peter Waddell, Scottish Falcon Books, Orkney Press, 1990, pp164-165.
[21] *The Romance and Tragedy of the Pioneer of Television*, S A Moseley and H J Barton-Chapple, Odhams Press Ltd, London, 1952, p129.
[22] *The Romance and Tragedy of the Pioneer of Television,* pp87-88.
[23] *Sermons Soap and Television*, John Logie Baird, Royal Television Society, 1988, p128.
[24] 'Who Really Invented Television', W Phillips, *Broadcast Magazine: 50 Years of Television, Golden Jubilee Issue*, 1986, p3.
[25] *The Romance and Tragedy of the Pioneer of Television*, S A Moseley and H J Barton-Chapple, Odhams Press Ltd, London, 1952, p250.
[26] J L Baird, British Patent Office, London, British patent *GB289104,* applied 15th October 1926.
[27] 'Engineer Extraordinary', R W Burns & A D Blumlein, *History of Technology. Engineering Science & Education*, 1995, p25.
[28] *Restoring Baird's Image*, Don McLean, Institution of Electrical Engineers, 2000.
[29] *Nature*, 5 February 1927, pp188-189.
[30] E G O Anderson, personal correspondence to Dr Peter Waddell, Strathclyde University, 1974.
[31] *Seeing by Wireless*, R Herbert, Published by Herbert, Surrey, sponsored by Quantel, 1996, p7.
[32] *Baird of Television*, R F Tiltman, Seeley Services & Co Ltd, London, 1930, p119.
[33] *The Romance and Tragedy of the Pioneer of Television*, S A Moseley and H J Barton-Chapple, Odhams Press Ltd, London, 1952, p167.
[34] 'The Birth of Modern Television', D C Birkenshaw, *Television*, November-December 1977, p33.
[35] *Sermons Soap and Television*, John Logie Baird, Royal Television Society, 1988, pp98-100.
[36] *Seeing by Wireless*, R Herbert, Published by Herbert, Surrey, sponsored by Quantel, 1996, p13.
[37] 'John Logie Baird: 50 Years of Television', M Exwood, *IERE History of Technology Monograph*, IERE, 1976, p7.
[38] For detailed Information, refer to sister publication by the same author *Images Across Space*, RSGB Publications, ISBN 978-1-874289-21-0.

[39] *Images Across Space*, pp39-43.
[40] Dr A H Sommer, private correspondence to the author, 6 August 1990.
[41] *Sermons Soap and Television, 2nd ed*, J L Baird, Royal Television Society, London, 1990, p140.
[42] *Sermons, Soap and Television,* p141.
[43] For detailed Information, refer to sister publication by the same author *Images Across Space*, RSGB Publications, ISBN 978-1-874289-21-0.

5

Colour, Light and Human Perception

This chapter reviews the knowledge that John Logie Baird applied to his colour television research and includes a non-mathematical introduction to the physics and human perception of colour supported by a number of illustrations which are reproduced in colour on pages C.1 to C.4.

THE PLANCK DISTRIBUTION CURVE of electromagnetic energy generated by the Sun is shown in **Fig 5.1** [also reproduced in colour on page C.1]. It illustrates the range of wavelengths present in comparison to the narrower band of light energy visible to the human eye.

The peak energy from the Sun, normalised in the graph to 1, is around 600 nanometers (nm) [1], which conveniently falls within the narrow 400nm to 700nm sensitivity region of the human eye, depicted in the illustration by the familiar rainbow spectra [2].

Colour vision and perception are partly explained with reference to the spectral composition and the additive nature of sunlight. Sir Isaac Newton (1672 - 1727) asserted that sunlight is not a discrete colour, nor a single wavelength, but comprises a combination of many wavelengths.

In 1704 Newton proved this theory by carrying out his now famous experiment involving a narrow beam of sunlight, a glass prism and a white screen to display the result. It had previously been observed that when a ray of light

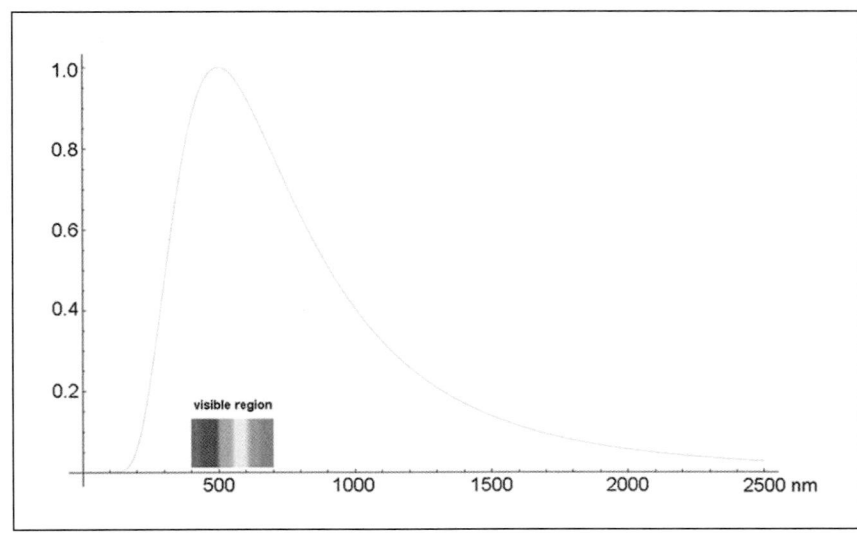

Fig 5.1: Planck energy distribution of the Sun (see page C.1 for a colour version)

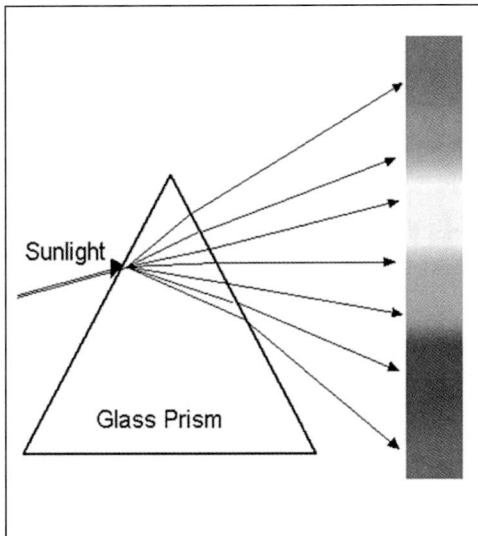 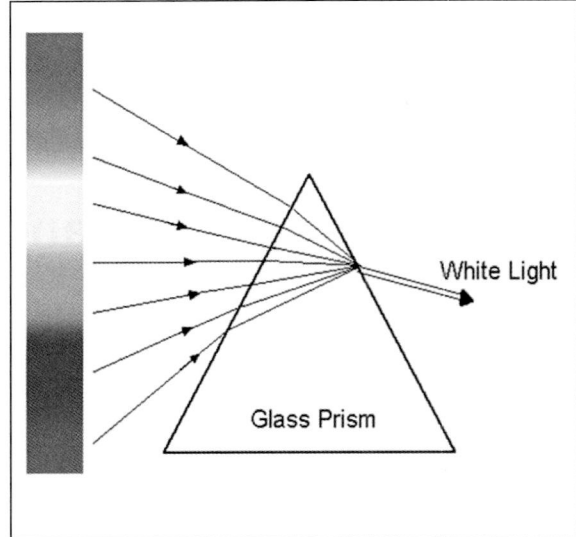

(left) Fig 5.2: Splitting white light into its spectral components

(right) Fig 5.3: Recombining the spectral colours to produce white light

(see page C.1 for colour versions of these pictures)

passes from one transparent medium to another, such as from air to water or from air to glass, the speed of the ray changes within the medium causing the ray to bend by a certain amount relating to its wavelength.

The effect, known as refraction, was first studied by Ptolemy around 140AD and mathematically redefined in 1621 by the Dutch physicist Willebrord Snell (1591- 1626).

Based on this knowledge, Newton allowed sunlight, formed into a narrow beam as it passed through a slit in window shutters, to fall upon a triangular glass prism in an otherwise darkened room. As shown in **Fig 5.2** [also reproduced in colour on page C.1], the light emerging from the prism is separated from the white source into the familiar spectral colours of the rainbow; red, orange, yellow, green, blue, indigo and violet.

Colour separation is caused by each colour of light ray having a unique wavelength and therefore tilting at a unique angle as it travels through the variable refractive index of the glass medium. Newton demonstrated that it is also possible by refraction to recombine the separated wavelengths and reconstruct white light by means of an identical prism as shown in **Fig 5.3** [also reproduced in colour on page C.1].

White light can be obtained not only from the sum of the known spectral colours, but also from primary colour combinations. According to the traditional trichromic theory [3] there are two main photoreceptor organs found on the retina of the human eye, known as rods and cones.

The rods are more sensitive to light than cones and provide a useful monochromatic grey-scale vision under poor ambient lighting conditions. Under brighter light, the cones, sensitive to three wavelength ranges (interpreted by the brain as red, green and blue) are activated to provide colour vision. The retina comprises more than a hundred million rods and perhaps five to ten million cones.

Fig 5.4 [also reproduced in colour on page C.2] illustrates that when three primary colours of light, red, green and blue are projected upon each other, the colours blend to provide the impression of white, but where they overlap on other

CHAPTER 5: COLOUR, LIGHT AND HUMAN PERCEPTION

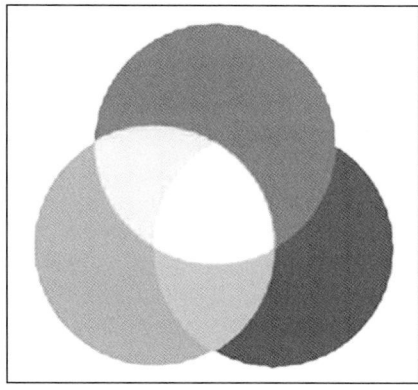
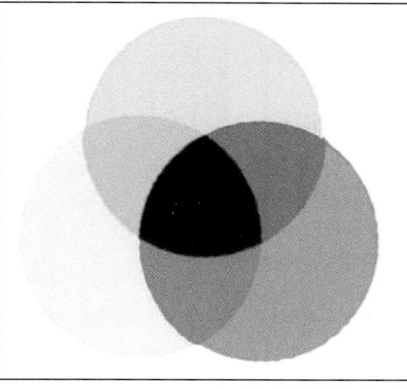

(left) Fig 5.4: Additive mixing of light

(right) Fig 5.5: Subtractive mixing of pigments

(see page C.2 for colour versions of these pictures)

primaries, the respective secondary colours of cyan, magenta and yellow are observed respectively. Therefore, by regulating the intensity and mixture of three primary colours, selected to match the sensitivity peaks of the eye, the full gamut of visible colours observed in nature may be produced.

This should not be confused with the 'subtractive' nature of colour pigments in the form of colour filters, paint, dyes or inks used by colourists and the printing industry. **Fig 5.5** [also reproduced in colour on page C.2] illustrates the result of mixing the three primary colours for pigments, cyan, magenta and yellow, which 'subtract' to produce secondary colours of red, green and blue.

In theory, at any location where the primary colours converge there should be an area of absolute subtraction where no colour exists, but in practice this often results in a poor representation of black [4].

It should be mentioned that modern television displays exploit the additive mixing nature of coloured light while only a few experimental colour television displays made use of subtractive colour mixing. Normally three primary colours of light, red, green and blue are additively mixed by the eye and brain, from a television or monitor screen to enable a wide spectrum of colours, including white, to be perceived.

It is important when selecting trichromic primaries that two should be selected from the extreme limits of the spectral light range while the third should be located near the centre. The wavelength and frequency range of the visible colour spectrum is shown in **Fig 5.6**.

Colour	Wavelength [nanometers]	Frequency [Terahertz]
Violet	400 to 420 nm	714 to 750 THz
Indigo	420 to 440 nm	682 to 714 THz
Blue	440 to 490 nm	612 to 682 THz
Green	490 to 570 nm	562 to 612 THz
Yellow	570 to 585 nm	513 to 526 THz
Orange	585 to 620 nm	484 to 513 THz
Red	620 to 700 nm	429 – 484 THz

Fig 5.6: Wavelength and frequency of the visible spectrum

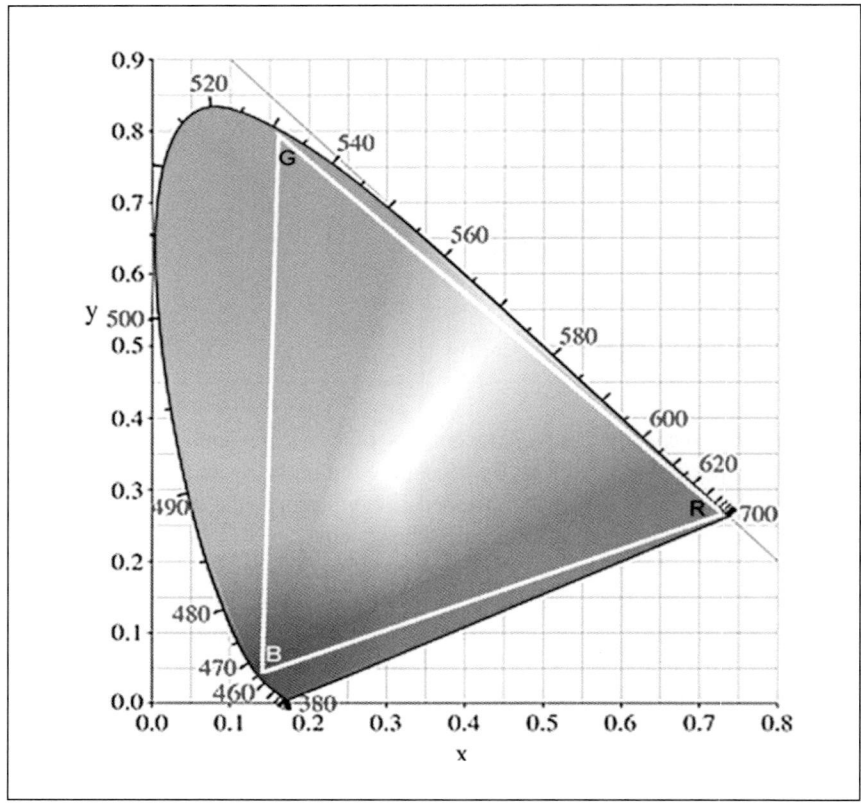

Fig 5.7: CIE (1931) colour space diagram (see page C.2 for a colour version of this picture)

Red is normally selected from the longer region of wavelength between 620nm to 700nm, 'blue' from the shorter region 400nm to 490nm and 'green' from the central region 490nm to 570nm.

Fig 5.7 [also reproduced in colour on page C.2] is the mathematically derived CIE colour space chromaticity diagram (dated 1931). The tongue shaped area scaled in nanometers around the boundary, describes the gamut of colours visible to the average person, but on the printed page the colours viewed are relative. The triangle shown in Fig 5.7 is drawn by plotting three wavelength points (R, G and B) on the boundary of the diagram, with red (700nm), green (530nm) and blue (465nm). The plotted points represent a set of trichromic primary colours while the area described by the triangle represents all possible secondary colours that may be reproduced from those primaries.

A full-colour image may be recorded on three (black and white) grey-scale photographic images by exposure using appropriate primary colour filters. The process shown in **Fig 5.8** shows the basis for Baird's colour television experiments.

A conventional (analogue) camera is loaded with black and white transparency film and mounted on a tripod to produce three black and white photographic slides of the same colourful scene. The experiment requires that the first photograph is exposed through a red primary filter, the second through a green primary filter and the third through a blue primary filter. Each resulting black and white transparency is an accurate recording of the intensity of the reflected light

CHAPTER 5: COLOUR, LIGHT AND HUMAN PERCEPTION

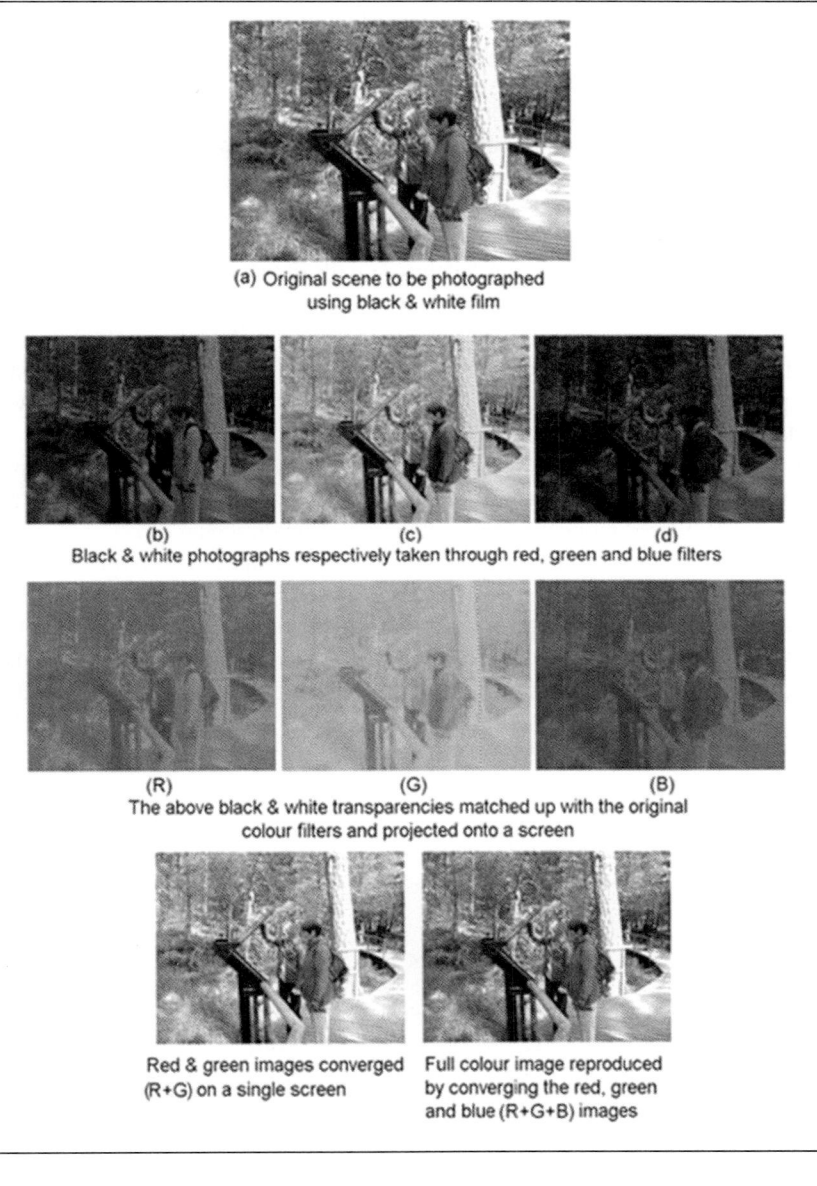

Fig 5.8: Obtaining and reproducing a colour image from black and white film (see page C.3 for a colour version of this picture)

(a) Original scene to be photographed using black & white film

(b) (c) (d)
Black & white photographs respectively taken through red, green and blue filters

(R) (G) (B)
The above black & white transparencies matched up with the original colour filters and projected onto a screen

Red & green images converged (R+G) on a single screen

Full colour image reproduced by converging the red, green and blue (R+G+B) images

from the scene, but only at the particular narrow band of wavelength of the primary colour filter used.

It can be seen that three monochromatic (black and white) slides, together with three colour filters, identical to those used in taking the pictures, carry all the information required to reproduce a full colour image of the original scene.

In Fig 5.8 [also reproduced in colour on page C.3], the scene to be photographed (a) is shown, followed by the images of the three resulting black and white transparencies, (b) was exposed through the red filter, (c) through the green filter and (d) through the blue filter.

Each image has a unique grey-scale relating to the intensity of the reflected light from the scene relating to the photographic filter used. The black and

73

white, grey-scale transparencies are individually placed, one in each of three projectors, correspondingly with a lens filter of the same wavelength used to produce the slide.

The three images are then projected side by side on a screen as shown in Figure 5.8: (R) is the red component, (G) the green component and (B) the blue component of the original colour scene.

To reproduce the original scene in full colour all that remains is to cause the three images to converge accurately on the screen causing the primary colour components to add. The bottom left image with red (R) and green (G) primary images converged, produces a very convincing effect. This image has a similar range of colours to the duochromic process, developed in 1922 by the Technicolor Motion Picture Corporation for the Hollywood film industry. Despite being limited in colour spectrum, the first Technicolor movies were acclaimed as a major technological breakthrough.

The final picture in Fig 5.8 is the full-colour image, virtually identical in spectrum to the original image (a). It was produced by converging the red, green and blue (RGB) filtered monochromatic images on a single screen.

This knowledge indicated to Baird that moving colour television images would also be obtained by a method or methods of encoding a monochrome television display (or displays) by means of a series of colour filters. Baird not only demonstrated colour television using red, green, and blue (RGB) trichromic primary

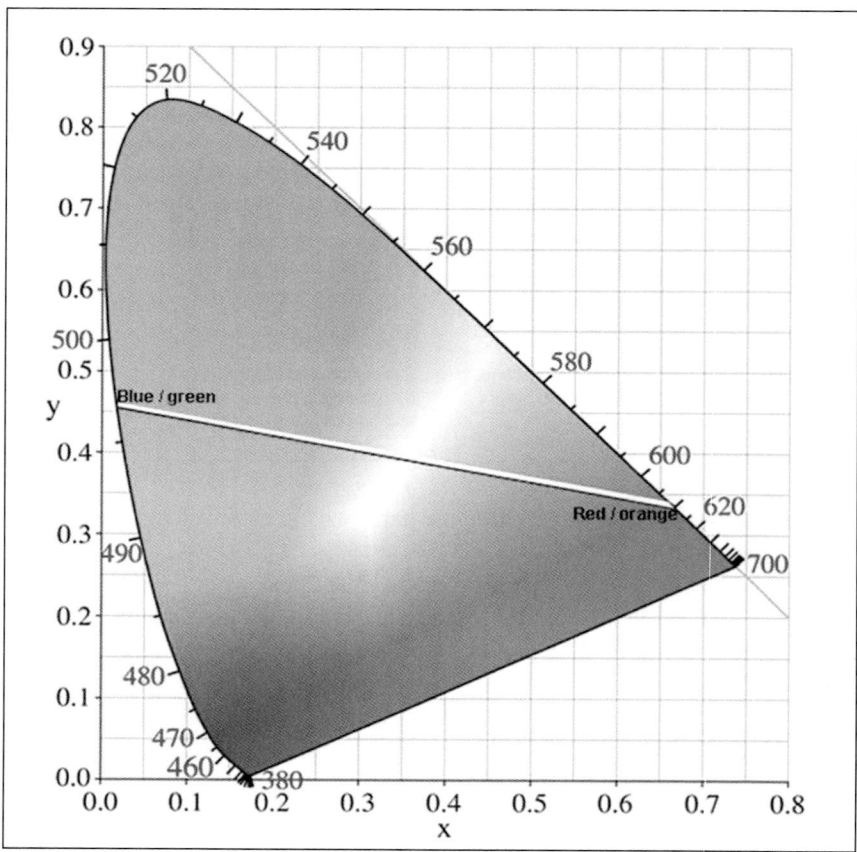

Fig 5.9: Baird selected (red/orange) and (green/blue) to achieve a wide range of colours (see page C.4 for a colour version of this picture)

colours, but also produced convincing results by using two carefully chosen primary colours.

Fig 5.9 [also reproduced in colour on page C.4] shows that the secondary colours obtainable by additive mixing, when using a duochromic system, are not described within an area but are limited to the range of hues that occur along a line between any selected two points on the CIE diagram. The graph shows a line travelling from red/orange (625nm) through pale yellow, pink,pink/white, light green/blue, and finally green-blue (495nm). Careful selection of two wavelengths of colour filter result in a picture with not only a better white, but also a more realistic range of colours than those present in the early Technicolor system circa 1921-1924.

It must be assumed that both Baird and the Technicolor designers were unaware of the human psychology, (only more recently understood), which enables images produced by a two-colour system to be perceived with a wider spectrum than is actually present on the screen. There is an inconsistency of human colour perception that assists in providing the impression that a wider range of colours is present than those actually existing in an image.

In **Fig 5.10** [also reproduced in colour on page C.4] a small block of colour approximating a wavelength of 495nm (Baird's blue/green primary colour), has been placed in the centre of two larger blocks of colour, green at 520nm and blue at 465nm. Although in both cases the wavelength of the centre colour is identical, it may be perceived as a shade of blue when compared visually with the surrounding green block or as a shade of green when compared with the surrounding blue block.

This simple experiment demonstrates how the brain will misinterpret a particular colour depending on the circumstances, which may be due to the presence of other colours, radiant light levels and other cues.

If human perception of colour did not operate in this way, the visual experience of the world would appear to be quite strange. Under a filament light source our vision would be biased red, while in daylight everything would be biased to blue. The brain continually corrects our vision by automatically readjusting the white balance based on the visual information present, to enable apparent colour consistency, but under certain circumstances the information processed will provide perception with results that may appear illogical as indicated by Fig 5.10.

Baird's experiments with duochromic images took place 36 years before the colour perception experiments conducted by Dr Edwin Land, founder of the Polaroid Corporation:

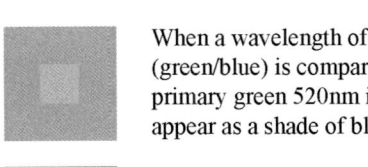
When a wavelength of 495nm (green/blue) is compared to primary green 520nm it may appear as a shade of blue.

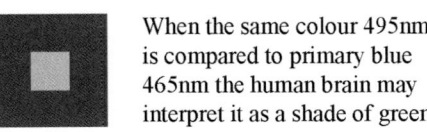
When the same colour 495nm is compared to primary blue 465nm the human brain may interpret it as a shade of green.

Fig 5.10: Colour interpretation relies on human perception (see page C.4 for a colour version of this picture)

"We have come to the conclusion that the classical laws of colour mixing conceal great basic laws of colour vision. There is a discrepancy between the conclusions that one would reach on the basis of the standard theory of colour mixing and the results we obtain in studying the total images." [5]

As part of the colour mixing demonstration previously described and shown in Fig 5.8, Land went through the process of photographing a scene three times in succession using three primary colour filters, and produced a full RGB colour image by adding together the images from his three slide projectors, each fitted with the appropriate monochrome slide and colour filter.

On returning later to dismantle the experiment he was surprised to find that it was partly dismantled with the blue component projector switched off, the red and green component projectors remained switched on and the green filter had been removed. The result was that the red image remained converged on the screen superimposed with a bright black and white representation relating to the green image. Logically, it would be expected that the red picture would appear pink and washed out, by the bright black and white image. Incredible although it may seem, Land and his students continued to perceive a colour image with a range of colours including the unlikely green and blue hues.

Later, in 1971, after considerable research, Land and McCann presented the 'Retinex' theory indicating that human colour vision is more complicated than the trichromic theory alone could explain.

Based on Land's work it is now understood that colour perception relies only to a smaller extent on the wavelengths of light being mixed at the retina, but to a much larger extent on the image processing that take place in the human visual cortex. By measuring the wavelength sensitivity of the cones in the average eye an interesting phenomenon is observed. The sensitivity chart of the cones shown in **Fig 5.11** indicate a considerable overlap of red and green sensitivity, which is a condition more significant than most observers would be aware.

"The cone cells of the human eye are sensitive to three wavelength ranges which the eye interprets as blue (narrow, with a peak near 419nm), green (broader, with a peak near 531nm) and red (also broad, with a peak near 558nm, which is actually more like yellow!)" [6]

When determining some red or green colours the brain is presented with data from both red and green cones, suggesting confusion with yellow:
"If 650nm photons hit your retinae, your brain will receive a mixture of

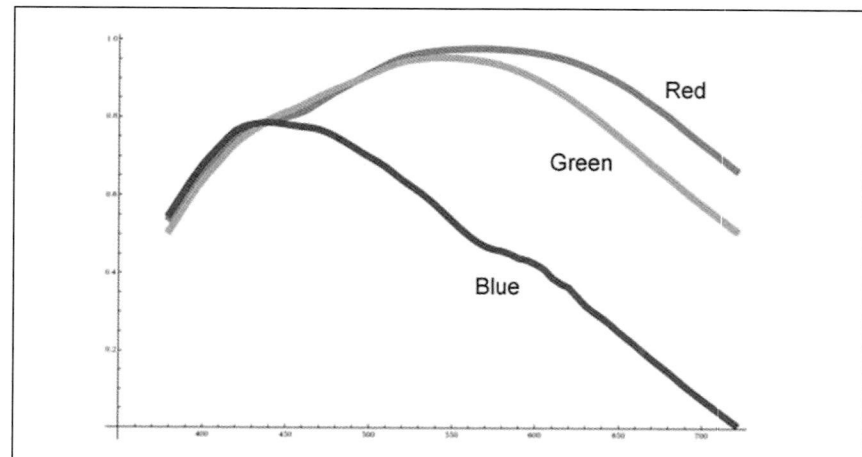

Fig 5.11: Spectral response of the eye against wavelength (Vos, J J Walvarem, P L 1971)

green and red signals, with more red than green but not too many of either. This will be interpreted as red. Similarly, 475nm photons will cause about equal numbers of blue and green signals, with only a few red; this will be interpreted as a sort of bluish-green. The number of signals for any one of these ranges depends on both the intensity of the light and the sensitivity at that wavelength. This leads to a vaguely disturbing contrast between sensation and perception: your eye sends only three kinds of signals to your brain, yet your brain 'constructs' the full color spectrum of 'reality' from them. Now you can understand why people can have violent disagreements over what color something is: no two people see exactly the same thing, yet we assume that color is something objective." [7]

This was the background to John Logie Baird's experiments with colour television and it will become apparent, from the evidence provided in the next chapter, that through the results of his intense investigations into colour scanning and reproduction, that he became familiar with and applied to his advantage, many of the psychological anomalies of the perception of human vision.

References

[1] One nanometer = 1 x 10^{-9} meters.
[2] Modified from an original diagram courtesy of Ken Koehler, University of Cincinnati, Ohio.
[3] Young-Helmholtz three-colour theory, *Encyclopædia Britannica,* 2010.
[4] Black produced by subtracting pigments of cyan (C), magenta (M) and Yellow (Y) is not colourless enough for the printing industry and a specially produced black ink (K) is provided. The four colour process is known as CMYK.
[5] 'Colour Vision and the Natural Image: Part 1', Dr E H Land, *Proceedings of the National Academy of Sciences*, 1959, pp115-129.
[6] Kenneth R Koehler PhD, Professor of Physics, Department of Mathematics Physics and Computer Science, Raymond Walters College, University of Cincinnati, Ohio.
[7] Text from Koehler, http://www.rwc.uc.edu/koehler/biophys/6d.html.

THE THREE DIMENSIONS OF JOHN LOGIE BAIRD

6

The Colourful John Logie Baird

The following is a history of colour television, from the low definition experiments of the 1920s, through the electronic and optical systems of the 1930s, concluding with a description of the first purpose-built all-electronic colour television cathode ray tube, demonstrated by John Logie Baird in 1944.

THE INVENTION OF THE TELEPHONE by Alexander Graham Bell on 3 June 1875 prompted widespread speculation among the scientific community on the possibility of developing a videophone carrying electrical representations of full colour images. Although this expectation was seriously challenged by the limitations of 19th century technology some rather interesting ideas began to emerge. In December 1889, Polumordvinov applied for a Russian patent describing a colour television system which used rotating cylinders with slits alternately covered by red, green and violet filters [1]. This was the first patent describing a 'sequential' colour television, predating by 37 years the first practical demonstration of monochrome television by John Logie Baird in 1926.

August 1904 saw Jaworski and Frankenstein apply for a colour television system patent, which used a red, blue and yellow sectored filter wheel [2]. In 1908, Johannes Adamian applied for a British patent [3], followed by the Andersons of Copenhagen, with a proposal to utilise a scheme of prisms to separate colour [4]. In 1923, the year that John Logie Baird is reported to have devoted his life to the development of television, Valensi applied for a French colour television patent proposing a combination of photocells and cathode ray tubes [5] and in 1923, Hammond Junior applied for a US patent for a monochrome, stereoscopic and colour television [6]. Despite these and other proposals for the sending and receiving of colour images, no practical application to the solution was known prior to Baird's demonstration of monochrome in 1926.

Baird disliked the tedium of the hours he spent in the boardroom of 'Television Ltd', and instead of engaging in the time-consuming discussion on policy and organisation he would be preoccupied with solving the problems of colour and 3D television. Moseley [7] wrote that Baird began colour work in 1927 while Barton-Chapple gave the year as 1928 [8]. In his autobiographical notes, Baird wrote:

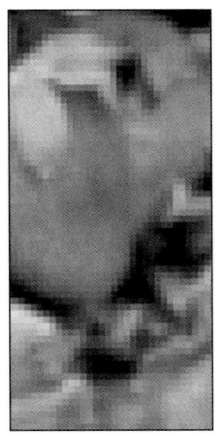

Fig 6.1: Emulation of a low-definition colour television image (see page C.5 for a colour version)

"I was intensely interested in the possibility of colour television, and early in 1927 I commenced experimenting. The strawberries came out particularly well and they were popular with the staff. The device was demonstrated at a British Association meeting in Glasgow in July 1928. This was the first occasion in which colour television was shown in public." [9]

Fig 6.1 [also reproduced in colour on page C.5] provides an impression of the reasonable quality of image that can be provided using a highly pixelated low-resolution colour image.

The above conflicting dates are not to be confused with Baird's first experiments with colour television. With only a few exceptions, Baird avoided revealing new inventions until patents were firmly in place. He believed that a certain amount of discretion was required to remain ahead of the competition.

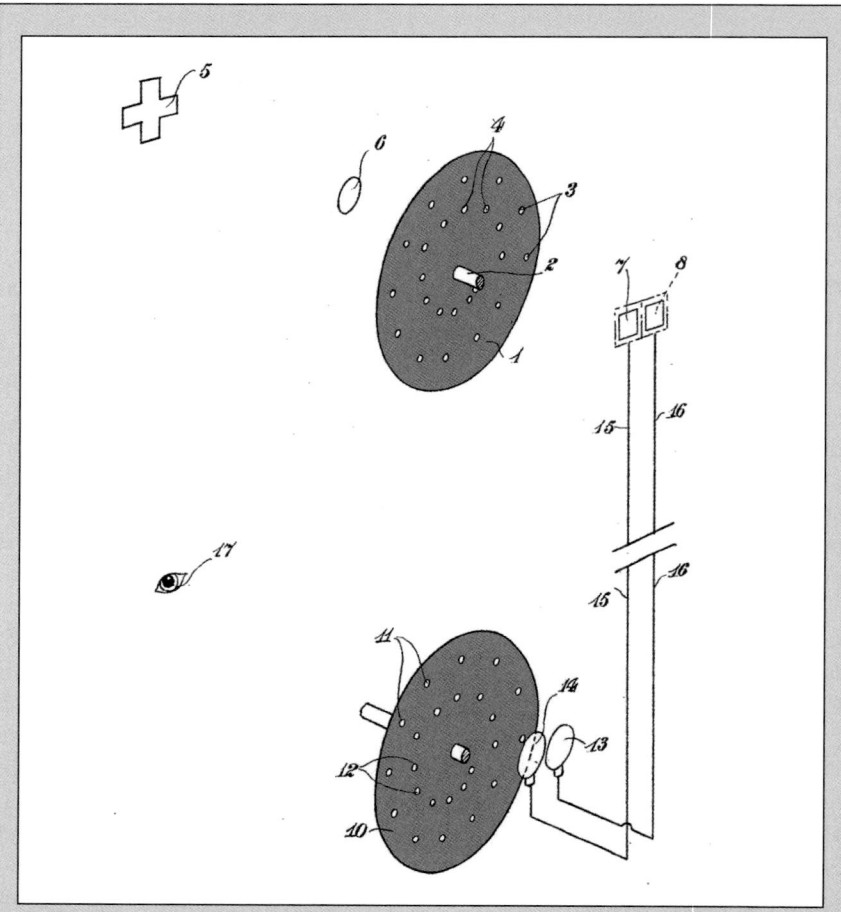

Fig 6.2: Abstracts from *GB266564*, a preliminary specification for the production of colour television

To transmit images in their natural colours two or more cells may be used each cell sensitive to one colour. These cells controlling correspondingly coloured lights at the receiver. Alternatively several adjacent images say two for example may be transmitted each through a different light filter, one image might correspond to say red the other to the blue, the blue image might thus be arranged to traverse the blue sensitive cell only and the red image the red sensitive cell. At the receiver the red and blue images are superimposed. To transmit in stereoscopic relief two images may be projected by stereoscopic lenses and this double image be viewed by the transmitter and transmitted to the receiver where it is viewed through a stereoscope.

CHAPTER 6: THE COLOURFUL JOHN LOGIE BAIRD

A patent search reveals an interesting anomaly that casts into doubt the date of the breakthrough in monochrome television as recorded by Baird in his autobiographical notes. Baird stated his breakthrough in resolving the first greyscale image was on 2 October 1925, yet on 1 September he had already lodged two provisional patents for colour television, *GB266564* [10] and *GB267378* [11]. This adds to the credibility that Baird was always at least one step ahead of the publicity released to the press.

One of these patents was for a two-colour low-resolution television system using the duochromic primaries red and blue. This was a dual patent, describing not only colour, but also a modification that would enable an early form of stereoscopic television (**Fig 6.2**). Throughout his television career, Baird would punctuate his activities with concepts for three-dimensional television with quite remarkable results. For that reason some Baird colour television patents which also have a partial interaction with 3D television are excluded from this section, to be treated separately in Chapter 8.

The second patent, dated 1 September 1925 (**Fig 6.3**), clearly indicates that Baird was applying the orange/red and green/blue duochromic colour theory as early as 1925. A clue to Baird's discretion lies in the fact that both patents were not completed until 1926 and took a further year before reaching the acceptance stage in March 1927. However, further research shows that a little early publicity had been released. The first was on 10 August 1926 in an article that appeared in *The Morning Post* entitled 'Living Pictures by Wireless':

"Mr. Baird has now transmitted by wireless a moving image, showing gradations of light and shade and even a little colour - red and blue."

The second article relating to colour television experiments appeared in *The Sphere*, on 28 August 1926, but gave very little detail. These early articles serve to show that prior to September 1925, Baird was aware that a process using primary colour filters could be applied to his embryonic monochrome television system. Baird was mindful of the competition that existed in America and careful to avoid further public disclosures of colour until after 1 March 1927, by which time the patents had been accepted and published.

To obtain colour effects the source . . . may be a neon lamp giving a red glow discharge at eg 200 to 250 volts and a blue arc discharge at eg 800 to 850 volts. A neon lamp with mercury electrodes or having, mixed with the neon, mercury vapour or helium or other gas giving a bluish spectrum is suitable. Two light-sensitive cells may in this case be used at the transmitting station, one, covered by an orange-red screen, being in circuit with the [first] amplifier . . . and the other, covered by a green blue screen, being in circuit with the [second] amplifier . . . and if a light-sensitive cell is used as described above for rendering the [first] amplifier . . . inoperative, a green blue screen is placed between the source . . . and this cell. Alternatively two cells responding differently to lights of different colours may be used at the transmitting station. The sources . . . may also differ in colour instead of differing in intensity.

Fig 6.3: Abstract from Baird's patent *GB267378*

Fig 6.4: Abstract from Baird's patent GB312560

Television apparatus for colour reproduction comprises at a transmitter, two or more exploring-discs each exploring an object and each associated with a separate light-sensitive cell and appropriate light filter, and at a receiver, two or more light sources each associated with a separate exploring device and light filter to reproduce an appropriately coloured image.

The novelty of a further colour television patent *GB312560* [12], taken out on 30 November 1927 (**Fig 6.4**) provides an insight into the many different ideas that were being considered by Baird. This patent, although not followed up, was designed to improve the image resolution by utilizing two or more exploring devices to deal with individual primary colours. This invention loosely followed the concept of *GB266564* as a parallel imaging concept with a multiplicity of primary colour images being presented simultaneously. At the receiving station a similar number of discs were to be used with appropriate coloured lamps and filters. The separate images were to be optically converged on a screen to produce a greater detail than otherwise possible by scanning with a single disc.

Baird then turned his attention to the concept of multiple interlacing discs designed to reduce perceived flicker. The starting point was *GB314591* (**Fig 6.5**) [13] [also reproduced in colour on page C.5], application date 4 January 1928, which introduced the concept of sequential scanning at low resolution, by scanning the object with spots of coloured light.

While this described a sequential red, green and blue, flying spot scanner disc a further patent on 5 June 1928, *GB321389* [14] disclosed a triple interlaced, field scanner providing 90-line monochrome or 30-line colour television image.

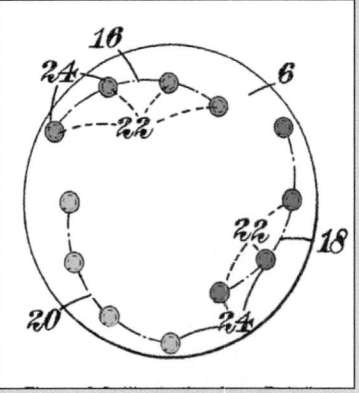

The object is explored by a spot of light of the same colour as itself. If the object is of varying colour, two or more light spots of different colour may be used, the colours chosen being those of the predominating colours of the object or those used in two-colour or three-colour reproduction. The periods of exploration by each light-spot may be separate but consecutive or may overlap. Several light-sensitive cells may be used, each sensitive to one of the colours used. At the receiving station, all the signals may modulate a source of light to reproduce a monochromatic or coloured image, or the signals derived from each coloured light-spot may modulate a correspondingly coloured light source so as to produce a coloured reproduction. The exploring disc at the transmitting or receiving station may have three separate spirals 16, 18, 20 of lenses covered by light-filters, the filters of each spiral differing in colour from those of the other spirals.

Fig 6.5: Abstract from Baird's British colour television patent GB314591 (see page C.5 for a colour version of the picture)

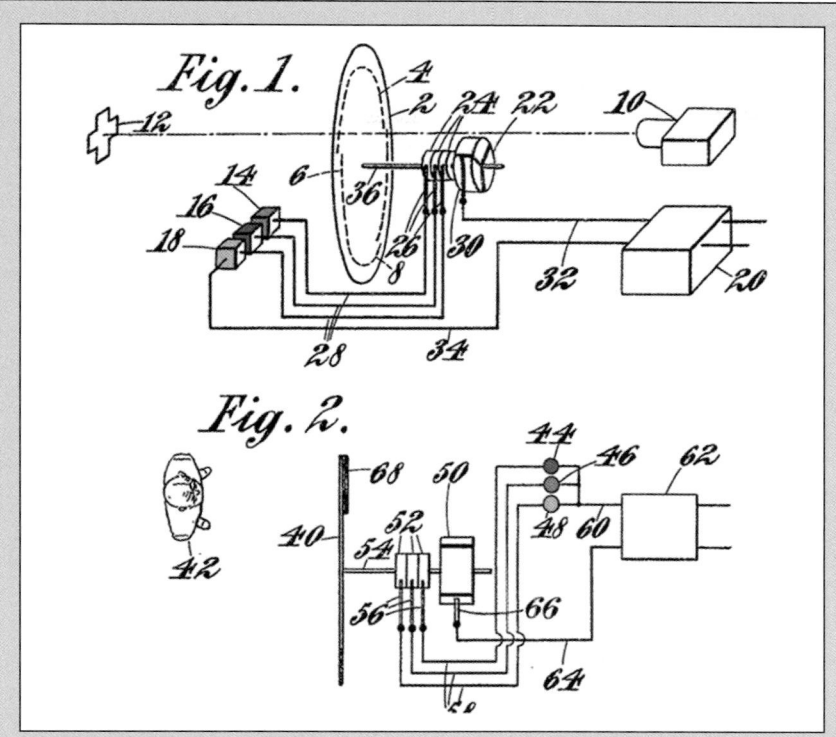

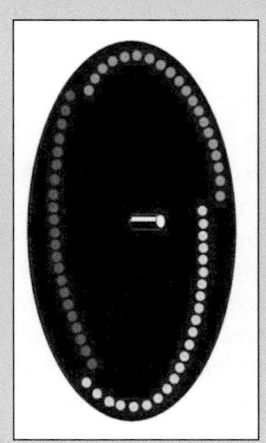

A television apparatus for reproducing images in natural colour comprises a plurality of light-sensitive cells of different colour sensitivity at the transmitter, a plurality of correspondingly coloured lamps at the receiver, and means for operating each cell successively with its corresponding lamp. The scanning disc 2, fig. 1, at the transmitter is provided with three sets of spirally arranged holes 4, 6, 8, light from each set completely traversing the object 12. Light-filters of blue, red and green respectively cover the three sets of holes, and a commutator 22 and slip rings 24 which rotate on the same shaft 36 as the disc 2 connect the particular light-sensitive device 14, 16, or 18 which is sensitive to the colour of the light falling on the object, to the outgoing amplifier 20. At the receiver a disc 40, Fig. 2, similar to the transmitting disc 2 rotates synchronously with the latter, a similar arrangement of slip rings 52 and commutator 50 serving to pass the incoming signals to the appropriate lamp 44, 46 or 48, which emits light of the same colour as that with which the object is being scanned at that instant. A translucent screen 68 may be placed adjacent the disc 40 to serve as a viewing screen for the image. The holes in the disc 40 may be provided with colour filters if desired. Overlapping of the exploration bands may be introduced by suitable modification of the scanning discs.

Fig 6.6: Baird's colour television patent *GB321389* (see page C.5 for colour versions of the pictures)

An artist's impression of an RGB colour sequential disc is shown in **Fig 6.6** together with the concept diagram from the patent [these pictures are also reproduced in colour on page C.5].

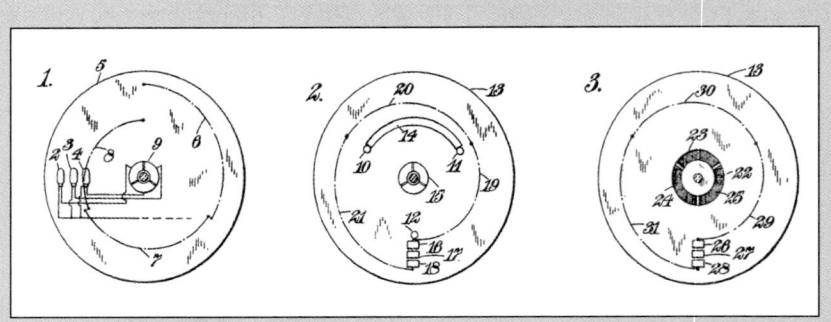

With reference to disc '1', each of the three lamps emits one of the colours for a three colour process and constitutes the source of illumination of an object of which an image will transmitted. Between the lamps and the object there is a rotating disc 5 carrying three sets of spirally arranged holes or lenses, each set situated a different distance from the centre of the disc compared with its companion sets, so that only one of the lamps effects exploration of the object at a time. By this arrangement the object is explored by coloured light spots. To ensure that the correct lamp is illuminated a commutator 9 is provided.

In disc '2', each of three lamps 10, 11, 12, emits one of the colours for a three-colour process. To illuminate the object as a whole, instead of with coloured spots, the rotating disc 13 carries a curved slot 14. The slot is of such a length as to subtend an angle of 120 degrees at the centre of the disc and of such a width that the emergent beam will illuminate the whole of the object. The three lamps are situated 120 degrees apart and positioned on the same radius as the slot. As the disc rotates the lamps will in turn illuminate the object as a whole. To ensure that the correct lamp is illuminated a commutator 15 is provided. The exploration of the object is carried out by light cells 16, 17, 18, each being allocated a set of spiral holes or lenses, 19, 20 and 21. The spirals are coterminous, forming a single spiral, arranged on the outer edge of the disc, further from the centre than the curved slot. Each set of spiral holes or lenses effects a complete exploration of the object by the light from one colour.

In the case of disc '3' a single lamp 22 emitting red, green and blue light, such as a mercury vapour glow discharge lamp may be utilised, but in this case the disc employs three curved slots 23, 24, 25 covered by the appropriate red, green or blue light filter positioned in front of the light source. The object is illuminated as a whole by light of each colour in turn. The three light sensitive cells 26, 27, 28 in conjunction with the three coterminous spirals 29, 30, 31 explore the object. In conjunction to the sets of lamps and or cells described, neutral density filters are chosen to equalise the effects that said colours exert upon the cell.

Fig 6.7: Abstract from Baird's colour scanning patent *GB322776*

The monochrome image was to be scanned at three times the normal speed to reduce image flicker, but despite this, colour flicker would be apparent at a frequency relating to a third of the scanning rate of a three-colour system and at half the scanning speed of a two-colour system. On 9 June 1928 Baird applied for colour television patent *GB322776* [15], **Fig 6.7**, describing three improved methods to selectively scan an image with primary colours of red, green and blue. A further novelty of the patent was correction of the photocell's spectral colour response by means of neutral density filters. The complexity involved in producing colour television led to yet another patent application on 20 June 1928, *GB319307* [16]. This dealt with problems associated with colour synchronisation, **Fig 6.8**, [see page C.6 for a colour version of the picture].

In July 1929 Baird learned that his earlier rival, Bell Telephone Laboratories in New York had demonstrated a three-colour television system employing three independent colour channels. The live pickup equipment consisted of three banks of cells with three primary colour responses. A flying spot scanned the object and a scanning disk served as the receiver to reconstitute the image. Goldmark wrote:

> *"It is interesting to note that while Bell Laboratories employed a three-channel system which obviously occupies three times the frequency spectrum over the corresponding black-and-white picture and requires three times the facilities, Baird, though similarly requiring three times the frequency space, employed rotating filters and was thus the first one to demonstrate the sequential, additive method of colour in television."* [17]

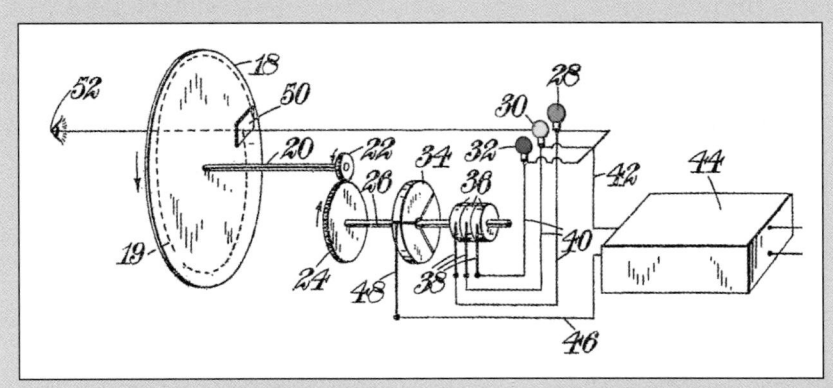

A television system for reproduction in colour comprises at the receiver an exploring-device 18 having a single set 19 of light apertures, means 28, 30, 32 for providing illuminations corresponding respectively in colour to the colours at the transmitter, and a means 34 for rendering each illumination operative in turn in synchronism with the operation of the exploring devices. The elements of each of the three sets of apertures in the exploring-disc 2 have light filters, not shown passing light of the appropriate colour. The exploring-disc at the receiver is rotated at three times the speed of that at the transmitter and the coloured lamps 28, 30, 32 are energised in turn through a commutator 34 on the shaft 26.

Fig 6.8: Abstract from Baird's patent *GB319307* showing the switching commutator (see page C.6 for a colour version of the picture)

Colour and 3D television became such a diversion that in 1933 John Logie Baird privately financed, equipped and staffed his own experimental laboratories in a converted stable adjacent to his home in Crescent Wood Road, Sydenham in London. Initially he had five members of staff: P V Revelley (Project leader), E G O Anderson, R Vince, A E Sayers and C F Oxbrow.

"The assistance of specialists. Experts in glass blowing who could produce cathode-ray tubes of unusual shape; and designers of photocells, who knew best how to develop panchromatic devices to collect the reflected light from a subject and convert it into an electrical signal, were frequent visitors to the inventor's private laboratory." [18]

In 1935, while development work continued at Baird Television, J L Baird was taking a serious interest in applications of the cathode-ray tube. On 12 December 1935 he applied for a futuristic patent *GB476195* [19] to display a large bright monochrome television image on a screen. The process involved projecting television images by using multiple monitors and lenses. Each cathode ray tube would display only a part of the full image to be projected on a screen.

Once again, Baird produced an invention that would become a practical reality of the future. The description of the concept illustrated in **Fig 6.9**, relates to a complete television picture constructed from a number of display panels or discretely projected images. This may be recognised as the forerunner to the 'video wall' display introduced commercially by Philips Lighting [20] in 1984.

On 9 April 1936 Baird patented (*GB473303*) an electro-optical scheme for converging and viewing a brightly projected colour image on a screen. A bright

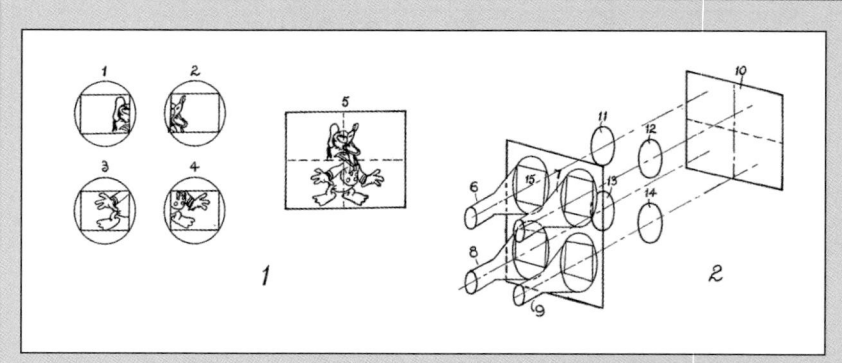

Fig 6.9: Abstract from Baird's video wall patent (1935) *GB476195*

Each of a plurality of cathode-ray tubes 6 .. 9 reconstitutes a portion of the received image 10. Masks 15 insure that the several partial images projected on the screen 10 are truly complementary and together form a completely contiguous image. The projecting lenses 11 .. 14 have screw adjustments for aligning the part images. All the cathode ray tubes are modulated by the same signal or they are severally modulated by separate signals corresponding with the separate zones of the image.

CHAPTER 6: THE COLOURFUL JOHN LOGIE BAIRD

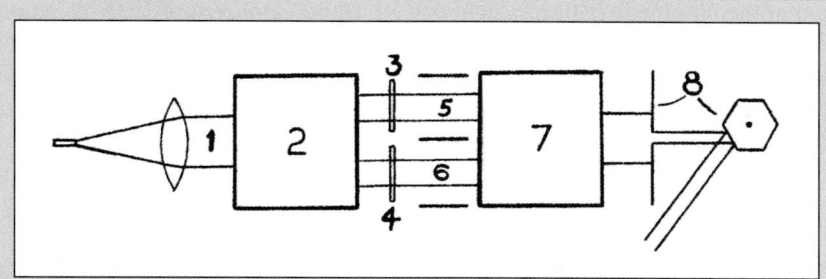

Fig 6.10: Abstract from Baird's optical switch patent *GB473303* for scanning or recombining colour images

In a television system for reception in colour a light beam 1 after passage through an Iceland spar or other double refracting prism 2 is split into an ordinary and extraordinary ray which are passed through two differently coloured filters 3, 4, e.g. red and blue, and thence through electro-optical cells 5, 6 which are modulated by image signals, corresponding to the red and blue component images, the coloured modulated rays being re-combined by a second prism or equivalent 7 and projected on to a screen by a scanning device 8. The cells 5, 6 employed may utilise the Kerr, Pockels, or Faraday effect.

beam of white light is optically processed by a combination of prisms, Kerr cells and a mirror drum to enable a televised colour image to be projected onto a large screen. **Fig 6.10** illustrates the scheme.

After the loss of the Baird Television laboratories in the devastating Crystal Palace Fire on the evening of the 30 November 1936, John Logie Baird continued his development work from the surviving South Tower and from his private laboratory at Sydenham. The decision to develop large screen theatre television for paying audiences gave Baird a commercial outlet for his private development of colour television. Baird had maintained an interest in television for the big screen since the date of his first television patent *GB222604* [21] in 1923.

"The first patent I took out described a television screen consisting of thousands of little lamps, something after the style of the publicity signs in Trafalgar Square where the lamps light up to form figures and letters."

The first showing of television on a Baird active screen to a paying audience was on 28 July 1930 at the London Coliseum. The event was rated 'top of the bill' and with three daily performances ran for three weeks, earning Baird Television fifteen hundred pounds [22]. The screen, which stood 2ft wide by 5ft high, (approx 0.6m x 1.5m) and consisting of 2100 bicycle lamps, produced a brilliant image with very little flicker due to the afterglow effect of the bulbs. On 1 June 1932 Baird surprised an audience at the London Metropole Theatre by presenting the finish of the Derby from Epsom Downs on a larger screen made from three adjacent active screens. This was named 'Zone Television' and had a total screen size of 9ft wide by 6ft high (approx 2.7 x 1.8m) with 6300 picture elements.

In a paper read before the Society of Motion Picture and Television Engineers (SMTE) Convention on 21 October 1947, Captain West described a theatre version of the intermediate film system that J L Baird employed in 1935. A remote television signal was relayed via an advanced centimetric microwave link to the theatre where it was received to be displayed on a small actinic cathode ray tube. A 35mm movie camera synchronised with the television signal photographed each frame as it appeared on the screen and after rapid development, fixing and drying, the film was projected to fill the large screen in the auditorium using the regular projector [23]. The microwave link was powered by split-anode magnetrons designed and fabricated by the Baird Television Company.

John Logie Baird was associated with two different theatre television projects, contributing to the 405-line large screen projector format at Cinema Television Ltd under the technical direction of Captain West and the innovative colour and monochrome television work carried out at his private laboratories.

On 9 April 1936 John Logie Baird, at least on paper, made the transition from mechanical to electronic scanning with British patent *GB473323* [24]. He described the use of a field sequential colour television system using a field synchronised filter disc. The disc rotated in front of a CRT flying spot scanner in the studio, which in combination with photomultiplier tubes picked up live events. The received colour images were reproduced on the screen of a standard monochrome television tube by viewing the screen through a rotating colour filter disc. The field sequential colour television system, although requiring mechanical rotating colours disc, offered an inexpensive solution, however, a working system of this type would not be revealed until 1940.

On 7 December 1936, Baird demonstrated a large screen, 120-line monochrome television picture on an 8ft by 6ft (approx 2.4m x 2m) screen at the Dominion Theatre, Tottenham Court Road in London, **Fig 6.11**. This was a joint John Logie Baird and Cinema Television project based on patent *GB418527* [25], utilising a mirror drum and a slotted disc with a fixed aperture slot.

At the receiver the process was reversed using a mirror drum, an arc lamp and a Kerr cell. **Fig 6.12** illustrates the basic concept with disc 'a' showing a single spiral slot passing a small vertical aperture to constitute the raster scan, while disc 'b' illustrates an alternative multiple spiral slot disc for the purpose of interlacing.

Fig 6.11: The Dominion Theatre, London where Baird demonstrated large-screen television

To generate the 120-line scan a mirror drum with 20 progressively tilted facets deflected the spot generated from an arc lamp, as it passed through the aperture of a disc with six spiral slots. This demonstration was the rehearsal for a colour television projection system using the same concept with a 12 spiral slotted disc. The large screen colour television image was first revealed in December 1937 at a press review held at the Dominion Theatre.

On 4 February 1938, Baird surprised an unsuspecting audience

COLOUR PICTURES

Fig 4.9: Baird televisor screen calibrated for ranging a reflected signal
(see page 55)

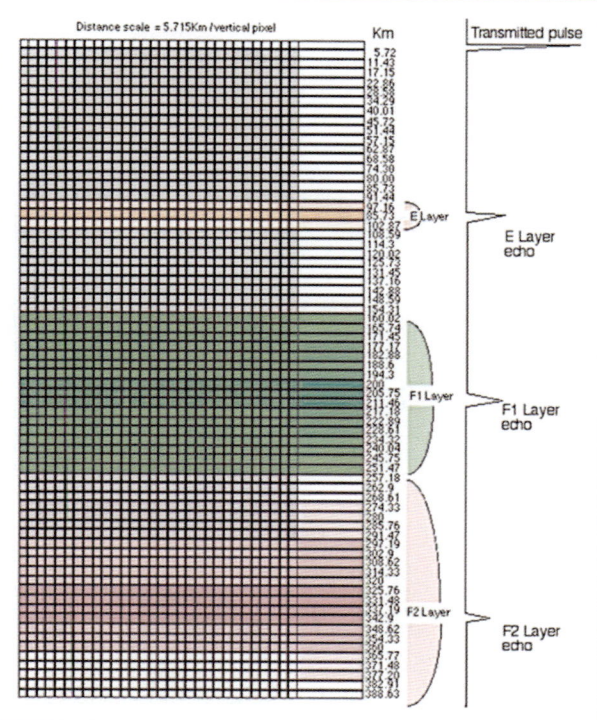

(below) Fig 5.1: Planck energy distribution of the Sun
(see page 69)

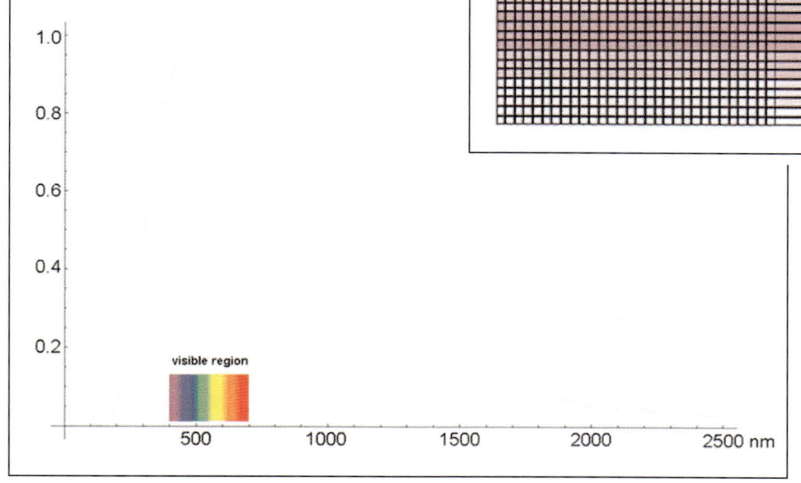

(below left) Fig 5.2: Splitting white light into its spectral components
(below right) Fig 5.3: Recombining the spectral colours to produce white light
(see page 70)

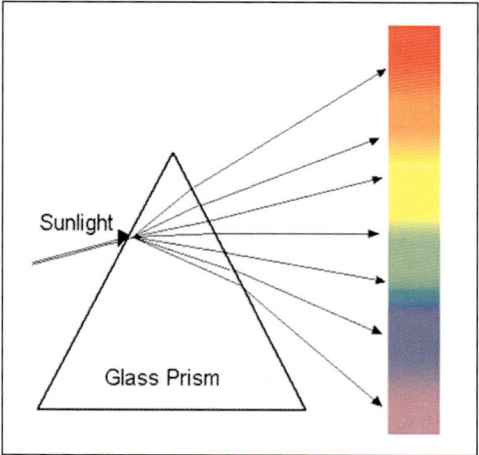
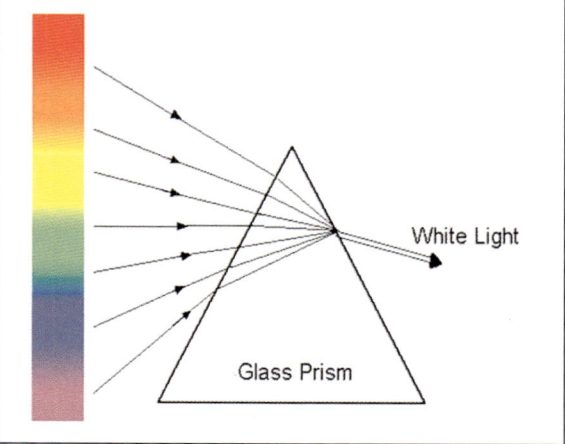

C.1

THE THREE DIMENSIONS OF JOHN LOGIE BAIRD

(left) Fig 5.4: Additive mixing of light
(right) Fig 5.5: Subtractive mixing of pigments
(see page 71)

Fig 5.7: CIE (1931) colour space diagram
(see page 72)

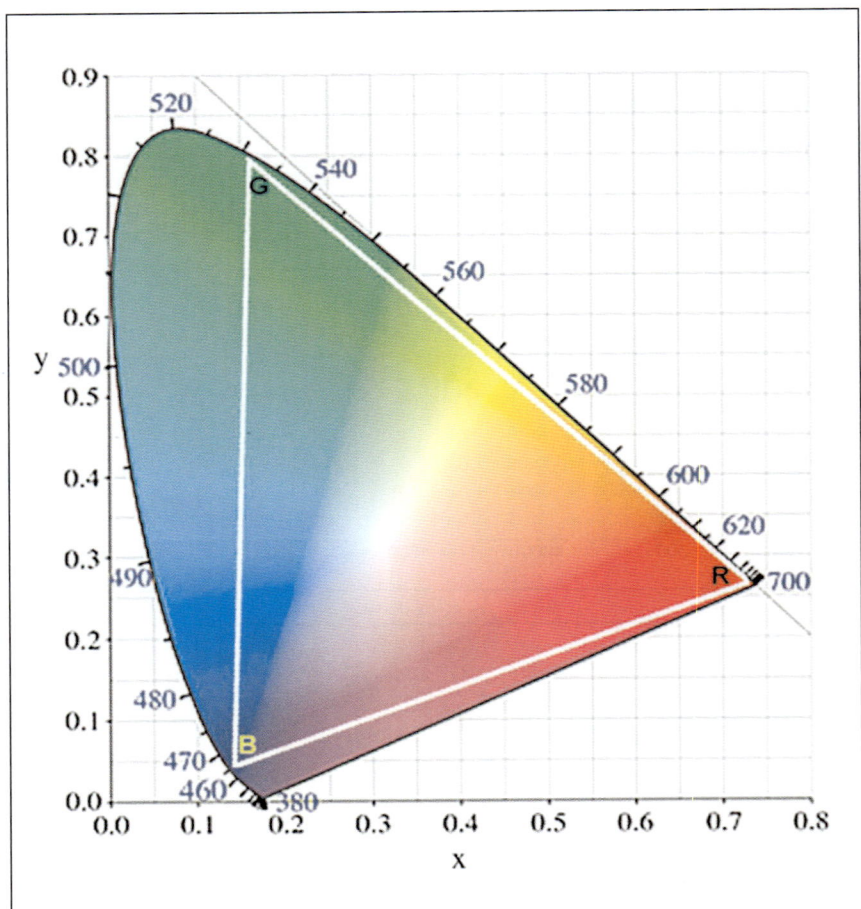

COLOUR PICTURES

(a) Original scene to be photographed using black & white film

(b)　　　　　　　　(c)　　　　　　　　(d)
Black & white photographs respectively taken through red, green and blue filters

(R)　　　　　　　　(G)　　　　　　　　(B)
The above black & white transparencies matched up with the original colour filters and projected onto a screen

Red & green images converged (R+G) on a single screen

Full colour image reproduced by converging the red, green and blue (R+G+B) images

Fig 5.8: Obtaining and reproducing a colour image from black and white film *(see page 73)*

THE THREE DIMENSIONS OF JOHN LOGIE BAIRD

Fig 5.9: Baird selected (red/orange) and (green/blue) to achieve a wide range of colours *(see page 74)*

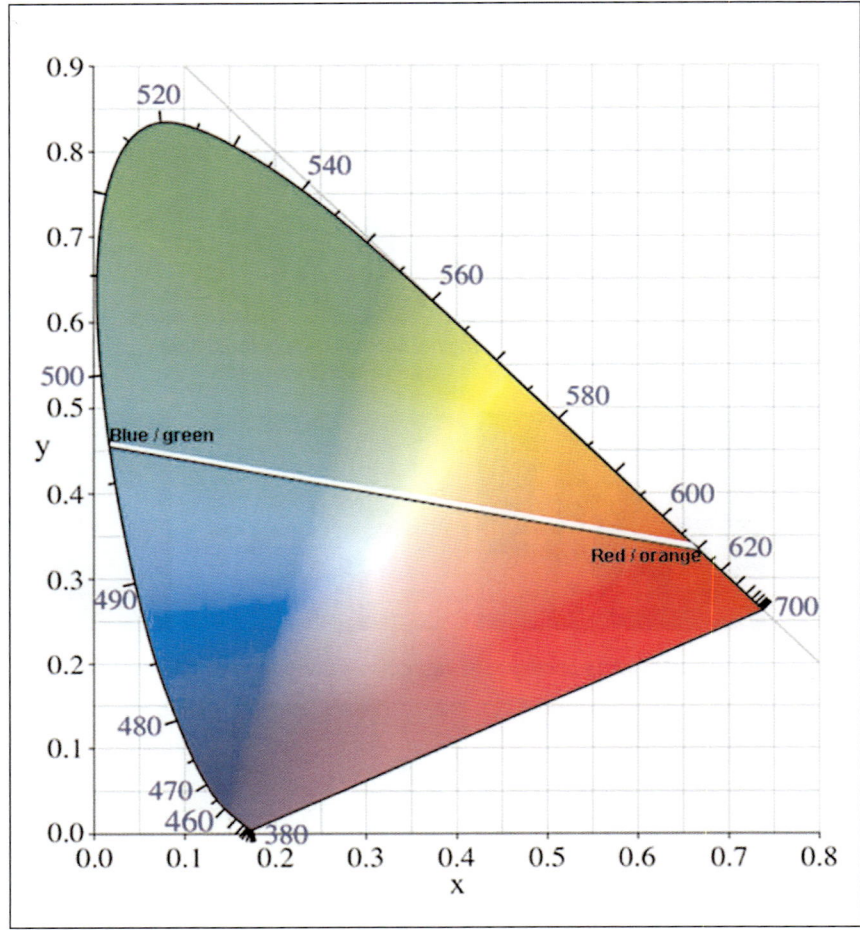

Fig 5.10: Colour interpretation relies on human perception *(see page 75)*

When a wavelength of 495nm (green/blue) is compared to primary green 520nm it may appear as a shade of blue.

When the same colour 495nm is compared to primary blue 465nm the human brain may interpret it as a shade of green.

C.4

COLOUR PICTURES

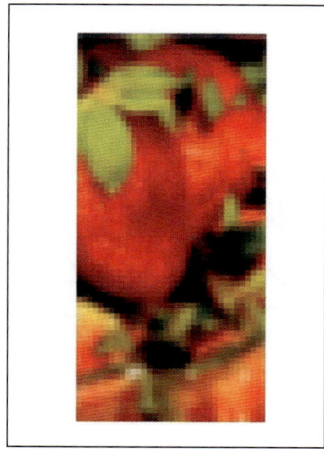

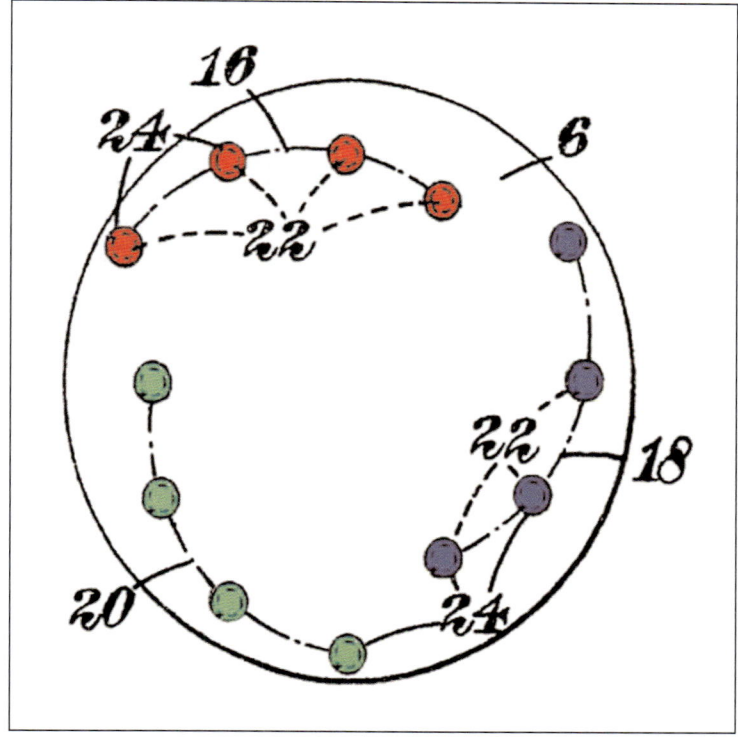

(above) Fig 6.1: Emulation of a low-definition colour television image
(see page 79)

(right) Fig 6.5: Part of Baird's British colour television patent GB314591
(see page 82)

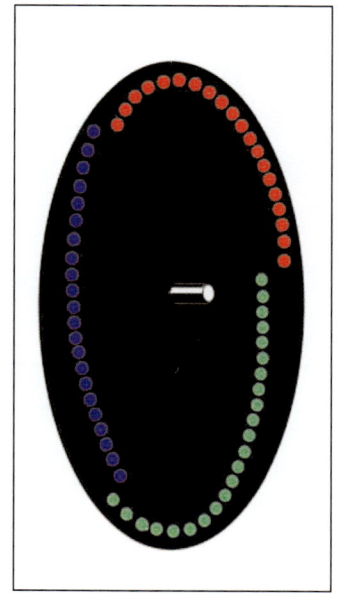

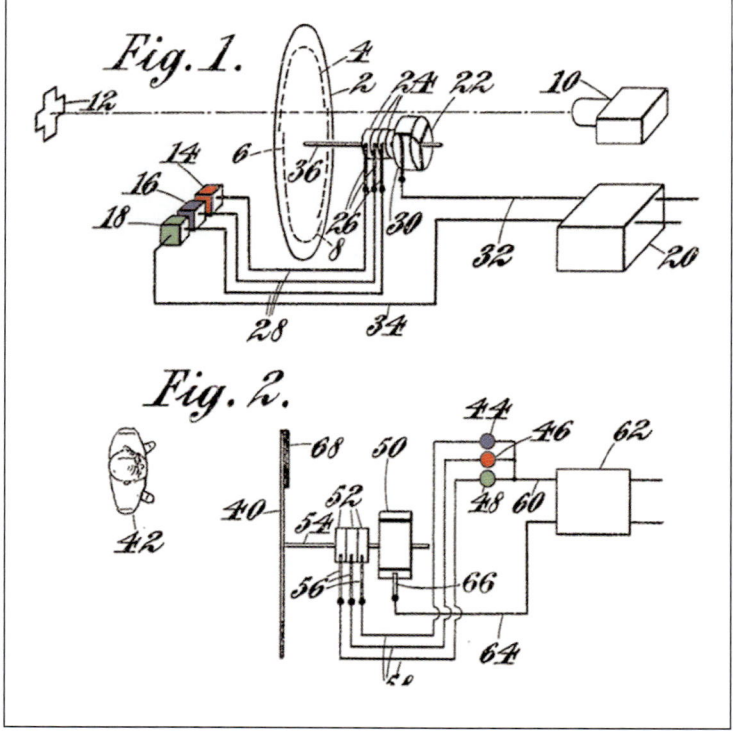

(above and right) Fig 6.6: Illustrations from Baird's colour television patent GB321389
(see page 83)

C.5

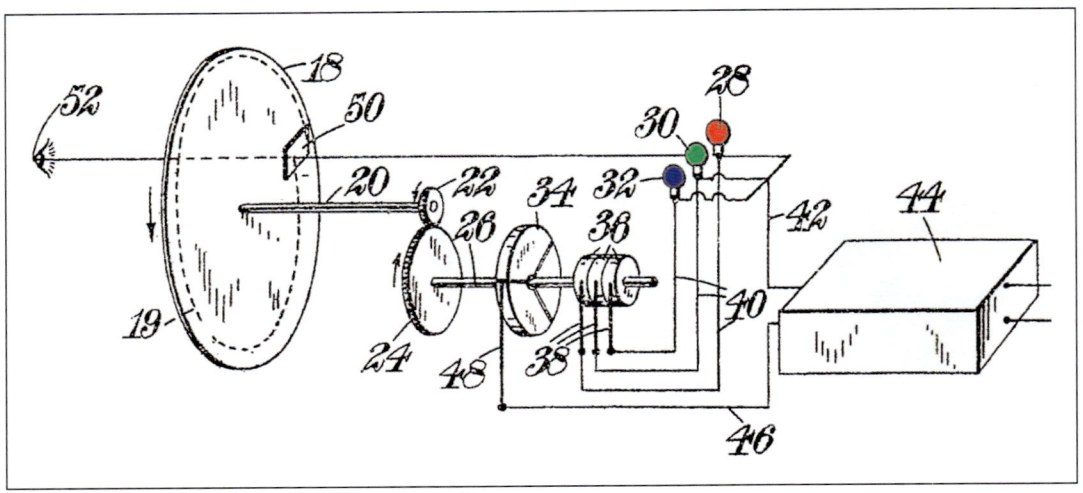

Fig 6.8: Abstract from Baird's patent *GB319307* showing the switching commutator
(see page 85)

(above) **Fig 6.14:** A Baird caesium-antimony photocell manufactured by Sommer at Cinema Television. Baird's patent GB532259 describes its use (1938)
(see page 91)

COLOUR PICTURES

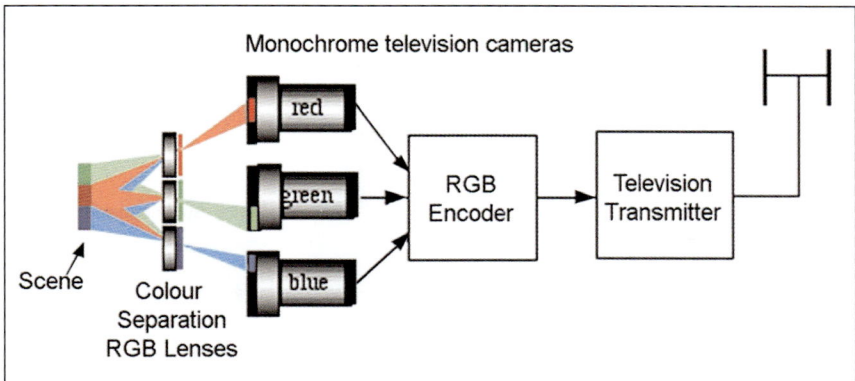

Fig 6.16: The three monochrome camera tube approach to colour separation *(see page 93)*

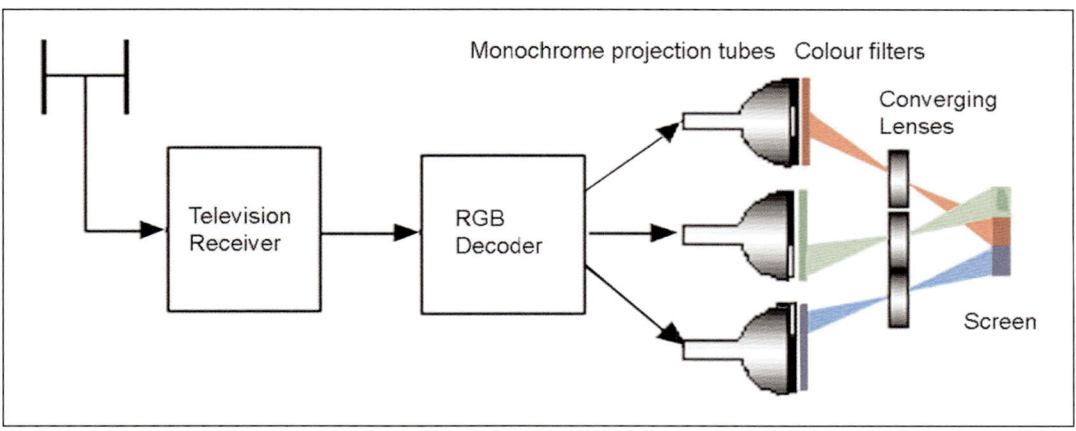

Fig 6.17: The three monochrome CRT colour television receiver *(see page 93)*

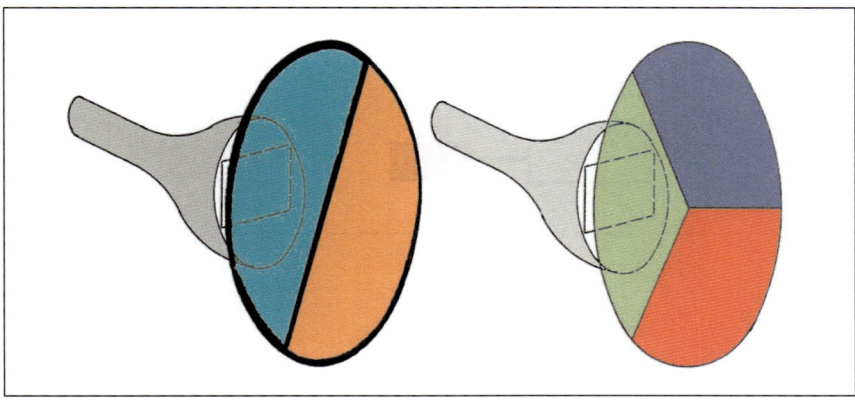

Fig 6.18: Colour filter discs for Baird's field-sequential colour television system *(see page 94)*

C.7

THE THREE DIMENSIONS OF JOHN LOGIE BAIRD

Fig 6.22: Baird's two-colour field-sequential transmitter (top), and receiver (below) *(see page 96)*

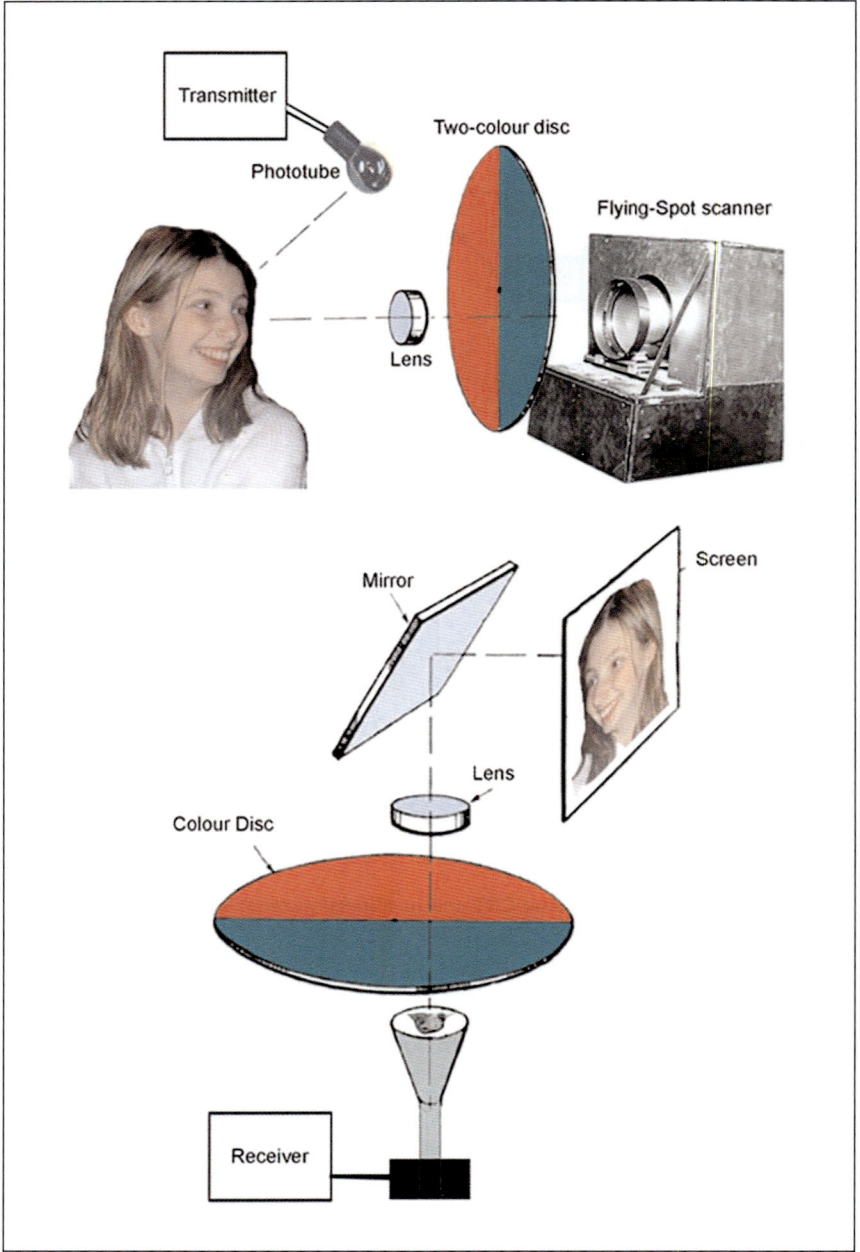

COLOUR PICTURES

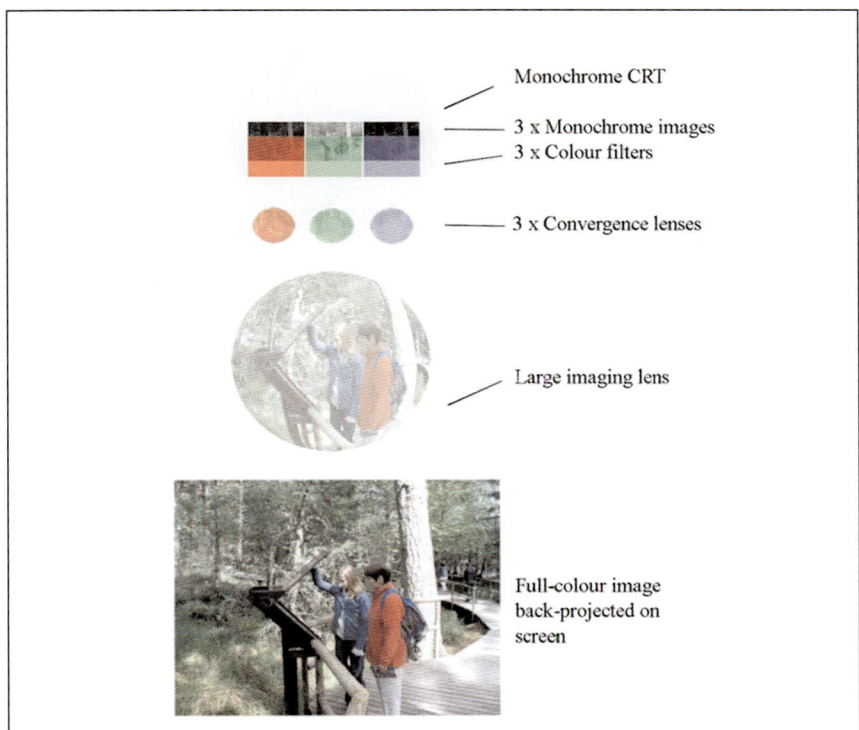

Fig 6.27: The receiver process described in *GB555167* *(see page 99)*

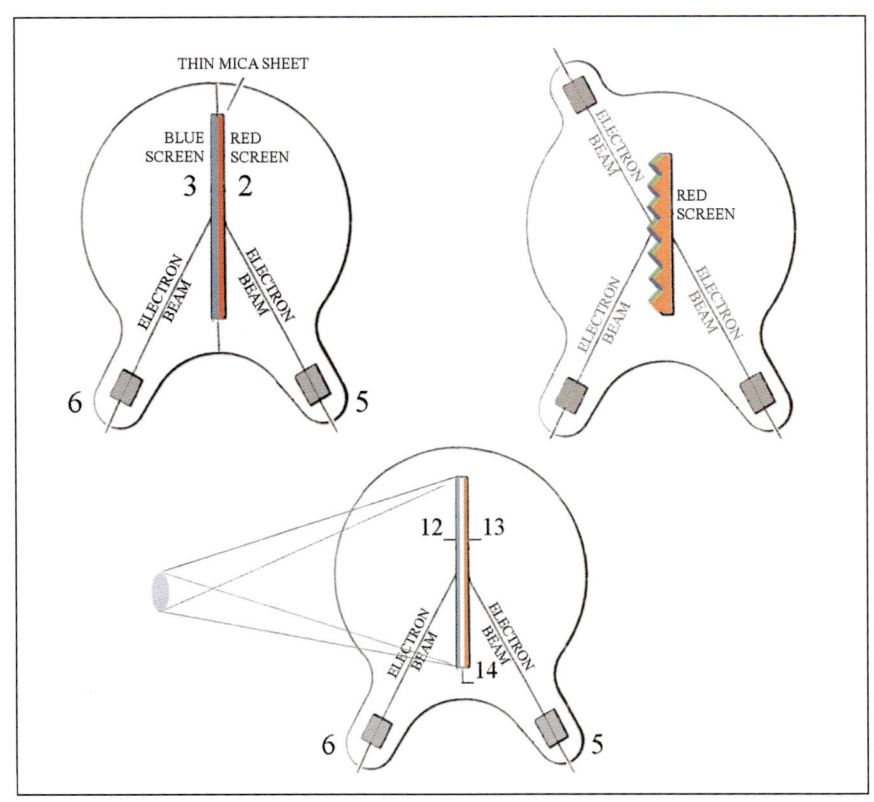

Fig 6.29: Illustrations from Baird's Telechrome patent *GB562168* *(see page 101)*

C.9

Fig 6.31: Illustration from Baird's patent GB562334 (see page 102)

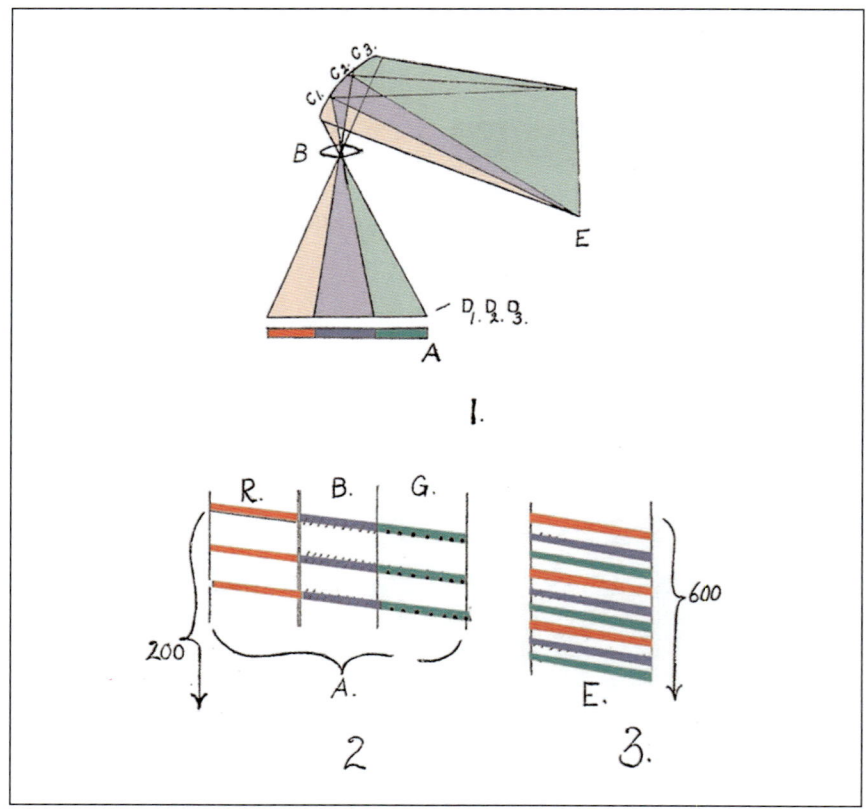

Fig 7.8: Kaplan's parallax colour tube (1947) (see page 112)

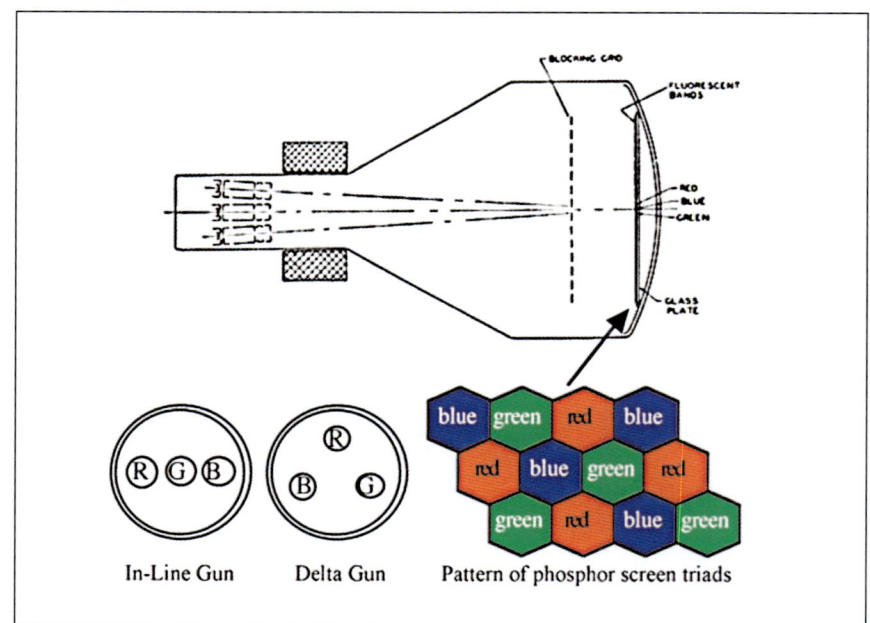

COLOUR PICTURES

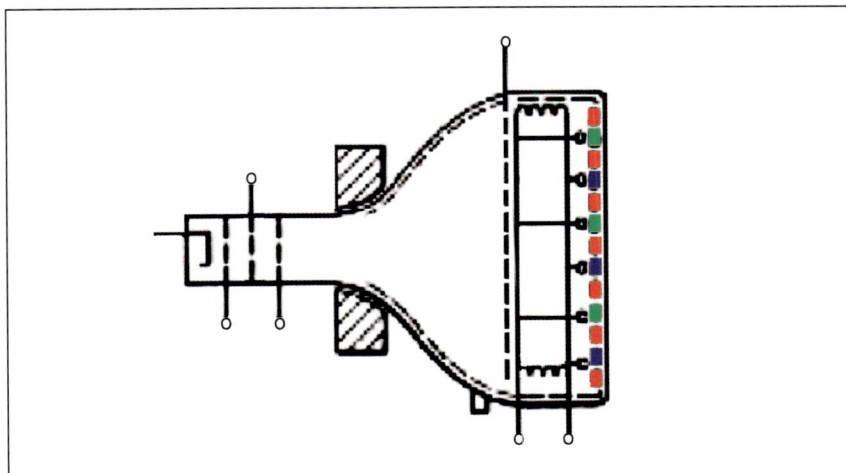

Fig 7.11: Lawrence Chromatron colour tube
(see page 115)

(below) Fig 7.12: The Sony Trinitron
(see page 116)

Vertically Flat Screen
Trinitron's cylindrical screen produces straighter lines, eliminating much of the image distortion found on a conventional spherical screen and producing a clearer, more lifelike picture. In addition, ambient light reflection is eliminated to provide a more pleasing screen image that is easier on the viewers eyes.

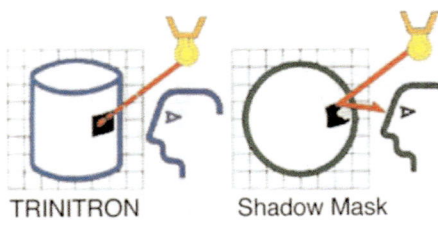

Aperture Grille
The Trinitron Aperture Grille, designed with long, unbroken slits, allows more light and colour to reach the screen, resulting in purer, more beautiful colour images. The mesh-like, darker screen of a conventional Shadow Mask creates a colour shift ("doming") distortion and prevents colour purity.

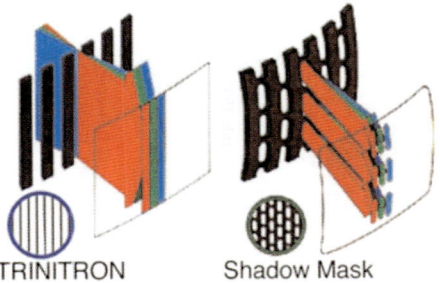

1-Gun, 3-Beam System
Trinitron's simple one-gun, three beam system provides more precise beam focusing. This avoids many problems faced by conventional three-gun systems. For example, the Delta version is troubled with complicated vertical and horizontal alignment and the In-line version requires either smaller lenses or a larger gun. Trinitron's one-gun, three-beam system creates clearer text and adds reality to the picture.

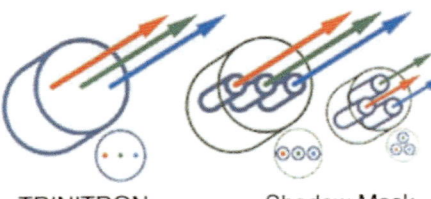

C.11

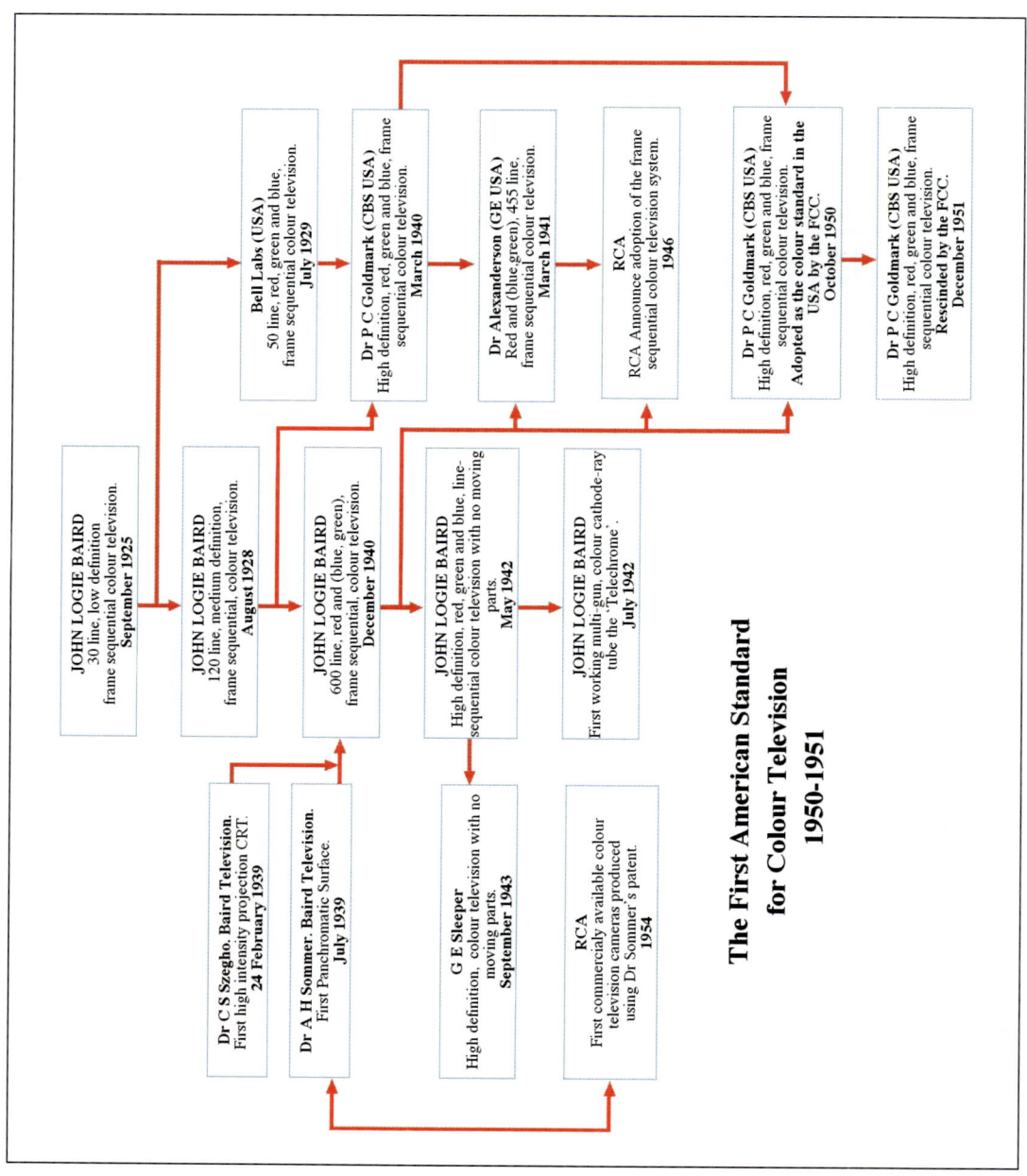

Fig 7.14(a): Baird's influence on the the evolution of frame sequential colour television (see page 120)

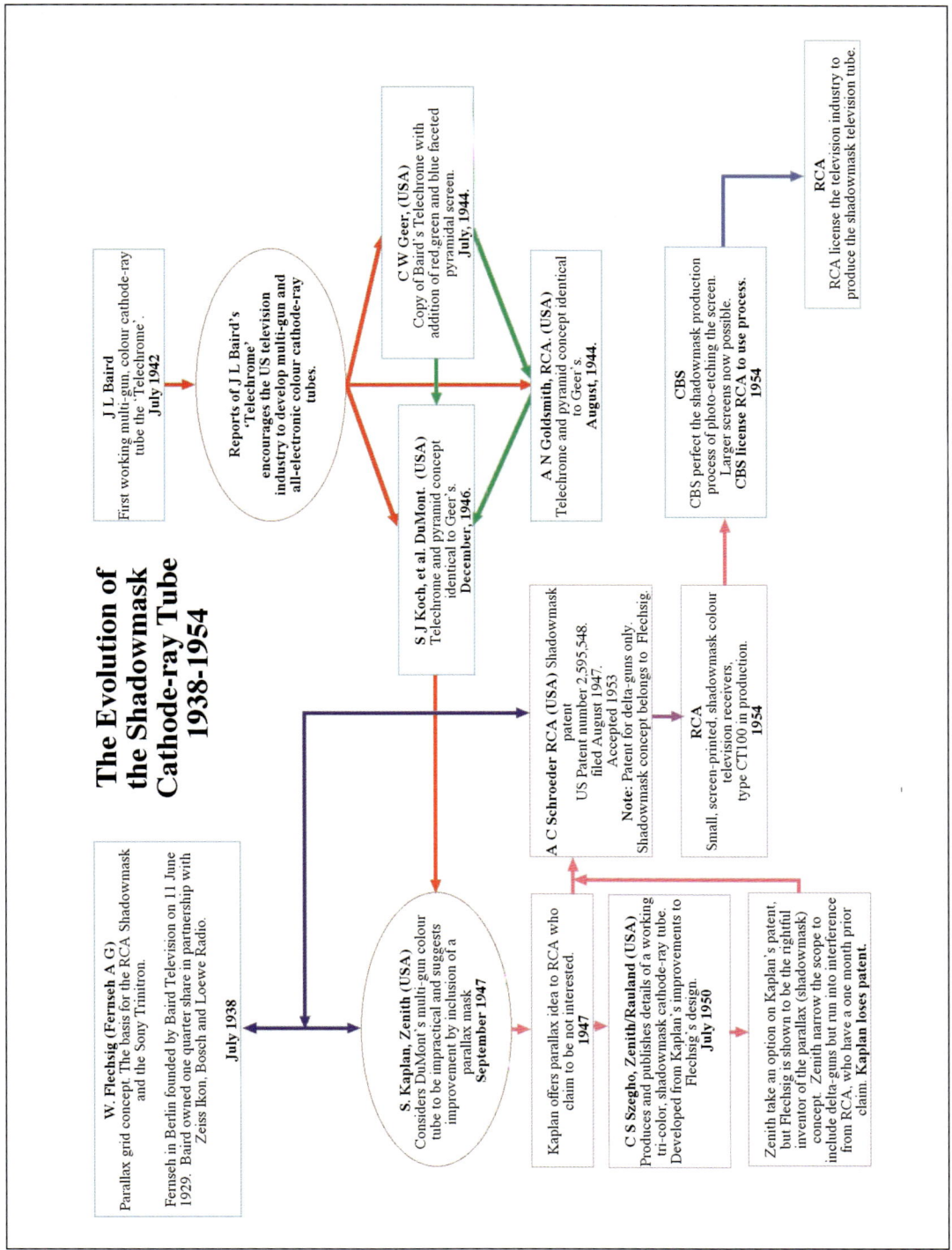

Fig 7.14(b): Baird's influence on the the evolution of the Shadowmask cathode-ray tube (see page 121)

THE THREE DIMENSIONS OF JOHN LOGIE BAIRD

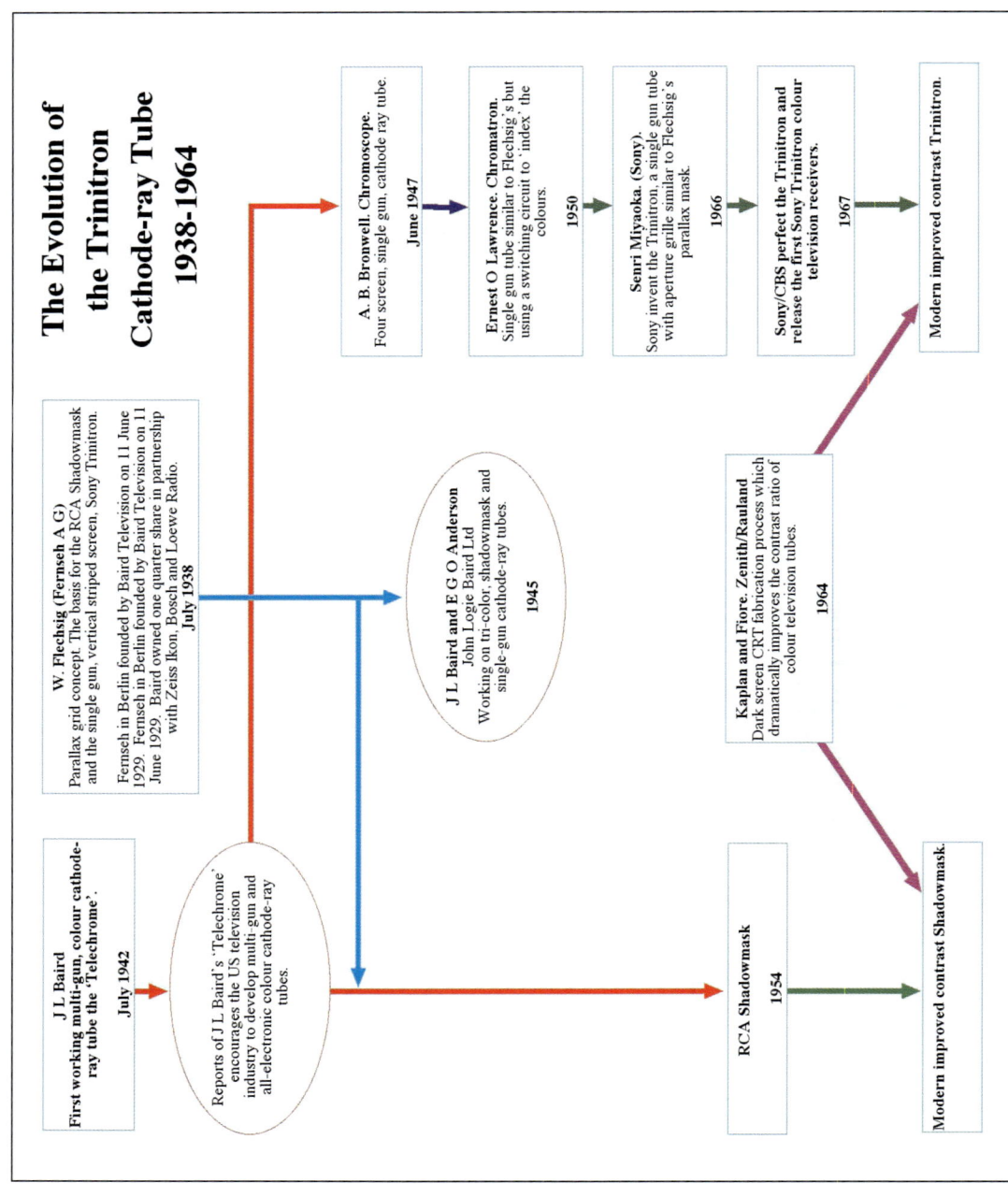

Fig 7.14(c): Baird's influence on the the evolution of the Trinitron cathode-ray tube
(see page 122)

COLOUR PICTURES

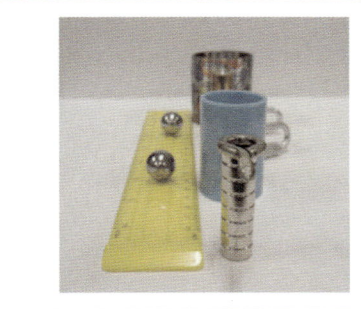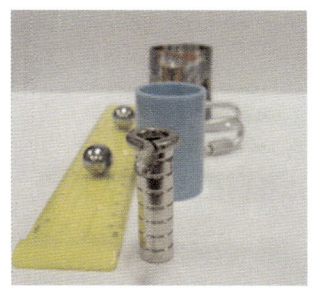

Fig 8.1: Example of a parallel viewed stereo-pair
(see page 126)

Fig 8.2: A colour anaglyph produced to be viewed with a red colour filter over the left eye and a blue colour filter over the right eye
(see page 126)

Fig A2.9: Colour bars photographed from the author's field sequential colour television display
(see page 172)

C.15

THE THREE DIMENSIONS OF JOHN LOGIE BAIRD

Colourised version of the portrait of John Logie Baird shown on page viii

CHAPTER 6: THE COLOURFUL JOHN LOGIE BAIRD

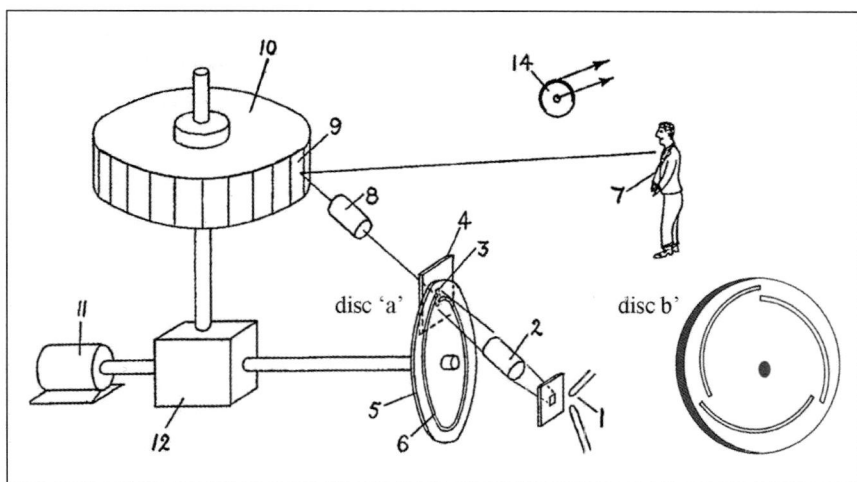

Fig 6.12: Spiral/ slot scanning arrangement described in *GB418527*

of three thousand people at the Dominion Theatre, when he demonstrated the 120-line colour television picture, six-times interlaced and projected on a 12ft by 9ft (3.7m x 2.7m) screen. During this event the pictures from Baird's studios were transmitted to the theatre on 37MHz from the remaining South Tower of the Crystal Palace.

"The colour camera mounted on a four wheel dolly could be trundled through a convenient gate near the studio for televising the red trolley buses in Anerley Hill. Never before had colour television been demonstrated in a theatre, or indeed transmitted by radio link - a double first for Baird." [26]

The two-colour, red/orange and green/blue process involved the use of a disc with 12 concentric slots. At the studio the slots were covered alternately with blue/green and red/orange colour filters to produce six alternate lines of each colour during a single revolution of the disc. A mirror drum with 20 facets, rotated at 6000rpm and the resulting vertical slices of light comprising the television image were reflected onto the rubidium cathode of an experimental photomultiplier cell.

At the receiver a similar arrangement was employed to provide the two-colour 120-line picture on a cinema screen. The rubidium photomultiplier tube was the work of Dr Sommer, a brilliant chemist and photoelectric specialist hired on 1 January 1936 to work with Dr Samson on the development of the all-electronic image dissector camera tube. Sommer's main task was to produce silver-caesium-oxide (Ag-O-Cs) light cells and photomultiplier tubes, which at the time, although being highly red biased and extremely insensitive, were the only useful photoelectric surfaces known. Sommer produced 10-inch (254mm) silver-caesium-oxide photomultiplier tubes for Baird's private laboratory, but due to their spectral response they were ineffective for colour television. Colour television required photomultipliers with a good response that ranged from red through green to blue, but no such device existed.

At Baird's request in 1938, Sommer began developing new experimental photo-surfaces by systematically adding or substituting other elements to the silver-

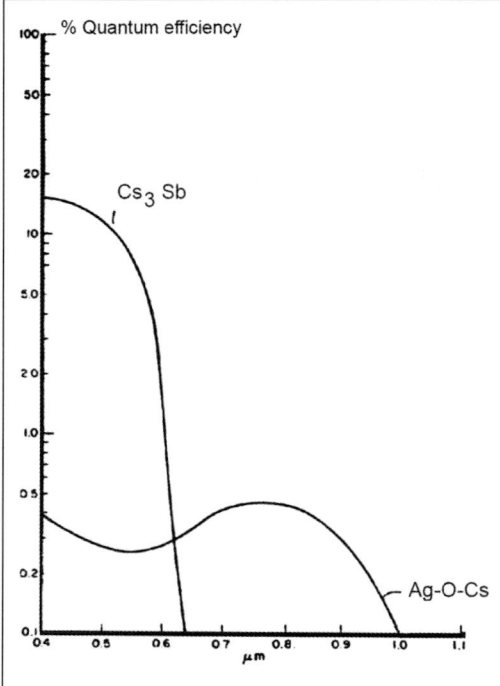

Fig 6.13: The improvement in blue response obtained by the caesium-antimony photosurface

caesium-oxide surface. One of the first experiments included rubidium, which rendered a slight improvement in the blue response. Although these tubes had very low efficiency, the improvement in spectral response over silver-caesium-oxide guaranteed the success of the 120-line colour demonstration.

On 1 November 1938 Sommer made a major breakthrough on a par with the importance of storage in television camera tubes that would revolutionise the sensitivity of all future phototube applications [27]. **Fig 6.13** shows the characteristic curve of the caesium-antimony photosurface, that became known as the 'daylight photosurface' exhibiting an almost 100-fold increase in blue sensitivity over Ag-O-Cs. This surface, later named S-11 was the most sensitive photosurface that existed and was exclusively used by Baird Television to enhance greatly the sensitivity of experimental monochrome television cameras.

The caesium-antimony photosurface was later used by Cinema Television for war work, which included a top secret British Air Ministry project to supply photocells for use in the nose cone of a special anti-aircraft rocket projectile known as the optical proximity fuse, the forerunner of the radar fuse. While this new photosurface was a giant leap forward in photoelectric technology, the peak in sensitivity within the blue region of the spectrum presented colour balance problems making it a poor choice for colour television. The original Ag-O-Cs photocells had resulted in a paper white response on a monochrome television screen for colours in the red and infrared region of the spectrum, while blue appeared black. The reverse situation occurred with the new surface making blue appear white and red to appear black.

In the process of systematically fabricating and testing new surfaces Sommer made another major breakthrough, providing Baird with the perfect solution to his problems, a panchromatic response photomultiplier with good quantum efficiency.

"The story of the Bi-Ag-O-Cs (Bismuth Silver Oxide Caesium) cathode is one of incredibly lucky events. At the time I was worried about the absence of red-response in all TV transmission systems, particularly when the red jackets of the Guards always seemed black on the tube! I tried unsuccessfully to combine the red-sensitive Ag-O-Cs (Silver Oxide Caesium) cathode with the blue-sensitive Cs-Sb (Caesium Antimony) cathode. Without any scientific justification I made a 'Friday Afternoon Experiment' and replaced Sb with its neighbour in the Periodic System, ie, Bi (Bismuth). By sheer luck I obtained a good panchromatic response in this first experiment and naturally continued with the experiments, and immediately applied for a patent in GB and USA. Since the War started soon afterwards, there was no great interest in using the cathode for camera tubes, but I continued (basically in

CHAPTER 6: THE COLOURFUL JOHN LOGIE BAIRD

my spare time because my main job was producing gas-filled phototubes for sound-film projectors). I had two incentives to continue: Baird's requirements for his colour system and some odd requirements by Woolwich Arsenal to improve their evaluation of gun flashes!" [28]

Fig 6.14 [see page C.6 for a colour version of the picture] shows a Baird caesium-antimony photocell manufactured by Sommer at Cinema Television.

The cessation of television broadcasting from Alexandra Palace at the onset of the Second World War and the subsequent reorganisation of the Baird Company led to Baird's contract being severed.

"I was in the middle of some extremely interesting and, I believed, important work on colour television and, rather than see it stopped, I continued this work at my own expense, keeping on two assistants and working in a private laboratory attached to my house at number three Crescent Wood Road. I also sent my name to the authorities and expected to be approached with some form of government work, but no such offer materialised...." [29]

Contrary to this denial, a level of evidence exists to suggest that Baird was involved in secret work involving radar and secret signaling for the Government during the war years. A research programme to identify John Logie Baird's clandestine projects during the Second World War is gathering momentum by a number of researchers, including the author, Prof Malcolm Baird, Dr Peter Waddell and others.

Baird continued:

".... In the meantime, the colour television made very good progress and I was able to show a six-hundred-line colour picture of very fine quality and

In the production of a photo-electrically sensitive surface with a special sensitivity approximating to that of the human eye, a layer of bismuth is evaporated on to a supporting surface such as the glass wall of an evacuated envelope, and a layer of silver is evaporated over the bismuth layer and oxidised by discharge. The composite layer is heated to between 100-150 degrees C. to reduce the silver-oxide and oxidise the bismuth. A layer of alkali metal, such as caesium, is evaporated on to the composite layer at a temperature between 150-200 degrees C. so as to reduce the bismuth oxide and form caesium oxide. It is preferable that only sufficient caesium is used to reduce the bismuth, and the composite layer is preferably finally baked and superficially oxidised till the desired sensitivity is obtained.

Fig 6.14: Abstract from Baird's patent *GB532259* describing the use of a Caesium-antimony photocell produced by Sommer (1938) (see page C.6 for a colour version of the picture)

suitable for broadcasting through the BBC, with very little alteration to their existing apparatus. With the European War raging, however, nothing could be done. My only hope was that I endeavour to keep going until after the War."

Dr Sommer's employment continued with Cinema Television throughout the war years. The former Baird Television Company's intellectual property rights to inventions were dissolved when the company was liquidated in 1940 and Cinema Television emerged. This enabled those named on the patents as 'joint inventor with the Company,' to become the owners of the IPR of the patents. In a letter to the author on 11 March 1991, Sommer described the circumstances surrounding the birth and development of the world's first panchromatic photocathode and how he secured its future:

"In 1947, I visited RCA for the first time and, ironically EMI encouraged me to tell them (RCA) about the new cathode. The reason was, that they felt they should reciprocate for the information they had obtained from RCA on various occasions and thought that it would be very clever to give them (as they thought!) useless information! RCA immediately started to work on the cathode, particularly with colour TV on the horizon, and had no problems reproducing my results. My cathode was used for the next twenty years and made colour television possible. EMI and RCA had patent exchange arrangements, so there was no problem in EMI using my US patent." [30]

Dr Sommer's development work at Baird Television is recorded in a small monograph published in 1946 [31]. Baird invited Sommer to work with him at his private laboratories in Sydenham, but he declined:

"I did not think it would be wise to leave the relatively secure employment of Baird Television to join him full time, but I continued to help him with his colour television project. I had just invented the first panchromatic photocathode (British patent number GB532259) which was very useful to him (It became the standard cathode for TV camera tubes in the 50s and was used in all Image Orthicons thereafter)." [32]

The wider spectral response of the panchromatic surface S-10 is compared with the original silver-caesium-oxide S-1 and the caesium-antimony S-11 photosurfaces in **Fig 6.15**. Sommer's discovery of the world's first panchromatic photo sensitive material gave John Logie Baird an exclusive advantage in colour television research.

In the late nineteen thirties it was generally believed in the USA that it would be a relatively short time before progressing from electronic black and white television to colour. Despite the methods developed by Baird and others, an inexpensive and practical solution proved to be more complicated than first anticipated.

Figs 6.16 and 6.17 [also reproduced in colour on page C.7] illustrate a basic theoretical colour television camera and receiver system. In Fig 6.16 the scene to be televised has been reduced for simplicity to a three-colour pattern comprising conveniently of primary colours; red, green and blue. It is known from the previous chapter that combinations of carefully selected red, green and blue primary colours of light mix by addition to produce the colours that occur in nature.

Lenses fitted with coloured separation filters are arranged to focus the red, green and blue components of the scene on correspondingly assigned monochrome television cameras. Each camera therefore produces a signal corresponding to the intensity of its associated primary colour in the scene.

Fig 6.17 illustrates a suitable receiver and colour display. Near-white cathode ray tubes are used to display the corresponding grey-scale pictures obtained from the cameras at the transmitter.

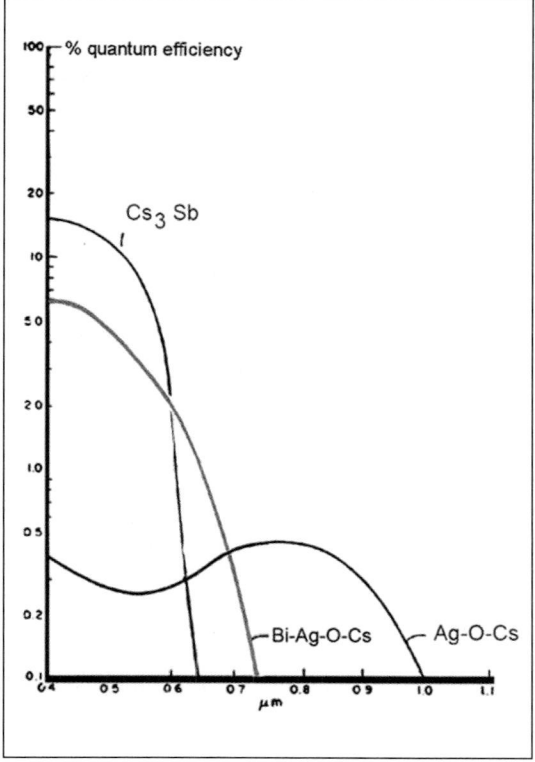

Fig 6.15: Panchromatic (Bi-Ag-O-Cs) photosurface characteristic

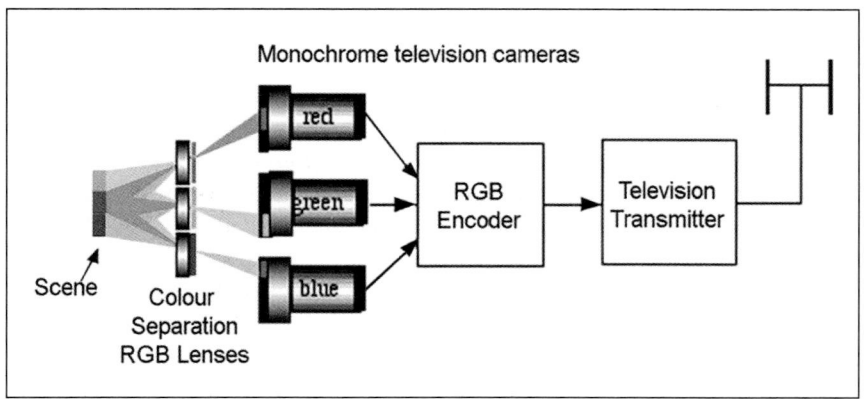

Fig 6.16: The three monochrome camera tube approach to colour separation (see page C.7 for a colour version)

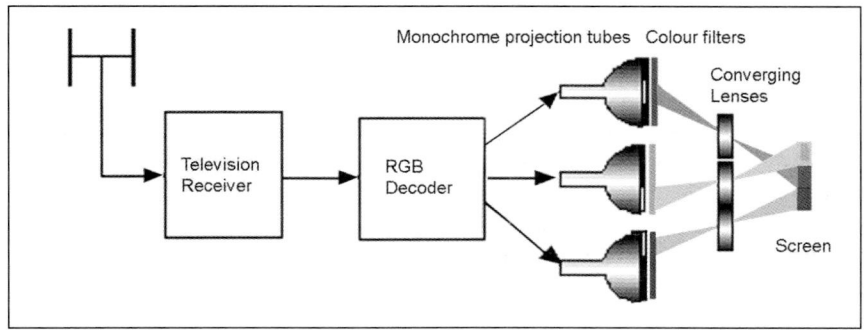

Fig 6.17: The three monochrome CRT colour television receiver (see page C.7 for a colour version)

Lenses fitted with coloured separation filters, converge the red, green and blue images to reassemble the original scene in full colour. While a quality colour television picture could be achieved in the laboratory by this process, the expense and complication precluded it from being a practical solution for domestic television. Evans and Little [33] reported that equipment of this this type was used to produce RCA's large screen colour television images demonstrated in 1947.

The possibilities of colour television attracted Dr Goldmark, an admirer of John Logie Baird's work, who left Britain to take up the position of Chief Television Engineer for the Columbia Broadcasting System (CBS). In 1940, based on Baird's rotating colour disc concept, Goldmark carried out the first experiments of a partial electronic colour television using cathode ray tubes for both scanning and receiving.

Goldmark gave a demonstration on 29 August 1940, of a 343-line system with a picture frequency of 60 frames per second and a colour frequency of 120 frames per second [34]. The Columbia system, which used a rotating red, green and blue, sequentially scanning filter disc was essentially the same as Baird's.

In the absence of a single cathode ray tube capable of producing the red, green and blue pictures simultaneously, Baird's returned to his field sequential CRT patent of 1936, *GB473323* [35]. This meant replacing mechanical scanning with a cathode ray tube flying-spot scanner, and by December 1940 he had developed the partial electronic field sequential system. The two-colour sequential scanning technique was designed to complement the twin-field interlaced, 405-line television system by arranging that alternate interlaced fields represented the blue/green and red/orange primaries of a colour picture. According to Parr:

> *"While it is essential to use the three primary colours for a true rendering of all colours, reasonably satisfactory results can be obtained by two colours which are a compromise between the three selected primaries - namely red-orange and green-blue.*
> *All three colours superimposed will, of course produce near-white. If the same experiment is made with the two-colour filters, you will see that 'white' is a pinkish white and that yellow is not obtained in a pure hue."* [36]

Fig 6.18 [also reproduced in colour on page C.7] shows disc filters for the two and three-colour field sequential process. Further Baird patents later supplemented this process. The first *GB508039* [37] on 24 December 1937 described

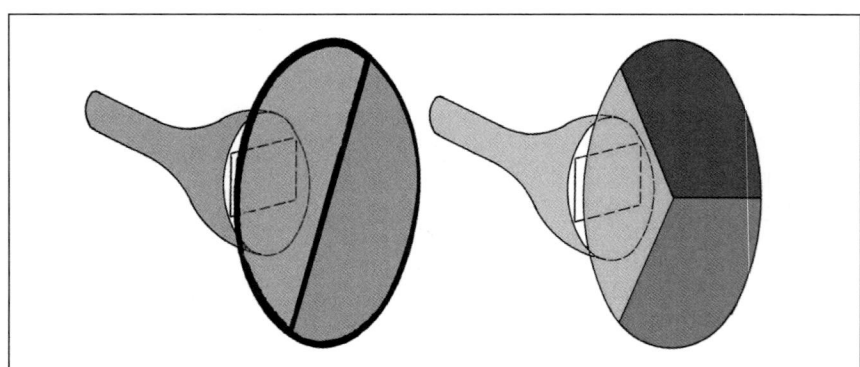

Fig 6.18: Colour filter discs for Baird's field-sequential colour television system (see page C.7 for a colour version)

the use of interlaced colour fields **Fig 6.19**. Another patent, *GB545078* [38] (**Fig 6.20**), on 7 September 1940, described a method of interlacing three fields of 200-lines to reproduce a 600-line colour television picture.

However, while the concept of rotating filters offered the only available practical solution, there were disadvantages. The addition of the rotating filters introduced mechanical inertia and an effect known as colour frequency flicker that became evident when there were large areas of the image consisting of either red or green.

Another disadvantage of the colour sequential system was that the image of a white ball moving at speed across the screen would break up into the primary colours. Some of these effects could be reduced at the expense of increasing the field rate to assist with persistence of vision, but the addition of mechanical components was considered a retrograde step. Baird derived his intensely bright electronically generated flying-spot from a large projection cathode ray tube developed by Dr Szegho for Cinema Television. The reflected light from the subject was picked up using one of Sommer's photomultiplier tubes fitted with the new panchromatic colour cathode. Colour separation at the transmitter and its recombination at the receiver being carried out by means of a rotating colour filter disc of the type described in *GB473323* (**Fig 6.21**).

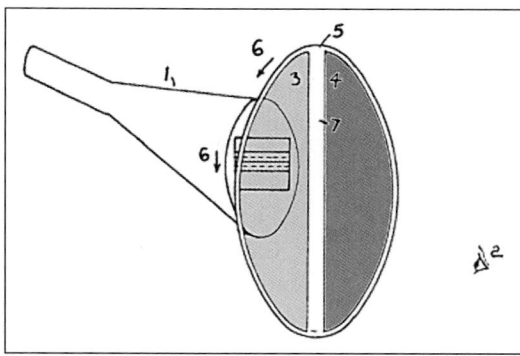

Fig 6.19: Interlacing described in patent GB508039

Each picture is scanned in three sets of 200 lines interlaced to produce 600 lines in all, and successive sets of 200 lines are passed through red and blue-green filters alternately whereby alternate lines of a complete picture of 600 lines are red and blue-green and of the next picture blue-green and red, the whole system being such that interlaced colour reproduction can be obtained by a receiver designed to receive the complete interlaced colour scans, and at the same time the same transmission can be received by a simple received designed to receive only 200 lines non- interlaced pictures.

FIg 6.20: Abstract from Baird's patent GB545078, describing a method of interlacing

In a colour-television system the screen 1 of a cathode ray tube at the receiver is viewed through a revolving disc 2 divided into coloured sectors or filters 4, 5, 6, preferably red, green and blue, the phase and speed of rotation of the disc being arranged so that the screen is viewed through the red filter when the red component is being reconstituted and so on. A disc fitted with only two coloured filters may be employed and, in all cases, co-operates with a correspondingly coloured and synchronised disc at the transmitter. Preferably the disc is rotated in a direction corresponding with the direction of the low frequency component of scanning, while reciprocating filters may be substituted for the rotating disc. If desired, the image of the screen 1 may be optically projected through filters on to a viewing screen.

Fig 6.21: Abstract from Baird's colour television patent GB473323

The discs were in perfect synchronization with the television field enabling, as the disc rotated at 1500rpm, a series of colour encoded monochrome television signals. This resulted in producing a signal that could be received as a series of alternately colour coded monochrome images on a conventional 'black and white' television receiver. The colour disc at the receiver being in synchronisation with the incoming television fields, was rotated at precisely the correct time to recreate by persistence of vision and colour addition the original colourful scene.

Baird designed this system as a bolt-on method of obtaining colour television for 405-line interlacing, where each even field of 202.5 lines would be encoded blue/green and every odd field with red/orange. Although recognising that a tricolor system would be the ultimate goal, Baird believed that a two-colour compromise would allow Britain to take the lead with colour television after the war.

Baird's demonstration of this new system was reported in the April 1941 edition of *Electronics and Television*. It was reported that he demonstrated a high quality 600-line colour television picture using cathode-ray technology coupled with blue/green and orange/red primary coloured filter discs. A quite remarkable 600-line colour photograph taken directly from the screen of Baird's colour 2.5ft x 2ft screen (approx 0.67m x 0.6m) showing a close-up of Miss Paddy Naismith, a visitor to the press demonstration, was published. Despite the image being generated from a two-colour, flying-spot scanner, the rich tones of her red hair, natural flesh tones and her floral garland were reproduced with excellent fidelity. **Fig 6.22** [also reproduced in colour on page C.8] illustrates the basic principle of the transmitting and receiving systems. The line structure of the resulting television image was hardly perceptible and the definition was similar to the former British analogue 625-line colour system.

Working independently from Baird Television, c/o Cinema Television, Baird applied for three patents on 23 October 1940. Two of the patents described a scheme to replace the mechanical filter disc, *GB545462* [39]

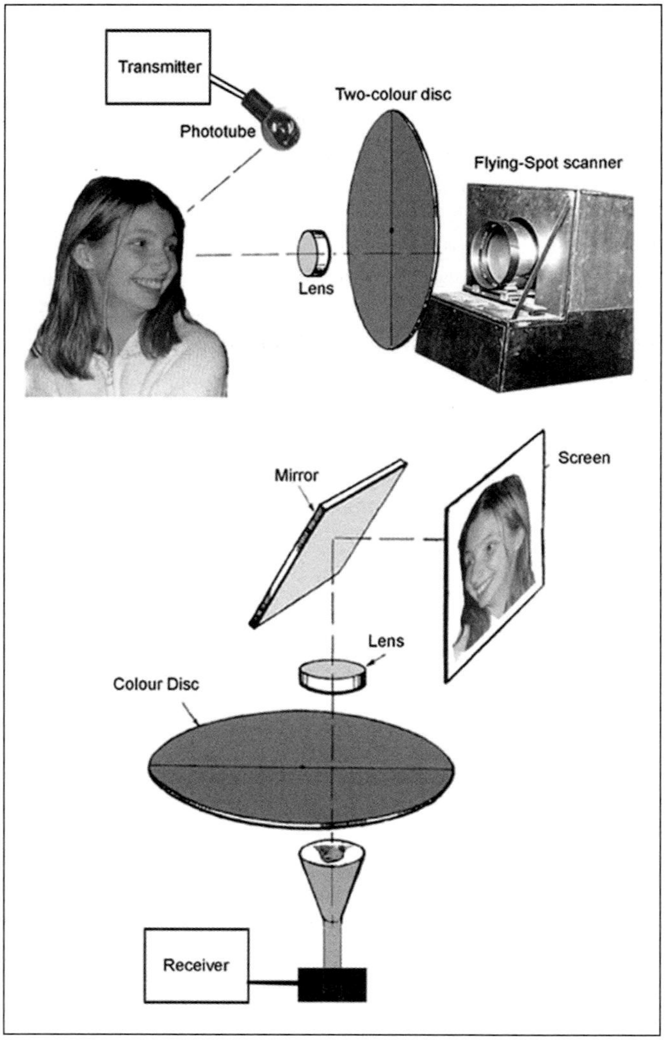

Fig 6.22: Baird's two-colour field-sequential transmitter (top), and receiver (below) (see page C.8 for a colour version)

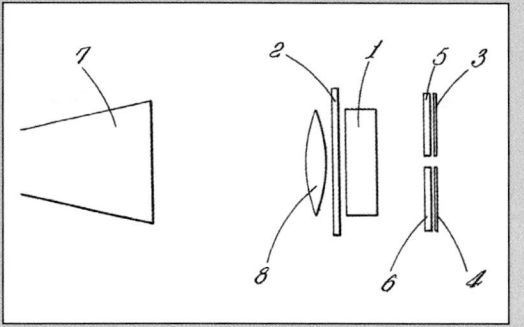

A transmitting or receiving apparatus for colour television comprises a Kerr cell 1 in association with a polarising prism 2 on one side and a plurality of differently coloured light filters 3, 4 and polarising prisms 5, 6 on the other side arranged to cover different portions of the Kerr cell, said prisms 5, 6 being polarised at different angles so that when voltages are applied to the Kerr cell to vary its polarising angle, the light passing through the cell will pass through a corresponding colour filter and will be cut off from the other filter or filters. In a modification, separate Kerr cells are used for the separate filters. In a transmitter this apparatus is placed against the lenses of the transmitter and different voltages are applied to the Kerr cell or cells at each alternate picture to pass red and blue-green alternately. At a receiver, the apparatus is placed against a lens projecting on to a viewing screen the image formed on the screen of a cathode-ray tube 7.

Fig 6.23: Abstracts from J L Baird's patents *GB545462* and *GB545491* describing the replacement of the rotating filter disc with an electro-optical device

and *GB545491* [40]. The concept electro-optically directed a television field to pass through the appropriate colour filter by means of polarising filters and an electro-optical switch, **Fig 6.23**.

The third patent *GB545603* [41] covered a less practical method of displaying a field sequential display. **Fig 6.24** illustrates this process, showing a cathode ray tube located within a set of rotating filters formed in the shape of a cylindrical drum.

As a further refinement to the two-colour rotating disc system described in GB423323 Baird applied for a Patent on 13 January 1941, *GB546470* [42], describing a modification to the shape of the junction between the filters on a rotating disc, **Fig 6.25**, to enable the colours to follow the scan lines accurately when a colour camera with a storage mosaic is used.

In March 1941 it was reported that Dr Alexanderson, a consulting engineer for General Electric (GE) demonstrated an identical system of field-sequential colour television to Baird's. This was similar to the detail of employing Baird's choice of duochromic orange/red and green/blue primary colours.

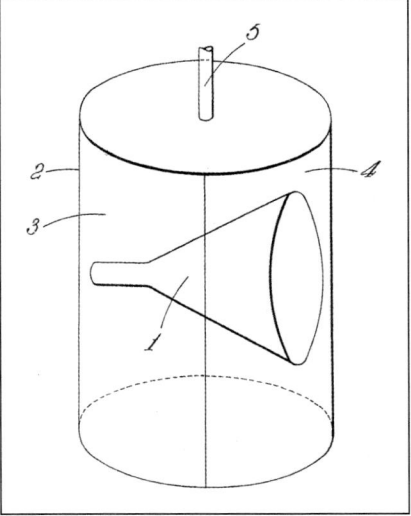

Fig 6.24: Cylindrical drum filter disc described in patent *GB545603*

Fig 6.25: Junction between filters is modified to perfectly match the storage matrix of an electronic camera device

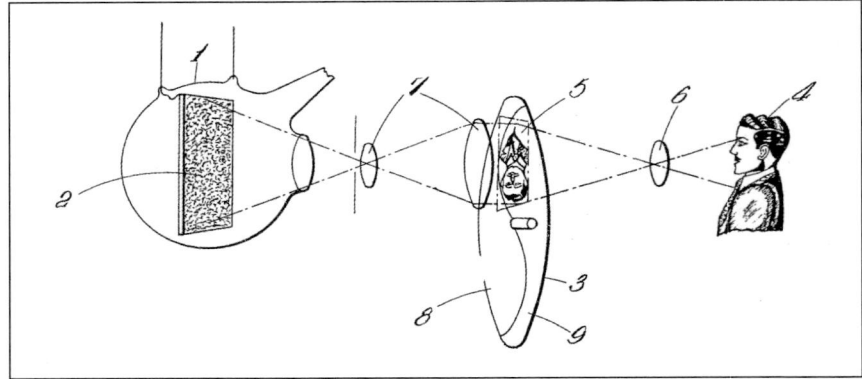

"The demonstration was staged at Dr Alexanderson's home, where he had installed a two-colour 24 inch revolving disc about a foot in front of the picture end of a cathode ray tube of his standard type receiver. As this whirled about at 1800rpm, its transparent field of orange-red and greenish blue reproduced the studio program in realistic colours." [43]

Inglis:

"Although the field sequential camera is not used for NTSC colour, it played an important role in the development of colour television. It was an effective vehicle for the promotion of colour in its early years, and it provided a test bed for research and development in colour. In addition, threat of its competition gave RCA and the rest of the industry the incentive to expedite the development of a compatible system.

With selected tubes, controlled conditions, and the skilled operation of Goldmark and his staff, it was capable of producing excellent pictures, judged by the quality standard of the day." [44]

On 1 June 1941, experimental field sequential colour television broadcasting was inaugurated from Station WCBW of CBS in the USA. On 20 June 1941, Anderson of the Radio Corporation of America (RCA), which held strong connections with (GE) also took an interest in Baird's colour television experiments and applied for a field-sequential patent to enable colour signals to be achieved from a monochrome iconoscope camera tube [45].

With no other income, Baird's financial situation was under considerable strain. He was using his dwindling savings to pay his remaining one full-time member of technical staff, Eddy Anderson and part-time glassblower, Arthur Johnson. The situation improved when Lord Inverforth and Sir Edward Wilshaw of Cable and Wireless met with Baird. It was decided that they would benefit from a mutual association with the famous inventor of the Baird system of television. Baird was appointed 'Consulting Technical Adviser' at a salary of one thousand pounds per annum for a period of three years on 1 November 1941.

In April 1942, Dr Goldmark (CBS) gave all due credit to the prior colour work of Baird including the one-hundred and twenty-line large screen colour demonstrations in 1938, and the cathode ray tube system with colour filters in 1939.

CHAPTER 6: THE COLOURFUL JOHN LOGIE BAIRD

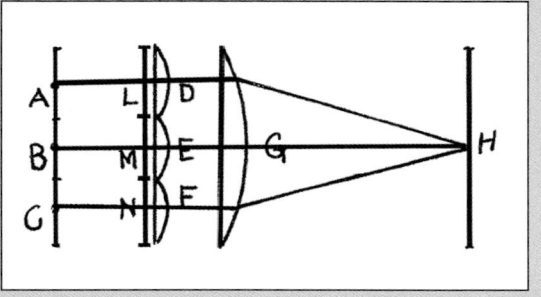

For producing television pictures in colour an optical system is used comprising lenses D, E, F, Fig. 1, side by side one for each primary colour, having their foci at the corresponding primary images A, B, C, and a lens G having its focus on a screen H and adapted to converge the parallel beams from the lenses D, E, F so that the projections of the primary images A, B, C are overlapped on the screen F. The colours may be produced by filters, L, M, N. An additional lens may be used between the lenses D, E, F and the primary images A, B, C when these pictures are small. The optical system may be used in the reverse direction in transmitters. The transmitter may be one using a scanning light spot producing separate fields in different colours by light filters or coloured powders in the cathode ray tube. Or, the transmitter may be of the flood lighting type producing separate images through coloured filters on a photo-sensitive surface.

Fig 6.26: Abstract from Baird's patent *GB555167* for a tri-band RGB cathode-ray tube

With the income from Cable and Wireless, Baird and his private staff continued with television projects and on 13 May 1942 applied for a futuristic colour television patent *GB555167* [46] requiring no moving parts. **Fig 6.26** describes an optical system to enable three images to appear positioned side-by-side on the face of a single cathode ray tube, to be projected and converged onto a screen as a full RGB colour image. The process is shown in more detail in **Fig 6.27** [also reproduced in colour on page C.9].

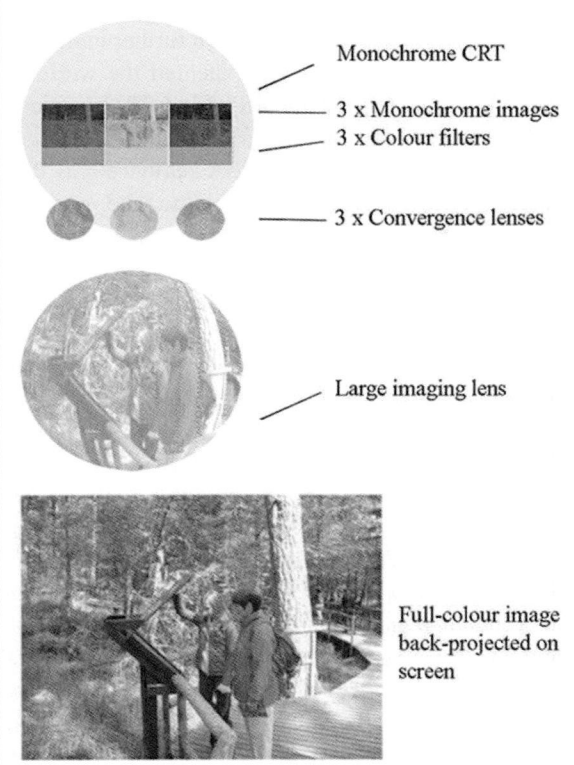

Fig 6.27: The receiver process described in *GB555167* (see page C.9 for a colour version)

99

Parr wrote in *Television*:

"As an alternative to the rotating filter disc it is possible to project a succession of coloured pictures on to the screen in sequence at such a rate that persistence of vision enables a complete coloured picture to be seen.
This is done in Baird's system by first reproducing the frames one above the other or side by side on the screen of a projection tube in the correct sequence. Lenses with their optical centres on the perpendicular through the centres of each of the images project the light onto a second large lens placed at a distance equal to the focal length from the matt receiving screen." [47]

On 11 July 1942 Baird patented a method of enabling the RCA iconoscope or EMI Emitron electronic camera or similar to generate side-by-side separate colour television images for the tri-band colour system (**Fig 6.28**).

Fig 6.28: Abstract from Baird's patent GB557992

A television transmitter is provided with three mirrors D, C, B set at different angles adjacent a single lens E and respectively fitted in front with red, blue and green filters to project three separate images F, G, H of an object A on to adjacent portions of a photo-electric mosaic K which is scanned in any suitable way.

John Logie Baird progressed much further in the field of colour television and on 25 July 1942 invented and patented the world's first all-electronic colour cathode ray tube, *GB562168* [48] (**Fig 6.29**) [the pictures are also reproduced in colour on page C.10].

This development made the USA heavyweights RCA, CBS and GE seriously pay closer attention. Incorporated in the tube, which he named the 'Telechrome' was a translucent colour-mixing fluorescent screen to enable colour images to be visualised. The 'Telechrome' (**Fig 6.30**) captured the imagination of the US television industry in their endeavours to produce the first commercially viable electronic colour television. On 16 August 1944 the London press were invited to see Baird's new invention of an all-electronic colour television, ready to be taken up by the radio industry and transmitting stations as soon as the war was over:

"The Telechrome as Baird called his new system, was demonstrated in his house in South London, among little heaps of broken glass, demolished pieces of furniture, and behind shutters which had been blasted into lopsided angles: Hitler's flying bombs, several of which had landed in the vicinity, had apparently tried to stop his work, but in vain; the intrepid Scotsman had made up his mind to stick to it, and he saw it through with the same unflinching determination that had been his characteristic feature a score of years before at Hastings." [49]

Baird made this new system of television entirely 'electronic' eliminating convergence lenses and revolving discs. The innovative Telechrome was the world's

CHAPTER 6: THE COLOURFUL JOHN LOGIE BAIRD

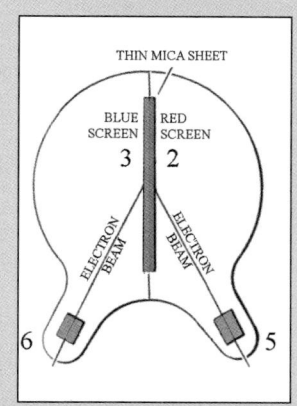 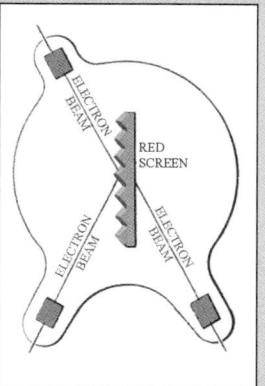 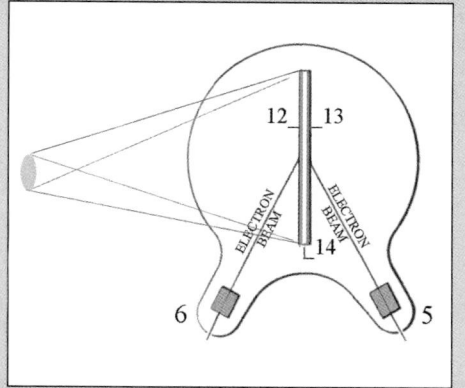

A cathode-ray tube for colour television Figure 1 [above left] has two screens 2, 3, each giving a different colour when scanned by an associated electron beam from guns 5, 6 and so placed that the coloured images are superposed when viewed from either side. In a modification for three colours, one side is divided by ridges into two screens 9, 10, Figure 2 [above centre], scanned by separate guns 6, 11. The transmitter has a corresponding tube comprising a transparent conductive plate or mesh 14, Fig 3 [above right], bearing on either side mosaic screens 12, 13 sensitive to a different colour and scanned in a suitable sequence by beams from guns 5, 6. In a transmitter of the light-spot type the tubes of Figs 1 and 2 may be used as light sources. According to the Provisional Specification, a light filter may be interposed between the screens so as to impart the desired colour characteristic to one screen when viewed through the filter.

Fig 6.29: Abstract from Baird's Telechrome patent *GB562168* (see page C.9 for colour versions of the pictures)

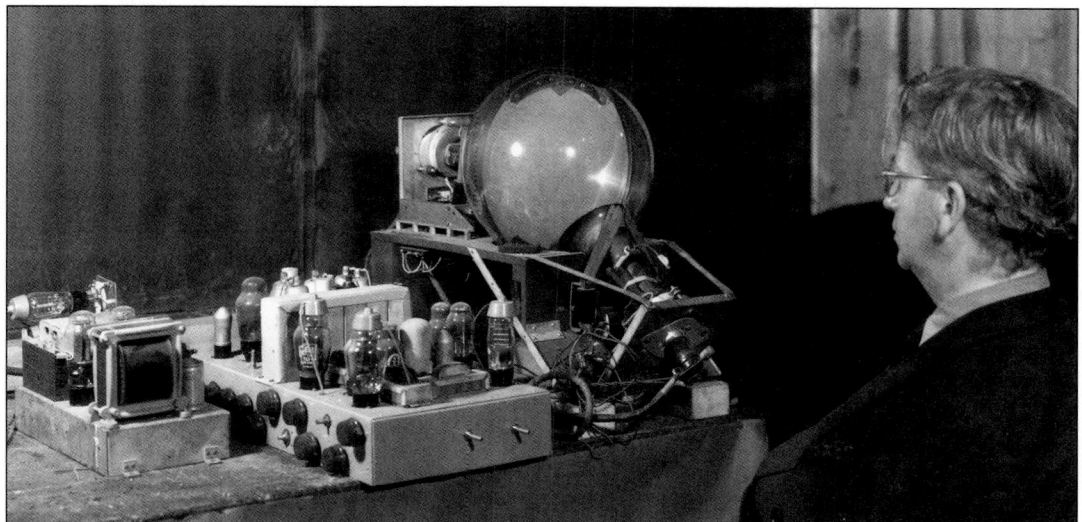

Fig 6.30: John Logie Baird working on the Telechrome colour television receiver. Photo courtesy of Science and Society Picture Library

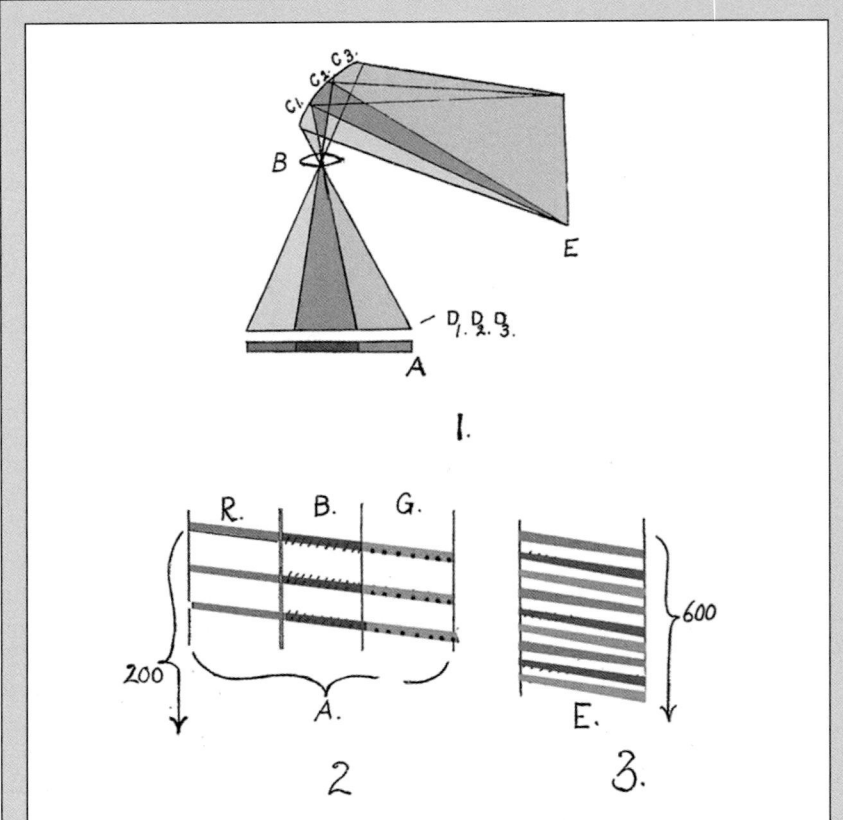

Fig 6.31: Abstract from Baird's patent *GB562334* (see page C.10 for a colour version of the picture)

A colour television apparatus comprises an electronic scanning system in which each frame is made up of successive lines of different basic colours, and successive frames are superimposed in such a way that individual lines are successively of different basic colours. In one arrangement, Fig 1, the screen A of a cathode-ray tube is focused by a lens B on to a scene or surface E to be scanned. Three colour filters (Red, Blue and Green) D1, D2, 'D3 are provided, one for each third of the screen A. Three mirrors C1, C2, C3 are slightly inclined to each other to superimpose each third of the screen A on the same surface E. The scanning of the screen A is by 200 lines according to the sequence indicated in Fig 2, the lines of each third being differently coloured by the filters D1, D2, D3. In the superimposed projection on the surface E, there will be 600 lines in alternate colours as indicated in Fig 3, the first line being Red. For the next. frame, the scanning ray is deflected the space of one of these 600 lines, so as to make the first line Blue. For the third frame, the ray is again deflected to make the first line Green. This sequence is then repeated. Modifications utilizing other numbers of lines and interlaced successions of frames are briefly mentioned. For two-colour reproduction, the cathode-ray tube, described in Specification 562168 may be used, opposite sides of different colours of a common screen being scanned in alternate lines.

first custom designed colour television cathode ray tube. The tube was constructed with two cathode-ray guns and instead of a conventional single fluorescent screen a centrally located transparent double-sided screen was used.

On the front face of the screen there appeared an image containing the orange-red colour components of a picture, and on its rear face the complementary colour components of blue-green were reproduced. The images produced separately by the two cathode ray guns were visually superimposed on the central screen where they were merged together and viewed as a colour television picture.

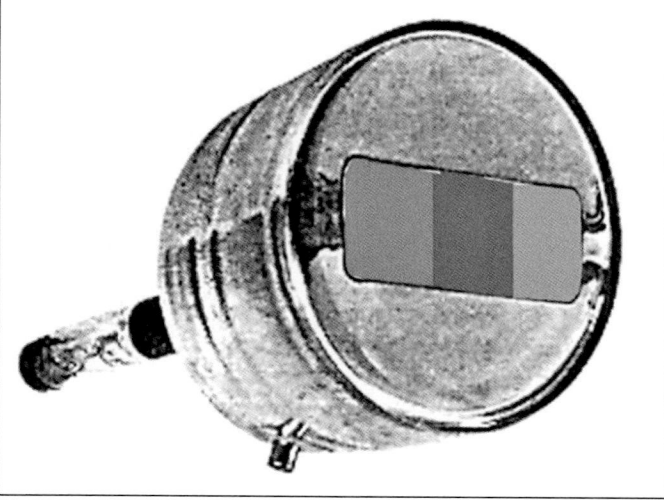

Fig 6.32: Sleeper's tri-band colour tube built by Rauland (1947)

Baird also included in his Patent a three-colour version of the tube (the middle picture in Fig 6.29) but it is not known if this version was constructed. In the three-colour tube, a third cathode-ray gun is added; the front face of the screen gave the red component while the back of the screen comprised ridges facing the blue and green guns respectively.

The application date of the Telechrome patent is 25 July 1942, which indicates that once again he was developing at least two years in advance of his demonstrations. The right hand picture in Fig 6.29 shows a further version of the Telechrome where it is to be employed as a storage camera device.

In his patent application *GB562334* [50], applied for on the 10 October 1942, Baird described a method of completely eliminating the colour flicker associated with frame sequential systems.

This patent described the line sequential system employed in the Telechrome during the 1944 demonstration. By scanning the red, green and blue primary images line-after-line instead of frame-after-frame, the colour frequency becomes very high and as a consequence colour flicker disappears at normal frame scan rates. The illustration of the process from the patent is shown in **Fig 6.31** [the picture is also reproduced in colour on page C.10]. The line sequential process of optically mixing the colours on the screen would result in colour quality similar to the analogue colour television standards: PAL, Secam and NTSC etc.

Baird's concept in *GB555167* of presenting three images on a monochrome CRT with converging and imaging lenses would appear again in the form of US patent *US2389645* [51] by George E Sleeper of Colour Television Incorporated, San Francisco, **Fig 6.32**.

The patent was accepted in 1945, and in 1947 only one year after J L Baird's untimely death. Szegho [52], formerly of Baird Television Ltd, who by that time was leading the Cathode-Ray Tube Division of the Rauland Corporation, fabricated Sleeper's copy of the cathode ray tube with its three coloured phosphor bands.

The next chapter shows that John Logie Baird's patents continued to be the catalyst for the development of colour television in the USA.

References

[1] Russian Patent No *10738,* A A Polumordvinov, applied 23 December 1889.
[2] DRP Patent No *172,376*, W Von Jaworski and A Frankenstein, applied 20 August 1904, issued 21 June 1906.
[3] British Patent *GB7219*, J L Adamian, applied 1 April 1908, issued 28 May 1908.
[4] British Patent *GB30188*, A C Anderson and L S Anderson, applied 24 December 1908, issued 22 September 1910.
[5] French Patent number *572,716*, G Valensi, applied 3 January 1923, issued 12 June 1924.
[6] US Patent *US1,725,710*, J Hammond, Washington DC, applied 15 August 1923, issued 20 August 1929.
[7] *John Baird: The Romance and Tragedy of the Pioneer of Television,* S Moseley, Oldhams Press Ltd, London, 1952, p106.
[8] *Television: Today and Tomorrow, 2nd ed*, H J Barton-Chapple and S Moseley, Sir Isaac Pitman & Sons Ltd, London, 1931, p12.
[9] *Sermons Soap and Television, 2nd ed*, J L Baird, Royal Television Society, London, 1990, p81.
[10] British Patent *GB266564*, Colour and Stereoscopic Television, J L Baird, applied 1 September 1925, issued 1 March 1927.
[11] British Patent *GB267378,* Colour Television, J L Baird, applied 1 September 1925, issued 1 March 1927.
[12] British Patent *GB312560*, J L Baird, applied 30 November 1927, issued 30 May 1929.
[13] British Patent GB314591, J L Baird, applied 4 January 1928, issued 4 July 1929.
[14] British Patent *GB321389*, J L Baird. applied 5 June 1928, issued 5 Nov 1929.
[15] British Patent *GB322776*, J L Baird, applied 9 June 1928, issued 9 Dec 1929.
[16] British Patent *GB319307*, J L Baird, applied 20 June 1928, issued 20 Sept 1929.
[17] 'Colour Television', Goldmark, Dyer, Pierce and Hollywood, *Proceedings of the IRE*, April 1942, p163.
[18] 'Baird's Colour Television 1937- 46', R Herbert, *Television: Journal of the RTS,* Jan/Feb 1990, p24.
[19] British Patent *GB476195,* J L Baird, applied 12 December 1935, issued 14 June 1937.
[20] http://www.lighting.philips.com.
[21] British Patent *GB222604*, J L Baird, applied 25 July 1923, issued 9 October 1924.
[22] *J L Baird: Sermons, Soap and Television,* Royal Television Society, London, p104.
[23] 'Development of Theatre Television', A G D West, *Journal of SMPE. New York*, August 1948, Vol. 51, pp127-168.

[24] British Patent *GB473323,* Baird Television and J L Baird, applied 9 April 1936, accepted 11 Oct 1937.
[25] British Patent *GB381898,* J L Baird, application date 30 May 1932, accepted 13 October 1932.
[26] 'Baird's Colour Television 1937- 46', R Herbert, *Television: Journal of the RTS,* Jan/Feb 1990, p24.
[27] Detailed Information is given in *Images Across Space,* RSGB, ISBN 978-1-874289-21-0.
[28] Dr A H Sommer, extract from a letter to the author, 11 March 1991.
[29] *Television and Me: Memoirs of John Logie Baird*, J L Baird edited by Malcolm Baird, Mercat Press.
[30] Dr A H Sommer, extract from a letter to the author, 11 March, 1991.
[31] 'Photoelectric Cells', Dr A H Sommer, *Methuen's Monographs on Physical Subjects,* Methuen & Co Ltd, London, January 1946, p92.
[32] Dr A H Sommer, extract from a letter to the author, 11 March 1991.
[33] 'Large-screen colour television projection', Evans and Little, *Journal of the Society of Motion Picture Television Engineers,* April 1955, pp169-173
[34] 'Review of Progress in Colour Television', G Parr, *Journal of the Royal Television Society, Vol.3*, London, 1942, p251.
[35] British Patent *GB473323,* Baird Television and J L Baird, applied 9 April 1936, accepted 11 Oct 1937.
[36] 'Review of Progress in Colour Television', G Parr, *Journal of the Royal Television Society, Vol.3*, London, 1942, p251.
[37] British Patent *GB508039,* Baird Television and J L Baird, applied 24 Dec 1937, accepted 26 June 1939.
[38] British Patent *GB545078,* Baird Television and J L Baird, applied 7 September 1940, accepted 11 May 1942.
[39] British Patent *GB545462,* J L Baird, applied 23 October 1940, issued 28 May 1942.
[40] British Patent *GB545491,* J L Baird, applied 23 October 1940, issued 28 May 1942.
[41] British Patent *GB545603,* J L Baird, applied 13 January 1941, issued 15 July 1942.
[42] British Patent *GB546470,* J L Baird, applied 23 October 1940, issued 4 June 1942.
[43] 'High Lights and Side Lights', G Bartlett and P C Caldwell, *General Electric Review*, March 1941, p185.
[44] *Behind the Tube: A History of Broadcasting Technology and Business,* A F Inglis, Butterworth 1990, p283.
[45] US Patent *US2297524,* E I Anderson, RCA, application date 20 June 1941, issued 29 September 1942.
[46] British Patent *GB555167,* J L Baird, applied 13 May 1942, issued 6 August 1943.
[47] 'Review of Progress in Colour Television', G Parr, *Television, Journal of the RTS, Vol.3,* 1942, p251.
[48] British Patent *GB562168,* J L Baird, applied 25 July1942, filed complete 23 July1943, issued 21 June 1944.

- [49] *Inventors' Scrapbook*, E Larsen, Lindsay Drummond Ltd, London, 1947, pp90-92.
- [50] British Patent *GB562334*, J L Baird, applied 10 October 1942, issued 28 June 1944.
- [51] Patent *US2389645*, G E Sleeper, Washington DC, filed 5 February 1943, issued 27 November 1945.
- [52] 'Color Cathode-Ray Tube With Three Phosphor Bands', C S Szegho, *Journal of the SMPTE*, October 1950, Vol.55, p367.

7

Colour from Cathode Rays

The history of colour television development in the USA began only a few years after the first demonstrations of television in 1925 by J L Baird in England, and C F Jenkins in the US [1]. The Association of Radio Manufacturers, formed in 1924, was renamed the Radio Manufacturers Association (RMA) in 1928, the forerunner of the Electronic Industries Association. The first RMA Television Committee included a representative from Baird's Company in England [2]. Following the Selsdon report in the UK, the RMA sponsored a National Television Standards Committee (NTSC) in cooperation with the Federal Communications Commission.

NEW SOURCE MATERIAL REVEALS that the Radio Corporation of America (RCA), General Electric (GE) and Columbia Broadcasting System (CBS) were not alone in considering John Logie Baird's work on colour television to be of value. On 11 July 1944, Geer disclosed an invention in US patent application *US2480848* [3] (**Fig 7.1**) that had a strong resemblance to all-electronic, two and three colour 'Telechrome' tube of Baird as described in the previous chapter.

Although Geer's application was preceded by J L Baird's corresponding US application, it was Geer who was awarded the US Patent because he could prove that he had the idea prior to the convention date of the English application. In 1944, J L Baird offered the Telechrome to the Rauland Tube Division, which had recently acquired Baird Television Incorporated of New York, but Rauland rejected the Telechrome with the argument that the viewing screen was inside and not on the face of the tube.

The surge of interest, which followed the announcement of the Telechrome suggested that Baird may have also attracted the interest of other American companies with a view to licensing its manufacture. Baird's main competitors in the US television industry then began producing their own versions of Baird's two and three gun Telechrome of 1942. Geer's patent, which cites Baird's

A transparent plate has on each side, one of the colours of a two colour system and is scanned with the two guns arranged opposite the sides, or this plate has on one side a line raster of triangular cross section where the two sides of the triangle are covered with the two colours of a three colour system, the third colour being again on the other side of the plate.

Fig 7.1: Abstract from Geer's patent *US2480848* (see also Fig 7.2)

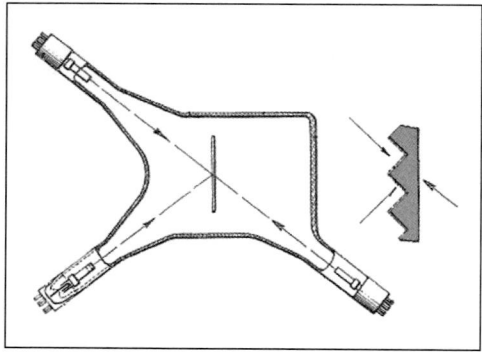
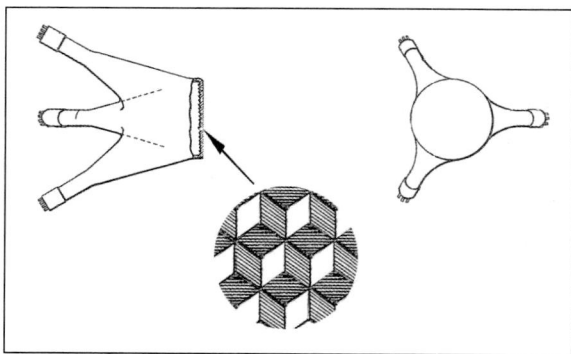

Fig 7.2: Geer's patent *US2480848* **(1944) based on Baird's (1942) Telechrome**

Fig 7.3: Geer's alternative pyramid screen enables the cathodes to be located at the rear

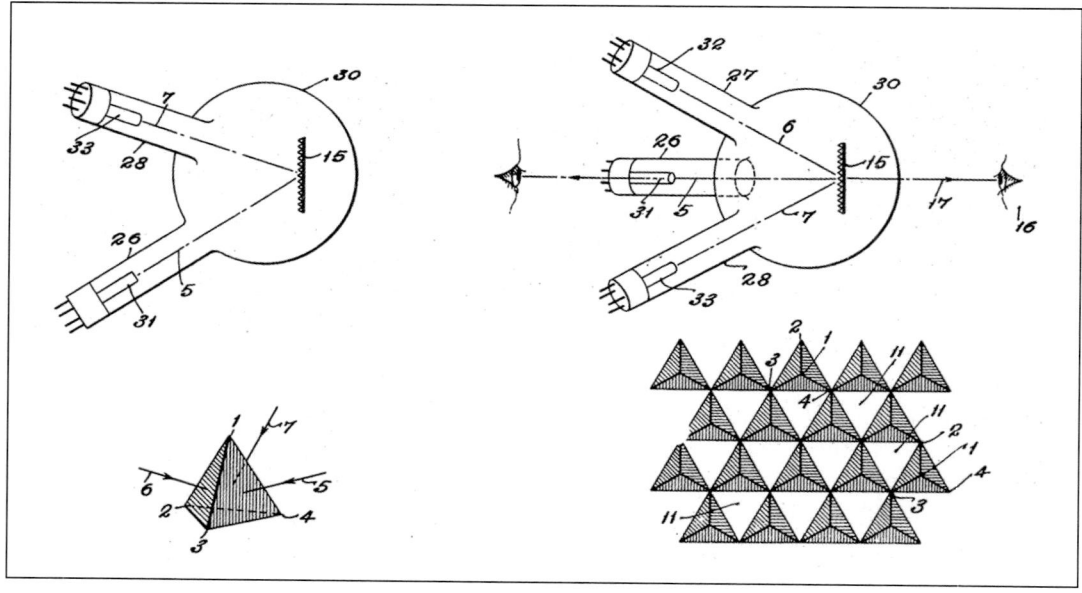

Fig 7.4: Goldsmith's three-cathode colour television tube *US2481839* **(1944)**

Telechrome, *GB395578*, provided a modification to enable the viewing screen to be on the front surface of the tube (**Fig 7.2**) but this resulted in one of the cathode-ray guns protruding from below the face of the tube.

In his patent, Geer describes an alternative configuration with the three guns positioned to the rear of the screen so that the scanning electron beam from each cathode reaches its corresponding red, green or blue phosphor target on the facets of small pyramid shaped sections illustrated in **Fig 7.3**.

Baird's Telechrome patent is cited in at least another two American patents including RCA and DuMont. The RCA patent by Goldsmith, *US2481839* [4], filed on 5 August 1944 is shown in **Fig 7.4**.

So apparently viable to the future development of colour television was Baird's three-cathode idea, that the US television industry could not easily let it go. On 26 December 1946, six months after Baird's death, DuMont applied for *US2544690* [5], **Fig 7.5**, which was not accepted until 13 March 1951.

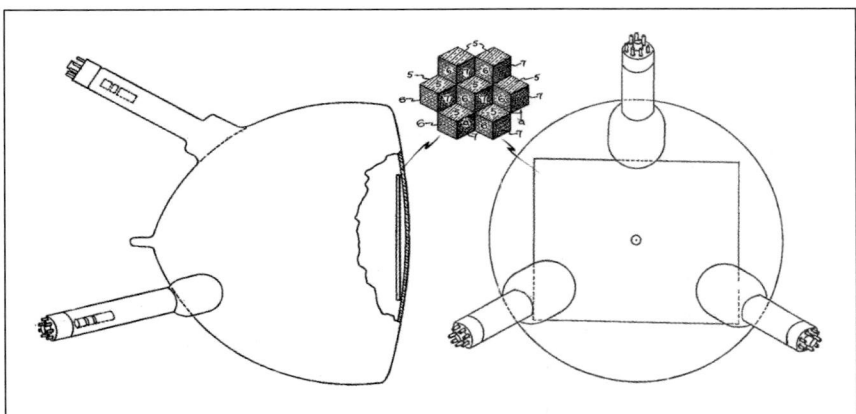

Fig 7.5: DuMont's three-colour cathode colour television tube *US2544690* (1946)

In theory the pyramid screen structure seemed an obvious solution to the problem of producing a front-viewing version of Baird's tube, but the complicated fabrication of the screen and the inability to guarantee colour purity, suggested that it may not have been as suitable a candidate for mass-production as it had first appeared.

The next major advance in the search for a practical colour television display was to combine the advantages of the multi-gun Telechrome, a direct view tube, with those of the tri-band indirect view tube of Baird. In other words to design a colour tube that resembles a monochrome tube, with a single electron gun, but employing a multiplicity of coloured screens, stripes or dots.

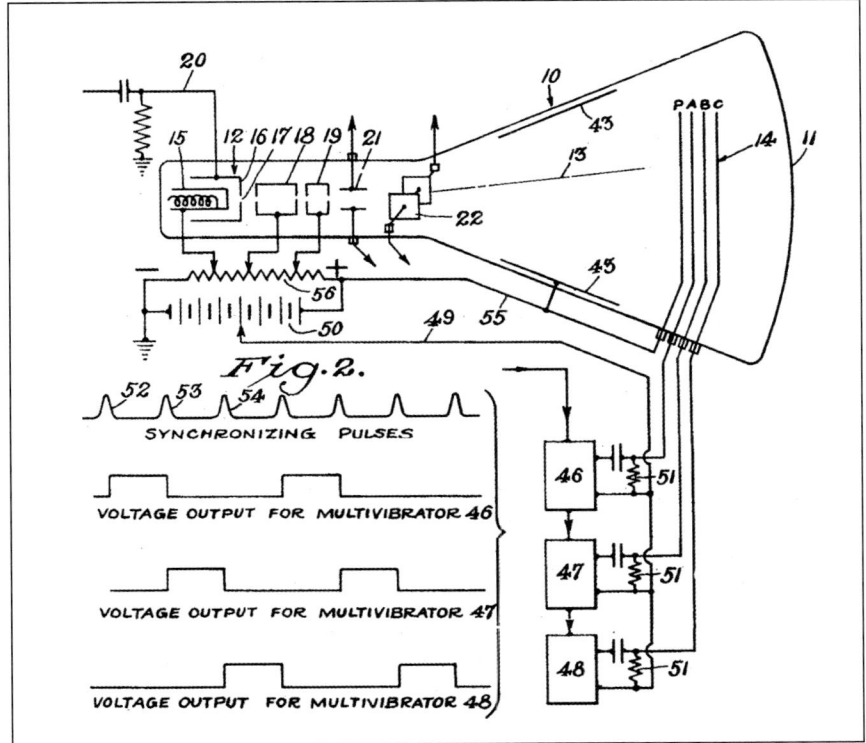

Fig 7.6: Arthur Bronwell's Chromoscope *US2461515* (1945)

One of the first published inventions of this type was the Chromoscope [6], *US2461545* [7], **Fig 7.6**. In his patent filed on 16 July 1945, Bronwell also cited Baird's Telechrome.

This novel approach used four partially transparent screens, three of which were coated with a colour rendering fluorescent material, red, blue or green, with the fourth screen held at a fixed potential. The screens were positioned one behind the other on the face of the tube to enable the additive method of colour mixing to take place. They were fabricated from a wire or grid structure and individually connected to an electrode to be electronically switched in sequence with the television field or line rates.

The Chromoscope was offered to the Rauland Corporation in 1947 and a few experimental tubes may have been produced, but as a solution for colour television the results were not favourable:

"We did not think much of his solution. It is only suitable for sequential systems and would cut down the available light to approximately one tenth. This tube cannot very well be used for direct view, even for projection it has the drawback that the grids do not lie in the same plane!" [8]

The Philips Electronics Company took John Logie Baird's concept of 'Theatre Television' to an advanced stage of development and, in February 1949, demonstrated a monochrome projection system as a visual aid for medical training to more than 200 spectators at the University Hospital of Leiden, Germany [9].

Also in 1949 the possibilities of colour in television and its application in medicine interested the American Medical Association. Nurses and doctors were first shown the application of colour television in medical teaching at the University of Pennsylvania, where the equipment developed by CBS went through its final tests. Pictures of an operation were shown on a row of three field-sequential colour television monitors manufactured by the Zenith Corporation.

Life magazine reported:

"The credit for getting colour television out of the laboratory and into practical, if still limited, use goes to the Columbia Broadcasting System, which has perfected a method of colour transmission, and the pharmaceutical house of Smith, Kline & French, which sponsored and helped develop its medical application." [10]

The main demonstration of the apparatus took place in Atlantic City's Convention Hall where fifteen thousand members of the American Medical Association watched the first large scale demonstrations in full colour.

According to *Life* magazine:

"On 20 specially built receivers, 1,000 doctors at a time will see a Caesarean, an appendectomy, a bone graft and other operations as up to now only the operating surgeon and his assistant could see them."

One of these field sequential colour television monitors, manufactured by Zenith for CBS, and formerly part of Michael Bennett-Levy's private collection

CHAPTER 7: COLOUR FROM CATHODE RAYS

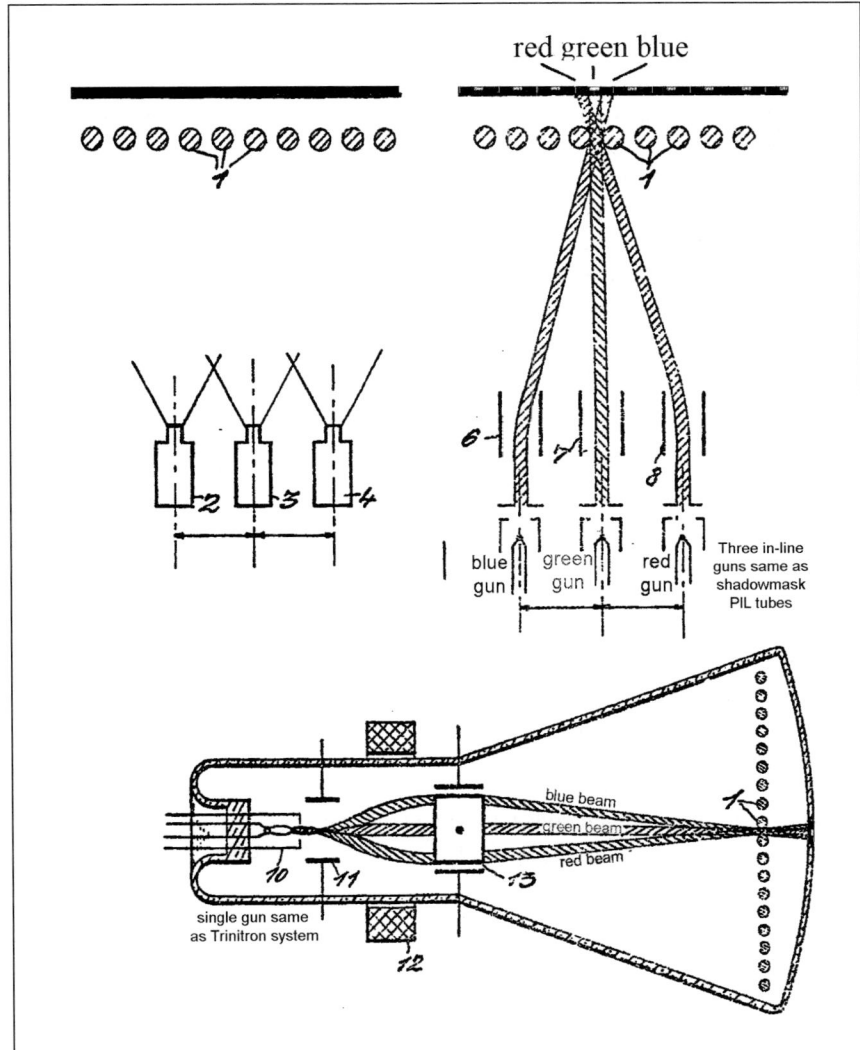

Fig 7.7: Werner Flechsig's colour picture tube (1938)

of prewar television receivers [11] may now be seen as part of the collection of the National Museum of Scotland in Edinburgh.

Probably one of the most important colour television disclosures relating to the future of colour television was the 1938 German patent, *DE736575* [12], **Fig 7.7**, by Werner Flechsig of the Fernseh company. Baird International Television originally formed the Berlin based Fernseh AG (translated 'Television Ltd') registered on 3 July 1929. Baird Television owned equal shares with Zeiss, Bosch and Loewe Radio until Hitler came to power.

Margaret Baird wrote:

"Hitler gave orders that our interests as a British concern must cease and that Fernseh must become wholly German. David Loewe who had wanted us to take over the Loewe interest and gain control was expelled from Germany because he was a Jew." [13]

In his autobiographical notes Baird wrote:

"Major Church however fixed up an arrangement whereby we retained an affiliation which gave us rights arising from our joint development in the British Empire; and in addition we were paid in cash the full value of our shares in the company." [14]

Flechsig had the foresight to describe not only the single gun, triple beam concept and parallax grid (aperture grille) that Sony would later employ successfully in the Trinitron cathode ray tube, but also the 'in-line' triple cathode arrangement eventually brought to perfection by RCA. Of particular importance was the parallax grid screen concept that would greatly influence American and Japanese development. In 1947, Kaplan of Rauland presented his advanced parallax grid concept and this appears to be the first disclosure of the practical parallax grid, shadowmask with a tri-colour screen as illustrated in **Fig 7.8** [the picture is also reproduced in colour on page C.10].

In the same year Goldsmith of RCA proposed an impractical version of Flechsig's concept, but simultaneously A C Schroeder [15], also of RCA, developed the format that would become the basis for the first practical tri-colour shadowmask tube. Early in 1950 RCA's Harold B Law, under the direction of Edward W Herold, decided that Schroeder's tube was the most promising and development work began. Law was able to demonstrate a working RCA shadowmask tube by March 1950 [16].

Inglis wrote:

"Development work then concentrated on tubes using this principle, and two working prototype receivers with shadowmask tubes were produced in the

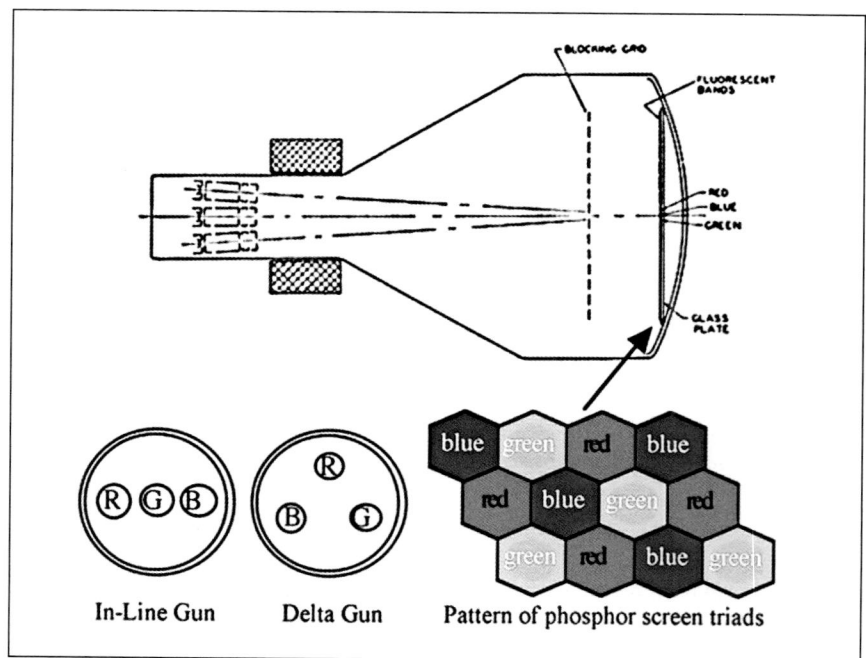

Fig 7.8: Kaplan's parallax colour tube (1947) (see page C.10 for a colour version)

In-Line Gun   Delta Gun   Pattern of phosphor screen triads

CHAPTER 7: COLOUR FROM CATHODE RAYS

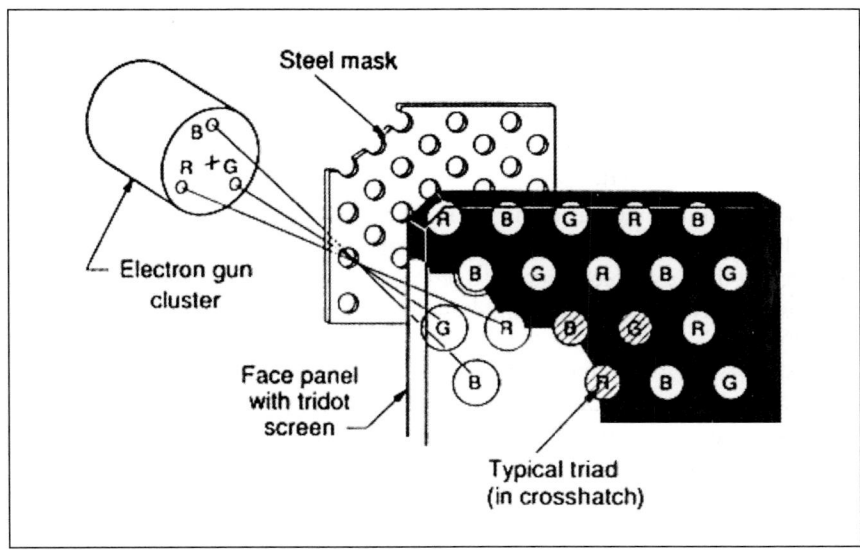

Fig 7.9: Shadow-mask principle

incredibly short time of two months. They were first demonstrated to the FCC on 23 March and to the press on 29 March 1950. The picture quality of the prototypes was surprisingly good, and although neither CBS nor the FCC would admit it at the time, the demonstrations sounded the death knell for the impressive CBS frame sequential colour television system." [17]

In July 1950, following RCA's demonstration of the shadowmask principle, illustrated in **Fig 7.9**, Szegho published an article describing a working version of Kaplan's experimental tri-colour parallax tube:

"The cathode ray tube had three electron guns and a blocking grid so placed between the guns and the fluorescent screen that as a result of parallax effects the grid permitted electrons from only one gun to strike any particular colour portion of the screen." [18]

Variations of the parallax grid concept are the RCA shadowmask and the Sony Trinitron aperture grille:

"There was, however, an enormous gap between the prototype and a reliable commercial product that could be manufactured economically. RCA was almost on the right track, but years of hard work lay ahead before it would have a manufactureable consumer product." [19]

The problem resulted from the technique used to deposit phosphor dots on the screen. To make tube components interchangeable the dots had to be printed with precision on a flat glass plate using an expensive silk-screen process. The plate could then be mounted with equal compatibility to any cathode ray tube assembly, but this greatly limited the screen size and caused the pictures to be very dim. With these inherent problems, RCA put the CT100 shadowmask colour television into production in 1954:

"To RCA's great embarrassment, the solution came from the Hytron Division of CBS, and its embarrassment was not mitigated by the fact that the solution was based on a photographic method of locating and depositing the phosphor dots that had been developed by Harold Law at RCA Laboratories in 1948." [20]

Although having the basic patents for the shadowmask process RCA had to pay a licence fee to use CBS's patented photographic etching process to enable them to put their own tubes into production. David Sarnoff was furious at CBS's success and gave his staff a severe dressing down:

"Soon afterward, Frank Folsom, RCA's president, convened a meeting of the top corporate executives. Shaken and grim, he said, 'Gentlemen, we must catch and surpass CBS. If we do not, I can assure you that I shan't be here next year and I will probably be joined by others.'" [21]

Early in 1951 it was reported that Rauland had an improved experimental tricolour picture tube based on Kaplan's parallax grid theory. In this disclosure [22] it was revealed that two versions were under consideration, the R-6073 a single electron gun type and the R-6074, a three-delta gun type. Both designs utilised a shadowmask containing about four hundred thousand holes, geometrically corresponding to one hundred and twenty thousand triangular groups, of three phosphor dots. Anderson, Baird's remaining technical assistant during the war years, stated that in 1945, he assisted Baird in developing single gun colour tubes with screens fabricated with red, green and blue vertical stripes. Each strip was a thin wire coated with a fluorescent material of the red, green or blue type and these were connected to related circuits that would electronically switch the beam to converge and illuminate the correct colour. The strips were aligned vertically to form the viewing screen of the CRT in the order of fluorescent coating, red, green and blue:

"The Strip Tube was very nearly the Trinitron of Sony!" [23]

Anderson also revealed that Baird was experimenting with the basis of the shadowmask concept. An extract from Baird's wartime diary with a drawing he dated 25 Jan 1944 illustrates the basic RGB, three-gun shadowmask process, **Fig 7.10**.

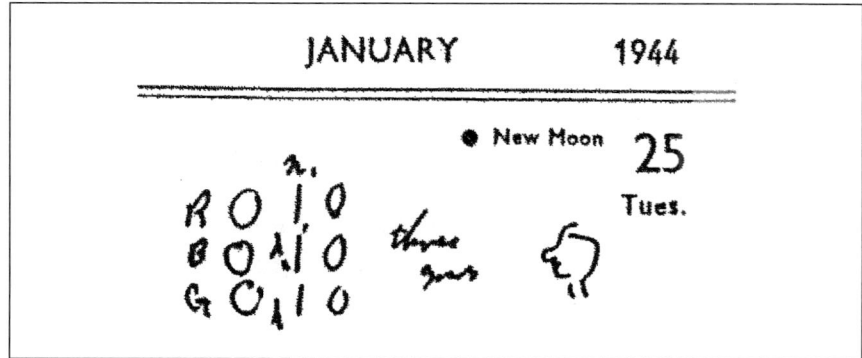

Fig 7.10: Extract from Baird's diary

CHAPTER 7: COLOUR FROM CATHODE RAYS

Anderson continued:

"Work was thought out by making a normal CRT with resist of RGB coated on the outside of the tube. This was very nearly the modern colour screen on the shadowmask tube. Only a drawing office design away from the shadowmask."

Other colour tubes based on the principles of Flechsig's 1938 concept were developed. The most notable was the Chromatron shown in **Fig 7.11** [also reproduced in colour on page C.11] invented in 1950 by Dr Ernest O Lawrence the Nobel Prize-winning inventor of the cyclotron [24].

This was developed by Chromatic Laboratories of the Paramount Pictures Corporation and produced colour television pictures with three times the brightness of the shadowmask. The process was very similar to Flechsig's single gun concept and Baird's 'Strip Tube' design. A single beam of electrons was consecutively switched to red, green and blue phosphor stripes via an 'aperture grille' of fine wires.

By 1961 the Chromatron was the only competing tube of any merit to challenge the RCA shadowmask, but the manufacturing processes had proved difficult and it had not reached a successful stage of mass production.

In 1961 Masaru Ibuka (founder of the Sony Corporation), looking for a unique product saw a demonstration of the Chromatron and was immediately impressed [25]. In 1962 Sony obtained a license for the development of the Sony Chromatron and for two years they struggled with the concept, which proved to be very difficult to accurately repeat in production.

By September 1964 Sony introduced a 17-inch Chromatron colour television and sold thirteen thousand sets in Japan. The Chromatron was a disaster; servicing and replacement cost were high and the development alone cost seven hundred thousand US Dollars a year. The Chromatron project was abandoned

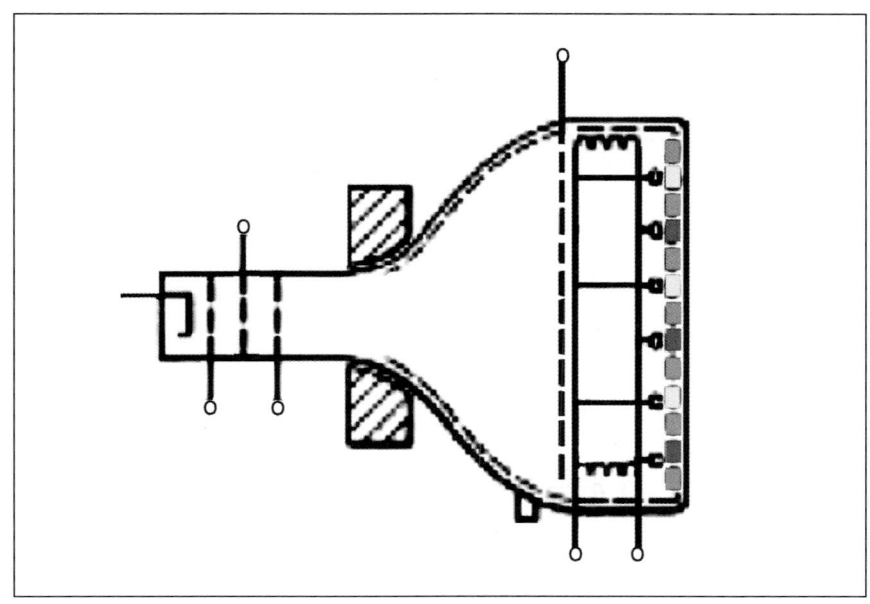

Fig 7.11: Lawrence Chromatron colour tube (see page C.11 for a colour version)

and work began on an in-house tube design. This decision seriously left them behind their competitors who had opted to obtain a license from RCA. This put a great deal of pressure on the development team, but a breakthrough was on the horizon. In 1966 a young Sony engineer Senri Miyaoka made a discovery:

"Towards the end of 1966, while running some experiments, Miyaoka made a mistake that produced some interesting results. Using the single gun, characteristics of the Chromatron, and three cathodes, he produced a blurred picture - but it was a picture, and by a new concept." [26]

Lyons continued:

"By February 1967 they had eliminated the blur and produced a decent picture with the new gun; many of the engineers thought it was already better than the shadowmask. Now they had to decide whether to use the shadowmask screen, with its colour dots, the Chromatron grid, or some new system."

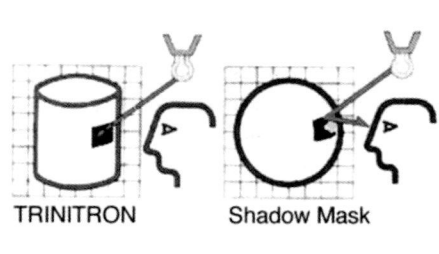

Vertically Flat Screen
Trinitron's cylindrical screen produces straighter lines, eliminating much of the image distortion found on a conventional spherical screen and producing a clearer, more lifelike picture. In addition, ambient light reflection is eliminated to provide a more pleasing screen image that is easier on the viewers eyes.

Aperture Grille
The Trinitron Aperture Grille, designed with long, unbroken slits, allows more light and colour to reach the screen, resulting in purer, more beautiful colour images. The mesh-like, darker screen of a conventional Shadow Mask creates a colour shift ("doming") distortion and prevents colour purity.

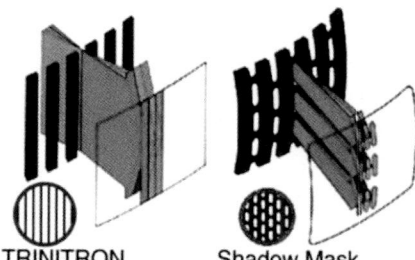

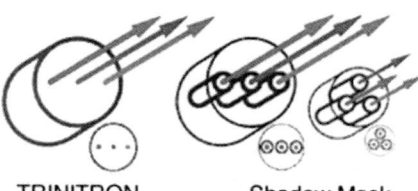

1-Gun, 3-Beam System
Trinitron's simple one-gun, three beam system provides more precise beam focusing. This avoids many problems faced by conventional three-gun systems. For example, the Delta version is troubled with complicated vertical and horizontal alignment and the In-line version requires either smaller lenses or a larger gun. Trinitron's one-gun, three-beam system creates clearer text and adds reality to the picture.

Fig 7.12: The Sony Trinitron (see page C.11 for a colour version)

The result of a series of experiments involving several systems was the aperture grille of vertical strips, which they produced by the new process of photo-etching. To compensate for the expansion and contraction of the aperture grille, which threatened to produce serious colour registration problems, two small tungsten wires tensioned the grid onto the glass faceplate. Finally the tube face was changed from the conventional spherical shape to a cylindrical one. On 16 October 1967, the Trinitron tube, **Fig 7.12** [also reproduced in colour on page C.11] was successfully demonstrated.

In 1967, 'CBS-Sony' was formed and cross licensing agreements reached between Sony and Philips over patent rights for audio cassettes. In 1967 both CBS and Philips put one million US Dollars into Sony, which helped finance the Trinitron venture. David Sarnoff, the President of RCA, would not have been pleased to discover that CBS had joined forces with Sony to become their major competitor.

By October 1958 the Philips Company having developed a theatre television system to include colour, installed a closed circuit system capable of projecting very large pictures (2.70m x 3.60m), for the Faculty of Medicine at the University of Marseilles [27]. The Philips system used a three-tube television projector with red, green and blue images positioned side-by-side. Colour convergence was achieved by mechanical adjustment. In the operating theatre the colour television camera incorporated three camera tubes, dichroic mirrors for colour separation.

After the introduction of the RCA Shadowmask colour display tube and its rival the Sony Trinitron, a number of refinements were made to improve their performance. An example of this was the introduction in 1972 of RCA's PIL (precision-in-line) tube with its slotted shadowmask and striped screen. The advantage of the PIL tube was a reduction in the amount of convergence circuitry that was essential for sets fitted with delta-gun tubes [28].

In 1982, Toshiba introduced the FST (flat-square-tube) screen, this was followed by Zenith in 1986 with their FTM (flat tension mask) [29] tube. However, by far the most important advance in colour CRT performance came in the 1960s. This major breakthrough was to improve the contrast ratio and more than double the brightness of all Delta gun, PIL and Trinitron colour cathode ray tubes manufactured thereafter.

The problem with contrast ratio was explained as follows: The black content of a television picture is related to the physical colour of the screen of the cathode ray tube and its reflective properties. The colour of non-radiating phosphors is usually white and a picture viewed on such a colour television in a totally dark room will appear to have a good visual black content. The range of visual contrast possible lies between the colour of the non-illuminated areas of the screen and the brightest illumination from the glowing phosphors. When the same picture is viewed in brighter conditions the contrast ratio is poorer and the perceived black appears grey.

Various techniques have been applied to mask this deficiency. One method was to trade-off contrast with brightness by fitting a transparent neutral density filter in front of the screen. This process nominally improved the contrast, where the reflected light travels twice through the filter while light from the image, only travels once. A further improvement was introduced by coating the faceplates with an anti-glare material to reduce reflected light.

Fig 7.13: Comparison between Samuel Kaplan's Chromacolor tube and the standard Shadowmask

COMPARISON OF PRE-DECEMBER 1968 TUBE TO THE CHROMACOLOR TUBE			
	STANDARD SHADOWMASK TUBE	CHROMA-COLOUR TUBE	% IMPROVEMENT
SCREEN OUTPUT	54.3	61.9	14
GLASS TRANSMISSION	42	80	--
BRIGHTNESS	22.8	49.5	117
CONTRAST RATIO AT 20FT CANDLES	7.3	9.5	31

The real solution to the contrast ratio problem came in August 1964 when Fiore and Kaplan of the Rauland Division of the Zenith Radio Corporation were awarded a US patent for a 'Dark Surround' tube [30]. Later in the same year, Zenith introduced its 'Chroma-Color' shadowmask tube based on this principle, and in 1969 a paper described the concept [31].

The new and more efficient tube comprised smaller phosphor dots interspersed with a light absorbing pigment such as black. The electron beam diameter was increased to increase the phosphor light output. In conventional shadowmask tubes the dots touch each other and the beam size has to be reduced to a smaller diameter to avoid colour purity problems. When comparing the performance of the newer Chroma Colour tube against the earlier conventional tubes there is a 117% increase in brightness and a 31% improvement in contrast over a conventional tube, **Fig 7.13**.

The dark surround mask manufacturing process, developed by Szegho and Kaplan at Rauland-Zenith, in 1968, was later applied to every colour television tube manufactured in the world. Evidence exists to show that John Logie Baird used a similar technique in 1931 when he addressed the problem of poor contrast ratio on his early active cinema television screen that utilised 2700 small lamps. Instead of using silver reflectors he painted the interior of each lamp cell or housing black [32].

The author was fortunate at the time of researching this book to communicate with Samuel Kaplan (then almost 82 years of age in the USA). His testimony has been included as a means of clarification and support to the work of this research. Kaplan began by describing his role in the birth of the shadowmask concept:

"In early 1947 I came across an article in Electronics *about an all-electronic colour television system by a Du Mont Engineer describing an all electronic colour CRT system (three faceted screen) and I thought that it was unlikely to succeed because of manufacturing difficulties. Shortly thereafter I remembered that a stereo photography parallax mask could give stereo photographs, separation on two or more planes, so why not use a parallax mask in colour television tubes to ensure each of the three guns in the single neck*

CHAPTER 7: COLOUR FROM CATHODE RAYS

illuminate only its proper colour (dot or strip) and not the others (being blocked by the metal between the slits). The more I thought about it the better I liked it, and then I contacted a competent patent attorney and filed my application in early September 1947. I contacted RCA, who said that were not interested. I found out later that they had the same idea!" [33]

The above shows that Kaplan's ideas for the shadowmask tube were stimulated as a result of the American industries obsession in reproducing the Telechrome after Baird's death in 1946.

Kaplan further explained that his idea for a shadowmask tube could not be developed without financial support. He contacted Zenith Electronics, a leading television manufacturer, but at that time they had no picture tube plant. He also contacted the Rauland Corporation a manufacturer of television tubes for Zenith. Dr Szegho (head of research at Rauland), also liked the idea but had the same problem of obtaining funding:

"I also contacted Motorola another well known local radio and television manufacturer who also had a well equipped research lot (sic) for television tubes, mainly for military radar displays (shades of the cold war!). Dr Schlesinger liked my idea but said that a tricolour screen could not be made to the necessary standards. RCA meanwhile indicated that they had an electronic colour tube - the shadow mask three gun tube - the same as I had conceived. It turned out that the shadow mask tube was invented by a German, Flechsig, who filed in 1938, he also filed it in France a year later."

In the interim Zenith decided to take an option on Kaplan's patent application, even though there was a question about obtaining it. This was because Kaplan had certain improvements enabling Zenith to narrow the scope (delta gun configuration as well as in-line), but in prosecution Zenith and Kaplan ran into an RCA patent interference. RCA also had the delta guns and won the Zenith-RCA interference action by filing about one month earlier.

In the interim, Zenith acquired the Rauland Tube Division as their main supplier of monochrome cathode ray tubes. Kaplan explained that the patent granted to Werner Flechsig for the basic shadowmask concept was upheld and that he must be recognised as the true inventor. A litigation by Frank Grotty (senior patent attorney for Zenith-Rauland), against RCA on illegal monopoly of patent licensing followed. The successful outcome saved Zenith a fortune in royalties and ironically Sam Kaplan became an RCA shadowmask licensee for free.

Kaplan also described the circumstances that followed the breakthrough in colour tube contrast enhancement:

"The dark screen surround was invented by myself and Joseph Fiore. The patent issued to Kaplan and Fiore was licensed to the industry, but a newcomer in the field Toshiba of Japan decided to attempt to invalidate the patent and they succeeded. For example, they sued the Sylvania Company, rather than Zenith direct, and Sylvania presented a poor defence, after all they had to pay Zenith royalties (RCA also paid)".

THE THREE DIMENSIONS OF JOHN LOGIE BAIRD

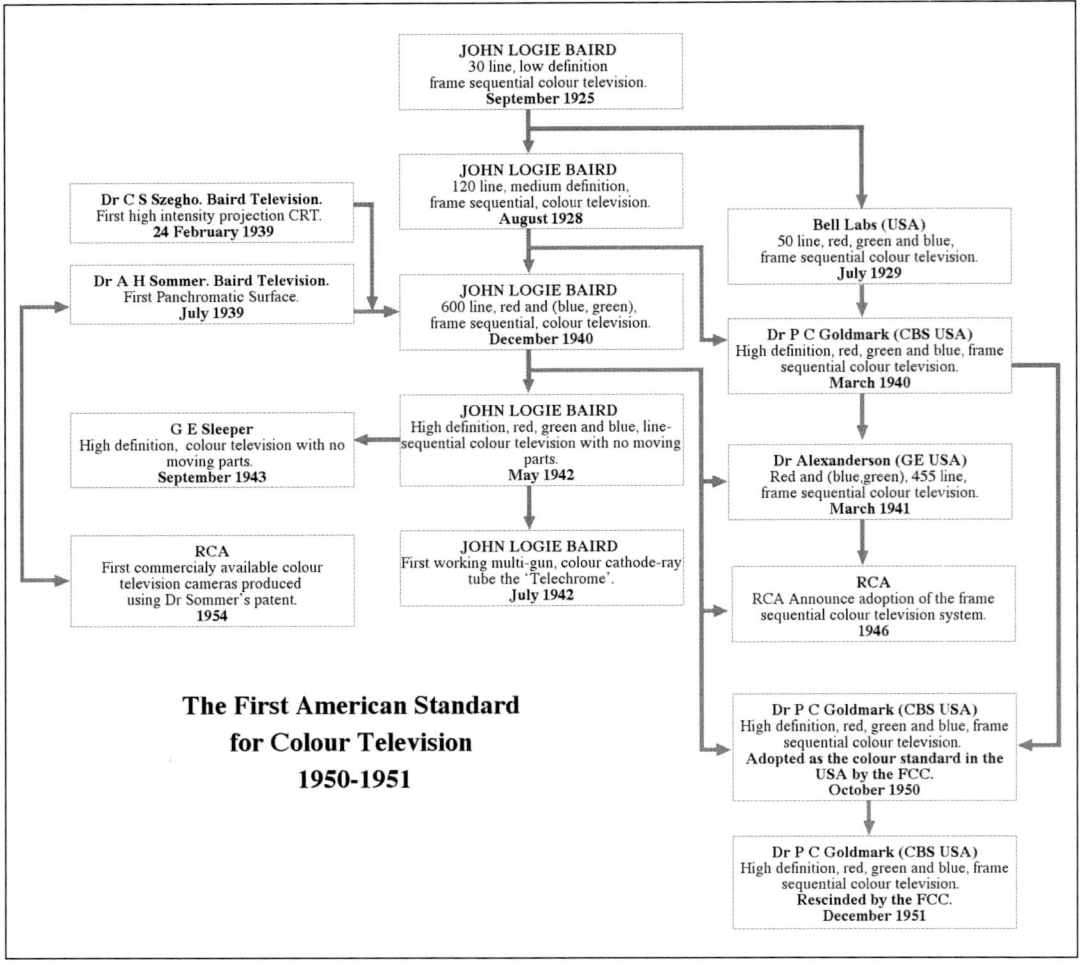

Fig 7.14(a): Baird's influence on the the evolution of frame sequential colour television (see page C.12 for a colour version)

Zenith Electronics was unable to defend its intellectual property rights for the dark screen surround. The Zenith Patent Department called it a miscarriage of justice.

Few history books record Flechsig as the inventor of the system used by Kaplan and RCA to produce a working colour television tube, and while Kaplan and Fiore invented the dark screen mask their names continue to remain in virtual obscurity.

The interaction of John Logie Baird's contributions to the development of modern colour television display systems, and how they inspired and linked with other developments from the American television industry, is shown diagrammatically in **Figs 7.14(a), (b) and (c)** [these pictures are also reproduced in colour on pages C.12 - C.14].

In summing up the colour dimension of Baird's life it is useful to review his achievements as the pacemaker for the development of modern colour television industry:

CHAPTER 7: COLOUR FROM CATHODE RAYS

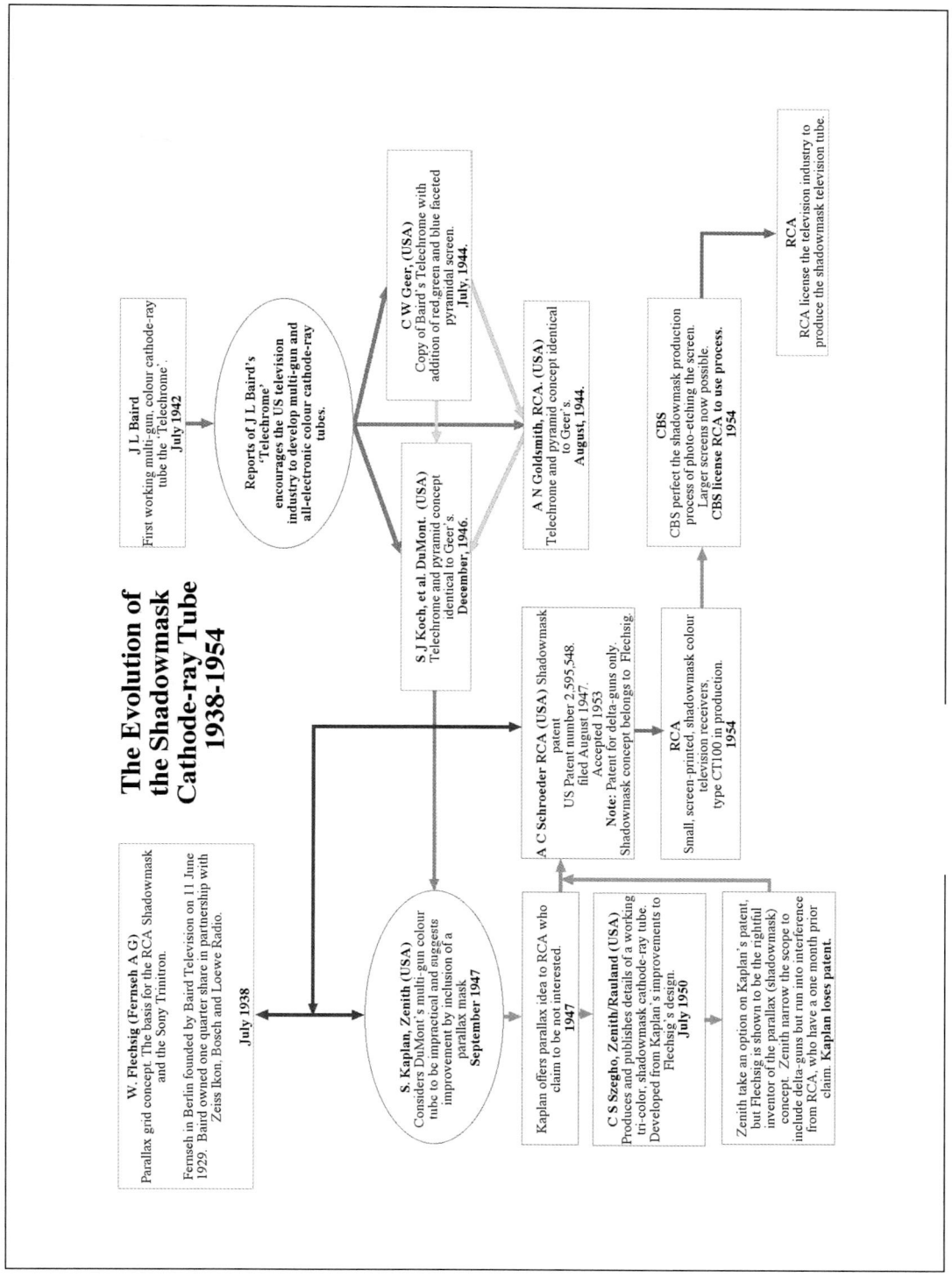

Fig 7.14(b): Baird's influence on the the evolution of the Shadowmask cathode-ray tube (see page C.13 for a colour version)

THE THREE DIMENSIONS OF JOHN LOGIE BAIRD

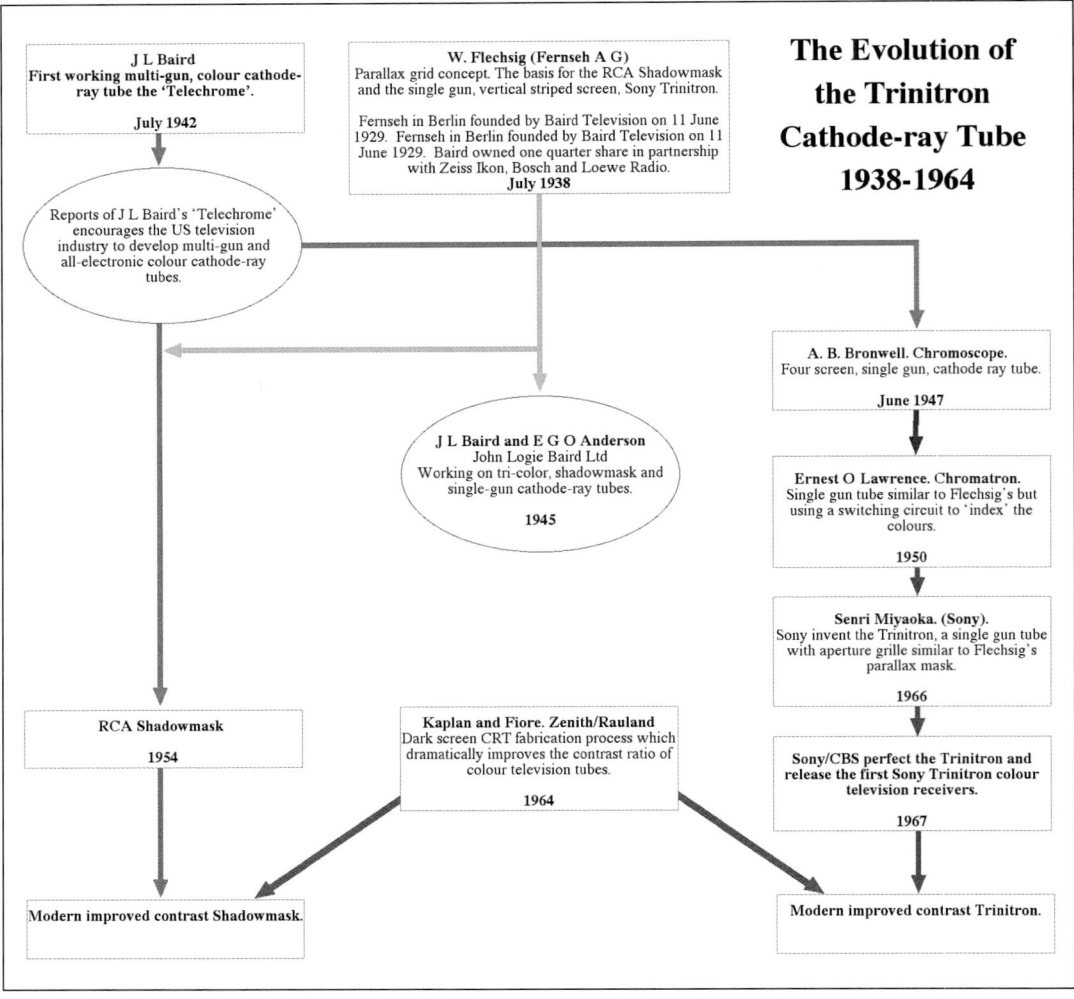

Fig 7.14(c): Baird's influence on the the evolution of the Trinitron cathode-ray tube (see page C.14 for a colour version)

1) Colour television in 1926 based on his 1925 patents.
2) Patented a video wall television system in 1935.
3) Demonstration of 120-line large screen colour in 1937.
4) Triple-interlaced 600-line colour picture in 1940.
5) Requested the development of panchromatic television surfaces at Baird Television, which was a technique exploited by RCA in the first commercial electronic colour television cameras.
6) In 1941 invented, produced and demonstrated the first working multi-gun cathode ray tube for colour television, the Telechrome.
7) The first line-synchronous colour television.

The next chapter will show that Baird developed the world's first stereoscopic 3D television systems, including a futuristic holographic system of colour television.

References

[1] 'Color Television-Part1. US', P C Goldmark, J N Piore, E R Dyer & J M Hollywood, *1944 Proceedings of IRE*, April 1942, p163.
[2] 'Another unsung Hero: E N Rauland', G E Hausske, *Journal of the Antique Wireless Association*, Vol 33, No1, May 1992, p1.
[3] Patent *US2480848*, C W Geer, filed 11 July 1944, issued 6 September, 1949.
[4] Patent *US2481389*, A N Goldsmith, filed 5 August 1944, accepted 13 September 1949.
[5] Patent *US2544690*, S J Koch et al, DuMont, filed 26 December 1946, issued 13 March 1951.
[6] 'The Chromoscope. A New Colour Television Viewing Tube', A B Bronwell, *Electronic Engineering*, June 1948, p150.
[7] Patent *US2461515*, A B Bronwell, filed 16 July 1945, accepted 15 February 1949.
[8] *Reciprocal Report number 15*, C S Szegho, Rauland Corporation, 1947.
[9] 'Colour Television in Medical Teaching', W A Holm, F H J van der Poel, *Philips Technical Review*, 1958/59, Vol.20, p327.
[10] 'Surgery in Colour Television', Anon, *Life* magazine, June 1949, p75.
[11] M Bennett-Levy, Private Television Collection.1996/97, Monkton House. Old Craighall. Musselburgh. Scotland.
[12] German patent *DE736575*, W Flechsig, applied 12 July 1938, issued 13 May 1943.
[13] *Television Baird*, M Baird, Citadel Press, Haum, Cape Town, 1973.
[14] *Television and Me*, J L Baird, edited by Malcolm Baird, Mercat Press, Edinburgh, p109.
[15] *US2595548*, A C Schroeder, Picture Reproducing Apparatus, filed 24 February 1947, issued 6 May 1952.
[16] 'A History of Color Television Displays', E W Herold, *Proceedings of the IEEE*, Vol 64, No 9, September 1976, p1333.
[17] *Behind the Tube. A History of Broadcasting Technology and Business*, A. F Inglis, Butterworth, 1990, p278.
[18] 'Experimental Tri-Color Cathode Ray Tube', C S Szegho, *Tele-Tech*, Vol 9, July 1950, pp34-35.
[19] *Behind the Tube. A History of Broadcasting Technology and Business*, A. F Inglis, Butterworth, 1990, p279.
[20] *Behind the Tube. A History of Broadcasting Technology and Business*, A. F Inglis, Butterworth, 1990, p279.
[21] *Behind the Tube. A History of Broadcasting Technology and Business*, A. F Inglis, Butterworth, 1990, p279.
[22] *Reciprocal Report No.20*, C S Szegho, Rauland Corporation to Cinema TV Ltd, 1951, Section 7, pp2-8.
[23] E G O Anderson, personal correspondence to Dr Peter Waddell, Strathclyde University, 1974.
[24] *The Sony Vision*, N Lyon, Crown Publishers, 1976, New York, p132.
[25] *The Sony Vision*, N Lyon, Crown Publishers, 1976, New York, p132.
[26] *The Sony Vision*, N Lyon, Crown Publishers, 1976, New York, p138.

[27] 'Colour Television in Medical Teaching', W A Holm, F H J van der Poel, *Philips Technical Review*, 1958/59, Vol.20, p327.

[28] 'The Development of Colour Tubes', E Trundle, *Television* magazine, IPC Publications, London, June 1986, Vol 36, No 8, Issue 428, p500.

[29] *Behind the Tube. A History of Broadcasting Technology and Business*, A. F Inglis, Butterworth, 1990, p281.

[30] Patent *US3146368*, J P Fiore and S H Kaplan, issued 25 August 1964. See also *US3365292*, J P Fiore and S H Kaplan, issued 23 January 1968.

[31] Patent *US3146368*, J P Fiore and S H Kaplan, issued 25 August 1964. See also *US3365292*, J P Fiore and S H Kaplan, issued 23 January 1968.

[32] *US1957815*, J L Baird, applied 15 January 1931, issued 8 May 1934.

[33] S Kaplan, private correspondence with the author, 15 November 1996.

8

The Third Dimension & Holographic TV

Monocular vision is vision from a singular point of view, an object or a scene observed or recorded in two-dimensions (2D) using a single eye, or a single television or photographic camera, either still or moving. Although sourced from a single viewpoint, 2D images may also be observed with normal stereoscopic vision resulting in a flat picture on a flat surface with no perceived parallax. Under such circumstances the brain has insufficient information to enable the viewer to visualise an object or a scene in stereoscopic, three-dimensional relief. Despite this, psychological cues may be present or intentionally added to an image viewed on a flat 2D display screen to assist in the estimation of depth: Evaluation of relative sizes of known objects in the distance or foreground; lighting and shading effects including highlights; tonal or colour differences; virtual reality; rotation of an object on a flat screen to observe a strong impression of solidity and camera panning to record parallax in a moving film or a television scene. It will be recognised that some of the above cues also assist stereoscopy in registering the impression of depth in the real world including: observing around or the rotation of a physical object and right-left head movement to increase the parallax. These are only some of the many visual cues that exist to assist in the estimation of depth. However, the estimation of depth is not the same as stereoscopic perception of images in three dimensions.

TRUE 3D VISUALISATION REQUIRES binocular vision and relies on the horizontal spatial separation of the eyes to register two slightly different two-dimensional viewpoints. The brain fuses these two 2D images containing light, shade, colour, height and width information and processes the differences between the left and right eye image to provide a perception of depth.

With both eyes functioning correctly, stereoscopic awareness of the world allows the observation of objects with apparent solidity. The effect of closer objects appearing boldly in front of more distant objects continues into the distance of a given scene, until eventually it becomes less effective at a point where stereoscopy is of less value to human behaviour. A photographic stereo-pair can be produced by suitably aligning a camera or cameras to produce exposures comprising equivalent left and right-eye viewpoints of a scene or an object. The horizontal separation of each camera lens is normally the same as the distance between the average human's eyes, however the stereoscopic effect may be enhanced for distant objects by increasing the separation. The opposite is true for enhancing the stereoscopic effect of closer objects, in which case the distance between the lenses is reduced.

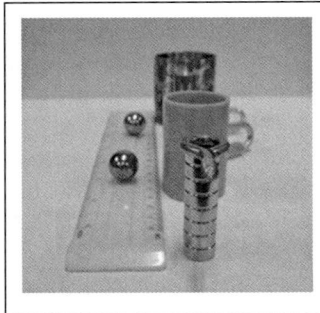 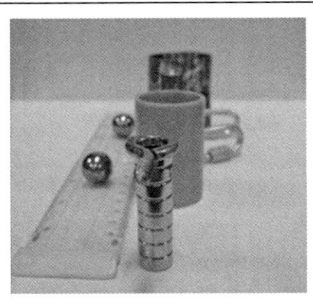

Fig 8.1: Example of a parallel viewed stereo-pair (see page C.15 for a colour version)

The modern 'stereoscope' is an optical binocular device for viewing a stereo-pair in relief, enabling the left eye to focus on the left-eye image and the right eye on the right-eye image. The same effect may be achieved by a number of other methods including viewing a stereo-pair of pictures placed side-by-side, with the right view on the right and the left view on the left. If they are positioned an optimum distance apart they may viewed with the naked eye by parallel viewing, **Fig 8.1** [also reproduced in colour on page C.15]. It should be noted that parallel viewing is a learned technique that does not involve cross-focusing the eyes and may take a little perseverance to achieve.

Other methods of viewing stereo-pairs in relief include devices using combinations of prisms and mirrors. Historically a popular method is to combine them in the form of a single image known as an anaglyph. This is a picture with the left-eye view printed in green or blue and superimposed on the right-eye image printed or tinted in red. Glasses fitted with respective colour filters enable the brain to discriminate between the left and right eye anaglyph images.

The red filter allows the red tinted image through to the left eye and blocks the green or blue image and the reverse is true for the right eye with the green or blue filter. The earliest 3D anaglyph movie shown to a paying audience was in 1922. Since then the technique has been used for special effects in theme parks, theatre productions, publications and occasionally television. Red and blue coloured anaglyphs were seen as a compromise for producing colour 3D pictures while red and green anaglyphs were mostly used for monochrome. A colour anaglyph is in **Fig 8.2** [*shown only in colour*, on page C.15]. This was produced to be viewed with a red colour filter over the left eye and a blue colour filter over the right eye.

There are three types of stereoscopic viewing spectacles, coloured-filter, cross-polarised and liquid crystal, (LCD) active shutter. This chapter will show that the modern 3DTV systems employ concepts using both the passive polarised and active shutters methods first patented by John Logie Baird in 1941.

On 3 April 2010, eighty-two years after Baird's first demonstration of stereoscopic colour television, British Sky Broadcasting ('Sky') launched the first commercial 3DTV service. At the transmitter a dual camera unit produces a stereo-pair, side-by-side on a single high definition (HD) 1920 x 1080 pixel frame. Stereo-pairs received by a 3DTV display are dealt with differently depending on the technology employed by the receiver manufacturer.

In the case of a 3DTV receiver that requires 'active shutter' glasses, the side-by-side images are electronically separated and stretched to fit the 1920x1080 format. The left and right frames are then switched alternately to appear on the screen. The 3D Blu-ray format will comprise the full 1920 x 1080 definition. To enable viewing of the alternating frames in 3D, a wireless synchronising signal from the television activates the liquid crystal shutter glasses, alternately blocking the left eye from the right eye image and vice versa at a minimum frame speed of 120Hz.

CHAPTER 8: THE THIRD DIMENSION & HOLOGRAPHIC TV

Some other 3DTV receivers use passive glasses. The left-eye and right-eye images are present together on the screen, separated into sets of odd and even lines, with each set having a different angle of polarisation. The double stereo image is resolved as a full 3D picture by wearing passive cross-polarised glasses. Athough the 3D polarising technique reduces vertical definition of each stereo frame by 50% the fused image received by the brain, enables cumulative perception of 1080 pixels of resolution. A prospective user would be advised to experience both systems.

In the case of a glasses-free 3D display, the stereo-pair is split into a series of thin pixel-sized vertical strips with the odd numbered vertical strips from one viewpoint being interlaced across the screen alternately with the even numbered strips from the other viewpoint. The screen is viewed either, through a lenticular screen designed to focus the left-eye image to the left eye and the right-eye image to the right eye, or viewed through a parallax barrier, which uses thin vertical slots between bars to block the left eye from seeing the right-eye view and vice versa. There may be more than two viewpoints interlaced in sequential order to increase the horizontal image parallax over a wider viewing angle. Lenticular imaging is a concept covered in more detail in the following descriptions of Baird's stereoscopic and three-dimensional television patents.

It is likely that Baird first experimented with three-dimensional television prior to his first dual colour [1] and stereoscopic patent on the subject (1 September 1925). It may be recalled that *GB266564* [2] described a two-channel system employing a twin spiral of scanning holes, where each set of holes was covered with a primary colour of either red/orange or blue/green. In the colour television description, the red and blue spirals scanned the scene or object from the same viewpoint and two photocells produced the red and blue television signals. At the receiver a synchronised scanning disc recovered the parallel red and blue images, and an optical device converged them to reproduce the colour television picture.

Baird indicated that a slight modification to the system would enable an image to be sent and received in stereoscopic relief. In the stereoscopic description it was arranged that instead of both spirals scanning the object from a single viewpoint, one spiral would optically resolve the object from a left-eye point of view and the other spiral from a right-eye perspective. At the receiver an optical stereoscope enabled the viewer to observe the images in colour and 3D. Alternatively, the coloured stereoscopic images could be superimposed on a screen as an anaglyph to be resolved by viewing through red/orange and blue/green filter glasses.

On 22 December 1926, Baird applied for a patent [3] to describe a method of producing and visualising stereoscopic or pseudo-stereoscopic images directly, without the requirement of a viewing device such as a stereoscope or coloured filter glasses. Three-dimensional images produced to be observed directly, fall into the category of 'auto-stereoscopic'. In the patent (**Fig 8.3**), Baird indicated that images produced by this invention relate to the production of photographs and the like and a system of television in relief.

This may be achieved in the case of a stereo-pair of photographs, by slicing each image into a number of thin vertical strips and removing every odd numbered strip from one picture and replacing them in the recomposed picture with the appropriate odd numbered strips from the other. There may be more than two viewpoints and therefore the slices would be juxtaposed in the order of the

Fig 8.3: Abstract from Baird's 3DTV patent *GB292365*

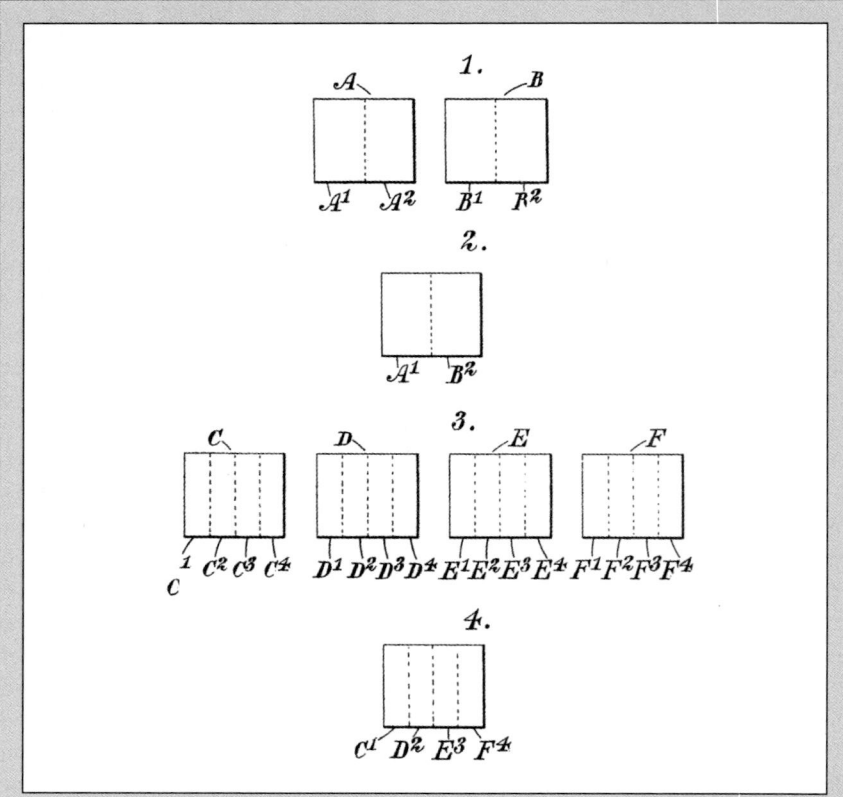

This invention comprises a method of providing a single representation giving an impression in relief, of an object, consisting in dividing vertically into two parts, each of the two reproductions or images as normally reproduced for stereoscopic purposes, and associating the right-hand part of the right-hand image with the left-hand part of the left-hand image. There is thus provided a single representation of the object which when viewed in the ordinary manner (without a stereoscope or like device) produces upon the observer an impression of solidity or relief in the representation. The two said parts may be each a complete half of a reproduction, in which case they are juxtaposed to provide the complete representation. Such an arrangement however, may result in showing a line of demarcation between the two parts and this effect is reduced or obviated by a modified method, according to which there are prepared a plurality of reproductions from a series of viewpoints spaced apart horizontally, and a corresponding plurality of vertical strips, one from each of the reproductions are associated to form the complete representation.

sequence of the viewpoints shown in Fig 8.3. In the patent description Baird indicated that while the stereoscopic or a pseudo stereoscopic effect could be obtained by this method, without the use of an optical system. Viewing the completed image through a plano-convex lens large enough to allow both eyes to view the image simultaneously enhances the effect. It is further described that the completed image, made up from the sets of the odd and even slices of an object, viewed from two or more parallax viewpoints or aspects, may also be produced by means of interlaced vertical scanning.

A modern technique for producing a single viewer, auto-stereoscopic display using a similar concept, is the parallel barrier or lenticular 3D screen recently exploited by television and game manufacturers including Sharp and Nintendo respectively. The technique is also used for the manufacture of 3D posters and the following description will assist in the understanding of the animated display process.

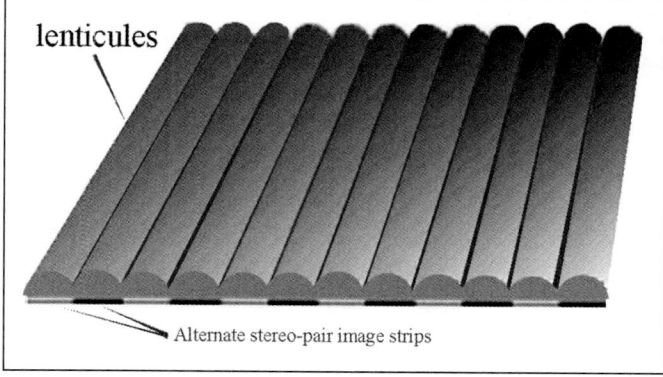

Fig 8.4: Section of lenticular panel

A photographic colour film is exposed through a special lenticular screen comprising several hundred vertical semi-cylindrical lenses per centimetre positioned side-by-side across the picture, **Fig 8.4**.

Each lens or 'lenticule' exposes a thin vertical slice of the photograph from a particular point of view in the same order that Baird described for a multiplicity of viewpoints. The completed photograph is viewed through the array, which acts as a parallax barrier directing the multiplicity of parallax left and right-eye images to the appropriate eyes. The result is that the observer views an auto-stereoscopic 3D lenticular picture with varying levels of horizontal parallax, depending on the number of viewpoints registered, without the aid of optical devices.

Despite the intensity of work on related projects including fibre-optics, night vision, video discs and spanning the Atlantic with television, Baird found time to return to the subject of stereoscopic television on 11 July 1928 with a further disclosure in patent *GB321441* [4] (**Fig 8.5**).

This stereoscopic patent, illustrated in Fig 8.5, describes a two and three colour, flying-spot scanner requiring only a single channel of communications to convey the signal to the receiver. The stereoscopic colour system uses a double 30-line scanner first demonstrated on 9 August 1928. The disc alternately scans the subject from two positions at the transmitter relating to left and right-eye views, and at the receiver images are viewed in colour and relief through a prismatic stereoscope. Referring to this concept at a later date in the development of 3DTV Baird recalled:

"This arrangement gave satisfactory results within the limitations of the definition and the only disadvantage was the necessity of using a stereoscope to view the received picture. Having demonstrated the possibility of such pictures, no further development was undertaken at the time owing to pressure of work in other directions and the subject was left until a more convenient time occurred." [5]

Baird was still deeply involved in the development of colour television in 1931, when he returned to the subject with a radical 3D concept [6]. It should be understood that a stereo-pair although recorded from two viewpoints relating to

Fig 8.5: Abstract from Baird's 3DTV patent *GB321441*

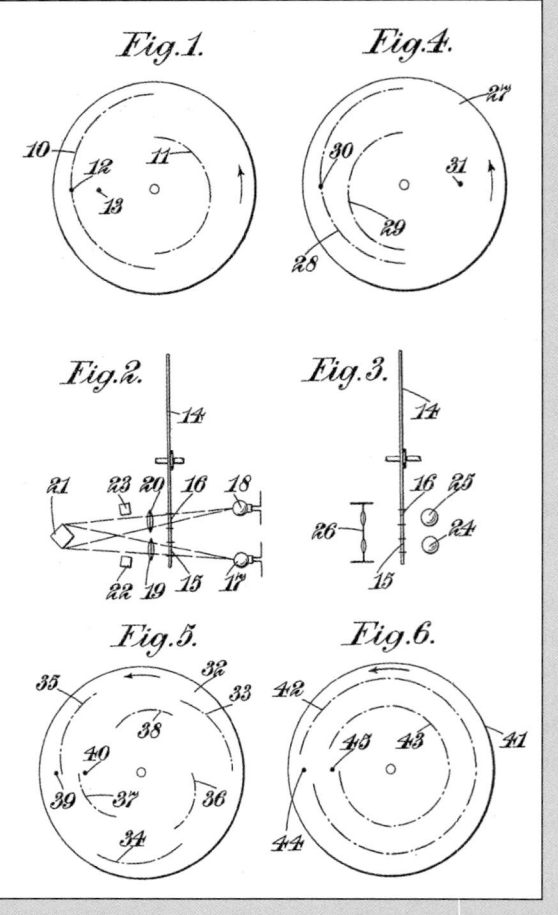

In an apparatus for stereoscopic television, two aspects of an object or scene are transmitted successively through a single communication channel and reproduced in juxtaposition, at such a rate as to be within the persistence of vision of the observer, which involves the transmission of some 12 images of each aspect per second. The scanning disc 14, Figs. 2 and consists 3 is provided with two sets of holes 10, 11, Fig. 1, lying on spirals whose "operative points" are 12 and 13 respectively, and which are spaced any desired distance apart, eg the spacing of human eyes.

In the transmitter, Fig. 2, the light sources 17, 18 scan the object 21 alternately, and from a different angle, through the spirals 10, 11, whose fields are represented as 15 and 16 respectively, and light from the sources, reflected from the object, is received by the light-sensitive cells 22, 23 respectively.

At the receiver Fig. 3, glow discharge lamps 24, 25 in conjunction with an identical scanning disc 14, produce two images in alternation, which are in juxtaposition and are viewed by a stereoscope 26. Exaggerated stereoscopic effects may be obtained by using scanning discs of the type shown in Fig. 4, in which 30, 31 are the widely spaced "operative points" of the spirals 28, 29 respectively.

Colour effects may be obtained by reproducing one picture in red and the other in blue by using a filter of appropriate colour with each spiral on both scanning discs. For a three-colour process the disc 32, Fig. 5, is preferably used in which the three outer spirals 33, 34, 35 have an "operative point" 39, and the three inner spirals 36, 37, 38 have an "operative point" 40, the spirals of each group being provided with blue, green, and red colour filters respectively, arranged in any desired sequence.

CHAPTER 8: THE THIRD DIMENSION & HOLOGRAPHIC TV

> *In a modification, at the receiver a screen of paper, gelatine, and the like is mounted upon a member, such as an iron body, adapted to be displaced in accordance with the depth signals, so that the screen is displaced according to the depth of the image. Alternatively, the screen may consist of a transparent enclosure filled with light-diffusing particles held in suspension, eg smoke, colloidal sulphur in water or dilute hydrochloric acid, fluorescence or chlorophyll.*

Fig 8.6: Abstract from the first volumetric imaging patent, *GB373196*

the horizontal separation of the eyes, provides only three-dimensional information from one particular viewing position. There is not enough information in a stereo-pair to visualise around the scene or object, and therefore a stereogram is not fully three-dimensional in the wider sense.

It will be shown that Baird not only developed the first stereoscopic television images, but also invented the first known system of its kind to transmit and receive the image of a three-dimensional object that occupied a real volume of space at the receiver. This concept liberated the observer from cumbersome viewing devices such as stereoscopes or coloured glasses and was a completely radical approach to the subject.

GB373196 [7] (**Fig 8.6**) is the premier patent for an area of research more recently known as 'volumetric' or 'cubic' imaging, which at the time of writing continues to remain under development. This type of image which contains the 'whole picture' and although not being produced by the same technique as the more familiar static laser hologram, is a form of dynamic holographic television image. Baird refers to such images as luminous solid and phantom 3D images.

In this patent specification Baird explained:

> *Whereby instead of the usual flat representation of an object or scene an observer at the receiving end of the system sees a representation of the same in three dimensions, giving an impression of the solidity or depth of the object or scene transmitted.*

The description stated that systems have previously been described to reproduce an object viewed from two discrete points separated substantially by the distance of separation of the human eyes. With regard to this patent, having first referred to the prismatic stereoscope, stereo-pair juxtaposed strip combination, anaglyph and other methods, Baird wrote:

> *Now these systems are open to the objection that the image is reconstituted in one plane and the stereoscopic effect is obtained by an optical system that can be used by only one observer at a time.*

The core of Baird's 'new' invention related to a television or like system wherein a three dimensional object at the transmitting station is separately analysed from two directions, substantially at right angles, and at the receiving station reconstituted in three dimensions as an image of a physical object constructed from light.

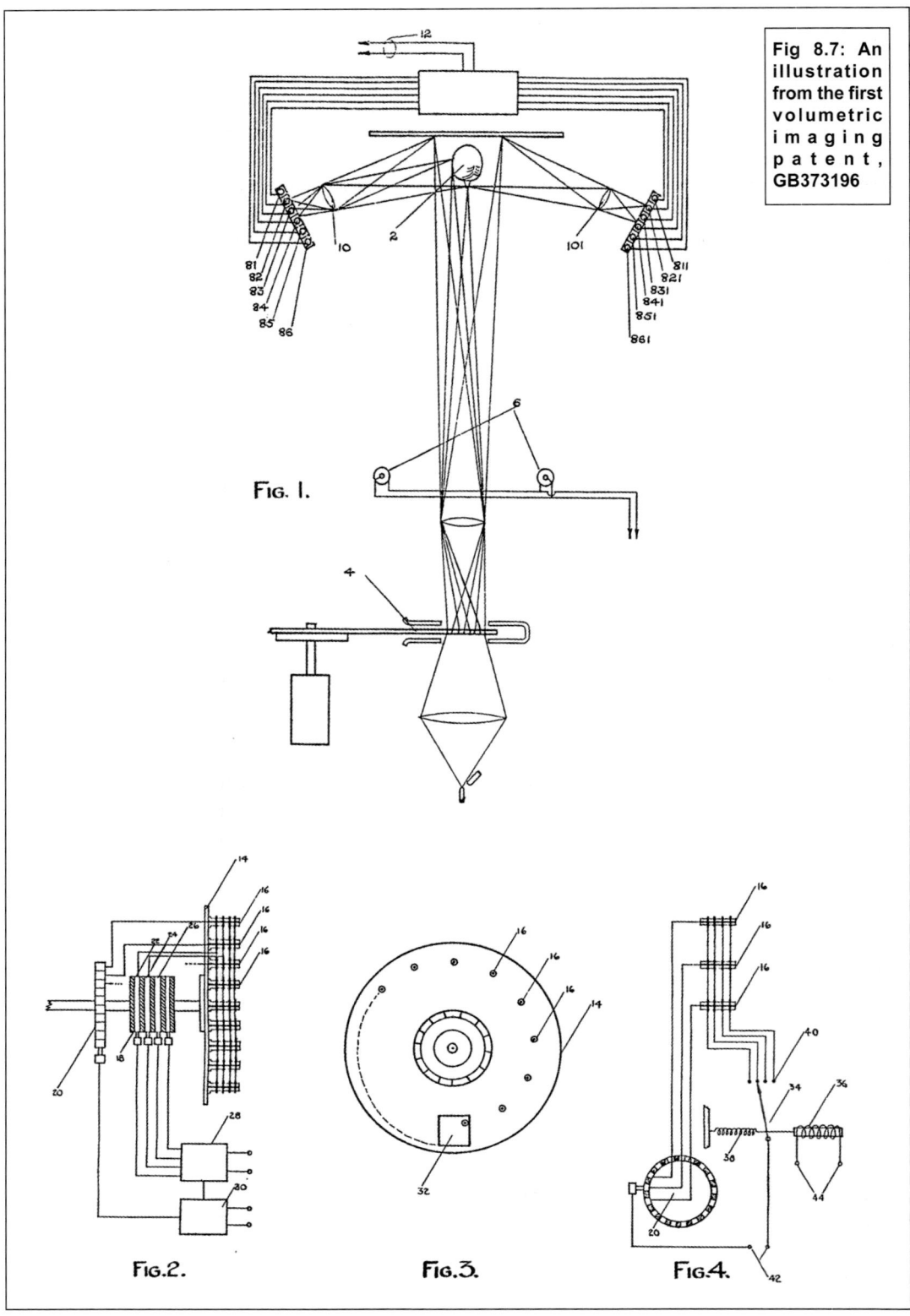

Fig 8.7: An illustration from the first volumetric imaging patent, GB373196

CHAPTER 8: THE THIRD DIMENSION & HOLOGRAPHIC TV

At the transmitter a number of photocells, carefully positioned, collected diffused light from the front to the rear of the object as it was scanned by a flying-spot. The outputs from the photocells relating to each depth plane were in parallel and used to activate one of a set of six relays. Each plane as it is scanned would activate a corresponding relay, which sent a unique level of current to the receiver, relating to the particular depth plane currently being scanned. Signal line 12 in the patent (see **Fig 8.7**) therefore carries serial information to the transmitter relating to the depth plane currently being scanned. The depth signal was transmitted separately from the television signal and used at the receiver to activate a two-dimensional screen at corresponding depth in the display. The multiplicity of images at various depth planes, together with persistence of vision gave the illusion of a physical replica of the original object reconstructed from points of light. Various depth screens were suggested including a transparent enclosure filled with a light-diffusing medium or a dynamic screen registering light spots at different positions within a volume of space.

While the process of mechanical scanning may be considered primitive by modern-day standards, the dynamic holographic concept of the patent is very advanced. Baird considered it a major breakthrough not to be revealed until it could be properly exploited, which he later achieved using higher-definition electronic scanning.

The Second World War and subsequent reorganisation of Baird and Cinema Television by the major shareholder and head of Gaumont British Picture Corporation, Isidore Ostrer, resulted in Baird's contract as President of Cinema Television being terminated.

Baird television Ltd did not conventionally go into liquidation, but continued to exist with staff levels reduced from around 300 to 30 under a Board of Trade agreed merger with Cinema Television Ltd. Many Baird employees were either called up, volunteered to join the armed services or were retained as employees vital to the war effort, while others simply lost their jobs. In the Companies Court, London, a scheme was sanctioned providing for the merger of Baird Television with Cinema Television, to preserve its goodwill and retain the technical staff until television transmissions were resumed in this country.

Subject to a sanction of the Board of Trade, it was proposed to retain the name Baird. Cinema Television continued to carry the Baird Television name and goodwill as 'Cinema Television Ltd (Incorporating Baird Television Ltd)' until at least 1942 when Ostrer sold the entire Gaumont British concern to the Rank Organisation [8]. Cinema Television exists today as Cintel International a highly successful world supplier of post-production equipment for the film industry and is no longer related to the Rank Organisation.

As previously indicated, Baird decided to continue with his colour and three-dimensional television development from his private laboratories, and was fortunate in receiving the continued support from his colleagues at Cinema Television. It is also known that Baird made informal visits to the factory during the war years, but due to the secrecy of the war work and his lack of a security pass, he was under the escort of A G D West [9]. With the advent of higher definition television and the assistance of his extended research group, Baird was able to dispense with the mechanical flying-spot scanners replacing them with high-intensity cathode-ray projection tubes, designed, built and supplied by Dr Szegho.

Baird also had unique access to the only large panchromatic photomultiplier tubes in existence, invented on his behalf by Dr Sommer. As Baird makes reference, in a number of his patents, to the use of matrix floodlight systems, it is reasonable to assume that storage cameras were also at his disposal. Although unlicensed for commercial exploitation, experimental Baird and Cinema Television storage camera tubes were manufactured under the direction of Dr Samson.

Former Baird engineer and diarist, Gilbert Tomes worked for Baird and Cinema Television from 1935 until 1945, before establishing the British nuclear support industry, Twentieth Century Electronics, appropriately renamed Centronics for the 21st Century. He made reference to Baird Television iconoscopes and included photographs in his memoirs [10]. **Fig 8.8** shows various types of Baird manufactured tubes.

With the advantage of Cinema Television's high intensity projection tubes and other electronic technology at his disposal, Baird returned to the subject of colour and stereoscopic television at his own expense. He arranged for his family to move out of London to a place less likely to be a target for German bombers and after some consideration chose Bude in Cornwall. Despite the personal danger, Baird continued to work from his private laboratories at Crescent Wood Road with his three assistants.

Baird revisited colour and stereoscopic television employing the principle of anaglyphs viewed through special coloured filter glasses. The main purpose was to produce a high quality three-dimensional, anaglyph system suitable for use with the 405-line BBC television standard. His first idea was to add colour and 3D, but limitations of the anaglyph viewing system made the system more viable for monochrome television.

For example, if the right eye view were red and the left eye view blue, a portion of the scene having a natural red colour disappeared when viewed through the blue (left) filter and thus lost relief. Colour rendering could thus only be achieved by this method at the expense of stereoscopy and vice versa.

With the cessation of television broadcasting during the Second World War eliminating any immediate requirement, he decided to take his research further in the hope that at the end of hostilities he could take the lead in television tech-

Fig 8.8: Baird manufactured matrix camera tube of the Zworykin type undergoing visual inspection (left), and (right) large quantities of Baird iconoscopes, image dissectors and other cathode-ray tubes ready for testing in 1939

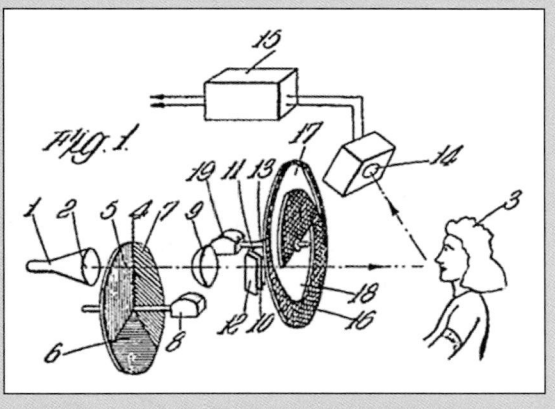

Fig 8.9(a): Abstracts from Baird's patent *GB552582*, including Fig 1 from the drawing for Baird's auto-stereoscopic active shutter 3DTV (1941)

> *A stereoscopic television transmitter, Fig 1, comprises a cathode-ray tube 1 producing a scanning spot on a single field area of its screen 2, a single lens 9 for focusing the light on to the required spot of the scene or object 3, a multiple mirror or like device 10, 11, 12, 13 close to the lens on the side remote from the cathode-ray tube for dividing the light into two beams scanning the objects from points virtually spaced apart to correspond to right and left eye views, a shutter 16 arranged to rotate between the mirrors and the object to obscure the respective beams alternately, and means such as a photo-electric device 14 for converting the light into electrical impulses and transmitting the impulses by wire or wireless.*
>
> *If colour reproduction is desired, a rotating colour screen 4 is provided between the cathode-ray tube 1 and the single lens 9.*

nology. The future did not appear certain, but he would continue until either he attracted a supporting income, or in the worst case, for as long as his £15,000 in savings would allow.

On 11 July 1941, financed only from his savings, Baird produced an electronic stereoscopic colour television patent *GB552582* [11], **Fig 8.9(a)**, which although dispensing with any form of mechanical scanning, required a rotating colour filter disc, and an alternating left-eye image, right-eye image shutter disc. Baird's mechanical shutter disc carried out precisely the same function as the wireless LCD shutter glasses used by modern 3DTV manufacturers. He gave a brief description of rendering the 3D system in colour and with typical foresight, he alluded to the addition of multiple sets of active shutters, which relate to the use of multiple active LCD shutter glasses (**Fig 8.9(b)**).

In completing his vision of future 3DTV, Baird provided an alternative to the active shutter system by describing the use of passive cross-polarised viewing glasses operating in harmony with a rotating left-eye and right-eye polarising shutter disc (Fig 6 from the patent specification, shown in **Fig 8.9(c)**). It was also noted that the system was not limited to the flying-spot scanner, but also functions with an iconoscope or an electron type camera.

Perhaps the most striking part of this 1941 patent was that it offered several methods of viewing the picture in relief and included an active shutter disc for auto-stereoscopic viewing. By suggesting a multiplicity of active shutter discs for multiple viewers, Baird not only alluded to a multiplicity of discrete active shutter glasses, but also removed the requirement when used with a single viewer.

In summary, the methods covered by *GB552582* are:

Fig 8.9(b): Abstracts from Baird's patent *GB552582*, including Figs 2, 3 and 4 from drawing for the auto-stereoscopic active shutter 3DTV (1941)

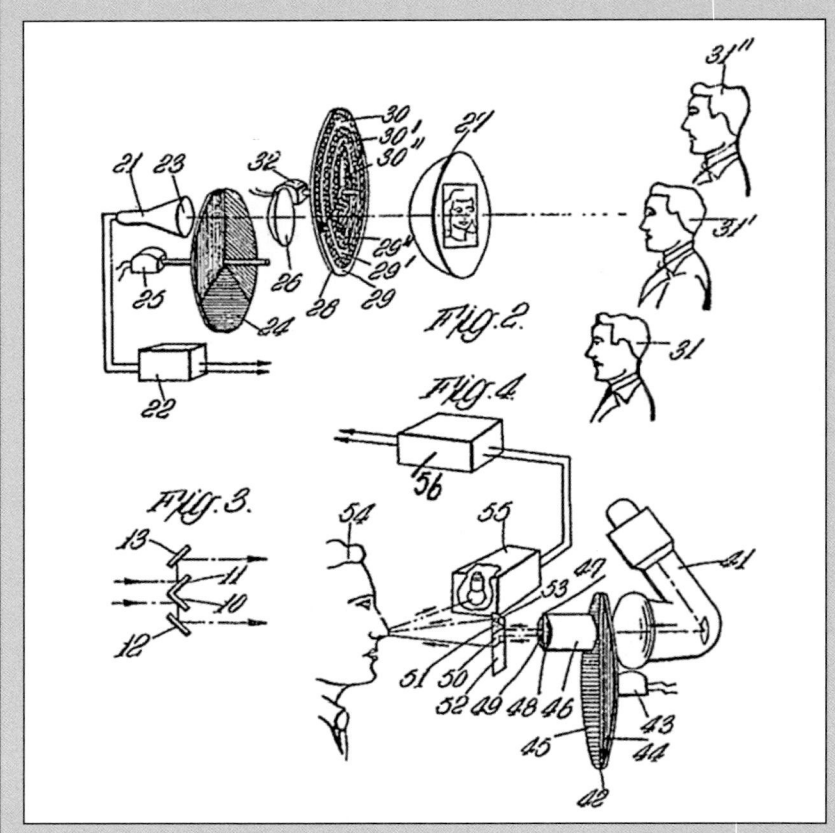

In a receiver, Fig. 2, of generally similar arrangement, the object is observed by a person 31 through a field lens 27 focusing a pair of slots in a rotating shutter 28 on the eyes of the observer.

Several pairs of slots may be provided for several observers 31, 31i, 31.ii In a modified transmitter, Fig. 4, half the lens 46 between the cathode-ray tube 41 and the multiple-mirror device 50, 51, 52, 53 is covered by a fixed screen 48 of one colour and the other half by another fixed screen 49 of another colour and a two-colour rotating screen 42 cooperates with the fixed colour screens to produce the alternate shuttering of the two stereoscopic beams.

1. Multiple active shutter discs allude to viewing with active shutter glasses.
2. Single viewer: auto-stereoscopic, without glasses.
3. Multiple viewers: auto-stereoscopic without glasses.
4. Multiple viewers: passive cross-polarised glasses.
5. Multiple viewers: passive anaglyph colour filter glasses.

Baird's financial situation was only moderately improved when Cable and Wireless appointed him as 'Consulting Technical Adviser' at a salary of one thousand pounds per annum for a period of three years from 1 November 1941.

CHAPTER 8: THE THIRD DIMENSION & HOLOGRAPHIC TV

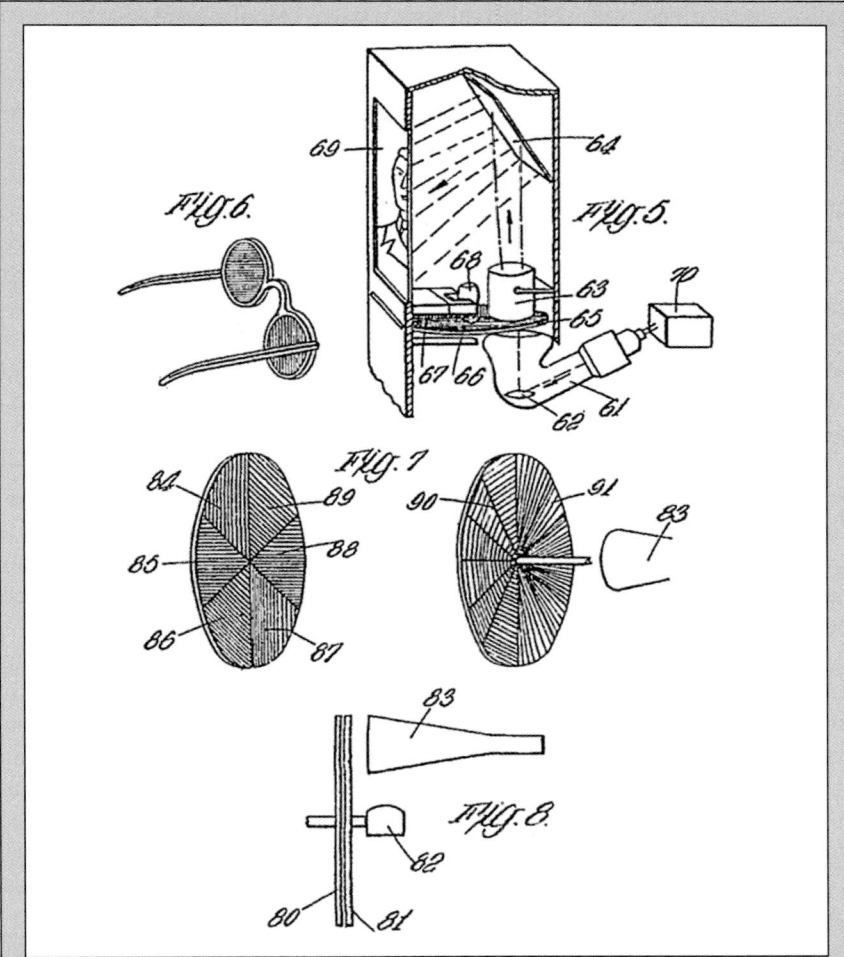

Fig 8.9(c): Abstracts from Baird's 3DTV patent *GB552582*, including Figs 5, 6, 7 and 8 showing alternative multiple viewer passive polarised spectacles (1941)

The shutter devices of Figs. 1, 2 and 4 may be replaced by shutter devices depending on the use of light polarisers. The transmitters, Fig. 1 and 4 may be made to operate with the light beams in reverse direction, that is to say, the object may be flood lighted and the light from the object proceed to a photo-electric mosaic or electron-camera type of transmitter. The specification also describes the use of receivers in which the image appears on a screen and is viewed by means of coloured or polarised glasses in front of the observer's eyes.

On Thursday, 23 December 1941, Baird gave a demonstration of his most recent system based on the above patent to the press showing stereoscopic television in colour without glasses. The 3D display presented full-colour, red, green and blue images of the left and right-eye views of a scene, in rapid alternate sequence, the sequential stereoscopic images being optically directed to the respective eyes of the viewer with an active shutter built into the optics of the display. A 500-line image of 'Eustace,' a manikin used in the studio for experiments

137

and demonstrations, was published showing a stereo pair of a 'head and shoulders' of the manikin, dressed in a vivid red jacket, green cravat and blue headband.

To the reporters this was a landmark in television technology, but although a few articles later appeared in technical magazines it was completely overlooked by the newspapers, where events in Europe took precedent.

In photographs supporting a few magazine articles, Baird is seen rigidly looking into a field lens viewing an auto-stereoscopic image, but the fixed position of the viewer was seen as a disadvantage. The reporters could not possibly have known that they had witnessed the basis for the future of commercial 3DTV and games consoles. This singularly important demonstration included the important elements required for multiple and single viewer systems. The familiar passive cross-polarised 3D spectacles, shown in Fig 8.9(c), are currently available for viewing movies produced for the 3D market and for some models of stereoscopic television.

Baird took another giant leap forward in the development of 3D television by combining the elements of his 1932 cubic (or volumetric) television patent *GB373196* [12], with electronic cathode-ray tubes. On 26 August 1943, Baird applied for what could now realistically be recognised as his most important patent, the detail of which, due to Baird's untimely death in 1946, remained obscure for almost fifty years.

The full details of his breakthrough remained virtually unknown until the author completed a search and analysis of Baird's patents in 1994. On compiling the 176 British patents issued to John Logie Baird during his television career covering two decades from 1928 until 1948, the author took particular interest in *GB573008* [13], which comprised fourteen pages of text and five sets of drawings. The application took Baird one year to complete from six consecutively modified provisional specifications. The patent was finally accepted on 1 November 1945, only seven months before Baird died on 14 June 1946. This patent is valuable not only by its contribution to technology, but also in providing the reader with an insight into John Logie Baird's highly creative thought process. The patent is in seven stages, with six provisional specifications leading to the complete specification. There is much to be gained by chronologically interpreting from the most relevant stages through to the completed specification (see **Fig 8.10**).

The first Provisional specification, number *13887/43* [14], described a method of improving the depth of focus in a television scene by means of a photoelectric 'range finder' device that was also claimed to be useful for determining the range and speed of moving objects (**Fig 8.11**).

To simplify a complicated description, an example was given relating to the process for switching the focus between the two individual planes of the back-

Fig 8.10: Abstract from Baird's patent *GB573008*

A television apparatus comprises a photoelectric range-finding device, whereby a signaling current is produced related continuously to the distant or range of the point of the object being viewed by the transmitting apparatus. This signal may be used to control the focusing of the scanning apparatus, or may be sent to a receiving apparatus to produce an image in three dimensions.

CHAPTER 8: THE THIRD DIMENSION & HOLOGRAPHIC TV

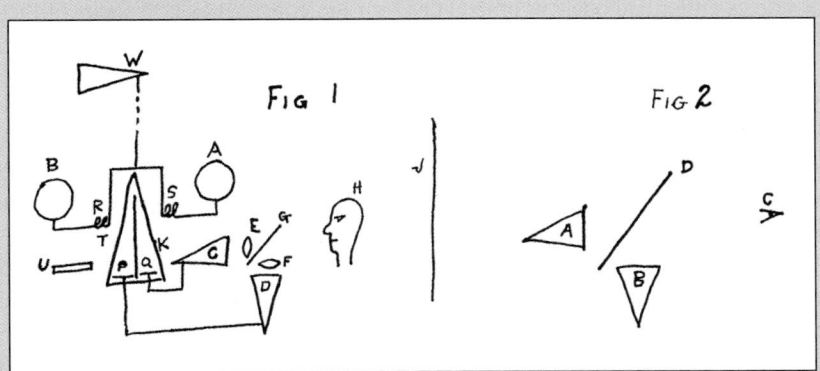

Fig 8.11: Baird's description of a means to improve depth of focus in his provisional specification No 13887

In television a wide aperture lens is very desirable - such lenses have little depth of focus and to obviate this, I use the following device:

Two photocells 'A' and 'B,' form the basis of a range finder and are placed within the proximity of a scene. If 'cell A' is nearest to the scene, then for distant objects the current from 'A' will be nearly equal to that of 'B.' For nearby objects 'A' will give a greater current than 'B.' The difference of output between the cells [15] is dependent upon the distance separating 'cell A' from the point on the scene upon which the light spot falls at any given instant. Although the difference in current was describes as a means of correcting the focus of the transmitter at two points, it was clarified that the system would measure over a multiplicity of depth planes.

ground and the foreground. It was assumed that background objects at distance of 'F' from 'cell A' would give an output difference between (A and B) of 'M' and that foreground objects at a distance of 'G' from 'A' would give an output difference between (A and B) of 'n'. As the light spot traverses the screen, it falls first on perhaps background objects giving a current 'M', which causes the spot of the transmitter to focus sharply upon the background. If the light spot next falls upon an object in the foreground the current difference then immediately changes to 'n,' and is used to bring foreground objects into sharp focus.

With reference to ('Fig 1') shown in Fig 8.11, the operation of controlling the focus of the spot of light on the object or scene is as follows. At the transmitter two flying-spot cathode-ray tubes 'C' and 'D' are associated with a pair of fixed focus lenses separated by a beam splitter comprising a half silvered mirror. Lens 'E' is pre-focused on near objects and Lens 'F' is pre-focused on distant objects. Depending on the distance from the transmitter at which the light spot falls on the object, the most appropriately focused spot from one of the two flying-spot cathode-ray tubes is selected.

The control signals 'M' and 'n' therefore alternately switches between one or other of the two light spots to maintain sharp focus of the spot throughout the scan. The switching device is a special purpose cathode-ray tube 'K', where the electron beam 'T' is pulled from or towards electrodes 'P' and 'Q', by the signals 'M' or 'n' via the magnetic deflection field coils 'R' and 'S'. A permanent magnet 'U' sets the beam to the desired zero position while the current from either 'P' or

'Q' selects either cathode-ray tube C or D for the consistent sharp focus of these two planes.

Unexpectedly the subject of the specification diverts from being solely related to a means of sharpening the focus in a television scene to that of a device for transmitting and showing depth. Baird indicated that the analogue variations in current corresponding to the distance of an object from the transmitter may be used to cause corresponding received images to be thrown upon screens at corresponding distances from the viewer's eyes.

'Fig 2' from the specification, shown in Fig 8.11 illustrates a form of 3D receiving display upon which cathode ray tube 'A' reproduces only an image of the background objects and cathode-ray tube 'B' reproduces only an image of the foreground objects. The images are superimposed to the observer's eye 'C' at the desired distance apart by means of a half-silvered mirror 'D'. The receiving tubes are controlled by the signal from the transmitting tubes so that each receiving tube acts in synchronism.

In summing-up, Baird suggested a number of modifications including replacing the two cathode-ray tubes, associated lenses and beam splitter with a rotating screen, or by employing a double-screen tube such as the Telechrome to reproduce the scene. It was also indicated that negative lenses could be used to emphasise depth and that colour could be added by any one of a number of techniques. At the end of this stage it appears that although the variable analogue range finder signal is able to determine a number of depth plane within the scene, the display systems suggested are only viable for reproducing a 3D image consisting of the foreground and the background. It will be shown that Baird moved on to resolve this limitation in his second provisional specification.

The second provisional specification No *15851/43* [16] directly relates to improvements in three-dimensional television. Baird makes direct reference to the depth signal derived from the previous specification, the intensity of which varies in accordance with the distance from the transmitter at which the spot lands on the object. It would appear that after experimenting with the previous concept and subsequently viewing the resulting image of the depth signal against time as a single two-dimensional line display on a cathode-ray tube, Baird made an interesting discovery. The pattern on the screen showing the repetitive two-dimensional depth pattern for each line of the scanned image in a completed frame appeared similar to a plan view of the object at the transmitter.

A simulation of the hemispherical surface of a human head being horizontally scanned using a conventional raster pattern, is illustrated in **Fig 8.12**, with parts of the object closest to the scanner being seen closest to the lower edge of the screen. The contour lines representing the part of the face below the nose

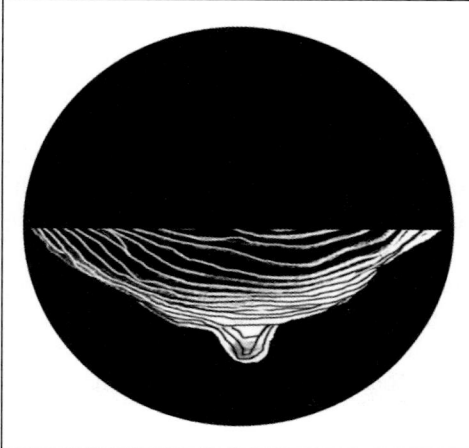

Fig 8.12: Plan view of scanned image

are also present on the cathode-ray tube but are obscured by the more prominent part of the plan view, objects near at hand being displaced on the scene from the more distant objects. The picture is made up of a succession of superimposed plan views of the scene, with the spot being modulated by the light and shade of the scene and displaced laterally by its distance from the transmitter.

If now the screen is received with its edge presented to the observer, that is, with the screen approximately horizontal, so that the observer looks along its surface, near objects will be represented by light spots of the part of the screen near and more remote objects on parts of the screen being remote from the observer as in Fig 8.12. With a stationary screen, the plan views given by each scanning line sent out by the transmitter are superimposed, but if the screen is displaced vertically by a distance approximately equal to the width of a spot by each line, vertical scanning is provided and the observer will see a picture in relief, **Fig 8.13**.

The effect of moving the screen vertically may be obtained optically. For example a revolving mirror drum or an oscillating mirror might be used with the observer viewing the screen by reflection. The depth of the images are limited by the depth of the screen, but the effect of depth can be increased by placing a negative lens in front of the edge of the screen. The device can also be used at the transmitter to give depth of focus by adapting the method described in co-pending provisional specification No *13887/43* of 1943.

The third provisional specification, 16413/43 [17], described a more convenient method of displacing the image on the screen. The specification indicated that in co-pending provisional specifications Nos *13887* and *15851* (1943) means are described for producing images in three-dimensions by providing at the receiver a screen which is viewed at a very acute angle, the width of the screen giving the depth of the picture and the vertical component being obtained by moving the screen (or an image of the screen) vertically. A convenient method of providing this movement is to make the screen in the form of a single turn of a spiral, the picture being projected upon the flat face of the spiral so that as the spiral revolves the screen is moved vertically by a distance equal to the pitch of the spiral.

Fig 8.13: Persistence of vision enables a 3D image when the screen is vertically displaced at the frame rate of the scanner

Baird also dedicated this specification to methods of providing colour, by indicating that three screens, each one-third of a complete spiral, could be arranged to act in sequence, each screen being coated with a different coloured powder. The spirals rotating within a cathode-ray tube, or the colours achieved by attaching coloured filters to the receiving screens, or viewing the

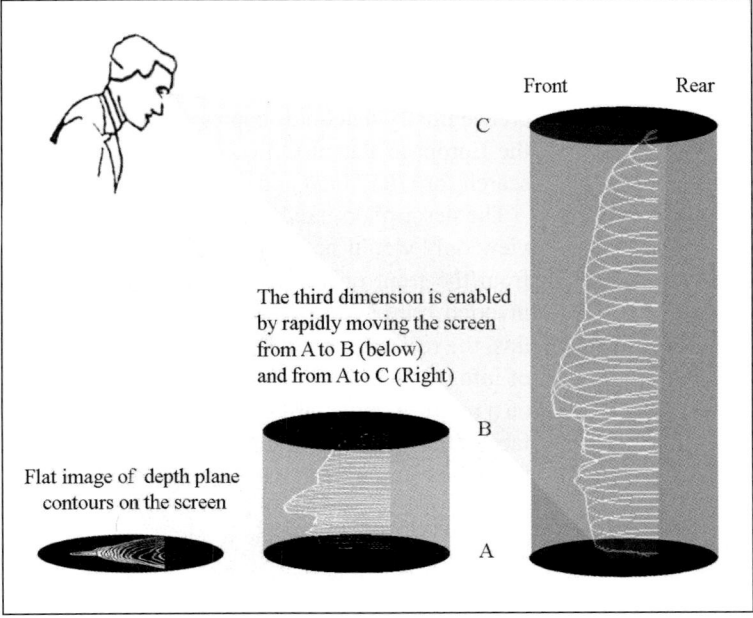

picture through revolving colour discs or by using two cathode-ray beams, one giving a red picture on the face of the spiral and the other a blue picture on the opposite face. Baird explained that the spiral or spirals could be external to the cathode-ray tube in the form of projected images.

This specification is summed up with a description of a method to overcome problems encountered when a negative lens is used at the receiver, to enhance the apparent depth of the images. The negative lens achieves this impression by increasing the perspective, but in doing so there is a noticeable reduction in the stereoscopic effect that contradicts the improvement. To obviate this, Baird exaggerates the perspective at the transmitter with a positive wide-angle lens.

The fourth stage, described in provisional specification *16581/43* [18] is less relevant to the completed specification and will be dealt with only briefly. The purpose of this disclosure is to describe an alternative method of measuring depth using a process that appears to be more academic than practical.

The specification described an alternative means of finding the distance of the light spot by utilising two lenses. One lens projected the spot upon the scene while the other, separated from it, projected an image of the spot on a screen. Baird explained that the angle made between the light beam issuing from the first lens and the light beam entering the second lens is proportional to the distance of the spot. The spot on the screen will therefore move up or down during any one horizontal scanning line where the scene contains objects at different distances. The up or down movement provides a graph on the screen showing the distance away of the spot at any instant. The less practical part of the spec related to the manner in which the movement of the spot on the screen was to be converted to an electrical signal by fabricating the screen from a grid of photo-electric detectors.

The fifth stage in the process is provisional specification No *19825/43* [19], which partly deals with the mathematics required in measuring the depth of the object or scene with the greatest accuracy.

If you were thinking that Baird worked on the 'hit and miss' principle you would be pleasantly surprised by the mathematical calculations provided in this specification. It is not the purpose of this book to delve into the mathematical complexities of Baird's work but readers are advised to visit an appropriate reference library that holds copies of patents, visit the UK Patent Office or log onto the European Patent Office [20]. In the case of the latter, carry out a number search for *GB573008* and download the original patent.

The description indicated that by using the devices so far described, the front view only would be seen, the effect being of, for example, a head illuminated from the front only and if the observer moved to the back of the receiver an inverted front view would be seen instead of the rear of the head. To overcome this, the rear view would be generated separately at the transmitter and both sets of information would be sent to a receiver. In the case of viewing the 3D image on a rotating single turn spiral screen, the front view would be projected onto one face of the screen while similarly the rear view would be projected onto the other face. To overcome the transparency effect of the 3D image it is arranged that the rotating screen consist of an opaque material, which will also behave as a shutter occluding the rear view of the image from the front, and the front view of the image from the rear. A 3D image produced in this way will have a viewing angle of 360 degrees in the horizontal plane, and 75-80 degrees in the vertical plane.

CHAPTER 8: THE THIRD DIMENSION & HOLOGRAPHIC TV

> *A modified form of the device could be used for X-rays, the body being scanned by an X-ray beam, for example, by passing the X-rays through a revolving lead disc with a spiral of holes. Then when the X-rays strike, for example a metal body, secondary rays could be detected by a fluorescent screen viewed by a photocell, these screens replacing the photo cells in the devices described above.*

Fig 8.14: Part of Baird's provisional specification No *19825/43*

> *The invention relates to improvements in television and is concerned with apparatus for the production in three-dimensions of phantom visual replicas of objects or scenes. It is to be understood that this invention does not refer to stereoscopic television (which is also sometimes referred to as three-dimensional) as in the latter type of television only two-dimensional images are received and the three-dimensional effect is entirely an optical illusion produced by presenting different flat pictures to the right and left eyes of the observer.*
>
> *In the present invention the image itself has actually three-dimensions does not appear upon a flat screen, but within a three-dimensional space. Such an image is hereinafter called a three-dimensional image.*

Fig 8.15: Abstract from the completed specification in Baird's patent *GB573008*

> 1. *At the transmitter a range-finding device produces and transmits a signal varying with the distance from the transmitter of the point of the scene being scanned and this signal is transmitted in addition to the signal giving the luminosity of the point. A number of suitable range finding devices are hereinafter described.*
> 2. *At the receiver the screen is viewed at an angle, being preferably approximately horizontal. Upon the screen are reproduced a succession of contour lines. The screen moves up and down in a vertical direction and the image is built up of a succession of adjacent contour lines. The image appears within a three-dimensional space, two of whose dimensions are the length and breadth of the field produced upon the screen and whose height is the vertical up and down motion of the screen. An optical device is also hereinafter described whereby the illusion of infinite depth may be given to the scene reproduced.*

Fig 8.16: Abstract from the completed specification in Baird's patent *GB573008*

In summing up, Baird indicated that the 3D system could also be used with a floodlight system such as a storage mosaic camera, provided it cooperated in synchronization with a flying spot camera acting as the range finder. Another interesting part of this specification lies in the paragraph shown in **Fig 8.14**.

The sixth provisional specification No *2370/44* [21] deals with range finding by the superimposition of a grid and is less relevant to the completed specification which is shown in **Fig 8.15**. In the completed specification *GB573008* [22], Baird described the invention as having two outstanding features (**Fig 8.16**).

The diagrams shown in shown in **Fig 8.17,** numbered from (Fig 1 to Fig 6) and (Fig 7 to Fig 11) are the ones that accompanied the completed specification. The

THE THREE DIMENSIONS OF JOHN LOGIE BAIRD

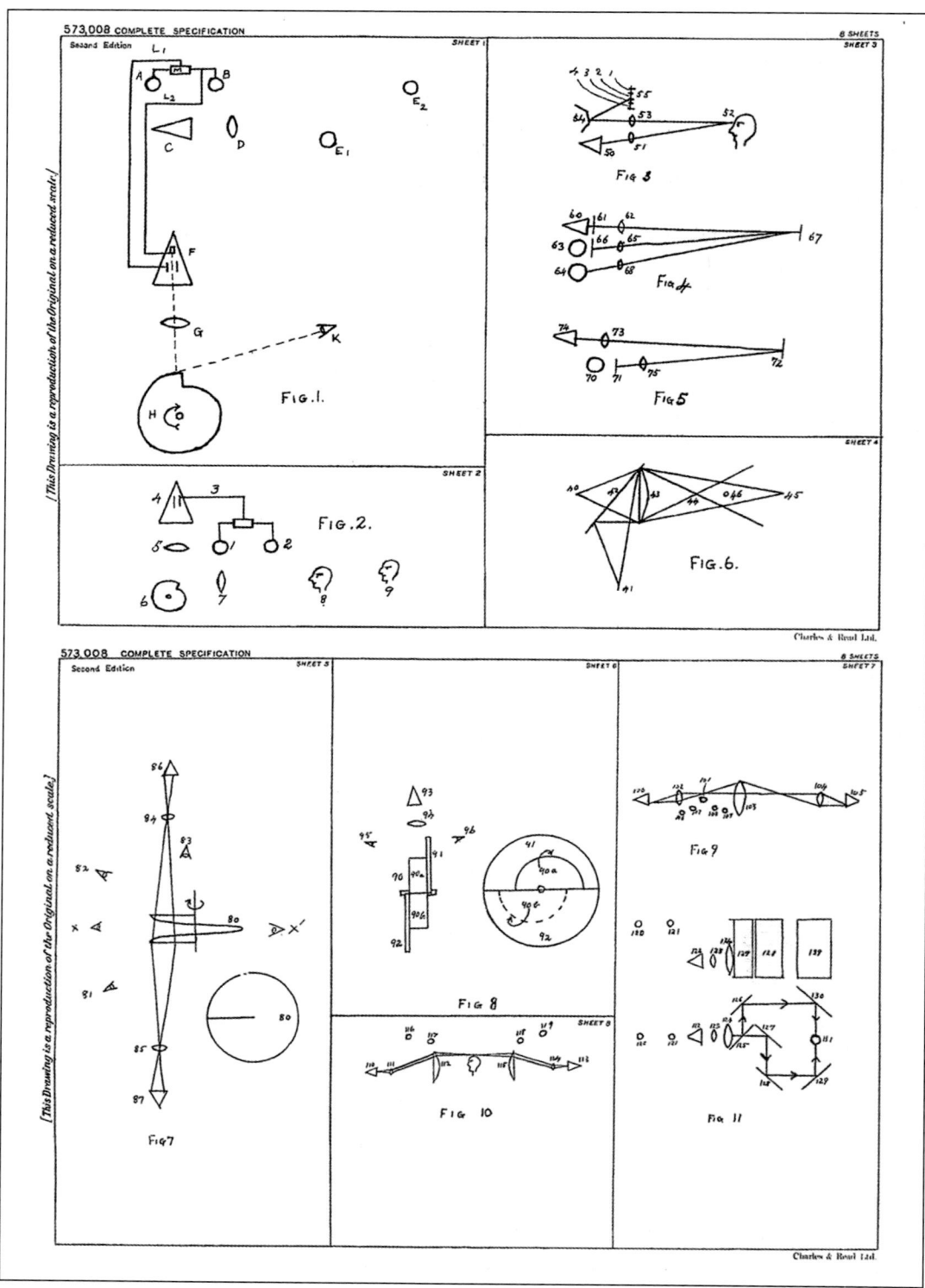

Fig 8.17: Figures 1 to 11 from completed specification *GB673008* (see pages 8.20 & 8.23 for detailed captions)

CHAPTER 8: THE THIRD DIMENSION & HOLOGRAPHIC TV

items labelled A and B are photoelectric cells; C is a cathode-ray tube at the transmitter, the fluorescent screen of which is traversed by a scanning spot and D is a lens causing an image of the scanning spot to traverse the objects E1 and E2.

The line L1 carries a current proportional to the output from A divided by the output from B; the reason for this is to vary the ratio as much as possible to collect detailed depth information. The line L2 carries current proportional to the output from B.

The current in L1 is fed to a pair of plates and deflects the scanning beam of the cathode-ray tube F at the receiver in a direction at right angles to its direction of line scanning. The current from L2 is fed to the gun of the cathode-ray tube F and modulates the intensity of the cathode-ray beam. The lens G projects an image of the spot on the face of a revolving spiral-shaped drum H, which revolves once for each frame. In this way a succession of modulated contour lines at different vertical heights are presented to the viewer K, producing by persistence of vision the visual impression of a luminous solid object.

The details of Fig 8.17 are as follows:

Fig 1 is the diagrammatic representation of the principles involved.
Fig 2 illustrates means whereby the scanning spot remains sharply focused.
Figs 3, 4 and 5 illustrate forms of range finding devices.
Fig 6 is one arrangement of range finding cells.
Fig 7 shows one arrangement of the receiving apparatus, where both front and back views can be produced.
Fig 8 is an alternative arrangement of the receiving apparatus, where both front and back views are reproduced.
Figs 9 and 10 are arrangements of the transmitting apparatus where both front and back views are transmitted.
Fig 11 shows a modified arrangement of the transmitting apparatus whereby both front and back views can be transmitted.

Possibly the most important part of this specification is an improved method of range finding where the distance is computed by electronic analogue calculation. The output from photocell A is divided by the output from photocell B for accurate distance measurement. This is achieved electronically by processing the signals from A and B separately through logarithmic amplifiers, which are then algebraically subtracted (logA - logB). This is achieved by inverting signal logB and passing the two signals through an electronic summing junction that normally produces a signal relating to the addition of the two signals, but as the logB has been inverted it is dealing with (logA ± logB) which is processed as (logA - logB). After subtraction the resulting signal is passed through an antilogarithmic amplifier producing the quotient of A/B, which is an accurate measurement of the depth signal.

This remarkable Baird invention is a concept much more advanced than stereoscopic television and may be the basis for a future holographic commercial television system. A simplified description of the holographic system showing the reconstruction of the front of a human head is shown in **Fig 8.18**.

Variations on Baird's theme are the subject of research worldwide. Baird's concept of producing solid images in colour within a volume of space also had

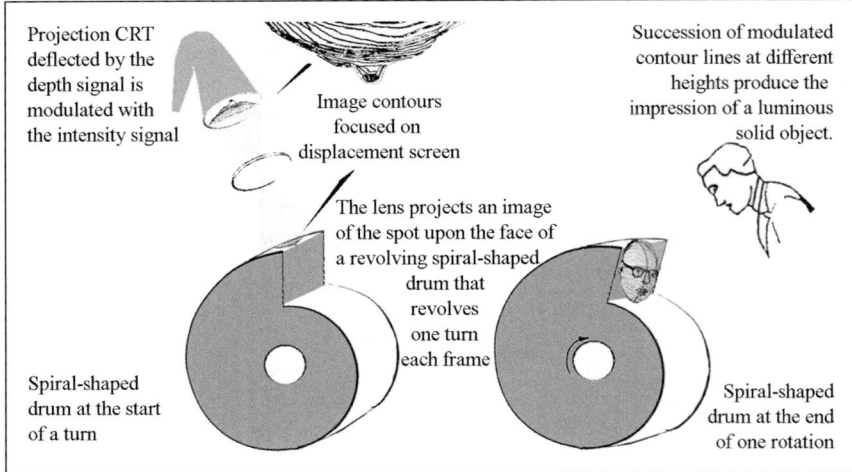

Fig 8.18: Dynamic holographic television concept described in the completed specification *GB573008*

possibilities for military, aviation and other applications. On 31 October 1943, Ferril [23] of the Sperry Corporation in the US patented a three dimensional Stereo Indicator proposed for air traffic control, which suggested using a cathode ray screen viewed on a mechanically oscillating mirror.

On 2 May 1945, Parker and Wright [24] proposed a 3D radar system in a patent on behalf of the Admiralty Signal Establishment. This employed a rotating stepped mirror for depth creation that was similar to Baird's equiangular rotating screen. A more recent concept based on the principle of volumetric imaging is the varifocal lens. This first appeared in 1956 in a paper by Dyson [25] describing a 'Common-Path Interferometer for Testing Purposes'. The device used a partial reflecting target and a triplet lens in conjunction with a parabolic reflector.

This was followed in 1958 by a 'Solid Image Microscope' article by Gregory [26], and in 1961 Muirhead [27] described a varifocal mirror consisting of a thin membrane of aluminised mylar in tension over the face of a loudspeaker using acoustic vibration to vary its focus. The basis of Baird's methodology for generating phantom images was further exploited in an article in September 1962 that discussed the use of vibrating mirrors and spiral screens to create three dimensional volumetric radar displays [28].

In March 1967 another development in the process of volumetric imaging was the suggestion that a birefringent material may be used as a means of replacing moving or vibrating screens [29].

In the same year Traub [30] published an article based on Muirhead's method for generating cubic images using a rapidly vibrating mirror. The varifocal theme of producing virtual three-dimensional images continued in 1968, with Rawson [31] followed by King and Berry [32] in 1970. King and Berry were troubled with 'out-of-focus' images later to be addressed by Gerritsen and Horowitz [33].

In June 1977 Hamasaki, Nagata, Higuchi and Okada, published a paper [34] on 'Real-Time Transmission of a three-dimensional image using Volume Scanning and Spatial Modulation'. This used a varifocal mirror and a vidicon camera at the transmission path, and a cathode-ray tube with a varifocal mirror at the receiver. It was confirmed that the required bandwidth of a suitable transmission line is n-times the bandwidth required for two-dimensional television, where 'n' is the number of sectional images per three-dimensional image. In March 1981, Dean Morelli [35] wrote of his three dimensional imaging system for real moving objects and described a moving test bed with laser depth sectioning, with an image dissector camera and a cathode-ray tube display on a vibrating table. The photographed results were good but Morelli concluded:

The number of scanned cross sections was 125 with 20 lost during retrace. At the very slow longitudinal scan rate of 13Hz the flicker was severe although the image appeared unbroken.

In 1987 Bernard M Ciongoli followed this with a method of three-dimensional imaging called Triology [36], in which a series of two-dimensional images were synthesised into a three-dimensional image, by means of a two-dimensional cathode-ray tube, rotating lens and a viewing screen. More recently the author's version of a volumetric imaging system was developed and constructed using computers, acousto-optical devices and lasers [37].

References
[1] The colour system is described in Chapter 6, see also Fig 6.2.
[2] Patent *GB266564*, JL Baird, applied 1 September 1925, accepted 1 March 1927.
[3] Patent *GB292365*, JL Baird, applied 22 December 1926, accepted 22 June 1928.
[4] Patent *GB321441*, JL Baird, applied 11 July 1928, accepted 11 November 1929.
[5] *Electrical Engineering*, JL Baird, February 1942, p620.
[6] *Images Across Space*, D Brown, RSGB Publications, UK (ISBN 976-1-874289-21-0), pp87-89.
[7] Patent *GB373196*, JL Baird, applied 18 February 1931, accepted 18 May 1932.
[8] 'Current Topics', *Wireless World*, December 1940, p500.
[9] *John Logie Baird: a life*, Antony Kamm and Malcolm Baird, NMS Publishing Ltd, 2002, p319.
[10] *'Jumbo's Diaries': The memoirs of Gilbert Tomes*, Never published but printed in a small quantity for family and friends.
[11] Patent *GB552582*, JL Baird, application 11 July 1941, accepted 15 April 1943.
[12] Patent *GB373196*, JL Baird, applied 18 February 1931, accepted 18 May 1932.
[13] Patent *GB573008*, JL Baird, application date 26 August 1943, completed 25 September 1944, accepted 1 November 1945.
[14] Provisional specification No *13887/43*, John Logie Baird. lodged on 26 August 1943.
[15] If the distance between A and B = x and the distance of the point on the scene on which the spot falls at the instant under consideration from cell A = y then the difference of the amount of light falling upon cell A and that falling upon cell B = $y^2 - (y-x)^2$; thus the difference of output between the cells is dependent upon the distance separating cell A from the point on the

scene upon which the light spot falls at any given instant. This difference in current is used to alter the focus of the transmitter.

[16] Provisional specification No *15851/43*, John Logie Baird, lodged on 27 September 1943.

[17] Provisional specification No *16413/43*, John Logie Baird, lodged 7 October 1943.

[18] Provisional specification No *16581/43*, John Logie Baird, lodged 11 October 1943.

[19] Provisional specification No *19825/43*, John Logie Baird, lodged 26 November 1943.

[20] http://gb.espacenet.com (note that website addresses often change).

[21] Provisional specification No *2370/44*, John Logie Baird, lodged 9 February 1944.

[22] Patent *GB573008*, J L Baird, application date 26 August 1943, completed 25 September 1944, accepted 1 November 1945.

[23] US patent *US2361390*, T M Ferril Jr, Stereo Indicator, filed 1 November 1943, issued 31 October 1944.

[24] Patent *GB592367*, E Parker, and Charles Seymour Wright, 3D Radar System, applied 2 May 1945, issued 16 September 1947.

[25] 'A Common-Path Interferometer for Testing Purposes', J Dyson, *Journal of the Royal Society of America*, May 1957, vol 47, No 5, p386.

[26] 'A Solid Image Microscope', R L Gregory, *Nature*, 1958, vol 182, p1434.

[27] 'Variable Focus Length Mirors', J C Muirhead, *Rev Sci Instrum*, 1961, vol 32, p210.

[28] 'Editorial: Space /Aeronautics, 3D Display', *Electronics*, September 1962.

[29] 'Digital Control of Focal Distances', H J Caulfield, *Applied Optics*, March 1967, Vol 6, No 3, p549.

[30] 'Stereoscopic Display using Rapid Varifocal Mirror Oscillations', A C Traub, *Applied Optics*, 1967, vol 6, p1085.

[31] 'Computer Generated Movies using a Varifocal Mirror', Eric G Rawson, *Applied Optics*, Aug 1968, vol 7, No 8, p1505.

[32] 'Varifocal Mirror Technique for Video Transmission of Three-Dimensional Images', M C King and H D Berry, *Applied Optics*, September 1970, vol 9, no 9, p2035.

[33] 'Experimental Demonstration of Optical Sectioning by Spacial Filtering and its Possible Use in Transmission of Three-Dimensional Pictures', H J Gerritsen and B Horowitz, *Applied Optics*, April 1971, vol 10, No 4, p852.

[34] 'Real-Time Transmission of a 3D Image using Volume Scanning and Spatial Modulation', J Hamasaki, Y Nagata, H Higuchi and M Okada, *Applied Optics*, June 1977, vol 16, no 6, p1675.

[35] 'Real-time Volume-scanning Three-dimensional imaging system for real moving objects', Dean Morelli, *Optics Letters*, March 1981, vol 6, No 3, p105.

[36] 'Innovations, patents, processes and products', Bernard M Ciongoli, *IEE Spectrum*, December 1987, p20.

[37] Patent *GB2307373*, D Brown, Three-dimensional imaging system, Glasgow, 21 May 1997.

9

The Final Chapter

The Television Committee chaired by The Rt Hon The Lord Hankey was appointed in September 1943, with the following terms of reference:

"To prepare plans for the reinstatement and development of the television service after the war with special consideration of:
a) The preparation of a plan for the provision of a service to at any rate the larger centres of population within a reasonable period after the war;
b) the provision to be made for research and development;
c) the guidance to be given to manufacturers, with a view especially to the development of the export trade."

The committee members were:
   Sir Stanley Angwin: Engineer-in-Chief, General Post Office.
   Sir Edward Appleton: Secretary, Department of Scientific and Industrial Research.
   Sir Noel Ashbridge: Deputy Director-General, British Broadcasting Corporation.
   Sir Raymond Birchall: Deputy Director-General, General Post Office.
   Professor J D Cockcroft: Air Defence Research and Development Establishment, Ministry of Supply.
   W J Haley Esq: Director-General, British Broadcasting Corporation.
   R J P Harvey Esq: Assistant Secretary, Treasury.
   Secretary: J Varley Roberts, Esq: General Post Office.
   Assistant Secretary: Miss C. Kennedy: General Post Office.

ON 25 MARCH 1944 LORD HANKEY wrote to Baird asking for his views on the future of television [1], indicating that he may have seen from the press that he was presiding over a committee, considering the future of television in this country. Hankey asked for a statement of Baird's views in writing in advance to circulate to committee members.

The particular points of reference were: (1) War research and the technical development of television; (2) Re-opening of television service on the 1939 standard of definition; (3) Television research; (4) New television service of radically improved type; (5) Home market and (6) Overseas market.

Baird replied on 14 April 1944 by first clarifying the situation that he was a free agent doing independent television research and not tied to any company or organisation. He wrote:

Dear Lord Hankey,

I have given some thought to the points mentioned in your letter of the 25th March, and before giving my views would like to make my present position clear. In 1939 I was president of Baird Television Ltd. When war broke out this company was put into liquidation by Gaumont British Ltd, which acquired its assets and transferred them to Cinema Television Ltd, a company controlled by them. With the liquidation of Baird Television Ltd, my contract terminated and I ceased to have any connection with Gaumont British Ltd., or Cinema Television Ltd. I then commenced research as a private individual in my own private laboratory at my own expense. My laboratory and my patents since 1939 are entirely my own property. I was approached in 1941 by Cable & Wireless Ltd, (who control the Marconi Co) and accepted a position with them as Technical Adviser. This position does not affect my laboratory or my patents, which are entirely my personal concern. The views given here are therefore those of a private individual carrying out research in his own private laboratory, and deal largely with this work.

In reply to the first point of reference on 'War research and the technical development of television', Baird wrote that he assumed this would relate to two aspects of such research, the first being war applications for television and the second being the justification of television research during wartime that had no war application.

He stated than in first instance he had direct experience of two devices, the first being his invention of an infrared device that could render directly visible objects in darkness. Of this device he indicated that since his first demonstration of such a unit, much development had taken place by 1939 on apparatus operating on this principle both by the Baird Company and others abroad, who were actively engaged on devices of this kind. The second device he related to was the secret installation of television transmitters in aircraft for aerial reconnaissance for the French Air Ministry, details of which may found in a sister publication by this author, *Images Across Space*, published by the RSGB [2].

Baird continued by describing his work on colour and stereoscopic television and indicated that although there was no apparent wartime application, he felt that it was of importance to the National Effort. In terms of colour television, he pointed out that his demonstrations had been given considerable publicity both in this country and abroad and were in advance of anything shown in the USA, where the only colour television had been by the revolving disc method first shown by Baird in 1939.

Baird indicated that in his view, the publication of his work had a stimulating and beneficial effect upon research by opening up fresh fields of investigation. He wrote that, he had reason to believe that several British companies had taken up research on colour television since his demonstrations and publications. In the USA, he continued, the Columbia Broadcasting Company and General Electric, had during the current war, given demonstrations of the Baird disc system of colour television, which Baird clarified in his reply had since been superseded by non-mechanical systems.

He also made it clear that his were the only demonstrations of stereoscopic television shown, anywhere in the world. He proudly wrote that his stereoscopic

CHAPTER 9: THE FINAL CHAPTER

demonstrations had attracted considerable attention from abroad resulting in enquiries from the USA, Russia, Australia, India, South Africa, Canada, Brazil and the Argentine. In his view, Baird wrote, this stimulation of interest in British television was a further justification for Wartime Television Research and will assist in opening future foreign markets. Baird concluded the first of Hankey's point of reference with:

Television will be a large post-war industry, and anything done now to help it forward will assist in providing post-war employment.

In replying to the second point relating to the 'Re-opening of Television Services on the 1939 Standard of Definition', Baird was of the opinion that broadcasting should resume as early as possible using a system capable of being received on pre-war receivers and with the guarantee of at least three years service. Baird indicated that in his opinion, he would review and test all improved systems and select the most substantial improvement. Before commencing simultaneous transmissions of an improved system in parallel with its older counterpart, it would, in his opinion, be realistic to notify the trade and the public that transmissions of the old system would cease after a reasonable time.

This has direct parallels with the procedures adopted in Britain by the Government for the transition in 1985 from 405-line (VHF) monochrome to the 625-line (UHF) monochrome system leading to the adoption of the PAL colour system in 1967 and more recently the change over from analogue to digital television completed in 2011. Baird indicated that this would ensure that owners of sets would be given three years use of their equipment before their sets became obsolete.

Baird was emphatic that the broadcasting monopoly in Britain should cease:

In answering this question I have assumed that the BBC retains its monopoly, but in my view this monopoly should cease. Broadcasting should be on the system used in Australia, where there is both an Australian Broadcasting Company similar to the BBC, and also separate independent broadcasting companies run by private enterprise, and obtaining their revenue from sponsored programmes. In 1938 I visited Australia as a guest of the Australian radio convention and had an opportunity of inspecting this system in operation. It appeared to give complete satisfaction and in my view is much to be preferred to a monopoly. Under the Australian plan television systems could be developed side by side (with proper safeguards to the public) with far greater elasticity than is possible under a monopoly.

I may add that a monopoly of broadcasting of any sort appears to me most undesirable. This question is, perhaps, outside the scope of the enquiry, but should you desire it I will be happy to give my views at length.

In response to the subject of Television Research, Baird offered an opinion on the lines along which television research should proceed.

There are in my view, two branches of research. The first relates to Wireless (or cable) transmission. This has at present limitations, which restrict range, over which the image can be broadcast. The transmission also limits the detail in the received picture. Research to find means to overcome these limitations is of first importance. A partial solution of the range problem is already available by using relay stations and the possibility of extending the broadcasting service to the continent and ultimately to the USA should be investigated.

The second branch relates to improving the quality of the image. The present standard is incapable of giving an image equal to the cinema. A standard of the order of 1000 lines should be aimed at. The television should have colour and stereoscopic depth.

On the fourth point of reference 'New Television of Radically Improved Type', Baird replied:

The ideal television service should show a picture in colour and stereoscopic relief and should operate on an International Standard of the order of definition represented by 1000 lines in conjunction with an International world-wide television broadcasting service.

Before this position can be arrived at considerable time must elapse and, in my view, one or more intermediate steps are inevitable.

The first step, and the one in which I am immediately interested is, is the introduction of colour and stereoscopic relief and I would like to make clear the position in regard to these branches of television.

Baird indicated that the disadvantages of rotating discs had led him to develop a colour system with no moving parts. The point here being that Baird had again taken the lead, leaving the Americans with a copy of his outdated mechanical technique. Baird described the first of his non-mechanical systems, which reduced to practice a single cathode-ray tube system that displayed three coloured images side-by-side on the face of a cathode-ray tube. He did not go into the finer detail of this method, the most important being that this was the development of the first ever line synchronous colour television, removing all visible colour flicker without the requirement of raising the frame rate by three times. He pointed out that this development while having the advantage of a non-mechanical system had the disadvantage of indirect viewing. This, he indicated had led him to develop a purpose built, direct view, double-sided, colour cathode-ray tube with high brightness.

On the subject of stereoscopic television, Baird dedicated a section of his reply to this, informing Lord Hankey that he alone had developed a number of systems:

1. The first has the advantage of simplicity and can be used with a two colour system with no alteration to the receiver. The audience must, however, wear coloured glasses.
2. In the second system no glasses are required the picture being received directly. It is restricted to a few persons sitting in a fixed position.
3. In the third system this restriction is overcome and the picture can be received from any position, no glasses being required.

CHAPTER 9: THE FINAL CHAPTER

Although he offered no details in relation to the above 'third system' of auto-stereoscopic viewing, which has been an anomaly for some time, it can now be revealed that Baird was referring to his breakthrough in holographic television. Baird next gave an opinion on which of the various facets of his research would provide the most practical results:

The best results have been given by a three colour system showing stereoscopic relief without the aid of glasses; while a three colour system such as this gives better colour rendering than a two colour, and stereoscopy without glasses is as much to be preferred to the use of glasses.

A two colour system gives so many practical advantages, and so greatly simplifies the apparatus both for colour and stereoscopy that, in my view, the most practical apparatus, which I have developed given the present transmitting conditions - is a two colour 600 line stereoscopic system.

Baird pointed out that this system had the following advantages:

1. Colour pictures sent out can be received on pre-war television sets without alteration as monochrome pictures.
2. No change is required in present wireless transmitters and only small change in in studio equipment.
3. Pictures in stereoscopic relief can be received on colour receiving sets without any alteration to these sets.
4. No revolving discs or moving parts necessary.
5. Definition can be increased to 600 lines or the present 405 lines used.

If the BBC decided to add colour and stereoscopy on this system, attachments could be fitted to their existing apparatus so that items in colour or stereoscopic relief could be introduced into the programmes, possibly in the first place as experimental interludes.

In reply to the question of the Home Market, Baird replied briefly in three stages:

1. This has already been discussed and I understand plans have been made to extend the service immediately after the war.
2. The second point is a cheaper receiver. This will be provided by mass production.
3. The third point is the provision of improved programmes. The abolition of the present BBC monopoly and the establishment of free competition would be in this respect.

On the final point of reference relating to the Foreign Market, Baird indicated in his reply that the largest potential foreign market appeared to be Russia.

He had several visits from the Russian Embassy, including Mr. Sobielioff, the Russian Minister, and his technical advisers. They informed Baird that television in Russia was far behind what we he was able to show them and they had expressed the greatest interest:

They also informed me that they hoped after the war to introduce television into Russia on a very large scale as they had a high opinion of its educative value. I should add that they were particularly interested in the stereoscopic picture shown. They have apparently been conducting a good deal of research on stereoscopic cinematography, but have, so far, not attempted to apply it to television.

Baird wrote that, in the pre-war days the European market was to a considerable extent dominated by Germany. The Fernseh Company (which was founded by Baird Television Ltd and which Baird Television Ltd had an arrangement for exchange of technical information and patents) had a powerful hold on the industry and produced apparatus of a high technical standard. After the war, however, Baird continued, the market may be open in France and the rest of the continent:

To encourage export it appears to me highly important that an International Standard of Transmission should be adopted, also that every effort should be made to establish an International System of Broadcasting whereby the programmes radiated from London could be received throughout the continent of Europe (and ultimately the World) and in the same way foreign programmes received in London from stations abroad.

In summing up Baird wrote:

I hope these views will be of interest and will be very happy to give any further information, or any assistance, which lies in my power.

On 16 August 1944, the London press were invited to see Baird's new inventions of all-electronic colour and facsimile television, which he believed were ready to be taken up by the radio industry and transmitting stations as soon as the Second World War was over. Interestingly, he chose not to reveal his breakthrough in holographic television, but demonstrated the Telechrome as the world's first all electronic colour and stereoscopic television by showing two colour dynamic anaglyphs.

Baird had made his new system of television entirely 'electronic', eliminating all lenses and revolving discs. With the Telechrome, both colour and stereoscopic pictures could be made to appear directly on the screen of his purpose-designed cathode ray tube. According to Larsen [3], the journalists who had braved the 'doodle-bugs,' coming all the way from Fleet Street to see Baird's new inventions, were highly satisfied. For stereoscopic viewing Baird provided them with coloured glasses, the left and right eyes looking through blue and red glasses respectively to view the anaglyphs in relief.

Larsen wrote:

The Telechrome as Baird called his new system, was demonstrated in his house in South London, among little heaps of broken glass, demolished pieces of furniture, and behind shutters, which had been blasted into lopsided angles: Hitler's flying bombs, several of which had landed in the

CHAPTER 9: THE FINAL CHAPTER

vicinity, had apparently tried to stop his work, but in vain; the intrepid Scotsman had made up his mind to stick to it, and he saw it through with the same unflinching determination that had been his characteristic feature a score of years before at Hastings. [4]

The Report of the Hankey Committee, was published on 29 December 1944 after being held as a secret file. It was not presented to Parliament until March 1945, although there was a leak in 1944, which made its existence known. It was reported that consultation had been made with representatives from: General Electric Ltd, Marconi-EMI Television Company Ltd, Scophony Ltd, Standard Telephones and Cables Ltd, Board of Trade, British Broadcasting Corporation, British Film Producer's Association, Ministry of Education, Radio Industry Council, REP Joint Committee and J L Baird Esq.

The Report highlighted that while radiolocation owed much to research work on television, war research had made no discovery of a fundamental character of much value to television. The greatest influence the war had on the future of the industry was in enhancing the skills of radio engineers and greatly increasing the number of skilled men and women able to work in the future design, development, manufacture and maintenance of television. It was reported that the television service in Great Britain prior to the war only available in the London area, was new and improvements were expected. It was assumed that the definition of 405-lines, while providing an adequate small image in the home was inadequate for the large screen where definition in the order of 1000-lines would be required.

The Committee reached the conclusion, based on evidence provided, that the correct course of action would be to re-open the service on the basis of the 405-line system rather than wait for the development of a new system as a result of research. This decision was based on the high degree of reliability that the Alexandra Palace transmitter had achieved, providing good entertainment value in the home. It was considered that, with minor refinements, the system would be improved further.

Another consideration was that a good deal of research and development would be required before a markedly improved service could be put into operation. The wartime demands on scientists and materials would delay the formation of a new service to the point that a gap of some years would be likely to elapse without television, which would seriously damp interest and retard commercial interest.

The Report recommended that vigorous research work be carried out to produce a radically improved system as soon as staff can be made available, if Britain is not to lag behind the rest of the world in the television field. The Committee had given a great deal of consideration to the evidence provided by John Logie Baird and this was evident when they reported on the provision of an improved television system:

We have taken a great deal of evidence on this important question and there is strong support for the view, with which we agree, that the aim should be to approach the cinema standard. We think that television definition should eventually be of the order of 1,000 lines and that the introduction of colour and stereoscopic effects should be considered.

Margaret Baird recalled:

The war was coming to an end. John's health was failing. Soon the journey of eight hours to Bude was too much for him, and early in 1945 I went to Bexhill to find a house. We picked Bexhill because the children's school, Sandown, was returning there and also because John wanted to live at the coast but near enough to London to control a new company he was forming. [5]

With the promise of a new commercial age for television dawning at the end of the Second World War, it seemed that everything Baird had worked towards would now become a reality, but he was very ill. He suffered a stroke early in February 1946 and was very frail, but he was as determined as ever to continue. He drew up plans for the formation of a new company with the help of close friend and schoolboy chum, the actor and producer, Jack Buchanan. In 1945 they registered the new company, 'John Logie Baird Ltd', with headquarters at 4 Upper Grosvenor Street, London W1, and a main factory at 466 Alexandra Avenue, Rayners Lane, Harrow, Middlesex [6]. This small but advanced industry produced some of the most elegant post war period 405-line receivers, including the Lyric and the impressive Grosvenor (**Fig 9.1**), featuring an enormous 30 inch direct view cathode-ray tube with a 28 inch picture.

Baird was fortunate to have the continued support from his reliable and capable assistant E G O Anderson, who had served him as ultimately his only remaining loyal employee in his private laboratories during WW2. There is still

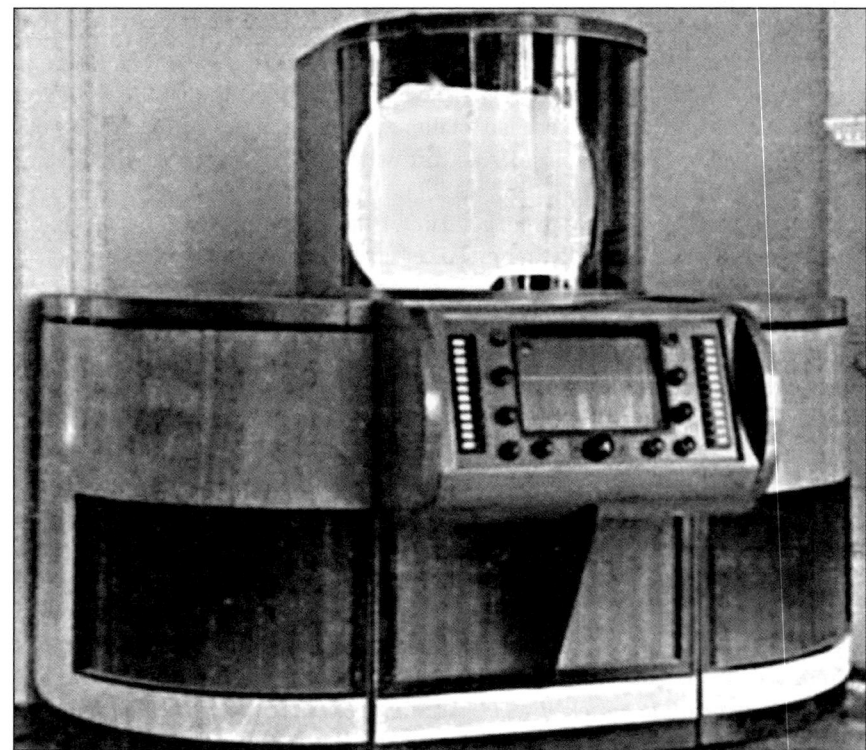

Fig 9.1: The impressive Grosvenor receiver

CHAPTER 9: THE FINAL CHAPTER

a question regarding why Anderson was not called up for duty, but would remain, working in a reserved occupation, a category only relevant to employees in an occupation considered to be important to the War effort. Even then, he was only able to retain Anderson financially, partly though his own savings boosted slightly by earnings from the three year appointment at Cable and Wireless at a salary of one thousand pounds per annum from 1 November 1941.

Kamm and Baird wrote:

"As Baird tried to regain strength in the spring of 1946, his attentions were focused on the resumption of television on 7 June, and the televising of the Victory Parade on the following day. His new company had arranged to feed the BBC television pictures of the parade to projector receivers at the Classic Cinema in Baker Street and the News Theatre in Agar Street. At the Savoy Hotel the programme would be seen on a 5-foot projection screen, and also on the 28-inch screen of the Grosvenor receiver, strategically placed in a room surrounded with lots of mirrors." [7]

On 14 June 1946, only one week after the resumption of television, John Logie Baird died peacefully in his sleep.

The business of 'John Logie Baird Ltd.' continued for two years under the directorship of Jack Buchanan, a successful entrepreneur owning 'Jack Buchanan Productions Ltd' and holding directorships in another five companies. In 1948, in an endeavour to release assets, Buchanan formulated a plan to rescue the failing Scophony Company which had been a sister company to Gaumont's Cinema Television Ltd.

The recovery plan was not to be a take-over but a merger involving two companies of which Jack Buchanan was the Principal Shareholder. The name Baird was retained in the new title and an agreed financial package was awarded to Buchanan in return for his release from directorship. The merger took place on 26 November 1948 and Scophony-Baird Ltd was officially born on 12 February 1949. It must be assumed that John Logie Baird's experimental colour and three-dimensional apparatus was part of the assets acquired by the new company. Scophony-Baird continued to operate from the Scophony locations at Lancelot Road, Wembley and from a factory at Wells in Somerset.

Scophony-Baird Ltd proudly advertised television sets as 'Baird' receivers. Advertisements in trade journals around 1950 depict the late film star, Terry Thomas demonstrating a 'Portable Baird Television Receiver'. The company appeared at Stand 39 at the 17th National Radio Exhibition, from 6 to 16 September 1950, held at Castle Bromwich, Birmingham, and Stand 50, Earls Court, Wembley in 1951. As well as displaying 'new' Baird products the company presented an exhibition of early television apparatus relating to the work of John Logie Baird.

There is some irony in the fact that Scophony-Baird was acquired in 1951 as a going concern by EMI Engineering Development Ltd, who recognised there was a market advantage to be gained by trading under the name 'Baird'. On 14 August 1952 Scophony-Baird at Lancelot Road became EMI's 'Baird Television Ltd' and Scophony-Baird at the Penleigh, Wells in Somerset became 'EMI Engineering Development Ltd' from 1951.

In 1968 SE Laboratories, which had been acquired by EMI, set up a branch on the Wells site, which then became EMI (SE Labs), a major EMI centre for magnetic recording technology. In 1970 SE Labs was relocated to an independent site near the Wookey Hole caves in Somerset, and eventually changed name to Thorn/EMI Datatech. This company was rated as a leading supplier of industrial magnetic recording systems in Europe. Datatech was sold to Penny and Giles, and EMI Engineering Development Ltd on the original Penleigh site became the 'Computer Systems Division, Thorn/EMI Electronics Ltd' [8].

The 'new' Baird Company at Lancelot Road appeared on Stand 33 at the National Radio Show, Earls Court in September 1952, but this time exclusively under the name 'Baird'. This was the last time that the 'Baird Museum' exhibits were put on public display. On 19 March 1954 this Baird Television Company was sold to A W M Hartley to become Hartley-Baird Ltd.

Margaret Baird related to events in 1960:

"I had letters from Mr A W M Hartley and Mr Perring-Thoms to say that he had sold the trade-name Baird to Mr Perring-Thoms' vast Radio Rentals group, which had decided to manufacture television sets. It appeared that when the directors were discussing a name for the subsidiary company someone said, facetiously: 'What we want is a name like Baird'. Mr. Perring-Thoms traced the name to the possession of Mr Hartley and bought it." [9]

The 'Baird' trade name enabled Radio Rentals to establish yet another 'Baird Television Company'. Margaret Baird continued:

"On the 13 June, 1961, I was taken to Bradford to see the new Baird factory, then with nearly one-thousand employees." [10]

After the death of Perring-Thoms on 31 July 1964, the Radio Rentals Group and the 'Baird' trade name was further acquired by Sir Jules Thorn of Thorn Electrical Industries who continued the business of manufacturing Baird television receivers. EMI once more came into the equation by regaining ownership of the 'Baird' trade name by their merger with Thorn in 1980. Radio Rentals and the Baird trade name then became the property of 'Radio Rentals (Thorn/EMI Rentals).' For many years the Baird badge could be found on rental video recorders and television receivers marketed by the Radio Rentals Company.

In August 1992 Thorn/EMI honoured the memory of John Logie Baird by naming their headquarters in Reading, 'Baird House' [11]. In 1996 Thorn and EMI demerged, the Baird trade name becoming the property of Thorn (UK) plc until 2008, although for many years prior to this it was not used commercially. In 2009 the Brighthouse group began retailing LCD television receivers under the name Baird, which continues to the present day.

Some of the early Baird Museum artifacts found their way to the 'Hartley Automotive Group' subsequently 'Shrewsbury Engineering' and finally Shrewsbury Technology Ltd [12]. When the company fell into the hands of a receiver, the items were donated to the National Media Museum in Bradford [13].

CHAPTER 9: THE FINAL CHAPTER

This book has shown that Baird made a significant contribution not only to the development of television, but also to other information sciences and their resulting technologies. The three dimensions of his life that were initially under consideration were; monochrome, colour and 3D television, but this list slightly extended during the preparation of the manuscript.

Despite ailing health and poverty, John Logie Baird did more to advance the early development of television than any other individual. He planted the seed, which has grown into a multinational, trillion dollar, video and communications media industry. Television is the fundamental display technology for every computer, games module, digital photo frame, digital camera, mobile telephone, web cam, flight and other simulators, to name a few.

New evidence provided by this research has shown that Baird discretely applied the flying-spot scanning technique to enable the first demonstrations of television. It has also been shown that John Logie Baird is the inventor of the advanced image enhancing circuit that was used extensively, albeit at higher frequencies, in every video-recorder and almost every modern analogue television receiver and associated video equipment.

This book has also described how two factions of society emerged with diverse views on the history of the birth of television. A number of Baird Television Company management errors compounded a situation, which may have been responsible for the partially damaged reputation of John Logie Baird. There were and remain those who revere John Logie Baird for his achievements, while others give him no credit whatsoever.

It has also been shown that Baird was the driving force whose determination led a skeptical radio industry in America and Britain to recognise that there was potentially a world market for television. By the mid-thirties the competition which Baird attracted was driven ruthlessly by the American radio industry, RCA, General Electric, Bell Laboratories, CBS and ultimately RCA's subsidiary company in Britain, EMI-Marconi.

The Marconi Company approached John Logie Baird in 1931 with a view to a merger, but in contrast to Baird's wishes, in 1932 the Baird Television board of directors voted against it. Marconi had previously merged with the Gramophone Company in 1929 and Columbia in 1931 to form Electrical and Musical Industries (EMI). It was not until 1934 that the seriousness of the error became apparent and inevitably Baird Television was faced with the impossible task of trying to supersede storage television cameras in 1936.

The demise of Baird Television was brought about by the unresolved problem of live-broadcasts, and the Selsdon Committee's edict that EMI-Marconi should be allowed to compete for a British standard based on American technology. EMI-Marconi claimed that the Emiscope, based on RCA's Iconoscope, was an all-British development. Prior to the system tests at Alexandra Palace, Baird had developed an operational in-house portable electronic storage camera, but the patent laws prohibited its use. In retaliation, Baird Television and Captain A G D West imported their own 'back-door' American technology. A licence was obtained for the Image Dissector camera from Philo T Farnsworth, which Baird Television developed as the Baird Electron Camera. While this device proved to be successful in the US for live television, for reasons outside the remit of this research it was not as successful at Alexandra Palace.

When EMI-Marconi gained the contract to supply the BBC with television transmitting apparatus in 1937, the legend of John Logie Baird began to develop. In his autobiographical notes Baird wrote:

"The success of their work was brought home to me when I overheard a conversation over the phone between an elderly lady (who I thought knew all about me) and a friend - 'Yes we have a Mr Baird staying with us', silence, then, 'You know! Mr. Baird, the Television Scientist', silence - the friend did not know! 'Oh! You must know, he invented a television nearly as good as Marconi's'. [14]

It has also been shown that Baird had many important supporters who spoke highly of his achievements: Ramsay MacDonald; Professor Taylor Jones of Glasgow University; Sir Oliver Lodge; Captain Leonard Plugge, owner of IBC Radio Luxembourg; Peter Goldmark of the Columbia Broadcasting System (CBS); Edward W. Herold, head of the development team working on the RCA Shadowmask; Donald Flamm, owner of WMCA and founder of the Voice of America, who for over 20 years until his death in 1998 campaigned to bring Baird's achievements to the American public's attention; Sir Edward Wilshaw, former President of Cable and Wireless, and Perring-Thom's Radio Rentals Group.

Another supporter was Sir Jules Thorn of Thorn Electrical Industries whose products carried Baird's name on Radio Rentals products. The Thorn Group honoured Baird's memory by creating a new Baird Television industry in the 1960s. Thorn merged with EMI to carry Baird's trade name to the Thorn/EMI Rentals Group, who appropriately named their new headquarters in Reading 'Baird House.'

J L Baird was an oracle who, despite being plagued by ill health, was far ahead of his time. In 1928 he described videodiscs and dual track stereophonic recording techniques. In the same year he fully disclosed in articles, the principles of scanning radar, colour television, fibre optics, Telecine, and more. The red, green and blue active colour television display and dark surround mask described in 1938 were later put into production by Sony and other television manufacturers.

Other advanced concepts have included theatre television, three dimensional television and volumetric imaging, of which he also suggested an X-ray, 3D body scanner for medical purposes. It has been shown that Baird demonstrated the first all-electronic colour television tube, the Telechrome in 1944. Dr A H Sommer, working at Baird Television made the breakthrough of inventing the first truly panchromatic photocathode material for electronic colour television cameras, initially as a requirement for John Logie Baird's colour television project in 1939 to later be exploited by RCA.

In addition, this book has shown that Baird unwittingly contributed to the development of the modern RCA Shadowmask and the Sony Trinitron display tube. These devices can be traced back to the work of Flechsig, an employee of Fernseh AG, Baird's German television interest, until Hitler came to power. Sam Kaplan of Zenith, substantially improved the development of the RCA Shadowmask and the Sony Trinitron while working with former employee of

CHAPTER 9: THE FINAL CHAPTER

Baird Television, Dr C S Szegho in the USA. The Sony Trinitron is shown to be related to the Chromatron work of Lawrence, and that CBS and Philips injected a total of two million dollars into Sony, which helped with in its production.

In conclusion, John Logie Baird was an engineer, businessman, inventor and manager with a profusion of ideas, which constantly brought him to face and overcome many new challenges. As an inventor he was a man of vision, so far ahead of his time that the supporting technology was not often ready to properly enable his advanced concepts. As an engineer and manager he moved himself and his company forward into the electronic age, yet the general public continued to associate his name with an outdated mode of television.

The 'Baird Cathovisor' cathode-ray tubes designed and manufactured by Baird Television were among the finest available in the world. Patents taken out by Baird and his company covered a wide range of electronic devices, many of which the company manufactured including advanced photomultipliers, specialised electronic television cameras and Telecine film scanners.

John Logie Baird was the pacemaker for almost all of our modern visual communication systems. Without his ability for publicity, his entrepreneurial skills, and remarkable achievement as a prolific inventor, there would have been little upsurge of interest in the development of radar and infrared, before and during the Second World War. Radar is basically the technology of television. The outcomes of the Battle of Britain could have been radically different without the intelligence obtained from the Chain Home Radar Stations situated around the coast of Britain. Infrared night-vision equipment used by the military for seeing the enemy in the darkness of night, and by fire departments and other services for locating warm bodies under rubble, was first anticipated by Baird's 1928 Noctovision apparatus.

Surprisingly, mechanical television continues to be used under the name of FLIR (Forward Looking Infrared) and was fitted to all-weather fighters and specifically the PAMIR system from Karl Zeiss Optics, fitted to German Air force Tornados. This apparatus, which may now be superseded by more superior modern equipment, originally used a 720-line output from a frame store that was slightly lower than analogue broadcast television quality.

Baird's evidence presented to the Hankey Committee of 1944 was largely accepted in the published report, which indicated that after a few years of resuming the old standard, television should work towards a system along the lines of 1000-line colour with stereoscopic effects. However, after Baird's death in 1946, the BBC continued to use the outdated 405-line monochrome standard for many years longer than the recommendation suggested.

The BBC resumed television on 7 June 1946 and it took a further 18 years before an improved 625-line monochrome system was introduced. This was made possible by the launching of a UHF service featuring BBC2 on 20 April 1964.

Another improvement was made with the introduction of the PAL colour system on 15 November 1967. This was followed by the Freeview digital television service, although not considered 'high definition,' when it was first introduced by the BBC, on 10 December 2004. Baird's vision of a 1000-line colour system had to wait a total of 60 years until 27 May 2006, when the BBC introduced the

HDTV terrestrial system of 1080-pixel resolution. As far as his vision of a 1000-line colour, stereoscopic service was concerned, on 3 April 2010, 82 years after his first demonstration of stereoscopic colour television, British Sky Broadcasting (Sky) launched the first commercial 3DTV service.

Baird has been shown to be the father of modern 3D Television having patented in 1941 early versions describing both the polarised glasses and active shutter system only recently introduced. He also patented the world's first holographic three-dimensional television system, where holographic means the 'whole picture' which is a technology still to be fully developed and realised. Such a system could revolutionise the future of domestic television by producing an RGB, three lasers system, forming a spectacular 360 degree viewable volumetric image in the centre of a room.

While this book has been based on the most definitive academic research available on the life and contributions of John Logie Baird with regard to modern television and other related technologies, aspects of his work specifically during the Second World War and relating to 'secret signalling' and 'radar' remain under investigation by various researchers including, Dr Peter Waddell, Prof Malcolm Baird and the author.

References

[1] Letter from the Privy Council Office, Whitehall, dated 25 March 1944. From Lord Hankey.

[2] *Images Across Space: The Electronic Imaging of Baird Television*, Douglas Brown, ISBN 9781 8742 89210, RSGB.

[3] *Inventors' Scrapbook*, E Larsen, Lindsay Drummond Ltd, London, 1947, pp90-92.

[4] *Inventors' Scrapbook*, E Larsen, Lindsay Drummond Ltd, London, 1947, pp90-92.

[5] *Television Baird*, M Baird, Haum, Cape Town, 1973, pp157-160.

[6] Scophony-Baird Ltd, Archived Records, London, Companies House Search Room.

[7] *John Logie Baird: a life*, Antony Kamm and Malcolm Baird, NMS Publishing Ltd, Edinburgh, p363.

[8] *Thorn/EMI. The History of Computer Systems Division*, company document. Thorn EMI Electronics Ltd, Wells, Somerset.

[9] *Television Baird*, M Baird, Haum, Cape Town, 1973, pp157-160.

[10] *Television Baird*, M Baird, Haum, Cape Town, 1973, pp157-160.

[11] *Thorn/EMI. The History of Computer Systems Division*, company document. Thorn EMI Electronics Ltd, Wells, Somerset.

[12] Shrewsbury Technology Limited, Company No 00566892, Companies House Search Room, London.

[13] Baird items in Museum, J Trenouth, former Curator of the National Media Museum, Bradford.

[14] *Television and Me: The memoirs of John Logie Baird*, John Logie Baird, edited and introduced by Malcolm Baird, Mercat Press Ltd, Edinburgh, 2004, p127.

Appendix 1: Experimental Phonovision Device

A few experimental Baird type television video discs have survived the experimental period and the author was fortunate in locating one such example and transferred it to tape. As synchronisation was not present on the original 78rpm recording, the author passed the replayed signals through an active second order bandpass filter tuned to the line frequency in the hope of obtaining a pseudo synchronising pulse from the image content. The 78rpm disc, contained a series of still pictures some of which yielded a regular enough signal to construct a line frequency pulse from the filter.

THE BLOCK DIAGRAM OF the experimental process is shown in **Fig A1.1**. It should be noted that the Nipkow disc was replaced with the oscilloscope screen shown in **Fig A1.2**.

Although this regular pulse obtained from the tuned bandpass filter always represented the line frequency, its phase in relationship to the original line-sync was incorrect. Using the pulse as a timing marker, it was used to trigger a fast pulse from a monostable circuit. This in turn was used to trigger a variable delay time pulse from another monostable, both being constructed from a 556 dual timer integrated circuit.

A variable resistor controlling the length of the resulting monostable pulse could therefore be manually adjusted until its leading edge reached the next image line accurately. As this stretched delay pulse regularly timed out (at the

Fig A1.1: Block diagram of the scheme used by the author to recover phonovision images

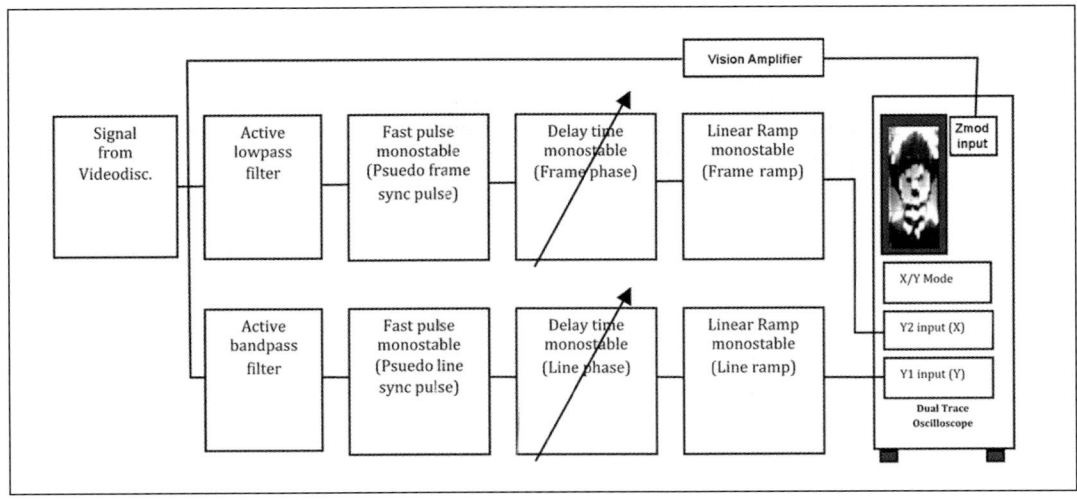

(left) Fig A1.2: A 30-line image reproduced from audio cassette on an oscilloscope

(right) Fig A1.3: A 30-line image drifting horizontally

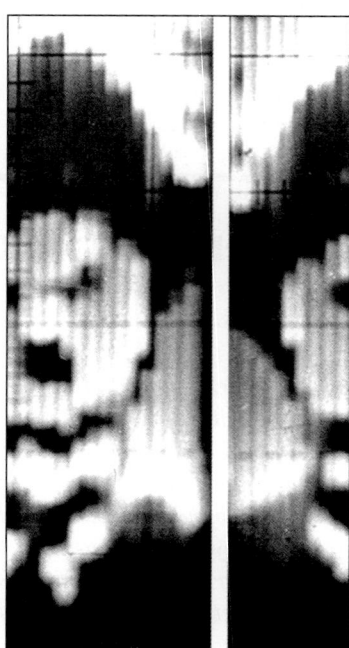

correct line frequency and phase) one half of a further 556 timer IC, wired as a 2.67mS linear ramp generator, was triggered. This repetitive ramp driven by the image data was connected to the 'Y' input of a real-time oscilloscope and adjusted to produce a blank 30-line vertical raster on the screen.

To produce an image on the raster the unfiltered signal from the videodisc was amplified and to avoid producing a negative image was phased correctly, before being used to drive the 'Z Mod' input of the oscilloscope. Provided the signal is large enough to meet the requirements of the Z Mod input it will modulate the oscilloscope's beam current, and in this case result in the formation of an image on the screen. To obtain the correct contrast and brightness the Z Mod input level should be adjusted together with the oscilloscope's intensity control respectively.

Adjusting the free-running horizontal time base on the oscilloscope enabled a series of vertically locked images slowly running from side to side on the screen, **Fig A1.3**. These images were vertically positioned on the screen by fine adjustment of the variable monostable phase control. The final step was to introduce an active low pass filter followed by another 556 to generate a variable delay pseudo horizontal sync pulse from picture information. When this low frequency signal was applied the oscilloscope's horizontal timebase the horizontally drifting image behaved in an erratic and unpredictable manner.

Fortunately, the oscilloscope had an X/Y mode switch to optionally bypass the time-base and select the 'Y1' input for the 'Y' channel and the 'Y2' input for the 'X' channel as shown in Fig A1.1. This involved replacing the oscilloscope's built-in timebase with an external frame linear ramp generator, constructed from the remaining monostable and adding a further 556 integrated circuit.

With the horizontal ramp locking the frame, and the vertical ramp locking the line, images became clearer and more stable. The block diagram of the setup is shown in Fig A1.1, and a recovered image from the 78rpm disc is shown sync-

APPENDIX 1: EXPERIMENTAL PHONOVISION DEVICE

locked in **Fig A1.4**. The original picture for this televised image is most likely a cartoon sketch of the late Charlie Chaplin (Sir Charles Spencer Chaplin, KBE, 1889 - 1977) the English comedian and film director from the silent film era. Unfortunately, the oscilloscope graticule is ever visible on the image due to the scale being etched on the inside of the cathode-ray tube screen and therefore could not be removed.

The author completed the Phonovision project in 1986 using the analogue and digital devices available at the time and without the aid of a computer. The equipment was first constructed for an exhibition at the Collins Gallery, University of Strathclyde in 1988, to celebrate the life and work of J L Baird. The equipment then successfully toured Britain for a further two years. Since that time, Donald McLean [1] (G2TV) has published extensively on the digital extraction and reproduction of historic images from the dawn of television history.

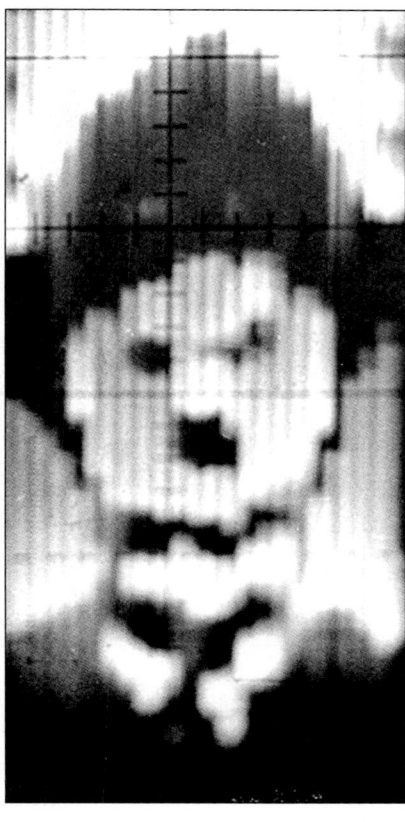

Fig A1.4: Stabilised image of Charlie Chaplin

References

[1] *Restoring Baird's Image*, Don McLean, Institution of Electrical Engineers, 2000.

THE THREE DIMENSIONS OF JOHN LOGIE BAIRD

Appendix 2: Colour from a Monochrome TV!

When I started writing this manuscript, the publisher asked if I could include a few pages describing my recent project for displaying colour television pictures on a monochrome television screen.

I had recently constructed this device using a field synchronised RGB tricolour rotating disc to assess the colour quality of Baird's invention on modern digital television pictures. The device takes a normal recorded or broadcast signal and encodes it to be displayed as three sequential colour fields, arriving one after the other, on the face of a monochrome screen. A tri colour filter disc is then synchronised perfectly with the colour information enabling the viewer to see the red field through the red colour filter, the green field through the green colour filter, and so on. As each field time is only 20mS the eye and brain work together to enable a spectacular full colour image when viewing the monochrome screen through the virtually invisible rotating colour filter disc.

To cover this in detail in only a few pages was a real challenge without reverting to the inclusion of a full constructional article, which was clearly out-with the remit of this book. Instead I have included the following overview using block diagrams, based on a recent presentation given to the Lomond Radio Club, (formerly Helensburgh Radio Club) where a full working demonstration of field sequential colour television, sourced from a Freeview box and a DVD player was given.

THIS DIGITAL FIELD-SEQUENTIAL PROJECT requires that RGB, composite video and the right and left audio signals are available from the SCART socket on a digital set top box or any other video equipment that you may choose to use. The full universal wiring is not always obtainable from SCART sockets, although two very inexpensive devices tested, a DVD player and a basic digital set top box costing under £20, both had the full wiring. The auxiliary SCART socket on the digital set top box was fully equipped with the RGB signal while the main TV SCART socket had only the video and the pair of audio signals connected. It is also worth checking that SCART leads also have the full universal wiring as many low cost connecting leads are wired for video and audio alone. **Fig A2.1** shows the connection details.

Once the signal wires have been identified from the connector, remove the SCART plug from one end of the lead, and common the grounds by connecting them together on an electrical connector block. Connect and appropriately label the output wires required for the project as shown in **Fig A2.2**.

THE THREE DIMENSIONS OF JOHN LOGIE BAIRD

Fig A2.1: SCART socket, numbered from the front, and its pin connections

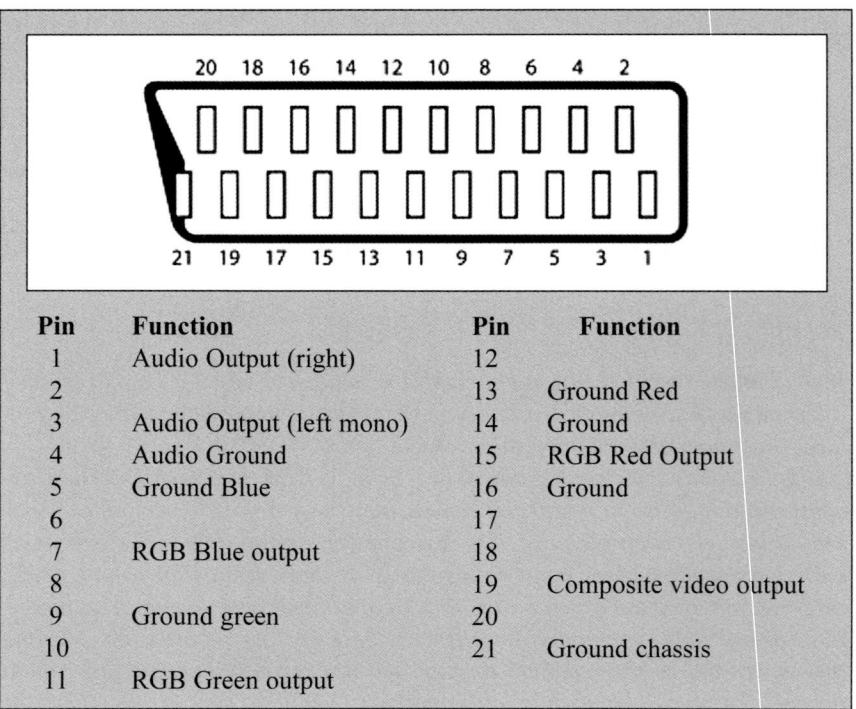

Pin	Function	Pin	Function
1	Audio Output (right)	12	
2		13	Ground Red
3	Audio Output (left mono)	14	Ground
4	Audio Ground	15	RGB Red Output
5	Ground Blue	16	Ground
6		17	
7	RGB Blue output	18	
8		19	Composite video output
9	Ground green	20	
10		21	Ground chassis
11	RGB Green output		

Fig A2.2: Outputs required from the universal SCART lead

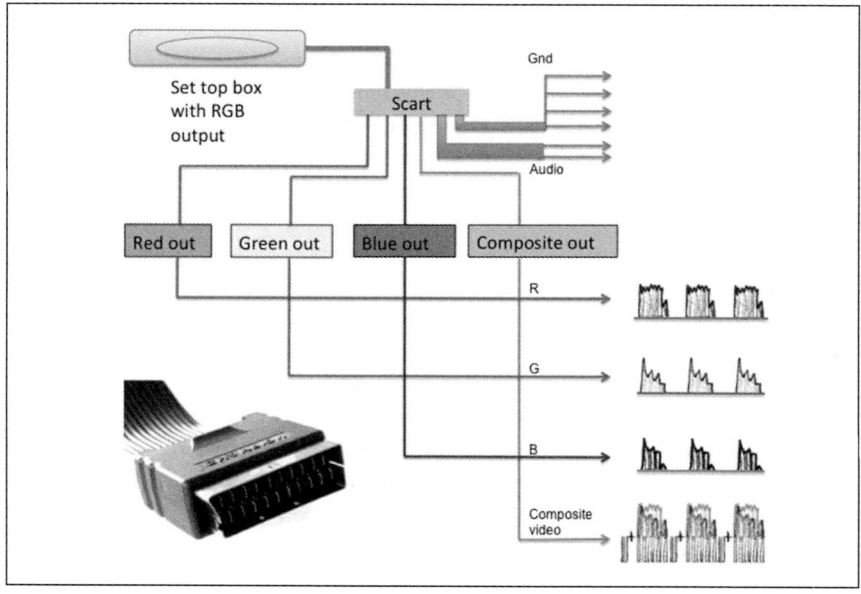

For the next stage a number of devices are required:

- 12 volt power supply.
- CMOS sync separator integrated circuit LM1881 or similar.
- CMOS 4066 quad analogue gate or equivalent.
- CMOS decimal counter similar to a 4017.

168

APPENDIX 2: COLOUR FROM A MONOCHROME TV!

Wire up the decimal counter to reset automatically on the edge of the third count (0=1; 1=2; 2=3) and give each of the outputs a colour designation; 1=Red; 2=Green; 3=Blue. Label the outputs accordingly. Feed the sync separator with the field signal from the composite video, SCART pin 19.

Wire the field output signal from the LM1818 sync separator to drive the clock input of the decimal counter 4017, which will be the RGB sequential control line.

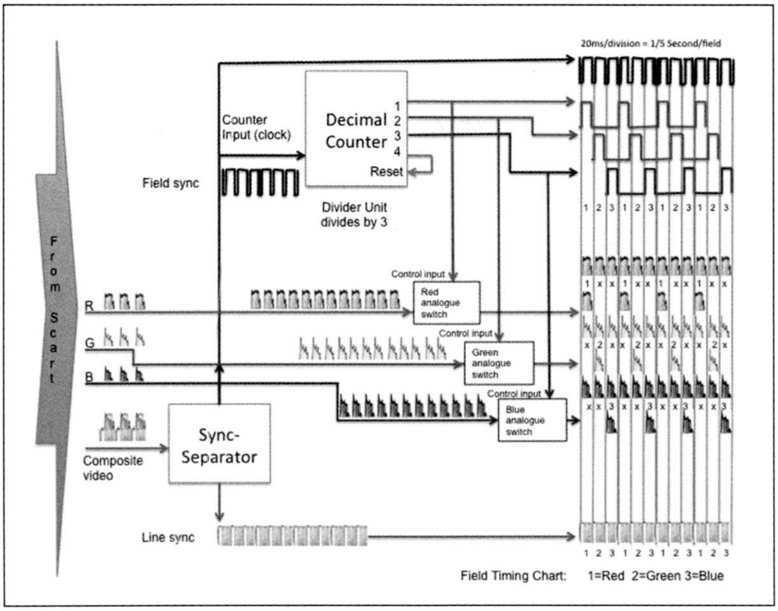

Fig A2.3: Sequential timing is organised by the decimal counter from the field syncs provided by the sync separator

Three of the analogue gates in the (4066) quad device are individually related to one of the colour channels and should be wired to either the Red, Green or Blue signal lines, (R, G or B) from the SCART connector block and labelled accordingly. The output lines from each of the utilised analogue gates are then electronically coupled together.

The Red control output (1) from the decimal counter, which will switch to a high level once every three fields, is wired to control the analogue gate appropriately connected to the Red signal. Therefore on every count of (1) there will be only a Red field present at the common output.

During the second count (2) the Green control output from the decimal counter is wired to control the corresponding analogue gate to enable a green field through to follow the Red field. Therefore by the same process the Blue field (3) will complete the cycle. The timing chart in **Fig A2.3** shows the analogue gates are switched in RGB sequence with two complete fields of each colour signal being gated out. This enables, by sequential switching via the analogue gates, that fields of the other colours replace them at the appropriate time.

The common output line will be coded with the monochrome representations of fields in the repetitive pattern of Red, Green, Blue, Red, Green, Blue, etc. However, before the field sequential vision signal can be connect to the normal video input of a monochrome monitor, it must be reunited with the line sync pulses from the original composite video signal.

The timing chart and block diagram in **Fig A2.4**, illustrate that the RGB and synchronising signals from the LM1881 recombine to produce the sequential television signal.

With the synchronising pulses reunited with the vision signal the video signal appears, (at least as far the television monitor circuit is concerned) identical to a normal monochrome signal. It may be necessary to amplify the field sequential video signal before connecting to a monitor. The next task, which may be the most

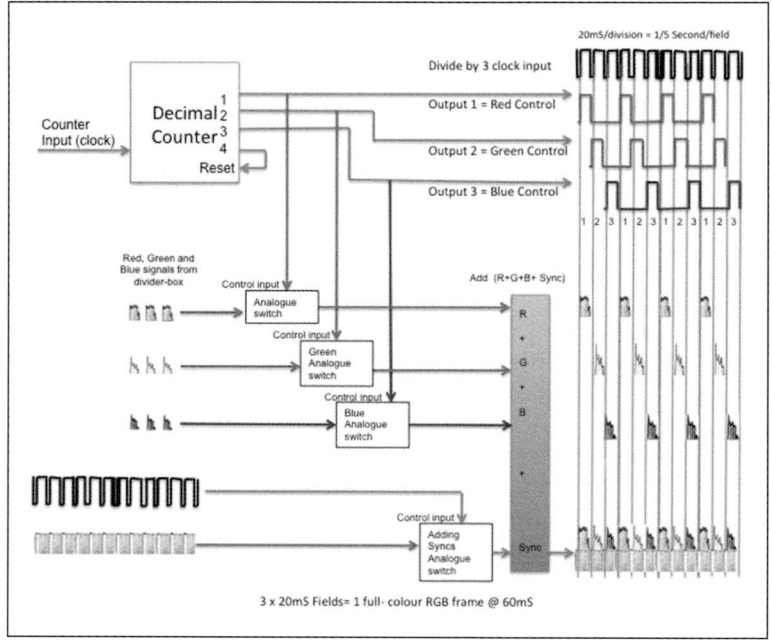

difficult, (now that broadcast analogue television no longer exists in Britain) is to identify an inexpensive, small monochrome, battery/mains adapter type analogue television receiver, which must provide a video and audio input socket.

The set shown in **Fig A2.5** is a small 5in (13cm) screen monitor, which provides an example of the type and size required. Having identified and obtained a small monochrome monitor with an A/V input, a suitable 12 volt DC motor will need to be identified, that runs comfortably and quietly at a speed of around 1500rpm. One rotation of the field synchronous disc, which will need to be as light as card, will take the time of three television fields, which equates to 1000rpm. It is important that the motor spindle is of a size to enable a suitable coupling to the disc. The disc can be fixed to the spindle by using small coupling parts designed for turning the gears or wheels of popular plastic technical kits.

Fig A2.4: Field sequential waveform constructed by sampling signals in correct RGB order and combining with sync pulse

A suitable motor should be mounted with its spindle close to the top left-hand corner of the monitor screen, in a position to be able to turn the disc, and covering most of the screen area. In the unit shown in Fig A2.5, the most distant corner of the screen from the motor spindle purposely remains uncovered, to show that the screen behind the colour image is actually monochrome.

Fig A2.5: Tri-colour disc and 5-inch monitor. Note the coupling method

At this stage it is important to ensure safety at all times, as the disc will rotate freely at a speed of 1000rpm. The disc should be extremely lightweight with no hard materials used in its construction. In operation the disc must be enclosed in a box with an acrylic window in front. It must be understood that this project is undertaken at the reader's risk and no responsibility will be taken by the author or the publisher.

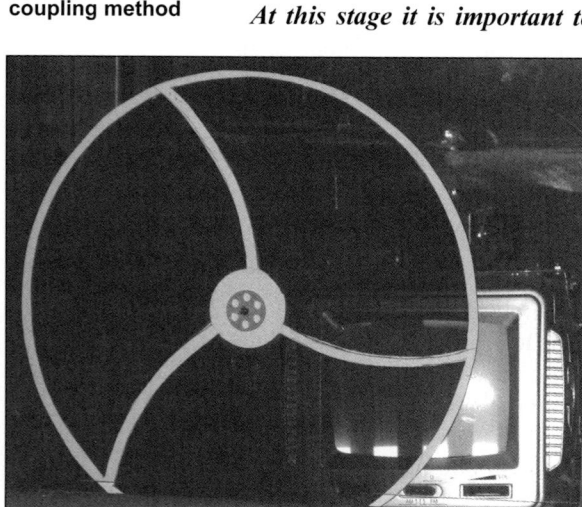

The sequential disc was made from a large sheet of 150g stiff card by drawing a circle with a 6in (15cm) radius and then another 5.75in (14.5cm) radius. The position of the thin spokes can be identified by marking at intervals 120 degrees around the circumference and drawing a line that intersects with

APPENDIX 2: COLOUR FROM A MONOCHROME TV!

the centre hole of the disc.

The photograph in Fig A2.5 shows curved spokes that more accurately follow the line scan across and down the screen, but straight spokes as shown in **Fig A2.6** are probably good enough to get the project going. It will be seen that the curved spokes are of the same radius as the disc. Suitable primary red, green and blue filter gels, intended

- *A lightweight disc comprising three 120 degree segments, each covered with a different primary coloured filter, red, green and blue is fitted on the spindle of a 12 volt DC electric motor.*
- *The time for one rotation of the disc will have to be precisely equal to the time of three sequential fields (60mS). This is equivalent to one rotation in 16.67 seconds or 1000rpm.*
- *The disc must rotate in front of a black and white monitor screen at the correct speed, but to ensure that a colour picture is accurately recovered the phase or rotational position of the disc must also be in time with the red, green and blue, grey-scale fields, as they appear on the screen.*

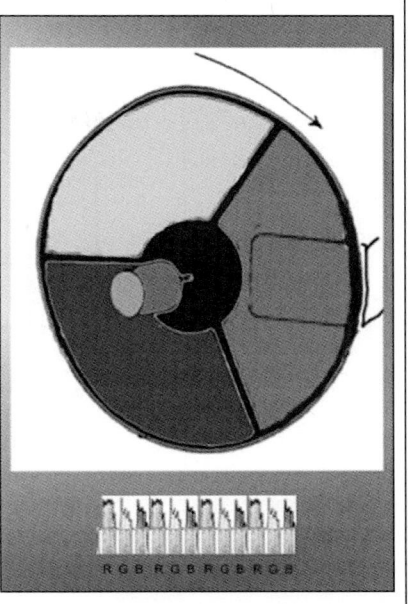

Fig A2.6: The disc may have straight or curved spokes but they must be strong to support the circumference of the disc

for theatre and studio lighting, can be obtained from sources on the internet. Once cut to size the filters should be carefully super-glued to the disc and to itself. When this hardware is completed, the electronic tachometer required to synchronise the disc to the Red sequential field will need to be fitted. This is a black timing line drawn from the centre to the circumference of a small white disc fixed on the inside of the hub (**Fig A2.7**).

As the timing line travels past a stationary reflecto-optical device it produces a pulse, to synchronise the motor speed and position with the correct colour of

- *The system needs to know when one of the colour filters is directly in front of the screen. The disc therefore has to send a signal once each rotation to tell the electronics where it is.*
- *The device used is called a tachometer and consists of a white disc with a black timing line (rotating with the disc) and a reflective opto-electronic detector. The line coincides with the start of the red filter.*
- *The variable pulse from the tacho is compared to the reference red control pulse.*

An electronic comparator is used to speed up or slow down the motor until the pulses are in time with each other. When that occurs the motor speed and phase are locked to the incoming red field and a stable full-colour image is seen.

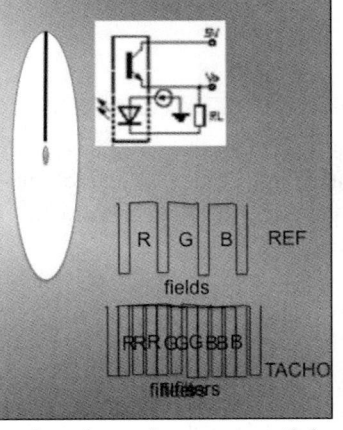

Fig A2.7: White disc with a black timing mark in conjunction with a reflective opto-electronic reflector generates a tachometer pulse once per cycle

171

THE THREE DIMENSIONS OF JOHN LOGIE BAIRD

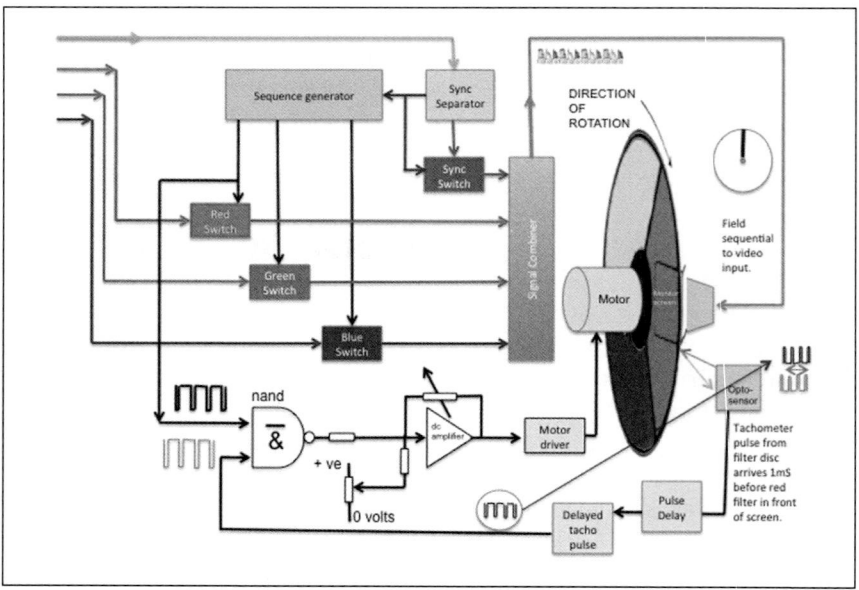

Fig A2.8: Complete operational block diagram

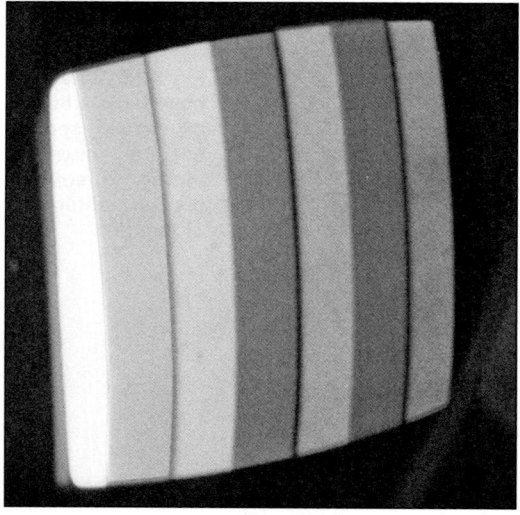

Fig A2.9: Colour bars photographed from the author's field sequential colour television display (see page C.15 for a colour version)

monochrome picture as it arrives on the screen. It is important that the opto-switch and timing-line are positioned to ensure that the pulse is produced to coincide with the centre of the blue filter.

It will be arranged that the motor turns clockwise when viewing the screen to enable the red, green and blue filters to pass the screen in that order. A power FET is used to drive the motor from the 12v supply with an adjustable bias on the gate to set the motor speed slightly faster than 1000rpm.

Also feeding the gate of the FET is a signal from the output of a simple comparator that will drive the speed down to lock the red filter to the Red field on the screen. The comparator may comprise a two input nand gate with one input receiving the red control pulse from the divider and the other receiving the pulse from the tachometer. To enable a level of control over the phase locking angle of the disc and to offer a fine adjustment, see **Fig A2.8**, the timing mark should be positioned to produce the pulse before the red filter is required to reach the screen.

The pulse is used to drive a monostable circuit of the 555 variable delay type, which in turn generates the pulse for the comparator. When adjusted correctly the unity will remain in lock and from switch-on will take only two or three seconds to produce a stable colour filter.

Fig A2.9, shows the colour bars reproduced from the author's field-sequential screen [also reproduced in colour on page C.15].

Appendix 3: Bibliography

Abramson, A: *The History of Television 1880 to 1941*. North Carolina: McFarland, 1987.

Adamian, J: *British Patent Office. GB7219*. 1908, applied 1 April 1908, issued 28 May 1908.

Anderson, AC and Anderson LS: *GB30188*. British Patent Office, London. Convention date 24 December 1908, issued 22 September 1910.

Anderson, EI: Radio Corporation of America, Washington DC: *US2297524*, US Patent Office. Application date 20 June 1941, issued 29 September 1942.

Anderson, EGO: Personal correspondence to Dr Peter Waddell, Strathclyde University. 1974.

Anon: 'Surgery in Colour Television'. *Life* magazine. August 1951.

Baird, JL: *GB289104,* British Patent Office, London. Applied 15 October 1926, accepted April 16 1928.

Baird, JL: *GB314591*, British Patent Office, London. Applied 4 January 1928, issued 4 July 1929.

Baird, JL: *GB476195*, British Patent Office, London. Applied 12 December 1935, issued 14 June 1937.

Baird, JL: 'Best Letters of the month'. *Television*. July 1928.

Baird, JL: *Electrical Engineering*. February 1942.

Baird, JL: Washington DC: *US2006124*. US Patent Office. Filed 31 December 1926, granted 25 June 1935.

Baird, JL: *GB381898*, British Patent Office, London. Applied 30 May 1932, accepted 13 October 1932.

Baird, JL: *GB473323*, British Patent Office, London. Applied 9 April 1936, accepted 11 Oct 1937.

Baird, JL: *GB222604*, British Patent Office, London. Applied 26 July 1923, issued 9 October 1924.

Baird, JL: *GB230576*, British Patent Office, London. Applied 29 December 1923, issued 19 March 1925.

Baird, JL: *GB266564*, British Patent Office, London. Applied 1 September 1925, issued 1 March 1927.

Baird, JL: *GB267378*, British Patent Office, London. Applied 1 September 1925 issued 1 March 1927.

Baird, JL: *GB269658*, British Patent Office, London. Applied 20 January 1926, issued 20 April 1927.

Baird, JL: *GB270222*, British Patent Office, London. Applied 21 October 1925, issued 21 April 1927.

Baird, JL: *GB312560*, British Patent Office, London. Applied 30 November 1927, issued 30 May 1929.

Baird, JL: *GB319307*, British Patent Office, London. Applied 20 June 1928, issued 20 Sept 1929.

Baird, JL: *GB321389*, British Patent Office, London. Applied 5 June 1928, issued 5 Nov 1929.

Baird, JL: *GB322776*, British Patent Office, London. Applied 9 June 1928, issued 9 Dec 1929.

Baird, JL: *GB555167*, British Patent Office, London. Applied 13 May 1942, issued 6 August 1943.

Baird, JL: *GB562168*, British Patent Office, London. Applied 25 July 1942, filed complete 23 July 1943, issued 21 June 1944.

Baird, JL: *GB562334*, British Patent Office, London. Applied 10 October 1942, issued 28 June 1944.

Baird, JL: *GB292365*, British Patent Office, London. Applied 22 December 1926. accepted 22 June 1928.

Baird, JL: *GB321441*, British Patent Office, London. Applied 11 July 1928, accepted 11 November 1929.

Baird, JL: *GB373196*, British Patent Office, London. Applied 18 February 1931, accepted 18 May 1932.

Baird, JL: *GB473323*, British Patent Office, London. Applied 9 April 1936, accepted 11 Oct 1937.

Baird, JL: *GB508039*, British Patent Office, London. Applied 24 Dec 1937, accepted 26 June 1939.

Baird, JL: *GB545078*, British Patent Office, London. Applied 7 September 1940, accepted 11 May 1942.

Baird, JL: *GB545462*, British Patent Office. London. Applied 23 October 1940, issued 28 May 1942.

Baird, JL: *GB545491*, British Patent Office, London. Applied 23 October 1940 issued 28 May 1942.

Baird, JL: *GB545603*, British Patent Office, London. Applied 13 January 1941, issued 15 July 1942.

Baird, JL: *GB546470*, British Patent Office, London. Applied 23 October 1940, issued 4 June 1942.

Baird, JL: *GB552582*, British Patent Office, London. Applied 11 July 1941, accepted 15 April 1943.

Baird, JL: *GB555167*, British Patent Office, London. Applied 13 May 1942, issued 6 August 1943.

Baird, JL: *GB562168*, British Patent Office, London. Applied 25 July 1942, filed complete 23 July 1943, issued 21 June 1944.

Baird, JL: *GB562334*, British Patent Office, London. Applied 10 October 1942, issued 28 June 1944.

Baird, JL: *GB573008*, British Patent Office. London. Applied 26 August 1943, completed 25 September 1944, accepted 1 November 1945.

Baird, JL: *Provisional Specification No 13887/43*, British Patent Office, London. Lodged on 26 August 1943.

Baird, JL: *Provisional Specification No. 15851/43*, British Patent Office, London. Lodged on 27 September 1943.

APPENDIX 3: BIBLIOGRAPHY

Baird, JL: *Provisional Specification No 16413/43*, British Patent Office, London. Lodged 7 October 1943.

Baird, JL: *Provisional Specification No 16581/43*, British Patent Office, London. Lodged 11 October 1943.

Baird, JL: *Provisional Specification No 19825/43*, British Patent Office, London. Lodged 26 November 1943.

Baird, JL: *Provisional Specification No 2370/44*, British Patent Office, London. Lodged 9 February 1944.

Baird, JL: *Sermons, Soap and Television, 2nd Edition*. Royal Television Society, London, 1990.

Baird, JL: *US1957815*, US Patent Office, Washington. Applied 15 January 1931, issued 8 May 1934.

Baird, JL: *US2006124*, US Patent Office, Washington DC. Filed 31 December 1926, granted 25 June 1935.

Baird, Malcolm: *Television and Me*. Mercat Press Ltd, Edinburgh. 2004.

Baird, Margaret: *Television Baird*.Haum, Cape Town. 1973.

Bartlett, G and Caldwell, PC: 'High Lights and Side Lights'. *General Electric Review*, March 1941.

Bennett-Levy, M: Private Television Collection.1996 - 97. Scotland.

Bidwell, S: *Nature*, 4 June. 1908.

Birkenshaw, DC: 'The Birth of Modern Television'. *Television, 50th Anniversary issue, 1927-1977*. November-December 1977.

Bronwell, AB: *The Chromoscope, A New Colour Television Viewing Tube*.

Brown, D: *Images Across Space: The Electronic Imaging of Baird Television*. Radio Society of Great Britain, London. 2010.

Brown, D: *Images Across Space: The Electronic Imaging of Baird Television*. Middlesex University, Teaching Resources. London. 2009.

Brown, D: *GB2307373*, Three-dimensional imaging system, UK Patent, Glasgow, 21 May 1997.

Brown, D: *US5949389*, Imaging System, US Patent Date of Patent 7 September 1999.

Burns, RW: 'JL Baird: Success and Failure'. *Proceedings IEE*. Volume 126, No 9, September 1979.

Burns, RW: 'A D Blumlein Engineer Extraordinary'. *History of Technology. Engineering Science & Education.* February 1995.

Caulfield. HJ: 'Digital Control of Focal Distances'. *Applied Optics*, Vol 6, No 3, March 1967.

Ciongoli Bernard M: 'Innovations, Patents, Processes and Products'. *IEE Spectrum*, December 1987.

Cooney, Charles: *US1506975*, Boot and shoe having inflated air cushion inserted in the sole and heel thereof. 21 Aug 1922, issued 2 Sept 1924.

Daily Express: 8 January 1926.

DuMont, AB: 'Experimental CR Tubes'. *Electronics*. March 1947.

Dyson, J: 'A Common-Path Interferometer for Testing Purposes'. *Journal of the Royal Society of America*. Vol 47. No 5. May 1957.

Editorial: 'Living Pictures by Wireless'. *The Morning Post*, 10 August 1926.

Editorial: 'Sending Pictures Through the Air', *The Sphere*, 28 August 1926.

Editorial: 'Sony's new Plasmatron'. *What Television*. London. September 1995.

Editorial: 'Current Topics', *Wireless World*, No 1062, December 1940.

Editorial. 'Space /Aeronautics. 3D Display', *Electronics*, September 1962.

Ekstrom, A: *32220*, Swedish Patent Office. Applied 24 January 1910, issued 3 February 1912.

Evans and Little: 'Large-screen colour television projection', *Journal of the Society of Motion Picture Television Engineers*, April 1955.

Exwood, M: *John Logie Baird 50 Years of Television*. IERE History of Technology Monograph, London, January 1976.

Faber, M: *A History of Communications*, Prentice Hall International Inc, London: 1964.

Ferril, TM Jr: *US2361390*, Stereo Indicator, US Patent. Filed 1 November 1943, issued 31 October 1944.

Fink, DG: *Colour Television Standards / NTSC*, New York, McGraw Hill, 1955.

Fiore, JP and Kaplan, SH: 'A Second Generation Color Tube Providing more than Twice the Brightness and Improved Contrast'. *IEEE Trans. Broadcast and TV Receivers*, October 1969, Vol BTR-15, No 3.

Fiore, JP and Kaplan, SH: *US3146368*, Washington DC: US Patent Office. Issued 25 August 1964.

Fiore, JP and Kaplan, SH: *US3365292*. Washington DC: US Patent Office. Issued 23 January 1968.

Flechsig, Dr W: *888065*. Paris: French Patent Office. Applied 11 July 1938, issued 16 June 1941.

Flechsig, Dr W: *736575*. DRP Patent Office. Applied 12 July 1938, issued 13 May 1943.

Gardner, JE and Hineline, HD: *GB225553*. British Patent Office. Applied 28 November 1923, issued 28 May 1925.

Geddes, K and Bussey, G: *Television the first fifty years*. National Media Museum. 1986.

Geer, CW: *US2480848*, Washington DC, US Patent Office. Filed 11 July 1944.

Gerritsen, HJ and Horowitz B: 'Experimental Demonstration of Optical Sectioning by Spacial Filtering and its Possible use in Transmission of Three-Dimensional Pictures'. *Applied Optics*, Vol 10, No 4, April 1971.

Goldmark. Dyer, Pierce, and Hollywood: 'Color Television - Part 1'. *Proceedings of the IRE*, April 1942.

Goldsmith, AN: *US2481839*, Washington DC: US Patent Office. Filed 5 August 1944, issued 13 September 1949.

Gregory, RL: 'A Solid Image Microscope'. *Nature*, 1958. Vol 182, p1434.

Hamasaki, J, Nagata, Y, Higuchi, H and Okada, M: 'Real-Time Transmission of a 3D Image using Volume Scanning and Spatial Modulation'. *Applied Optics*, Vol 16, No 6, June 1977.

Hammond, JH: *US1725710*, US Patent Office, Washington DC. Applied 15 August 1923, issued 20 August 1929.

Hartley-Baird Ltd: The microfiche file, Search Room, Companies House, London.

Hartwich: *Colour Television Servicing Handbook*, Vol 1, London: Philips Technical Library, 1965.

Hausske, GE: 'Another Unsung Hero: EN Rauland'. *The Old Timer's Bulletin*, Journal of the Antique Wireless Association, Vol 33, No 1, May 1992.

APPENDIX 3: BIBLIOGRAPHY

Herbert, R: *Baird Newsletter*, No 2, November 1983.

Herbert, R: 'JL Baird's Colour Television 1937 -46'. *Television*: Journal of the Royal Television Society, Jan/Feb 1990.

Herbert, R: *Seeing by Wireless*. Published by Herbert, sponsored by Quantel. 1996.

Herold, EW: 'A History of Color Television Displays'. *Proceedings of the IEEE*, Vol 64, No 9, September 1976.

Holm, WA and Poel, FHJ van der: 'Colour Television in Medical Teaching'. *Philips Technical Review*, 1958/59, Vol 20.

http://gb.espacenet.com.

http://www.lighting.philips.com.

http://www.localhistory.scit.wlv.ac.uk/Museum/Engineering/Electronics/history/ValveEra.htm.

Hubel, DH: *Eye, Brain, and Vision*. New York: Scientific American Library, 1988.

Inglis, AF: *Behind the Tube; a history of broadcasting technology and business*. Butterworth. 1990. USA.

Jaworski, W von and Frankenstein, A: *172376*, DRP Patent Office. Applied 20 August 1904, issued 21 June 1906.

Kamm, A and Baird, M: *John Logie Baird: a life*. NMS Publishing Ltd. Edinburgh, 2002.

Kaplan, S: Private correspondence. Letter to author. 15 November 1996.

Kaplan, S: Private correspondence. Letter to author. 14 December 1996.

King, MC and Berry, HD: 'Varifocal Mirror Technique for Video Transmission of Three-Dimensional Images'. *Applied Optics*. Vol 9, No 9, September 1970.

Koch, SJ, et al, DuMont: *US2544690*. Washington DC: US Patent Office. . Filed 26 December 1946, issued 13 March. 1951.

Koehler, Kenneth R: Professor of Physics, Department of Mathematics.Physics and Computer Science. Raymond Walters College, University of Cincinnati, Ohio: http://www.rwc.uc.edu/koehler/biophys/6d.html.

Land, EH: 'Colour Vision and the Natural Image. Part 1', *Proceedings of the National Academy of Sciences*. 1959.

Larsen, E: *Inventors' Scrapbook*. London: Lindsay Drummond Ltd. 1947.

Lefroy, Lieut Col: Air Ministry File No 43A. London: Kew. Public Records Office. Reference: AIR2 269/S/24132.

Leman, J: *TV is King*. BBC Scotland. April 1994.

Letter from Lord Hankey, Privy Council Office, Whitehall to John Logie Baird 25 March 1944.

Letter: Cable and Wireless, Archives, 29 October 1941.

Letters to the Editor: 'Opinions', *The Glasgow Herald*. March/April 1994.

'Living Face by Wireless', *The Morning Post*, 25 January 1926.

Lyons, N: *The Sony Vision*. Crown Publishers. New York. 1976.

McLean, D: *Restoring Baird's Image*. Institution of Electrical Engineers. 2000.

Miceli, Alfonso: *US1516395*, Shoe Attachment. US patent. Applied 14 November 1923, accepted 18 November 1924.

Morelli, D: 'Real-time Volume-scanning Three-dimensional imaging system for real moving objects'. *Optics Letters*. Vol 6, No 3, March 1981.

Moseley, SA: *John Baird, The Romance and Tragedy of the Pioneer of Television*. London: Oldhams Press Ltd. 1952.

Moseley, SA and Barton-Chapple, HJ: *Television Today and Tomorrow, 2nd ed.* London: Sir Isaac Pitman & Sons Ltd, 1931.

Muirhead, JC: 'Variable Focus Length Mirrors'. *Rev Sci Instrum*, Vol 32. 1961.

Nipkow, P: *30105*. Electrisches Teleskop. German Patent Office. Berlin. 1884.

Nuttall, TC: 'The Development of a High Quality 35mm Film Scanner'. *IEE convention on 'The British Contribution to Television', 28th April-3rd May 1952.*

O'Neil, W: 'Baird's Clear Vision of the Future'. *New Scientist*, Vol 128, No 1747, 15 December 1990.

Parker, E and Charles-Seymour: *GB562367*, 3-D Radar System. Applied 2 May 1945, issued 16 September 1947.

Parr, G: 'Review of Progress in Colour Television'. London: *Journal of the Royal Television Society*, 1942, Vol 3, 1942.

Phillips, W: 'Who Really Invented Television?', *Broadcast Magazine: 50 Years of Television, Golden Jubilee issue.* International Thomson Publishing Ltd. 1986.

Polumordvinov, AA: *10738*, Russian Patent. Applied 23 December 1889.

Rawson, Eric G: '3D Computer Generated Movies using a Varifocal Mirror'. *Applied Optics*. Vol 7, No 8, Aug 1968.

Register of Defunct Companies, Macmillan Publishers Ltd under licence from the International Stock Exchange. 1990.

Rignoux, GPE: *FR390435*, Paris: French Patent Office. Applied 20 May 1908, issued 5 October 1909.

Russell, Dr A and Crookes, WR: *Nature*. 23 November 1926.

Russell, Dr A: *Nature*, 3 July 1926.

Schroeder, AC: *US2595548*, Picture Reproducing Apparatus. Washington DC: US Patent Office. Filed 1947, issued 1953.

Scophony-Baird Ltd: Archived Company Records. London: Company House.

Shrewsbury Technology Limited: Company No.00566892. London: Companies House Search Room.

Sleeper, GE: *US2389645*, US Patent. Filed 5 February 1943, issued 27 November 1945.

Sommer, Dr AH: 'Photoelectric Cells'. *Methuen's Monographs on Physical Subjects*. London: Methuen & Co Ltd, January 1946.

Sommer, Dr AH: Private correspondence to the author. 11 March 1991.

Sommer, Dr AH: Private correspondence to the author. 6 August 1990.

Swinton, C: 'Secret memo on television'. Letter to Major HE Wimperis, RAF, 23 January 1928. London: Kew. Public Records Office. Air Ministry File. Call number: Air 2: File no. 269/S/24132.

Szegho, CS: 'Experimental Tri-Color Cathode Ray Tube'. *Tele-Tech*, Vol 9, July 1950.

Szegho, Dr CS. and Kaplan, SH: *US3615462*, Processing Black Surround Screens. US Patent. Applied 6 November 1968, issued 26 October 1971.

Szegho, Dr CS: 'Color Cathode-Ray Tube With Three Phosphor Bands'. *Journal of the Society of Motion Picture and Television Engineers, Inc*. Vol 55, October 1950.

Szegho, Dr CS: *Reciprocal Report Number 15*: Rauland Corporation (previously Baird Television Incorporated) to Cinema Television Ltd (Incorporating Baird Television Ltd). 1947.

APPENDIX 3: BIBLIOGRAPHY

Szegho, Dr CS: *Reciprocal Report Number 17*: Rauland Corporation (previously Baird Television Incorporated) to Cinema Television Ltd (Incorporating Baird Television Ltd). 1949.

Szegho, Dr CS: *Reciprocal Report Number 20*: Rauland Corporation (previously Baird Television Incorporated) to Cinema Television Ltd (Incorporating Baird Television Ltd). 1951. Section 7.

Technical Editor: 'Technical Notes'. *Television*. April 1928.

Television Ltd: Closed records of the company. London: Kew. Public Records Office. Call Number 206588.

Television: Journal of the Television Society. July 1928.

Thorn/EMI: *The History of Computer Systems Division*. Company Document, Thorn EMI Electronics Ltd, Wells, Somerset, UK.

Tiltman, RF: *Baird of Television*. London: Seeley Service & Co. 1933.

Tomes, GAR: *Jumbo's Diaries: The memoirs of Gilbert Tomes*. Unpublished.

Traub, AC: 'Stereoscopic Display using Rapid Varifocal Mirror Oscillations'. *Applied Optics*. Vol 6, 1967.

Trenouth, J: Baird items in Science Museum. Former Curator National Film Museum, Bradford.

Trundle, E: *Beginner's Guide to Videocassette Recorders*. London: Butterworth & Co (Publishers) Ltd. 1983.

Trundle, E: 'The Development of Colour Tubes'. *Television*. IPC Magazines. Vol 36, no 8, Issue 428, June 1986,

Turner, LW: *Electronics Engineers Handbook. History of Electronics. 4th edition*. Newnes-Butterworth. London. Section 2, 1976.

Valensi, G: *572716*, French Patent. Applied 3 January 1923, issued 12 June 1924.

Waddell, P and McArthur, T: *Vision Warrior*. Orkney: Scottish Falcon Books, 1990.

West, AGD: 'Development of Theatre Television'. *Journal of SMPE*. New York. Vol. 51, August 1948.

Wood, K Sir, 'The Right Hon'. MP. Postmaster General: *The Report of the Television Committee*. London: His Majesty's Stationery Office, January 1935.

Young-Helmholtz: 'Three-colour theory'. *Encyclopædia Britannica*, 2010.

THE THREE DIMENSIONS OF JOHN LOGIE BAIRD

Index

See also the Contents list on page iii and the List of Illustrations on page iv.

Abramson, Albert 33
Admiralty, British 44, 146
Adamian, Johannes 79
Air Ministry, British 33
Air Ministry, French 150
Alexanderson, Dr E F W 97, 98, 120
American Baird Company 66
American Medical Association 110
American Telephone and Telegraph Company
    (ATT) 39, 43
Anderson, Ian 45, 46
Anderson, Edward G O 53, 86, 98, 114, 115,
    .... 121, 156, 157
Anderson (of RCA) 98
Andersons (of Copenhagen) 79
Angwin, Sir Stanley 149
Antique Wireless Association Museum 53
Appleton, Sir Edward 149
Apollo 11 1
Argyll Motor Company 14, 27
Ashridge, Sir Noel 149
Association of Radio Manufacturers 107
Atlantic City Convention Hall 110
Australian Broadcasting Company 151

Bain, Alexander 3
Baird Development Company 41, 45, 63
Baird Incorporated 66
Baird International 41, 45, 111
Baird Laboratory 54
Baird, Prof Malcolm 91, 157, 162
Baird, Margaret 21, 28, 54, 156, 158
Baird's Speedy Cleaner 18, 27
Baird Television Ltd 23, 41, 56, 58, 60, 61,
    .... 62, 63, 64, 65, 67, 88,
    .... 92, 100, 133, 150, 154,
    .... 157, 158, 159, 160, 161
Baird Undersock 15, 16
Bakewell, Frederick 3
Barton-Chapple, H J 40, 79

Bell, Alexander Graham 7, 12, 46, 79
Bell Telephone Laboratories 34, 41, 43,
    .... 85, 120, 159
Bennett-Levy, Michael 110
Berengaria (liner) 54
Berry, H D 148
Bidwell, Shelford 21
Birchall, Sir Raymond 149
Blumlein, Alan 32, 48, 49
Board of Trade, British 133, 155
Bosch 111, 121, 122
Boon-Danahar boxing contest 64
Bragg, Sir William 42
British Association 52
British Broadcasting Company / Corporation
    (BBC) 39, 40, 41, 44, 45, 47, 56, 59,
    .... 60, 63, 64, 92, 149, 153, 155, 161
British Film Producers Association 155
British Sky Broadcasting ("Sky") 126, 162
British Thomson-Houston 39
British Westinghouse 39
Broadcasting 23
Broadrip, Captain 27
Bronwell, Arthur 109, 110
Brooks, Roger 26, 27
Browne, C O 48
Brunel, Isambard Kingdom 7
Buchannan, Jack 3, 156, 157
Burns R W 41
Buxton Hydro 18

Cable and Wireless 66, 98, 99, 136, 150, 160
Campbell-Swinton, A A 42
Caselli, Abbe Giovanna 3
Centronics 134
Chaplin, Charlie 49, 50, 165
Church, Major 112
Cinema Television Ltd 62, 64, 65, 66, 88,
    .... 91, 92, 95, 133,
    .... 134, 150, 157

Cintel International 62
Ciongoli, Bernard M .147
Clapp, Ben 28, 53, 65
Classic Cinema, Baker St 157
Clayton, Harry 64
Clyde Valley Power Station,
    Rutherglen 14, 15, 16
Cockroft, Prof J D 149
Coia, Emilio 47
Columbia Broadcasting System (CBS) 94, 98,
    . . . . .107, 110, 113, 114, 116,
    . . . . .117, 120, 121, 150, 159, 160
Companies Court, London 133
Cooney, Charles 22
Crookes W R 52

Daily News 23
D'Albe, Dr Fournier, ...22, 23
Day, Wilfred Ernest Lytton 23, 26, 27, 28
Dept of Scientific and Industrial Research 149
Derby, The 23, 46, 47, 67
Dominion Theatre. London 64, 88, 89
DuMont 108, 109, 118, 121
Dyson, J 146

Eckersley, Peter, Captain 39, 48
Eddison, Thomas Alva .6
Ekstrom, A 35
Electronic Industries Association 107
Electronics magazine 118
Electronics and Television 96
EMI (see also Marconi-EMI) . . .48, 49, 62, 63, 92,
    . . . . .100, 157, 158, 159, 160
Eustace 137
Exwood, M 42, 44

Farnsworth, Philo T 41, 58, 61, 62, 159
Federal Communications Commission
    (FCC) 107, 113
Feril, T M Jr 146
Fernseh AG 64, 111, 121, 122, 154, 160
Ferranti 60
Fessenden, Reginald Aubrey 8, 9
Fiore, Joseph P 117, 119, 120
Flamm, Donald 160
Fleming, J Ambrose 9
Fleschsig, Werner 111, 112, 115, 119,
    .120, 121, 122, 160
Folsom, Frank 114
Forest, Lee de 9
Frankenstein, A 79
Fournier, A 14
Friese-Green, William .6

Gables, Ed 53

Gardner, J E 35
Gaumont British Film Company 56, 63, 64, 65,
    .66, 133, 150, 157
Geer, C W 107, 108, 121
General Electric Ltd 39, 97, 107, 150,
    . . . . .155, 159
GEC, British 39
German Broadcasting House 45
Gerritson, H J 146
Glasgow and West of Scotland Technical
    College 14
Glasgow University 14
Goldmark, Dr Peter Carl 94, 98, 120, 160
Goldsmith, A N 108, 112, 121
Gramophone Company 159
Gray, Dr Frank 34, 43
Great Eastern, The 7
Greer, Sir Harry 64
Gregory, R L 146
Grotty, Frank 119

Haley, W J 149
Halley's Industrial Motors 14
Hamasaki, J 146
Hammond Jr 35, 79
Hankey, Lord 2, 149
Hart, John 36
Hart, Robert 53
Hartley Automotive Group 158
Hartley A W M 158
Hartley-Baird Ltd 158
Harvey R J P 149
Helensburgh Radio Club 167
Herold, Edward W 112, 160
Hertz, Heinrich 7
Higuchi, H 146
Hineline H D 35
Hitler, Adolph 111, 160
Homer, William George 6
Horowitz, B 146
Hutchinson, Oliver George, Captain . . .18, 27, 28,
    . . . . .29, 41, 42, 45, 46, 53
Hutchinson's Rapid Washer 18, 27

Ibuka, Masaru 115
Image Telegraph 3
Inglis, A F 112
Institution of Electrical Engineers (IEE) 28
International Congress of Electricity 3
Inverforth, Lord 98
Ives, Herbert E 43

Jaworski, Werner von 79
Jazz Singer, The 41
Jenkins, C F 107

John Logie Baird Ltd 2, 156, 157
Johnson, Arthur 98
Jones, Professor Taylor 160

Kamm, Antony 15
Kaplan, Sam H 112, 113, 114, 117, 118,
. . . . .119, 120, 121, 160
Kennedy, Miss C 149
Kinematograph Weekly .23
King, M C 146
Kirke, H L 44
Koch, S L 121

Lance, T M C 56, 65
Land, Dr Edwin 75, 76
Larchfield School 11
Larsen, E 154
Law, Harold B 112, 114
Lawrence, Dr Ernest O 115, 122, 161
Lefroy, Lt Col 33
Life magazine 110
Lodge, Sir Oliver 160
Loewe Radio 111, 121, 122
Lomond Radio Club .167
London Coliseum 87
Loxdale, Norman 25, 26
Luger, Len 53

Marconi, Guglielmo 7, 8, 49
MacDonald, Ramsay .160
Marconi Co / Marconi-EMI (see also EMI)
. . . . .39, 41, 47, 53, 58, 60,
. . . . .62, 150, 155, 160
McArthur, Tom 21
McCann, John J 76
McLean, Donald .51, 165
McGee J G 48
Metropole Theatre, London 87
Metropolitan Vickers 39
Miceli, Alfonso 22
Mills, Victor 34
Ministry of Education 155
Ministry of Supply 149
Mitchell-Thomson, Sir William 59
Miyaoka, Senri 115, 122
Moon, Captain 65
Moreli, Dean 147
Morning Post, The 18, 43, 81
Morse, F B 7
Moseley, Sydney A 44, 45, 47, 49, 56, 79
Motorola 119
Motor News, The 22, 27
Muirhead, J C 146

Nagata, Y 146

National Media Museum, Bradford 158
National Museum of Scotland, Edinburgh 111
National Telephone Company 12
National Television Starndards Committee
(NTSC) 107
Nature 28, 42, 53
News Theatre, Agar St 157
Newton, Sir Isaac 69, 70
Nintendo 129
Nipkow, Paul 7, 8, 23, 163

Odhams Press 23
Okada, M 146
Osmo Boot Polish 16
Osram 33
Ostrer, Isidore 56, 63, 64, 65, 66, 133
Oxbrow, C F 86

Pantelegraph 3
Paramount Pictures Corporation 115
Parker, E 146
Parr, G 94, 100
People's Friend, The 15
Perring-Thoms, Percy 158, 160
Perskyi, Constantine 3
Phenakistiscope 6
Philips Electronics Company . . .110, 116, 117, 161
Philips Lighting 86
Phillips, William 47
Physical Society, The 28
Planck 69
Plateau, Joseph 6
Plugge, Capt Leonard 160
Polumordvinov, A A 79
Post Office, British 39, 40, 42, 53, 149
Ptolemy 70

Queux, William Le 23

Radio Communication Company 39
Radio Corporation of America (RCA) 39, 41,
. . . . .42, 56, 58, 62, 92, 94, 98,
. . . . .100, 107, 108, 112, 114, 115,
. . . . .117, 118, 120, 121, 122, 159, 160
Radio Industry Council 155
Radio Luxemburg 160
Radio Manufacturers Association (RMA) 107
Radio Rentals Group 158, 160
Radio Times 23
Rank Cintel / Cintel International 62, 133
Rank Organisation 133
Rauland Corporation 66, 100, 110, 112,
. . . . .114, 119, 121
Rawson, Eric G 134
Reith, John 39, 44

REP Joint Committee .155
Reuters Press 53
Revelley, P V 86
Rhumer, Eric 13
Rignoux, Georges .14, 35
Roberts, J Varley 149
Robinson, F H 23
Roentgen Society 42
Roop, Ram 17
Royal Institute of Great Britain / Royal
   Institution .28, 36, 42
Royal Technical College 44
Russell, Dr Alexander 28, 52, 53

Samson, Dr 58, 89, 134
Sarnoff, David 114
Sava, Anatas Botzaritch 47
Savoy Hotel 157
Sayers A E 86
Schlesinger, Dr 119
Schroeder, A C .112, 121
Science Museum, London 28, 29, 32, 60
Scophony Ltd 64, 155, 157
Scophony-Baird Ltd .157
SE Laboratories 158
Selfridge, Gordon 26
Selsdon Committee / Report 59, 60, 107
Selsdon, Rt Hon Lord .59
Sharp 129
Shaw Robert 36
Shrewsbury Engineering 158
Shrewsbury Technology Ltd 158
Sleeper, G E 103, 120
Smith, Kline and French 110
Smith, Willoughby 7
Snell, Willebrord 70
Sobielioff, Mr 153
Society of Motion Picture and Television
   Engineers (SMTE) 87
Society of Telegraph Engineers 7
Sommer, Dr Alfred H 58, 89, 90, 91, 92
   . . . . .95, 120, 134, 160
Sony Corporation 113, 114, 115, 116,
   . . . . .117, 122, 160, 161
Sperry Corporation 146
Sphere, The 81
Springfield Electrical Works 14
Stampfer, Simon Ritter von 6
Standard Telephones and Cables Ltd 155
Stock Exchange, London 42
Stookie Bill 27, 28, 34
Strathaird 64

Stroboscope 6
Sullivan, Edmund J 46
Sylvania Company 119
Szegho, Dr Constantin S 58, 64, 65, 66,
   .94, 100, 113, 117,
   . . . . .119, 120, 121, 133, 161

Taynton, William 27, 33, 34
Technicolor Motion Picture Corporation 76
Television Advisory Committee (Selsdon)
   . . . . .59, 159
Television Committee (Hankey) 149, 155, 161
Television magazine 42, 47, 100
Television Ltd 27, 39, 41, 43, 44, 56, 79
Televisor 28, 33
Thomas, Terry 157
Thorn Electrical Industries 158, 160
Thorn/EMI Datatech .158
Thorn, Sir Jules .158, 160
Tiltman, Ronald F 18
Times, The 17, 18, 22, 23
Tomes, Gilbert 64, 134
Toshiba 117, 119
Traub, A C 146
Trundle E 31
Twentieth Century Electronics 134

Unversity Hospital, Leiden 110
University of Marseilles 117
University of Pennsylvania 110
University of Strathclyde 165

Valensi, Georges 79
Vince, R 86
Voice of America 160
Vowler and Company .45

Waddell, Dr Peter 21, 91, 162
Wells, H G 11
West, Capt A G D 23, 56, 65, 87, 88, 133, 159
Western Electric 39
Westinghouse Company 39, 42
Wilshaw, Sir Edward 98, 160
WMCA 160
World Fair, London 3
Wright, Charles Seymour 146

Zeiss .111, 121, 122, 161
Zenith Corporation . . .110, 117, 119, 120, 121, 160
Zoetrope 6
Zworykin, Vladimir 41, 42, 56, 57, 62, 134

THE BRISTOL IDEAS BOOK OF WALKS

By
Robin Askew
Eugene Byrne
Melanie Kelly
Amy O'Beirne

Edited by
Andrew Kelly

 BRISTOL BOOKS

Bristol Books CIC, The Courtyard, Wraxall, Wraxall Hill, Bristol, BS48 1NA
www.bristolbooks.org

The Bristol Ideas Book of Walks
By Robin Askew, Eugene Byrne, Melanie Kelly, Amy O'Beirne
Edited by Andrew Kelly

Published by Bristol Books 2024

ISBN: 978-1-909446-44-1

Design: Joe Burt
Cover Illustration: Till Lukat

Parts of this book were originally published in 2016 for the Bristol800 programme, which was managed by Bristol Cultural Development Partnership (BCDP – later renamed Bristol Ideas). A revised edition was published in 2017 to coincide with the Festival of the Future City, another BCDP initiative. In 2019, the Homes for Heroes 100 Book of Walks, looking at four council estates in the city, was published.

Text copyright: Walks 1, 3, 4, 5, 9, 10, 11, 12 devised and written by Melanie Kelly (BCDP); Walk 2 devised and written by Amy O'Beirne (BCDP) and developed further by Melanie Kelly; Walk 6 devised and written by Robin Askew; and Walks 7 and 8 by Eugene Byrne. The views expressed in this publication are those of each author.

Images: Every effort has been made to contact copyright holders of material reproduced in this book. The names of the copyright holders of images are provided with the captions. All maps are © 2023 Bristol City Council.

The authors have asserted their rights under the Copyright, Designs and Patents Act of 1988 to be identified as the author of this work. This book may not be reproduced or transmitted in any form or by any means without the prior written consent of the publisher, except by a reviewer who wishes to quote brief passages in connection with a review written in a newspaper or magazine or broadcast on television, radio or on the internet.

Printed on paper from a sustainable source.

A CIP record of this book is available from the British Library

Please note. There's lots to see in Bristol but cities change, and Bristol is no exception. Details provided in this book were correct at time of publication. Please take care on walks, paying particular attention to pavement conditions, car traffic, scooters, bikes and skateboards at all times. Bristol Books, the authors and past partners of Bristol Ideas, are not responsible for any problems encountered by people taking these walks.

Introduction	5
Walk 1: Commerce and Public Life by Melanie Kelly	8
Walk 2: Bristol and Romanticism by Melanie Kelly and Amy O'Beirne	20
Walk 3: Brunel's Bristol by Melanie Kelly	39
Walk 4: Art and Culture by Melanie Kelly	54
Walk 5: Nature in the City by Melanie Kelly	72
Walk 6: Bristol's Rock Music History by Robin Askew	85
Walk 7: Bristol's Urban Myths by Eugene Byrne	101
Walk 8: Unbuilt Bristol by Eugene Byrne	122

Council Estate Walks

Council Housing in Bristol Since 1919 by Melanie Kelly	141
4 June 1919: 4 June 2019	149
The Centenary Poem: Vanessa Kisuule	150
Walk 9: St Pauls by Melanie Kelly	153
Walk 10: Hillfields by Melanie Kelly	171
Walk 11: Sea Mills by Melanie Kelly	186
Walk 12: Knowle West by Melanie Kelly	202
Bibliography and References	219
Acknowledgments	221

Introduction

Bristol Ideas has organised many walks about ideas and the city covering arts, culture, nature, religion, Brunel and the Romantic poets. This book brings these together with a series of walks prepared for 2019 for our Homes for Heroes 100 celebration on the history and future of council housing, with new walks commissioned for this book on urban myths, rock music and unbuilt Bristol. All walks have been updated.

This is the fourth in a series of books providing routes around the city that allow residents and visitors to explore Bristol's past in the context of the present and with an eye to the future. We begin the series in the vicinity of the city centre where Bristol began as a trading post on the banks of the River Avon and then move out to look at the history of some of Bristol's council estates.

Walk 1 starts at the site of Bristol's old castle, once the seat of royal power, and ends at City Hall, the seat of today's directly elected local government, with a look at the role of merchants and the church in the city's early history along the way. Walk 2 starts at the top of Park Street and ends at St Mary Redcliffe church, following in the footsteps of some of those who encouraged critical debate and challenged the status quo in the eighteenth and nineteenth centuries. Walk 3 is a circuit that begins and ends at Temple Meads station, looking at the contribution to the city's development made by Brunel's engineering projects, as well as the changing fortunes of the city's docks. Walk 4 demonstrates Bristol's rich and varied artistic and cultural life, starting at the studios belonging to the BBC on Whiteladies Road and ending at Temple, for centuries the location of an annual fair that combined commerce with entertainment. Walk 5 begins on Clifton Down and ends on Brandon Hill, demonstrating the importance of nature and green space in the city.

Walk 6 looks at Bristol's music history especially in the 1960s and 1970s. We have always believed that urban myths are an important part of the culture of a city and Walk 7 looks at some of these Bristol Yarns. Walk 8 is slightly unusual as it is about unbuilt

Bristol and there is nothing to see. We have provided images for what might have been in the city and the whole volume is illustrated throughout.

The second section of this new book incorporates updated editions of four walks made in 2019 looking at four of Bristol's council estates. Walk 9 covers St Pauls. Walk 10 takes you around Hillfields. Walk 11 looks at Sea Mills. And Walk 12 looks at Knowle West. Each walk covers the history of the estate, the types of housing there, and testimonies from residents. The section has a special introduction which provides more information about Bristol and council housing.

Details about public transport – where relevant – are provided in each walk.

As with any modern city, Bristol is in a constant state of redevelopment, and you may find your way disrupted by road works and diversions. You may therefore need to improvise a little before picking up the suggested route again. Note is made in the text where roads, paths and pavements may have uneven surfaces that could make walking or using a wheelchair more challenging. On Walk 1 there is an optional short cut to avoid climbing steep steps. The most difficult walk is probably Walk 5 due to the hills, raised pavements and unavoidable steps when you reach the park on Brandon Hill.

The suggested time is always described as one taken at a leisurely pace. Although specific points of interest are mentioned in the text, you are encouraged to take the time to linger and to look around you. This is particularly true for city residents who may feel they already know these areas.

Look up and the mundane modern ground floor of a frequently passed shop front may reward you with an unexpectedly exotic upper storey. Observe overlooked details: an unusual archway above a door, perhaps; remnants of lettering from an old advertising sign; or a pane of coloured glass in an otherwise plain window. Stop to look at the views that open up before you – or are shyly revealed through a gap between two buildings. Look down at the stones, concrete and slabs beneath your feet. Be aware of the people around you: Bristol is proud to be a multicultural city where 91 different languages are spoken by residents and where visitors are welcomed from around the world. Listen to the sounds of voices, birds, traffic, the wind in the trees and – occasionally – silence. Smell food, coffee, flowers, water. Touch stonework, grass and wood. On a guided walk led by author Will Self in 2015 as part of Bristol's Festival of the Future City, participants were encouraged to liberate their senses. The key to exploring a city is to take nothing for granted and to be endlessly curious.

The maps in this book are from the award-winning Bristol Legible City (BLC) way-finding information system, which includes maps and on-street signage designed to improve people's understanding and experience of the city. The BLC walking maps are free of charge and widely available. They can usually be found at visitor attractions, libraries, hotel receptions, travel arrival points and many other sites. Use them with the confidence that you can wander off the known track, following your instincts, but soon find your way back to familiar ground when you are ready. Note that the scale of the maps means that, in some places, only an approximation of the route and location of the site of interest can be shown in this book. However, the detailed written directions will keep you to the right path.

You'll find a lot of useful information about what's on in the city on the website of Visit Bristol, including details of guided tours (visitbristol.co.uk/things-to-do/sightseeing-and-tours). There are many other walking guides and city histories available from the Bristol library service and local bookshops (some of those used for the research of this book are listed at the back).

The details of visitor attractions were correct at the time of writing, but please check the relevant website before making a trip. The religious buildings referred to in these walks are all of Christian origin, but you can learn about some of Bristol's many other faith communities at bristolmultifaithforum.org.uk. Bristol Ideas and the Festival of Ideas have run a number of walks on sacred Bristol (of all faiths and other sacred sites), as well as other walks on women and Bristol, Bristol and the trade in enslaved Africans, Old Market and many others. This is the final Bristol Ideas publication. We hope that others will add to what we have done with new walks covering these and many other areas of the city, its history, life and work.

Walk 1:
Commerce and Public Life

Bristol was originally established hundreds of years ago to serve as a trading centre for the region. Throughout this walk you will learn about Bristol's trading past and the influence of trade and commerce upon the city, as well as visit sites connected to local politics and civic and public life. It features architectural points of interest dating from the twelfth century to the present day.

The walk begins and ends on level ground with a number of hills to climb up or down along the way. There is an optional section for those who are not able to manage steep steps. Most of the pavement surfaces are of good quality. Allow at least 45 minutes, not including stops for refreshment or for visitor attractions.

The Walk

The walk begins at the lamp-post at the top of the ramp to the ferry landing in Castle Park (1), looking out across the water.

Bristol came into being as a trading centre because it was perfectly situated on a well-drained and easily defended knoll on a navigable waterway. It was built at the junction of the Rivers Frome and Avon; was within reach of abundant grazing and agricultural land, building materials and fuel; and enjoyed a generally mild climate. It was close to the old Roman roads of the Fosse Way (linking Lincoln to Exeter) and the Via Julia (from London to the port of Sea Mills via Bath), and to the post-Roman defensive Wansdyke earthwork. Bristol was originally known by the Saxon name of Brigstowe or 'Place of the Bridge'. Around 1240, a major engineering project was begun to divert the course of the River Frome. This was to increase the number of quaysides and provide deeper berths for trading vessels and took seven years to complete. The first significant extension to the town took place around 1248 when the existing wooden bridge across the Avon was replaced with a stronger stone one and the area south of the river – the direction you are looking now – was incorporated. Since 1809, the river at this point has been non-tidal and part of William Jessop's Floating Harbour (see Walk 3).

Goods imported to Bristol in the early thirteenth century included cloth, tin, timber, wool and madder (a plant used in dyeing). Fish would have come from Iceland and wine from Northern France. By the early sixteenth century, Bristol merchants were trading further afield, including Northern Europe, Spain, Italy and the Middle East, and

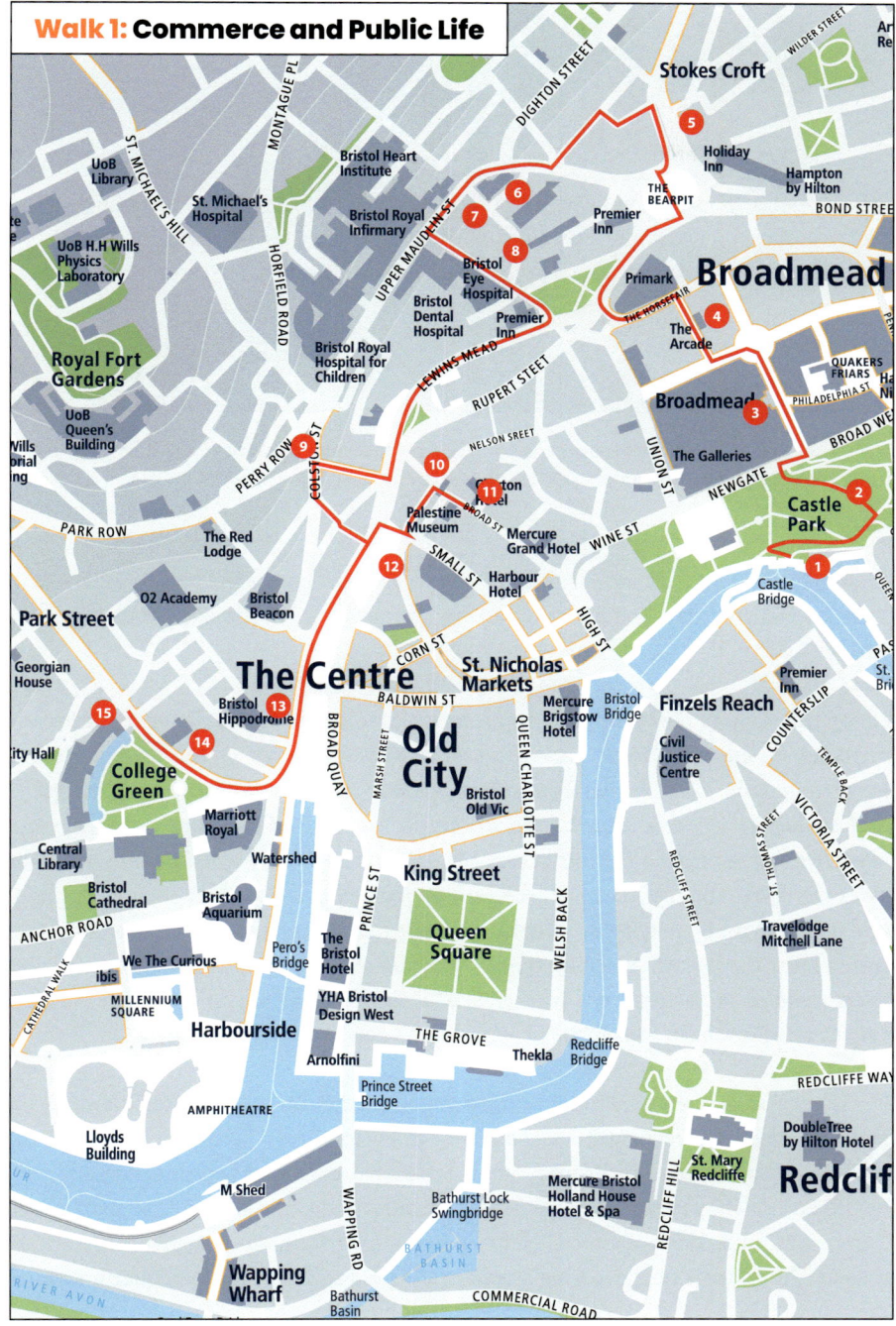

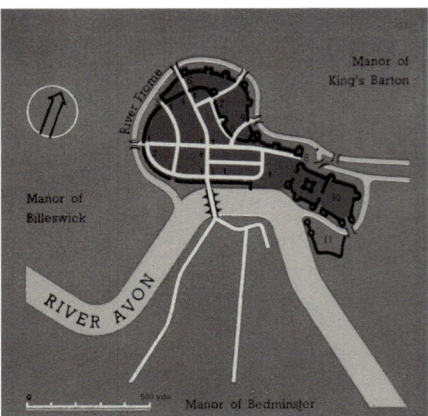 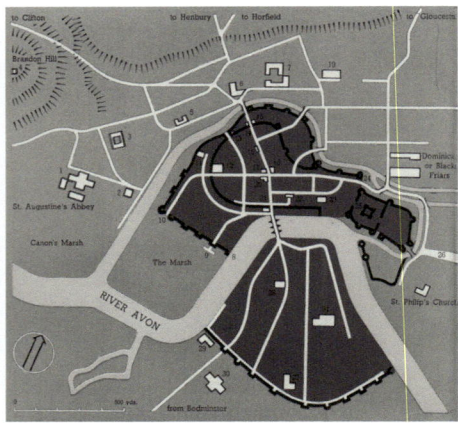

Maps showing Bristol's city walls and the changing course of the Frome c1200 and c1248 (from *English City: the Growth and the Future of Bristol* published in 1945 by J S Fry and Sons Ltd). The bridge across the River Avon is also shown.

beginning to explore the possibilities of what was then termed the New World (North America and Canada). You'll learn more about the city's international trade on other walks.

Go up the slope to the main path and turn right. You will pass on your left a bronze drinking fountain by Kate Malone decorated with images of the cod that were once caught by local fishing boats. John Cabot, the Venetian explorer, navigator and cartographer, who sailed from Bristol in 1497, hired crews who already had experience of venturing far out across the Atlantic in search of fishing grounds. His ships were fitted out by Bristol merchants. By travelling west rather than east, Cabot had hoped to find a profitable alternative route to the commercial riches of Asia, but instead he 'discovered' Newfoundland. The path you are following runs parallel to what remains of the southern wall of Bristol Castle. Continue to the Bristol Legible City Castle Park map board (2).

There was a castle at Bristol by 1088, built on the orders of Bishop Geoffrey de Montbray who had been awarded the manor of Barton Regis (which included Bristol) by William the Conqueror. It was strengthened and extended between 1122 and 1147 by the Earl of Gloucester – William's illegitimate grandson. Gloucester supported the Empress Matilda, the daughter of Henry I, in her long-running struggle for the British throne with Henry's nephew, King Stephen, and the castle was the headquarters of her campaign. When Matilda's son Henry became king on Stephen's death, he claimed the castle as his own; his son, John, later made the people of Bristol pay for its upkeep. Among its many prisoners was Princess Eleanor of Brittany who died here in 1241 having been held captive for 40 years: King John feared she or her heirs would claim the throne.

In later years the castle's military significance was greatly reduced and it fell into disrepair. It was partly restored during the Civil War (1642-51) – as the second biggest city in England at the time, the control of Bristol was a prime objective for both the Parliamentarian and Royalist forces – but was dismantled after the war, along with the city's defensive walls, on the orders of Oliver Cromwell. The land on which it once stood was used for housing. Following the destruction caused by the Bristol Blitz in the Second World War, there were ambitious redevelopment plans for the site, but it remained largely derelict until this park was opened in 1978. A few remnants of the old castle can still be seen.

Take the path that passes the bandstand to exit the park at Newgate. Cross at the pedestrian crossing on your right and go straight ahead into Merchant Street. Continue to The Merchant Tailors' Almshouse (3) on your left.

The Guild of Merchant Tailors received its charter in 1399. By 1604, the guild was using tenements on this street (then known as Marshall Street) as an almshouse for its impoverished members. The building you see before you replaced the old tenements and was completed in 1701. The guild closed in 1824 when its last surviving member died.

During the medieval period, craftspeople could only practise their trade if they were a member of a guild. It usually took seven years of apprenticeship before a person was received as a journeyman and, if he wished to set up his own business, he had to submit a test piece (or 'masterpiece') for appraisal by the guild officers' panel. Across the country, the decline of the old guild system was hastened by the Industrial Revolution, when mechanisation replaced some forms of skilled labour and mass production developed. However, Bristol was slower to industrialise than other major cities (like Birmingham and Manchester), partly due to the conservatism of the local merchants who preferred to focus on traditional trading for as long as it remained profitable.

Continue up Merchant Street. To your left is the Galleries Shopping Centre, which forms part of Bristol's Shopping Quarter, along with Broadmead and Cabot Circus. Before the destruction caused by the Second World War, Bristol's shops were still concentrated on the medieval heart of the city around Corn Street, Broad Street, Wine Street and High Street (see Walk 2). Turn left on Broadmead to reach John Wesley's

Chapel, The New Room (4) on your right.

This is the world's oldest Methodist chapel. It was completed in 1739, but was considerably extended and partially reconstructed in 1748. John Wesley, a Church of England minister, had been invited to Bristol to preach to the poor by George Whitefield. The large gatherings were initially held out of doors. Wesley organised his followers into religious 'societies' that would meet in each other's homes for Bible study and prayer. He decided an indoor meeting place was needed to avoid the disapproving attention of members of the Anglican church, including the Bishop of Bristol. A new worldwide religious movement was born; one that particularly appealed to the poor and working-class, though many were put off by its strict rules of behaviour. The New Room was also used as a dispensary and school room. A major programme of renewal was finished in 2017.

The New Room: Visitors welcome. Normal opening hours: Mon-Sat 10am-4pm. There is no admission charge. There is a retiring collection for the Friday lunchtime concerts. www.newroombristol.org.uk 0117 926 4740.

If The Arcade is open (on your right) you can take this to The Horsefair where you cross the road and turn left. Otherwise continue along Broadmead and turn right into Union Street. From the bottom of The Horsefair bear round to your right, without crossing the road, to St James Barton Roundabout. Go down the slope to the subway. Go straight across The Bearpit and exit by the opposite subway (headed Stokes Croft, Kingsdown). Turn left then right up the slope. You should now be standing with buildings to your left and the road to your right. Follow the road round to your left, under the building that was once used as offices by the now disbanded Avon County Council, and into Stokes Croft (5).

Unofficially referred to as The People's Republic of Stokes Croft, this was a once thriving inner-city suburb, which suffered from decades of neglect and poor planning decisions. In recent years, it has re-emerged as an unconventional cultural quarter noted for its street art (much of which has been authorised by the buildings' owners, if not always by Bristol City Council) and its independent shops. As is often the case, there is a fine balance in meeting the needs of all residents – new and old – including those seeking a less idiosyncratic form of regeneration, those wanting a radical alternative and those at risk of being priced out of their rented homes as property values increase.

Turn left into Cherry Lane and then right up Barton Street. Turn left into Charles Street (look out for number 4, once home to Charles Wesley, the Methodist minister and writer of hymns, and younger brother of John) *and right into Marlborough Street. Cross Marlborough Street at the pedestrian crossing and stop by the University of Bristol Dorothy Hodgkin Building (6).*

Among the Nobel Prize winners who have studied or worked at the University of Bristol is Dorothy Hodgkin, who won the Nobel Prize for Chemistry in 1964 for her work on X-rays. She was the university's chancellor from 1970 to 1988. The university was officially founded in 1909 when it received its royal charter, but it had its origins in a number of pre-existing colleges including the Merchant Venturers' Technical College dating from 1595. Winston Churchill was appointed the university's third chancellor in 1929, a position he held until his death in 1965. Winifred Lucy Shapland was appointed registrar in 1931, making her the first female registrar of a British university. Both the University of Bristol and the University of the West of England play an influential role in the development, economy and culture of the city.

Marlborough Street becomes Upper Maudlin Street as it curves to the left. Continue to The Old Building of Bristol Royal Infirmary (7) which will be on your right.

This is one of the oldest infirmaries in the country. It was founded in 1735 with the support of Paul Fisher, a wealthy merchant. Throughout the city you will see evidence of public acts of charity on the part of local business people from a time when there was no state provision for the poor, the unemployed and the sick. Our reaction to this benevolence is influenced by present-day judgments on how some of their wealth was acquired – views that were rarely shared by their contemporaries. Much of the economy of the city in the eighteenth century, for example, was dependent upon slave labour in the colonies (see Walk 3).

Opposite the Old Building is a newer building, opened in 1912. The fundraising campaign was led by Sir George White, who had been elected president of the infirmary in 1906. Bristol has been an international centre for aviation since 1910 when White's British & Colonial Aeroplane Company opened its first factory here. British & Colonial was later known as the Bristol Aeroplane Company and was a major local employer. White was also the chairman of the Bristol Tramways & Carriage Company. In 1895, under his guidance, Bristol became the first city in Britain to have electric trams that were approved by the Board of Trade. In addition he introduced Britain's first motor taxi service to Bristol.

Cross the road to turn left down Lower Maudlin Street (a steep hill), passing the University of Bristol Dental Hospital and the Bristol Eye Hospital, and turn left into Whitson Street for the West Entrance of St James Priory (8).

This former Benedictine priory is the oldest surviving building in Bristol. It was founded in 1129 by Robert Fitzroy, Earl of Gloucester. The Benedictine monks who lived here in the medieval period cared for the poor and the sick. The priory was a major landowner and had a significant influence upon the city's commercial affairs, as well as its parishioners' spiritual well-being. The week-long St James Fair was held every Whitsuntide in the churchyard and was one of the most famous in the country. The last took place in 1837, by which time the trading component had been eclipsed by entertainments deemed unsuitable by the church. The priory was dissolved during the reign of Henry VIII and many of its buildings were demolished. Only the west end survived as a smaller parish church, and this is still used for worship today.

> **St James Priory:** Normal opening hours for visitors: Mon–Fri 10am-4pm and Sunday 7.30-9.30am. There is no admission charge, but donations are welcome. There is a café on site. www.stjamesprioryproject.org.uk 0117 933 8945.

Return to Lower Maudlin Street and continue walking down the hill. For a closer view of the exterior of the church, either go up the steps to your left and along St James' Parade or follow the road round to your left and take the sloping path up through St James' Park. Then return to the bottom of Lower Maudlin Street and, using the pedestrian crossing, enter Lewins Mead. Head left on Lewins Mead towards the city centre.

If you are unable to manage steep flights of steps, continue through Centre Gate Passage and as far as the pedestrian crossing that will take you across the road to the front of Electricity House, the large Art Deco building. Otherwise, turn right up Christmas Steps. Until 1669 this was an unpaved path and until 1774 it went by various names, including Knifesmiths' Street because of the many cutlers who worked here. Turn left on Colston Street to Foster's Almshouse (9).

Here is another example of a building associated with philanthropy. The almshouse was founded in 1483 by the merchant John Foster, who served as a mayor of Bristol. With the exception of the fifteenth-century chapel dedicated to the Three Kings of Cologne (on the left), the original building was demolished and rebuilt in 1702. It was rebuilt again between 1861 and 1883, at which time the chapel was restored. In 2007 the building was sold by Bristol Charities for private development with the money raised paying for a new purpose-built property in Henbury that provides retirement, sheltered and almshouse accommodation.

Continue a little way along Colston Street and go down the steep and narrow stairway on your left. Cross the road and take the next flight of steps down (Zed Alley) to Colston Avenue. Use the pedestrian crossing to your left to cross to Electricity House, formerly used as offices and showrooms for the South Western Electricity Board and converted into apartments. Its designer, Sir Giles Gilbert Scott, also designed Battersea Power Station and the iconic British red telephone boxes. This is where you will rejoin the walk if you missed out Christmas Steps. Bear round the side of the building and into Quay Street. Look out for more examples of street art. Cross to St John's-on-the-Wall (10) on your right.

This is the only survivor of five medieval churches that were built over the gates in Bristol's defensive walls. The church was founded in the late fourteenth century by William Frampton, another merchant and Bristol mayor. His effigy lies on his monument in the chancel with a sculpture of his dog at his feet. No longer used for regular worship, the building is run by the Churches Conservation Trust. It is opened regularly by volunteer stewards throughout the year, but please call the Bristol office on 0117 929 1766 before your visit.

Go through the arch under the church and up Broad Street for a look at Edward Everard's Printing Works (11), on your left.

Bristol was once famous for its glass and soap manufacture, its pottery, its wine merchants and its brewers. Tobacco, sugar-refining and chocolate – on which much of the city's eighteenth-century prosperity was built – continued to play an important role in Bristol's economy well into the twentieth century. By the early 1900s, newer businesses in metalwork, printing and packaging, and boot and shoe manufacture had developed. Before you is the colourful façade of the printing works owned by Edward Everard, which was completed in 1901 (the works themselves were demolished in the 1970s). It is in the flamboyant Art Nouveau style. The figures depicted on the marble tiles are the printers Gutenberg (misspelt) and William Morris, the Spirit of Literature and a woman holding up a lamp and a mirror to represent Light and Truth. Significant contributors to Bristol's economy today include aerospace, the arts, media and technology companies, 'green' businesses, higher education and legal and financial services, but what many feel makes the city special is its diverse range of self-employed individuals and independent businesses working outside of the mainstream.

Go back down Broad Street and retrace your steps to the pedestrian crossing on Colston Avenue. Turn left and go down the avenue until you are opposite The Cenotaph (12).

Around the city there are many memorials – including crosses, tablets and plaques – commemorating those who were killed during the First World War, but Bristol was among the last of the major cities in Britain to build a civic monument to the dead. The Cenotaph was unveiled before a crowd of 50,000 people on 26 June 1932. In the preceding years there had been considerable debate about what form a memorial should take; whether it might be of practical benefit to the living, like a hospital, or a structure designed for remembrance ceremonies. There were also debates about where it should be located. Under pressure from the local British Legion, a fundraising campaign backed by the regional press raised sufficient money to pay for this monument. A design contest for local architects was launched in January 1931 and won by Harry Heathman and Eveline Blacker.

Among those who had protested against the war was Walter Ayles, the Independent Labour councillor for Easton and founder member of the No-Conscription Fellowship. He was imprisoned for refusing to serve in the military. He was released from prison in 1919 and elected MP for Bristol North in 1923. A blue plaque for Ayles is on his home in Ashley Down.

You will pass St Mary-on-the Quay, begun in 1839 and designed by local architect Richard Shackleton Pope, who worked with Isambard Kingdom Brunel (see Walk 3). As its name suggests, it would have originally been set close to the dockside when the waters of St Augustine's Reach (to your left) had yet to be covered (see Walk 2). As you continue down the avenue, note the empty plinth where the statue of the slave-trader Edward Colson was until it was toppled in 2020. The plaque – which contains the words 'Erected by citizens of Bristol as a memorial of one of the most virtuous and wise sons of their city AD 1895' *– remains. Colston Avenue becomes St Augustine's Parade (13), where protesters once marched in support of the Bristol Bus Boycott.*

After the Second World War thousands of people from around the Empire and Commonwealth were encouraged to come to Britain to fill the many jobs needed to rebuild the country. When the state-owned Bristol Omnibus Company refused to employ non-white drivers and conductors – with the connivance of local trade unionists – a bus boycott was organised by Paul Stephenson, Bristol's first Black social worker, with Guy Bailey (who had been rejected as a bus conductor when he applied and which sparked the boycott), Prince Brown, Audley Evans, Roy Hackett, and Owen Henry. The boycott was successful and – a month after conceding – the company hired Sikh graduate Raghbir Singh as Bristol's first bus conductor of colour. He was soon joined by four other conductors: Mohammed Raschid and Abbas Ali from Pakistan, and Norman Samuels and Norris Edwards from Jamaica. The campaign helped to end employment discrimination on racial grounds by the bus company and to pave the way for the Race Relations Act of 1965. Stephenson was made the first Black Honorary Freeman of the City of Bristol.

Among the boycott's supporters was Tony Benn, who was first elected Labour MP for Bristol South East in 1950. In November 1960 his father died and, as his heir, Benn became Viscount Stansgate. As a hereditary peer, he could no longer be an MP. A by-election was held in May 1961 and he was re-elected, even though he was officially disqualified. On 31 July 1963, the Peerage Act gave hereditary peers the right to renounce their titles. Benn was the first to do so and won a new by-election on 20 August 1963. He held the Bristol South East seat until June 1983 when the constituency was abolished. There is a bust of Benn in City Hall.

....

Go past the Hippodrome (*see Walk 4*) *and into College Green. Stop at The Lord Mayor's Chapel* (14) *which will be on your right.*

....

This chapel is all that remains of the Hospital of St Mark, which had been founded in 1220. The hospital provided food and care for the poor until the dissolution of the monasteries in 1539. It was then purchased by the Bristol Corporation (the forerunner of the city council) for the use of Queen Elizabeth Hospital School for Boys. It was later occupied by Bristol Grammar School and, in 1885, the Society of Merchant Venturers established its School and Technical College on the site. From 1687 to 1722, the chapel was used as a place of worship by Huguenots who had fled persecution in France. When they moved to a new building in Orchard Street, the corporation decided to make the chapel its official place of worship, having fallen out with Bristol Cathedral. It became known as the Mayor's Chapel.

Bristol has had an elected mayor – rather than a royally appointed one – since 1216, albeit that those citizens eligible to vote were few in number for centuries. The first mayor whose name is engraved in the wall of City Hall is Roger Cordwainer, but he is likely to have been appointed by King John. The first to be elected was probably Adam le Page in 1216. The mayor was Bristol's leader and represented the city to national government. In 1899, Queen Victoria granted Bristol special privileges and its leader was then referred to as the Lord Mayor. The Lord Mayor is elected annually by the city's councillors. The Lord Mayor's official residence is the Mansion House in Clifton Down.

Bristol's first female Lord Mayor was Florence Brown, a former tobacco stripper and shop steward at the Wills factory, who was elected in 1963.

Between 2012 and 2024, Bristol had a mayor directly elected by the people. The Lord Mayor carries out civic and ceremonial engagements and chairs council meetings, while the elected mayor headed the city's government. A referendum in 2023 voted to abolish the post of elected mayor in favour of a committee system.

> **The Lord Mayor's Chapel:** Normal open hours for visitors: Thurs-Sat 10.30am-4pm. 0117 903 1450.

....

Continue past the chapel. Cross Unity Street. Go past the pedestrian crossing for a view of the artwork by Banksy on the wall to your right (see illustration on page 71) then return to the crossing to reach City Hall (15).

....

This is Bristol's fourth Council House (renamed City Hall in 2012). The decision to replace the third one (which you will see on Walk 2) was first made in 1897. Construction began in 1935, was interrupted by the Second World War and completed in 1952. The gilded unicorns on the roof feature on the city's coat of arms, along with the castle and a sailing ship (see also Walks 7 and 8). The city motto is 'By virtue and industry'.

Bristol became an independent county in 1373. This was, in part, in return for money paid to King Edward III by local merchants and other wealthy citizens to support the war against France. From this point Bristol could have its own law courts, stronger local government and a clearly defined boundary that remained largely unchanged for 450 years. Bristol was formally granted city status in 1542 when it was made a Bishopric.

Since the 1850s the city's population has grown dramatically. This is not only because of people moving here, but also because former independent villages and suburbs like Clifton and Bedminster have been absorbed within the spreading city boundary and new housing estates have been built on the outskirts, including Hillfields and Hartcliffe. The current population is around 479,000. A recurrent topic of debate is the extent to which Bristol should work regionally with the neighbouring local authorities of Bath & North East Somerset, North Somerset and South Gloucestershire. A devolution for the West of England was agreed by three of the four councils in July 2016 and the West of England Combined Authority was created in 2017. There have been two elected mayors since then: Tim Bowles (2012-2021) and Dan Norris (from 2021).

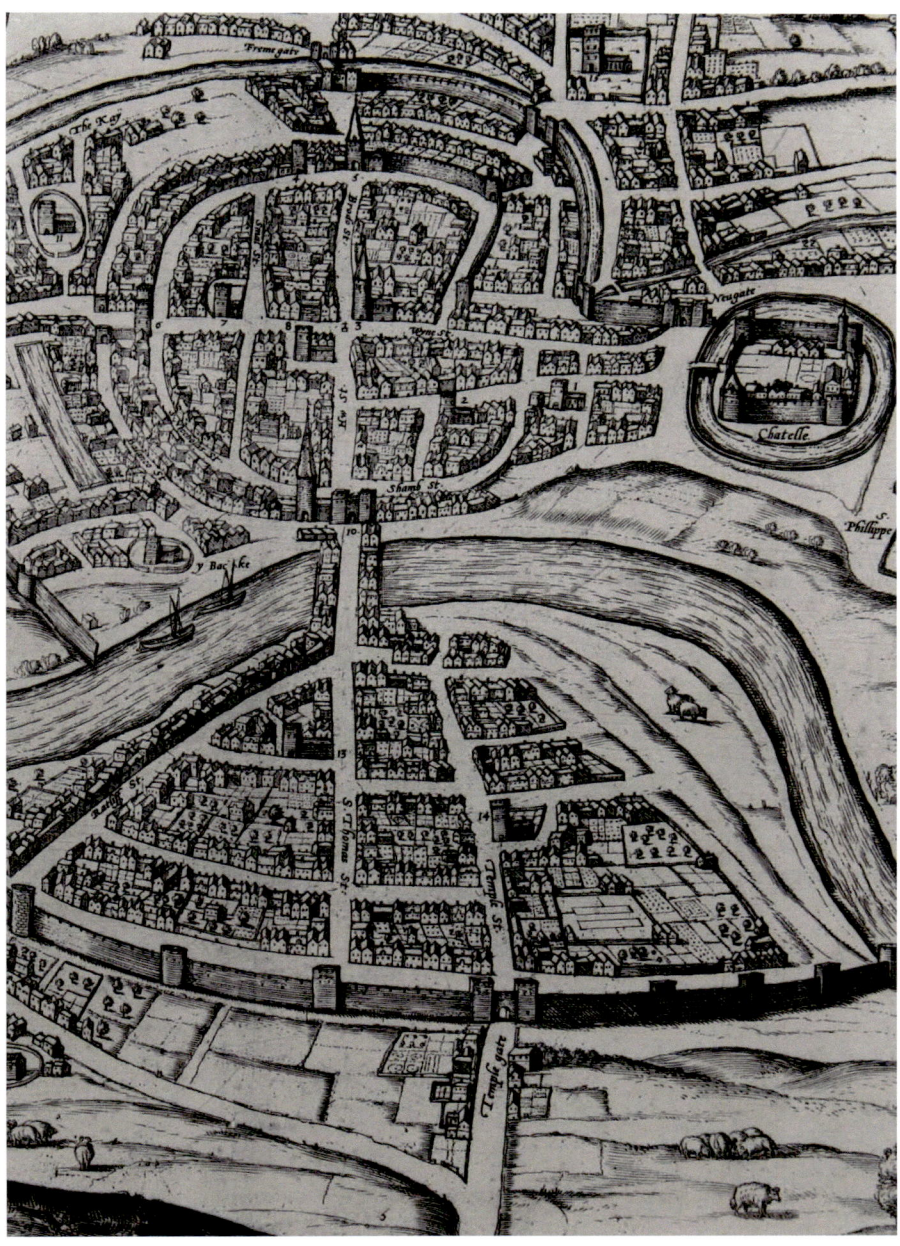

Detail from Joris Hoefnagle's plan of Brightstowe, 1581 (Bristol Culture M5277).

Walk 2:
Bristol and Romanticism

Romanticism – a period roughly bookended by the years 1780 and 1830 – marked a time of revolution; medical and scientific progress; the beginning of democratic politics; and the wide discussion of ideas.

Bristol was central to this movement. It was a city of political and religious dissent and unconventional views; it was home to newspapers, publishing companies, coffee houses, meeting rooms and lending libraries, providing fertile ground for debate; and it produced and attracted a series of uniquely talented writers and thinkers.

This route from Park Street to St Mary Redcliffe enables you to walk in the footsteps of some of the key figures of Romanticism; to learn where they lived, worked, visited, lectured and wrote poetry; and to find out more about the ideas they argued and debated. These include: the Bristol-born boy poet Thomas Chatterton (1752-1770), an icon of neglected genius and the inspiration of the Romantics who followed him; the Devon-born Samuel Taylor Coleridge (1772-1834), who fostered critical debate with his celebrated series of lectures; and Bristol-born Robert Southey (1774-1843), the radical poet and playwright who became a pillar of the establishment. Coleridge and William Wordsworth (1770-1850), from Cumberland, collaborated on the first edition of *Lyrical Ballads*, which was produced by Bristolian publisher Joseph Cottle (1770–1853) and is now considered a landmark of English Romanticism.

This walk is mainly level with fairly steep declines down Park Street, Hill Street and St George's Road and a short climb up College Street. Allow around an hour-and-a-half to complete the route and longer if you wish to include time for the many attractions and opportunities for refreshments along the way. Other points of interest, unrelated to the Romanticism theme, are also included.

The Walk

The walk begins at the top of Park Street (1) *on the right-hand side. Over half of the buildings on this street were damaged or destroyed by bombing in the Second World War but, unlike other areas of the city centre, when they were rebuilt in the 1950s their character remained much the same as before. This was thanks to the efforts of city architect Nelson Meredith.*

In 1798, Wordsworth and his sister Dorothy came to Bristol to see *Lyrical Ballads*

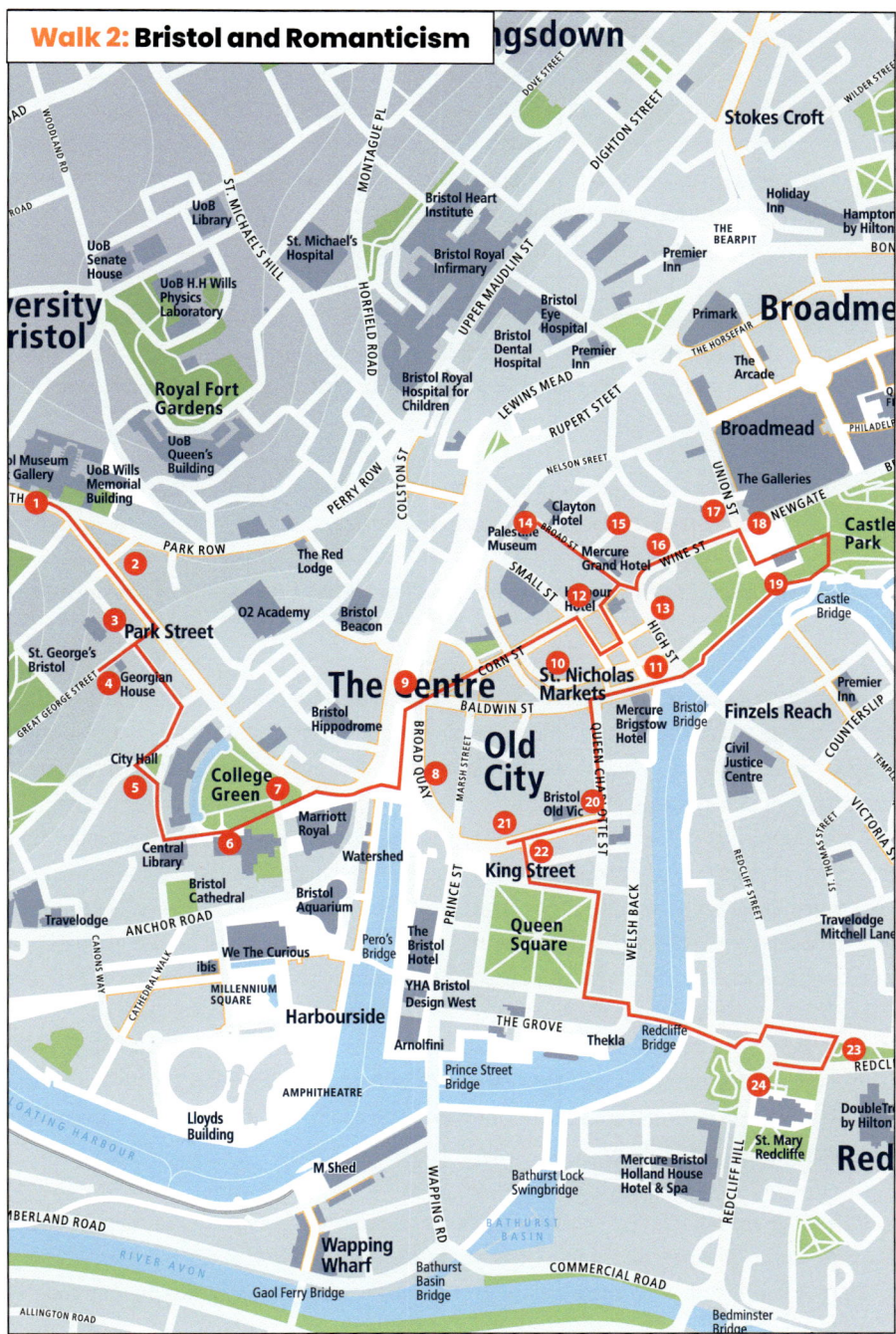

through the press. They stayed with Cottle in Wine Street (which you will see later in this walk). While they were here, they took a trip to Tintern in the Wye Valley. By the late 1700s, the abbey there had become a popular destination for tourists travelling in search of the picturesque. On 13 July 1798, as they walked down Park Street on their way to Cottle's house, Wordsworth composed the last passage of 'Tintern Abbey', a poem which encapsulates his philosophy of nature.

Wordsworth later wrote:

I began it upon leaving Tintern, after crossing the Wye, and concluded it just as I was entering Bristol in the evening, after a ramble of four or five days, with my sister. Not a line of it was altered, and not any part of it was written down till I reached Bristol.

'Tintern Abbey' was the last poem to be written for the original *Lyrical Ballads* and it was probably at Cottle's home that it reached the page.

Walk down the hill until you are opposite 60 Park Street (2) on the other side of the road.

Born in Fishponds, Hannah More (1745-1833) was one of the most influential women living in England in this period. She was a playwright and poet, but is now better known for her religious and political writing, her philanthropy, her educational campaigns on behalf of the poor and her passionate support of the abolitionist movement. In 1762, she and her sisters established an Academy for Young Ladies in specially-built premises on this site, then number 43, following the success of their previous school in Trinity Street. The school concentrated on 'French, Reading, Writing, Arithmetic, and Needlework', with each sister taking responsibility for a particular part of the curriculum. Cottle's sisters were educated here and it is likely that Sarah, Edith and Mary Fricker, the women who married, respectively, Coleridge, Southey and the Bristol poet Robert Lovell (1771-1796), were too. The Mores retired from the school in 1790. More was a patron of the poet Ann Yearsley ('the Bristol Milkwoman'), who also wrote against the trade in enslaved Africans, but Yearsley eventually found More's attentions too demanding (see Walk 5).

Continue down Park Street, pausing opposite number 52 (formerly number 47). This was the home of Mary Estlin who was secretary of the Bristol & Clifton Anti-Slavery Society. Ellen Craft, a runaway slave from America, stayed here during her tour of England when she and her husband spoke at public meetings about their experiences.

Hannah More presenting Ann Yearsley to Mrs Montague (Special Collections, University of Bristol Library Restricted HAe).

Turn right into Great George Street, noting the building on the opposite corner, the former home of New-York-born Henry Cruger, a Bristol MP, a US Senator and a Merchant Venturer. Cross Hill Street and continue to St George's Bristol (3).

St George's Church was completed in 1823 and was the city's first building in the Greek Revival style. Its architect, Robert Smirke, designed the opera house at Covent Garden and the British Museum. With the congregation dwindling, the building was rescued from redundancy in 1976 by a group of local music enthusiasts, founders of the St George's Music Trust. It is now one of the country's leading concert halls, noted for its superb acoustics (www.stgeorgesbristol.co.uk 0117 929 4929).

Cross the road to The Georgian House Museum (4).

John Pretor Pinney was a wealthy sugar merchant who owned plantations on the

Caribbean island of Nevis. He moved into the newly-completed six-storey townhouse at 7 Great George Street in 1791. William and Dorothy Wordsworth stayed here between 21 August and 26 September 1795. It was during this time that Wordsworth was introduced to Cottle, Southey and Coleridge and it is likely that some early meetings between Coleridge and Wordsworth took place at Pinney's house, though probably not the first. 'Coleridge was at Bristol part of the time I was there,' Wordsworth wrote in October 1795. 'I saw but little of him. I wished indeed to have seen more – his talent appears to me very great.'

The house's last private owner, Canon R T Cole, presented it to Bristol Corporation in 1938 to be used for the display of Georgian furniture. It has been restored and is open to visitors, showing what life was like above and below stairs in the city in the eighteenth century. Pinney's plantations were worked by enslaved people, but Coleridge, Southey and Wordsworth all wrote against the transatlantic slave trade.

The Georgian House Museum: Normal opening times: April to December: Sat-Tues 11am-4pm. No toilet or café on site. Admission is free. www.bristolmuseums.org.uk/georgian-house-museum 0117 921 1362.

Return to Hill Street. Turn right and go down to St George's Road. Turn right. Continue past the mini-roundabout to the pedestrian crossing outside Brunel House. This building was originally a hotel that was intended to form part of Isambard Kingdom Brunel's integrated passenger service between London and New York (see Walk 3). Cross the road here, keep straight ahead and continue across the second pedestrian crossing to College Street (5).

Today College Street is mainly occupied by the rear of City Hall (see Walk 1) and a car park. However, in 1795, 25 College Street was the home of Coleridge, Southey and George Burnett, the three originators of a movement they called Pantisocracy. Coleridge and Southey met in Oxford and this scheme, to emigrate to America and found a utopian commune-like society in the wilderness, developed during their long discussions. The name for the proposed community came from the Greek 'pan-socratia', meaning an all-governing society. The community was to consist of 12 men and 12 women who would support themselves by farming the land. Coleridge and Southey thought that no more than three hours of labour would be required each day, and so planned for the remaining time to be devoted to study, liberal discussions and educating their children. Members of the community were to be allowed their own opinions in matters of politics and religion, but land would be held in common, belonging to everyone.

At the end of College Street cross over Deanery Road to the Central Library, considered one of the city's finest buildings. It opened in 1906. Go through the old abbey archway on the left-hand side for a view of the rear of the building. Its architect, Charles Holden, appears to have picked up some of the new aesthetics coming from mainland Europe that had influenced Charles Rennie Mackintosh in Glasgow. Come back through the archway to see the statue of Rajah Rammohun Roy, the Indian philosopher who died during a visit to the city in 1833. He was staying at the home of Lant Carpenter and his daughter Mary, a campaigner for educational reform (see Walk 4). Continue to Bristol Cathedral (6).

In the 1840s, Cottle decided that Bristol should inaugurate a project to honour Southey, who had been Poet Laureate from 1813 until his death. He initially wanted a monument to be built but the money raised fell short and the committee that took over the management of the campaign downgraded the project to a bust. This was created by E H Baily in 1845 and is installed in the Seafarers' chapel area of the cathedral. Founded as an Augustinian abbey in 1140, the cathedral boasts some of the most important medieval architecture in the UK. Look out for the Norman stone carving in the Chapter House, the medieval stained glass preserved in the cloister, the brightly coloured Eastern Lady Chapel and the lofty arches and vaults which distinguish Bristol Cathedral as being of a medieval hall church design.

> **Bristol Cathedral:** Visitors welcome. Normal hours, excluding special services and events: Mon-Fri 8am-5pm; Sat-Sun 9 am-3pm. Admission is free. www.bristol-cathedral.co.uk 0117 926 4879.

From the cathedral, turn right and walk along the side of College Green (7).

When Coleridge arrived in Bristol in early August 1794, he came to Lovell's house on College Green in search of Southey. Lovell had been disowned by this rich Quaker family for marrying Mary Fricker earlier that year. When Coleridge reached the house, he found himself in the midst of a lively family party; Southey, Lovell, Mary and Sarah Fricker were all there.

Another of the houses on the green was home to Elizabeth Tyler, Southey's aunt and Edith Fricker's employer. Southey spent a large part of his childhood here and often stayed with his aunt when he was not at university in Oxford, so he frequently saw

Edith. Southey proposed to Edith in 1794 and the two intended to emigrate to America along with the other members of the Pantisocracy scheme. However, on 17 October 1794, all thoughts about moving to America were cast into doubt when Southey's aunt found out about the plan to emigrate, as well as Southey's secret engagement to Edith, whom she referred to as 'a mere seamstress'. She threw Southey out of her house without his coat, though it was cold and raining heavily, and told him that she wished to have nothing more to do with him or his family.

Pause outside the Bristol Marriott Royal Hotel. This is built in limestone in the Italianate style and was designed by W H Hawtin in 1864. The extension to the east of the site was built during renovations in the early 1990s. The statue of Queen Victoria in the turning circle outside is by Joseph Boehm and commemorates the Queen's Golden Jubilee. While you stand here, note the pretty Art Nouveau upper storeys on number 38 College Green, across the street. This is the former Cabot Café, which was designed by the Bristol architects LaTrobe and Weston (1904). The ground floor originally had grand Mackintosh-style doors and windows, which have been lost. Continue down the hill. Cross Canon's Road and then St Augustine's Parade to the fountains on the Centre Parade (8).

At the time of the Romantics, where you are standing now was the northern section of St Augustine's Reach, a human-made water channel dug in the thirteenth century during the diversion of the River Frome. It was built to increase the capacity of the docks, but was covered over in the 1890s when there was a need to provide more space for road traffic. The water is still there beneath your feet.

Bristol's centre was originally near Bristol Bridge, at the crossroads you will see later on this walk. When people refer to the Centre today they usually mean here, the former site of the Tramways Centre, the hub for the city's old tram routes. The area was redeveloped in the 1990s in an effort to overcome congestion problems and to provide a more clearly defined public space. Critical reaction to the scheme by some people was less than enthusiastic, but there had also been little affection for how it looked before.

Keeping the fountains to your right, walk towards the stand of trees ahead of you and the statue of Edmund Burke (9).

Edmund Burke, the Irish philosopher and politician, was the MP for Bristol between 1774 and 1780. In his speech to the electors of Bristol on 3 November 1774, Burke said:

Parliament is a deliberative assembly of one nation with one interest, that of the

Broad Quay, Bristol, attributed to Philip Van Dyke, c 1760 (Bristol Culture K514).

whole; where, not local purposes, not local prejudices, ought to guide, but the general good, resulting from the general reason of the whole. You choose a member indeed; but when you have chosen him, he is not a member of Bristol, but he is a member of Parliament.

Burke is widely remembered for his opposition to the French Revolution. Wordsworth read Burke's 1790 book *Reflections on the Revolution in France* in spring 1791, and attacked Burke in *Letter to the Bishop of Llandaff* (1793).

Walk back towards the fountains, and cross Broad Quay using the first pedestrian crossing on your left. Turn left, crossing Baldwin Street. Continue straight ahead along pedestrianised Clare Street. Continue into Corn Street, an area once noted for its eighteenth- and nineteenth-century commercial and legal offices. These were lavish buildings designed as visual statements of confidence to reassure customers. Those that survive have mostly been converted into shops and restaurants. Much of their grandeur has been lost at street level, but look up to the upper storeys to get a sense of their former opulence. On your left, you will pass The Commercial Rooms, built in 1810 to provide convivial spaces where the local bankers, lawyers and merchants could meet over coffee. Its designer was the 24-year-old Charles Busby. Its first president was John Loudon McAdam, surveyor of the Bristol Turnpike and inventor of the road construction method known as macadam. Where Corn Street is pedestrianised, continue to The Exchange (10) on your right.

This was originally a meeting place for merchants, designed by John Wood the Elder and built between 1741 and 1743. To make room, the old hall of the Coopers' Company was demolished. The Coopers were paid £900 and provided with a new site on King Street, which you will see later. Wood had transformed nearby Bath with his designs for Prior Park, Queen Square, North and South Parade and the Royal Mineral Hospital. He would later design The Circus. The brass 'nails' outside the building are historic relics of the tables on which the merchants once conducted their business (there is an information board by the door giving details). The decorative façade depicts products from the four corners of the world, illustrating the global trade in which the merchants were engaged. Look at the clock above the entrance. This was first installed in 1822 and later given two minute-hands, which can still be seen. One hand shows the old Bristol time, which, with the coming of the railway and the need to synchronise train schedules across the country, was adjusted to London time, indicated by the other hand, just over ten minutes ahead. The building was converted into a corn exchange in 1872 and now provides an entrance to St Nicholas Market. There are plenty of places to visit and get something to eat in the market and there is an outside street food market on Tuesdays and Fridays.

If The Exchange is open, enter the market and walk straight through, taking note of the courtyard roof overhead, an addition from 1870 when it was finally conceded that it might be better to conduct business undercover, paying particular attention to the decorated cornices around the interior of the market. Exit the building on the left hand side and turn left to **The Rummer Hotel (11)** on the corner of All Saints Lane. If The Exchange is closed, walk past the entrance and turn right down the lane. Note this is quite narrow, which may make it awkward for wheelchair users and those with pushchairs.

In late 1795 or early 1796, a group of friends met with Coleridge at The Rummer Tavern to persuade him to start a new radical periodical. Entitled *The Watchman*, it would contain news, parliamentary reports, original essays, poetry and reviews, and Coleridge would be its editor, publisher and chief contributor. Its motto was 'That All may know the Truth; and that the Truth may make us free'.

Having attracted 250 subscribers in Bristol alone, the first issue of *The Watchman* went out on 1 March 1796. Coleridge and Cottle spent four hours arranging, counting, packing and invoicing the copies for the 150-or-so London and provincial customers. The journal was issued every eighth day (to avoid tax) and survived until 13 May 1796, when the tenth and final issue appeared.

Return to Corn Street via All Saints Lane. The church that gives the lane its name dates back to the eleventh century. Opposite the entrance to the lane is the former West of England Bank and South Wales District Bank. This is now the Harbour Hotel (12).

This building was designed by W B Gingell and T R Lysaght and built between 1854 and 1857 in an extravagant Venetian style using Bath and Portland stone. The sculptured frieze on its façade is by John Evan Thomas who also worked on the Houses of Parliament. On the ground floor the sculptures depict the five main towns where the bank did business: Newport, Bath, Bristol, Exeter and Cardiff. On the first floor are female figures representing Peace, Plenty, Justice, Integrity and other elements considered conducive to making money in this period. The bank collapsed in 1878.

On this site once stood The Bush Tavern, Bristol's leading coaching inn. It was used by Burke for his political campaign headquarters. Before Coleridge found Southey at Lovell's house on College Green on that day in August 1794, he came here. Coleridge had just arrived in the city having been on a walking tour to Wales. Southey had come to Bristol shortly before him and was busy recruiting friends to their Pantisocracy scheme, including Lovell.

Turn right to the end of Corn Street. You are now at the crossroads that once marked the medieval city centre, where the four principal streets – Corn Street, Broad Street, Wine Street and High Street – meet. On your right, at High Street Corner, is 49 High Street (13).

In the days that followed Coleridge's arrival in Bristol, Lovell and Southey introduced him to a city strong in political radicalism. Coleridge met Cottle, whose shop stood on this site. Cottle considered Pantisocracy an 'epidemic delusion' but acted as a patron for the poets and offered Coleridge a guinea-and-a-half for every 100 lines of poetry he produced. In April 1795 he published *Poems on Various Subjects*, Coleridge's first major collection. Cottle also commissioned and printed *Lyrical Ballads*, although he disliked the idea of a joint volume and the plan of anonymous publication.

A red plaque on the building reads:

On this corner site from 1791-1798 Joseph Cottle (1770-1853) bookseller, publisher and poet. The first effective publisher of the poems of Coleridge, Southey, Lamb and Wordsworth (some of whose works were written here).

Turn left down Broad Street, noting on the corner the Old Council House – the city's third, according to records, and now the Bristol Registry Office – and continue to The Guildhall, also on the left-hand side of Broad Street near the end of the street (14).

This building was completed in 1846 and designed by local architect Richard Shackleton Pope, who is closely associated with Brunel's work in the city (see Walk 3). The sculptures of leading Bristolians on the front are by John Evan Thomas. It replaced an earlier Guildhall dating from the medieval period. Until the mid-sixteenth century a Guildhall served as the central meeting place for a city's most important guildsmen as well as its civic leaders (often one and the same).

Crop failure in 1794 and the effects of the war with France resulted in national scarcity, which, by the end of 1795, led to popular protests. In London George III's coach was attacked by crowds throwing stones and crying 'Bread! Peace! No Pitt!'

A meeting was held in Bristol at the Guildhall on 17 November 1795 to congratulate the king on his escape from the attack, but this also attracted a large number of people who were against the war. One voice repeatedly called out 'Mr Mayor! Mr Mayor!' in an attempt to be heard. That voice was Coleridge's, arguing that although the war had been costly to the rich, they still had a great deal; 'but a PENNY taken from the pocket of a poor man might deprive him of a dinner'. *The Star*, a London newspaper, published an account of the Bristol Guildhall meeting and reported Coleridge's speech as 'the most elegant, the most pathetic, and the most sublime Address that was ever heard, perhaps, within the walls of the building.'

Cross the road for a clearer view of the building then turn back up Broad Street to The Grand Hotel (15).

The White Lion Inn once occupied this site. Between 28 October and 24 November 1813, Coleridge gave a series of twice-weekly lectures on Shakespeare in the inn's Great Room. The first lecture had to be cancelled when, in the coach at Bath, Coleridge changed his mind about coming to Bristol and decided to escort a lady to North Wales instead. He turned up a couple of days later, agreed on another time, and was then 'only' an hour late for his audience. Cottle wrote that 'the lectures gave great satisfaction'.

The present-day hotel was designed by Foster & Wood and completed in 1869. It has an Italian Renaissance design reminiscent of the buildings of Venice. The ground floor, which projects out onto the street, was originally occupied by shops.

Continue up Broad Street to Christ Church (16).

This church, designed by local architect William Paty, was built in 1786, replacing the medieval church that once occupied this site. Southey later wrote, 'I was christened in that old church, & at this moment vividly remember our pew under the organ'. Southey also wrote that when he was young he enjoyed the Quarter Jacks – two figures over the entrance that strike the quarter hours: 'I have many a time stopt for a few minutes with my satchel on my back to see them strike. My father had a great love for these poor Quarter Boys who had regulated all his motions for about 20 years.' The Jacks had been carved by Paty's grandfather and were retained for the new building. The organ, reworked, was also reinstalled. The Quarter Jacks were taken down in 2013 for redecoration and found to be in such a poor state that they could not be allowed to be used outside again. The church is doing its best to get replacements made and re-sited in their original place, but this is a long process and will take years to achieve.

Turn left into Wine Street. This area suffered considerable bomb damage during the first Bristol Blitz on 24 November 1940, which led to the loss of around a quarter of the medieval city, as well as the Dutch House (a landmark five-storey timber-framed building dating from 1676 on the corner of the High Street) and St Peter's Hospital (the site of which you will visit later in this walk). Where the side wall of Christ Church abuts the end of the Prudential Buildings you will see a plaque commemorating the house where Southey was born (17).

In August 1774, Southey was born above his father's shop at 9 Wine Street, a linen draper's identified by the sign of a golden key. Southey called his place of birth 'Wine Street below-the-Pump', referencing the pump which divided the street.

In a letter in March 1804, he wrote:

when I first went to school I never thought of Wine Street & of that Pump without tears, & such a sorrow at heart – as by heaven no child of mine shall ever suffer while I am living to prevent it! & so deeply are the feelings connected with that place rooted in me, that perhaps in the hour of death they will be the last that survive.

The pump on Wine Street by Charles Bird from Picturesque Old Bristol, 1886 (Bristol Reference Library BL10F).

Cottle moved into a house on Wine Street on 7 March 1798, and moved his shop to 5 Wine Street later that month. The shop (since destroyed) was larger than his previous premises but in a less prominent position; 35 years later, Wordsworth recalled that the move had been financially disastrous. Among the many other buildings lost on Wine Street during the war was the former Corn Market. By late February 1795, Coleridge had organised a series of public lectures here. Entrance to the lectures was charged at one shilling per head and the money collected was intended to help fund Coleridge and Southey's emigration to America.

The lectures attacked Pitt's government and condemned the war against France. Coleridge dealt well with hecklers. On one occasion, some men who disliked what they heard began to hiss. Coleridge responded instantly: 'I am not at all surprised, when the red hot prejudices of aristocrats are suddenly plunged into the cool water of reason, that they should go off with a hiss!' After the second lecture it was felt necessary to move the third to a private address.

....

Continue along Wine Street to the pedestrian crossing and cross to Castle Park (18).

....

This area is also sometimes referred to as Castle Green and Coleridge gave the third lecture of his 1795 series at a house somewhere near here. Further lectures by both Southey and Coleridge were to follow; Coleridge delivered one notable speech attacking the transatlantic slave trade, and at the end of June he was to begin a series of six lectures at the Assembly Coffee House, on the quayside, comparing the English Civil War and the French Revolution. A prospectus for these lectures has survived but it is not known for certain whether he actually delivered them.

....

Behind the bomb-damaged ruins of St Peter's Church is The Castle Park Physic Garden (19), supported by Jo Malone London and St Mungo's, the national homeless charity. It opened in 2015.

....

The garden is close to the site of St Peter's Hospital, which was destroyed in the war. Sometime in 1798, Wordsworth wrote 'The Mad Mother'. It is possible that the subject of this poem is Louisa, the Maid of the Haystack, who lived for a time at the hospital. In 1776, a young, well-mannered girl entered a house at Flax Bourton asking for milk. After leaving, she wandered through the nearby fields and slept under a haystack for four nights. Local women fed her and offered her a bed in their houses but she refused them. The women then clubbed together to purchase the haystack for her. The girl was eventually taken to St Peter's Hospital, but she returned to the haystack where she lived for four more years. The locals continued to feed her and gave her the names 'Louisa'

The back of St Peter's Hospital from the Floating Harbour, 1820, Hugh O'Neill (Bristol Culture M2702).

and 'The Maid of the Haystack'. Hannah More became involved in her care in 1781 and had her taken to the Henderson Asylum at Hanham; she continued to pay for her keep there until Louisa's death in 1800.

If you can manage steps, walk through the garden, along the side of the church, turn left and then right, passing the linear ponds of Beside the Still Waters by Peter Randall-Page (1993). Continue down the steps then turn right to go down to the waterfront and right towards Bristol Bridge. If you are unable to manage steps, return to the entrance of the church and take the sloping path to your left down to the waterfront and towards the bridge. Cross the road ahead of you via the pedestrian crossing into Baldwin Street. Continue to Queen Charlotte Street to your left. Turn down here, cross Crow Lane and continue to King Street where you turn right to the Bristol Old Vic (20). (Note that road surfaces in this area are cobbled and can be uneven underfoot.)

This is the oldest continuously working theatre in the English-speaking world and celebrated its 250th birthday in 2016. It has been home to the Bristol Old Vic company since 1946. In 2012 a major refurbishment of the historic Georgian auditorium was completed. The redevelopment of the front-of-house was completed in 2018. (www.bristololdvic.org.uk 0117 987 7877).

The Coopers' Hall – which had replaced the demolished premises on Corn Street – became part of the theatre complex in the early 1970s, providing a new two-tiered foyer space. The Coopers Company, which included many local wine merchants, had long since gone into decline and its hall had been used for exhibitions, Baptist missions, warehousing and auctions since the late eighteenth century.

By 1784, the craze for balloon flights had reached England and ascents, with or without people on board, were taking place in almost every large city, including Bristol. High balloon ascents prompted advances in meteorology and drew people's attention to the formation and beauty of clouds. Poets and writers, including Coleridge and Wordsworth, saw ballooning as a symbol of hope and liberation. In January 1784, Michael Biaggini exhibited an air balloon at the Coopers' Hall for three days. He charged a 2s 6d (12.5p) entrance fee, and the balloon, around 30ft/9.14m in circumference, attracted much public interest. For an extra 2s 6d, Biaggini allowed those who were interested 'to see the method and process of filling the balloon with inflammable air'.

Continue along King Street to the building on your right, set back from the road behind a paved courtyard. Currently occupied by a restaurant, this was once Bristol Library (21).

The Bristol Library Society, founded in 1773, charged an entrance fee and an annual subscription of one guinea per member until 1798, when the fee increased to four guineas. In 1798, the library had around 200 members and held 5,000 books, as well as providing custody of 2,000 books belonging to the city. You were not allowed to become a member if you owned a lodging-house, inn, tavern, coffee house, place of public entertainment or circulating library. The library was made free to the public from 1856. Coleridge, Southey, Lovell and Cottle all valued the library and used it frequently. Southey was library member number 278 and Coleridge number 295.

Furnishings from the library, including the ornately carved over-mantle from the reading room's fireplace, can be seen in the Bristol Room in the Central Library. The building was taken over by the War Pensions Office during the First World War.

Go back to King William Avenue, on your right. Turn here, cross Little King Street and enter Queen Square. Turn left to 2 Queen Square (22).

This was once the home of Josiah Wade, a radical Bristol tradesman who became

King Street, 1825, Thomas L Rowbotham (Bristol Culture M2509).

a principal supporter of *The Watchman*, Coleridge's political journal. Coleridge stayed with Wade from late October to late November in 1813 while he was presenting a series of lectures on Shakespeare and Milton in the city. Coleridge intended to begin a further series on 7 December but, on 2 December, a physical and mental crisis, induced by opium and alcohol, overcame him.

Continue clockwise around the square, exiting at Bell Avenue to your left, the pedestrian path between numbers 24 and 26. There is an information board marking the Brunel Mile to your right. Continue straight ahead, crossing Welsh Back to the left-hand side of Redcliffe Bridge. As you cross the bridge, look to your left to the brick-faced former Western Counties Agricultural Co-operative warehouse (1909-12), a Grade-II listed building which was converted by the Bristol Churches Housing Association for social housing in 1997. Continue straight ahead when you leave the bridge. Cross Redcliff Street by the pedestrian crossing and continue along Portwall Lane, which marks the old city boundary, keeping the car park to your right. Cross Phippen Street and turn

***right to the Chatterton House** (23).*

This was constructed in 1749 as the master's house for the adjoining Pile Street School, which was founded around 1739. Chatterton was born here in 1752 and subsequently educated at the school, where his father was master. In the 1930s, when the surrounding buildings were demolished to make room for Redcliffe Way, part of the façade of the school was attached to the house. The council-owned building has been used as a museum and currently houses a restaurant.

Chatterton left Bristol for London in April 1770, allegedly disappointed with his lack of recognition at home, and died shortly afterwards of arsenic poisoning. His early tragic end – now thought to have been an accident – has led to the romantic legend of the boy genius destroyed by a philistine world, a legend enhanced by Henry Wallis' famous portrait of the penniless young man lying dead in his London garret.

Re-cross Phippen Street and continue down to Redcliffe Way. Turn right and head to the pedestrian crossing which will take you to St Mary Redcliffe (24).

Queen Elizabeth, on a visit to Bristol in 1574, is said to have declared this to be the 'fairest, goodliest and most famous parish church in England'. Parts of the structure date back to the twelfth century. The Canynges, a Bristol mercantile family, were among the most high-profile of the church's early patrons, paying for major building projects in the fourteenth and fifteenth centuries. The Canynges Society, founded in 1848 to raise funds for essential restoration work, is still active on the church's behalf, having been revived in 1927. The imposing spire, which was truncated after being struck by lightning in 1446, was rebuilt to its full height of 292ft/89m in 1872.

It was on 4 October 1795 that Reverend Benjamin Spry married Coleridge and Sarah Fricker in a quiet ceremony at St Mary Redcliffe. Their marriage was witnessed by Mrs Fricker and Josiah Wade. On 14 November 1795 Southey married Edith Fricker, with Cottle and his sister, Sarah, acting as their witnesses. Cottle also paid for the ring and marriage licence. The marriages of Coleridge to Sarah and Southey to Edith were intended as a prelude to emigration. Southey's friend George Burnett also intended to join the Pantisocracy scheme, and proposed to Martha Fricker, one of the younger Fricker siblings. Martha turned him down.

St Mary Redcliffe was where Chatterton claimed to have discovered poems written by a fifteenth-century monk named Thomas Rowley. The poems were hailed as a magnificent find and experts were unstinting in their praise. However, the Rowley poems were found to have been the work of Chatterton himself. Both Coleridge and Wordsworth wrote about Chatterton; Wordsworth in 'Resolution and Independence' and Coleridge in 'Monody on the Death of Chatterton'.

St Mary Redcliffe: Normal opening hours for visitors: Monday-Saturday 8am-5pm, Sunday 12pm-4.30pm. Sunday services are now held at 8am, 9.30am, 10.30am and 5.30pm. Visitors wishing to view the church but not attend the service are not admitted at these times. The Arc Café is located in the undercroft. www.stmaryredcliffe.co.uk 0117 231 0060.

Walk 3:
Brunel's Bristol

Isambard Kingdom Brunel (1806-1859) was one of the most versatile, audacious and inspirational engineers of the nineteenth century and Bristol is home to some of his finest work.

As well as building the Great Western Railway (GWR) and designing the Clifton Suspension Bridge, Brunel led two major shipbuilding enterprises in Bristol that transformed ocean-going travel. Less well known is that he was engaged as consulting engineer for the Bristol Docks Company, working on a number of projects, the most significant being the redesign of the South Entrance Lock and his plans for dealing with the recurrent problem of silt in the Floating Harbour. The ss Great Britain is a familiar city landmark and popular visitor destination, but there are also substantial remains of Brunel's docks work that can still be viewed today, as this walk reveals.

The circular route includes Brunel's station at Temple Meads, the ss Great Britain, Underfall Yard and the Brunel lock plus a view of the Clifton Suspension Bridge. It also provides an insight into the history of the city's docks. Allow at least two hours to complete at a leisurely pace, not including time for visiting attractions or stopping for refreshments along the way. This is mainly a level route, but extra care may be needed alongside the Floating Harbour, where surfaces can be uneven, and some of section 4 may not be suitable for wheelchair users or those with pushchairs.

The Walk

The walk is divided into four sections with optional routes suggested at various points.

Section 1

The walk begins beneath the clock tower at Temple Meads station (1), which is at the heart of the Temple Quarter Development Zone. This is the main entrance to the station.

It is a common error to believe this station was designed by Brunel. In fact, work on the station began in 1865, six years after Brunel's death, and was completed in 1878. The station originally served the Bristol & Exeter and Midland railway companies. Its architect was Brunel's colleague Sir Matthew Digby Wyatt, who had co-designed Paddington station. The arched iron roof was designed by Francis Fox, whose father,

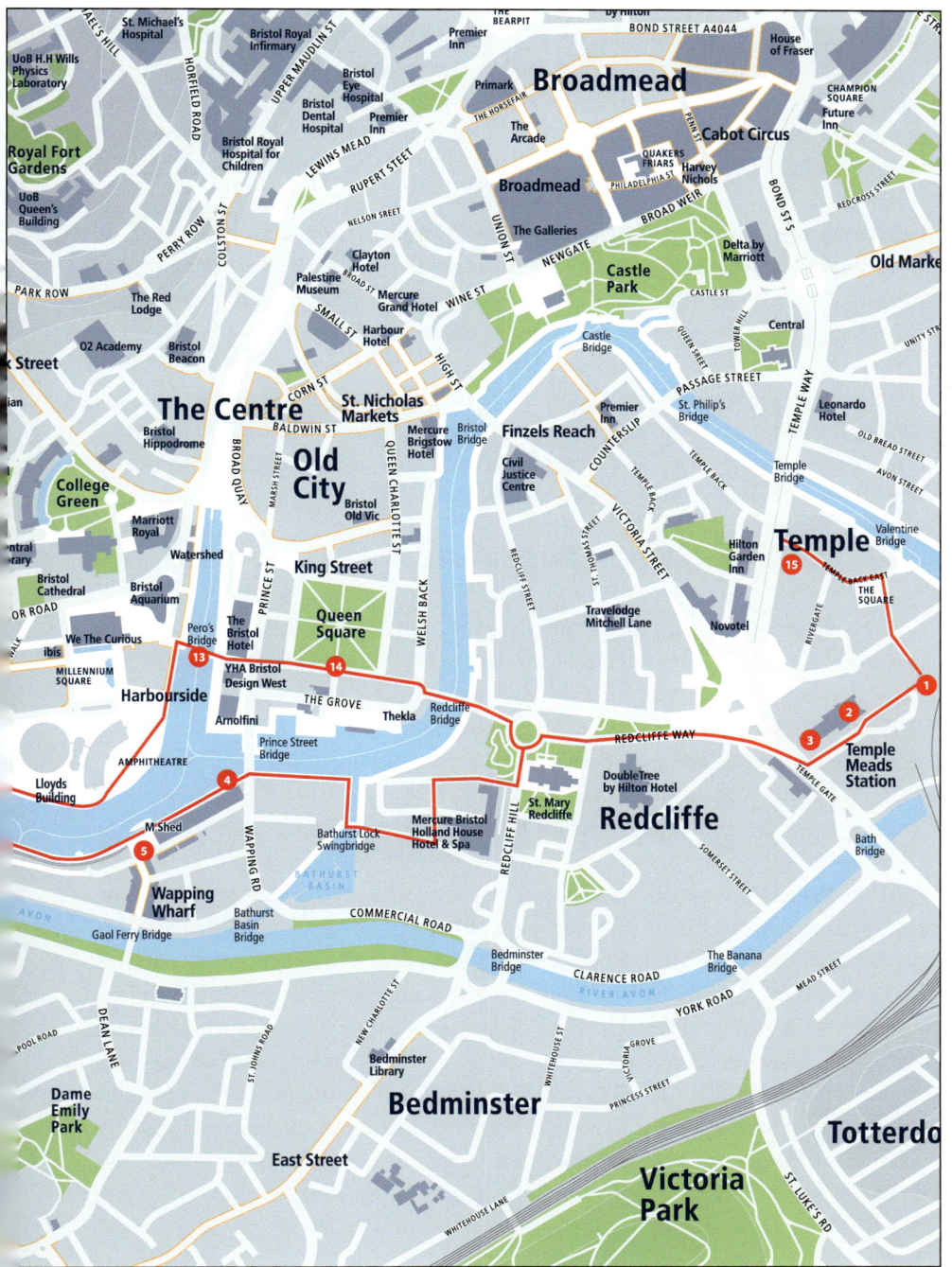

Sir Charles Fox, had constructed the roof at Paddington and was also involved in the design of the Crystal Palace in London. The station has a neo-Gothic exterior in pink stone with Bath stone dressings and a red-brick interior. A plaque at the entrance commemorates Emma Saunders (1841-1927) who founded the Bristol and West of England's Railwaymen's Institute.

Walk down the right-hand side of the station approach road. Across the road, on the left-hand side, you will pass some Jacobean-style offices built in the 1850s for the Bristol & Exeter Railway. On your right is the passenger shed (2) built by Brunel for the original GWR station, which is situated at the end of the road.

A beautiful hammer-beam roof spans the 72ft/22m of the shed, which is now used as a conference, exhibition and entertainments venue. You may be able to step inside the entrance to take a look. The beams are largely decorative as most of the roof's weight is supported by the iron columns along the aisles.

Near the entrance to the passenger shed is the statue of Brunel by John Doubleday.

Although intended as a serious tribute the figure is considered comical by many because of its height and stance. It was unveiled in May 1982 outside the head office of Bristol & West off Broad Quay, moved to Osborne Clark solicitors in 2006 – as part of the Brunel200 celebrations (www.brunel200.com) – and installed here in 2021. Take a moment to read the information boards about Brunel and the station.

Continue down to Temple Gate to see the front of Brunel's station (3).

In 1833 Brunel was appointed the chief engineer of the GWR, although he had no previous experience in railway construction. The line's promoters in Bristol were facing stiff competition from the docks at Liverpool and needed to enhance the transport and communication facilities offered by the city. Brunel became personally involved in every aspect of the enterprise. He negotiated with the clients and landowners; devised the route; secured finance; drew up the specifications for the carriages and locomotives (designed by Daniel Gooch); found radical solutions to civil engineering problems encountered along the way; and recruited, motivated and managed staff. He even designed the lamp-posts and livery. The London-Bristol section of the route was completed in 1841. Brunel had insisted on using his broad gauge (7ft/2.14m) system instead of the standard gauge (4.7ft/1.43m) used by Robert and George Stephenson. The broad gauge system was more comfortable and allowed for faster travel than the

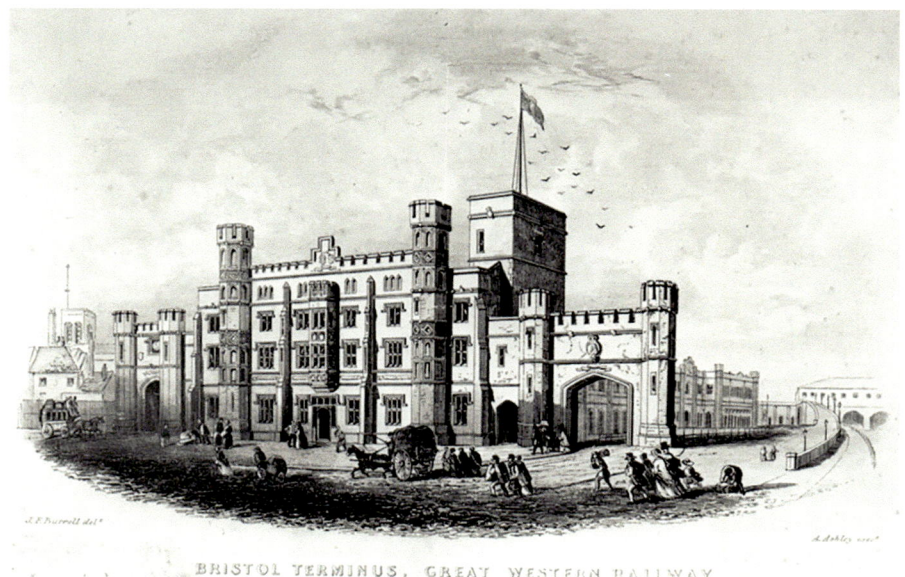

Bristol Terminus, Great Western Railway (Bristol Culture J20)

narrower gauge. However, in 1846 the government decided in favour of the standard and all new lines were built to that width (the GWR would complete its conversion to standard in 1892).

The Bristol station is thought to be the first true railway terminus, with trains and people all inhabiting the same integrated space beneath a single roof. The booking hall was at ground level and passengers reached the platforms on the first floor by climbing an internal staircase. The track was supported on brick vaults. The front of the station has a three-storey entrance in Bath stone in a hybrid revival style reminiscent of a Tudor mansion. It was designed by local architect Richard Shackleton Pope in consultation with Brunel. The right-hand wing was removed in 1878 to make room for the road. The station closed in 1965 and now houses offices and meeting rooms for a variety of organisations.

Cross Temple Gate to the Holiday Inn opposite using the pedestrian crossings. Turn right. Continue to follow the road round to your left. Cross at the bottom of Redcliff Mead Lane and continue along Redcliffe Way. On the opposite side of the road is the Chatterton House. On your left you will pass St Mary Redcliffe church (see Walk 2). Cross at the bottom of Redcliff Hill. Turn left up the hill then right into Redcliffe

ss Great Western (private collection).

Parade East.

This is the highest point on the walk so take time to look out at the view of the Floating Harbour from the car park on your right. Directly below you is Redcliffe Wharf, where the replica of John Cabot's ship, the Matthew, was built to mark the 500th anniversary of his voyage to Newfoundland. Across the water is Harbour House, a restaurant housed in what is thought to be Bristol's earliest surviving transit shed. Built around 1865, this is an iron-framed building that would originally have been open-ended to allow the swift unloading of goods headed for the warehouses or for transportation.

Turn left into Jubilee Place, which becomes Alfred Place. Turn right into Guinea Street (*a fairly steep downward hill*), **passing the former Bristol General Hospital** (*founded 1832*). **At the bottom of the hill, cross the road and then the pedestrian bridge over Bathurst Basin. Turn right along Trin Mills then left to Merchants Quay, which will bring you to M Shed** (4) **on the corner of Wapping Road.**

You are now on Prince's Wharf where Brunel's oak-hulled paddle steamer, the ss Great Western, was launched on 19 July 1837 before sailing to London for fitting out (there is a commemorative plaque on the side of the museum). The ship was built at the yard of William Patterson for the Great Western Steamship Company. On her return trip to Bristol, fire broke out in the boiler room and Brunel was injured when he fell 18ft/6m from a burning ladder. When the ship left Bristol on 8 April 1838 for her first voyage to New York, 50 of those who had purchased tickets cancelled their booking as they considered her too risky a venture.

For centuries Bristol had produced a variety of ocean-going and coastal vessels. The last ship to be built here was MV Miranda Guinness, launched in 1976. The term 'shipshape and Bristol fashion' refers to Bristol's reputation for building ships that were strong and seaworthy. It also refers to the need to stow everything well to withstand the River Avon's unusually high tidal range in which the water level can drop as much as 40ft/12.3m, leaving vessels stranded on mud banks twice a day.

Brunel had envisaged an integrated passenger service between London and New York via Bristol. On St George's Road at the back of City Hall is Brunel House (see Walk 2). Like the station, this was designed by Pope in consultation with Brunel. It was completed in 1839 and has been much altered since then, but it retains its four-storey Greek revival façade. It was originally the Royal Western Hotel and intended as a stopping off point for travellers who had come to the city by the GWR before they embarked for their transatlantic voyage. The ss Great Western did have a successful career as a transatlantic liner, making 74 crossings, but the service ran from Liverpool rather than Bristol. She was later purchased by the Royal Mail Steam Packet Company, operating out of Southampton on the Caribbean run. She was broken up at Millbank in London in 1857.

On the wharf outside M Shed are travelling electric cranes built by the Bath company Stothert & Pitt, a steam crane, a steam railway, two tugs and a fire-boat, all in working order and regularly brought to life as part of M Shed's programme of activity. M Shed was created out of a 1950s transit shed that had previously housed the city's Industrial Museum.

M Shed: Entry to the museum is free, but there is an admission charge for some special exhibitions. Normal opening hours: Tuesday to Sunday, 10am-5pm. www.bristolmuseums.org.uk/m-shed 0117 352 6600.

Continue along the Harbourside. Take special care when crossing over the tracks of the Bristol Harbour Railway. Pause between the first and second electric crane and look out across the water. To your left is the Lloyds Amphitheatre, which is used as an open-air concert venue, and to your right, Bristol Cathedral. Between the two is a view of Cabot Tower on Brandon Hill. The tower was built to commemorate the 400th

anniversary of Cabot's voyage (see Walk 5). On this walk you may see the replica of Cabot's Matthew if she is in the city (tel 0117 927 6868 for details of her schedule).
Continue past M Shed and walk as far as the Fairbairn steam crane (5).

This crane was completed in August 1878 and is capable of lifting 35 tons. Prior to its construction, there had been no crane in the docks capable of lifting more than around three tons, a serious commercial disadvantage when Bristol was hoping to attract vessels with heavy loads. As you continue on your way, you will see on either side of the harbour how the area has been redeveloped for leisure, commerce and residence since the turn of this century. From the Middle Ages to the eighteenth century, Bristol was considered the second most important port in Britain. However, the winding tidal waters of the River Avon had always been a challenge to navigate and the inconveniences of the harbour were accentuated as ships became so much larger. After a period of decline, new facilities built at Avonmouth and Portishead secured a degree of recovery. It was the strategic importance of the docks, along with the presence of the Bristol Aeroplane Company, which made Bristol one of the most heavily bombed British cities in the Second World War. Although the old City Docks were closed in 1975, the Port of Bristol continues to be a major international gateway. In 1839, Brunel had proposed a large floating pier at Portbury, near Portishead, for the transatlantic passenger service but his plans were not developed.

Prince's Wharf becomes Wapping Wharf as you pass the crane. There are information boards marking the 200th anniversary of the opening of the Floating Harbour in 1809 at various points along the waterfront, including by Brunel's Buttery. Continue to ss Great Britain (6).

Brunel's ship was launched on 19 July 1843 from Bristol's Great Western Dockyard, where she now lies. The honours were performed by Prince Albert, who had travelled to the city from Windsor by the GWR. The ship set new standards in engineering, reliability, speed and ocean-going comfort. She was the first iron-hulled, screw-propelled steamship to cross the Atlantic. The difficulties of navigating in and out of the harbour, along with the high charges incurred for using the docks (originally raised to cover the construction costs of the Floating Harbour), meant she never operated a service from Bristol. Her transatlantic first voyage was from Liverpool in July 1845. On 23 September 1846 she ran aground at Dundrum Bay in Ireland. It took nearly a year to refloat her, by which time the Great Western Steamship Company was bankrupt. She was bought by Gibbs, Bright and Company and converted to sail. Between 1852 and 1876 she made 32 voyages to Australia and is thought to have transported the forebears

Launch of the ss Great Britain, 1843, by Joseph Walter (ss Great Britain Trust).

of around 250,000 modern-day Australians.

Following storm damage off Cape Horn in 1886, she struggled to the Falkland Islands where she remained as a storage hulk. She was deliberately scuttled in Sparrow Cove in 1937 after a failed salvage attempt, but returned to Bristol on 19 July 1970 in an epic operation that entailed transporting her across the Atlantic upon a barge. She has undergone extensive conservation work. In 2005, construction of an innovative glass sea was completed at the ship's water line, which provides the roof to an airtight chamber to prevent any further corrosion of her hull. Also on the site is the Brunel Institute, which houses one of the world's finest maritime collections. It is a collaborative venture between the ss Great Britain Trust and the University of Bristol.

> **ss Great Britain:** Tickets include free unlimited return visits for a year (excluding group tickets, schools, or venue hire guests). Open every day, except 24 and 25 Dec and the second Monday in Jan. Check website for opening hours. Access to the Brunel Institute is free. The institute has separate opening times to the ship. www.ssgreatbritain.org 0117 926 0680.

Section 1 ends here. If you do not wish to continue with the next section you can either retrace your steps or use the Cross Harbour Ferry and then continue down the other side of the harbour (see section 4 for further details).

> **Number 7 Boat Trips:** Cross Harbour Ferry runs daily except 25 and 26 Dec. www.numbersevenboattrips.com 0117 929 3659.

Section 2:

From the ss Great Britain, retrace a few steps and turn right to cross the car park. Walk up Gas Ferry Road. On the corner of Caledonian Road are administrative offices belonging to the award-winning animation company Aardman Animations Ltd (7), creators of Wallace and Gromit. A more recent HQ (completed in 2009) is next door.

These offices are not open to the public. Aardman's founders, Peter Lord and David Sproxton, began their animating partnership at school. They moved to Bristol in 1976 where they produced their first professional production, creating Morph for the children's art programme *Take Hart*. They have been based on Gas Ferry Road since 1991. The site includes studios housed in a former warehouse where bananas imported from the Caribbean were once ripened. Aardman's main production facilities are at Aztec West. In 2013, more than 80 giant Gromit sculptures decorated the streets of Bristol for ten weeks in the Gromit Unleashed trail. This was one of the highest-profile charity arts trails the country has ever seen. It was followed by a Shaun the Sheep trail in 2015, Gromit Unleashed 2 (2018) and there's plans for a new Gromit trail in 2025.

Continue to the end of Gas Ferry Road. Turn right into Cumberland Road. Continue to Spike Island Artspace (8).

The area of the city called Spike Island was formed as part of the development of the Floating Harbour and the creation of the tidal New Cut, which runs parallel to Cumberland Road. It was once known for its shipyards, warehouses and busy quays. Although dock-related business still takes place, it is increasingly thought of as a place of cultural activity. Spike Island Artspace is housed in a former tea-packing factory.

> **Spike Island Artspace:** Gallery free. Normal hours: Wednesday to Sunday, café is open Tues-Sat 10am-5pm, Sun 10am-4pm. Café: weekdays, 8.30am-5pm; weekends, noon-5pm. www.spikeisland.org.uk 0117 929 2266.

Turn right up Mardyke Ferry Road and continue straight ahead. Just before you reach Cumberland Close, turn right along a footpath that will lead you to Bristol Marina.

Turn left and left again. Continue along the waterfront to Underfall Yard (9), pausing outside The Cottage Inn for a view of the terraces of Clifton and a tantalising glimpse of Brunel's bridge.

In the early nineteenth century, the engineer William Jessop was engaged by the Bristol Dock Company to create a non-tidal harbour. This was needed to combat continuing problems associated with ships being stuck in the mud at low tide, limiting the time available for loading and unloading goods at the quaysides. Jessop's solution was to contain the water in the harbour behind lock gates so ships could remain afloat at all times. The Floating Harbour was opened in 1809. Part of the project included building a dam at Underfall Yard with a weir that allowed surplus water to flow into the New Cut.

Brunel was called in at a later stage to deal with problems of silting. Among the measures he introduced were the replacement of the dam with sluices that controlled the flow of water through the Floating Harbour and allowed dredged mud to be washed away. A Brunel-designed dragboat, for scraping mud from the sides of the harbour, remained in operation until the early 1960s. An information board at the entrance to the yard provides further details and you can see where the sluices are housed just beyond this point.

Most of the buildings and engineering installations you see were constructed between 1880 and 1890 under the direction of John Ward Girdlestone. Today's yard tenants include businesses building classic boats and working on leading-edge fibre composite applications. Income from the tenants and the slipway is put back into maintaining the yard and buildings. The Harbour Master and the Docks Engineer are also based here.

The Underfall Yard Visitor Centre: Admission is free. Check www.underfallboatyard.co.uk for updates.

Take care as you go through to the other side of the yard as this is a working area. You will pass information boards along the way.

Turn right towards the visitor centre, then left, left and right to exit by the gate next to the Avon Scout County Sailing Section facility. Turn left and continue to the Nova Scotia Hotel (10). (If the yard is closed when you take this walk, you can reach this point by taking the path on the left of the entrance out to the Cumberland Road and turning right down Avon Crescent.)

You have now reached the end of section 2. You can either continue with section 3 (a circuit) or go to the start of section 4. Other options include:
- Catch the ferry service from the Nova Scotia stop to the city centre. Bristol Ferry Boats: Daily except 25 Dec. www.bristolferry.com 0117 927 3416
- Join a Bristol Packet docks tour from the Wapping Wharf stop near ss Great Britain. This runs 8.00am-5pm every day during summer and school holidays. www.bristolpacket.co.uk 0117 926 8157

Section 3

Opposite the Nova Scotia you will see a car park. Cross the road (take care as there is no pedestrian crossing). Follow the slope down to the waterside path. Pause to look across the water of the Cumberland Basin. To your right is Junction Lock Bridge; to your left is the Plimsoll Bridge. Both can be swung open to allow large vessels to pass through. Continue along the path, which will become uneven in the vicinity of the Plimsoll Bridge and may not be suitable for wheelchair users: extra care will be needed on foot. Cross the concrete footbridge on your right then go under the left-hand arch of the Plimsoll Bridge. The bridge, which was opened in 1965, is named after Samuel Plimsoll (1824-1898), a politician and campaigner for safety at sea who was born in Redcliffe. To your left is Brunel's South Entrance Lock, which is no longer in use. Continue round until you reach Brunel's wrought-iron tubular swing bridge (11), which is now set on the quayside.

This bridge was once part of the rebuilt South Entrance Lock, opened in 1849, and its innovative design later informed Brunel's triumphant Royal Albert Bridge at Saltash, which is still in use today. Brunel had previously modified Jessop's North Entrance Lock after the ss Great Britain was trapped in the Floating Harbour following her launch in 1843. She remained trapped for 18 months.

Continue to the furthest point of the island where you will be rewarded with a stunning view of the Clifton Suspension Bridge, spanning the Avon Gorge. On your right is Thomas Howard's lock, completed in 1873, which is still in operation. The North Entrance Lock was sealed when this new lock was opened. Across the water is Hotwells, a place which once attempted to rival Bath with its healing spa water facilities.

Turn round and walk back towards the South Entrance Lock, keeping to the right-hand side of the island. Information boards giving details of the swing bridge and plans for its conservation are by the railings. Cross over the static tubular bridge

Aerial view of the Cumberland Basin in the 1930s, before the current road scheme was built (Bristol Culture PBA 345).

ahead of you, a near-replica of Brunel's designed by Howard, pausing to look down to your right at the curved, masonry walls of the lock's entrance. At low-tide, when the New Cut is reduced to a muddy trickle, you will appreciate the necessity of creating the Floating Harbour.

When you reach Cumberland Road, turn left and go under the bridge arch. The flyover above you carries Brunel Way. Taking care of the traffic (there is no pedestrian crossing), cross the road and walk towards the large red-brick structure ahead of you, 'B' Bond (12).

This is a former Wills tobacco warehouse dating from 1908 and is one of the earliest large buildings to be built on a reinforced concrete frame. By the 1670s, about half of Bristol's ships were engaged in the tobacco trade and by the mid-eighteenth century a number of tobacconist shops had been set up in the city, concentrated around the Castle Street area. In 1786, Bristol tobacconist Samuel Watkins took on a new partner, Henry Overton Wills (1761-1826), who had recently arrived from his hometown of

Salisbury. This was the foundation for the Wills tobacco manufacturing company, which became one of the city's biggest employers.

Tobacco was first grown as a commercial crop in the British colonies of the Caribbean and North America in the early seventeenth century. A triangular transatlantic trade developed. On the first leg of a typical voyage, a ship would sail from Bristol to the African trading centres of the Gold Coast, Angola and the Bight of Benin, laden with manufactured goods. The goods were exchanged for enslaved people who had been captured from across West Africa in inter-tribal wars and in raids on villages. On the second leg the slaves would be transported to the colonies in appalling conditions, chained below-deck. Those who survived the journey would be sold in private sales, at auction or in a free-for-all 'scramble'. The ship would then load up with local goods (mainly sugar, but also tobacco, coffee, rum, cocoa and tropical woods) and return home on the third and final leg of the journey. The transatlantic slave trade was abolished by the British government in 1807 but the Emancipation Act, freeing slaves in the colonies, was not passed until 1834. The government paid compensation to plantation owners and mortgage holders. Most of the money was reinvested in new engineering and manufacturing ventures including canals and railways. No compensation was paid to those who had been enslaved. Brunel's opinion on slavery is unknown.

....

Make a circuit of 'B' Bond. This now houses Bristol Archives (https://www.bristolmuseums.org.uk/bristol-archives/) (www.bristolmuseums.org.uk/bristol-record-office) and the environment centre Create (www.createbristol.org). Both often host free exhibitions. Re-cross Cumberland Road, turn right and return to the Nova Scotia.

....

You have reached the end of section 3. You can now choose one of the options at the end of section 2 or continue to section 4.

Section 4

....

Cross Junction Lock Bridge, taking note of the row of dock cottages to your right. Turn right and follow the path that runs by The Pump House (originally built to house the machinery that operated the sluice gates). Your route is signposted as Harbourside Walk. Along the way you have an excellent view of ss Great Britain. Opposite the ship is a memorial to Samuel Plimsoll. The Plimsoll line on a ship's hull indicates the legal limit to which the ship may be safely loaded under various conditions.

If you need to avoid the steps at Porto Quay, turn left up Gas Ferry Road then right through the archway in the wall. Keep taking right-hand turns until you reach Hannover Quay where you turn left. This is where you will pick up the route if you have taken the Cross Harbour Ferry.

Continue along Hannover Quay and cross the Lloyds Amphitheatre. The GWR's docks service once terminated in nearby Canon's Marsh, now the site of Millennium Square (see Walk 4). Continue to Pero's Bridge (13).

This footbridge is named after Pero Jones who came to Bristol from the Caribbean as a personal slave of the merchant John Pinney (see Walk 2). The bridge was opened in 1999. It was designed by the artist Eilis O'Connell with the engineering company Arup. The central span can be raised for shipping. The horn-shaped sculptures act as counterweights for the lifting section and the bridge is sometimes referred to as The Horny Bridge.

Cross the bridge. Walk straight ahead, cross Prince Street at the pedestrian crossing and enter Queen Square (14).

The square was the focus of the Bristol Riots of 1831 in which Brunel, who was in the city to supervise work on the Clifton Suspension Bridge, served as a special constable. Business confidence in the city plummeted following the riots, which contributed to delays in the construction of Brunel's bridge (see Walk 5). The rioting was in protest at the House of Lords rejecting the second Reform Bill and also prompted by discontent at the corruption of city officials.

Keep straight ahead and, where the path forks on Bell Avenue by The Hole in the Wall, keep to the right. Cross the road at the end of The Grove and go to the right-hand side of Redcliffe Bridge. Cross the bridge, taking note of the view you now have of Redcliffe Parade, where you were earlier, and of the cliff that gives this area its name. Bear round to your right to the Quakers' Burial Ground (there is an information board by the entrance). Retrace your steps back to the station from the pedestrian crossing at the bottom of Redcliff Hill. When you reach the station, you may wish to make a short diversion through the station to Temple Back East (15) to see the former offices of the law firm Osborne Clark, facing Temple Way.

Osborne Clark has an interesting Brunel connection. One morning in 1833, Jeremiah Osborne, who was the GWR's solicitor and a founder of the firm, rowed Brunel down the Avon to survey the river banks when he was planning the route of the railway. In 2006, the statue of Brunel you encountered near the start of the walk was moved here before being installed at Temple Meads station in 2021. Osborne Clark held an exhibition looking at Brunel and the company in 2006.

Walk 4:
Art and Culture

The arts and culture are important contributors to Bristol's unique identity and way of life, as well as its economy. They are valued in their own right and for the diverse means of expression they offer to a wide range of individuals and communities. They also make the city a particularly attractive and stimulating place for visitors and for those who choose to study, work and live here.

This walk can only give you a brief glimpse of what the city has to offer culturally, but it will take you from the studios of the BBC to the site of one of Bristol's oldest fairs with a variety of performance and exhibition spaces in between, as well as some examples of interesting architecture and design. Look out for street art along the way (you'll see more on Walk 1).

Allow at least an hour to take this walk at a leisurely pace, not including stops for refreshments and visitor attractions. There are some fairly steep downward slopes in the first half, but no climbs of note. The pavement and road surfaces may be uneven around the Harbourside and on the setts in Queen Square, where extra care will be needed.

The Walk

The walk begins outside the BBC (1) *on Whiteladies Road.*

This is the BBC's regional television centre for the West of England, but it also plays a significant national and international role. National radio programmes have been produced at these studios since the 1930s. *Any Questions?*, the weekly live topical debate programme, has been made here since 1948. It was created by Frank Gillard, a former BBC war correspondent, who also set up the BBC Natural History Unit in Bristol, makers of Sir David Attenborough's ground-breaking *Life on Earth* (it is estimated that 40 percent of the world's natural-history films have links to studios in Bristol, including those of the BBC). Other Radio 4 programmes produced in Bristol include *Poetry Please*, the world's longest running poetry request show. Programmes are also made here for BBC Radio 3 and for BBC Radio Rural Affairs, which moved to Bristol in 2012. Network television programmes from Bristol include *Antiques Roadshow* and *Flog It!*.

In 2022, BBC Bristol moved to new studios in Finzels Reach in the centre of Bristol, although still retains the base at Whiteladies Road. The BBC also has studios and

Royal visit for the opening of the School of Architecture, 1921 (from the RWA Permanent Collection).

post-production facilities at Paintworks, the creative business quarter near Arnos Vale Cemetery on the Bath Road.

Bristol is world-famous for its contribution to animation, with many companies based in the city including, most significantly, the Academy Award-winning Aardman Animations (see Walk 3). Founders Peter Lord and Dave Sproxton's early work included the Gleebies, created in the 1970s for the BBC's *Vision On*, the innovative children's programme that was partly filmed in Bristol.

Other broadcasters in Bristol include Ujima Radio 98FM, a Community Interest Company supplying listeners with news, discussion and music with a particular focus on celebrating African and Caribbean cultures (www.ujimaradio.com). Production companies include Tigress, specialising in wildlife, adventure, science, features and documentary projects.

Keeping to this side of the road, walk down to the Royal West of England Academy (*RWA*) (2).

The Bristol Society of Artists held its first public exhibition at the Bristol Institution for the Advancement of Science, Literature and the Arts on Park Street in 1832 (the building is now home to the Freemasons). In 1844, the society was incorporated into the newly-founded Bristol Academy for the Promotion of Fine Arts. The local artist Ellen Sharples was an enthusiastic Academy member and gave it a substantial financial

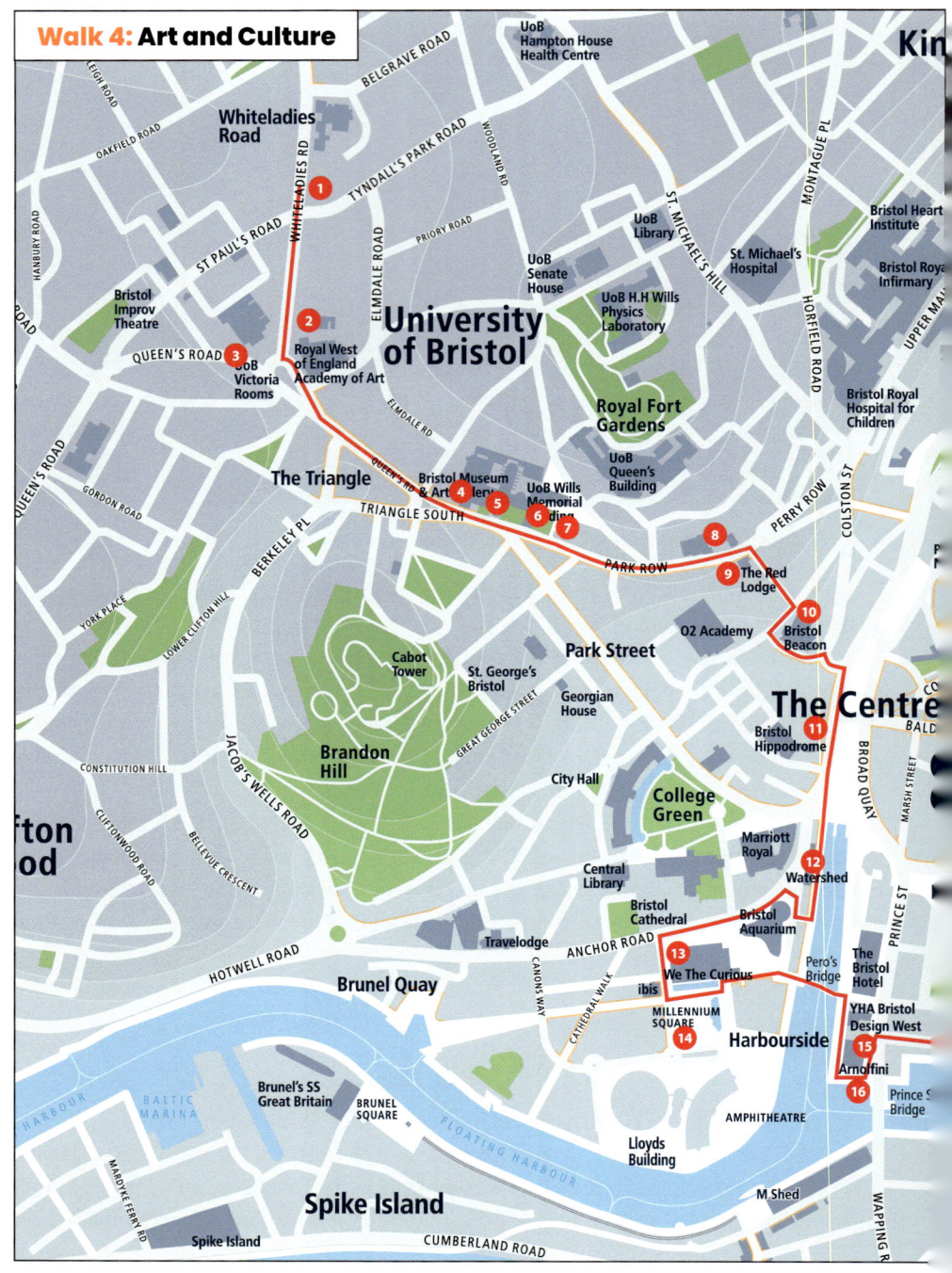

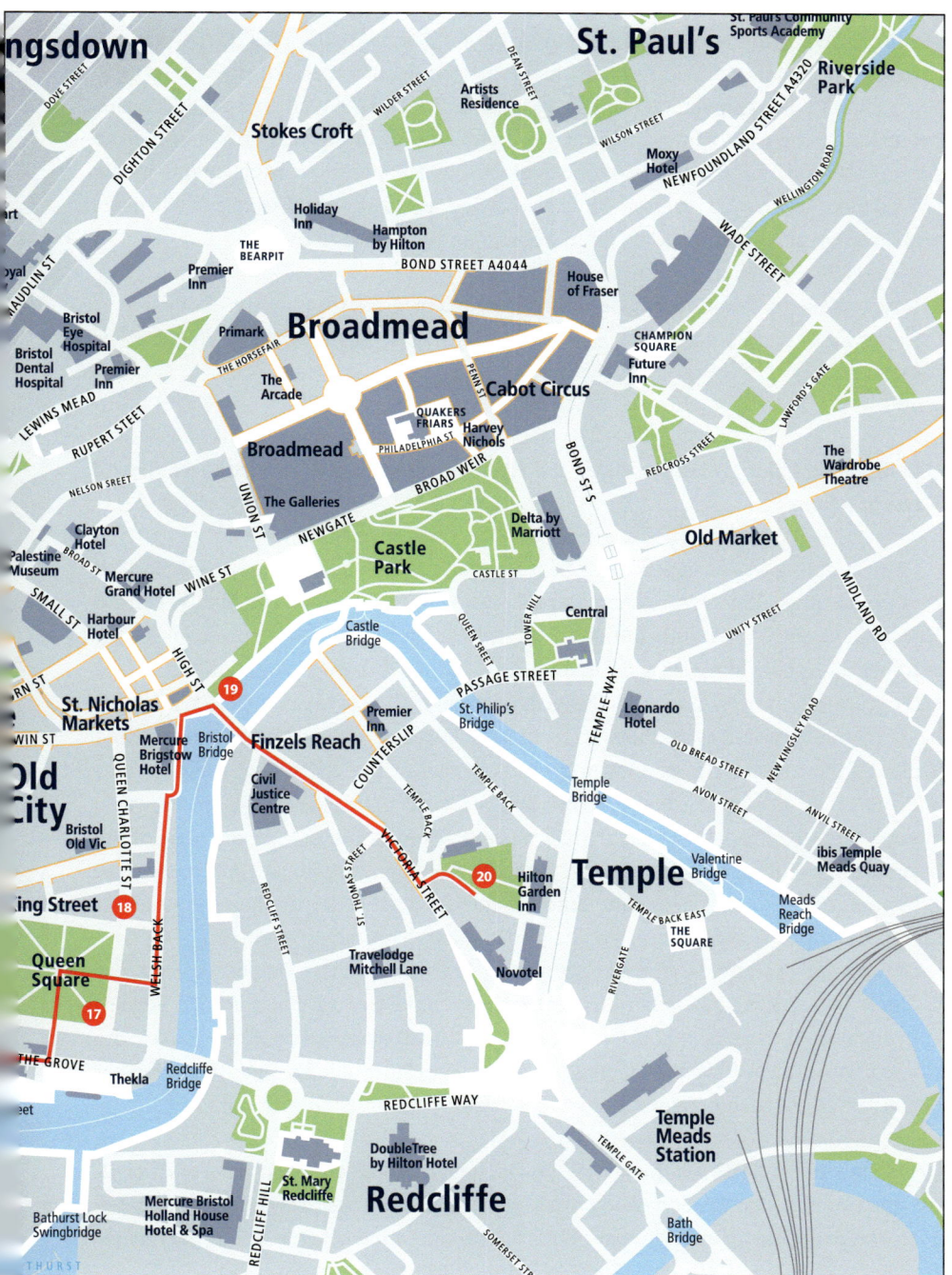

gift in 1845, with additional funding coming on her death in 1849. This money was put towards the building of a permanent home for the Academy, which opened in 1858. The gallery's patrons included Isambard Kingdom Brunel (see Walk 3). In 1913, a major extension to the front of the building was completed and King George V granted the academy its royal title. It is one of only five royal academies of art in the country. The RWA School of Architecture was officially opened in 1921 by HRH Prince of Wales. It was taken over by the University of Bristol in 1963 and closed in 1983.

> **Royal West of England Academy:** Normal opening hours: Tues-Sun 10am-5pm, closed Monday. Also open Bank Holiday Mondays. There is an admission charge for most exhibitions. www.rwa.org.uk 0117 973 5129.

Use the pedestrian crossing outside the RWA to cross to the University of Bristol Victoria Rooms (3).

This building was designed by Charles Dyer and completed in 1842. It was paid for by a group of wealthy Conservatives who considered the Assembly Rooms at Clifton (see Walk 5), their previous haunt, insufficiently exclusive. It became a place for music, readings and political meetings – among those known to have performed here in the early years were Jenny Lind (the 'Swedish Nightingale') and Charles Dickens – but it never really succeeded as a public venue. In 1920, it was purchased by Sir George Alfred Wills (of the tobacco company which was then one of the city's biggest employers) and presented to the University of Bristol for use as the Students' Union. It was a cinema from 1920 to 1931. The main hall was destroyed by fire in 1934 and little of the original interior remains. From 1964 to 1996, the building was used as a conference and exhibition centre before becoming home to the university's Department of Music. It includes a 700-seat auditorium, a recital room and two recording studios.

Above the entrance, the pediment sculpture by Jabez Tyley depicts Athena, the Goddess of Wisdom, in her chariot, accompanied by the Graces. The statue of Edward VII which stands outside marks his death in 1910. It was designed by Henry Poole, who worked with Edwin Rickards in creating the extravagant, neo-Baroque fountain that forms part of the memorial and symbolises Bristol's relationship with the sea.

Cross back to the RWA and turn right down Queen's Road. On the corner of Queen's Avenue you will pass Beacon House, which was opened by the University of Bristol in 2016 and provides quiet study space for students and a public café. It was the former Habitat store. Cross University Road and stop at Browns Restaurant (4).

This building was built to house an amalgamation of the privately-financed Library Society (see Walk 2) and the Bristol Literary and Philosophical Institution. Its design, by John Foster and Archibald Ponton, was partly inspired by the Doge's Palace in Venice and is similar to Bristol Beacon, which you will see later on this walk. It opened on 1 April 1872. In June 1893, ownership was transferred to Bristol Corporation when it became the Bristol Museum and Reference Library. The books were removed to the new Central Library in 1906 (see Walk 2), increasing the available space for the museum collection. Much of the interior was destroyed by bombing during the Bristol Blitz and the surviving collection was moved to the adjacent art gallery. Rebuilt after the war, the building was used by the University of Bristol as the Senior Common Room and later as a refectory.

Continue to Bristol Museum and Art Gallery (5).

Funding for this building came from Sir William Henry Wills, later Lord Winterstoke. The lead architect was Sir Frank Wills, who designed many of the buildings used by the family business. Among these was what is now called the Tobacco Factory in Bedminster, South Bristol, a model of urban regeneration that houses a café, living and work space, and one of the country's most respected theatre venues. The first section of the art gallery was opened on 20 February 1905. In 1925, when it became obvious that more space was required, Sir George Alfred Wills (William's nephew) paid for a substantial extension to be built at the back, with the design again handled by the family's architectural firm. It was completed in 1930.

One of the main permanent galleries is devoted to the Bristol School of Artists, an informal group that was active in the early nineteenth century and held its first group exhibition at the Bristol Institution in 1824. It includes paintings by Edward Bird, Samuel Colman, Francis Danby, Samuel Jackson, Rolinda Sharples (daughter of Ellen) and Edward Villiers Rippingille. Bristol is now famous for its street art. Its best-known – and most elusive – practitioner is the multi-talented Banksy who began as a graffiti artist with the Bristol DryBreadZ Crew in the early 1990s. He increasingly used stencils, which allowed him to work more quickly, and some of these have become familiar Bristol landmarks, including the naked man on Park Street (see page 71). In 2009, more than 300,000 people visited the free exhibition *Banksy vs The Bristol Museum*. One Banksy sculpture was left behind: the Angel Bust – or the paint-pot angel – which is currently on display.

Bristol Museum and Art Gallery: Normal opening hours: Tues-Sun 10am-5pm. General admittance is free; special temporary exhibitions are sometimes charged for. www.bristolmuseums.org.uk/bristol-museum-and-art-gallery 0117 922 3571.

Continue to the University of Bristol Wills Memorial Building (6).

This city landmark was built in honour of Sir Henry Overton Wills III and paid for by his sons. It was designed by George Oatley. Construction began in 1914 but was soon interrupted by the First World War and not completed until 1925. Among those who worked on the project was a plumber called Harry Patch, who became the longest lived British survivor of the horrific fighting that had taken place on the Western Front during the war. He was awarded an honorary degree from Bristol University in 2005 and died in 2009.

If you get the opportunity, take a look inside to see the carved stone vaulted ceiling and the double stone staircase that leads to the oak-panelled Great Hall, which is used for graduation ceremonies and a range of public events and often substitutes for parliament in films. Former students of the university associated with the arts and culture include Julia Donaldson, Sarah Kane, Matt Lucas, Chris Morris, Simon Pegg, Tim Pigott-Smith, Mark Ravenhill, David Walliams and Emily Watson.

Bristol-born Allen Lane was the founder of Penguin Books and he revolutionised book publishing in the 1930s. Between 1965 and 1969, he donated approximately 7,800 Penguins to the University of Bristol from his personal library, forming the basis of the Penguin Archive in Special Collections at the Arts and Social Science Library on Tyndall Avenue. Among the authors published in the current Penguin Classics series is the award-winning Angela Carter, who graduated from the University of Bristol in 1965. Her novels *Shadow Dance* (1966), *Several Perceptions* (1968) and *Love* (1971) are sometimes referred to as *The Bristol Trilogy*.

Continue down Park Row, past the University of Bristol Centenary Garden, and down to the figure of Nipper the Dog (7) above the doorway of the building on the corner of Woodland Road.

In the early twentieth century, the Gramophone Company Ltd began using a painting called His Master's Voice as its official trademark and in its advertising. The painting shows a dog listening to a phonograph. It was painted by Francis Barraud using his dog, Nipper, as the model. Nipper had originally been owned by Francis' brother who was the stage-set designer at the Prince's Theatre on Park Row. The advertising campaign was so successful that the Gramophone Company changed its name to His Master's Voice, later HMV.

The Prince's Theatre was destroyed by German bombing on 24 November 1940. It had opened in 1867 and initially focused on presenting serious drama, but became famous for its pantomimes; George Bernard Shaw was said to be a fan. It was originally

Building the University of Bristol by Reginald Bush, 1922 (Bristol Culture M4309).

called the New Theatre Royal – its manager, James Henry Chute, also owned the lease on the Theatre Royal on King Street, now home to Bristol Old Vic (see Walk 2), and the Theatre Royal in Bath – but its name was changed in 1884. This was partly prompted by the bad publicity that still lingered from the tragedy of Boxing Day 1869 when 14 people had been killed as the audience surged into the pit and gallery to take their places before the show's start. It had been Bristol's principal theatre before the war, but remained a bomb site until 1967 when it was sold to Western Motors. The land was later used to build accommodation.

Cross Woodland Road. Continue to the University of Bristol Theatre Collection (8).

In 1946, the University of Bristol established the UK's first university Department of Drama. The university's Theatre Collection was founded in 1951 to serve as a research resource for members of the department and the local community. It has since expanded to become a fully accredited museum of national importance. It also has a large live art archive.

> **University of Bristol Theatre Collection:** Normal opening hours: Tues-Fri 9.30am-4.45pm. No admission charge, but donations welcome. www.bristol.ac.uk/theatre-collection 0117 331 5045.

1946 was also the year in which the Bristol Old Vic Theatre School was opened by Laurence Olivier as a training school for the Bristol Old Vic Company. In 1954, Dorothy Reynolds and Julian Slade wrote the hit musical *Salad Days* for the school. With the money made from this production, the school could afford to move to bigger premises on Downside Road in Clifton. These were officially opened in 1956 by Dame Sybil Thorndike. Among those who trained at the school are Stephanie Cole, Jeremy Irons, Daniel Day-Lewis, Pete Postlethwaite, Miranda Richardson, Patrick Stewart and Mark Strong.

Continue to the pedestrian crossing that will take you across the road to The Red Lodge Museum (9).

Campaigners for educational reform to come from Bristol include Mary Carpenter, who opened a series of schools for girls and the poor in the city. Her pioneering Reformatory School for Girls was housed in one of two late-sixteenth century lodges built in the grounds of the Great House, a mansion belonging to Sir John Young (now the site of Bristol Beacon). The reformatory was closed by 1919. In 1920 the building became an annex for the city art gallery, with the support of Sir George Alfred Wills and the Bristol Savages, an artists' club formed in 1904 and still active today. The Bristol Savages is now called Bristol 1904 Arts.

> **The Red Lodge Museum:** Normal opening hours: Monday, Tuesday, Saturday and Sunday from 11am-4pm. Entry is free. www.bristolmuseums.org.uk/red-lodge-museum 0117 921 1360.

Turn right and go down Lodge Street (a steep hill). Cross Trenchard Street and turn right to the rear entrance of Bristol Beacon (10) (Continue to the corner, keeping Bristol Beacon on your left, go down Pipe Lane then turn left.

The Elizabethan Great House was purchased by the Colston Hall Company in 1861 and demolished, making room for the new concert hall which opened in 1867. It was designed by local architects John Foster (whose work you saw earlier in the walk) and Joseph Wood in a style called Bristol Byzantine (you'll see another example of this later). The interior was seriously damaged by fire in 1898 and again in 1945. Bristol Beacon was purchased by Bristol Corporation in 1919 and has been managed by Bristol Music Trust since 2011. It is currently the largest concert hall in the city.

As part of a major redevelopment scheme, Bristol Beacon's new foyer was completed in 2009. Performers at the concert held to mark the re-opening included Bristolian drum & bass star Roni Size and his band Reprazent (winners of the 1997 Mercury Prize for their debut album *New Forms*), and the award-winning Bristol-based jazz musician Andy Sheppard and his 100-strong Saxophone Massive Choir.

Colston Hall's name was controversial for many years. By 1710, Edward Colston had established Colston's School in the Great House. Colston was a Bristol-born merchant and MP who usually lived in Surrey but maintained close ties to his home city. In 1680 he became an official of the Royal African Company. At that time, the company held the British monopoly on slave trading. The monopoly was broken in 1698 following intensive lobbying by Bristol's Merchant Venturers. From then until the slave trade was abolished in 1807, up to ten percent of Bristol's trading voyages were slaving trips. The transatlantic slave trade was a systemised and brutal form of slavery on a scale not seen before or since and was based upon a form of racist ideology that championed white supremacy.

Colston was a major benefactor to the city through his donations to good causes, but the source of his wealth meant that some performers refused to appear at the concert hall (most notably Massive Attack) and aroused wider controversy in the city leading to the toppling of the Colston statue in 2020 (you can still see the empty plinth – see Walk 1 for the wording that remains on this) and changes to the names of schools and roads. In 2024, new research was published by Richard Stone (University of Bristol) which found that though the society did not own plantations or enslaved Africans, it benefitted from income from running the port of Bristol. The report also found that around half of Bristol's slaving voyages received investment or were managed by individual members of the society. These transported 250,000 enslaved Africans to North America or the Caribbean with an estimated 44,000 of them dying during the journey. On publication of this research, The Society of Merchant Venturers commented: 'The modern Society of Merchant Venturers is deeply sorry for the historic part this organisation played in the unimaginable suffering that resulted from this abhorrent trade in human lives'. (Find out more about the transatlantic slave trade in Walk 3.)

Next to Bristol Beacon was Lesser Colston Hall, which opened as the Little Theatre in 1923. It was the home of the Rapier Players from 1935 to 1963 and was then used by the Bristol Old Vic. The building was converted into Colston Hall's bar in 1987. It is now The Lantern Hall.

Go down Colston Street to St Augustine's Parade where you turn right. Stop at Bristol Hippodrome (11) (*www.atgtickets.com/venues/bristol-hippodrome*).

West End shows that have been premiered here include *Guys and Dolls* (1953) with Sam Levene and Stubby Kaye, *The Music Man* (1961) with Van Johnson and the Disney-Cameron Mackintosh production of *Mary Poppins* (2004). The theatre opened in 1912. Its owner was Oswald Stoll and it was designed by Frank Matcham, the most eminent theatre architect of the day. A fire in 1948 engulfed much of the backstage area, but fortunately the damage to the auditorium was mainly limited to that caused by the smoke and water. There was once another Hippodrome in the city. This was the Bedminster Hippodrome, which opened in 1911. The ornate glass 'Hippodrome' sign outside the current Hippodrome was originally at the Bedminster Hippodrome. It presented music hall acts and other live entertainment, but its owner, Walter de Frece, was repeatedly refused a drama licence. In 1914 he sold the theatre to Stoll who reopened it in 1915 as a cinema. The People's Palace in Baldwin Street, which had opened in 1892, had been converted from a music hall to a cinema in 1912. Many of Bristol's grand picture palaces of the past have been lost, but in 2016 the Whiteladies Picture House, a Grade-II listed building dating from 1921, re-opened after a 15-year break. Since then Bristol has lost more cinemas, including two multiplexes.

Use the pedestrian crossing to cross to the Centre Parade then turn right to

In the Gallery by Alexander J Heaney, c1928 (Bristol Culture Mb408).

Harbourside. Continue to Watershed (12) (www.watershed.co.uk), which is open seven days a week until late in the evening.

Watershed opened its doors in 1982 and declared itself to be 'Britain's First Media Centre'. It is the leading film culture and digital media centre in the South West. It advances education, skills, appreciation and understanding of the arts, with a particular focus on film, media and digital technologies. One of the events that takes place here is Encounters Short Film and Animation Festival, which promotes the short film as a way of developing the next generation of film-makers and animators, and is one of the world's best known and most respected showcases for emerging talent (the festival paused in 2024 but aims to be back in 2025). Watershed is also home to The Pervasive Media Studio, a creative technologies collaboration between Watershed, the University of the West of England and the University of Bristol. It is a multi-disciplinary lab where artists, creative companies, technologists and academics work on commercial and cultural projects.

Another important organisation in Bristol supporting digital technologies is Knowle West Media Centre in South Bristol, which was founded in 1996 and provides a range of ways for people to get involved in community activism, education, employment and local decision-making through the arts.

....

Turn right at the side of Watershed (extra care may be needed) then left into Canon's Road, which becomes Anchor Road. The red-brick building with the high curved wall to your left is the Bristol Aquarium and the former IMAX cinema. Look at the series of blue plaques commemorating some local engineering achievements along its length. Continue to We The Curious, the arts and science centre which was the first major project of Bristol Ideas (13).

....

> **We The Curious: Normal opening hours:** Tues-Sun 10am-5pm during term time; Bank Holidays and Bristol school holidays open every day 10am-6pm. There is an admission charge. www.wethecurious.org. 0117 915 100.

This building was developed from a 1906 railway goods shed and is filled with interactive exhibits and activities. It is a National Lottery Millennium project that opened in 2000 as part of the regeneration of Bristol Harbourside. It had two components: a science centre and natural history media attraction. Wildwalk and the linked IMAX cinema closed in 2007, although the IMAX cinema now shows occasional films and hosts festivals. The science centre was renamed We The Curious in 2017.

We The Curious was founded by Bristol Ideas in 1993. It was one of the venues used by the Bristol Festival of Ideas (www.ideasfestival.co.uk), which was run by Bristol Ideas. The festival aimed to stimulate people's minds with an inspiring programme of discussion and debate throughout the year. Speakers included scientists, artists, politicians, journalists, historians, novelists and commentators covering a wide range of topics. The festival took place between 2005 and 2024.

....

Walk past We The Curious and continue along Anchor Road. On your left you will pass a sculptural tribute to the Bristol-born physicist Paul Dirac, **Small Worlds** *(2000) by Simon Thomas, which was sponsored by the Bristol-based Institute of Physics Publishing. Turn left into Millennium Square. You will pass on your left the We The Curious planetarium. Stop at the statue of Cary Grant (14) by the Millennium Square community garden. It was unveiled in 2001 and is by Graham Ibbeson. Other statues nearby include one of the boy poet Thomas Chatterton (see Walk 2) by Lawrence Holofcener (2000).*

....

Cary Grant, the epitome of old-style Hollywood charm and sophistication, was born Archibald Leach in Horfield, Bristol, in 1904. While still at school he became an assistant at the Bristol Hippodrome and at the age of 14 he joined Bob Pender's Knockabout Comedians as an acrobat. He travelled with the troupe to America in 1920 and decided to stay. He appeared in vaudeville and Broadway plays and musicals, and was signed by Paramount Pictures in 1931. When he was ten, his father told him that his mother had gone away on holiday and had died. In fact she had been put into a mental institution, something he did not discover until he was in his 30s, after which he made regular trips back to Bristol to visit her. Rubble from buildings destroyed during the Bristol Blitz was used as ballast in American cargo ships during the Second World War. It was later incorporated into the foundations of East River Drive in Manhattan; Grant unveiled a plaque in New York's Bristol Basin commemorating this in 1974. The first Cary Grant Comes Home festival was held in Bristol in 2014 and continues every two years. (www.carycomeshome.co.uk).

From Cary Grant, walk straight across the square, past the big screen on the side of We The Curious and through the Aquarena (2000), a water sculpture by William Pye. Continue across Anchor Square. Keep walking straight to cross Pero's Bridge (see Walk 3) and turn right to the Architecture Centre and The Architect café/bar (15) (again, extra care may be needed because of the uneven surface along Narrow Quay).

The Architecture Centre is an independent, not-for-profit organisation that champions better buildings and places for everyone. It seeks to inform and inspire people about the possibilities of good design and encourage everyone to get involved. It was opened in 1996 and was the first purpose-built architecture centre in the country. The Architecture Centre is now closed to the public but its café/ bar called The Architect is open and a handy place to stop for refreshments on this walk.

Continue to Arnolfini (16).

Arnolfini was founded in 1961 and moved to its current location – a former tea warehouse dating from the 1830s – in 1975. It is one of Europe's most important centres for the contemporary arts. Arnolfini's resources and facilities are shared with the University of the West of England, which also collaborates on programming and education workshops, seminars and events, as well as academic research.

Arnolfini: Normal opening hours: Tue-Sun and Bank Holiday Mondays 11am-6pm. Entrance to the galleries and building is free. www.arnolfini.org.uk 0117 917 2300.

In the early 1990s, the award-winning British choreographer Matthew Bourne was a regular artist-in-residence at Arnolfini with his ground-breaking company Adventures in Motion Pictures. He developed and premiered several dance performances here including *Highland Fling*, an alcohol-soaked update of *La Sylphide*, which premiered in the spring of 1994. Bourne later choreographed the famous all-male version of *Swan Lake*.

....

Continue to the end of St Augustine's Reach to view the statue of John Cabot by Stephen Joyce (1985). Turn left and left again and then use the pedestrian crossing to cross into The Grove (another street where the pavements may be uneven in places). **Go past Mud Dock and cross at the pedestrian crossing into Grove Avenue. Continue to the centre of Queen Square to view the statue of William III (17).**

....

This Grade-l listed statue by the immigrant Flemish sculptor John Michael Rysbrack dates from 1736 and is considered an outstanding example of the artist's work. The king is depicted on horseback as a triumphant Roman emperor. During the Second World War it was moved to Queen Mary's temporary home at Badminton for safe-keeping. The statue is controversial to some given William III's connections to the trade in enslaved Africans and there have been calls for its removal.

Queen Square is regularly used for events. In 2003, Massive Attack – who refused to appear at Colston Hall – played a concert to an audience of 20,000 in an evening marking Bristol's bid for European Capital of Culture 2008 (the bid was ultimately unsuccessful, though Bristol was shortlisted). Among the support acts were Goldfrapp, featuring Bristol-born musician and composer Will Gregory. Massive Attack was formed by Robert Del Naja, Grantley Marshall and Andrew Vowles in the late 1980s. They had all been members of The Wild Bunch, a group of Bristol DJs, musicians and sound engineers based in St Pauls.

Other outdoor venues for concerts and large-scale entertainments in the city include Ashton Court Estate, which is used for the annual Bristol International Balloon Fiesta. A 20,000 capacity arena – originally planned for the city centre – is being developed on the site of the former Filton Airfield's Brabazon hangar.

....

Take the path to your right and exit the square via Mill Avenue to Welsh Back. Turn left and continue to The Granary (18) on the corner of Little King Street, which is currently a restaurant downstairs with apartments above.

....

The style of architecture sometimes described as Bristol Byzantine emerged in the 1850s and was mainly used for industrial buildings such as warehouses and factories.

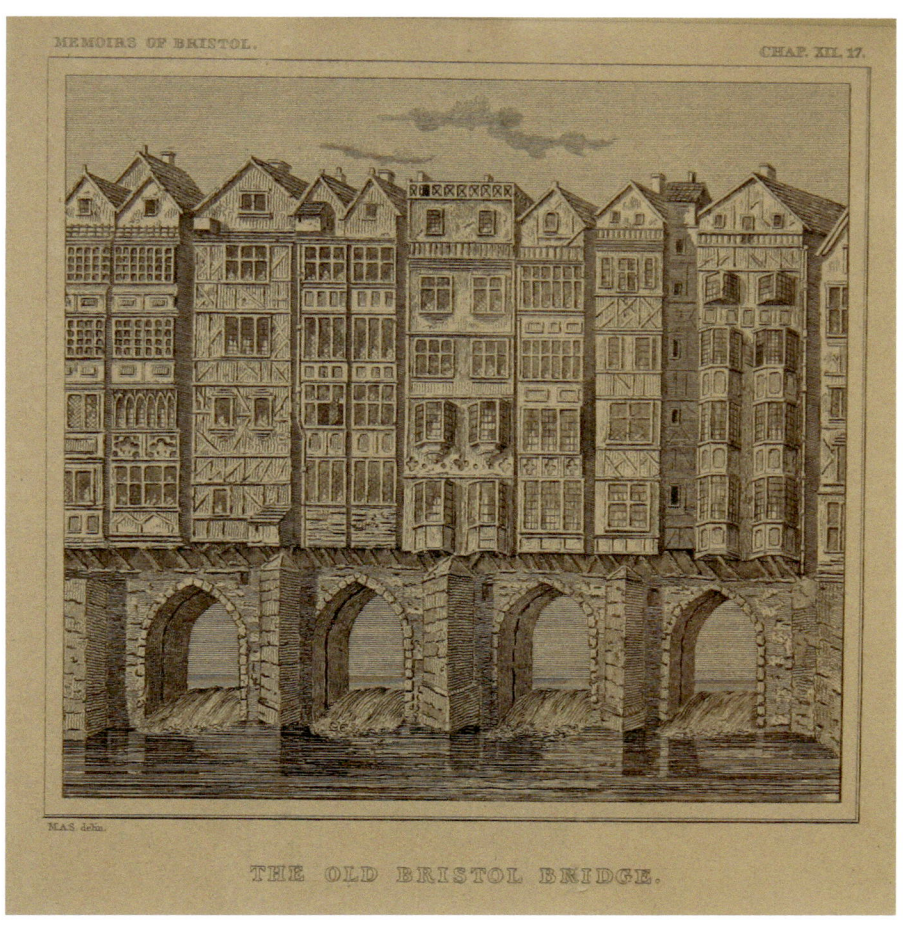

The old Bristol Bridge (Bristol Culture K4785).

This is considered the best surviving example. It was built in 1869 to the designs of the local architectural firm Ponton and Gough for the company Wait, James and Co (the dynamic Wait became Mayor of Bristol that same year). The machine-made red and buff bricks came from the Cattybrook brickworks in Almondsbury. It is a beautiful building, but also works with machine efficiency; the open brick-grilles were required to dry the grain stored here, while the multiple arches transferred the load from the floors above into the piers positioned between them. The port-holes on the ground floor were originally needed for the chutes that brought the sacks of grain down to the transport wagons. It is a reminder that this area was once a busy working dockside. Walk 6 has more details of the music history of The Granary.

Continue along Welsh Back. When you cross the end of King Street take a look up to see The Old Duke, a locally famous jazz club, on the right. Opposite is The Llandoger Trow, long rumoured to be the inspiration for the Admiral Benbow tavern in Robert Louis Stevenson's **Treasure Island** (*a novel which includes scenes set in Bristol, though there is no record of Stevenson ever visiting the city*) *and also where Daniel Defoe may have met the marooned sailor Alexander Selkirk, the inspiration for Robinson Crusoe. Continue past the Merchant Navy Memorial* (*on your right*) *to Baldwin Street. Turn right then right again so you are on Bristol Bridge* (19).

At one time the bridge (see page 69) that crossed the River Avon at this point was an impressive structure lined with shops and houses that fetched some of the highest rents in the town. It was constructed in 1248 and dismantled in 1761.

Cross the bridge then use the pedestrian crossing to the other side of the road and turn right. Notice the curved building on the corner of Bath Street, the former Talbot Hotel (*c1873*)*, which has an attractive arched entrance and multi-coloured brickwork. Cross Counterslip and continue down Victoria Street. Go past Temple Street and turn left into Church Lane to Temple Gardens and Temple Church* (20) (*also known as Holy Cross Church*)*.*

The area of Temple was given to the Knights Templar in 1145 by the Earl of Gloucester (see Walk 1) and is considered to be Bristol's first suburb. The Knights were soldier-monks who guarded pilgrims travelling to the Holy Land. Their order was abolished in 1307 and Temple was awarded to the Knights of St John. It was the site of one of Bristol's great fairs (another was St James, see Walk 1). Alongside the traders, entertainers of all kinds – jugglers, minstrels, tumblers, bear-keepers, strolling players – would flock to the fair to play to the vast crowds that gathered there. It was a lively occasion for business and pleasure.

Temple was once the centre of Bristol's weaving trade and the guild had its chapel in Temple Church. The church was severely damaged during the Bristol Blitz and its bombed-out shell is now a listed monument, owned by English Heritage. The tower is not leaning because of the bombing: it had already started to tilt as the result of subsidence when it was being rebuilt by the Knights of St John in the fourteenth century. In 2015 it provided the setting for Sanctum, one of six Arts Council England Exceptional Fund projects that formed part of the programme when Bristol was European Green Capital. Bristol-based art producers Situations invited Theaster Gates, one of the most sought-after American artists of his generation, to produce his first UK public project in the

Banksy on Park Street, viewed on this walk and Walk 1 (Visit England).

city. He chose the ruined Temple Church for the creation of an innovative temporary performance space. Performances ran here continuously for 552 hours and around 1,000 artists (many of them local) took part. The schedule was developed by MAYK, who produce MayFest, Bristol's unique festival of contemporary theatre.

Walk 5:
Nature in the City

The beautiful green spaces within and surrounding the city are among Bristol's greatest assets. In addition to The Downs and Brandon Hill, which mark the start and end of this walk, there is Castle Park (see Walk 1) and Queen Square, visited on three of the other walks. Blaise Castle Estate, Ashton Court Estate and Arnos Vale Cemetery are a short bus-ride from the city centre. A little further out is the countryside of the South Cotswolds and rural Somerset.

Avon Wildlife Trust manages and cares for 3,000 acres of nature reserves in Bristol and the surrounding area, most of which are open to the public for free all year round (www.avonwildlifetrust.org.uk). The Trust is a member of the Bristol Natural History Consortium, which manages the annual Festival of Nature (www.bnhc.org.uk). The festival gives people of all ages the opportunity to explore, enjoy and get close to the natural world. Another consortium member is Wildscreen, a charity that uses the world's best wildlife photographs and films to promote a greater understanding of the natural world (www.wildscreen.org). The Wildscreen Festival is the world's most influential and prestigious wildlife and environmental film-making event. Bristol is also home to a range of organisations and companies exploring green initiatives and championing the natural environment, including Sustrans, the Soil Association, the Environment Agency and the BBC Natural History Unit.

Allow at least two hours for this walk. The most convenient refreshment stops are in Clifton Village. There's also a cafe at the Observatory, which has stunning views from the tower. This is potentially the most physically challenging of the walks as it entails some steep climbs and descents. The pathways and pavements are narrow and uneven in places. Steps are avoided wherever possible, but this is not possible at the point at which the route enters Brandon Hill (there are level entrances to the park elsewhere).

The Walk

The walk begins at the Merchants' Hall (1) on The Promenade on Clifton Down Road.

References to the Guild of Merchants in Bristol date back to the thirteenth century. The Guild was granted a royal charter by Edward VI in 1552, which gave its members a monopoly of seaborne trade from and to Bristol. In 2007, the Merchant Venturers joined with the Lord Mayor of Bristol and other civic representatives in signing a statement regretting Bristol's role in the transatlantic slave trade (their involvement in the trade, which was abolished in 1807, is referred to in previous walks and new research is noted in Walk 4). Today, the Merchant Venturers' main objectives are: to contribute to the prosperity and well-being of the greater Bristol area; to enhance the quality of life for all, particularly for the young, aged and disadvantaged; and to promote learning and the acquisition of skills by supporting education. They are also stewards of various buildings, open spaces and charitable trusts. Membership is by invitation.

The manor of Clifton – then known as Clistone – is mentioned in the Domesday book (1085). It was purchased by the Merchant Venturers in 1676. It remained a small hamlet of scattered farms and houses until the 1700s when it began to be developed to provide an escape from Bristol's inner-city congestion and pollution. Wealthy city-dwellers were attracted by its clean air and the views across the Avon Gorge. The population rose from around 450 in the early eighteenth century to nearly 4,500 in the 1801 census. Visitors were also attracted by the warm springs here and at nearby Hotwells, coming in the summer season before moving on to the more fashionable Bath spas in the winter. Clifton was incorporated into the city of Bristol in 1835.

Find a safe place to cross Clifton Down Road then turn left, keeping to the footpath that runs parallel to the road. As you head towards Clifton Village, on your left you'll have the grand mansions of The Promenade; on your right is Clifton Down (2).

In the mid-nineteenth century, the Society of Merchant Venturers – the owners of Clifton Down – joined forces with Bristol Corporation to promote The Clifton and Durdham Downs (Bristol) Act, which was passed in 1861. The corporation was given permission to purchase Durdham Down (to the north-east) and the combined Downs were henceforth preserved as a whole 'for ever hereafter, open and unenclosed' for public use. This meant the spread of the encroaching suburbs was successfully curtailed in this part of the city and Bristol's 'green lung' was safe from development. The Downs cover 422 acres and are a Site of Nature Conservation Interest. The University of Bristol maintains the ancient right for the land to be used for public grazing by regularly grazing sheep here. The management of the Downs continues to be a joint venture between the Merchant Venturers and Bristol City Council.

Continue along Clifton Down Road. Cross Observatory Road and continue on the path that is closest to the road. You'll see ahead the spire of Christ Church. On your right, on Clifton Green, you will pass an obelisk dedicated to the politician William Pitt and a sarcophagus, which serves as a memorial to the men of the regiment of General Sir William Draper who were killed during the capture of Manilla in 1763. There's a good

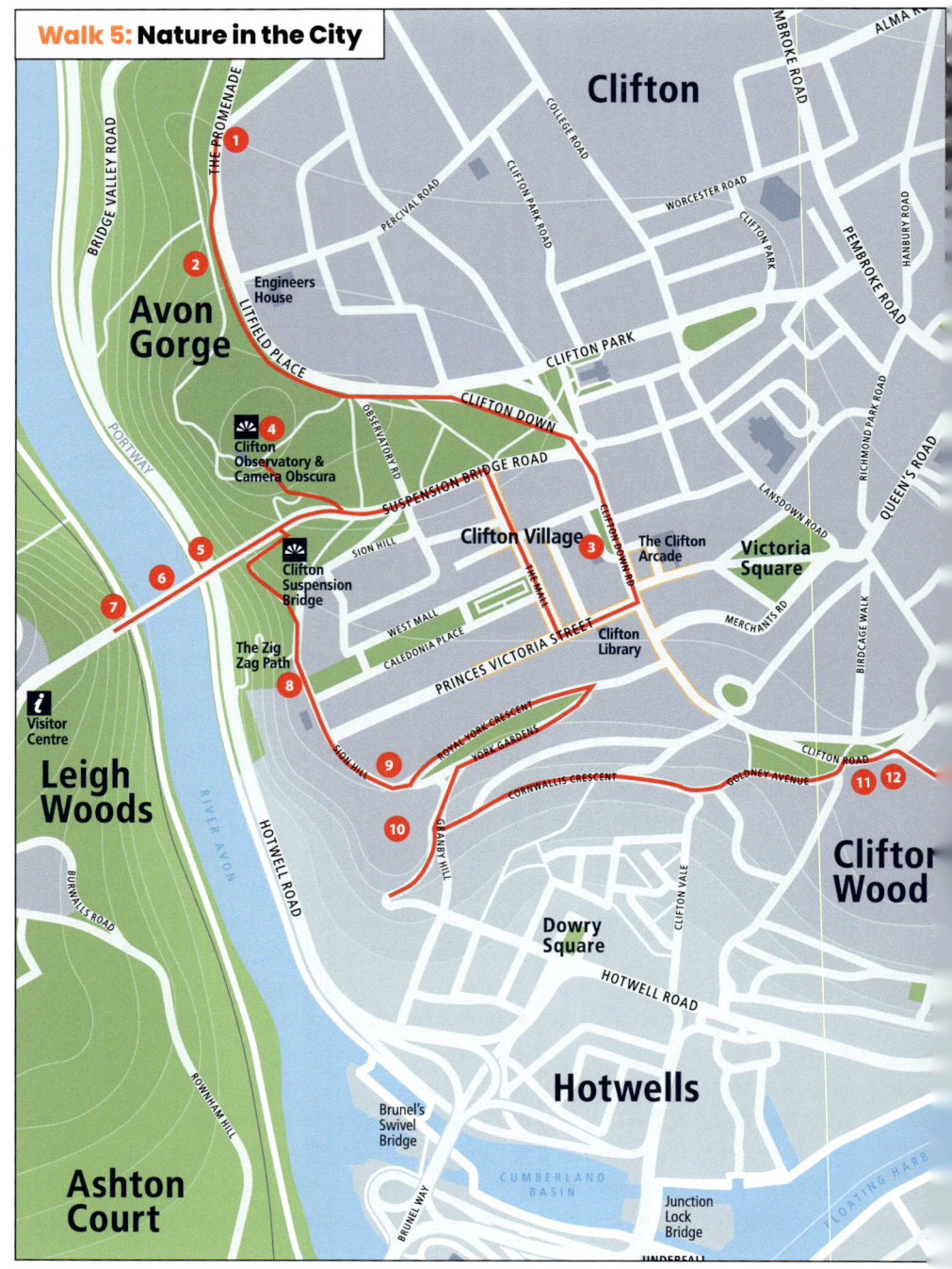

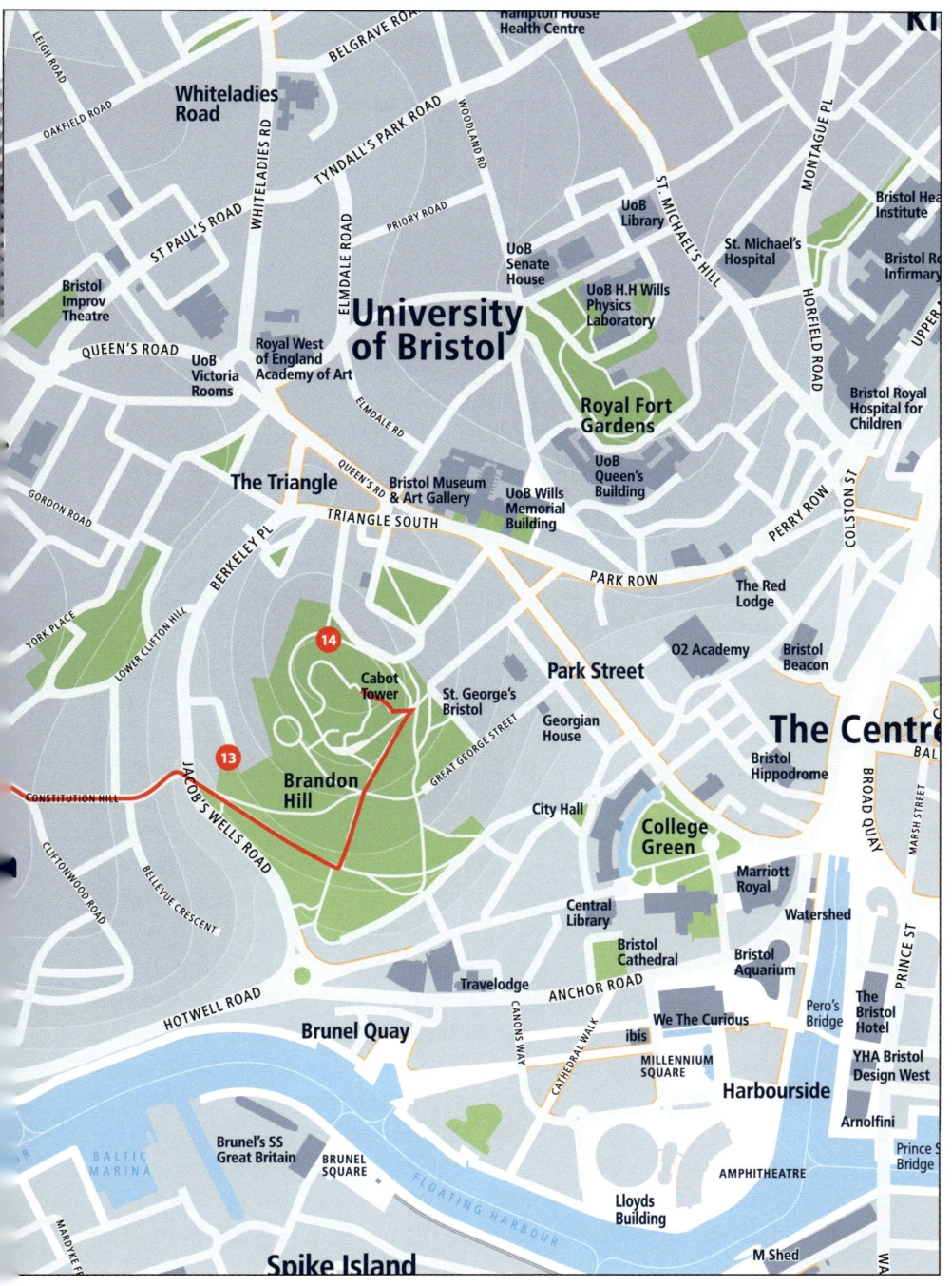

information board at the bottom of the green about both memorials. Cross at Beaufort Buildings and Portland Street then go up the driveway that leads to The Rodney Hotel. Stop at Number 3, Rodney Place (3), where there is a particularly attractive plaque installed by the Clifton and Hotwells Improvement Society.

This was once the home of Dr Thomas Beddoes, a leading figure in the scientific life of Bristol, a philanthropist and a political radical. His Pneumatic Institute in Dowry Square in Hotwells was a centre for research into diets, drugs and inhalable gases.

In 1798, Beddoes invited Humphry Davy (later inventor of the Davy lamp used by miners for safety) to take up the post of superintendent. Davy experimented with the effects of nitrous oxide (laughing gas) and its use as an anaesthetic in minor surgery. As well as experimenting on himself, he tested the effect of the gas on friends and acquaintances, asking them to record their experiences. The Bristol-born poet Robert Southey (see Walk 2) wrote to his brother: 'Davy has actually invented a new pleasure, for which language has no name. Oh, Tom! I am going for more this evening! It makes one strong and happy! So gloriously happy!' When Davy left Bristol for London in 1801 to join the Royal Institution, Beddoes converted his institute into a charitable dispensary, the Preventive Medical Institution for the Sick and Drooping Poor. His grave is in the Strangers' Burial Ground on Lower Clifton Hill.

Continue along the driveway to rejoin Clifton Down Road. Turn right into Princess Victoria Street then right into The Mall. At Caledonia Place, cross the road to read the information board giving the history of Mall Gardens (once private but now open to the public). It is worth making a circuit before continuing along The Mall. Opposite the entrance to West Mall is the former Clifton Assembly Rooms and Hotel, which was completed in 1811 and was later in competition with the Victoria Rooms on Queen's Road (see Walk 4). It was designed by Francis Greenway, who was transported to Botany Bay for forgery in 1814, and went on to design many government buildings in Australia. At the end of The Mall turn right and use the pedestrian crossing to cross at Beaufort Buildings then turn left, heading towards the bridge. The information board by the remains of the old drinking fountain gives further details about The Downs, including where to find the peregrine watch point. Where the path forks by the lamp-post, take the right-hand path up Observatory Hill to the Observatory and Camera Obscura (4), which is built on the site of an Iron Age camp.

This building was originally a windmill for corn that was later converted for the grinding of snuff. It was left derelict following a fire in 1777 and was converted by William West into an artist's studio in 1828 (he was a member of the Bristol School of

Artists, see Walk 4). West initially installed a telescope in the tower but replaced this in 1829 with a camera obscura, now one of only two still open to the public in England. He also built a tunnel through to St Vincent's Cave (familiarly known as the Giant's Cave) in St Vincent's Rocks.

Clifton Observatory and Camera Obscura: Normal opening hours: April-October 10am-5pm, November-March 10am-4pm. There is an admission charge. www.cliftonobservatory.com 0117 974 1242.

Take a final look at the view from this excellent vantage point before retracing your steps back to the lamp-post that you passed on your way up. Then turn right and continue to the tablet commemorating Isambard Kingdom Brunel on the wall by **Clifton Suspension Bridge** (5). In the flower bed beneath the tablet are examples of some of the rare and beautiful plants from the Avon Gorge and the Downs that make this one of the UK's top botanical sites. The information board tells the story of how Brunel took steps to save the threatened 'autumn squill' during the construction of the bridge, an early example of plant conservation in the face of development. There is another information board giving details of some of the native flora further along the footway overlooking the steep rocks you can see to the north of the bridge known as St Vincent's Rocks.

A scene from the Camera Obscura showing the Clifton Suspension Bridge (*The Newcomers,* BBC, 1964).

....

From the tablet commemorating Brunel, cross the road carefully to the footway on the south side, cross the bridge and stop at the viewing point by the far pier. Look down to the row of buildings on the Portway below you, on the other side of the river. The first – the curved red-brick building with white pillars – is The Colonnade (6).

....

The Colonnade was built in 1786 as an addition to the visitor facilities offered at Hotwell House. The spa was already in decline by the 1790s and Hotwell House, which had contained the pump room and accommodation, was demolished in 1822. The Colonnade is all that remains of the original complex. Ann Yearsley (see also Walk 2) opened a lending library in the building in 1793. She was one of only a few working-class

Sketching party in Leigh Woods, c1830, Samuel Jackson (Bristol Culture K2761).

women of the time to gain recognition as a writer, and her success was thanks in part to the patronage of Hannah More, whose home you'll see later in this walk. Yearsley was also one of many prominent Bristol women who campaigned against the transatlantic slave trade and her numerous verses include 'Poem on the Inhumanity of the Slave Trade'.

A little further ahead on the right hand side of the road is the Clifton Suspension Bridge Visitor Centre, which is open every day, excluding Christmas Day, Boxing Day and New Year's Day. Entry is free (www.cliftonbridge.org.uk).

Return along the bridge on the north side to the half-way point, then stop to look down into the Avon Gorge (7).

The steep limestone walls of the gorge provided the docks at Bristol with natural protection from the prevailing south-westerly winds, as well as from maritime invaders attempting to travel up the river from the Severn Estuary. However, among the challenges faced by ships' pilots were the Avon's unusually wide tidal range, its strong currents and the unreliable winds around the river bends. This is why newer facilities were developed at Portishead and Avonmouth (see Walk 3).

The trees to your left are part of Leigh Woods, a diverse woodland managed by the

Front View of Bristol Hotwells and St Vincent's Rocks, 1793 (Bristol Reference Library 393).

National Trust (www.nationaltrust.org.uk/leigh-woods). It was a favourite destination for the Bristol School of Artists. In his elegy on Thomas Chatterton, the poet Samuel Taylor Coleridge imagined his subject roaming the wooded sides of 'Avon's rocky steep' where 'the screaming sea-gulls soar' (see Walk 2).

....

*Before setting off again, look at the entrance to the Floating Harbour to the left of the river (see Walk 3). Cross the bridge and go through the old turnstile at the far end. Take the path to your right, back in the direction of the bridge. Note the pretty balconies on the houses on Sion Hill opposite. The path sweeps round to the left and then takes you down to The Lookout (**8**).*

....

While recuperating in Clifton from injuries sustained in an accident at the Thames Tunnel, Brunel learnt of a competition to build a bridge across the Avon Gorge. In 1754, William Vick, a wealthy Bristol wine merchant, had left £1,000 in his will with instructions that the money should be invested until it reached the sum of £10,000, an amount he felt would pay for the construction of a stone bridge across the gorge. The bridge would be free to travellers and would link the hamlet of Clifton and the then-private estates of Leigh Woods. As the proposed bridge would seem to serve little economic purpose, it is uncertain what Vick's motive was in leaving these instructions.

By 1829, Vick's legacy had reached £8,000 and a committee was set up to decide how to fulfil his dream. It was soon realised that a stone bridge would cost in the region of £90,000. An iron suspension bridge would be cheaper, but would still require tolls to cover its cost and maintenance. On 1 October 1829, a competition was announced with a prize of 100 guineas for the winner. The judging proved shambolic but eventually, on 16 March 1831, Brunel was declared the winner. The foundation stone of his bridge was laid on the Clifton side of the gorge on 21 June 1831. Work was soon halted as business confidence in the city fell and it did not resume until 1836. Financial difficulties and contractual disagreements led to further long delays in construction and the bridge was not completed until 1864, five years after Brunel's death and as a memorial to him. Although built for pedestrian and horse-drawn traffic, the bridge was so ingeniously constructed that it is now capable of carrying around four million cars a year, and has become a major route to the motorway network. Brunel's original design included Egyptian sphinxes on top of the piers, but these did not make it to the final version.

The first civilian to cross it on foot was Mary Griffiths, a 21-year-old barmaid from Hanham, who dashed under the barrier just before it was opened. She said:

> I was determined to get as near and to see as much as I could. So I got right up to the gate that had been erected across the entrance to the bridge. When the signal was given and the gate opened I began to run. I had not run very far when I heard my uncle shout 'run, Mary, run!' and I turned round and saw a young man behind me. I could see he was trying to beat me across so I tucked up my long dress and ran for dear life - and I beat him by a few yards!

Continue towards the Avon Gorge Hotel. You will pass the entrance to the Clifton Rocks Railway, a funicular from the 1890s which closed in 1934, but can be viewed by pre-booked groups or on open days (www.cliftonrocksrailway.org.uk). Next to the hotel is the entrance to the Clifton Spa Pump Room, which was built in 1894 at the request of the Merchant Venturers to exploit the warm, allegedly healing Clifton waters. In the 1920s it was converted to a cinema and then a ballroom. Go past the hotel and cross the road opposite the White Lion before continuing down Sion Hill (this means you'll avoid the steps down from the high pavement further on). Stop at Royal York Crescent (9).

Construction of this crescent commenced in 1791. The financial collapse linked to the Napoleonic Wars, which began in 1793, bankrupted many of the city's merchants along with the property speculators who had invested in the development of Clifton, bringing building projects like this one to an abrupt halt. Work was finally completed

The Proposed Suspension Bridge from Rownham Ferry, c1836, Samuel Jackson (Windsor Terrace can be seen on the right) (Bristol Culture K1374).

in 1818. In the 1840s, when Georgian architecture had fallen out of fashion, many of the houses were converted into flats. Among famous past occupants of this street is the author Angela Carter, who was a student at the University of Bristol. The crescent was designed by William Paty, whose design for Cornwallis Crescent you will see later in this walk.

If you can manage steps, climb up the flight to your left so you can walk along the pavement directly in front of the terraced houses. If you can't manage steps, you can go along the lower-level road used by cars. At the other end, either take the steps down or turn the corner to enter York Gardens (the gardens themselves are private) and walk back in the direction of Sion Hill. At the end of York Gardens the road bears to the left. At the junction with Cornwallis Crescent, Granby Hill and Windsor Place, cross over to Windsor Place which will take you to Windsor Terrace (a private road with a cobbled surface). Stop at Number 4 (10).

The writer, campaigner and philanthropist Hannah More lived at this house from 1829 until her death in 1833 (see also Walk 2). Her charitable work included providing educational, spiritual and financial help to impoverished miners and agricultural workers in Somerset. Work began on the terrace in 1790 at the height of the short-lived

– and ultimately financially disastrous – building mania. Its west end is supported by a massive human-made cliff that was paid for by William Watts, a Bristol plumber who, in 1782, made his fortune patenting a new process for producing high-quality spherical shot. Construction stopped in 1793 but it was eventually completed in 1811.

Go back up the hill to the junction and now take Cornwallis Crescent, ignoring Hensman Hill on the left. Just before the crescent drops down to the right, go up Goldney Avenue (you'll need to switch to the left-hand side further up). Turn right on Regent Street to Goldney Hall (11).

This is now a University of Bristol hall of residence, but was originally built for Thomas Goldney II around 1720. Its attractive gardens were designed by Goldney's son. In 1737, work began on a grotto decorated with rock crystal from the Avon Gorge and an assortment of fossils, shells and corals from around the world. It was developed over 25 years. The Goldneys were Quakers but their fortune partly came from gambling, privateering (a licensed form of piracy) and the manufacture of cannons.

Opposite the hall is a narrow green, on the other side of which is Clifton Hill (12). Ann Yearsley's home was on Clifton Hill.

Yearsley had been taught to read and write by her mother, a milkwoman who trained her daughter to follow her in the same occupation. By May of 1784, she and her family had fallen into destitution. They were rescued from near-starvation by local charitable individuals including Hannah More. More organised the publication of a volume of Yearsley's poetry, *Poems, on Several Occasions*, paid for by subscriptions from her literary and wealthy friends. This was published in June 1785. More than 1,000 subscribers are listed. 'Clifton Hill' was the title of the last and longest poem in the publication.

From Goldney Hall, turn right down the very steep Constitution Hill (you'll need to cross to the left-hand side as the pavement runs out on the right). At the bottom of the hill, on the left hand corner, you'll find an information board about the washhouses of Jacobs Wells. Find a safe place to cross Jacobs Wells Road, then go up the steps to Brandon Hill Nature Park (13).

In 1980, the Avon Wildlife Trust partnered with Bristol City Council to transform five acres of urban parkland on Brandon Hill into this haven for wildlife (www.avonwildlifetrust.org.uk/reserves/brandon-hillnature-park). Brandon Hill is a prominent green oasis that provides an essential fuelling station for migrating

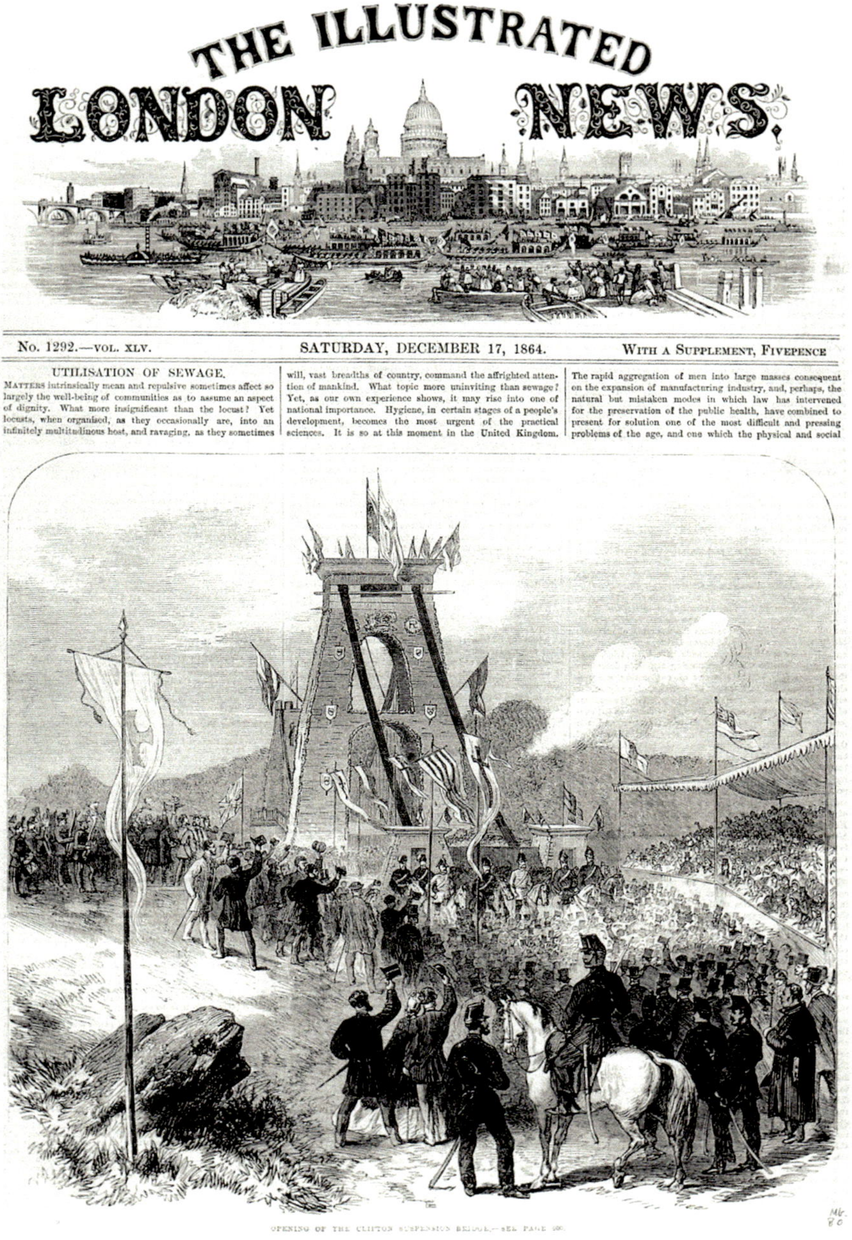

Cover of The Illustrated London News marking the opening of the Clifton Suspension Bridge (University of Bristol Library).

birds. It shelters flocks of redwings and fieldfares escaping the freezing conditions in Northern Europe in winter. In spring the wildlife pond is full of frogspawn and toads. In summer cowslips, oxeye daisies and knapweed help to attract butterflies and bees. Foxes and pipistrelle bats come out in the early evening and there are finches, tits, thrushes and warblers in the woodlands. It demonstrates the important contribution made to our environment by urban conservation projects.

Brandon Hill is thought to be the UK's oldest public park. It was given to Bristol's town council in 1174 by the Earl of Gloucester and sub-let to farmers as grazing land until 1625 when it became a public open space. Citizens of Bristol still have the right to dry their clothes here and beat their rugs. The hill was of strategic importance in the defence of the city and the remains of a fort and earthworks can be found to the west and south of Cabot Tower.

The hermit Lucy de Newchurch lived here in an anchorite cell in the 14th century, writing that she was 'tired of this world' and wanted peace and quiet.

Queen Elizabeth 1 visited Bristol in 1574 and complained that her ruffs had become dirty. They were washed at Jacobs Hill by washerwomen and hung out to dry on Barton Hill.

....

From the information board, take the right-hand path to follow the first half of the nature trail. From the board telling the story of Brandon Hill (point 3 on the trail) continue up to Cabot Tower (14).

....

This tower was commissioned to commemorate the 400th anniversary of John Cabot's voyage to Newfoundland and was paid for by public subscription. In Italian, Cabot's name was Giovanni Caboto and in Venetian it was Zuan Chabotto. The tower was designed by William Venn Gough and is constructed from reddy-pink sandstone and cream Bath stone. The coats of arms on the sides include that of the Society of Merchant Venturers, with whom we started this walk. The tower was closed to the public in 2007 after cracks appeared when the supporting ironwork corroded. It re-opened on 16 August 2011 after the vital work to make it safe again was completed. The climb up the narrow winding spiral staircase will reward you with panoramic views across the city.

Cabot Tower: Cabot Tower is open every day except Christmas Day, Boxing Day and New Year's Day from around 8am to 6pm. Entry is free. www.bristol.gov.uk/museums-parks-sports-culture/brandon-hill.

Walk 6:
Bristol's Rock Music History

Bristol might not have a place in musical history to match that of Liverpool or London, but local audiences got to see some amazing shows by artists of national and international renown during the sixties and seventies. Many of these took place at venues that no longer stage live music, such as the Granary, Corn Exchange and Victoria Rooms. This walk takes you back in time to visit some of the key sites where eventful shows took place between the early sixties and late seventies. It should take around two hours to complete.

The walk begins at 32 Welsh Back, the site of the Granary (1). **This is near Queen Square, in Bristol City Centre.**

Take a moment to admire the magnificent architecture of this Grade II listed building, which was built in 1869 and has been described as the best-preserved example of the Bristol Byzantine style. The ground floor is currently run as a restaurant, with luxury apartments above. But before gentrification hit the city docks, this was home to the city's most legendary, riotous rock club from 1969 until 1988 – its heyday being in the 1970s. As you passed through the entrance, you'd find the Granary Roll of Honour, listing all the great artists who played here, from the pioneers of prog rock (Yes, Genesis, King Crimson, Barclay James Harvest) to the great metal bands of the late seventies and early eighties (Iron Maiden, Motörhead, Def Leppard). Having climbed the stairs to the first floor, you'd find yourself in the club itself, which had a balcony stretching round half of the venue to the DJ booth above the stage.

The Granary began its life as a music venue as a jazz club under the guidance of Acker Bilk, opening seven days a week from 8 October, 1968. But as the popularity of jazz waned, the early part of the week became particularly quiet. This provided a perfect opportunity for a quartet of hairy young men who called themselves Plastic Dog. These guys (Al Read, Mike Tobin, Ed Newsom and Terry Brace) were at the centre of Bristol's musical counterculture in the late 1960s and were looking for a new venue to stage live music, having outgrown the Dugout Club. Initially, they hired the Granary on Monday nights. The first big non-jazz event staged here was by the great Chicago bluesman Muddy Waters and his band on 15 November 1968. The only thing that didn't go according to plan that evening was Muddy's entrance. Legend has it that as he emerged from the dressing room to greet the hero-worshipping hordes, the blues giant banged

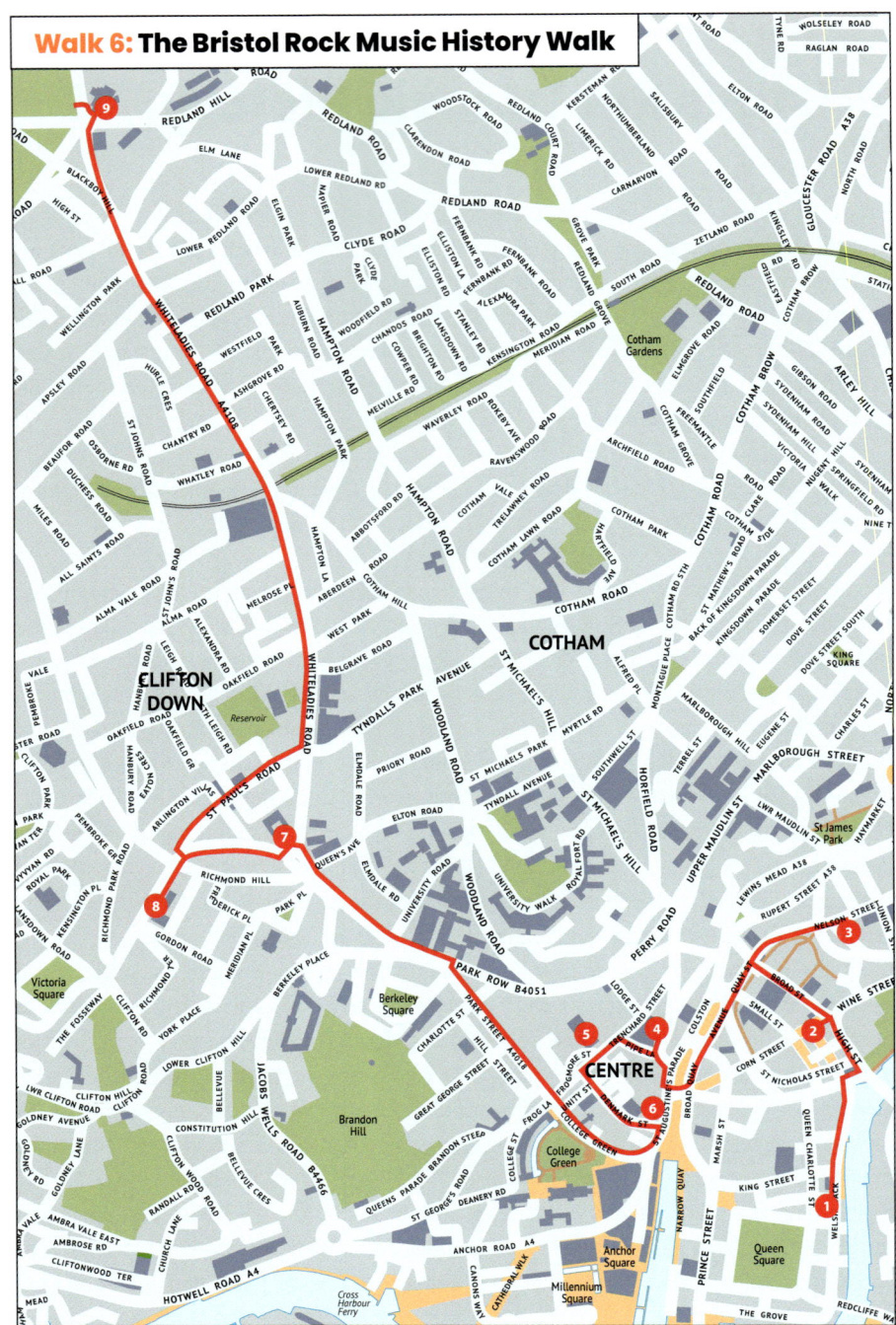

his head on the low beam above the door, dislodging his toupee.

When US all-girl group The Runaways played here in 1976, ticketless punters resorted to scaling the outside walls in a desperate attempt to get in through the windows. In 1978, John Cougar Mellencamp performed a show at the Granary in his earlier, poppier incarnation as Johnny Cougar, long before he became a multi-million selling US heartland rocker. Everybody wanted to play the venue but not all artists were an instant hit. The club actually lost money putting on Dire Straits in June 1978 as the meagre takings didn't cover the band's £125 fee. But big names were often eager to grace the tiny stage. After Led Zeppelin split, Robert Plant played an invitation-only show here in May 1981 with his band The Honeydrippers. A month earlier, Motörhead returned at the height of their fame to play a benefit show for the local chapter of the Hells Angels.

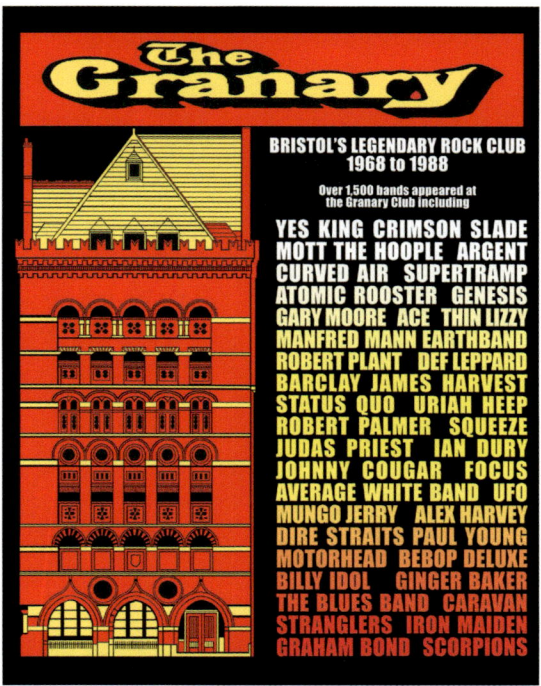

The Granary souvenir T-shirt design. Now a sought-after collectors' item. (The Granary archive).

....

Walk along Welsh Back in the direction of Broadmead. This will mean the river stays on your right-hand side. Cross Baldwin Street and turn right to follow High Street, which curves round to the left past St Nicholas Market, with Castle Park on your right, until the road starts to curve. Turn left into Broad Street, and then quickly left into the pedestrianised Corn Street, where you'll find the Bristol Registry Office on your right. On your left after a few hundred metres you'll find the entrance to the historic Corn Exchange (2).

....

Today this is a market, but back in the 1960s it was a live music venue that played host to many of the stars of the future, including the Rolling Stones (their first Bristol

The entry sign at the Corn Exchange.

show was in December 1963), The Who, Rod Stewart and The Kinks.

There were two key music promoters here. Tuesday nights were taken by a fella who styled himself 'Uncal' Bonny Manzi, who ran the Bristol Chinese R&B Jazz Club. Although the club was hung with Chinese lanterns and advertised 'fly lice' there was nothing Chinese about it. Bonny was a Cockney gent from a showbiz family who originally ran a string of jazz clubs around the country. But after the bottom fell out of the jazz market he switched to R&B, which he preferred to pop because he perceived it to have a higher standard of musicianship. Wednesday nights were taken by Freddy Bannister, who went on to organise the Bath Festivals of Blues and Progressive Music as well as the legendary Knebworth Festivals in the 1970s. Freddy staged the more rock and pop-oriented gigs at the Corn Exchange.

There were plenty of great shows at the Corn Exchange. When gospel/blues pioneer Sister Rosetta Tharpe played the venue in the autumn of 1964, a huge crowd of ticketless punters gathered outside hoping to hear her play. Roger Daltrey of The Who was once walloped by a bouncer for breaking the 'no girls backstage' rule. Pete Townshend wrote *My Generation* on the same day of that gig (19 May 1965). Eric Clapton played his very last show with The Yardbirds here in 1965, just weeks before his 20th birthday. Freddy Bannister secured something of a coup in bagging The Byrds for one of the dates on their brief UK tour in August 1965. They arrived by train from London accompanied by DJ Annie Nightingale. Alas, although The Byrds looked fabulous, everyone agreed the show was awful. Other notable gigs included Pink Floyd's first Bristol show, with Syd Barrett in 1967, and the original shock rocker Screamin' Jay Hawkins, whose spring 1965 show opened with him climbing out of a coffin to perform *I Put a Spell On You*.

Walk back the way you came and turn left into Broad Street. This will take you past the Grand Hotel, where Mick Jagger was allegedly denied entry to the restaurant in January 1964 because he was improperly dressed. After passing through St John's Gate at the end of Broad Street (Bristol's last remaining city gateway), turn right into Quay Street, which turns into Nelson Street. You'll pass Rough Trade on your left before crossing Bridewell Street and arriving at a nightclub on your right. This has had multiple names over the years (including Papillons, The Works, Top Rank, Baileys, Romeo and Juliets, Odyssey and The Syndicate) and currently trades as SWX (3).

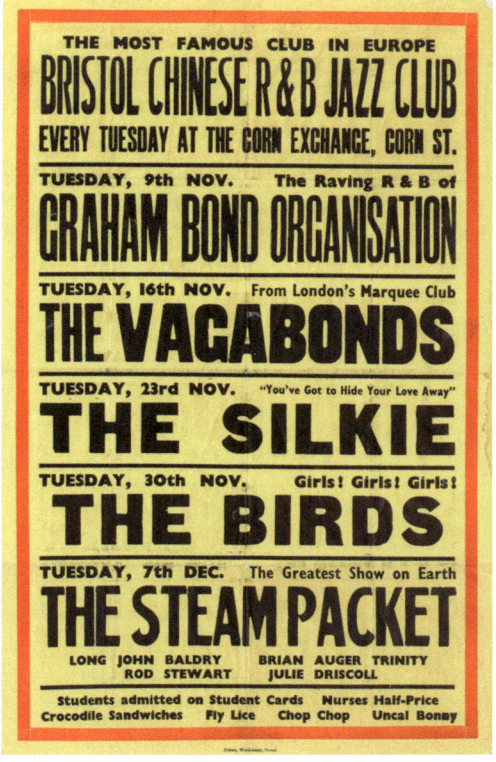

Bristol Chinese R&B Jazz Club poster for Nov/Dec 1964.

Between 1966 and 1974, this club was called the Top Rank. It hosted plenty of live music, including shows by The Kinks, Cream, Sweet, Status Quo, Peter Frampton, Can, Humble Pie and many of the great progressive rock acts, including King Crimson, Yes, Genesis, Gentle Giant.

Now retrace your steps, back down Nelson Street, which will turn into Quay Street. Follow this road all the way until you meet the junction with Baldwin Street. Turn to your right to cross the Centre and then St Augustine's Parade, and then turn left into Colston Street. Immediately on your right, you will find Bristol Beacon (4). Previously known as the Colston Hall, this was one of the most important venues on the UK concert touring circuit.

The Beatles played here three times. On the last occasion, in November 1964, the Hall's manager Ken Cowley erected an iron ring of security around the venue to ensure no one without a ticket got to see the Fab Four. This was swiftly breached when 17 schoolgirls snuck in through a back entrance left open for cleaners and hid in a toilet. Only when one of them coughed was their enterprising ruse rumbled. Matters got worse for Mr Cowley when a prankster opened a lighting vent above the stage and showered the band with flour during the climax of *If I Fell*. Many great bands performed in Colston Hall during this period and there's some wonderful stories, as the section below illustrates.

Who appeared at Colston Hall?

Colston Hall had some bizarre bills, the oddest of which was probably the Walker Brothers/Jimi Hendrix/Englebert Humperdinck/Cat Stevens package in April 1967, and there were usually two shows a night to maximise revenue. Timings were tight, with lower billed acts given as little as eight minutes to strut their stuff, while the headliners frequently got just 20 minutes. Occasionally, this led to trouble.

When the Small Faces played here with Canned Heat in 1968, a full-blown riot ensued and the police were called. 'Fighting end to pop show' was the headline in the following day's *Bristol Evening Post*. While Canned Heat played during the second house, the Small Faces had been enjoying themselves very much backstage. So much so that they eventually took to the stage a few minutes before the curfew. Halfway through their number one hit *All or Nothing*, the house lights went up. All hell promptly broke loose. The highlight of the ensuing fracas came when a bouncer picked up diminutive keyboard player Ian McLagan and hurled him off the stage into the stalls.

When Led Zeppelin played their second show at the Colston Hall in January 1970, there was nearly another riot. Heavy snow had delayed the arrival of John Bonham and Robert Plant, so everything started late. The hall's management moved to impose the 11pm curfew, cutting short the gig during the second encore and leading to a chorus of boos from the audience.

There had also been ructions the previous month, when Delaney and Bonnie played Colston Hall in December 1969. This time it wasn't the fault of jobsworth hall management or artist tardiness. D&B were touring with friends, one of whom was widely known to be Eric Clapton. It had also been strongly rumoured that George Harrison would be joining them on stage, in what would have been The Quiet One's first post-Beatles show. But no matter how loudly the punters yelled for Beatle George, he failed to appear and the promoter threatened the disgruntled audience with arrest if they didn't all leave before the second show. It later transpired that George had been present all along, hiding backstage while the audience rioted. He'd been alarmed by the presence of the press and didn't want to be the centre of attention.

The *Bristol Evening Post* reported 'nasty moments' when Black Sabbath headlined in February 1972, principally when 'some hysterical kids clambered on the lip of the circle and looked in danger of plunging down and later when the bouncers formed an on-stage wall between audience and stars.' When the tired and drugged-out heavy metal pioneers returned on their tenth anniversary tour in 1978, they were blown away by their exciting young American support act, Van Halen, who ushered in an entirely new kind of metal for the 1980s.

Bob Marley's final show in Bristol, with the Wailers in June 1976, is often trumpeted by Beacon management in their eagerness to move the conversation on from slavery. But on this hot and sweaty midsummer night there was a heavy police presence and an atmosphere that the *Bristol Evening Post*'s reviewer described as 'edgy'. It seems that significant numbers of Bob's fans declined to pay for tickets to see him.

AC/DC played their first Bristol show here a few months later, incurring the displeasure of officials when the audience leapt from their seats to rock. Interestingly, this was also a rare AC/DC show at which lead guitarist Angus Young failed to bare his backside. Apparently, such was the Colston management's disapproval of his antics that the paranoid guitarist became convinced there were members of the Bristol Vice Squad planted in the audience ready to haul him off to the clink at the slightest flash of a pallid, scrawny Antipodean buttock.

In May 1966, Bob Dylan pitched up for the first date on his ill-fated English tour, just ten days before the infamous 'Judas!' incident at the Manchester Free Trade Hall. The show comprised two full sets: an unaccompanied, seven-song acoustic one and an electric one for which Dylan was backed by The Hawks (who later became The Band). Many reviewers objected to their folk hero embracing rock. The *Western Daily Press* called it a 'noisy, blaring, ear-splitting disaster'.

In October of that year, the Rolling Stones provoked a now-familiar teen girl frenzy as excited young ladies hurled themselves at the stage, only to ricochet off the ample belly of a portly, amiable bouncer who'd been positioned to intercept them. For many of those present, the real stars of the show were support act, the Ike and Tina Turner Revue. 'This was pop at its very best,' enthused the *Bristol Evening Post*.

The huge success of a sold-out show by Jethro Tull and Ten Years After in May 1969 provoked something of an agonised post-mortem in the local press, especially as a recent show by the previously bankable Herman's Hermits had been an under-

The Rolling Stones (left to right: Charlie Watts, Keith Richards, Bill Wyman, Mick Jagger, Brian Jones) backstage at the Colston Hall, 1964 (*Bristol Post*).

attended disaster. 'Underground music' had usurped those smartly dressed beat boom acts and the hacks were keen to find out what this meant. The Hall's manager, Ken Cowley, took a suitably progressive approach, describing underground music as 'maturer, deeper. Pop is better for it.' The *Bristol Evening Post* noted that while 'hippies arouse deep feelings of hatred among the respectable public,' Colston Hall was welcoming them with open arms. 'They certainly look very odd,' conceded Mr Cowley, 'but they don't come to take part in the frenzied fan worships we used to see not so long ago. They are extremely polite.'

Many of the biggest names of the 1970s started out in lowly positions on packed Colston Hall bills. David Bowie was at the bottom of the bill for two packages in 1969. At the first of these (supporting his old chum/rival Marc Bolan's Tyrannosaurus Rex) he didn't sing a note, performing a mime act instead; at the second (supporting his former schoolmate Peter Frampton's Humble Pie) he was heckled. When Bowie returned on the third leg of the Ziggy Stardust tour in June 1973, he was a huge star and the Hall's manager hired 40 bouncers to keep order. Just two weeks later, Bowie dramatically retired Ziggy at the Hammersmith Odeon in front of DA Pennebaker's cameras.

Elton John had a particularly unfortunate experience. After naughty young

Nick Mason and Roger Waters of Pink Floyd at the Colston Hall, unveiling what was to become 'Atom Heart Mother' as part of the University of Bristol's Timespace Arts Festival, 7 March 1970 (*Bristol Post*).

Elton played here in November 1971 on the *Madman Across the Water* tour, he was summoned for a stern dressing down by the starchy suits for performing a jig on the Hall's piano. When he showed up again in March 1973 on the ironically named *Don't Shoot Me I'm Only the Piano Player* tour, he was issued with a warning before his performance: put one foot on our municipal piano, chummy, and your show will come to a premature end. By November of that year, Elton was a huge star and turned up for a sold-out show with his own concert grand piano. 'He can have a bath in it for all we care,' sniffed a Colston Hall official.

The shows with the biggest demand for tickets in the early seventies were by the Rolling Stones and Pink Floyd. Local TV news footage depicts queues snaking all the way up Colston Street in the driving snow as punters hoped to get their hands on the 75p tickets for the Stones' show in March 1971. Why such eagerness? This nine-date jaunt was the Stones' first UK tour since 1966 and also marked the band's first live gigs since the Altamont debacle in December 1969. Everyone also knew that this was the last chance to see the Stones before they went into tax exile in the south of France to avoid the start of the new financial year. Unfortunately, it turned out that they were pretty awful on the night.

Pink Floyd had previously played here in March 1970 as part of the University

Queen (left to right: Freddie Mercury, John Deacon, Brian May, Roger Taylor) at the Colston Hall, 17 November 1975. Photograph: *Bristol Post*.

of Bristol's Timespace Arts Festival, during which they unveiled their quadraphonic sound system. In February 1972, they returned to run through the work-in-progress that would become the mega-selling *Dark Side of the Moon* for only the eighth time on stage, a full year before its official release. They certainly pushed the boat out on the sound and lighting front. The band turned up at the Colston Hall with nine tons of equipment, having insisted that the venue be available from 8am. Their elaborate lightshow and surround sound system took six hours to set up and a further four hours to take down again. A guarantee was demanded that the stage would be able to take the weight of all their gear and 16 local roadies were recruited to help haul it out of the trucks.

With no UK arena circuit at the time, bands surfing huge waves of popularity in the 1970s occasionally played two nights at Colston Hall. On 6 and 7 May 1974, Status Quo opened their *Quo* tour with a two-night stand in Bristol. The extravagantly staged 104-date *The Lamb Lies Down on Broadway* tour by Genesis reached the Colston on 29 and 30 April 1975. That same year (17 and 18 November), Queen's *A Night at the Opera* tour came to town just as *Bohemian Rhapsody* took up residence at number one in the singles chart for the first time. The national music press were out in force to cover these shows. A few years later, Canadian prog-metallers Rush sold out two

The programme for the Jimi Hendrix Experience/Pink Floyd show at the Colston Hall, 24 November 1967.

nights at the venue on their *Hemispheres* tour.

As for the venue's most extraordinary shows, it is hard to beat the 50-piece jazz-prog combo Centipede organised by Bristol musician Keith Tippett and featuring members of King Crimson, Soft Machine and Spooky Tooth. The unwieldy combo played just four shows, including this hometown one in March 1971. The best 1960s package arrived here in November 1967, and was headlined by The Jimi Hendrix Experience, with support from The Move, Pink Floyd and The Nice (featuring Keith Emerson).

....

Retrace your steps along Colston Street, which turns into Pipe Lane, and go around the left-hand side of Bristol Beacon. Bear left when you reach the junction with Frogmore Street and continue down the steep hill. At the bottom, you will find the O2 Academy music venue on your right-hand side (5). This was formerly an ABC cinema, which formed part of a large entertainment complex known as the New Bristol Centre.

....

The block of student accommodation to the left of the O2 Academy was previously home to a nightclub and live music venue, originally named the Locarno. It was here that Jimi Hendrix made his Bristol debut on his first UK tour in February 1967. There's a great *Daily Mirror* photograph from this show, taken from behind Hendrix to show the audience. Although the Summer of Love was just months away, Bristol's teens looked very square and conservative. That's partly because, in common with many other Bristol venues, the Locarno had a strict dress code which persisted until the arrival of progressive rock in 1970.

After his show, Hendrix retired to the Heartbeat club above the Locarno, which was to become his favoured hangout whenever he came to Bristol.

One of the most extraordinary Locarno shows was by the Ike and Tina Turner Revue on 17 October 1966. Ike and Tina had stayed in the UK after the Stones tour to play a short run of headlining gigs. These included a hugely ambitious plan to play the Bath Pavilion and Bristol Locarno on the same evening. Ike had agreed that they would use borrowed equipment at the Pavilion, beginning their show at 8.20pm and coming off stage at 9pm. That would give the entire Revue just enough time to pile into a van and charge over to Bristol, where their own equipment had been set up, for a 9.40pm start. But when Ike turned up in Bath, he ruined these carefully laid plans by insisting on using their own gear for both shows. This caused such a delay that the Revue didn't appear on the Locarno stage until 10.20pm.

The Who played the Locarno twice in 1967 and a new series of weekly progressive rock shows was announced in 1970. Bottom of the bill at the second of these on 7 June was Kevin Ayers and the Whole World, which featured a young chap named Mike Oldfield on bass. In July 1972, Hawkwind turned up on the first of a short run of Electric Mecca gigs at the venue which offered two acts for less than a pound. Their support band was a bunch of art-rockers who called themselves Roxy Music. When the Hawks returned to the venue in November of the same year, they had a more unusual support act: local folkie Fred Wedlock, nine years before his novelty hit *The Oldest Swinger in Town*. Brave Fred had agreed to step in at the last minute after tour support Fat Mattress failed to turn up. By all accounts, he went down rather well.

Between those two shows, David Bowie arrived on the third Bristol date of his *Ziggy Stardust* tour, with the young Thin Lizzy in support. After their soundcheck, Bowie and the Spiders trooped off to the Centre's naff Polynesian-themed Bali Hai bar in full regalia, where the unwary might have found them sitting round on bamboo chairs, surrounded by plastic palm trees and fake totem poles, being served cocktails by Bristolian-accented waitresses wearing grass skirts.

The New Bristol Centre was also home to a rather more upmarket venue called the Mayfair Suite. This is where the *Western Daily Press*'s *Teenpage* held its Teenpage Ball on

1 November 1969. The Ball was a glorified beauty contest with a few pop acts, including a local band contest, but they pushed the boat out in booking an unlikely hard rockin' headliner: Deep Purple.

Cross the road and walk with the Hatchet Inn (see Walk 7) to your right until you reach the junction with Denmark Avenue. Follow this road, which becomes Denmark Street until you reach St Augustine's Parade. Turn left and after a couple of hundred metres you'll come to the historic Hippodrome Theatre (6).

These days, the Hippodrome – known as The Hippo – tends to focus on theatrical shows with long runs. But this wasn't always the case. In the past, individual rock shows were often slotted in to its busy programme.

Eddie Cochran perished in a car crash outside Chippenham on the way back to London after a gig at the Hippodrome in April 1960. His passenger and tour mate Gene Vincent hobbled away with a broken leg. In December 1969, The Who performed their legendary rock opera *Tommy* pretty much in full at the Hippodrome in an ecstatically received show. Dutch progressive rock band Focus enjoyed a massive career boost after appearing on *The Old Grey Whistle Test* and were at their commercial peak when they played to a packed Hippodrome in January 1973. But the biggest band of 1973 was Rod Stewart and The Faces. When tickets went on sale in March for the first of their two Hippodrome shows that year, the box office queue snaked halfway round the Centre.

In November 1973, Neil Young played his one and only Bristol show at the Hippodrome. Unfortunately, it was a disaster. Neil was eager to flog his downbeat, drug-themed *Tonight's the Night* album, while the increasingly annoyed audience just wanted to hear 'Cinnamon Girl'. The support act that night was the little-known The Eagles (also playing their only Bristol show), who'd just released their *Desperado* album.

The Hippo secured another coup on in May 1974 when Steely Dan turned up on their *Pretzel Logic* tour. By all accounts they were magnificent, but the entire UK tour was abandoned shortly afterwards when Donald Fagen fell ill. Paul McCartney and Wings played here twice, the second show in September 1975 being part of the epic *Wings Over the World* tour, which stretched into 1976. And since the two-hour Hippodrome gig was the second show on that tour, Bristol audiences were among the first to hear Beatles songs performed live by McCartney for the first time in nearly a decade.

In August 1976, Ritchie Blackmore's Rainbow played their first ever UK show at the Hippodrome. 'It was the noisiest, most deafening, ear-splitting show I've seen in years and the fans were in ecstasy,' grumbled the *Bristol Evening Post*'s veteran pop correspondent. Peter Gabriel turned up on his first post-Genesis tour in October 1977. Naturally, he took advantage of the ornate Hippodrome architecture to go walkabout in the circle and stalls, occasionally popping up unexpectedly from the darkness. The following night, Hawkwind unveiled their most ambitious production: the giant, multi-coloured, pulsating Atomhenge set, which stretched the width of the stage and towered over the audience. And when Judas Priest played here on 12 November 1978, they introduced the leather-clad look that would dominate heavy metal for the next decade. In April 1979, Kate Bush played the Hippodrome on her only UK tour. A low-quality audience bootleg recording of the show can be found on the internet.

With the Hippodrome on your right, retrace your steps along St Augustine's Parade and follow the road round to your right onto Park Street, with College Green on your left. Walk up Park Street.

You are now following the route of the 3 March 1969 Bristol University Rag Week's torchlight procession from Colston Street to the Victoria Rooms, where a happening ensued. We'll come to that soon. Glance across the road as you continue up Park Street and you'll see number 77 is now a men's hairdressers. Its 1960s occupants had little need of such services as the first floor was home to the Plastic Dog collective, who programmed live music at the Dugout club (just round the corner at 52 Park Row) and, later, the Granary. They also produced Bristol's very own alternative music magazine – *Dogpress* – which started out as a typewritten gig list but later grew to incorporate such of-their-time features as *Groupie of the Month*.

At the top of Park Street, cross the road onto Queens Road and continue past the Wills Memorial Building and City Museum on your right. Carry on along this road past Beacon House until you reach a zebra crossing. Cross the road here and you reach the Victoria Rooms (7).

These days it houses the University of Bristol's Department of Music, but back in the 1960s it housed the Students' Union and was frequently used as a live music venue. That 1969 happening mentioned earlier was headlined by the newly four-piece (Roger Waters, David Gilmour, Nick Mason, Richard Wright) Pink Floyd, with radio DJ John Peel in attendance. Other notable shows here included Bristol prog heroes Stackridge (whose June 1970 gig got them signed to MCA Records), Sparks (July 1971), German experimentalists Can (May 1975) and Supertramp, who unveiled their masterpiece *Crime of the Century* here in December 1974. Search on YouTube and you'll find that Edison Lighthouse shot the promo film for their 1970 number-one hit *Love Grows (Where My Rosemary Goes)* on the steps of the Victoria Rooms.

A selection of covers of *Dogpress*, the Bristol alternative music magazine, 1969-1971 (The Granary archive).

Follow Queens Road around past the front of the Victoria Rooms, so you pass the building on your right hand side. When you reach the roundabout, take the first exit, cross Richmond Hill and continue until you reach the Bristol University Students' Union at number 105 (8).

Opened in 1965, this houses the Anson Rooms on its first floor, reached by a spiral staircase. Few shows take place here now but plenty of groovy gigs happened in the past. In January 1967, The Move smashed up a TV set with an axe, also whacking holes in the stage and prompting a row over who would pay for the damage – estimated at £25-£50. In March 1968, Arthur Brown managed to set off the fire alarm with his smoke bombs and flaming helmet routine at the End of Rag Ball, causing the venue to be evacuated. The Who delivered a suitably incendiary performance in December of that same year, right in the middle of an 11-day student sit-in at Senate House, at the climax of which Keith Moon kicked over his brand new drum kit. During her short-lived pop career, Marsha Hunt taunted her student audience for their appearance and dress in October 1969. And in March 1972, Dutch progressive rock band Focus headlined a show, supported by a chap named David Bowie.

Retrace your steps back to the Victoria Rooms on Queens Road, cross over at the zebra crossing and take the left fork along St Pauls Road until you reach the junction with Whiteladies Road. Cross over the road so you reach the Royal West of England Academy and continue up Whiteladies Road, with the BBC on your right. This is quite a hike up an increasingly steep hill, but eventually you will reach a fork. Follow Westbury Road to your right. Cross the roundabout, taking the second exit. At the top, you will see the Spire private hospital on your right (9).

There was a nightclub on this site from the 1950s. To an earlier generation of rug-cutters it was The Glen. But by the late sixties it had become Tiffany's, which was famed for its plastic palm trees. This was later home to a regular Wednesday live music night which went by the very seventies name of Boobs, whose logo was a silhouette of a naked woman visibly feeling the effects of the cold. The night began on 1 March 1972 with local heroes Stackridge. On 15 November of that year, you could have paid 50p to see Status Quo. In May 1973, Bob Marley and the Wailers played their first Bristol show here, followed by German avant-gardists Faust in June. Punk acts such as The Clash played here, too, in the late 1970s and one of the last shows at the venue was by Motörhead in October 1978, who were on the cusp of hitting the big time.

Cross the road and you'll come to the open space of The Downs.

Modern live music events here are corporate, fenced and ticketed affairs, but back in 1971 this was the site of Bristol's very first free festival - and one of the country's first events of this nature. The Bristol Free Festival was organised by a group of local anarchists with a generator supplied by the West Coast chapter of the Hells Angels. Three local bands (Wisper, Flash Gordon, Magic Muscle) played. Naturally, the police turned up. But they just took a quick look and went away again. They would not be so lenient when faced with such unbridled hippy merry-making in future.

Head back to Whiteladies Road to get buses and trains back to the city centre and transport stations. Head to Rough Trade Records near Broadmead for more music and check out the programmes at Bristol's music venues. And do read Robin Askew's book: *The West's Greatest Rock Shows 1963-1978* **(Bristol Books, 2024) for more stories.**

Walk 7:
Bristol Urban Myths

Bristol has more than 1,000 years' worth of history. But when you drill into the detail beyond all the business of the rise of the port, the trading with Europe, the Civil War, the transatlantic trade, Brunel, the Blitz and all the rest, you encounter some tales which are only half-true, and plenty more which probably or definitely aren't true at all.

We'll look at some of these tales today, some old and some more recent. Some of the more recent ones might be described as urban legends, a term coined by American academic Jan Harold Brunvand in the 1960s to highlight the fact that modern industrial urban communities have their myths and legends just as much pre-industrial ones did. Where people once passed on stories in fields or a village alehouse, they then exchanged them in factories, offices and bars, and nowadays online or with our phones as well.

We will start, though, back in the distant mists of time, with the story of how Bristol got here in the first place. Well, one version of it anyway…

The walk is mostly on level ground. There are a few gentle slopes and three sets of steps, two of which can be avoided with short detours. Allow at least 90 minutes, or a little longer for stops for refreshment, relief and distraction.

The Walk

Start in Queen Square (1)

Until it was laid out and houses built in the early 1700s, this was a part of Bristol known as the Town Marsh. The works carried out over the centuries to create the port of Bristol and, later, its Floating Harbour, easily obscure the fact that central Bristol (the low-lying meeting point of the rivers Avon and Frome) used to be very marshy and muddy.

By the 1500s, if not sooner, there arose a fairy tale as to how the city came into being. It was all thanks to two giants, Vincent and Goram, brothers who competed for the hand of the fair maid Avona. She declared she would choose whichever of her suitors managed to drain the swamp. So they set to digging a channel which would take the water away to the sea. The tale goes that this is how the Avon Gorge came into being.

Vincent was industrious and sensible but Goram drank too much and fell asleep on the job. When he woke, he found he had lost the contest and, in his despair, he threw

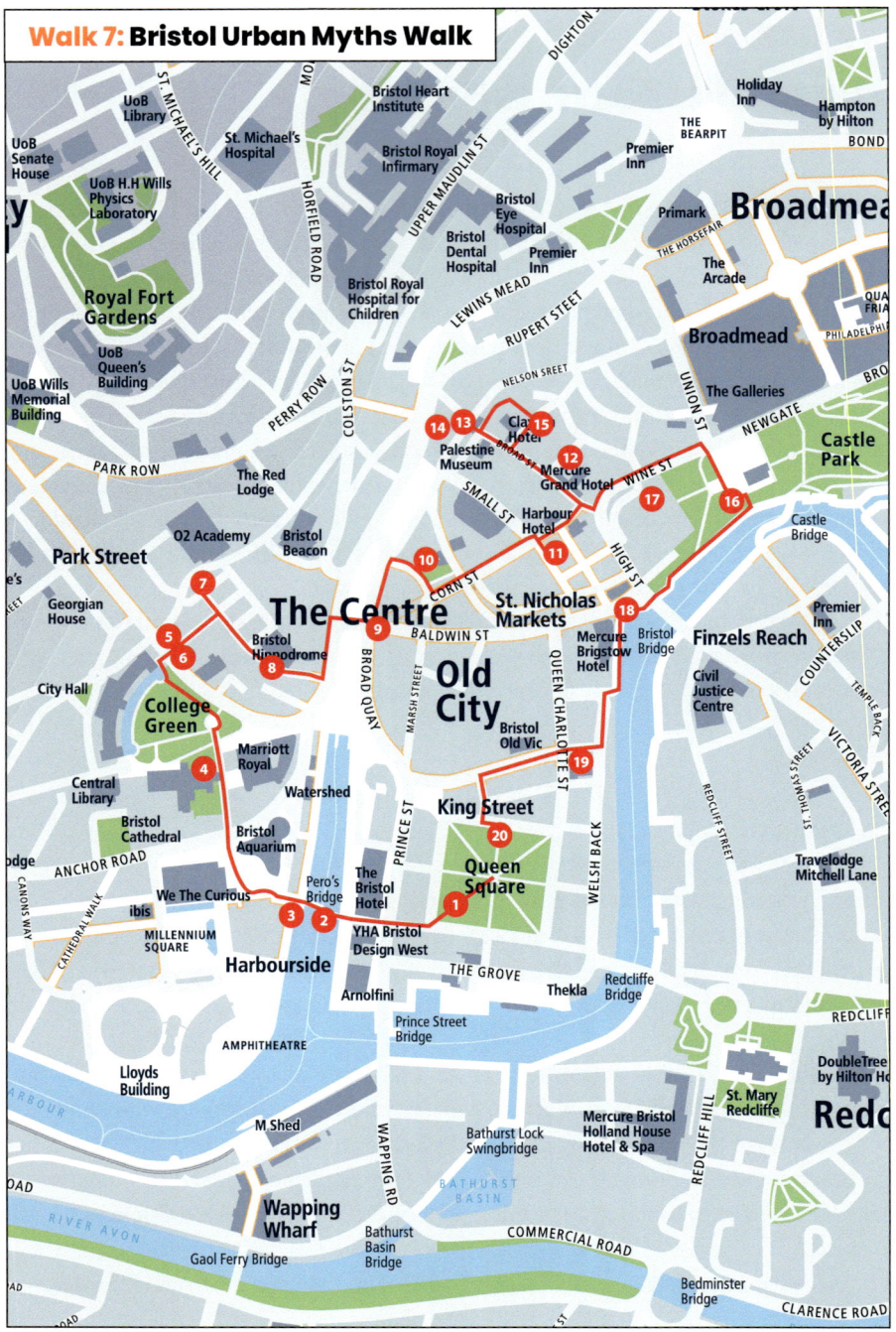

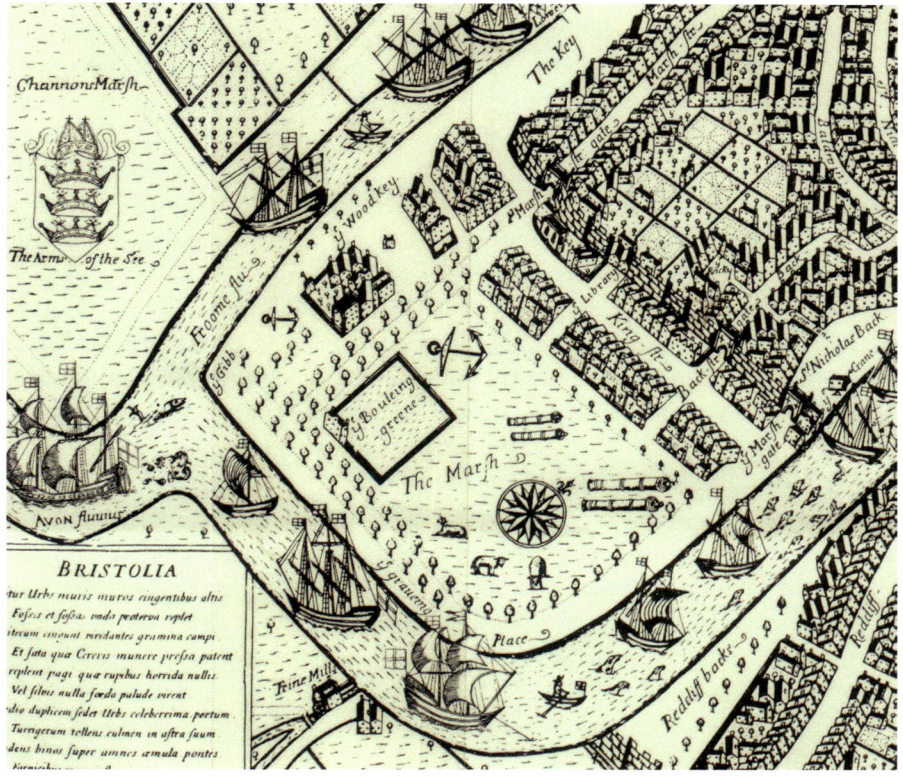

'The Marsh' from a late-1600s map of Bristol before it was Queen Square.

himself into the Bristol Channel where bits of his giant corpse can still be seen sticking out of the water as Flat Holm and Steep Holm.

Other versions of the story are available but, at heart, it is a dull fable. Work hard, stay off the ale and you'll be a winner. Until more recent decades, when Bristol acquired its reputation as a place of music, art and creativity, Bristol was known to locals and outsiders alike as a serious, hard-working place where the pubs closed earlier than they did in a lot of other towns. Blame that on nonconformist religion and the Victorian/Edwardian temperance movement, or blame it on the terrible example of what happened to poor Goram.

While we're here ... Everyone admires the equestrian statue of King William III in the centre of the Square, which generations of mothers used to tell their children comes to life at night, when no-one's looking, to go for a ride around town.

There's also the tale that the sculptor, the Flemish John Michael Rysbrack, was a closet Jacobite – a supporter of the Stuart dynasty which had been ousted when the

Protestant William III and his wife Mary deposed the Catholic James II. William III died in 1702 after falling from his horse, which supposedly tripped on a molehill (more likely it was felled by stumbling into a mole's burrow); Jacobites from then on drank the health of 'the gentleman in black velvet'.

So the little hump that the horse's rear right hoof rests on might be a secret allusion to the molehill. Rysbrack was a Catholic and therefore open to suspicion of Jacobitism, but he worked on commissions for all sorts of people. It is more likely that the little hump is just there to add a sense of movement to the horse.

....

Go to the south-western corner of the Square and along Royal Oak Avenue. Cross Prince Street at the pedestrian crossing and follow the lane straight in front of you to Narrow Quay and to Pero's Bridge (2). Stop at the bridge.

....

This pedestrian bridge, designed by sculptor Eilis O'Connell, opened in 1999 and was named for an African 'servant' – in reality an enslaved man – of sugar merchant John Pinney who lived in Great George Street, and whose home is now the Georgian House Museum (see Walk 1). The bridge opening coincided with a major exhibition at the city museum, *A Respectable Trade?*, which attempted to give a clear account of Bristol's part in the transatlantic trade in enslaved people. An important part of the exhibition was its attempt to challenge many of the myths surrounding Bristol's role, one of the most notorious being that pub cellars around the city, and even Redcliffe Caves, had once served as 'warehouses' for kidnapped Africans. This never happened.

Another curious tale to be challenged was the racist legend as to how many people of African-Caribbean heritage ended up living in St Pauls. The story, apparently current in the 1960s and 1970s, was that a wealthy sugar planter had retired to a big house in St Pauls, along with his Black 'servants' and, disliking his neighbours, took posthumous revenge on them. After he died, his will stipulated that his servants should be freed and that they should inherit his house. They invited all their friends and relatives over from the Caribbean – and this was why there were so many Black people in St Pauls. If you know anything about the real history of St Pauls, you'll see that this tale makes no sense whatsoever.

....

Now cross Pero's Bridge, stop at the other end of the bridge by the Bordeaux Quay, near the Pitcher & Piano. (3).

....

This is the spot where the Colston Statue was thrown into the harbour in June 2020. It is also one of many areas around the docks where real people have jumped into the water down the years.

There are drowning tragedies in the docks. Sometimes people fall in accidentally; or jump in to cool off or as an act of drink-fuelled bravado and get into trouble as a result. Sometimes people jump in to try to end their own lives. Around 2017-18, there was a widely-believed story going around that a serial killer was deliberately pushing people into the water. It wasn't true. The story was identical to equally baseless claims from a few years previously that someone was doing the same thing in Manchester. Successive coroners' reports and statements from the police that state that a number of drownings were accidents had little effect.

On a happier note, we could mention the Bristol Crocodile here. Sightings of this notional amphibian first arose around 2014, with a few witnesses swearing that they had definitely seen it, although usually in the Avon rather that the docks. You can believe them or not, as you wish.

Now continue walking away from the harbour and go through Anchor Square, turning right at We The Curious and cross Anchor Road at the pedestrian crossing. Close by on your right is a lane with five flights of steps taking you past the Cathedral along Trinity Street. Go along here and into the middle of College Green. Beware of cyclists, scooters and skateboarders.

If you do not wish to use the steps, carry on along the Anchor Road footpath towards the Centre, and then double back towards College Green. Stop on College Green to admire the Bristol Cathedral (4).

In the late eighteenth and early nineteenth centuries, one of the senior clergymen at Bristol Cathedral was Dr Frederick William Blomberg, a man on whom favours had been showered by the Royal Family. He had been chaplain and private secretary to the Prince of Wales (later Prince Regent, later George IV), and had been appointed to many well-paid sinecures within the church, although he did very little work in any of them. He was appointed Prebendary of Bristol in 1790, at the age of just 28. Dr Blomberg was a case study in ecclesiastical venality, and he did it with the backing of the Royals. How had he managed it?

One version of the story went: Blomberg's father, an army officer, had secretly married a young woman against the wishes of one or both of his parents. Two children resulted from this furtive union but, alas, their mother died and Major Blomberg sent them to a house in rural Dorset to be brought up in secrecy. The American War of Independence saw the Major recalled to his regiment and he was killed in action, orphaning the two children that no-one knew about - aside from whoever their guardian was. But then, Major Blomberg's ghost appeared to one of his fellow officers. The ghost told him about the children, where to find them, and how to ensure that they inherited

Bristol Cathedral in the 18th century, as imagined by Bristol artist FG Lewin in 1922. The clergy there were sometimes more worldly than spiritual...

his property.

When Queen Charlotte (wife of King George III) heard about this, she insisted that young Frederick William be sent for so that he could be brought up with her own children. Perhaps Blomberg's great good fortune in amassing ecclesiastical offices was the result of his being brought up as a virtual member of the Royal Family.

You can believe this if you like. Or you may prefer to believe that Frederick William Blomberg was in fact an illegitimate son of George III to whom he bore a strong physical resemblance.

....

Now cross College Green to the bottom of Park Street and take the pedestrian crossing near Bristol City Hall. Turn left and walk about 10m to a spot near the Banksy mural of the naked man hanging out of a window (5).

....

There are many tales concerning Banksy and his famous murals. The one you are looking at appeared in 2006 and it was spattered with blue paint three years later, just before the hugely popular Banksy exhibition was due to open at Bristol Museum & Art Gallery up the road. The blue paint might have been the work of some arty type(s) either doing something edgy and subversive, or it was the work of some attention-seeking idiot(s), according to taste. Others claimed it was the work of Bristol Rovers fans (the paint is a Rovers-type shade of blue, and Banksy is a Bristol City fan). Others claim the mystery splatterer was Banksy himself making some sort of artistic statement, or simply being mischievous. But never mind all that. Somewhere near here, Spring-Heeled Jack attacked in Bristol.

Spring-Heeled Jack was one of the most important and interesting folkloric figures of nineteenth century Britain. 100 years ago, everyone knew who Spring-Heeled Jack was while few know of him nowadays.

This bogeyman attacked people, sometimes to rob them, sometimes to molest young women. His most distinguishing features were his strange attire and his ability to leap extraordinary distances – due, some said, to powerful springs in the heels of his boots. From the 1830s to the 1890s, there were sightings of him all over the country including London, Liverpool, the Midlands, Scotland and a few in Bristol. One attack took place here in 1841 when 14-year-old Ellen Hurd was walking home. A newspaper report tells us:

> She was seized hold of by a tall man, dressed in a rough greatcoat who, with many imprecations if she uttered a syllable, took out a sharp pair of scissors and, in a minute, cut off the whole of the hair at the back part of her head. The poor girl, as may well be imagined, was dreadfully frightened. This, we suppose, is one of the freaks of the unmanly ruffian who, under the name of 'Spring-Heeled Jack,' has been lurking about in bye-places lately, and annoying and frightening females. We sincerely trust that the police may pitch upon him and that the springingness of his heels may be put to a severe test on the treadmill.

As the report suggests, there was no Spring-Heeled Jack as such, but a lot of copycat attacks, and the press was happy to perpetuate the myth of a superhuman criminal. Although there doesn't seem to be anything superhuman about Bristol's Jack. Six months later, poor Ellen Hurd was up before the magistrates on charges of theft, but the court was told that the attack had so disturbed her that she had spent some time in an asylum and the case was dropped. The newspaper account of the case suggested that other young women had been attacked in Bristol in a similar manner and that Bristol's Jack was still at large.

....

Turn to your right to go down towards the Centre and take the first turning on your left into Unity Street. Stop a little way down and turn to admire the unicorn at this end of City Hall (6).

....

The story goes that one of the reasons why architect Emanuel Vincent Harris chose to place unicorns at either end of what we used to call the Council House was that one

Allegedly not just a decorative flourish but a rude gesture to one of the architect's enemies (Eugene Byrne).

of his professional enemies lived on Unity Street. So the first thing the man would see as he emerged from his home each morning would be a unicorn's arse.

Continue along Unity Street to the first turning on your left, Denmark Avenue. Wander down to take a close look at the door of the Hatchet Inn (7).

The Hatchet Inn is Bristol's oldest still-functioning pub, dating back to at least 1606. It has got plenty of ghosts and, in the late 1700s and early 1800s, was associated with several of Bristol's famous prize-fighters (a wall plaque round on the right-hand side from the door commemorates this). But the most gruesome legend associated with the Hatchet is that its main door, underneath the tar and paint, has a covering of human skin. Why on earth any respectable landlord would want to do this is a bit of a mystery. But that's the story.

Go back down Denmark Avenue which becomes Denmark Street until you come to the stage door of the Hippodrome on your left (8).

Generations of Bristolians have hung around here hoping to get a glimpse of, or an autograph from, showbiz figures who performed at the Hippodrome down the years (see Walk 6 for more details). Now imagine it is April 1960 and loads of excitable teenagers are waiting here to see some rock 'n' roll idols.

One of the most bizarre stories in the Hippodrome's long history concerns Eddie Cochran, who, although American, is probably better remembered in Britain than in his home country. In April 1960, he played several nights here on a bill which also included Gene Vincent. After their final show on 16 April 1960, Cochran, Vincent and Cochran's girlfriend, Sharon Sheeley, hired a taxi to take them to London to get a plane back home. However,

The door at The Hatchet Inn (Eugene Byrne).

the car crashed into a lamp-post near Chippenham. Vincent was injured, Sheeley and the driver were unhurt, but Cochran was thrown from the car and died at St Martin's Hospital, Bath, the following day. He was 21.

There are numerous legends around the death of this teen idol, although an incontrovertible fact was that the Eddie Cochran single which topped the charts after he died was called *Three Steps to Heaven*. Before he died, Cochran was said to have been depressed and dwelling on the deaths of his friends Buddy Holly and Ritchie Valens in a plane crash a year previously. He was supposed to have had a premonition of his own death.

One of the first officers to arrive at the site of the car crash was Wiltshire Constabulary police cadet David Harman, who then taught himself to play Cochran's guitar, which was being kept in the evidence room at the police station. This would launch his own musical career and he went on to stardom in the 1960s as Dave Dee of Dave Dee, Dozy, Beaky, Mick and Titch. Earlier on in the tour, the guitar had been carried for Cochran by adoring young fan, Mark Feld, who later achieved fame as Marc Bolan – and who would also die in a car accident.

In Bristol there remains a persistent legend that, when he had been in the city, Cochran fathered a love-child. This seems unlikely given that his girlfriend stayed close to him but you never know. An investigation by *Venue* magazine some years ago contacted the head of Cochran's fan club in the US, who said he had been approached by an English woman claiming she had had Cochran's child.

Carry on to the end of Denmark Street, turn left onto St Augustine's Parade and continue past the Hippodrome to take the pedestrian crossing and stop by the corner of Baldwin Street and the pedestrianised Clare Street (9).

In August 1714, Queen Anne died and was succeeded by King George I, the first of the new Hanoverian dynasty. This was at a time when political rivalries between the Whigs and Tories could be violent and savage. From the time of Cromwell's rule, the country spent over a century in a state of incipient civil war. These political divisions were particularly bitter in Bristol. The arrival of an obscure German family on the throne was a cause of celebration for Whigs but most definitely not for Tories. In Bristol, there was a discovery which, according to accounts from the time, was seen as a political omen – the appearance of an immense cobweb.

Eddie Cochran; teen idol whose death sparked some strange tales.

A cooper living in Baldwin Street had invited his friends over for an afternoon, proposing that they sit in the summerhouse in his garden (they had gardens in central Bristol back then, often big ones). According to local legend, this summerhouse had been a meeting place for Bristolians taking part in the Rye House Plot to overthrow King Charles II some 31 years previously.

As the cooper and his party entered the pavilion, they found an enormous black cobweb 34ft wide. He swore that it had been swept clean just a few days previously, and many people saw it as a terrible sign connected with the new regime. The house was swamped by sightseers and the cobweb was soon destroyed by souvenir-hunters grabbing pieces of it. But what of the spider that could make a 34ft web? There were never any sightings of it. You might not expect any reports though; anyone who had seen it would surely have died of fright. Or was it all just a load of nonsense put about by one faction or another?

Continue along Colston Avenue, which is to the left of Clare Street, in the direction of the Cenotaph and the Colston statue plinth. Take the first turning on the right into St Stephen's Avenue (not St Stephen's Street). The entrance to St Stephen's church (10) is close by. If it is open, go in if you like.

There are many tales associated with this church, not least because of its longstanding associations with the Society of Merchant Venturers. If you go in, you'll see their coat of arms in one of the stained glass windows. What you will also see is the tomb of Edmund Blanket and his wife, Mrs Blanket. Local legend has it that the blanket was invented by a Thomas or Edmund Blanket of Bristol, an idea widely ridiculed as an unfounded bit of nominative determinism. Some feel that Witney in Oxfordshire, which for centuries was the centre of English blanket-manufacturing, invented the blanket.

We can't prove that a Bristolian named Blanket invented woollen bedclothes. But we can't prove he didn't either (even some Witney histories say it might have been a Bristol invention). Blanket's family originally came here from Flanders. It is likely that 'Blanket' was an anglicisation of their original name (Blanquet? Blanchette?) and they were involved in the cloth-weaving industry which was so important to medieval Bristol. The Blankets made local cloth production more efficient, and while the local weavers guild disliked their industrial-style methods, they had the backing of King Edward III himself. It is possible that they produced an off-white/grey woollen cloth (known in French as 'blanchet') and that this became a popular form of bed-covering.

When you come out of the church, take a moment to admire the little churchyard which now functions as a miniature park. More on Bristol's tiny graveyards later.

Continue along St Stephen's Avenue and turn left into Clare Street and continue along it when it becomes Corn Street. Don't follow the curve of the road round to the left but instead keep going straight on and to the pedestrianised area. Just past the Exchange (St Nicholas Market) and the Corn Street nails you'll see a turning on the right called All Saints Lane. Go along here until you come to the church on your left (11).

This church, which is closed to the public, is the last resting place of Edward Colston, although we don't know exactly whereabouts he's buried. Colston died at his home in Mortlake, Surrey, on 11 October 1721 at the age of 86, and his body was taken to Bristol to be interred here at All Saints in a grand ceremony in the pouring rain. Some years later, a spectacular monument in the church was completed featuring a massive

The Blanket tomb in St Stephens (Eugene Byrne).

reclining effigy of Colston. It was carved by John Michael Rysbrack, who also made the William III statue in Queen Square.

In the 1840s, during building work, workmen found what was assumed to be Colston's tomb. The body was well-preserved and over the coming days it was looked at by various local dignitaries. It is said that pieces of the corpse's clothing (he was dressed in shirt, stockings and underdrawers), and possibly even pieces of the corpse itself, were taken by these visitors as relics or souvenirs. This is probably the basis of strange rumours that various bits of the man who made his fortune dealing in human beings are ceremonially venerated by the Society of Merchant Venturers.

In truth, they were almost certainly not Colston's remains. He was supposed to have been buried with his sister Ann, although she was not in this tomb. Furthermore, the body they were looking at had a full set of teeth in good condition. Few 85-year-old men nowadays have such good teeth, never mind the early 1700s.

Return to the junction with Corn Street and turn left. Take the next left onto Broad Street. Stop near or opposite the Mercure Bristol Grand Hotel (12).

During the Second World War, the Grand Hotel was the check-in desk for Britain's only civilian airport. Civilian flying was stopped and airports were closed in 1939 on government orders. Just one remained open, as it was needed for diplomats, celebrities, business people and, well, spies. This was Bristol Airport, which back then was at Whitchurch – much of the former airport is now Hengrove Park. Initially, there were only a few flights each week, and all to just one destination – Lisbon, the capital of neutral Portugal. There were later flights to Shannon in neutral Ireland, connecting with flights from the United States.

Passengers reported to the Grand Hotel where, once their papers and passports were checked, they would be taken to a bus with blacked-out windows so they couldn't see where Britain's only airport was. All sorts of strange and fascinating people came and left via Whitchurch and it is a fair bet that some of them were spies. The spy connection might explain the loss of BOAC Flight 777 from Lisbon to Whitchurch on 1 June 1943, which has been the subject of all sorts of conspiracy theories ever since.

There was an unwritten agreement between the British and German air forces that they would not attack civil airliners, but Flight 777 was shot down by German aircraft over the Bay of Biscay with the loss of all 17 passengers and crew. Why did this happen? Probably because, as one of the German aircrew stated after the war, they never got the memo and assumed that all British aircraft were fair game. This, though, did not satisfy everyone. Some claimed the Germans shot the plane down because they thought Prime Minister Winston Churchill was on it. Another explanation was that the Germans believed that RJ Mitchell, designer of Britain's Spitfire fighter aircraft, was on board which seems a little absurd because Mitchell had died in 1937.

However, one of the passengers was actor Leslie Howard, whose most famous role was that of Ashley Wilkes in *Gone With the Wind*. Howard had also starred in a popular 1942 British film *The First of the Few* in which he played... RJ Mitchell. Could it be possible that a German agent in Lisbon saw 'the Spitfire designer' getting on the plane and alerted the German air force?

Continue along Broad Street to St John's Arch (13) at the end.

This is St John's Arch and to either side at the top of the arch are a pair of statues representing Brennus – or Brennius – and Belinus, the two mythical sons of a mythical king of Britain. Brennius is supposed to be the founder of Bristol. Like Rome, Bristol was built on seven hills and, like Rome, it has a founding myth involving two brothers. In fact, we go better than Rome, having *two* founding myths with two brothers, the previously mentioned Vincent and Goram being the other ones. The Brennus and Belinus story is too ridiculous to dwell on. There they are holding orbs with crucifixes, even though their legendary rule pre-dated Christianity and even the Roman invasion.

Brennus and Belinus at St John's Arch (Eugene Byrne).

Go through St John's Arch and stop just on the other side to Electricity House (14).

The big grey building across the road to your left is Electricity House. For many years this was the HQ of the South West Electricity Board (SWEB), and on the ground floor was a showroom featuring all manner of electrical appliances. When Sir Clive Sinclair launched his C5 battery-powered three-wheel vehicle in the mid-1980s, the story goes that two were sent to Bristol for display in the showroom. According to legend, the Sinclair C5s were never put on display, because fun-loving SWEB staff used them as dodgems in the underground car-park during their lunch-breaks.

While we're here ... A little further to your left along Quay Street they used to have the head of an executed terrorist. Here's the story:

On 22 January 1777, three ships in Bristol's harbour were deliberately set on fire. The fire failed to catch in two of them but one was damaged. It was later found that all three had been daubed with pitch overnight. This was before Bristol had a floating harbour. The docks were crowded with wooden ships which simply settled on the riverbed when the tide was out. If the fires had caught, several ships close by might also have been destroyed because it would have been impossible to move them out of harm's way at low tide.

Three days later, a warehouse in Bell Lane was destroyed by fire while two attempts were made to burn a sugar warehouse. Evidence of attempted arson was found at a number of other buildings and Bristol was thrown into complete panic. At a time when buildings, never mind ships, contained so much wood, fire was a real and constant fear. It remains a miracle that there never was a Great Fire of Bristol. A huge reward was offered for the capture of the arsonist who, some weeks later, was arrested in Lancashire. This was John (or James) Aitken, also known as John (or Jack) the Painter, a rapist and small-time crook who had fled to America and there become 'radicalised'. Returning to Britain, he decided he could help the cause of American independence through what we would now call terrorism.

Believing that British power could be destroyed by arson in the Royal Navy's lightly protected dockyards, Aitken had set fires in Portsmouth as well as Bristol. He was hanged from a ship's mast in Portsmouth on 10 March 1777 and so ended the first terrorist in Bristol's history, who was acting in support of… the United States!

For decades afterwards, a curious tradition persisted in Bristol about John the Painter. A warehouse was built on Quay Street around the time he was hanged, and the builder had bought part of the ruins of an old Abbey in Keynsham and took a corbel from it with the carved head of a saint and put it on the new building. Many Bristolians believed that this stone carving was the actual skull of John the Painter. The building is no longer there and the head's fate remains unknown.

***

Go back through St John's Arch and immediately take the first turning on the left into Tower Lane. Follow the lane until you see a turning on the right. Take this, passing the Bank pub. Over to your left you will see a small graveyard with a little set of steps leading up to it. Go into St John's in the Cemetery (15).

***

This yarn could easily apply in any of the old churchyards in central Bristol, but as this one has been restored and opened to the public in recent years, you might not have visited this wonderful little space yet so we'll mention it here.

In the past, parliamentary elections could be rumbustious affairs, with massive bribery, corruption and violence. It was not unusual for candidates in contested parliamentary elections to spend upwards of £2,000 on the campaign, which in modern money could be translated as £1 million. Well over half of this would be spent on bribing voters with drink.

In Bristol, candidates' agents would also scour the city for the widows of freemen. Freemen or Burgesses of Bristol were the privileged citizens. You couldn't carry out a trade in the city without being one, and you usually had privileged access to the welfare provided by local charities. Before the reforms of the nineteenth century, only freemen were allowed to vote. You could only become a freeman by being the son of one, serving an apprenticeship under one or by paying a very large sum of money. But there was one other way; you could marry the widow of a freeman. At election times these ladies were in great demand.

On several occasions in the 1700s, these women were induced with cash bribes, to marry supporters of one party or the other, who would then cast their votes accordingly. Because the law on marriage was vaguer than now, it was considered completely acceptable for these sham marriages to then be dissolved in a bizarre ceremony. The marriage vows at the time plainly stated that man and woman were joined together in holy matrimony 'until death us do part'. So once the election was over, 'man and wife' would go to a churchyard and stand on either side of a grave. They would then solemnly declare that death had indeed parted them.

While we're here ... St John's Graveyard, like every other parish graveyard in central Bristol, was seriously overcrowded by the early nineteenth century, when the city's population was rapidly expanding. There were horror stories of bodies buried inches beneath the soil, of hands and feet sticking out of the soil, of dogs chewing and carrying off body parts. These stories may have been exaggerated, not least by the directors of the new Bristol General Cemetery Company at Arnos Vale, where for the first 15 years – it opened in 1839 – business wasn't nearly as brisk as the investors had hoped. But then the old parish graveyards were closed on government orders and Arnos Vale became, for a couple of decades, Bristol's main burial place. The directors reported healthy profits for many years.

....

Go back down the steps, turn left and carry on back to Broad Street. Alternatively, if the gates are unlocked, go through these and through to Tailors Court. Either way, go to Broad Street and turn left. Carry on to the top of Broad Street and turn left onto Wine Street. Carry on along here until you get to the pedestrian crossing. Head over to the ruins of St Peter's church. Go to the square near the church overlooking the river and the riverside path (16).

....

St Peter's Hospital, as it came to be known, was seriously damaged in the Blitz and later demolished, but until then it was an important landmark building in Bristol. It was situated in front of you, between the church – also gutted by fire in the same air raid – and the water's edge.

The site had previously been the home of Thomas Norton, a medieval alchemist who claimed that he'd discovered the elixir of life but that it was (rather conveniently) stolen from him by a woman. A later version of the tale has the wife of merchant William

Canynges as the thief. In a Victorian work of fiction possibly based on local legend, she demanded that her husband stop giving so much money to St Mary Redcliffe church because she was now going to need all his cash as she was going to live forever. Supposedly, she was struck down for this by the Virgin Mary herself and found that Christ's mother had also dried up the liquid in the elixir bottle.

In the 1600s, Robert Aldworth, a wealthy sugar merchant, built a very elaborate house on the site. It was a sugar refinery, later a mint, and later still was supposedly the first workhouse in England. By the 1800s, it was always overcrowded because it was housing impoverished adults and children, some of whom were suffering from various mental illnesses.

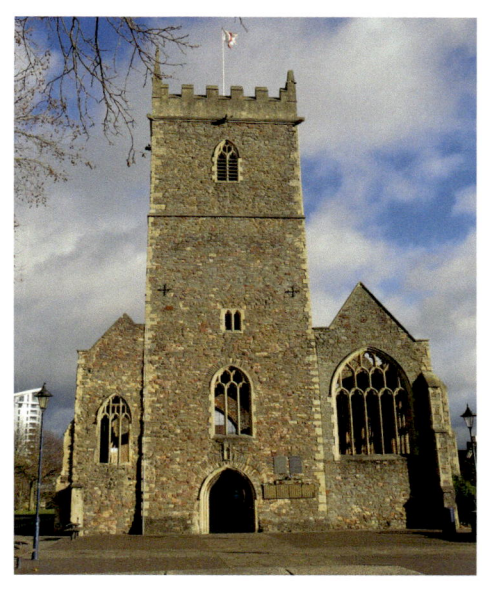

St Peter's, Castle Park, ruined in the Blitz, along with the nearby St Peter's Hospital (Eugene Byrne).

Life in the workhouse was meant to be hard to encourage people to support themselves, and not be dependent on their social betters (as they saw themselves) who had to carry the oppressive (as they saw it) burden of having to pay the poor-rate. The regime was usually hard – a basic diet, a lot of work, and men and women were kept separate. The workhouse was dreaded by the poor and so it wasn't surprising to find that, in the 1830s, a rumour arose that this workhouse was deliberately trying to kill its inmates to save money.

In 1832, Bristol was hit by a terrifying new disease, cholera. Hundreds died across the city. In St Peter's Hospital, about 600 people were sharing 300 beds. 208 of them contracted the disease and 94 died. Cholera was new and little understood. At the time, the medical profession was baffled. Many people despised doctors anyway because they only attended those who could afford to pay, and many doctors had connived in the theft of their loved ones from graveyards in order to dissect them.

Sometimes doctors pronounced cholera patients dead without realising that sometimes they might suffer a 'cold stage' in which they appeared dead but were not. Because of this, the inmates of St Peters almost rioted when they believed that some of their fellows were being buried alive. Meanwhile, evangelical Christians were distributing tracts in Bristol saying that the epidemic had been caused by the moral degeneracy of the poor.

While we're here... As a child, you were made to eat up your carrots because you were told they're good for your eyesight, right? You might also have been told they're especially good for your night vision. It is a myth. Consuming huge amounts might have a marginal effect, but not one you'd notice. This myth was turbocharged by the Blitz. Following one of the night raids on Bristol in early 1941, the German bombers were heading home when one was shot down over Hampshire by a Filton (or Weston-super-Mare) -built Bristol Beaufighter piloted by F/Lt John Cunningham of 604 Squadron. This was the first time a German raider had been shot down at night.

The UK government's propaganda machine claimed that 604 squadron had been training for night interception by eating carrots. Between the wars, increasing scientific knowledge of the properties of vitamins had suggested that carrots were indeed good for your eyes. The press dubbed Cunningham 'Cat's Eyes Cunningham' because of his allegedly excellent night-vision. He eventually shot down 20 German planes, 19 of them at night.

But the carrots thing was all nonsense. The Government put out the story because they didn't want the Germans to get wind of the real reason why Cunningham and other RAF night-fighter pilots were finding German raiders in the dark – they had developed radar sets small enough to fit into aircraft. Misleading the British public was collateral damage and the public fell for it, growing loads of carrots in gardens and allotments because they were desperate to be able to see better in the blackout. It was a particularly urgent matter in Bristol because the central part of the city was a working harbour with no safety rails anywhere.

....

Back opposite the memorial tablets to the dead of the Bristol Blitz on the church wall is a pathway that runs almost parallel to the river. This leads to the derelict buildings by Bristol Bridge. Follow this path to the ruins of St Mary le Port Church (17).

....

St Mary le Port, also ruined by the Blitz, dates back to the late Middle Ages. In Tudor times, it included a chapel dedicated to St Wilgefortis, usually known in England as Saint Uncumber. Her cult was extremely popular as she was the patron saint of unhappily-married women.

St Wilgefortis, the legend goes, was a princess somewhere in Europe who had converted to Christianity and taken a vow of chastity. Her pagan father was angered as he had promised her in marriage to a pagan king, so she prayed to God to spare her from this ghastly prospect. Her prayers were answered. She woke up the following morning to find she had grown a copious beard.

So it was here that unhappily married women in Bristol came to pray to St Uncumber to intercede on their behalf, usually bringing an offering in the form of a small bag of oats. Sir Thomas More, himself later martyred by Henry VIII, wrote that nobody knew why one should bring oats, unless it was to 'provide a horse for an evil husband to ride to the devil upon.'

Take the path down to the riverside pathway. There are steps, but these can be avoided by just walking on the grass. Go left onto the riverside path to the pedestrian crossing at Bristol Bridge and cross over to the other side of the street (do not cross the bridge). Stop by the entrance to Welsh Back (18).

The tomb of John Whitson in the crypt of St Nicholas (not open to the public) (Eugene Byrne).

The church across from you is St Nicholas. Its walls are pock-marked by WW2 bomb damage. In its crypt is buried Tudor merchant John Whitson who lived nearby. He's the subject of a local legend, our own Dick Whittington if you like. Whitson was a poor but honest lad born in Bristol or Gloucestershire in the 1550s, and was apprenticed to a Bristol wine merchant who died, leaving everything to his widow.

'He was a handsome young fellow,' wrote John Aubrey, his step-grandson in his book *Brief Lives*. 'And ... his mistress one day called him into the cellar and bade him broach the best butt in the cellar for her; and truly he broached his mistress, who after married him.'

Who seduced whom? Who knows? Whitson went on to become a wealthy and successful merchant. He also became the Sheriff of Bristol, mayor of Bristol, Master of the Society of Merchant Venturers and an MP. He married three times, being twice widowed, and had three daughters. He outlived them all, and when he died he left most of his money to local charities, including for the founding of Red Maids school.

Now walk along Welsh Back, keeping the river to your left, to King Street, the second turning on your right. Turn down King Street and stop by the Llandoger Trow (19).

The Llandoger Trow has more ghosts than Bristol has living people. It is also said to have been where Daniel Defoe got the inspiration for *Robinson Crusoe*. Some will tell you it was the model for the Admiral Benbow pub in Robert Louis Stevenson's *Treasure Island* but then so is The Hole in the Wall on The Grove. All of these tales may well have been embellished by the public relations people of Berni Inns in the 1950s or 1960s, when the Llandoger was one of their most prized outlets.

But the weirdest story is that it was moved. According to a couple of sources on the internet, including a compendium of interesting facts about Bristol compiled for the otherwise respectable *Time Out* website, the pub has wheels under it. This apparently was something to do with it having to be slightly moved when Bristol's docks were converted to a floating harbour in the early nineteenth century.

When this story was written up in the *Bristol Times* section of the *Bristol Evening Post* some years ago, a reader wrote in to point out that an ancient half-timbered pub in Exeter had indeed been moved on wheels and rails in 1961. There are plenty of photos of this event and it does bear a passing resemblance to the Llandoger, so maybe that's where it came from.

Turn left onto Queen Charlotte Street and return to Queen Square (20).

On the southern side of the Square is a blue plaque commemorating Woodes Rogers, sea captain, privateer and pirate hunter. With navigator William Dampier and a very unpromising crew, he led a privateering expedition of two Bristol ships in the early 1700s. They circumnavigated the globe, preying on Spanish and French ships during the War of Spanish Succession. Along the way, the ships put ashore on the Pacific Island of Juan Fernandez and rescued a man described by Rogers as 'clothed in goat-skins who looked wilder than the first owners of them'.

Alexander Selkirk had been marooned on the island six years previously and survived on crawfish, cabbage, turnips, pimentos and goats. The goats were his only company, apart from the dozens of cats, which had reached the island and bred as a result of shipwrecks. He passed his time chasing goats, carving his name in trees, reading psalms and singing and dancing with his cats and goats.

Afterwards, Selkirk came to Bristol and stayed at Rogers' house in Queen Square where, the story goes, he met journalist and writer Daniel Defoe and this was the beginning of the tale of *Robinson Crusoe*. Except that he may never have met Defoe at all. There's also a story that when staying here, Selkirk was in the habit of walking around the square in his goat skins during the evenings when wealthy Bristolians were out for a stroll, in order to cadge beer money off them.

We started this tour with one of Bristol's foundation myths. So let's end with

Lot and his daughters and wife flee the destruction of Sodom (and Gomorrah), as imagined by Gustave Doré. But what's the Bristol connection?

another one that's very different.

William Comyns Beaumont (1873-1956) was a 'British Israelist', that is, he believed that the events described in the Bible had not taken place in the Middle East but in the British Isles. This eccentric belief can be dated back to the seventeenth century, but enjoyed some currency in Beaumont's time because how can it possibly be that all the action in the sacred book at the centre of your Christian beliefs happened in places where none of the people are white and none of them spoke English? He spent years working out how Biblical geography translated to Britain, but claimed he had his eureka moment when crossing the Clifton Suspension Bridge one day in the 1920s.

It all fell into place. The Bristol Channel was the Sea of Galilee, Jesus was born in Wales, the Garden of Eden was at Glastonbury, Edinburgh was the site of Jerusalem and Ireland was ancient Eqypt. London was Damascus. And Bristol? Bristol was Sodom.

Now make up some Bristol yarns of your own and see if you can get anyone to believe them.

Walk 8:
Unbuilt Bristol

Bristol's long history has seen the rise and fall of numerous imposing buildings and other structures. Should you want to find out about the history of any of them, there are plenty of books, academic papers and online resources that will tell you all about the history of the city's built environment.

But few books or academic papers ever note that for everything that gets built, there's a wealth of proposals for things that never made it beyond the drawing board or, in some cases, the back of an envelope.

This walk will look at just a few of the proposed, but never built, structures for the central area of Bristol. This tour can't invite you to look at anything; instead, you're being asked to try and imagine things that aren't there.

The walk is for the most part on level ground, with a few gentle slopes. There is one significant set of steps but this can be avoided with a short detour. Allow at least 90 minutes, or a little longer for stops for refreshment, relief and distraction.

The Walk

Start in Anchor Square, or near the entrance to We The Curious (**BS1 5DB**) ***This is where The Orbit Millennium Project* (1)** *would have been built.*

What's the point of looking at the stories of structures that were never built? Well, if you're seeking some intellectual purpose in this adventure, perhaps we can say that looking at the things that didn't happen can sometimes tell us as much about the past as looking at the things that did. People's values and assumptions change; over time, they can change a great deal and sometimes they change quite quickly.

Nowadays, for instance, we value nature and wildlife much more than our forebears. If you told one of our Victorian city fathers who was proposing a railway scheme that it might harm the wildlife, he'd have considered you a candidate for the Stapleton Asylum. Nature, as far as he was concerned, was to be conquered.

Likewise, if you were to tell a city father or mother in the 1950s and 1960s that roads and cars were a bad thing, they'd have thought you almost as deluded. A councillor of the political left – and many on the right, too – would have demanded to know why the working man and his family should be denied the same convenience and leisure provided by the car that the well-to-do enjoyed. If you told them that Victorian buildings

shouldn't be demolished to make way for modern ones, they'd have thought you a crank. Bright, clean new buildings of glass, steel and concrete represented progress, while fussy, overly-ornate Victorian ones just got in the way.

This walk will highlight some of the ways in which values and priorities in Bristol's built environment have changed. But to begin with, we'll look at something which was very much of its time, and one which many would doubtless still like to see happen.

We The Curious, Anchor Square, Millennium Square and all the buildings in the neighbourhood are at the heart of the transformation of the former City Docks into what we now call Harbourside. This transformation happened between the 1980s and 2000s and, if inclined, you could spend years studying the details of the debate – a sometimes bitter argument – that took place between developers, planners, councillors, amenity societies and citizens over the best use of this vast stretch of central Bristol real estate that was freed up by the closure of the City Docks.

The Orbit Millennium Project was just one tiny proposal among many bigger ones. It was, at the time, and remains, a hugely attractive idea. The project was conceived to celebrate the turn of the millennium – a scale model of the solar system. The idea came from Martin Rieser, a senior lecturer in electronic media at UWE, and from Bristol architect Mike Richards of Inscape architects.

The sun would be here by We The Curious and would be 840mm in diameter. Earth would be 100 yards from the sun – near this side of Pero's Bridge – and about the size of a pea. Mercury would be over on this side of the Cathedral, and Mars would be in Millennium Square. Jupiter, the largest planet, would be the size of a grapefruit. Pluto would be a mile and a half away on the Downs to the north of the Suspension Bridge, and would measure less than a fifth of a centimetre.

Each planet would be positioned to correspond to where they would rise on the first day of the year 2000. Because so many were tiny, each scale model would be in a window inside a three-sided obelisk six metres high. There would be nine obelisks, all finished in various materials, such as granite, marble, glass, bronze and stainless steel, and each would be designed by an artist. Each obelisk would also have a laser on top, so that on special occasions – starting with 31 December 1999 – the sky over Bristol would be lit up by a laser display.

At the centre of the scheme, the sun would be covered by a glass dome which would also contain an orrery – an interactive moving display of the solar system. Sensors would pick up the movement of people around it to create music patterns. This 'Music of the Spheres' was to be composed and programmed by Bristol composer Edward Williams.

It would have fitted perfectly with the At-Bristol exploratory (now called We The Curious), which was being built then, as well as being part of the wider development

City Hall suggested municipal buildings and war memorial.

around Millennium Square.

Everyone loved it. One local councillor even complained that there wasn't going to be a planet in his ward and could they do something about this please? Alas, the scheme failed because the bid for Lottery funding was rejected. The whole thing would have cost less than £1.5m but there were worries about the expense of the lasers, which weren't really necessary anyway.

....

Now go to Anchor Road and cross at the pedestrian crossing. Close by is a lane with five flights of six steps taking you past the Cathedral along Trinity Street. Go along here and into the middle of College Green. Beware of cyclists and scooters. You will arrive at City Hall (2). If you do not wish to use the steps, carry on along the Anchor Road footpath towards the Centre, and then double back towards College Green.

....

The story of what was originally known as Bristol's new 'municipal building', later the Council House and now City Hall, is long and complicated. They were talking about a new building in Victorian times to replace the Council House (now the Registry Office) in Corn Street. As Bristol got bigger and the responsibilities of the Corporation grew, the old Council House was not big enough. Nor was it grand enough for a city like Bristol, which was being shamed by the magnificent town halls being built for the industrial cities of the Midlands and North of England.

Various locations were considered and several different plans produced, but it was the usual old Bristolian story of delay, dithering, argument and indecision. At the end of the First World War, the plans also got entangled with the saga of Bristol's war memorial. One proposal for the new municipal building on this site from 1919 suggested that it should also function as a memorial to those Bristolians who had lost their lives in the war. This was only one of a number of suggestions for a memorial. There was a big debate over what form it should take. Should it be a monument or should it be something of use to the living – such as a hospital, concert hall, art gallery or even a grand bus shelter in the city centre? Another suggestion was to re-build Bristol Bridge, or at any rate to decorate it with sculptures symbolising victory and peace.

The city fathers argued and argued about it. In the end they opted for the cenotaph on the Centre. This was nothing like the grand schemes which had been put forward to begin with, and it was not unveiled until June of 1932, almost 14 years after the war had ended. Meanwhile, work on the new municipal building finally started on this site in the late 1930s. Construction was suspended again by the Second World War and it wasn't finished until the 1950s. The grand vision of architect Emanuel Vincent Harris included removing trees and levelling the ground on College Green; the flat expanse of grass you're standing on now wasn't always like that.

Harris's finishing touches to the building included the unicorns at either end (see also Walks 1 and 6) , but one other pair of additions never materialised, as we shall now see.

....

Now walk to the fountain at the Cathedral/Central Library side of City Hall. This will take you to where The Goddess of Sex Fountain would have been (3).
....

Take a good look at the fountain. Has it ever struck you as odd that it is bare plumbing? Just pipes with water squirting out? This is City Hall's shameful little secret; the fountains were never properly finished. On top of these pipes should be a 30ft tall bronze statue of a naked woman. Emanuel Vincent Harris planned on topping off the fountains with huge sculptures which, from a distance, would be framed by the great archways behind them. Harris engaged the famous sculptor Alfred Hardiman, who proposed two bronze nudes called Night and Day. Night was modelled on a Greek Goddess. Day, who would be rising out of the fountain at the Park Street end, was either going to be another goddess, or maybe a god. Perhaps Apollo. We don't know.

Hardiman made a half-scale maquette of Night, which was exhibited at the Royal Academy in 1945 and won a silver medal from the Royal Society of British Sculptors the following year. Harris approved the designs but then the plan ran into fierce controversy. There were letters in the local press expressing dismay that a massive naked woman was going to be positioned right next to the Cathedral.

Exposed plumbing in front of Bristol's City Hall (Eugene Byrne).

Said one letter:

> A sensually suggestive idol … a shameful flaunting of human nakedness in the face of God! The war was sent by God as retribution for the ungodliness of men, but heedless of this the Christian citizens of Bristol are now contemplating the erection, in full view of the Cathedral, of a modern goddess of sex!

But there were letters of support, too:

> As an ex prisoner-of-war, I have to tell you that if you had been incarcerated as I was, you would have been delighted to see College Green once more, even if it had a statue of a dustbin on it. The Council deserves praise for commissioning this work, though I do wish they would remove that statue of Queen Victoria holding a bomb from College Green.

Another wrote:

> I must take issue with the old women and prudes who object to the statue which I find very beautiful, apart from her hairstyle. May I suggest that those who find the

statue objectionable should wear gas masks when they are passing.

The Cathedral's Dean and Chapter came out against the statue as well, but Harris and the council stood their ground and insisted the statues would be built. And then … nothing. The statues were never cast and Hardiman died in 1949. Perhaps the council simply preferred to bury the controversy and save some money at a time of great austerity. It is also possible that in the postwar years, it would have been difficult to get hold of the enormous amount of bronze needed. And so the plumbing on the fountains remains exposed to this day.

Now go back along College Green towards the Centre and take the pedestrian crossing opposite the Tesco Express. Take the path into the Centre, along St Augustine's Parade, past the Hippodrome and use the pedestrian crossing by Dominions House, opposite to Baldwin Street. Find a convenient place to stop on the traffic island, clear of cyclists and scooters. This area could have had trams running through it (4).

We all call this area the Centre, although pedants will inform you that the historic centre of Bristol is elsewhere (at the junction of Corn, Wine, High and Broad Streets). It became the Centre because it was originally the Tramways Centre, the heart of the city's Victorian network of horse-drawn trams. The system was electrified in the 1890s and, for a while, Bristol had one of the most technologically advanced and widely-envied public transport systems in the country. The trams, along with suburban railways, made Bristol's massive growth in the second half of the nineteenth century possible, as people could now easily and cheaply travel long distances to their workplaces.

By the 1930s, the system was unpopular and outdated. The upper decks of the trams were open to the elements and new motor-buses were more flexible. For example, they could cater for the new council estates without new tramlines having to be laid. The trams were being phased out by the late 1930s and disappeared completely after the main power line to the system was severed by a German bomb in 1941.

By the 1970s, the growing numbers of private and commercial vehicles were making traffic congestion a problem, and Bristol started talking about a revived tram or light rail system as a solution. The first serious and viable proposal came in the 1980s with a company called Advanced Transport for Avon (ATA), with a system modelled on London's Docklands Light Railway. Much of it would use existing or redundant rail lines. The first route would go from here to Portishead, and later ones would go to Filton, Emersons Green, Yate and Bradley Stoke, Hartcliffe and Weston-super-Mare. The costs were projected at £230m but these would have been met by the private sector.

ATA was led by Richard Cottrell, an extrovert former broadcaster and Tory MEP for

Lovely trams in Bordeaux that we can't have (Eugene Byrne).

Artist's impression of the proposed 1980s Avon Metro.

Bristol. He called it 'a free gift to the people of Avon'. Part of the idea behind the scheme was that it would be paid for by 'planning gain' – the presumed increases in the value of land near the Metro stops. But there were problems. Cottrell was a Conservative politician and some of the company's other members were aggressively Thatcherite in tone. Bristol's Labour-run council and Bristol South MP Dawn Primarolo were concerned that the local public transport system would be in the hands of the private sector. The council also raised difficulties about the key part of the system, right here in central Bristol.

The scheme did, however, enjoy some support from Avon County Council but, to cut a long story short, the property bubble in the wider world burst and the company went under. And there ended the first proposed scheme, a perfectly sensible one. And we have been talking about trams, light rail systems and even an underground system ever since.

Take the pedestrian crossing onto Colston Avenue and walk northwards, away from the harbour, along Colston Avenue which turns into Quay Street until you come to the archway of St John's Church. Go through the archway on your right and take the immediate turning left onto Tower Lane. Walk along Tower Lane until you get to the Bank Tavern on your right. The next project we will discuss is streets in the air (5).

You have just been following the line of part of the old city walls. This is a very ancient part of Bristol but much of what you have passed will have been things built from the 1960s onwards, including a few of the last vestiges of late 1960s/early 1970s plan to build streets in the sky.

In the mid-1960s, everyone could see the volume of traffic in Bristol was constantly growing. And now there would also be motorways which would bring business here from London and allow families to take holidays in the South West. There were also plans to build a new bridge over the Severn. Something had to be done about the traffic and the city planning department dusted off a proposal originally suggested by the Bristol Architects Forum in 1959.

This was to keep all the traffic on the roads, while pedestrians would move around on pathways on an upper level. The system would cover all of central Bristol, from Lewins Mead to College Green. It would go up as far as Park Row, taking in the recently-opened New Bristol Centre. There would be walkways across the Centre, with a piazza in the middle of it. You could walk from the Bristol & West Building (now the Radisson Blu Hotel) or the Colston Hall (Bristol Beacon), as far as Broadmead without encountering any cars or having to cross a road. It was also wheelchair and pram-friendly. To use this network, you wouldn't have to use steps; it would all take advantage of topography.

Castle Street before the German air raid (*Bristol Post*).

Your route would not merely take you over bridges; for much of your journey you would be going along sidewalks next to the first or second floors of new office buildings and apartment blocks. There would be shops, pubs and cafes, too. It seems ambitious and expensive but on paper, at least, it was feasible. At the time there was a major office building boom going on – we now use those offices to accommodate students – and any developer wanting a piece of Bristol's property market would be required to put their part of the system into place. Many Bristolians hated the idea of making the streets of Bristol a no-go area for pedestrians, and a lot of developers lobbied furiously to water down the conditions.

In the early 1970s, the council commissioned a report from Ove, Arup and Partners, which concluded that it wouldn't make the traffic move any faster, would increase noise and air pollution and shops on the upper decks would probably never be sufficiently profitable. And it fizzled out. But here you can see some of its remains. Near here, there used to be a pedestrian bridge across Rupert Street leading to windblown plazas atop office buildings opposite. The plazas are still there but the vision is long gone. The terraces on the first floor of the Beacon Tower and Bambalan bar on the Centre are actually the vestiges of a pedestrian walkway which was built-in for this very purpose when the buildings were constructed.

Turn right by the Bank Tavern and go along John Street to Broad Street. Turn left into

Castle Street after the German air raid (*Bristol Post*).

Broad Street and carry on to the junction with Corn Street and High Street. Stop on High Street by the Bristol Cameras shop. The site you are looking at now is where you would have found The Dutch House (6).

Across the road, the site of the derelict Bank of England building was the location of Bristol's famous Dutch House. Its frontage was roughly where the traffic island in the middle of High Street now is. It was built in the late 1600s, and over its history was used for offices, shops and homes. It was a much-loved landmark at the centre of the old city and featured extensively on postcards and railway posters.

It was seriously damaged in the first major night-time air raid on Bristol, on 24 November 1941 and the decision was made to pull it down. Ever since, there have been periodic calls for it to be rebuilt on the same site. In 2012, for instance, mayoral candidate and architect George Ferguson said that Bristol should rebuild some of the architectural treasures lost in the war. A new Dutch House, he said, could either be a faithful reconstruction or it could be a contemporary equivalent.

Proceed along High Street, with Castle Park to your left-hand side and take the pedestrian crossing over to Castle Park (7). Go into the park and find a decent vantage point on the grass before you get as far as the ruins of St Peter's Church.

For centuries, there was a castle here. When it was pulled down on Oliver Cromwell's orders in the seventeenth century, many of the stones were used in what was the city's first great speculative building boom. You're standing in what used to be a dense area of streets, where shops, pubs, homes and businesses sprang up.

It all ended on 24 November 1940 with the first major German air raid on central Bristol, which destroyed or damaged much of Castle Street, Wine Street and the surrounding area. The burned-out shell of St Peter's Church was left standing as a permanent memorial. There have been disputes about what to do with the area ever since. In the initial postwar plans, the area was zoned for culture/amenity purposes. There was to be a park, but also some sort of civic centre or arts centre or exhibition hall.

At one point in the late 1960s, the Arnolfini, which was then an art gallery in Queen Square, came up with plans for an arts centre on the site. There would be a gallery and exhibition space, a cinema, a performance space, craft workshops and a bar and restaurant. The latter would also offer fashionable *cordon bleu* cookery courses for the aspiring middle-classes. This failed because the council couldn't find the money for its share of the plan – the landscaping – and so the Arnolfini moved to Bush House, its present home on Harbourside.

In the late 1950s, the council went against its own guidelines and decided to lease some of the land at the Bristol Bridge end for two office buildings. Despite considerable protest, the Norwich Union and Bank of England buildings were built. They have now been derelict for decades and condemned by many as an embarrassing eyesore. There have been several attempts to replace them. In the 2000s, a local firm proposed a mixed-use development of offices, shops and flats. Bristol City Council, which owns the land, was behind the plan but, because it would involve taking up some more of what was now established parkland, it met with furious opposition. Thousands signed a petition opposing the development. Proposals were batted to and fro before this plan failed. In 2023, a new proposal was put forward by property developers MEPC, with plans by Bath architects Fielden Clegg Bradley. Work on this scheme was expected to start late in 2024.

While we're here, take a moment to admire the curvy Castle Bridge. We might have had something very different in its place, namely the **Mobius Bridge** (**8**).

Across the water is Finzels Reach, a large development on a former industrial site taking in much of the old Courage brewery – George's brewery before that. The original development ran into financial troubles some years ago, and one of the casualties of the first stage was a bridge for pedestrians and cyclists which would have been quite remarkable. This was the so-called Mobius Bridge, a giant Mobius Strip which would have crossed from Finzels Reach to Castle Park. It was designed by a London architect named Julian Hakes. This design also inspired him to produce a line of women's shoes, which look as though they're made of orange peel. Google 'Hakes Mojito' and see for yourself.

Go back towards Bristol Bridge and cross by the bridge (*but do not cross the bridge*) **onto Baldwin Street. Turn left onto Welsh Back and continue, taking the second right onto King Street. Go down King Street until you reach the Old Vic and stop across the road from its entrance. This is where there would have been a huge tower (9) built by the General Post Office.**

Jerry Hicks GPO Sketch (*Bristol Post*).

In 1971, the General Post Office (GPO), which was then responsible for the nation's telephone services as well as the mail, proposed a building behind the Theatre Royal (now Bristol Old Vic). It would house a telephone exchange and microwave aerials as well as offices. This would be a tower some 305ft tall. That's half as high again as the Bristol & West tower – the Radisson Blu – and four times as wide. It would have been 100ft higher than the tallest commercial structure in Bristol at the time, the DRG building (aka the Robinson Building, now One Redcliff Street), and taller than the spire of St Mary Redcliffe. We do now have a taller building, Castle Park View. The GPO building would have been almost as high.

The plans sparked protest from Bristol Civic Society and others, who said it was grossly inappropriate. It would loom over the historic streets of Welsh Back, Queen Square and King Street. A particularly effective episode in the campaign against it was the publication in the local press of a simple line-drawing of the massive bulk of the proposed building dwarfing King Street and the Theatre Royal. The drawing was by Jerry Hicks, a local art teacher and Civic Society activist; in an article in *Architectural Design* magazine he satirised the design as part of a sinister plot to make Bristol as terrible a place to live in as possible:

The destruction [of King Street] will be achieved visually by dwarfing its scale to a point where it appears ridiculous. Immediately behind the Theatre Royal ... will loom a gigantic edifice hundreds of feet high full of GPO machinery and 16 men.

Campaigners also talked of flying a balloon at 300ft over the site so that passing Bristolians would see exactly how tall it was going to be. The design was rejected by the council's Planning Committee and in April 1972 the Post Office announced the project would be abandoned.

....

Walk a little further along King Street, then turn left onto King William Avenue and continue to Queen Square. Should plans have gone through, this is where Bristol Central Railway Station (10) would have been.

....

Bristol's main railway station is in the wrong place. This was as true in Victorian times as it is today. Through the second half of the nineteenth century, a number of rail companies proposed putting the main station in central Bristol. These all failed for various reasons.

The first important station proposal came in the early 1860s; it was going to be right where you are standing now. The Bristol and Clifton Railway Company would build a line from Temple Meads east to west through the city, through Redcliffe, across the water at Welsh Back, across Queen Square on arches and a low embankment. This embankment would accommodate a new main passenger station for the city. The railway would then continue out along the line of Anchor Road to terminate at another station at the bottom of Brandon Hill. The idea had a lot of support, and not just because the new passenger station would be in a better location. It would link the City Docks to the rail system for the first time, and horse-drawn tramways on quays would mean that goods could be moved quickly from ships to trains.

The idea was killed off by vested interests, and one of the most bizarre lobbying stunts in Bristol's history. Railways coming into the docks threatened the profits of some and the livelihoods of many more – merchants, shipmasters, traders, warehouse owners, dock workers and lightermen. Opponents enlisted the support of a now-forgotten group of local workers, the women who made a living taking in laundry and then drying the washing on Brandon Hill.

The railway would need Parliamentary approval. A newspaper account of some of the proceedings tells us:

> A bill now before Committee of the House of Commons for constructing a line of railway between Bristol and Bath, has to encounter a curious kind of opposition from a large body of 'laundresses' and washerwomen, who ventilate their linen on

George V Memorial (Eugene Byrne).

the line at Clifton, and who contend that any accommodation it may confer for the conveyance of their cargoes of clean clothes between Brandon-hill, Bristol, and Bath, will be counterbalanced by the destruction of their drying grounds, and the incessant scattering of sparks and 'blacks', or what the French call 'London snow', from the funnels of the locomotives.

The washerwomen who made a living around Jacobs Wells and Brandon Hill were obviously fearful that steam locomotives were going to affect their livelihoods. The Parliamentary bill was defeated, partly at least because of the testimony of 'three respectable washerwomen' who lived on Jacobs Wells Road.

From Queen Square, go through Thunderbolt Square towards the city centre, crossing over Prince Street at the pedestrian crossing. Go along Broad Quay and left onto Narrow Quay so you are by the harbour. Stop at the fountain with the brick arch over it. Half of this is the King George V Memorial (11).

King George V, great-grandfather of Charles III, had been very popular in Bristol. He'd visited the city far more often than his father or his grandmother, and when he died in January 1936 it was felt that Bristol might have some sort of memorial to him. An appeal was launched and money was raised, some of it spent on a children's playground

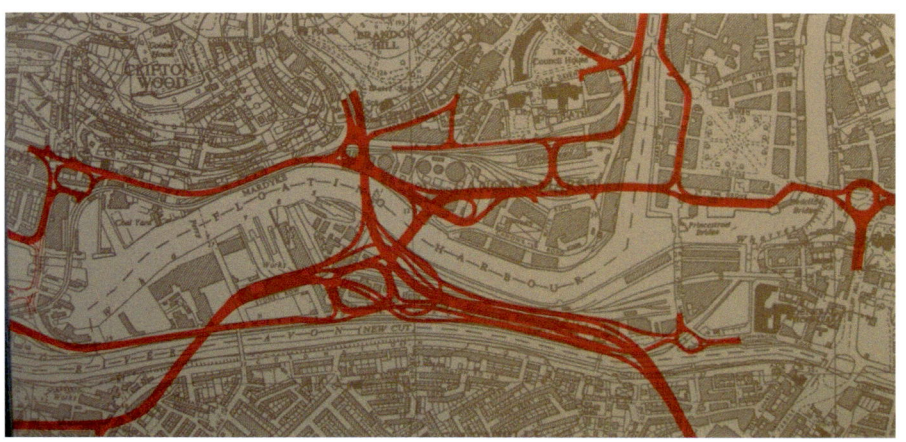

Casson Report Map.

in Barton Hill/St George. King George's Playing Field was opened in July 1938.

The rest of the money was to go on a monument. This would fit nicely with council plans in the late 1930s to cover over more of the water in the Centre, and to landscape the area. The Cenotaph would be a feature at one end of this landscaped area, and the George V monument could be a feature at the other. This was to take the form of a 60ft tall stone 'pylon' topped by a 16-sided bronze lantern which would be lit at night, reflecting George's connections with the Royal Navy and navigation, and Bristol's own seafaring heritage.

Plans were postponed by the outbreak of war in 1939. There were intermittent discussions about it in the post-war years, with various alternative suggestions including, apparently, a sunken pool and fountains, or a 68ft metal obelisk. By the late 1960s, the memorial fund to which Bristolians had donated 30 years previously had reached about £8,000 and was becoming an embarrassment, due to the significant delays in the building of the monument.

In the end, a memorial was built using the donated funds, although one imagines that scarcely one Bristolian in a hundred can point it out. Well now you, at least, will be able to! On one side is the face of a lion, and on the other are two statues – a farm worker and coal miner – which were salvaged from the old Co-Operative Wholesale Society building that had stood near here until it was demolished in the 1970s.

The two figures were from a set of six representing different occupations that had stood over the CWS building entrance. Many Bristolians felt it would be a shame if they were lost and so two were incorporated into the memorial, which was paid for by the King George V memorial fund and unveiled by Lord Mayor George Maggs in May 1982. Whatever the fountain's artistic merits, it is the product of a bizarre compromise.

Continue along Narrow Quay, and stop by the Arnolfini (12).

Try and imagine a dual carriageway thundering across the docks right next to Bush House, the former dockside warehouse that has long been home of the Arnolfini. In the late 1960s, the old City Docks, owned by Bristol City Council, were losing money. It was awkward for large ships to come up and down the river Avon, and the docks weren't adapted to the newer and bigger container ships. Well over 90 per cent of the city's maritime trade was now going to Avonmouth.

The council decided to close the City Docks to shipping, a decision that was also prompted by talk of the government nationalising commercial docks. No Bristol council, Tory or Labour, would willingly hand over a massive part of the city centre to the government. But it was also a great opportunity. If you covered over or filled in the docks you could recover a huge amount of land for new developments. Most of all, the area could be reclaimed for the massive road schemes the council was planning.

The Bristol Civic Society, along with other amenity groups, particularly the Clifton & Hotwells Improvement Society, fought a furious campaign to save the docks, not as a working harbour but as a water feature which could be used for leisure, and as the backdrop to houses, offices, bars, restaurants and so on.

In the end, the council's plans were defeated and the docks were saved, partly because of the economic downturn of the 1970s, and partly by growing objections to the road schemes, which councillors could now see would affect their wards as well as central Bristol. The amenity groups and other campaigners were not responsible for stopping the council's plans for the docks – but they had won the argument. By the mid-1970s, almost everyone accepted that the docks were an asset.

At the same time, the Civic Society and other groups were campaigning against other developments, most notoriously a plan to build a hotel – or rather to extend the Avon Gorge Hotel – on the side of the Avon Gorge. These events, this civic revolt against the destruction of old buildings, and a cultural change in the wider country, changed things dramatically. By the time the money and confidence were available for big developments once more in the 1980s, everything had changed. The council had to be more open and responsive, and old buildings were no longer knocked down willy-nilly. We preserve any old Victorian pile now – or at we least stick a new building behind its old façade.

Stay where you are and look across the water towards the gap between V Shed and the Lloyds Amphitheatre. This is where The Harbourside Centre (13) *would have been.*

Two views of the Behnisch & Behnisch model of the Harbourside Centre (Behnisch & Behnisch).

Once the docks were saved from concrete, there were all manner of arguments from the 1970s to the 1990s about what the area should be used for, what should go where and how big it should be. In the end, though, there was broad agreement that this should be an area for residences, businesses and leisure. Arnolfini, behind you, was an early leader in the regeneration of the area, ensuring from the start that what we came to call Harbourside would be a cultural destination, too.

Between then and now, all the different proposals for structures around the City Docks would easily fill a big book. But the one that perhaps most people regret not happening was the Centre for the Performing Arts, aka The Harbourside Centre.

This was to be the focus of the regeneration of the Canons Marsh area in the 1990s

– a landmark concert hall and performance venue that would be Bristol's answer to the Sydney Opera House. The project had been put out to an international design competition in 1996. It was won by Behnisch & Behnisch of Stuttgart with a startling design famously compared to an exploding greenhouse. In 1997, Arts Council England awarded £5.4 million to get things moving. It was due to be completed by 2002 and would feature a 2,300-capacity concert venue, plus a smaller 450-seat auditorium. It was going to be home to the Bournemouth Symphony Orchestra and the Orchestra for the Age of Enlightenment. The centre would even have its own composer-in-residence and in-house choreographer and two dance companies. The project had the wholehearted support of the council and the local arts and business communities. Of its projected £89m cost, some would be raised in sponsorship, some would come from Bristol City Council, and some £58m would come from the Arts Council's Lottery Capital Programme.

In 1998, the Arts Council withdrew its support, citing a number of vague, ill-defined management problems. Effectively, they'd decided they could no longer afford it following new policies on Lottery money introduced by the new government. At the time of writing, the site remains vacant.

This ends the tour. Take a break at one of the cafés/bars on the Harbourside and have a drink or meal and imagine how the city might have looked or how you might make it better.

Five 11-storey blocks under construction at Hartcliffe, 1964 (Bristol Archives 40826/ HSG/56/2).

Homes for Heroes 100

Council Housing in Bristol Since 1919

Homes for Heroes 100 was a programme of coordinated community projects, special events and new publications marking the centenary of the Housing and Town Planning Act 1919 (sometimes referred to as the Addison Act). This Act led to the development of the first large-scale council estates in the UK. This was not just about building homes; it was also about creating new communities and changing the social fabric of the country.

Originally promoted as homes for the returning heroes of the war and as fresh starts for those displaced by slum clearance, in recent decades council housing – and the social housing that has partly come to replace it – has attracted an unwarranted stigma. Council house residents have been marginalised for generations and the culturally important heritage that lies within their estates is little-known. Through community-based heritage research, hands-on creative activities and showcase events, the Homes for Heroes 100 programme aimed to celebrate the council estate and its residents – past, present and future.

In November 1918, Prime Minister David Lloyd George delivered a speech in which he called for 'a country fit for heroes to live in' (frequently misquoted as 'homes fit for heroes'). During the First World War, the Tudor Walters Committee had been set up

Postcard of Forest Avenue, Hillfields Park estate in Fishponds, early 1920s (Bristol Archives 43207/4/17).

with the aim of improving the country's housing provision by ensuring higher standards of design and location. This had been prompted by the realisation that many recruits had health problems which were the result of inadequate living conditions at home. Housing became a major government responsibility in the post-war period because the private sector was unable to meet demand. Although widely accepted as a bold vision that would make a positive difference to the lives of citizens, it could also be argued that the council housing programme was at least partly driven by political necessity. It was a means of smoothing the transition to peace, driven by the government's anxiety to avoid an uprising in the wake of the Russian Revolution of 1917.

With the 1919 Act, councils across the country began to establish housing committees, guided by recommendations from central government. Before 1919, local authorities had supported a range of rentable housing opportunities for those on low incomes who were unable to afford private sector rents or to buy a home of their own. In Bristol, for example, an early instance of council housing was a municipal lodging house for 123 men built in 1905 on Wade Street, St Jude's, under powers granted by the Housing of the Working Classes Act 1890. Other small-scale pre-war developments included what were described as 'workmen's dwellings' built in 1901 in Baptist Mills, Easton, St Philips and St Werburghs. The West Block at 19A-19J Mina Road, St Werburghs, survives to this day. (An optional diversion on the St Pauls walk will take you there.)

However, nothing had previously matched the ambitious scale of state intervention

Communal kitchen in the Wade Street lodging house depicted by Samuel Loxton (Bristol Reference Library A47).

embodied by the new Act It is from this point that the term 'council housing' becomes widely recognised. These were homes where costs were shared between tenants, local rate payers and the Treasury. The first council-estate homes to be completed in Bristol were located on Beechen Drive (1919), and the first houses to be occupied were on the corner of Briar Way and Thicket Avenue (1920). You will pass both of these sites during the Hillfields walk.

Other estates that were largely completed during the early 1920s included Knowle, Shirehampton and Sea Mills. The first sod at Sea Mills was cut by Christopher Addison, the government minister who was largely responsible for the Act, in a ceremony that took place on 4 June 1919: an oak was planted in Sea Mills Square to mark the occasion. You will see the mature tree during the Sea Mills walk.

Housing supply still fell well below demand. Out of the original 5,000 homes planned in Bristol under the 1919 Act, only 1,189 permanent and 141 temporary dwellings were completed before work was suspended following Addison's resignation from government and the ending of the subsidy. The following passage is from a booklet entitled *Bristol House Famine Campaign, 1923* (Bristol Reference Library B21855 p3).

> The house famine must be relieved, and it can only be relieved by the building in the immediate future of several thousands of dwellings. Thus a very serious situation

has to be faced, and faced without delay.

Just pause for a moment and try and think of your own Home and then try and imagine what this shortage of houses means to the thousands of unfortunate people – men and women, youths and young women, and little children herded together, two, three, or even four families in one house, boys and girls occupying the same bedrooms, common decency, to say nothing of privacy, quite impossible. Try and imagine the babies born every day under these conditions, and the mothers at childbirth and after. Think of sickness and death when the whole family inhabit one or two rooms. Think of these things and the grave possibilities to all citizens of Bristol inherent in the housing shortage will be clear to you.

It was calculated that there was a shortage of 7,270 habitable dwellings in the city in the early 1920s.

Work did eventually resume on building new homes: first under Neville Chamberlain's Housing Act 1923, which favoured the subsidising of private builders erecting houses for sale, and then by John Wheatley's Housing Act 1924, which provided subsidies to build cheaper council homes for which lower rents could be charged. To reduce building costs, the size of council homes had already begun to shrink while estates had increased in density.

The Housing Act 1930 focused on the post-war inner-city slum clearance and relocation programmes. The quality of the new homes generally remained good, although council housing was increasingly associated with the very poor rather than the more affluent members of the working-class for whom it had originally been planned.

Many residents of the new suburban estates would have previously lived in homes deemed unfit for human habitation. The following is from The Housing Needs of Bristol in Relation to the Greenwood Act, February 1931 (Bristol Reference Library B22170 p25).

> Considering the conditions under which they had lived, with insufficient accommodation and lack of proper facilities for cooking, washing, drying and other domestic duties, without gardens to afford relaxation or to stimulate interest in pursuits other than ordinary daily toil, it is generally regarded as highly satisfactory that the response of tenants to improved conditions has been so marked. They take a real pride in their new homes and gardens are well cultivated.

Between 1919 and 1939, 21,985 homes were built privately in Bristol, of which 3,020 were state subsidised. Over the same period, 14,610 council homes were built of which 711 were built without state subsidy. (These figures are taken from *Housing Estates* by

Photo of housing clearance in St Philips from the 1934 Bristol Housing Report (Bristol Reference Library B14102).

Public opening of a pre-fab at Shirehampton (Bristol Aero Collection).

R Jevons and J Madge, Arrowsmith 1946). The newer council estates included ones in Bedminster, Horfield, St Anne's Park, St George and Southmead. As with the earlier estates, they lacked enough shops and services to meet the needs of residents for at least the first few years of occupancy, and they were inconvenient for those still employed in the city centre and dependent on costly and unreliable public transport.

(This is still often the case with new developments.) The residents gained a superior quality of home at an affordable level of rent, but the clearance programmes meant that they also often lost contact with a strong, settled community that had grown up over several generations.

The Second World War interrupted the house building programme. By the war's end it was estimated that 750,000 new homes were required to meet demand in England and Wales alone. This was the result of the ongoing housing shortage and slum clearances as well as the loss of so many dwellings from bombing raids.

Among short-term solutions was the construction of pre-fabricated aluminium housing in kit-form that was estimated to have a life of ten years. In February 1944, the Minister of Aircraft Production, Sir Stafford Cripps, brought together representatives of the UK's leading aircraft manufacturers to discuss how they could help solve the housing crisis, recognising that the industry possessed the necessary skills, technology and materials. This work was overseen nationally by the Aircraft Industries Research Organisation on Housing, which had been jointly formed by the Ministry of Aircraft Production and the Ministry of Works. Founder members included the Bristol Aeroplane Company. Other companies were also involved in pre-fab construction locally and nationally and it is estimated that nearly 157,000 pre-fabs were erected between 1945 and 1949, of which around 2,700 were built in Bristol.

In 1955, it was reported that 43 families per week were moving into new Bristol council homes. Some of the shortcomings of the estates of the inter-war period were repeated in those built at Hartcliffe, Henbury, Lawrence Weston, Lockleaze, Stockwood and Withywood after the war. They too had limited local amenities and public transport. They also suffered from the abandonment of the garden suburb principles – which had been a redeeming feature of the earlier developments – and other cost-cutting measures. The increasing use of pre-cast reinforced concrete construction resulted in major structural problems (sometimes referred to as concrete cancer) that would begin to come to light in the 1980s and eventually lead to the demolition of hundreds of council homes across the city.

The Right-to-Buy policy introduced with the Housing Act 1980 meant that councils had to sell homes to sitting tenants who wished to become owners at the same time as the building of new homes was being cut back, thereby drastically reducing availability. By the first decade of the twenty-first century, many local authorities were faced with major housing debt and restrictions on investment, bringing a complete halt to new building. Funding was available from government for improvement and regeneration, but this was often conditional on transferring ownership and/or management to not-for-profit housing associations (also known as Registered Social Landlords). These stock transfers had been facilitated by the Housing Acts of 1985 and 1988. This was

Photo of prize-winning front garden created by Mr Edgell at his home in Hartcliffe, from the 1955 Bristol Housing Report (Bristol Reference Library B14100).

Hartcliffe, March 1988 (*Bristol Post*).

Design for new housing at Challender Court, Henbury (Emmett Russell Architects).

accompanied by a shift from the term 'council housing' to 'social housing'.

Council housing had become a safety net for the most vulnerable in society, but more recently there are signs that it is becoming a tenancy of choice again rather than a tenancy of last resort. Bristol City Council is committed to delivering a renewed building programme. The Labour government elected in 2024 is also committed to a major programme of house building.

Homes for Heroes 100 aimed to raise awareness of the mistakes of the past that should not be repeated – including problems of social alienation, lack of essential facilities and limited availability – as well as the many positive outcomes that should be publicly recognised and aspired to in the future.

4 June 1919: 4 June 2019

On 4 June 2019, the centenary of Christopher Addison's visit to Bristol was celebrated with two events.

Cllr Paul Smith, dressed as Addison, reads extracts from the minister's speeches in support of the Housing Act 1919 at a birthday celebration for the Addison Oak, Sea Mills.

Tree-planting ceremony at Ashton Rise, South Bristol. This marked the start of a mixed tenure housing development built by Willmott Dixon where the proceeds of homes sold to the private market were used by Bristol City Council to construct new council houses (both photos: Evan Dawson).

The Centenary Poem: Vanessa Kisuule

Bristol City Poet Vanessa Kisuule read her new poem at the celebratory events that took place on 4 June 2019.

> Close your eyes
> Picture the house of your dreams.
> Is it nestled in the wooded ribs of a glade,
> laced by the gentle sound of the sea?
> Perhaps perched on a hill overlooking
> The twinkling lights of the city
> Looking down at those that
> live side by side and top to toe.
> When we imagine perfect homes
> They're tucked away on private acres
> Unsullied by the bonds of social living
> Yet we lament the rise of loneliness
> the sickness making graveyards
> of us long before our last breaths.
> The underclass is frowned upon,
> their livelihood a punchline, a
> cautionary tale told with half
> its chapters missing.
>
> Their grievances are many, but
> Too few of us listen.
> So many facts too often forgotten:
> the good faith that built these dwellings
> The rich communities that flower here,
> every family behind every window
> With a story as unique as a fingerprint
> These buildings once trembled
> with the soft glow of utopia.
> In the aftermath of war, a bold law was passed:
> An act to build homes that would last.
> Fertilised by green space and great hopes,
> Owned and enjoyed by those in need.

A young man hollowed out by
The horrors of combat could return
To a home 'fit for a hero',
A low earning mother could
raise her children in a house
with working lights and running
water.

This was not a given.
It still isn't. We have yet to make
good on this 100 year promise.
We've neither the space or
luxury to be islands, not whilst
Waiting lists for houses gets longer
And the life span of the homeless
Gets shorter. Let idealism
Gleam on the horizon once again
as it did in 1919, bolstered
by the lessons we've learnt
Let's meet the ever urgent
need for all of us to live
amongst and for each other
In a city where everyday living
Makes heroes of us all.

Vanessa Kisuule reading her poem beneath the oak at Sea Mills, 4 June 2019 (Evan Dawson)

The Walks

The walks in this book give you an opportunity to look at the development of three inter-war suburban council estates and to explore an inner-city residential area that is more than 200-years-old.

The routes are mainly on level pavements with no significant hills to climb or descend. Pavement surfaces are generally good and there are no high kerbs to negotiate. The walks will each take around an hour to complete at a leisurely pace, allowing time to linger at particular points of interest. As you will be walking through residential areas, please be respectful of other people's privacy, especially when taking photographs.

Please note that although we realise that the local authority would have been referred to as a corporation during much of its history, we are using the word 'council' throughout these sections.

Walk 9:
St Pauls

This walk through St Pauls examines the area's housing history, which dates back to the late 1700s. It considers how residential neighbourhoods have changed over time and how poverty and inequality have informed the city's housing stock.

Charting the rise of municipal and state-provided housing in this inner-city area, it includes a look at compact, low-rise apartment-block council housing of the 1960s, but may also take you past the tents and makeshift shelters belonging to people who are homeless. This contrast shows that the issues which first inspired the need for publicly-provided housing have yet to be resolved.

(Note that 'St Pauls' is sometimes written with an apostrophe and sometimes without. For consistency, we are not using the apostrophe in this publication.)

The starting point, Brunswick Square (1), is a short walk from Bristol Bus and Coach Station. The garden in the centre of the square provides a pleasant place in which to read the opening text.

In the late-eighteenth and early-nineteenth centuries, St Pauls was set to rival Clifton as a fashionable residential district for the city's wealthier citizens. Brunswick Square was laid out at a time when Bristol was starting to spread eastwards. An advertisement in the Bristol Journal on 15 February 1772 referred to uninterrupted views that could be enjoyed from a ground floor window at a rental property which adjoined the square.

The houses in the square were constructed on behalf of various speculators – such as Messrs Lockier, Macaulay and Co – who had interests in the African trade and Guyanan plantations. Construction was intermittent. Some houses were occupied by 1771 but the east side was not completed until the mid-1780s and the north side was not finished until some years later.

Brunswick Square's – and St Pauls' – elite status was short-lived. Industrialisation and an expanding urban population brought prosperity to a few but was accompanied by cholera, pollution and congestion for the many, compounded by poor drainage. By the 1840s, the wealthier residents abandoned the area for Clifton and elsewhere. Many single homes were converted into multi-occupancy lodging houses or commercial premises.

As conditions deteriorated, private landlords – many of whom were themselves struggling financially – resisted the city's attempts to impose sanitation reforms. By the

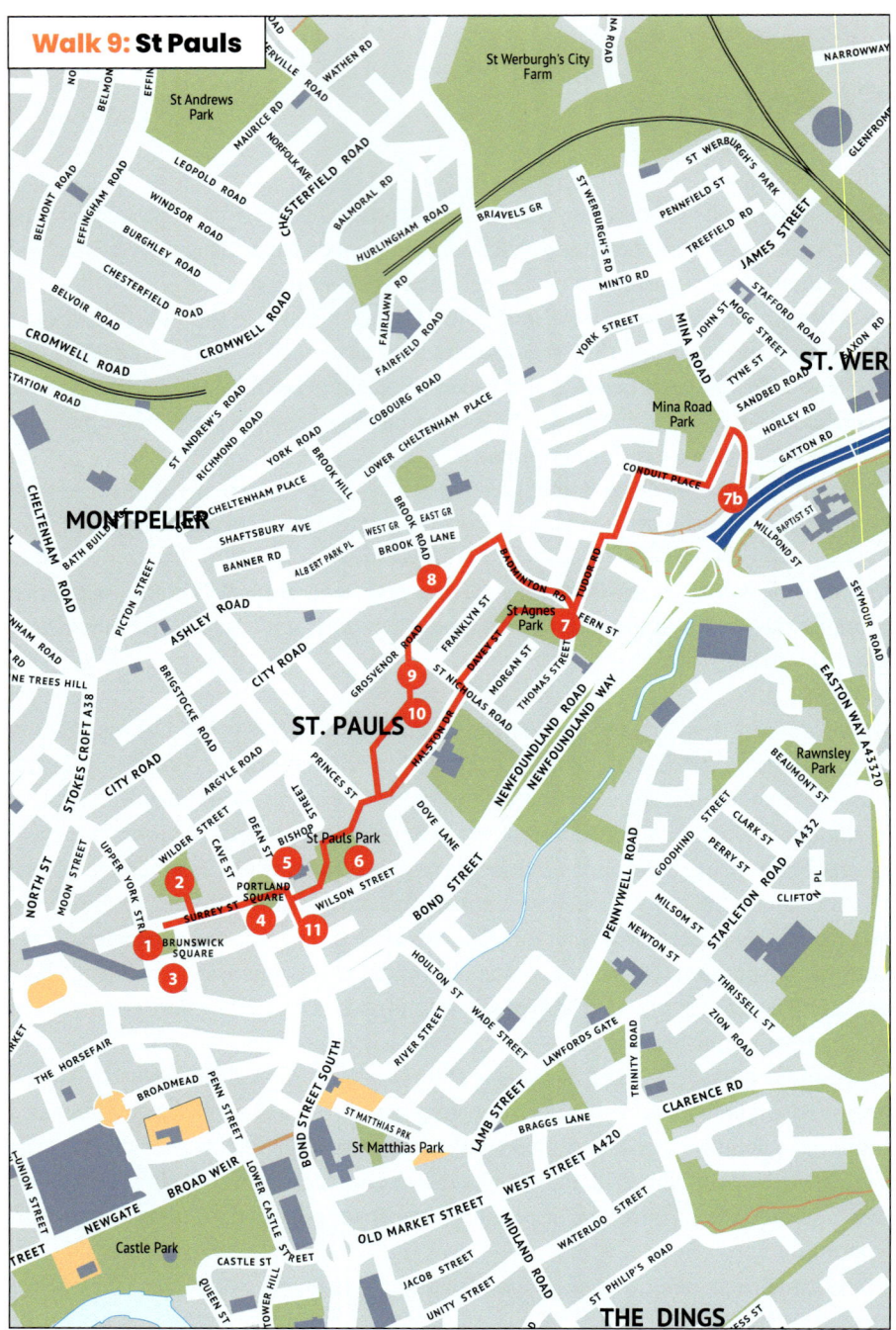

Aerial view of St Pauls taken in the 1930s (Bristol Archives 44819/3/224). You can see Brunswick Square, Portland Square, St Pauls Church and, at the top right, St Agnes Park. The area that would become St Pauls Garden Estate still comprises terraced streets in this image.

1880s, poor Jewish immigrant families fleeing Eastern Europe moved into many of the area's buildings, working as boot and shoe workers and cabinet makers, and living ten to a room in what were sweatshops.

A 1931 report, entitled *The Housing Needs of Bristol in Relation to the Greenwood Act* (Bristol Reference Library B22170), examined five areas of bad housing in the city, including St Pauls where over a quarter of the 117 houses surveyed were deemed to be overcrowded. It was reported that 'the odour of pickles does not appear to disturb the number of rats in this neighbourhood' (this was a food associated with Eastern European Jews). Vermin, leaking roofs, rotting floors and uninhabitable rooms meant many of those interviewed were anxious to leave for the sake of their children. The report cited the case of a husband and wife and their six children (two of whom were adult) living in one rented room who had been on the waiting list for council housing for a year.

When people from the Caribbean responded to the invitation to help rebuild Britain after the Second World War and began to settle in St Pauls, critics were quick to blame them for the deterioration of the area. However, St Pauls had already become run-down. That is why it was the first port of call for poor migrants looking for somewhere cheap to live, just as it had been 50 years previously. Directly after the war, St Pauls had attracted de-mobbed Poles, Ukrainians and people from the Baltic States who wanted to remain in this country. In addition to those from the Caribbean, the later 1940s and early 1950s also saw the arrival of Hungarians, Irish, Cypriots, Italians and South Asians, among many others. A *Bristol Evening Post* article on 21 March 1968 reported that 50 percent of the children in the primary schools of St Pauls and the neighbouring parish of St Agnes were first or second generation immigrants. Because one had to be resident in Bristol for a year before even getting onto the council housing waiting list, immigrants were largely confined to sub-standard private accommodation – and racism and xenophobia further limited where they might live. The vicar of St Agnes was quoted in the article saying: 'We are the last great reservoir of rented accommodation in Bristol'.

....

Facing the Unitarian Meeting Hall, on the north side of the square, take the path on your right that leads into Brunswick Cemetery Gardens (2).
....

An article in the *Bristol Evening Post* on 13 January 1956 referred to the Brunswick burial ground as the 'cemetery no-one wants'. It was created in the late-1760s and had been used as the cemetery of Lewins Mead Unitarian Church for much of its history. The most recent burial at the time of the article had taken place four years previously. In the *Western Daily Press and Bristol Mirror* on 5 August 1958, it was described as a 'white elephant' by the Rev Birtles, minister at Lewins Mead. Only a few burial spaces remained and these were mainly in plots belonging to families who no longer had a connection to the area. All the houses in the adjacent square were occupied by commercial firms with only a few residents remaining, among whom was the cemetery caretaker.

The last burial here took place in 1963. In the 1980s, the cemetery was reconceived as an informal public park. An extensive re-landscaping programme was completed in 2010. It forms part of a green corridor developed by Places for People, working in partnership with Bristol City Council. St Pauls is a densely developed and populated neighbourhood, but also has green spaces, like this one, providing recreation, tranquillity, nature conservation and visual interest.

Hew Locke's artwork 'Ruined' (2010), which is in the north-east corner, comprises a series of cast-iron grave markers. These are based on share certificates and other historical documents belonging to commercial companies that have either ceased to exist or been transformed as the result of economic and political change. Among the locally-linked companies depicted is that of WD and HO Wills Ltd, a family-run firm that grew from a small tobacconist shop on Castle Street in the late-eighteenth century to being one of Bristol's biggest employers, with extensive factories in Bedminster, Ashton and Hartcliffe, all of which have now closed.

Return to Brunswick Square. Remain by the Meeting Hall but stop to look across to the south-side terrace (3).

In 1832, Dr Francis Black opened a homeopathic dispensary on Upper Berkeley Place, which developed into Bristol's first homeopathic institute. By 1883, interest in homeopathy had grown so strong in Bristol that a group of patrons were able to raise sufficient funds to pay for a homeopathic hospital, which finally opened at 7 Brunswick Square in 1903. An operating theatre, lift and veranda were added in 1907. Further extensions were needed by 1911, a sign of its popularity. The hospital could accommodate 12 in-patients. Louisa Wills, a rich and enthusiastic supporter of homeopathic treatments, raised money for the endowment of a free bed to be offered to those who would otherwise have been unable to afford to stay.

During the First World War, two wards in the hospital were allocated for wounded soldiers. Walter Melville Wills, managing director of the family's profitable tobacco company and husband of Louisa, was appointed the hospital's president in 1916. He commissioned a new building as a gift to the city in remembrance of his son Bruce who was killed in action in 1915. This opened at Cotham Hill in 1925. The old premises were sold to the Bristol Maternity Hospital. They have since been demolished and a new building erected on the site.

By the 1980s, the remaining houses in the south-side terrace of Brunswick Square were in a particularly poor condition. The effects of bomb damage and neglect grew worse because owners postponed paying for repairs and renovations in anticipation of redevelopment schemes that failed to materialise. In the early-1970s, the local press carried articles about what they termed 'the Brunswick Square Battle', the campaign to stop the demolition of numbers 1 to 6 to make way for a multi-million pound development at St James Barton. Objectors included Bristol Civic Society, The Georgian Group and the Bristol Visual and Environmental Group, whose members argued that east central Bristol should be returned to residential use and that its unique character was just as important historically as that of the more esteemed Clifton. It was the work of such groups, combined with private and public investments, that gradually brought back some of Brunswick Square's original elegance. Where possible, the focus has been on the improvement and conservation of what already exists rather than demolition and redevelopment. However, the buildings are still primarily occupied by businesses rather than residents.

Vaughan collection postcard of the Bristol Homeopathic Hospital, 7 Brunswick Square, c1920, and John Trelawney-Ross' photograph of 1-6 Brunswick Square, c1980 (Bristol Archives 43207/9/45/4 and 45212/Of/3/47).

Turn left and walk along Surrey Street into Portland Square (4).

In 1787, an Act of Parliament gave the local council the powers to carve out a section of the existing parish of St James to create the parish of St Pauls. Portland Square was planned around the new church. The land was owned by John Cave (who campaigned in support of slavery), William Pritchard and the Dean and Chapter of Bristol, and plots were first advertised for sale to speculators in 1789.

One of the arguments put forward by supporters of the proposal to knock down part of Brunswick Square was that the square's architectural significance was diminished because the terraces were never completed. By contrast, Portland Square was completed by around 1820 on all four sides, although it also had a somewhat chequered history. Several of the original building contractors were bankrupted by the financial crisis of 1793 that followed the outbreak of war with France.

Like Brunswick Square, Portland Square changed from a place where Bristol's more prosperous citizens aspired to reside (members of the Wills family were residents, for example) to becoming the site of warehouses, industrial premises and commercial offices. Among the light-industrial companies based in or near here were manufacturers of boots and shoes, hardware and corrugated paper. A local writer recalled that as a child in the 1880s he associated the square with 'the appetising smell that came from one of the tall houses, beyond the church, where Packers had started chocolate making' (*Western Daily Press*, 8 October 1957). As the nineteenth century progressed, established

Vaughan postcard: Scout March, Portland Square, 23 July 1916 during a visit by Sir Robert and Lady Olave Baden-Powell (Bristol Archives 43207/36/9/3/9). Number 15 Portland Square was the headquarters of Bristol Boy Scouts from 1915 to 1933.

families moved out to the quieter suburbs and houses fell vacant or were converted into boarding houses. By 1906, Portland Square included the Working Jewish Girls' Club; by 1916 a hostel for women; and from 1939 to 1975 a Salvation Army men's hostel.

The square was partly rehabilitated after the damage and destruction of the Second World War but the collapse of the building boom in the early 1970s left the area with 'the highest concentration of derelict buildings in the city', according to an article by John Cornforth in *Country Life* (3 November 1983). Like other parts of St Pauls, Portland Square had been in decline for many years and was deteriorating rapidly, with many of its structures in a dangerous condition. It was designated part of a conservation area on 19 June 1974 and new permanent residents began to move in from the late 1980s.

Pausing in the circular garden at the centre of square, it seems opportune to remember that, for more than 200 years, many people have tried to make the best of things while living in St Pauls. In the survey of bad housing referred to earlier, investigators remarked how very clean most residents kept their homes under the circumstances, and how one elderly couple was in the process of repapering and painting the walls. Although many parents were torn between deciding whether they should spend what little money they had on food or on rent, their children were often still out playing and having fun.

For example, contributors to the *Western Daily Press* in the 1930s and 1950s recalled how Portland Square had once been a popular destination for novice cyclists. 'Boneshakers' could be hired for a few pennies from a shop on nearby Milk Street and many local children 'first experienced the pleasures – and sometimes pains – of learning to ride a bicycle' by going around the railings of the central garden in relative safety (6 March 1936 and 8 October 1957).

Cross to St Pauls Church (5) now the home of Circomedia.

The residents of Portland Square and its surrounding streets could not be accommodated at the church of St James, which had no spare pews available to buy or rent, hence the necessity of building a new church. In his book *Bristol's 100 Best Buildings* (Redcliffe Press Ltd, 2010), Mike Jenner writes that, at that time, pews, 'like parking spaces today, affected house prices and rents'. Daniel Hague, the architect responsible for developing Portland Square, was also commissioned to design its church, which was completed in 1794. Initially, it was considered an elegant structure but tastes changed and later commentators called it a monstrosity. The tiered tower was rumoured to have been designed by the vicar because it appeared so amateurish.

Dwindling congregations led to the closure of the church in 1988. It remained empty until 2004, when it reopened as the premises of a circus school.

Facing the church, follow the alley on your right into St Pauls Park (6).

This was originally the church's burial ground but it was designated a public park in 1935. Like Brunswick Cemetery Gardens, this park contributes to the St Pauls green corridor developed by Places for People. It contains play areas, innovative artworks and space for community events, and was awarded the Best Urban Green Space in the Local Government News Street Design Awards. The overall regeneration scheme, which cost almost £1million, received the Green Flag Award, the national standard for parks and green spaces in England and Wales. Places for People also manages more than 300 homes in the area and, in 2018, was granted permission to develop a vacant site between St Pauls Gardens Estate and Cabot Primary School (which you will pass shortly) for 230 homes plus shops and office space. A percentage of the homes are classified as affordable.

Keep to the left-hand path. Turn left just beyond the children's play area. At the exit, turn immediately right onto another path. When you reach the road (*the top of Prince's Street*), turn right and then take the right-hand path that will bring you out

in the parking area at the top of Halston Drive. Continue along Halston Drive and you will pass Cabot Primary School on your right. Cross St Nicholas Road, turn left, then right into Davey Street. Continue to St Agnes Park (7), another part of the award-winning green corridor.

In order to counter mid-Victorian Bristol's reputation as Britain's third most unhealthy city, reformers such as the Rev James M Wilson (headmaster of Clifton College) pressed for a series of 'parks for the people' and other urban improvements for those living in East Bristol. Wilson successfully petitioned the council to purchase an old orchard near St Agnes Church for use as a park which, by 1885, had become a popular local amenity.

It is worth noting, however, that St Agnes Park and the new speculative housing around it had displaced allotment holders who, before 1870, had squatted the area by installing chimneys in their tool-houses and converting them into dwellings. So long as they did not build a second storey, they had been left alone by the authorities. In all, some 500 of what were described as 'the roughest class of people' lived there breeding chickens, pigeons and pigs, growing vegetables and watching boxing matches and cock-fights. The newly-built terraces that replaced these homes were let to artisans and mechanics escaping the congested city centre. In effect, the very poorest Bristolians were moved on in the interests of the social class just above them.

In his memoir *A Family in St Pauls 1920-1940: Scenes from Childhood* (Redcliffe Press Ltd, 1985), Cecil Pope writes:

> St Agnes Park was nearby. There was a pond there with fish in it. The pond was raised up and you could see the fish close up. The Play Park was separate and all tarmac. Nothing in it. We preferred it out in the road for hoops. I had an iron hoop which was rusty with a wire handle... The hoop snapped one day but the blacksmith up in Conduit Place or Lower Ashley Road (I think) patched it. It always had a bump in it after that and bounced in an ugly unpleasant way and made a noise. Sometimes coming down Bishop Street it ran away and dashed across St Nicholas Road into the cobbled Davey Street right in front of a bus.

St Agnes Park, one of the best-funded of the city parks, remained an attractive destination for local people for nearly a century. However, by the 1970s, as conditions declined, it developed a reputation for being a dangerous haunt of drug dealers and their customers. It became overgrown and derelict until the beginning of the twenty-first century when improvement schemes – involving volunteers of the Friends of St Agnes Park – helped to make it a welcoming place once more.

Another important local organisation was St Pauls Unlimited Community Partnership (SPUCP, 1998-2017), which was funded by Bristol City Council until 2014. It was central to much of the area's regeneration, including urban renewal programmes which focused on social cohesion, green space and housing. SPUCP made a significant contribution to developing the St Pauls Neighbourhood Plan (adopted 2006), creating St Pauls Learning and Family Centre (seen later on this walk), improving local parks and minimising the impact of crime.

In an interview conducted by the Architecture Centre, long-term St Pauls resident Sue said:

> St Pauls Unlimited was a great organisation for the neighbourhood. It gave voice and agency to the local people with the council, businesses and developers. For years St Pauls was used as a bit of a dumping ground for people perceived to have social issues (ex-offenders, those with drug, alcohol or mental ill health issues) and so it had more problems than other parts of the city and a rather negative reputation. St Pauls Unlimited started to change the narrative of the neighbourhood and change people's perceptions, championing all the positive things about the area. It is such a shame that it had to close. There are still active residents, but it feels the community has ceased to have a voice in the same way and this is really worrying when developers are trying to buy up every spare bit of land for high end development. It really does feel like gentrification rather than community regeneration and there are times when I feel like I don't recognise my neighbourhood anymore.

When discussing regeneration, the less favourable term of 'gentrification' is sometimes used. This is a contentious issue. Gentrification can reverse decades of suburban flight and inner-city decline, but those who have benefitted most have often been small groups of young, educated and affluent people, thereby widening the gap between rich and poor. An article in *The Observer* on 11 December 2016 had the heading 'Riot flashpoint to housing hotspot: hipsters help to bring St Pauls back to life'. Many residents will feel safer if the streets are free from litter, vandalism, gap-sites, boarded-up shops and anti-social behaviour but, unless they are property owners or council tenants, their housing situation may become more precarious due to gentrification. This is because when house prices rise, private landlords will be tempted to raise rents or sell-up.

Cross the park. On your right you will pass a mural of Clifford Drummond, one of the co-founders of the St Pauls Carnival. This is part of the Seven Saints of St Pauls® Art and Heritage Trail, a project led by local artist Michele Curtis, which celebrates positive Black history. Keep left of the play area and exit into Fern Street. The mural of Barbara Dettering, another pioneering community activist and co-founder of the carnival, is opposite.

At this point in the walk you can either continue to the next location on the main route – Grosvenor Road (8) – or take a 20-minute diversion to see Bristol's oldest surviving council-built housing on Mina Road (7B).

If you are going to Mina Road, cross into Tudor Road. Turn left on Lower Ashley Road. Cross the road at the pedestrian crossing then turn right and left into Conduit Road. Near the end of Conduit Road, on your left, you will see Owen Henry House, named after one of the organisers of the Bristol Bus Boycott of 1963 (of which, more later), a co-founder of the St Pauls Carnival. His mural portrait is on the junction of City Road and Ashley Road.

Turn right into Conduit Place. Ashley Street Park will be on your left and you will pass Parkway Methodist Church on your right. Conduit Place leads into Rosebery Avenue. Keep on this road, which curves around to the left, to the end, then turn right into Mina Road. The tenement blocks you are looking for are on this side of the road, at the far end, by the exit from the M32.

....

These maisonettes are the only survivors of 73 municipally-funded council dwellings completed in Bristol before the First World War. The others were in Chapel Street, Braggs Lane, Fox Lane and Millpond Avenue. They were built to house those displaced by slum clearance programmes, and were referred to in the local press as 'houses for the industrial classes'. Their design was condemned by some. Members of the local Independent Labour Party, for example, dismissed them as 'brick-built barracks'. Tenants complained about cold floors, damp walls and chimneys that did not draw properly. The land at the Mina Road site was described as permanently boggy and was at risk of flooding from the River Frome (which now runs under the motorway). By the 1980s, the Mina Road homes were only being used by the council for short-term lets. In 1991, the council leased them to the local charity Solon Housing. Since 1998, the non-profit-making, volunteer-run Members In Need of Accommodation (MINA) Housing Cooperative has provided self-managed housing for a small group of single people living here. All that remains of the original Mina Road properties is this West Block; the East Block was demolished in 2007 as it was deemed beyond repair.

....

Retracing your steps, return to St Agnes Park to resume your walk along the main route. Before you reach the park, turn into Badminton Road and walk up to the junction with Ashley Road. Ahead you will see the old signage of Jenner and Co: Drapers and Milliners. Turn left at this junction, and then left again into Grosvenor Road (8).

....

Postcard from the Vaughan collection showing a view of Grosvenor Road, c1910 (Bristol Archives 43207/9/4/57).

Between the First and Second World Wars, Grosvenor Road was a thriving shopping street with a wide variety of stores that met all the community's needs. A nostalgic letter in the *Bristol Evening Post* on 27 September 1974 recalled some of these: Norris the baker; Watts the cycle shop where bicycles could be hired for three-pence for half an hour; Miss May's wool shop; Chamberlain's fish shop; Bartlett's greengrocers; Saxton's the grocer; and Browning's hairdressers, where regular customers had their own shaving mugs, marked with their name.

After the Blitz, many locals moved out of central Bristol and the demographic profile of the area changed to include a more transient and poorer population with higher rates of unemployment. The housing stock was left to deteriorate and by the 1970s homelessness grew, despite the presence of empty houses. Under the Housing Act 1974, the St Pauls Housing Action Area E was established with its lower boundary in Grosvenor Road. The area comprised 209 dwellings built before 1881. Of these, 96 were deemed in whole or in part unfit for human habitation, with 283 households being identified as lacking access to standard amenities. The 92 properties in multiple occupation contained a total of 340 households, indicating over-crowding. There was a mix of owner-occupied and privately rented properties. This was just one part of St Pauls that was of continuing concern to the council at the time. Although it may have been more effective to declare the whole of St Pauls a General Improvement Area in order to address long-standing housing problems, there was only funding available to focus on small, scattered housing zones.

....

Continue along Grosvenor Road. In the small park on your right is a bust of Alfred Fagon, the poet, playwright and actor. The location was chosen as Fagon reportedly regarded the corner of Grosvenor Road and Ashley Road as 'the heart of St Pauls'. In the 1980s, the council's Land and Administrative Committee backed a locally-led campaign to save this 'village green' and prevent a supermarket building on the site. Continue down the road and stop at the St Pauls Learning and Family Centre (9) on your left.

....

This centre provides a range of meeting rooms, learning and office space, art rooms, computer rooms, a woodwork space, a photographic dark room and a café. The library is on the ground floor. Since 2015, it has been run by The Ethical Property Company. Take the path on the café-side of the centre (to your left, when facing the main entrance), and look for the mosaic based on a diagram of the Liverpool-registered slave ship, the Brookes. This diagram became one of the first images to be widely distributed in abolitionist propaganda. In 1789, it appeared on 700 anti-slavery posters produced to illustrate the inhumane conditions in which enslaved African people were transported to the British colonies of the Caribbean to work on plantations. The economic prosperity that financed the early development of St Pauls was largely built on profits from the transatlantic slave trade and the sale of goods dependent on slave labour (tobacco, sugar, rum).

This image of suffering is in contrast with the celebratory 'All Our Tomorrows', the *bas relief* on the external end wall of the Learning Centre created in 2003 by Valda Jackson, a Jamaican-born artist resident in Bristol. With this commission, she became the first Black woman to produce a piece of public sculpture in the city.

....

Continue along the footpath that runs along the café-side of the centre. Cross Ludlow Close and enter St Pauls Gardens Estate (10). Use the path to the right of the wall that has the map of the estate on it. At the play area, turn right and, with the spire of St Pauls to guide you, walk to the enclosed park at the heart of the estate, where you will find more mosaics.

....

These council-built, low-rise concrete blocks of flats were completed in the late 1960s. The original plans show that the intention was to extend the estate further south and west, reaching to Portland Square. The estate is the oldest, modern public-sector

Bristol City Council Public Relations artist's impression of Bishop Street, St Pauls after redevelopment, c1960s (Bristol Archives 40826/PLA/36).

housing in St Pauls.

In *Portrait of Bristol* (Robert Hale, 1971), author Keith Brace writes of this area:

> A recent low-rise flat-development in the Bishop Street-Martin Street area behind St Paul's Church, replacing decrepit terraces, has brought residents back to or into the district, though, ironically, many of the nearly 200 flats were occupied in 1970 by people evacuated from a nearby high-rise block regarded as potentially unsafe. The flats in various groupings are clustered around grassy central spaces. Probably they will never fully replace the arms-akimbo, door-step camaraderie and bantering of the old streets. But one only has to look at the exposed backs of the remaining old houses round about, with their domestic rubbish, old bath tubs, broken bicycles, outhouses and outside lavatories, washed-up like flotsam around the end of a pier, to understand that the elbowing friendliness of the old way may be a small loss compared with the warmth and comfort of the new flats.

Another movement of immigrants into St Pauls came in the late 1990s/early 2000s, including refugees fleeing the conflict in Somalia, some of whom were housed on this estate. Somali is the third most commonly spoken of the estimated 91 languages in Bristol today and around 20,000 people of Somali heritage now live in the city. The first

Entrance to The Bamboo Club, c1970 (Bristol Archives 43845/Ph/3/6).

Bristol Somali Festival took place in 2015 to celebrate this well-established community.

Continue walking in the direction of the spire. You will pass the mural celebrating Full Circle, the youth and family project, on your left. Keep left and cross the small patch of green. You will exit near the end of the path you took from St Pauls Park earlier. Return to Portland Square. Turn left. At number 15, you will see a Bristol Civic Society plaque honouring race equality champion Batook Pandya. Cross Wilson Street and continue down St Paul Street. Stop at number 12, where there is a BBC Music Day blue plaque marking the site of The Bamboo Club (11).

Tony and Lalel Bullimore opened The Bamboo Club on 28 October 1966. It was the original headquarters of the Bristol West Indian Cricket Club but is best remembered for the musicians who performed on the top floor. They included Jimmy Cliff, Desmond Dekker, Ben E King, Bob Marley and Percy Sledge. Alfred Fagon was among those associated with the club's theatre workshop. It was one of the first music venues in the country to cater for the African-Caribbean community and was a cultural hub for St Pauls until 1977, when it was destroyed by fire.

An article on St Pauls in the *Western Daily Press* on 15 October 1969 included an interview with a Jamaican-born man named Leroy who had lived in Bristol for eight years. Leroy was employed as a bus conductor. He said:

> One of the chaps on my beat came to me one day and he says 'You know, Roy, I didn't have much to do with coloured people till I worked with you. But you seem a genuine sort.' He and his missus and I go to The Bamboo Club, and, man, did we have a good evening out.

Owners Tony and Lalel Bullimore pictured in The Bamboo Club's Orange Grove Restaurant, which served Jamaican food. (Bristol Archives 43845/ Ph/3/3)

In 1968, the first St Pauls Festival was held, led by residents. It was, in part, a celebration of community unity following the Bristol Bus Boycott. This had been the first Black-led campaign in Britain against the colour bar and it successfully overturned the open racial discrimination of the Bristol Omnibus Company, which would not hire non-white crews. The festival was an opportunity for local people to come together, enjoy themselves, learn about the different cultures to be found in the area and dispel negative stereotypes. It later evolved into the St Pauls Afrikan-Caribbean Carnival.

The walk ends here. If you continue down St Paul Street, turn right into Newfoundland Street then right into Bond Street, you will be able to find your way back to the bus station.

St Pauls/ Community Voices

The following comments were collected in projects run in St Pauls by the Architecture Centre. www.architecturecentre.org.uk.

Lisa: I lived in a rented house of multiple occupancy in St Pauls. It backed on to St Agnes Park and the house was always lively (sometimes too lively) and full of people. I wasn't used to city living back then and my bedroom overlooked the park – it was a real sanctuary for me, that view over the park – a kind of a green oasis with the grass, trees and beautiful stone church tower in the background – like a surprising little slice of middle England in the heart of the city! Don't get me wrong, it wasn't without its problems, especially after dark, but I think that view, and the access to green space on our doorstep, helped to keep me (a country girl at heart) sane during those lively years in my 20s. I also have happy memories of community events and parties that happened in the park around carnival time and throughout the summer – music, food and people of all ages and different cultural backgrounds all hanging out happily in the sun. Like many of the people living nearby, we only had a tiny yard outside our house, so the park was a haven to chill out in during those summers... with the noise of music and kids playing you could hardly even hear the grumble of traffic on the M32!

Michael: In the late '80s, my partner and I lived in a Housing Association flat on City Road. My son was born whilst we were living there and so it holds a special place in my heart. The area was pretty run down then, but very friendly and genuinely multicultural. At that time the housing association was helping long-term residents buy a home (outside of the area as it wanted to keep the inner-city properties) of their own through the Tenants Incentive Scheme. They gave us a pretty reasonable deposit to buy our first house over in East Bristol – not something that would happen nowadays! We had mixed feelings about leaving as we'd had happy times living in St Pauls – there was always lots going on, the Malcolm X [Centre] was a real hub and everyone knew each other. At that time lots of people (especially if they had/were planning kids) wanted to move out of the area to somewhere perceived as safer, with more space and into houses with gardens, rather than flats. I think we were really lucky – I'm not sure my children will ever get an opportunity to get on the housing ladder – I can't believe the house/rental prices in places like St Pauls now.

Sue: When my sons were younger, City Road was such a friendly, mixed community where everybody knew each other. There was a motorbike and car repair shop on the corner of Brunswick Street and City Road. It was a focal point. Not only could you get your car fixed but you could drop by, have a sit, a chat and a cuppa. It is now a barbers and an independent record shop. My neighbours were Afro-Caribbean as well as Ugandan Asians who taught me how to cook dahl, Italian families from Naples, a tall Hungarian chap always out with his huge St Bernard dog, along with a midwife, an artist and visiting Cuban musicians. Nowadays the population in parts of the neighbourhood is much more transient, with a constant turnover of tenants. They are nice people but we don't get to know each other and they don't emotionally invest in the area, as they don't stay for long – it's not their fault of course – the private rents are too high, so they move on to other parts of the city.

The row of houses I lived on had long gardens, which backed on to a green space with mature trees and even a family of foxes. It was a lovely place for my sons to play in. Green space like this is so important in the inner city for people's (especially children's) wellbeing. My neighbour only had a small back yard, so her children used to come across the road to play in our garden. As developers buy up more and more spaces in St Pauls, I really worry about the impact of the loss of green space in the neighbourhood. Bristol Churches and Knightstone Housing Associations owned many of the larger Victorian properties in the area so were able to rent out whole town houses to extended families. I remember one large Jamaican family who lived near Hepburn Road – they always had herbs like thyme growing out the front for cooking with their rice and peas. Most of the new housing in the neighbourhood now is tiny flats for single people, as I guess it's more profitable, but I don't feel it's better for residents and the community.

Member of TALO women's group: We moved into St Pauls nearly ten years ago as a young family with one small child and another one on the way. Before moving in, I was anxious about the area as, living in Bristol for most of my life, I was well aware that St Pauls always had a negative reputation and questioned whether it would be safe bringing up my children. I was contradicted instantly and found in St Pauls a peaceful and welcoming community from the moment we moved in. My first taste of this helpful spirit was in the communal (St Pauls Gardens) laundry room. I must have left the laundry room for less than half an hour to come back to my clothes all folded and ready to take home. When asked who was responsible for this good deed as I wanted to thank them, I was told that it was maybe 'so and so... but it wasn't a 'big deal as we all do this for each other'. I came to realise that this was the norm, helping each other with school runs, car sharing for groceries shopping or keeping an eye on all children whilst at the park. Ten years on, I do all these things and more for my neighbours and I hope that my children and other younger generations continue this culture of supporting each other without expecting anything in return.

Tara: I moved from the hustle and bustle of London to St Pauls eight years ago where I feel a sense of strong community. Neighbours are friends and we look out for each other. I feel very blessed to be a part of a rich and diverse community. I hope the needs for affordable and decent housing in St Pauls will be a top priority for the local authority.

Walk 10:
Hillfields

On 5 June 1920, more than 500 delegates from the Inter-Allied Housing and Town Planning Congress came to Bristol to see a range of new council-built housing, including the first homes to be built under the Addison Act.

The visitors were drawn from across the UK, the British Dominions (Canada, South Africa, Australia, New Zealand), Europe and the USA. They were led through the Hillfields Park Demonstration Area by groups of Boy Scouts and Girl Guides. On this walk, you will see some of the houses they inspected that day. These experimental housing blocks are in the northwest of the estate and were funded through the Local Government Board, the president of which was Christopher Addison. Following the Right to Buy Act, private owners have added personal touches to their homes but you should still come to recognise many of the original forms.

....

The number 6 bus will take you from Bristol city centre to the starting point – The Lodge Causeway Shops (1).

....

Alight at the Forest Road stop. The Hillfields Park housing estate (now generally known as Hillfields) is located on the edge of Fishponds on land once covered by the Royal Forest of Kingswood. Lodge Causeway is an ancient route that, in Saxon times, led from Fishponds to the Kingswood Lodge on Lodge Hill. By the seventeenth century, Fishponds had become a thriving village occupied by the families of locally-employed quarrymen and miners. During the General Strike of 1926, it was reported that some Hillfields residents were able to dig for coal in the shallow seams that ran under their back gardens. Fishponds was later associated with the aeronautical and automobile industries.

Hillfields represented a new approach to working-class housing that followed the recommendations set out by the Tudor Walters Committee during the First World War. These recommendations informed the Housing Act 1919. The middle-class suburbs had traditionally marked a clear physical division between work and home. Now this division was seen in a planned working-class community, with many residents commuting to the city centre for work. The new council estates aimed to be self-contained and self-sufficient, following the principles of the garden suburb movement. However, the Lodge Causeway shops were not completed until 1925/1926, so the first residents were often dependent on door-to-door hawkers charging higher than average

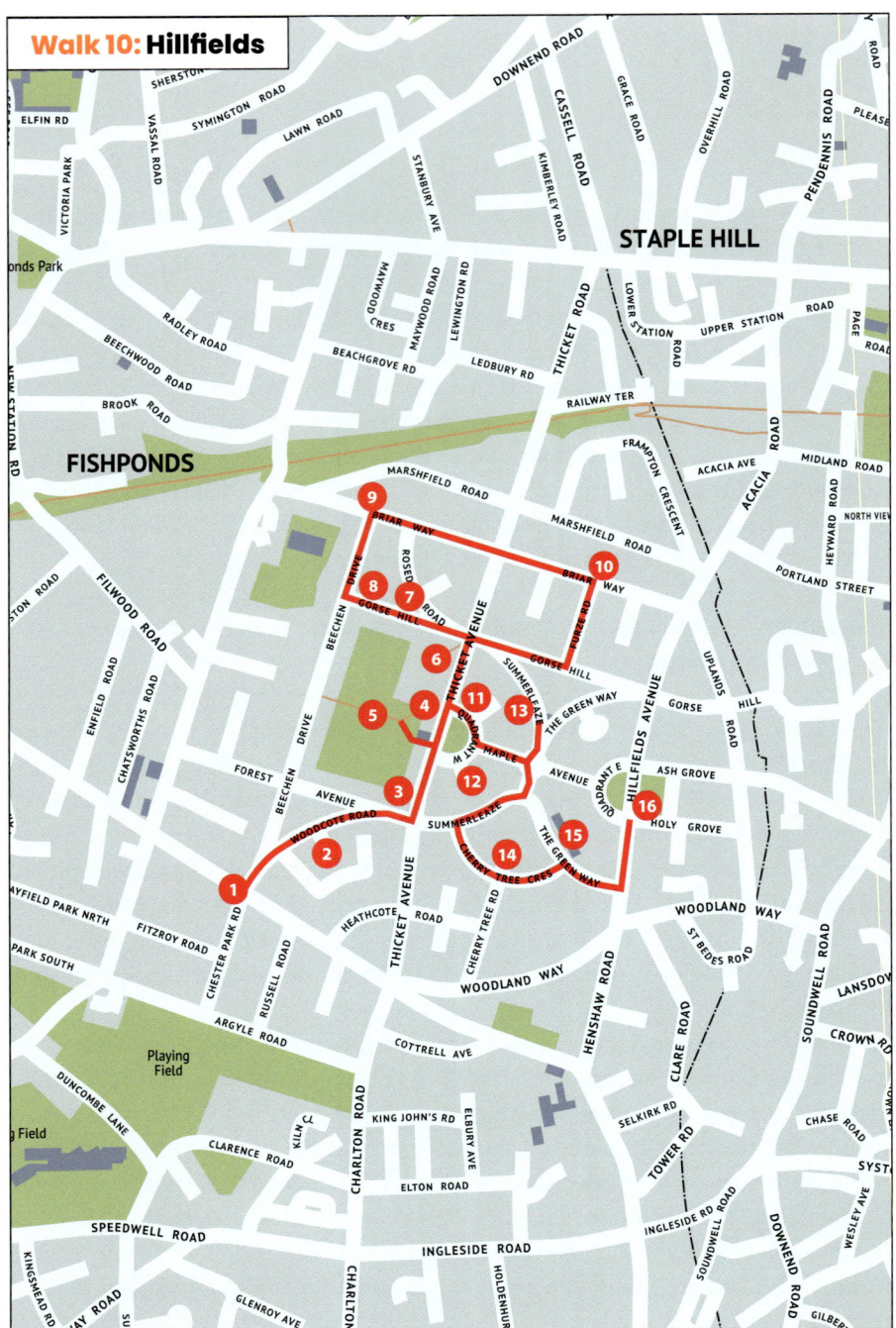

Aerial view of Hillfields taken in the 1920s/1930s (Bristol Archives 44819/3/176). You can see the crescents of Summerleaze and The Greenway, the east and west Quadrants linked by Maple Avenue, Market Square and the Recreation Ground.

prices, or deliveries from shops in Soundwell, Maple Park and Staple Hill. For many local people, Lodge Causeway eventually became the principal shopping street for the estate. Some shops were also later opened at Market Square, the final stop on this walk.

Walk up the hill to the junction with Beechen Drive. Turn left here then bear right into Woodcote Road (2).

The road sign may be hard to spot but it is the turning immediately after Woodcote Walk. The variety of homes found on the early post-war estates contrasted with the uniformity of working-class terraces built in the city centre. There was enough repetition of styles to give a sense of order, but also enough diversity to prevent visual monotony. Hillfields is exceptional because it has 30 individual types – the average is 15 – of which the majority were designed by winners of an architectural competition, rather than the City Engineer's Department. A few examples of these different types will be pointed out along the way. In the fork formed by the junction of Woodcote

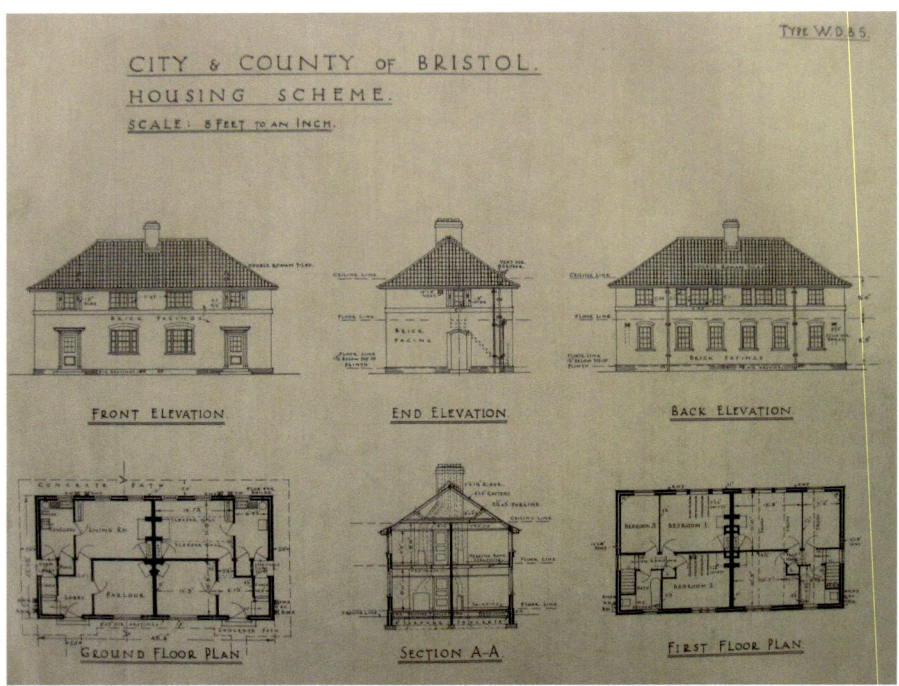

Architect's drawing of the three-bedroom WDB5 (Bristol Archives Red Label Plans 22/3/30). More than half the houses built under the Housing Act 1919 at Hillfields and Sea Mills had parlours compared with only 15 percent of those built at Southmead under the Housing Act 1924.

Road and Beechen Drive there is an example of the MGA2 by the architectural firm of Maidman and Greenen, one of the principal designers of Bristol council housing in the 1920s. This early version of its three-bedroom non-parlour semi-detached house was built between 1921 and 1925. On the left-hand side of Woodcote Road are some REA5s, designed by the architect SS Reay. There are more of these on the right-hand side along with some WDB5s designed by Benjamin Wakefield.

The WDB5 became one of the standard parlour-housing types in Bristol. Around 2,000 were built in the city between 1921 and 1939. A parlour could serve as a second living room – often kept for best – or provide ground-floor sleeping accommodation, perhaps for a former serviceman injured during the war and unable to manage stairs. In September 1920, more than 87 percent of the 676 applicants for houses on the Hillfields estate were from ex-service personnel, in keeping with the 'homes for heroes' campaign. These were designed as family homes and there was no provision for single people. Many of the first residents would previously have lived in overcrowded, insanitary conditions in the city centre.

At the end of Woodcote Road, the road joins Forest Avenue. Turn right up Forest Avenue then left into Thicket Avenue (3).

This is the only road within Hillfields that pre-dates the estate. It was previously a track running through the fields between St George and Staple Hill. Most of the land on which Hillfields was built was purchased by the council in early 1919 and comprised 128.5 acres. Additional plots totalling just over 31 acres were acquired soon after.

Work commenced at the site on 12 May 1919, before the drawings had been officially approved and before the Housing Act received royal assent (31 July 1919). At the peak of activity, 184 men were directly employed by the council on the initial layout work. Between 1919 and 1921, this entailed widening Thicket Road (previously a lane and later renamed an avenue) by 20 yards across a distance of 854 yards; completing 383 yards of new roadways; ballasting 6,200 yards of ground for further roadways; and laying 4,361 yards of pipe for soil sewers and 5,221 yards for storm-water (taken from the 1921 Bristol Civil Engineer's Report, Bristol Reference Library B6521). (A yard is approximately 0.92 of a metre.) During this same period, 15 contractors were engaged in house building; 134 homes were completed and occupied; and 530 other houses were either constructed or on course for completion soon.

By the time of the 1930 Bristol Housing Report, the estate had 788 parlour and 670 non-parlour homes built by the council and 22 privately-built parlour homes.

Continue on Thicket Avenue to the Baptist Church (4), on your left.

The church appears on the original plans for the estate. It was opened on 8 May 1929. In her book *Rough Hewn and Gentle Pride* (1994), recollections of her life on the estate in the 1930s and 1940s, Dolores Powell writes:

> Every child used to attend Sunday School in those days… We loved Sunday School as we were bathed and dressed in bonnets, white socks and satin dresses… My sister and I wore identical clothes and would walk hand in hand carrying small prayer books to Thicket Road Baptist Church which was situated on the edge of 'The Rec'… playing fields. Sunday School was held in a room under the main church and we had to descend steps into the basement

The house to the left of the church is a WSA2 designed by W H Watkins who was then better known for his work on public buildings including cinemas. This became a standard type of non-parlour Bristol housing.

Children on the Recreation Ground, c1922 (Know Your Place, Community Layer, submitted by Mr Baker, March 2012).

When facing the church, you'll see a path to your right. Take this path and walk past the play park to the Recreation Ground (5).

This Recreation Ground quickly proved to be a popular play area for children. Among the memories of early residents gathered by Jane Baker in a local history study of the late 1980s were those of the old blackthorn tree nicknamed Twisted Willie, which the children liked to climb, and the occasional visits from Pruett's fairground roundabouts. From the 1950s, the site was used by the Hillfields Park Community Association for the annual summer fete if the weather was fine. Enclosed open spaces with a clear, designated function were an important part of the design of a garden suburb. The focus was on providing amenities that supported the moral, cultural, religious and approved recreational needs of residents. Playing fields and churches were in; public houses were out. Bristol's Baptists were among those who successfully campaigned for the new estates to be alcohol-free. You'll find out more about garden suburbs in the Sea Mills walk.

To one side of the Recreation Ground was the community centre (6).

Before this centre was built by more than 40 volunteers in 1950/1951, there were few public meeting places on the estate other than those provided by churches. The Hillfields Park Community Association was formed in October 1945, but discussions about having facilities such as these had begun in the mid-1930s. Fundraising schemes led by the association included children's fancy dress parties held on the Recreation Ground. The official opening ceremony for the centre was on 29 September 1951.

Retrace your steps back to the church on Thicket Avenue and turn left. Turn left at the junction into Gorse Hill, crossing to the right-hand pavement. The first few houses on the left-hand side are examples of the BOA5, a three-bedroom parlour home designed by Austin Botterill, one of the first semi-detached council houses to be built in Bristol. Stop just past the junction with Rosedale Road to look at Mr Mitchell's semi-detached (7) on your right. You are now entering the demonstration area.

Many of the houses assessed by the visiting delegates in 1920 would become standard types used on estates throughout Bristol. This semi-detached property by the architect Arnold Mitchell is the only one of its kind to have been built. The MIA5 was spacious, with three good-sized bedrooms, a parlour and a downstairs toilet. However, higher quality meant greater construction costs and the necessity of charging dearer rents, putting such homes beyond the means of poorer tenants.

Generally, Hillfields and Sea Mills were designed for better-paid workers who could not only afford the rent and rates but also cover the costs of travel, furniture, food and other necessities. These tended to be more expensive outside of the city centre because of the lack of competition. Heating costs would also have been more because homes were larger. Residents were expected to keep high standards of housekeeping and garden maintenance and were closely vetted by council officers. The Hillfields Park Tenants Association was founded in March 1922 following a meeting of around 300 people to discuss rent reductions for the estate. The association had 20 members, five of whom were women. In addition to lower rents, they demanded a reduction in the rates, better public transport, the provision of more shops and other amenities, and the right to buy their homes (showing that some tenants were clearly better off than others). The council's Housing Committee was broadly sympathetic and in the autumn of that year the Ministry of Health approved the proposed rent revisions. However, with growing unemployment, many tenants still found themselves in arrears with rent, forcing them to sub-let rooms (a practice that was unauthorised). The need to build for those who were only able to pay low rents meant that cheaper construction methods were used and council homes became smaller.

Continue down Gorse Hill. The chimneys you see in the near distance relate to the ES & A Robinson packaging company, once one of the main local employers. The houses on your right are WSA5s designed by WH Watkins. Turn right into Beechen Drive (8) and stop at the plaque above the archway between numbers 84 and 86.

Housing Department photograph of volunteer labour building the Hillfields community centre out of a former army Nissen hut, c1950 (Bristol Archives 40307/1/128). The adjacent mess centre was built by council-employed labour at the same time.

This plaque marks the first properties to be completed in Bristol under the Housing Act 1919. It is a block of four three-bedroom parlour homes.

In 1968, campaigning journalist Barbara Buchanan wrote an article headlined 'Living in Bristol's first council house is "pure bliss" (says Angela, a pretty mother of three)' (*Bristol Evening Post*, 21 February 1968). Angela Piaseki lived at 82 Beechen Drive with her husband and their three young children. They had previously been living with Angela's parents-in-law in Ashley Hill. The family had been offered the council house in Hillfields in November 1967. The article reads:

> They... moved in straight away with the furniture they had stored and the wedding presents they had kept unopened and now it is pure bliss – a bedroom for Anthony [the youngest child], one for the twins, one for themselves, a sitting room, a parlour, a kitchen, a bathroom and a separate toilet. They don't, to use the familiar phrase, 'know themselves'.

Eveline Nicholas, another of Buchanan's interviewees, had lived at number 86 for 36 years, having first arrived with her husband and young son. Now widowed and with her son in a place of his own, she shared the house with her sister and brother-in-law.

Eveline remembered the area when it was still fields and number 82 had been a showhouse. She said: 'I think the Co-op furnished it and everyone crowded to see it. I think the furniture they put in was more modern than mine is now!'

Number 88 was one of 1,200 council homes in Bristol that had been bought by its tenant, well before the coming of Right-to-Buy. Audrey and Ivor Johnson had purchased it eight years previously. At the time of the interview, the house was technically overcrowded because ten people lived there, of whom two (the Johnsons' eldest son and his wife) were on the council waiting list. The parlour had been converted into a bedroom for Audrey and Ivor and their youngest son.

Continue along Beechen Drive. The last houses on your left before you reach Briar Way comprise a short terrace of three-bedroom non-parlour REA2s, designed by SS Reay. This is the only block of its type as it was considered too expensive. Cross Briar Way and stop at the Heathman Blacker HBA2 (9), **the short terrace facing down Beechen Drive.**

A total of 400 three-bedroom non-parlour HBA2s were built in Bristol, but this is the only block of this type in which the gables face the back garden rather than the street. Eveline Dew Blacker, Bristol's first female architect, co-designed with Harry Heathman several house types that can be found on local council estates. The firm later went on to design Bristol's Cenotaph (see Walk One). Blacker served her apprenticeship under the architect George Oatley, whose work you will see on the Sea Mills walk.

Turn right along Briar Way. The first block of houses on your right comprises two-bedroom WSA1s by WH Watkins. They can only be found in Hillfields because the visiting delegates thought them too small. Continue along Briar Way, crossing Thicket Avenue (*the first houses to be occupied on the estate are at this junction*) **passing Briar walk on your left and stop at the Heathman Blacker HBA1** (10), **facing Furze Road. This should take around five minutes.**

This block of four two-bedroom non-parlour houses designed by Heathman Blacker is the only one of its type. However, you will see three-bedroom terraces with similar gables fronting the street on many inter-war estates. Like the WSA1, the visiting delegates thought the HBA1 was too small and that council homes should provide for growing families with a minimum of three bedrooms.

Cross into Furze Road. At the bottom, turn right into Gorse Hill then left into Thicket Avenue. Stop at Quadrant West (11).

Furze Road depicted on a postcard from the early 1930s, looking towards the HBA1 block on Briar Way (Know Your Place, Community Layer).

The government review of spending in 1921/1922 (sometimes referred to as the Geddes Axe) brought a temporary halt to many of the building projects started under the Addison Act, though some houses were completed under existing contracts in Hillfields, including those around the Quadrant West. Those to the left are Heathman-Blacker HBA5s, semi-detached pairs of three-bedroom parlour homes of which 338 were built in Bristol in the 1920s, mostly in Knowle Park. Those on your right are Heathman-Blacker HBB5s, of which 62 were built, again, primarily in Knowle Park. During the Second World War, underground air-raid shelters were built on the Quadrant West green, some of which were assigned to pupils at Hillfields Park Primary School. Many residents would have had their own Anderson shelter in their back garden, a few of which survive to this day. Hillfields was on the flight path for German bombers heading to attack the Bristol Aeroplane Company at Filton, but did not itself sustain much damage.

Walk halfway around the Quadrant and turn into Maple Avenue (12).

Several of the houses in this street were built in 1923/1924 by the Robinson company to accommodate local employees. Occupants paid weekly instalments – deducted from their wages – towards the eventual purchase of their homes. Ground rent was paid to the council. The firm specialised in the manufacture of paper bags, waxed paper and cardboard boxes, as well as printing. Employees who lived in Hillfields but had to work at Robinson factories elsewhere in the city were issued with travel vouchers to use on the trams. With the Geddes Axe and the cancellation of the Addison subsidy, members of the Hillfields Park Tenants Association expressed concern that empty sites would be taken up by private developers and the estate would lose the quality associated with a garden suburb, but this was not the case. All the other houses here were built by the council. The trees that give this avenue its name are thought to have been imported from Canada.

Turn left into Summerleaze and stop at Hillfields Library (13).

The library was built in the 1950s on a site that is shown on the original plan for the estate as an open grassed area. It provided a hub for displays, memory-sharing events and hands-on activity during Local Learning's Hillfields Homes for Heroes project. It is usually open three days a week. Check the Local Learning website for details of what's on: www.locallearning.org.uk/hillfields-homes-for-heroes.

Return to Maple Avenue, cross over the roundabout and turn right onto Summerleaze Crescent. Turn left into Cherry Tree Crescent and stop at the entrance to Cherry Tree Close (14).

To your left is a group of one- and two-bedroom flats designed for sheltered independent living for people more than 60 years of age. It is one of 23 purpose-built schemes in the city managed by the charity Brunelcare. Applicants for homes here need to initially contact Bristol City Council's HomeChoice team, who will pass on referrals.

This was formerly the site of St Bede's Church, which was included in the original plan for Hillfields and opened in 1926. St Bede's was initially designated as a mission of St John's parish in Fishponds but became an ecclesiastical parish in its own right in 1929. It was abolished in 1962 and reabsorbed into the parish of St John. At the outbreak of the Second World War, the vicar of St Bede's, the Rev G H Dymock, was briefly detained by the authorities because he was rumoured to be a member of the British Union of Fascists and, therefore, a potential security threat. In 1934, Dymock had been among those who welcomed Sir Oswald Mosley and 500 of his 'blackshirts' to Bristol, attending a rally at the Colston Hall (now called Bristol Beacon). Residents recall that Dymock was a heavy drinker and that he campaigned – unsuccessfully – to have a pub built on the estate.

Continue along Cherry Tree Crescent which reaches The Greenway, stopping at Minerva Primary Academy (15).

Photograph of a classroom at Hillfields Park Primary School, 1933, submitted by Lorna Tarr to the Hillfields Homes for Heroes project (Know Your Place, Community Layer)

Hillfields Park Primary School (now the Minerva Primary Academy) opened in 1927. The site was set aside for the building of a school on the original plan of the estate but, as with the shops, there was a delay of several years so the children of the earliest residents had to travel elsewhere for their education. Until the community centre was completed, parts of the school were used in the evenings by the community association for 'Old Time' dancing and general get-togethers, and for meetings of the Hillfields Youth Club and Drama Group. The oak tree near the school entrance was planted in about 1932 and featured on the school crest until recently. In 2012, the school joined the Cabot Learning Federation, leaving the control of Bristol City Council. Since then, this modern structure has replaced the 1920s buildings that were demolished.

In 2008, ten pupils from the school represented Bristol at the 'Portrait of a Nation' showcase that formed part of the high-profile European Capital of Culture programme in Liverpool. 'Portrait of a Nation' engaged young people from 18 cities across the UK in celebrating their local, regional and national identities through arts projects. At Hillfields, Class 3/4M focused on the diversity of their community, exploring how they had all come together in this place with heritages that included Indian, Vietnamese, Swedish, Belgian, Somalian, Italian, Irish, Hungarian, Bangladeshi and Pakistani as well as British.

Turn right down The Greenway and continue to Hillfields Avenue. Turn left and stop at Market Square (16), which is on your right.

Despite its name, there was never a market in this square. The original estate plan shows shops on the corners, of which four were eventually built. The Green (as it is referred to locally) proved popular as a place for games, a counter-part to the recreation ground on the other side of the estate. By now, you may recognise the semi-detached WDB5s that make up most of the properties facing The Green. The central terrace comprises WDA5s, another type of three-bedroom parlour home designed by Benjamin Wakefield. Some of the houses around the square were built for Robinson employees to live in and purchase.

This is where the walk ends. There is a bus stop by Market Square, where you can catch the bus back to Bristol city centre.

Hillfields/ Community Voices

The following comments were collected in projects run by Local Learning in Hillfields. www.locallearning.org.uk

Philip: We lived in St Bede's Road in a three-bedroom, semi-detached house with a downstairs toilet as you went out the back door. It had a coal house just inside the kitchen and a larder in the kitchen. Although it had three bedrooms, my mother and father had six children; four brothers including myself and two sisters... Everything was very basic. Mother done the washing in a big gas boiler with a gas-ring underneath and ironed it with an iron that you had to put on the fire to get hot. That's how it was. Our garden was quite small, but my father had an allotment. You got Woodland Way, you got Uplands Road and then you've got Hillfields Avenue. In the back of all those houses there's a big area of ground which was converted into allotments and my father had an allotment there. The official gate was off Woodlands Road, but my mother and father were quite friendly with someone who lived on Woodland Way and they used to allow us to walk round the back of their house down the back garden and into the allotments... I used to go there digging, helping father grow vegetables; potatoes and anything like that.

Bryan: I live in Frampton Crescent. I was born there in 1932... I moved away for 20 years, but the greater part of my life has been at Frampton Crescent. One of the things I always remember about living on the crescent and living on the estate was the honesty: you never locked your doors and you could leave your push-bike outside the local shops overnight and it would still be there next morning, or even parked up against the kerb. There were no cars in those days or very, very few... At home, you got up and it was a fight for the bathroom... The bathroom consisted of just one bath and to fill that bath, down in the scullery in the corner was the boiler. You boiled the water up by gas – a lot of people don't realise, but all the houses have four chimneys, but only three fireplaces, the fourth chimney was the flue for the gas boiler. Having boiled your water up it was pumped by a large hand pump – backwards and forwards – to fill the bath up... The houses were well-built, and the windows had loads and loads of little panes. The house was heated by coal and you had the coal-man come round every week to see if you wanted coal. As you went into the porch on your left you had the coal house and on your right the toilet and then you had your main back door. After the war they were classed as outdoor toilets and that's why every council house had a door put on the porch... so the toilet was classed as indoors. Some houses had the toilet just inside the front door and there was always the joke that you could be sat on the toilet and your mail could drop in your lap or your morning paper.

Janet: I live in Forest Avenue now. I moved up to this area when I was a child and moved to Frampton Crescent and we lived in that house for about ten years 'til it got too small for us because we had a large family and we moved to Forest Avenue. There was my grandmother, my mum and dad and five children. We were a happy family. In Frampton Crescent, the street was lovely and there was lots of families. When we came up to this area we didn't feel lonely at all because there was such a lot of children around. You could make friends very easily in those days. We had lived in a very small terraced house in town. I think it had two rooms downstairs and a kitchen, but it had no toilet or bathroom. The toilet was out in the yard and we had to bath in a big tin bath in the kitchen. So when we moved out to Frampton Crescent as children we thought we were in heaven because we had a garden at the front. We could open the front door and go in and run out the back door into another lovely garden and there was also an allotment attached and it was lovely... Yes, we loved that house. It was good because there was so many friendly people around, all children playing together, going to school together, going out for walks together, playing in the street. My brothers used to sit on the kerb and play marbles and we could play hopscotch in the road, there wasn't any traffic about. It was good, it was good fun... We moved [to Forest Avenue] because we could make the little parlour into a bed-sitting room for my gran and the two boys had the small front bedroom, us three girls had the big front-bedroom and Mum and Dad had the back room, so we got there all right. But one bathroom between eight people: you can imagine us going to school in the morning all trying to get in the bathroom before somebody else. But it was fun. It was a good house. We appreciated it. I used to know everyone in the street almost, but I hardly know anyone now... I have some nice neighbours, I just don't see them very much.

Cyril: Our house on Gorse Hill was just red-brick with a garden. For people who moved from the centre of town it was quite good. Indoor bathroom, upstairs, and upstairs loo. I always thought there were two bedrooms. They're now saying that there can be three at a pinch – there's a parlour-type thing, downstairs, front room. My father always said it could be used as a miner's bedroom. There was quite a lot of miners around here from Speedwell Pit and to save going upstairs and disturbing the house they could use the front room because of their shift work. I don't know how far it's true.

Tina: I currently live on Beechen Drive, but I grew up on Holly Grove, just up by Market Square. We always had the [community centre] Hub. It was the playgroup, it was the youth club, it was somewhere I always went and thankfully that's still there. The park I think is great. I miss the wooden structure. When I was a kid there used to be a wooden bridge that you used to be able to launch a friend off if you stood just right, I used to love that… There used to be an event up on Market Square every year when I was a kid and I used to love that because it was local and there was a little fair and there were games and things to do. It's not something that goes on anymore which is a shame. I still feel a sense of community around here and it's a joy to be a part of. If I'm asked where I'm from I say Hillfields.

Walk 11:
Sea Mills

This walk will take you around one of the country's finest municipal garden suburbs, one that was praised by John Betjeman in a 1937 radio broadcast for its 'vistas of trees and fields and pleasant cottages'.

Among the characteristics of garden suburbs to be found in Sea Mills are: a planned layout of low-density housing and clearly defined streets; houses coherently grouped in symmetrical pairs with occasional short terraces to add variety; generous rear gardens and spaces between properties to maximise access to sunlight and circulating air; a pleasant green setting and attractive outlooks; tree-lined streets and deep grass verges with houses set well back from the road; and provision of allotments, recreational areas, shops, schools, places of worship and a library. Sea Mills was designated a Conservation Area in February 1981.

Bristol's earliest garden suburb was at Shirehampton. It was set up by the Bristol Garden Suburb Company Limited before the First World War. The company was founded by Elizabeth Sturge, a pioneer in education for women; her nephew Frederick Allen Sturge Goodbody; and Eliza Walker Dunbar, one of the first female physicians in the UK. They had been inspired by a visit to Bristol by Ebenezer Howard, one of the early leaders of the garden city movement. George Oatley was appointed the company's architect and in 1909 land was purchased for the building of the estate from Philip Napier Miles, the biggest landowner in the north-west of the city. The project faced financial difficulties. Only 44 cottages for rent had been built by the time war broke out and the scheme was never completed. Napier Miles was a keen supporter of garden suburbs and would go on to be an important figure in the development of Sea Mills.

....

If travelling from Bristol city centre, you can reach the starting point of the walk – Meadway (1) – by taking the number 4 bus to the Sea Mills Lane stop and then walking in the direction of the estate along Shirehampton Road. If you prefer to take the train from Temple Meads to Sea Mills station, walk up Sea Mills Lane then turn left into Shirehampton Road.

....

Meadway is marked by one of the last of the original 1920s road signs for the Sea Mills estate to survive. The name is set on a wooden board attached to a tall concrete post. You'll see another later in the walk. If you have travelled here from the direction of Bristol, you will have reached this spot by crossing the River Trym, which provides

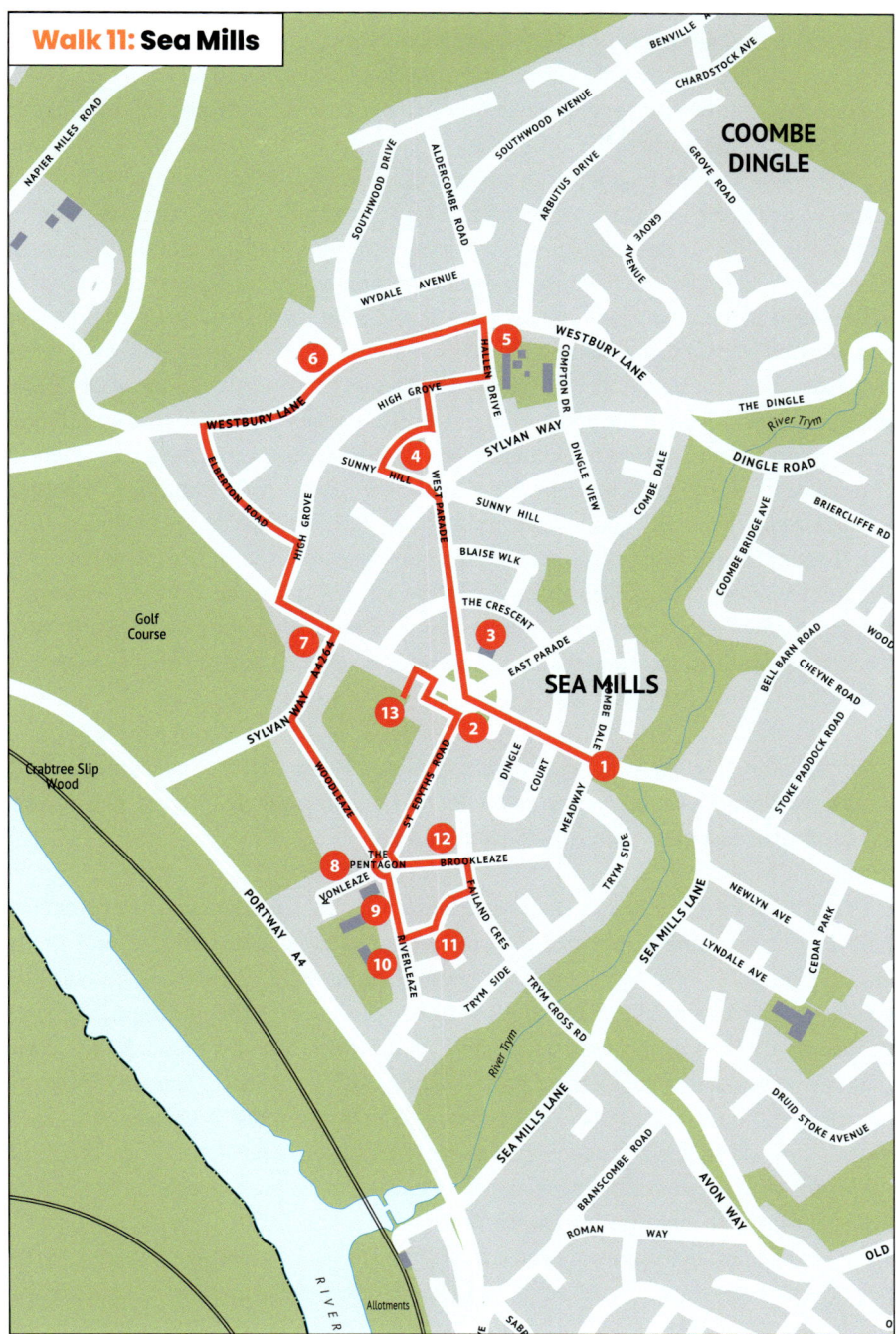

a natural buffer between Sea Mills and neighbouring Stoke Bishop. Until the late eleventh century, the river was navigable from the Avon as far as Westbury-on-Trym. Shifting land patterns have reduced its flow but it is rich in biodiversity and a valuable green asset for the area.

Coins and various fragments discovered at the mouth of the river have provided evidence of Portus Abonae, a Roman ferry station and military port that was located where Sea Mills now stands. It was abandoned by the Romans around the fourth century. If you walked up from the train station, you would have passed remnants of a dock built here in the eighteenth century. It was used for the fitting out of privateer and whaling ships, but was too far from the city centre to be commercially viable for trade and was abandoned in 1770. Although attempts to operate a dock at Sea Mills have proved fruitless, in the early years of the estate many local residents would have been employed at the docks in Avonmouth. Sea Mills tenants were charged the highest council rents in the city (a reflection of the high quality of the housing) so people needed steady work and good wages to live here. Other estates were expanded after the Housing Act of 1930 to accommodate those relocated from inner-city slums (see the Knowle West walk, for example). Sea Mills was left untouched by this, with little by way of new building until the infilling that took place after the Second World War. Therefore, the community remained relatively prosperous and relatively stable for years.

....

Continue along the left-hand side of Shirehampton Road. You will pass Dingle Close where the Olympic champion ice skater Robin Cousins grew up. Stop at The Square (2) by the Addison Oak and Sea Mills 100 Museum.

....

On 4 June 1919, the Bristol Housing Scheme was officially launched at a ceremony held here. Government minister Christopher Addison cut the first sod and Emily Twiggs, the Lady Mayoress, planted an oak tree. In his speech, Addison said that 'the scheme would be one that not only Bristol but the nation at large would be proud of' and he could 'not imagine a more glorious position than that of Sea Mills'. Later that day he spoke at a public meeting at Colston Hall (now Bristol Beacon) where he encouraged the council to provide 'houses with air about them' because, unless they did so, they would 'have to spend enormous sums annually on sickness'. At the time, Addison was president of the Local Government Board but he was appointed the first Minister of Health and Housing shortly afterwards.

Work on the Sea Mills estate commenced officially on 28 January 1920 and, at the height of activity, 185 men were employed by the council on laying roads and preparing the ground for housing. The only building contractor was William Cowlin and Sons. By

The bus stop at The Square, c1930 (Bristol Archives 43207/9/247).

the time of the 1921 Civil Engineer's Report, 58 homes were occupied; 1,200 yards of new roads had been completed; and more than 3,720 yards of sewage pipes laid.

This square, which comprises five greens bisected by axial routes, is the heart of the estate. It provides views to Kings Weston Hill in one direction and the trees of the Trym Valley in the other, and is edged by cottage-style, two-storey buildings with shallow-pitched roofs. The Tudor Walters Report on post-war council housing recommended that on the new estates there should be no more than 12 units per acre. At Sea Mills, there were fewer than 12 (in some places only eight), thus enhancing access to light and air as well as the sense of openness and space.

As part of the centenary celebrations, the volunteers of the Sea Mills 100 initiative converted the old telephone box which stands by the oak into a mini-museum with information boards, audio recordings and takeaway material. Opening hours are Mon-Sun 9am-5pm. See www.seamills100.co.uk for the latest news.

Cross to the other side of The Square (there's a pedestrian crossing on Shirehampton Road) and go to Sea Mills Methodist Church (3).

This landmark building provides a focal point for the square and was completed in 1930. It was designed by George Oatley, the architect of, among many other structures, the Wills Memorial Building for the University of Bristol (which you can see from the Northern Slopes on the Knowle West walk) and Bristol Homeopathic Hospital at

Building the Community Centre, late 1950s (photo submitted by Lisa Dicker to Sea Mills 100)

Cotham (which is referred to in the St Pauls walk). He was actively involved in Bristol's social housing movement and charitable programmes that aimed to alleviate problems faced by the poor. Eveline Blacker served as Oatley's apprentice for four years and his assistant for six. By 1919, she had set up an independent architectural practice in Bristol with Harry Heathman and together they designed several housing types for the city's inter-war estates. (See also the Hillfields and Knowle Park walks.)

Take West Parade to the Sea Mills Library (4). From the junction with The Crescent onwards, you will be passing a few of the more than 400 CNA2 type three-bedroom, non-parlour homes designed by the City Engineer's Department that were built in Bristol's inter-war estates. Alternating with these along West Parade are W H Watkins' WSB2s, a three-bedroom, non-parlour home. The significance of parlours is explained in the Hillfields walk.

The library is another landmark building on the estate. It was designed by C F Dawson, Chief Architect in the City Engineer's Department, and opened in 1934. This is one of Sea Mills' secondary centres and is located at the convergence of six roads with a clear line of vision in each direction.

There used to be a community centre behind the library. The Sea Mills Community

Children at the play group at the Community Centre, 1960s (photo submitted by Mary Wallis – who ran the group – to Sea Mills 100).

Association was founded in September 1945 and led a long grass-roots campaign to build a community facility for tenants. Residents were invited to purchase shares in the project for £1 each and enough money had been raised to begin construction by the late 1950s. Work was carried out over a period of five years by volunteers.

Continue past the old community centre site and turn right into Alveston Walk (there's another original 1920s road sign on the corner). Turn left into West Parade. The houses in this section of the Parade are MGB2s designed by Maidman and Greenen. This is a three-bedroom, non-parlour housing type of which 686 were built on the inter-war estates. At the end of West Parade, turn right into High Grove then left into Hallen Drive. Stop opposite what was once Sea Mills Infant School (5).

Land was reserved for a school on Hallen Drive in the original estate plans but it was not built until the 1950s. By then, the mixed infant and junior school on Riverleaze (which you will see later) had become too small to accommodate both age groups.

Continue on Hallen Drive and turn left into Westbury Lane, which has a mix of WDA5 and WDB5 three-bedroom, parlour homes designed by Benjamin Wakefield. The Red Bus Nursery, on the opposite side of Westbury Lane, is located in a former public

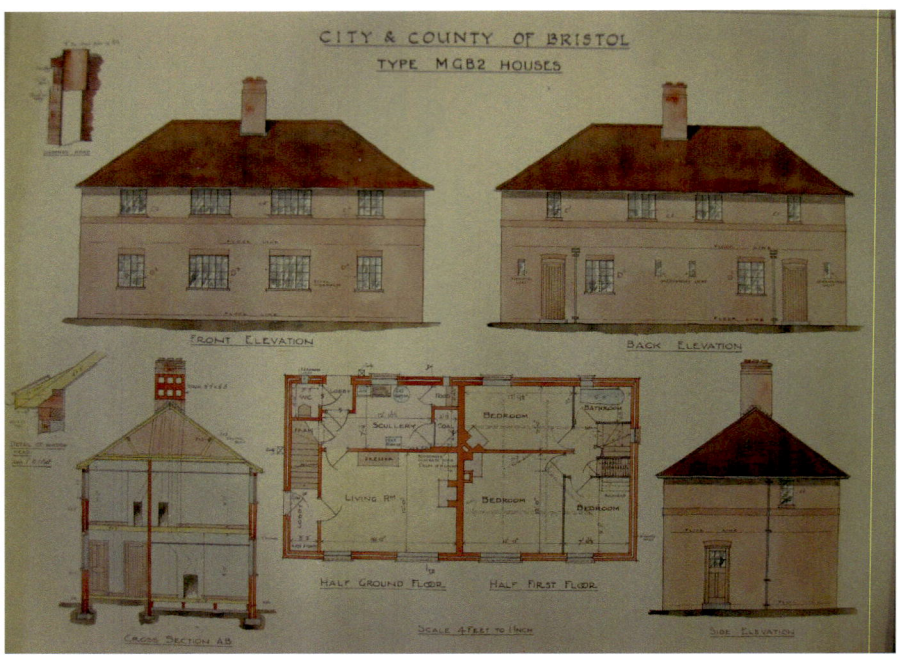

Original plan for MGB2 (Bristol Archives Red Label Plans 22).

house, the Progress Inn, which opened in 1936. In a perhaps naïve attempt to deter unruly behaviour on the part of council tenants, no pubs were allowed on the estate itself. Its building was controversial at the time with petitions gathered by both sides. However, as technically Sea Mills only occupies the left-hand side of this road, the pub could be built on the right-hand side in the neighbouring suburb of Coombe Dingle. Continue along Westbury Lane and stop opposite Haig Close (6) on your right.

These almshouses, positioned around a central green, were built for disabled ex-servicemen after the First World War on land donated by Napier Miles. Sixteen dwellings were completed between 1929 and 1930 with more added in 1936 and 1955. There is a mix of three and two-bedroom houses, three-bedroom flats and two-bedroom bungalows.

After Haig Close, take the next turning on your left off Westbury Lane into Elberton Road. Continue to High Grove. Turn right then left into Shirehampton Road. Turn right again into Sylvan Way (7) keeping to the left-hand pavement.

Image from The Builder (28 May 1920) showing construction of houses on Sylvan Way (Know Your Place, Community Layer).

The houses on Sylvan Way were the first to be occupied in Sea Mills, with residents moving in from August 1920. Within a couple of years, an article in the *Western Daily Press* remarked how Sea Mills was no longer a sleepy spot as every train between Bristol and Avonmouth now stopped there on weekdays, providing a regular commuting service to the new residents (30 June 1922).

Sylvan Way is the widest main route through the estate and has the deep grass verges characteristic of garden suburbs. It provides an attractive green gateway to the estate leading up from the Portway. Short terraces alternate with semi-detached houses. Their design is influenced by the simplicity of form of the Arts and Crafts movement – with minimal ornamentation and mainly flat fronts and backs – that might be considered unremarkable individually but collectively make a picturesque impression. Construction of the homes was partly financed through the selling of Bristol Bonds, as promoted in an advertisement in the *Clifton and Redland Free Press* on 22 April 1920 headed 'Do Your Share to Solve the Housing Problem'. Bonds ranged from £5 to £100 in value and could be redeemed after a minimum of five years at an interest rate of six percent per annum.

When the delegates of the Inter-Allied Housing and Town Planning Congress had completed their inspection of the Hillfields Demonstration Area on 5 June 1920 (see Hillfields walk), they came to Sea Mills to view some of the 250 Dorman, Long and Company (DorLonCo) homes that are unique to the southern half of the estate and include those on Sylvan Way. The houses were constructed on a steel frame. The outer walls are of cement and sand; the inner ones of breeze block. This was a nontraditional method that provided a quick alternative to brick-built homes when skills and materials were in short supply, but they were not cheap and the experiment dwindled from lack of funds after the Geddes Axe of 1921.

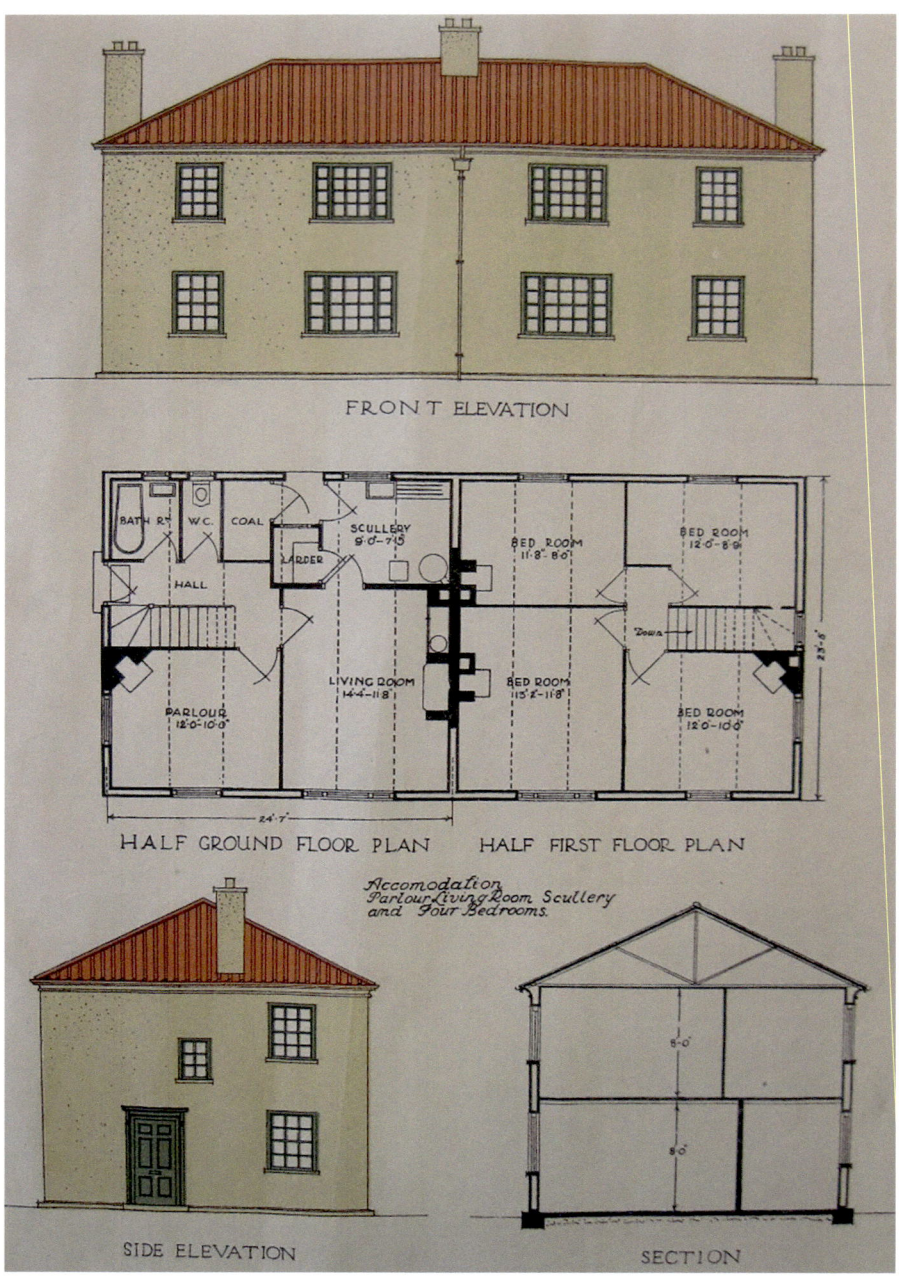

Original plan for a four-bedroom Dorman, Long and Company steel-frame parlour home (Bristol Archives Red Label Plans)

Evacuees from Sea Mills in Wellington, Somerset, 1941. They were accompanied by the Rev Wilson, vicar of St Edyth's, who is in the centre of the back row (submitted by Peter Hodgson – fourth from left in front row – to Sea Mills 100).

Another non-traditional method used in Sea Mills was for the Parkinson Pre-Fabricated Reinforced Concrete (PRC) houses built under the Housing Act 1924. They were finished in roughcast render with the exception of four red-brick pairs in Failand Crescent. Concrete proved less durable than brick and, by the 1980s, the council considered the Parkinson homes too expensive to repair, and demolished and replaced 132 of them. Locals campaigned to stop the demolition programme as there was no guarantee that people could move back again. It is a period remembered as one of uncertainty and sadness, although the campaign was ultimately successful and the council reverted to repairing the properties.

Continue down Sylvan Way. When you reach Woodleaze (on your left), look across at the pair of semi-detached houses on the opposite side of Sylvan Way. These are the only DLA6s – four-bedroom, parlour homes – to be built. Turn into Woodleaze. The houses immediately to your left and right are DLA2s, three-bedroom, non-parlour homes of which 100 were built. Alongside them are blocks of DLB5s. Continue to The Pentagon (8).

This is another of Sea Mills' secondary centres, marking a convergence of five roads. The pillar box is of interest. It was made for use in Scotland in the 1950s and therefore does not have the cipher for Elizabeth II that would normally be found on post boxes from this period. The Catholic Scots did not recognise Elizabeth I, hence the reluctance to refer to our current queen as being the second to have that name. Despite extensive research, it is not known why the box ended up in Sea Mills. It is an unusual sight in England.

Pupils from Sea Mills celebrate the coronation in 1937 (photo submitted by Jane Macfarlane to Sea Mills 100).

Adjacent to The Pentagon is St Edyth's Church (9).

This landmark building makes what is referred to in garden suburb planning as a 'terminal feature', one that arrests the eye when looking down the long straight stretch of St Edyth's Road from The Square. It is the focal point for The Pentagon. The foundation stone was laid in 1926 and the church was consecrated in 1928. Like the Methodist Church seen earlier, it was designed by George Oatley, who was knighted for his public service in 1925.

From the church, turn into Riverleaze and stop at Sea Mills Primary School (10) (formerly Sea Mills Junior School).

The first block of this school, designed by Alfred Oaten, was opened in 1928; the second in 1931. It uses the same red pennant stone that can be seen at St Edyth's Church. Until permanent buildings were erected, a repurposed wooden hall that had been moved from a nearby aerodrome had served as both a school and church for the estate. The separate school blocks were used for the infants and the juniors until the site became overcrowded. Once the new school on Hallen Drive – which you saw earlier in the walk – was built, the infants could move out. In 2009, the schools were merged again.

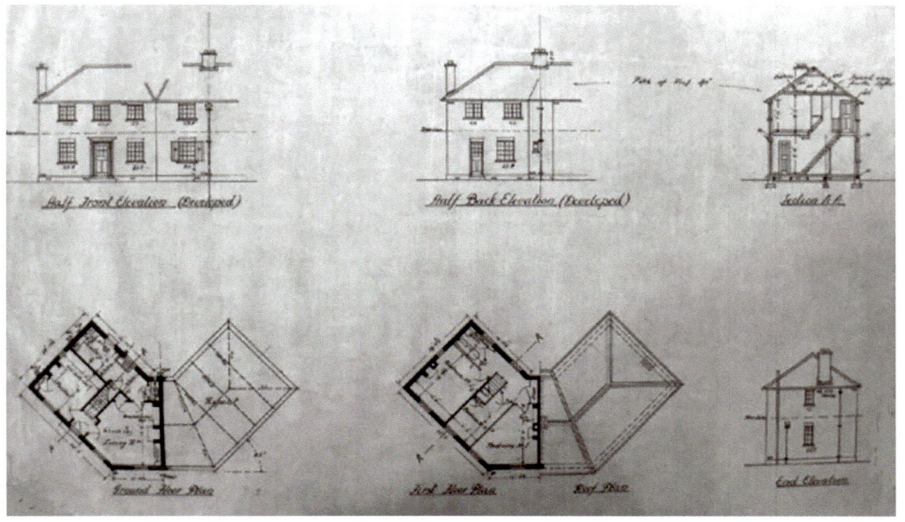

Original plan for CBA5 (Bristol Archives Red Label Plans 22).

Turn left into Bowerleaze. The houses on the corner are three-bedroom, CBA5-type parlour homes designed by the City Engineer's Department, of which only a few were built. On Bowerleaze you will pass CNA5s, another rarely built City Engineer's Department design for a three-bedroom, parlour home. Stop at Sea Mills Farm (11) on your right.

Sea Mills Farm, which was owned by the Napier Miles family, originally covered most of the land used for the Sea Mills estate. Napier Miles sold 205 acres to the council at the cost of £160 per acre on the condition that it would be used for no other purpose than the building of the garden suburb. Records suggest the land was purchased in segments while it was still being worked. The farmhouse dates from 1772, with some later additions. By the time the estate was being built, it was in a poor condition and the sitting tenant, Charles Pearce, insisted that repairs should be carried out at the council's expense. Pearce seems to have been persuasive as he was also compensated for the cost of fencing that would stop his cattle from straying beyond their fields and into the new development area. Research carried out by James Powell as part of the Sea Mills 100 project shows that the council only gained possession of the house when the longest surviving family member – Charles' youngest daughter, Alice – died in 1983.

Bowerleaze merges into Failand Crescent. Turn left along Brookleaze to The Pentagon. Turn right into St Edyth's Road (12).

The team of Sea Mills AFC 1932-1933 at the Recreation Ground (with kind permission of Sea Mills AFC). The ball is held by Bert Gill.

The houses on the corner are more of the council-built CBA5s. The rest of the housing on St Edyth's Road was built for private sale in 1923/1924. After an enthusiastic beginning in 1920, progress on the estate faltered following the cut to the housing subsidy in 1921 and did not pick up again fully until 1927. During the time of the slump, the council was left with empty plots and no money to build on them and so some land was leased to private developers. St Edyth's Road was an owner-occupied enclave in a municipal estate, making Sea Mills an early example of mixed development. Like Sylvan Way, this is a partially tree-lined street characteristic of garden suburbs, with houses placed well back from the traffic. The houses have bay windows and are finished in a rough-cast render combined with redbrick. Semi-detached homes alternate with terraces of three or four dwellings.

Continue to the end of St Edyth's Road. The shops on The Square are versions of the WDB5 design that you will also see on the Hillfields and Knowle Park walks. There were originally 12 shops, but some have been converted to residential use. Turn left into Shirehampton Road and then left into the Recreation Ground (13).

Napier Miles recommended to the council's Housing Committee that at least five acres of the new estate should be set aside for sport. Like the Recreation Ground in the Hillfields walk, this is an enclosed yet open space with a perimeter of houses. A cricket pavilion was opened in May 1923, which became an important venue for social activities. The Sea Mills Tenants Association was formed there, holding its first meeting in November 1927. It was destroyed by fire several years ago. Sea Mills Amateur Football Club was founded in 1925 and continues to play to this day, although the games now take place in Kings Weston.

The walk ends here. If you are in need of refreshment before continuing your journey, The Café on the Square (converted from a former toilet block) is recommended (see www.smci.org.uk/cafeonthesquare.htm for opening hours). The café is run by Sea Mills Community Initiatives, a charitable company set up in 2009 by the local churches, which has also developed a community garden on the estate. You can catch the number 4 back to Bristol city centre from the nearby bus stop. Alternatively, you could go to the library and catch the number 3 bus, which will also take you back to the centre.

Sea Mills/ Community Voices

These quotes from residents have been extracted from oral histories recorded by Sea Mills 100 volunteers.
www.seamills100.co.uk

Joan: I was born in a house just the next street away from where I live now. When my parents moved in it was a brand-new house. They moved in three weeks before I was born. It was the first time they had electricity or anything like that. They had been living in a tiny flat in Clifton. They were thrilled to bits to have a house… We grew up knowing all our neighbours and what their husbands did and what all the children did, and we all went to school together. I started school at five. I can remember it so well. It was a lovely school. Sea Mills school was then Infants and Juniors together. It was a lovely warm little [class]room and in the winter it had a coal fire. Everyone who went to that school will remember Mr Godwin. He was the caretaker. He was a very strict caretaker and in the winter he would come with the coal and that and light the fire in the babies' class… You used to come home for dinner. There was no such thing as school dinners. When you were small your mum would have to come and pick you up and take you back again. I remember well being in the Juniors over the other side of the playground. There was one teacher, a man teacher, and he was a wonderful teacher and I am sure his name was Mr Hunkin. All the teachers used to teach everything then, but he's the one teacher who got me interested in the world and geography. He's the one person that made me want to travel. I used to love his lessons. I couldn't wait for them. Before I had finished there the war started.

Stan: We lived in one of the concrete houses (DorLonCos) in Failand Walk. It was where you could have a hot bath and a cold shower all in one operation because the water used to condense on the ceiling and then drip cold into the hot water. One of the chief tools you took into the bathroom with you was a sponge on a long handle so you could wipe the ceiling dry so you could avoid the cold water dripping on you while you enjoyed a hot bath. All the houses in the road were the same, one sitting room downstairs, the kitchen with the obligatory larder and coal house, downstairs toilet with three rooms upstairs. The bathroom was downstairs.

Gill: Sea Mills library made a great impression on me. I didn't read before I went to school and when I went to school I didn't find it difficult to read but I loved going round to the library. I really, really did. I loved the fact that there was a children's section and the adults' section. When you went in you turned left and there was the children's section with lots of lovely books to enjoy. I would take them out with my ticket. I remember when I was 12 or 13 thinking I can go in the adults' section now and I remember that transition. I remember they had newspapers on boards that were slightly inclined so that you could read the newspapers.

Derek: The community was pretty friendly. In fact, I knew people everywhere in Sea Mills. I think because it was very mixed. There were people I knew who were dockers, or whose fathers were dockers. If you're a docker you were not on regular pay. They might be working today and not working tomorrow. There were a whole load of people you would meet, in the shops, in school, at church, in the scouts. What I remember about Sea Mills that I did not realise at the time was so good was that it was in an astonishingly beautiful setting. Not many council estates have hills rising on one side with Blaise Castle, which is marvellous walking country. Almost continuously it led to the golf course which was National Trust country which was great tobogganing. We had much more snow in those days. Then there were the woods and fields of Portway and on the opposite side was endless, endless Somerset, going forever. Yes, it was a landscape that came free, and it made a great improvement in life in general that you came out and looked around and saw trees and hills. The standard walk for the youth club was to meet on the Sea Mills Square around 2 o'clock on Sunday, set off and walk to Blaise Castle where there was a tea shop just beyond the big house and then turn round and walk back again. It was very healthy indeed.

Jane: My favourite haunts were with my bestie Sue. We used to fish. The damage I did to the Sea Mills environment was appalling when I was a kid! We used to fish for tiddlers with jam-jars in the stream behind where Sue lived in Trymside. I'd bring them home in jam-jars because I wanted to look after them, but they would always die because I did not get that they needed oxygen. They always died while I was in that lane between Failand Walk and St Edyth's Road. Oh no! So I used to run, and by the time I got home they were all dead. Once I got one tadpole, just one, and grew it into a frog in a goldfish bowl. When it was a tiny little frog, I took it to my uncle's pond.

Mervyn: The doctor's surgery used to be in a house in St Edyth's Road. On the outside, on the right-hand side of the front door, completely unlocked was a wooden box. That's where they put the prescriptions... We would put all the prescriptions in there, and people would rifle through all the different ones. You'd pick out your own one and off you went. We only had one incident of someone who was very angry and threw the whole lot over the garden. The waiting room had these benches which were probably a death trap because they were so old. It was small, like anyone's living room, and there was a hatch where someone was answering the phone behind. They could hear everything that was being said in the waiting room and they could probably hear everything that was going on in the back room.

Walk 12:
Knowle West

This walk will take you through an inter-war estate in South Bristol that owes its existence to the Housing Act 1930 and the government's commitment to a slum clearance programme.

Knowle West was constructed primarily to accommodate those who had been moved from parts of Bedminster and other areas of housing deemed unfit for human habitation in Bristol's city centre. Knowle Park – further to the east – was built under the more financially generous terms of the Housing Acts of 1919 to 1923. During this period, houses were of the highest quality, expensive to construct and aimed at more prosperous council tenants. A wide band of cheaper housing commanding slightly lower rents was built adjacent to Knowle Park under the Housing Act 1924.

In the introduction to his post about Knowle West on the Municipal Dreams blog (municipaldreams.wordpress.com), author John Boughton writes:

> If a single estate can be taken to encapsulate the social, political and planning history of council housing in this country it is probably Knowle West in Bristol. You'll find in it all the hopes and dreams, all the good intentions and unintended consequences that have marked the complex story of council housing over the last hundred years or so. And you'll find families and communities that have lived this story in all its complexity.

The new arrivals to Knowle West encountered broad streets containing a mix of semi-detached houses and short terraces laid out in accordance with garden suburb principles. There was access to green open spaces, as well as freedom from the industrial pollution that plagued the inner-city. Those who had been moved out of the slums had left behind the sociability (as well as the health hazards) of closely-packed, densely-populated dwellings, and lost touch with many of their former neighbours. Although some were unable to settle and took the first opportunity to leave, it is thanks to the commitment of those who remained that such a strong sense of community exists in Knowle West today.

Jayne Cogan, who has contributed to Knowle West Media Centre's *100 Years of Knowle West Style* (of which more later), says:

> I have memories of popping up to Filwood shops to get the shopping after school

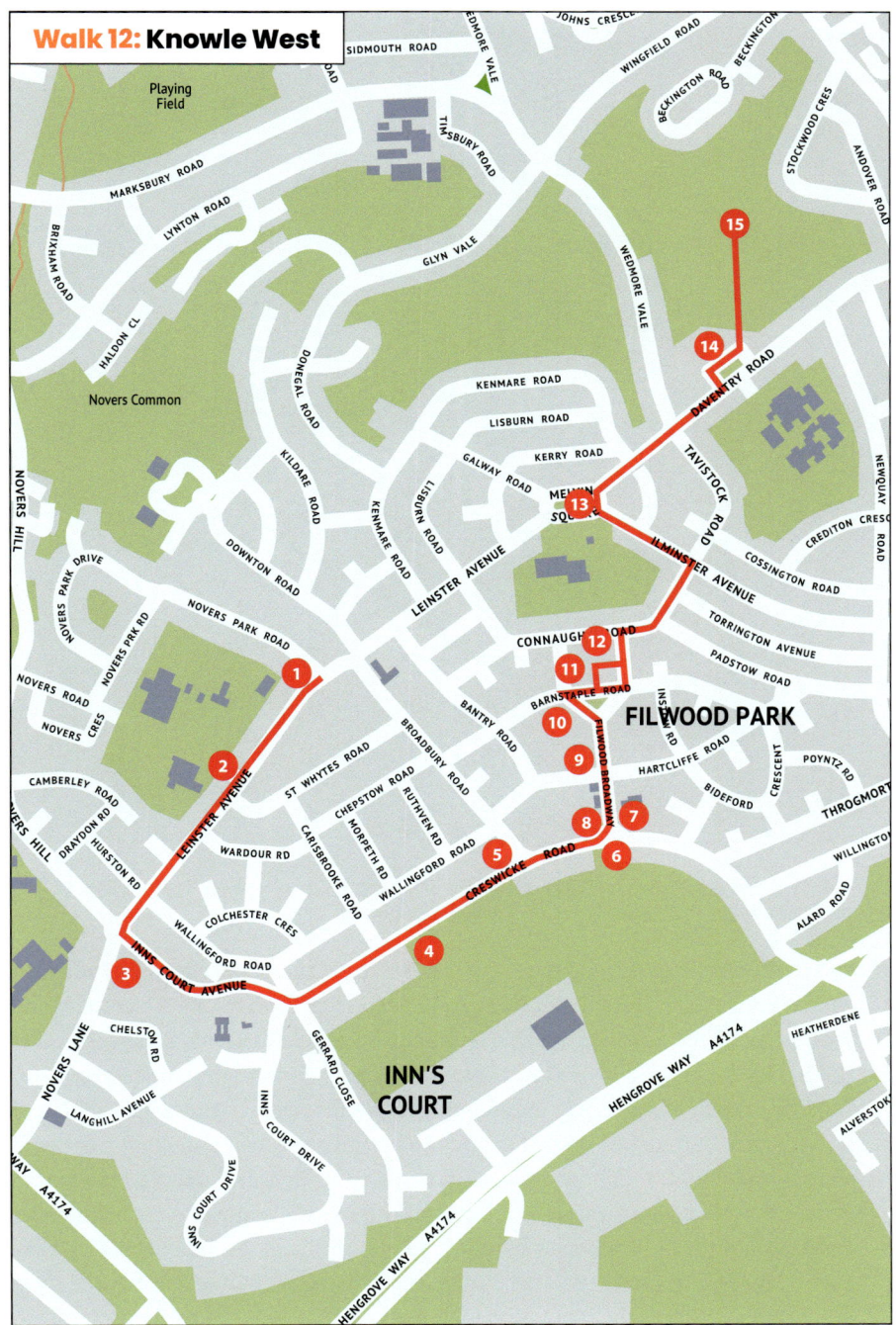

Plan of Knowle and Bedminster in the 1936 Bristol Housing Report (Bristol Reference Library B14103). This shows the area before the completion of the Knowle West housing.

and queuing outside the veg shop on Xmas eve to get the Christmas veg before they shut at dinnertime. I also remember a balloon coming down in Creswicke fields and me along with all the other local kids running down to have a look. I also remember the ice cream vans that used to come round, and going to get an ice cream for 10p! My grandad used to take me and my brother for walks in Creswicke fields and all over the old knocked-down houses nearby and show us the wild flowers (and weeds) growing. It's all been built on now. At 15 while still at school I worked at Jarman's off-licence two days per week on their sweet counter – and remember weighing out the old-fashioned sweets from the jars – and the kids and adults alike coming in for a '20p mix-up'. It was quite tough living in Knowle West at the time, but it never deserved the reputation it always had to outsiders. I'm always proud to tell people I come from Knowle West. It's full of spirit.

If travelling from Bristol city centre, you can reach the starting point of the walk – Knowle West Media Centre (1) – by taking the number 91 bus and alighting at the Broadbury Road stop. Cross Leinster Avenue, turn left and the Centre will be on your right.

Knowle West Media Centre (KWMC) is an arts venue and registered charity that supports people to make positive changes in their lives and communities by harnessing the power of technology and the creative arts. As part of Homes for Heroes 100, KWMC worked with the artists Lukus Robbins, George Lovesmith and Holly Beasley Garrigan on a project called '100 Years of Knowle West Style'. This celebrated the estate's distinctive culture and stories, giving residents the opportunity to decide what is remembered by future generations. Cheryl Martin, who has lived in Knowle West for 50 years and leads neighbourhood health and photography walks, provided invaluable background knowledge to both the KWMC project and the wider centenary programme. She believes it is important to know more about our history. She says: 'to understand where you're at, you have to know where you've come from.'

KWMC is an environmental construction, with the external walls made from straw bales and more than 100 solar panels on its roof. It was designed in partnership with local young people and was opened on 14 February 2008. Prior to this, the KWMC team was based in the former William Budd Health Centre, which was located on this site until its demolition in 2007. Budd was a Bristol surgeon who worked in the area of contagious diseases. His name lives on in the William Budd Health Centre on the nearby Knowle West Health Park.

Continue along Leinster Avenue, keeping Knowle West Media Centre on your right. Stop at the Knowle DGE Learning Centre and Sixth Form (2) on your right.

This was formerly the site of the South Bristol Open Air School, one of a number of purpose-built education establishments that were developed in the late 1930s to combat the widespread rise of tuberculosis among children. In 1938, some 1,035 children in Bristol were reported to be infected by the disease, partly as a result of the malnutrition prevalent among the poor during the economic depression of the 1930s. The schools were built on the concept that fresh air, good ventilation and exposure to the outside contributed to improved health. This was part of the wider philanthropic efforts of the era that aimed to improve the quality of life for families living in poverty. A 1939 survey showed that half the children in Knowle West lived below the poverty

Inns Court Farm, c1890s (Know Your Place, Community Layer).

line, with one in four heads of household unemployed or dependent on casual labour. The South Bristol Open Air School was the first of its kind to be built in the city and catered for 315 children transferred from existing local schools. It opened in January 1940 and closed in 1953.

Previously, this had been the site of the Novers Hill hospital for infectious diseases. In 1892, the Bristol Sanitary Committee purchased 13 acres to build a facility here for smallpox victims, which opened in 1897. It was also used for the treatment of those infected by measles, diphtheria, tuberculosis, scarlet fever and influenza. By 1933, the construction of the surrounding housing estate meant the site was no longer sufficiently isolated. Instead it became a convalescent facility for patients from Ham Green Hospital, before closing in 1937.

From here, continue to walk down Leinster Avenue until you reach the roundabout. Turn left into Inns Court Avenue and look across to the pastel-coloured houses on the edge of the Inns Court estate (3) on your right.

By 1946, there were approximately 26,000 people on the waiting list for council housing in Bristol. This demand was caused by the suspension of the much-needed house-building programme during the war and the damage to existing properties through bombing raids. Over the coming years, new large-scale estates were built in the suburbs, including Hartcliffe, while existing estates were extended wherever possible.

In the 1960s, Inns Court was developed on the edge of Knowle West. Its design marked a radical departure from the garden suburb principles of the rest of the estate. Instead of wide streets and open spaces, homes were constructed around small cul-de-sacs and enclosed pedestrianised areas. The planners' aim was to increase the sense of living within an intimate community. This is often referred to as a Radburn design, taking its name from the place where it was first used: the town of Radburn in New Jersey, USA. By the late twentieth-century, this type of design had been discredited. It was blamed for intensifying social isolation (cutting residents off from the surrounding streets) and the risk of crime. Following a review in 2009, Bristol City Council planned to demolish Inns Court and relocate residents, but this was successfully resisted by those who lived there. The chair of the residents' committee said at the time:

> I don't see there is any necessity for demolishing our homes. When they were built, we were told they would last for 100 years but now they are talking about taking them down. I have lived here for more than 30 years. My wife and I are happy here. We brought up our family here. If they demolished the estate and rebuilt it, it would devastate the community.

Evidence of a Roman settlement from the third and fourth centuries has been found in the Inns Court area. At the heart of the estate, there are also the remains of an early fifteenth-century manor house built by Sir John Inyn that was demolished in the nineteenth century, save for its medieval stair tower. The tower was incorporated into the Holy Cross Inns Court Vicarage, which was built in the 1950s and can be glimpsed from Inns Court Green off Inns Court Drive (a short optional detour). The building is now empty and the tower is on the English Heritage 'At Risk' register. The Knowle West estate was developed on fields that formerly belonged to Inns Court Farm. In the mid-1950s, an open barn still existed on Inns Court Green that was used as a milk-collecting point.

Continue along Inns Court Avenue, which becomes Creswicke Road once you've passed Inns Court Drive and Gerrard Close on your right. Stop at Filwood Fields (4) on your right.

This has been an area of play for generations of children in Knowle West. At one time, it was possible to walk from here to the hangars of nearby Bristol (Whitchurch) Airport. The airport opened in 1930 and closed in 1957 when the operation moved to Lulsgate. During the Second World War, it was one of the few British airports to continue running a regular civilian service.

After the war, temporary pre-fabricated housing was erected along Creswicke Road. Former Inns Court resident Bonita Croot shared her memories of the area with KWMC during the '100 Years of Knowle West Style' project. She says:

> Whilst living in the pre-fabs, approximately aged four, I can remember hearing bells and cheering around the area and neighbours going in to a neighbour's home a few doors away to watch Queen Elizabeth II crowned on 2 June 1953 from the only TV in the Walk! On Creswicke Road, near the pre-fab shops of Inns Court, there was a cow shed and sty where we would buy our fresh milk. As a child it was brilliant [living there], with friendly, helpful neighbours/kids. We used to go on long walks.

Continue along Creswicke Road, stopping at the junction of Broadbury Road (5) on your left. This is the first junction that you'll come to.

The houses on the corner of this junction are Maidman and Greenen MGC2s, one of the most widespread designs for three-bedroom, non-parlour semi-detached homes built on Bristol's inter-war estates. They are normally located at junctions such as this one. The short terrace on the right of Creswicke Road, facing Broadbury Road, is made up of HBB2s, a three-bedroom, non-parlour home designed by Eveline Dew Blacker and Harry Heathman. There are 620 houses of this type to be found on the later inter-war estates of Bristol, including Horfield and Southmead, as well as in parts of Knowle and Bedminster.

Continue along Creswicke Road until you reach Filwood Broadway (6) on your left. This is the first junction that you'll come to.

Filwood Broadway was designed as the central hub of Knowle West. In the early stages of the estate's development, there were few community facilities, in part because government subsidies only covered the cost of building homes. It was also felt moving people out of the unhealthy slums should be the council's first priority. As a result, thousands of people were relocated to 'a wilderness of houses' that were likened by some to 'barracks' (Martin J Powell in *Bristol: The Growing City*, 1986, Redcliffe Press, pp110 and 112). With the construction of the Broadway in the mid-to-late 1930s, residents were able to shop locally, having previously been dependent on the shops in Bedminster or door-to-door hawkers. However, many families still went to Bedminster for their weekly 'big shop' and a fish and chip van was among several mobile retail services making regular rounds of the estate well into the 1950s.

Some of the shops on the Broadway are now closed. There are several reasons for this: people increasingly using their cars to shop further away where there is more choice and more competitive prices; the growth of online shopping; the loss of major local employers and the reduction in household income for many; and, some argue, the unacceptably high business rates. As part of the council's Knowle West Regeneration Framework – a 20-year plan for meeting community aspirations that was launched in 2008 – this area is set for redevelopment. In both council consultations and their own 'grassroots' projects, residents have made it clear that they wish the Broadway to remain at the heart of Knowle West. There are many community-led initiatives currently working to create positive change in the area. In recent years children have also painted bollards and street furniture on the Broadway while a group of gardeners and makers constructed benches and planters for flowers. In 2023, Filwood Broadway was awarded £14.5m from the Levelling Up fund.

Opposite the library, which you'll see on your left, is the former site of Filwood Swimming Baths (7).

These were the first public baths to be built in Bristol after the Second World War and were large enough to host international events, accommodating up to 500 bathers and 450 spectators. They were opened on 14 July 1962 and closed in 2005. Jayne Cogan, who contributed her memories for the '100 Years of Knowle West Style' project, says:

> I mostly spent my leisure time at Filwood swimming pool from about the age of ten. We used to go about five times a week and stay in there as long as we could get away with... I got my first job there where I worked while at college. I have memories of Saturday and school holiday fun – swims when all the local kids would turn up in droves, and us lifeguards would have fun with them, squirting them with hosepipes and trying to knock them off the floats with the water... For me, Filwood swimming pool was an iconic place.

Turn left to walk past the library and continue along the Broadway to the Church of Christ the King (8) on your left.

This landmark church was completed in the 1960s. Prior to this, Mass had been held in the hall of the Catholic primary school around the corner on Hartcliffe Road (built 1938). Since the cross was stolen from the statue on the baptistery wall, some locals refer to the figure that remains as 'The Floating Jesus'. The sites of the church and school were acquired by the Bishop of Clifton during the early planning of the estate.

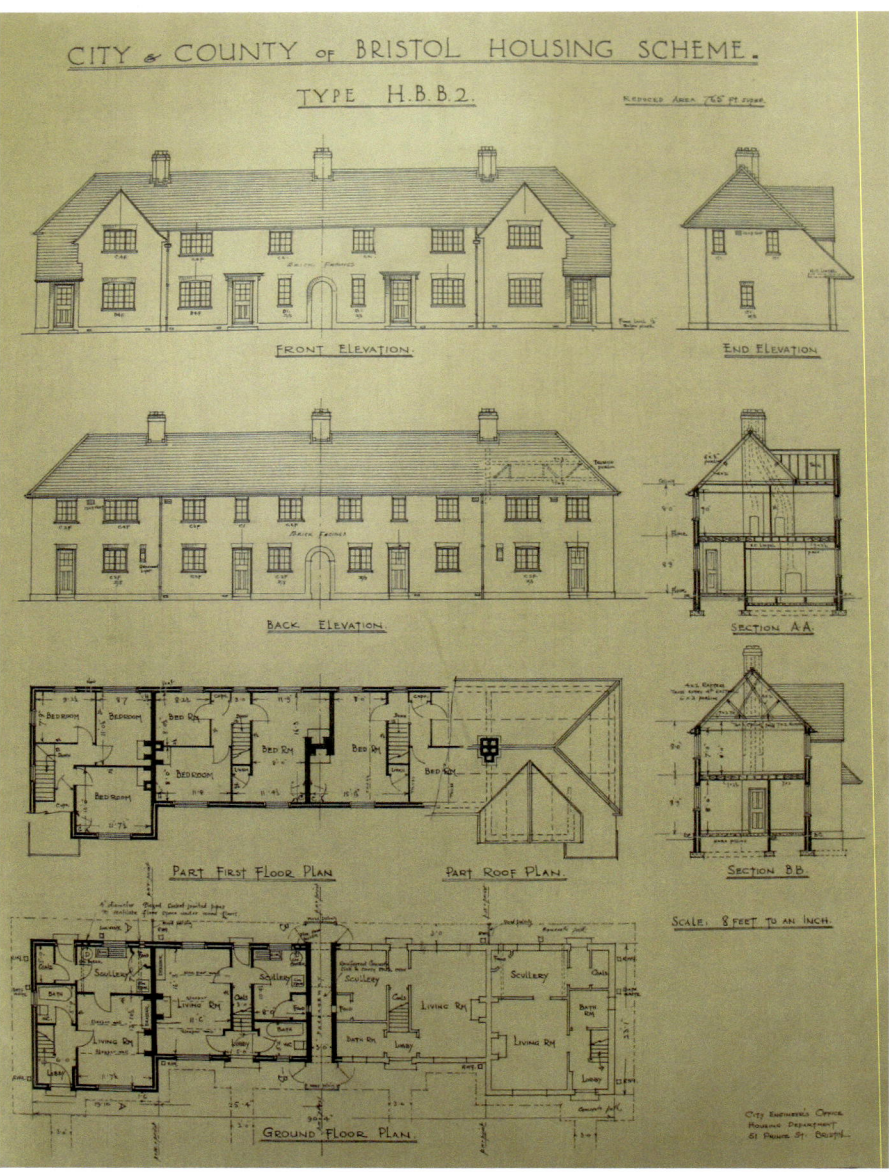

Original plan for HBB2 (Bristol Archives Red Label Plannns 22/3/4).

He also acquired the site further along at 2 Filwood Broadway that was used to build a convent for the Religious Sisters of Charity (opened in 1937). This became a small missionary community of the Lee Abbey Movement in the mid-2010s.

Filwood Swimming Baths, exterior, in the 1960s (Bristol Archives 40826/BAT/4).

Cross Hartcliffe Road and continue to the site of the former Filwood Broadway Cinema (9).

The license to build a cinema in Filwood Broadway was put out to tender in 1937 and won by Roy Chamberlain, who was granted a £7,000 mortgage by the council and a 99-year lease. The three-story Art Deco-style building opened on 20 October 1938 and had 1,000 seats. As stipulated in the council agreement, it had a separate rear entrance for lower-class patrons. Known locally as 'The Bughouse', boxing matches and other live events were staged here in addition to film screenings. In 1971, unable to compete with the attractions of television and other entertainments, the building was converted from a cinema into a bingo hall. It closed in the mid-1990s and remained empty until it was demolished in 2023 for new housing and retail. Some of the original cinema projectors and seating are now in the possession of M Shed, the museum of Bristol history on the Harbourside.

Continue along the Broadway, past a row of shops. The last of these was once the site of the Bristol Rediffusion showroom (10).

Bristol Channel was a community TV initiative set up by the Rediffusion cable network in 1973. Between December 1973 and March 1975, a team of volunteers and up to 16 staff members worked with organisations across the city, recording hundreds of hours of footage that was transmitted to around 23,000 Bristol homes, initially only at weekends but later also on weekday evenings. Among the groups involved was a large contingent of Knowle West residents who contributed 21 hours of material. They conducted street interviews, filmed around the community, and presented the Knowle West TV news bulletins. In 2014, the Knowle West TV stock was digitised. As well as being held by KWMC, footage is at Bristol Archives, alongside the administrative paper archive. Extracts of the show can also be found on the YouTube channel Knowle West TV. It is well worth a watch. Although the Rediffusion showroom moved from Filwood Broadway to Old Market in 1961, many locals associated the programme with this site.

Cross Barnstaple Road to the Filwood Community Centre (11), opposite the green that marks the end of Filwood Broadway.

Since 2010, the Filwood Community Centre (filwoodcentre.org.uk) has been managed on a day-to-day basis by the charity Community in Partnership Knowle West. The centre was built in 1937 at a cost of £16,000. It was arranged around a central lawn that could be used for outdoor entertainments. One side of the quadrangle was designed for the use of adults, the opposite side was for children and adolescents. It contained a dance hall/theatre; a fully-equipped gymnasium with changing and shower rooms; meeting and games rooms; reading and common rooms; a canteen; a skittle alley; and workshops. Its aim was to provide a homely club where the community could make friends and experience a range of leisure opportunities. It remains a well-loved community asset and a source of pride for the estate with many of the original design features and services still intact. KWMC has its origins here, beginning as a 1995 photography project that refurnished the dark-room facilities and taught photography skills to local people. If the centre is open, you can walk through the quadrangle and out to the community garden at the back.

Once you reach Filwood Community Centre, turn right, and then take the first left to Marwood Road. You will then reach the wood-clad We Can Make prototype home (12).

Over the summer of 2017, this eco-friendly Transportable Accommodation Module was built as part of the We Can Make housing initiative – an ongoing collaboration between local residents, KWMC and the architectural practice White Design. The programme aims to support people to develop the housing solutions they want and need. The house is fully plumbed and wired, has walls made of straw bales and has triple-glazed windows. More than 400 people stayed overnight to test it out. The first

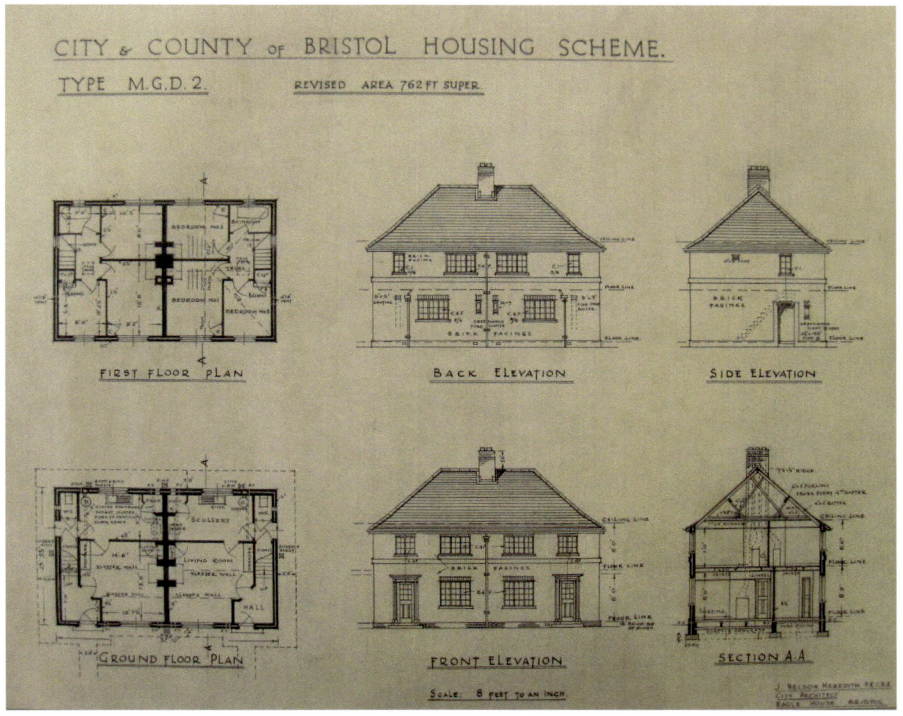

Original plan for MGD2 housing (Bristol Archives Red Label Plans 22)

two houses were built in 2022 and are now occupied. Melissa Mean, Head of Arts at KWMC, explains the reasoning behind the project:

> From our research and conversations with families in Knowle West, we've seen that the current competitive housing system doesn't work for many people – they are struggling to find the kind of home they need at a price they can afford. However, there's a keen interest in trying something new. Ninety percent of people we asked said they thought 'micro-plot' homes were a good idea for the neighbourhood and 73 percent thought they were a good idea for their street.

Knowle West residents were part of the construction team and worked with artists Charlotte Biszewski and Alex Goodman to ensure the house reflected the local area. They made curtains, cushions and tiles using natural dyes, and built furniture for the kitchen with support from the technicians at KWMC's manufacturing and making space KWMC: The Factory, which is based at Filwood Green Business Park.

The Venture Inn, c1950 (Bristol Museums, Galleries and Archives, Hartley Collection 259235).

Before leaving Marwood Road, look at the houses opposite. The first, coming up from the direction of Barnstaple Road, are CND2s designed by the council's City Engineer's Department and among the most common variety of three-bedroom, non-parlour homes built on Bristol estates from the mid-1920s to late 1930s. Next to them are Maidman and Greenen MGD2s, another common type of three-bedroom, non-parlour home.

From Marwood Road, turn right onto Connaught Road. Head towards Ilminster Avenue. Turn left and continue to Melvin Square (13).

Before the building of Filwood Broadway, this square was the only purpose-built shopping area for the estate. A once prominent feature was The Venture Inn, owned by the Bristol brewery George's, which opened on 16 December 1935. None of the other inter-war estates were provided with public houses so this was a significant exception. It also represented a new venture for the brewery (hence its name). It functioned as both a licensed pub and a neighbourhood social centre. There were separate entrances for the club room and for a café that provided soft drinks and light meals. Concerts, dances and meetings were held in the assembly hall that could hold up to 300 people. The licensed area was extended to cover the hall in 1938 to reduce overcrowding in the public bar and smoke room, and the need for people to spill out into the street. The building was demolished in 2006 and the Carpenters Place flats built on the site. Many locals have memories of times at The Venture; you can still hear the long-standing local joke that patrons would 'venture in and stagger out'.

George Campbell, who shared her memories with KWMC's '100 Years of Knowle West Style' project, says:

> An icon of Knowle West was the off-licence situated behind the Venture pub. Dad would send me up the 'Vench' where I dodged the small beer crates along the grey, dark alleyway which eventually took me to the main door. He'd also send me down Newquay Road to get a packet of ten Woodbines from the machine next to what I think was Harpers shop. Next to the fag machine was the chewing gum machine where every fifth turn you'd get a free packet of Beech-Nut chewing gum. The other icon was the rare trip to the Gaiety cinema [located on Wells Road, Knowle]. We'd make our way to Salcombe Road and weave the streets until we finally got to Broadwalk and the Gaiety. All this for a Western! We slept, ate and played out our Cowboy and Indian fantasies. There was another dance hall above the parade of shops at Melvin Square that is also fondly remembered.

There are now no pubs remaining within Knowle West, and community gatherings and social events circulate within the numerous social clubs in the area, of which there is a rich tradition. Eagle House Social Club in particular (situated on Newquay Road) has played a significant role in the music scene of the neighbourhood. The Eagles, a local band formed at the club in 1962, soon turned professional and featured on the soundtrack of the Kenneth More/Angela Douglas 'Swinging Sixties' film *Some People*, which was filmed in Bristol and promoted the Duke of Edinburgh Award scheme. In later years, Fresh Four, pioneers of the hip-hop-inspired Bristol Sound epitomised by their top-ten single 'Wishing On A Star', held their first gig at Eagle House. Knowle West resident band-member DJ Krust went on to be a leading light of the UK Drum and Bass scene. Trip-hop legend Tricky was also from Knowle West as his 2008 album *Knowle West Boy* attests.

From Melvin Square turn right into Daventry Road and stop at St Barnabas Church (14) on the left, slightly back from the road.

The church was completed in 1938. The houses on the corner of the green, nearest Daventry Road, are CBE5s, a semi-detached, three-bedroom, parlour home designed by the City Engineer's Department. Those next to them on the green are CNE2s, also designed by the City Engineer's Department and a variation on the CND2s seen earlier in the walk. The houses on the opposite side of Daventry Road are Benjamin Wakefield's

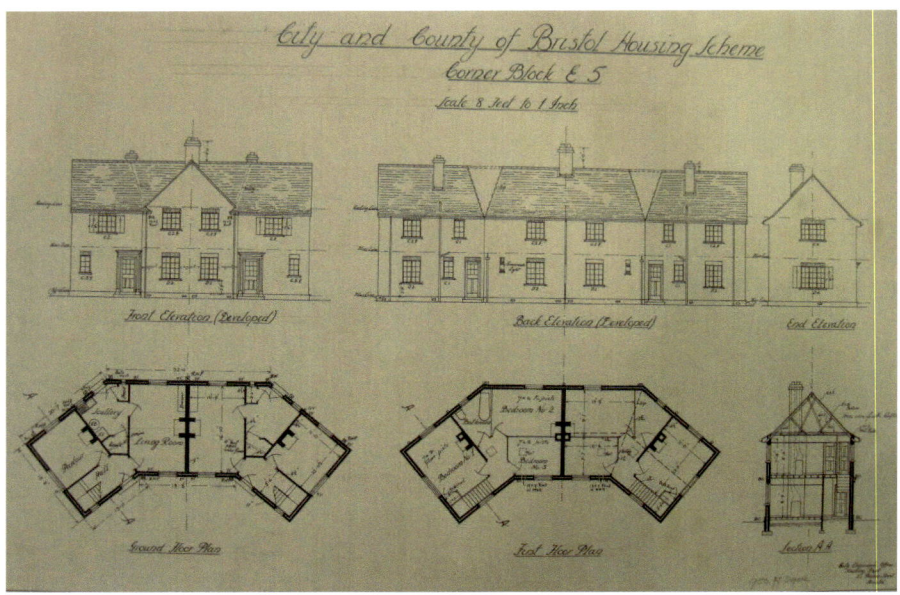

Original plan for CBE5 (Bristol Archives Red Label Plans 22)

WBD5s, the most common type of three-bedroom inter-war parlour home in Bristol. The design was used for private as well as council housing. There is a plan of a WBD5 in the Hillfields walk.

Facing the church, look to your right for the footpath that opens out to the Northern Slopes (15). Walk out to the viewing point from where you can see the top of the Wills Memorial Building and Cabot Tower in the city centre and the Clifton Suspension Bridge.

This area of woodland, wildflower meadow and scrubland, fed by two seasonal streams, gives you a sense of why Knowle West was dubbed 'the five thousand island forest' by the workers who built it. From this vantage point you can see how the estate – which comprises about 100 streets, 5,000 homes, and 12,000 people – is situated on a hill surrounded by green space. It is poised on the edge of the city and the rural landscape, somewhat isolated and detached from both. During the Second World War, nearly all of this area was turned over to food production as part of the Dig for Victory campaign. A temporary water store was built outside St Barnabas Church to facilitate this. On 3 January 1941, Knowle West was hit by German bombs, including one that landed but did not explode at Beckington Road at the other side of the Slopes. This might be the origin of the local nickname 'the Bommie' to describe this part of the estate.

Sharing her thoughts on the significant places in Knowle West, as part of '100 Years of Knowle West Style', Rianna Dyer, says:

> I spent my time when I was younger playing in the Bommie, taking the dog out with my mum to the Bommie, spending time with my aunties in what is now allotments... [They] used to have horses at the back of our house... We used to go swimming at Filwood swimming pool... I was devastated when they knocked it down.

This is where the walk ends. However, if you can manage walking on what might prove to be a steep and slippery path, it is worth continuing down into the Slopes and up the other side in order to look back at this viewing point.

The Park Centre (theparkcentre.org.uk/wp) opposite St Barnabas Church and the entrance to the Northern Slopes, has a community café and would be a good place to get some refreshments. Eggs from the local Springfield Allotments are sold here as well as handmade chutneys and jams. Filwood Chase History Society is also based at The Park Centre and can be visited to learn more about the rich history of the local area. If you need to return to Bristol city centre, you can either walk back to Melvin Square or forward to Broadwalk Square to catch the bus.

Bibliography and References

The walks devised in this book have drawn upon a number of previous publications including:

General
- Brown, Shirley and Dyer, Dawn, *100 Women of Bristol* (Bristol City Council, 2002).
- Foyle, Andrew, *Bristol* (Yale University Press, 2004).
- Gould, Peter, *Bristol Backs: Discovering Bristol on Foot* (Bristol Group Ramblers/Bristol City Council, 2002).
- Jenner, Mike, *Bristol's 100 Best Buildings* (Redcliffe Press Ltd, 2010).
- Little, Bryan *Bristol: The Public View* (Redcliffe Press Ltd, 2002).

Walk 1
- Byrne, Eugene and Gurr, Simon, *The Bristol Story* (BCDP, 2008).

Walk 2
- Cottle, Basil, *Joseph Cottle and the Romantics: The Life of a Bristol Publisher* (Redcliffe Press Ltd 2008).
- Mayberry, Tom, *Coleridge and Wordsworth in the West Country* (Alan Sutton, 1992).
- Sisman, Adam, *The Friendship: Wordsworth and Coleridge* (HarperPress, 2006).

Walk 3
- Kelly, Andrew and Kelly, Melanie (editors), *Brunel: In Love With the Impossible* (BCDP, 2006).

Walk 4
- Barker, Kathleen, *Bristol at Play* (Moonraker Press, 1976).

Walk 5
- Various plaques and information boards around Clifton provided details used in this walk and are the work of Clifton and Hotwells Improvement Society (www.cliftonhotwells.org.uk).

Walk 6
- Askew, Robin, *The West's Greatest Rock Shows 1963-1978* (Bristol Books, 2024).

Walk 8
- Byrne, Eugene, *Unbuilt Bristol: The City That Might Have Been 1750-2050* (Redcliffe Press, 2013)

Council House walks
In addition to resources cited in the text, the following material has been used in researching this publication.

- Malpass, Peter, and Walmsley, Jennie, *100 Years of Council Housing in Bristol* (UWE, 2005)
- Dresser, Madge, 'People's Housing in Bristol 1870-1939' in *Bristol's Other History* (Bristol Broadsides, 1983)
- 'Portland and Brunswick Square: Character Appraisal' (Bristol City Council, May 2008)
- *Sea Mills: Character Appraisal and Management Proposals* (Bristol City Council, January 2011)
- The websites Know Your Place Bristol (www.kypwest.org.uk) and Homes for Heroes Story Map (www.locallearning.org.uk/hillfields-history)
- Most of the local newspaper articles came from various collections of clippings kept at Bristol Reference Library.

Acknowledgements

We are grateful to many people and organisations for their help with this book – both in its original form, revised editions and in this latest version. The funding for some of these walks came from projects supported by Arts Council England, National Lottery Heritage Fund and Bristol City Council. We are grateful to them and to National Lottery players. We are also grateful for support to the Society of Merchant Venturers, the Guild of Guardians, Bristol and Bath Cultural Destinations, Business West, University of Bristol and University of the West of England.

We thank Robin Jarvis for providing guidance on the route for the Bristol and Romanticism walk, making additions to some of the entries and contributing some text. Lucy Prior also provided advice and comments on this walk. Marie Mulvey-Roberts provided advice on Angela Carter. The St Pauls walk is partly based on the current re-imagining of the 'Green link' route in St Pauls – a network of inter-linked green community spaces that have the potential to be re-activated through a local partnership project led by the Architecture Centre (our thanks to Georgina Bolton and Amy Harrison). The route for the Hillfields walk was devised by Peter Insole of Local Learning who, with Ruth Myers, led the Hillfields Homes for Heroes project. Peter also devised the route for the Sea Mills walk, with additional input from Mary Milton who led the Sea Mills 100 project. The route for the Knowle West walk was devised by Celia Turley, who led Knowle West Media Centre's Homes for Heroes 100 project, guided by resident Cheryl Martin and artist Lukus Robbins.

Thanks also go to: Ed Bramall, Debra Britton, Jane Duffus, Jacqueline Gerrard, Fiona Gilmour, Anna Keen, Alison Parry and Ruth Sharman for proof-reading and commenting on drafts; all the team at Bristol Ideas; Dawn Dyer from Bristol Reference Library for all her help in identifying research material; Madge Dresser for her guidance and research support, particularly for the St Pauls walk; Phil Sherborne and Jack Smith for advice on Legible City maps; and members of the Homes for Heroes 100 advisory group. We owe a great debt to Kate Sim Read for reviewing all walks and details here.

Our appreciation also goes to those who contributed to the image research and those who helped by proofreading and commenting on drafts of the text.

Thanks to Joe Burt and Bristol Books for the design and getting this book into print.

Finally, thank you to Bristol Ideas for providing a legacy grant to publish this book.

Notes

Notes

Notes

Notes

Notes

Notes

Notes

797,885 books
are available to read at

Forgotten Books
www.ForgottenBooks.com

Forgotten Books' App
Available for mobile, tablet & eReader

ISBN 978-1-332-16801-9
PIBN 10293023

This book is a reproduction of an important historical work. Forgotten Books uses state-of-the-art technology to digitally reconstruct the work, preserving the original format whilst repairing imperfections present in the aged copy. In rare cases, an imperfection in the original, such as a blemish or missing page, may be replicated in our edition. We do, however, repair the vast majority of imperfections successfully; any imperfections that remain are intentionally left to preserve the state of such historical works.

Forgotten Books is a registered trademark of FB &c Ltd.
Copyright © 2017 FB &c Ltd.
FB &c Ltd, Dalton House, 60 Windsor Avenue, London, SW19 2RR.
Company number 08720141. Registered in England and Wales.

For support please visit www.forgottenbooks.com

1 MONTH OF FREE READING

at

www.ForgottenBooks.com

By purchasing this book you are eligible for one month membership to ForgottenBooks.com, giving you unlimited access to our entire collection of over 700,000 titles via our web site and mobile apps.

To claim your free month visit:
www.forgottenbooks.com/free293023

* Offer is valid for 45 days from date of purchase. Terms and conditions apply.

English
Français
Deutsche
Italiano
Español
Português

www.forgottenbooks.com

Mythology Photography **Fiction** Fishing Christianity **Art** Cooking Essays Buddhism Freemasonry Medicine **Biology** Music **Ancient Egypt** Evolution Carpentry Physics Dance Geology **Mathematics** Fitness Shakespeare **Folklore** Yoga Marketing **Confidence** Immortality Biographies Poetry **Psychology** Witchcraft Electronics Chemistry History **Law** Accounting **Philosophy** Anthropology Alchemy Drama Quantum Mechanics Atheism Sexual Health **Ancient History Entrepreneurship** Languages Sport Paleontology Needlework Islam **Metaphysics** Investment Archaeology Parenting Statistics Criminology **Motivational**

NEPAL
(The Birth Place of Kalidas)

by
Pt. Muralidhar Bhattarai

Price (3.

Printed by Manohar Press, Varanasi, (India.)

FOREWORD

Shri Harsha, the celebrated poet of 12th century, has mentioned the name of Nepal in Naishadiya Charitam, as we read on १२/४३—12/43.

कृपानृपाणामुपरि क्वचिन्नते नतेन हा हा शिरसा रसादृशाम् ।
भवन्तु तावत्तव लोचनाञ्चला निपेय नेपालनृपालपालयः ॥

But at the time of Kalidasa Ist century B. C. it seems to me that Nepal was not in the same political form as she is at present. Then the Western Nepal was in Uttarakoshala and the Eastern was in Mithila, as is learnt from the Sarayumahatmya and Mithilamahatmya. It is vividly seen that Nepal in those days was divided into the several petty states, as we find in the above quoted stanza. नृपालपालय.

But the case of Kalidasa is different from that of Shri Harsha. Kalidasa was thoroughly acquainted even with an ordinary grass of Nepal, though he has not given the name of Nepal as Shri Harsha did. It was partially from his un-selfish nature and partially out of the non-existence of a particular kingdom known as Nepal separate from the other ones at the time of Raghu.

The Plots of the Poet's Works

The story of Kumarsambhava

The Gods annoyed by the demon Taraka tried to unite Shiva and Parvati, the daughter of king Himalaya in marriage in the hope of having a son from the father of the Universe as a commander for their army in the war against the demons. They had been successful in their scheme. Kumara, the son of Shiva, killed the demon and restored the kingdom of the heaven to the Gods. The plot is borrowed from the Skanda Purana.

Raghuwamsa

In this work the ideal characters of the kings of the solar lineage are depicted. The story is extracted from the Ramayan.

Vikramorvashia

In this drama the extreme affair of love between the king Pururaba and Urvashi, a celestial nymph, is described. This story is taken from the Mahabharatam.

Malvikagnimitram

This drama is a social one. Here the love

between Malavika and Agnimitra is expounded and the importance of music too is exhibited.

Meghadutam

In this Kavyam a Yachhya, who owing to the curse of his lord Kuvera had to stay for a year in the mountain of Ramagiri, being separated from his beloved wife, sends his painful message to her through the cloud. This plot also is taken from the Ramayana, where Ram sends his pathetic news through Hanumana to his wife Sitaji, in Lanka.

Ritusamhara

Here are beautiful poems describing the six-seasons. In this Kavyam the description of nature is unique.

The age of Kalidasa

Kalidasa did not refer to his birth place, date of birth and the king by whom he was supported. For the great men who have dedicated themselves to the service of mankind have very little time to look at themselves. This is the duty of his countrymen to trace out when and where he flourished. To decide his age the evidence of the Archaeological survey of India for 1909-10 will be undisputed proof. The coin found from the excavation of

Bhita near Allahabad has the inscription of a scene from the Shakuntala which is decided by the specialist to be of the Sunga period. The Sungas ruled from 187 B. C. to 72 B. C. So our poet must have existed in Ist century B. C. And in the last stanza of the Malavikagnimitra we find the line गोतरि अग्निमित्रे, where the present tense is used, therefore it can be boldly said that Kalidasa flourished at the time of Agnimitra, the son of Pukhyamitra. According to Hilbrade, a German scholar Kalidasa was before (अश्वघोष) Ashwaghosha.

Nepal

The Birth place of Kalidasa

Scholars of the east and the west alike differ from each other in their opinions regarding the time and place of the great poet of the world, Kalidasa. All of them seem to have endeavoured to bring into light the date and place of his birth. But very few of them seem to have succeeded even to a little extent in their efforts. Some of the learned men are of the opinion that the poet was born at Ujjayini. The first and foremost thing, on which they are firmly determined for their resolution of Ujjayini to be the birth place of our poet, is the great affection shown by the poet towards her. Some erudites ascribe Bengal to be the birth place of Kalidasa on the evidence his mentioning frequently the cultivation of paddy crops in his works. And there is not a small number of thinkers who believe Kalidasa to be an inhabitant of Kashmir on the ground of his exposition of the dance of ever green creepers in his literature. In this way a lot of people have tried to make Kalidasa their own, but not in a very successful manner.

The most popular critic Rajashekhar in his Suktimuktawali says that there had been three poets know as Kalidasa.

एको न जीयते हन्त कालिदासो न केनचित् ।
शृंगारे ललितोद्गारे कालिदासत्रयी किमु ।

Here I am concerned only with the author of Shakuntala, Malavikagnimitra, Vikramorvassiya, Raghuwansha Kumarssambhava, Meghaduta and Ritusambara and not with the name of Kalidasa at all. It can also be said that out of the many one or two Kalidas as could have belonged also to Kashmir, Bengal, and other provinces. But the thorough study of the above mentioned books of Kalidasa bears the testimony to the truth that the real Kalidasa was born nowhere but in Nepal. By the critical investigation through the literary works of Kalidasa, at the same time Ujjayini would be probed to be the place for his literary performances and the land of his father-in-law.

Similarity between Nepal and Ujjayini

Both, Nepal and Ujjayini have been the famous Hindu holy places from the Vedic up to the pressent time. Chhipra, in Ujjanini, is held upon as holy as the Gandaki, Koshi, and Bagmati of Nepal. In Nepal the temple of Pashupatinath has been the centre of Hindu attraction from time immemorial, so is the temple of Mahakalashiva in Ujjayini. Both of the places have been centers of hermitage for ascetisism to the warriors of Mahabharata. It is seen in the Mahabharata that Pandawas, the great heroes of Mahabharat had been frequently visiting Pashupatinath in Nepal and Mahakala in Ujjayini.

Kalidasa, moreover, praised now and then the quiet and calm region near the temple of Pashupatinath and Mahakala far from the din and dust of the world. Near the temple of Mahakala there is a temple of Saptarishis in Ujjayini. The traces of their Ashramas are found on the banks of Kali, Trishuli, and Koshi, in Nepal. There is a famous cave of Rishiswara near Palung on the road to Tribhuban raj path in Nepal. Mela is held on the occasion of Shivaratri in Pashupatinath temple and in Mahakala of Ujjayini. As in Nepal so in Ujjayini a grand temple of Harisiddhi is to be seen. In both the country Harasiddhi is said to be family diety of king Vikramadittya. As in Ujjayini there is a story of cutting his head to please the Goddess prevalent in Nepal too. Thousands of people go every year to see the head of Vikrama lying near the temple of Tara Layannya Puri in Nepal. As in Nepal there is a temple of Dakhinkali on the outskirt of the city in Ujjayini as on the top of the Farping hill in Nepal. As the Bhairabhgarh (fortress) of Kirtipur in Nepal, we have similarly a fort called Bhirahgadh at a distance of three miles from the city of Ujjayini. Kalabhairava is worshipped in both the countries as a terrible as well as a powerful God of war. As in Kageastami in Nepal the Bhairabastami is celebrated with great devotion by the people of Ujjayini. The Bhairavayatra of Ujjayini has very little difference with the Bhairavayatra of Nuwakot in Nepal. The worship of Ganesh in Nepal and Ujjayini is popular alike.

Even the images of the five headed Hanuman and sixteen-handed Ganesh are found in the same style in both the countries. As the Navagrahas are worshipped within the camp of Pashupatinath, so also exactly in Ujjayini, also there is a grand temple of Matsendranath in Ujjayini as in Nepal. Kautilya, the great Hindu economist also had mentioned the name of Nepal and Ujjayini along with the names of other cities. According to Bhagawata Purana Lord Krishna had gone to Ujjayini to acquire the knowledge of the Shastras from a Brahamin preceptor, Sandeepini by name. In the same book, in the same way it is seen that Shri Krishna came to Nepal and dedicated a temple of Shiva at the confluence of Wagamati and Visnumati to gain power to conquer the enemy, called Banasura. The Pradumneswara and the Gorkeshwar are the witness of the truth. In the Mahabharat the name of a king of Nepal and a king of Ujjayini are found, who fought against the army of the Kaurawas, in favour of the Pandawas.

Nepal for thousand of years had been closed and cut off from the world. No body could give attention in this direction and at the same time similarity between Nepal and Ujjayini had made the scholars confused to distingush the place wherein our poet was born; nor could they clearly understand his emotion in light of which he had expressed it in his works. Now Nepal also is out of the darkness in the history of the world. So we should widen our outlook.

There is a renowned story told in every roof of Nepal in respect of the early life of Kalidasa. About one thousand and two hundred years ago there was a wonderful throne near Pashupati temple. It lay buried under the ground, which was somewhat raised above the general level. For a long time it had been a play-ground of the cow-herd boys who went there daily to tend their cattle. They used to elect a man from their group and make him their king to rule over them for that day. The king was chosen in an election of a peculiar type. The candidates had to run a race from a fixed point to the place under which the throne was kept. And he who could reach first of all was made a king to rule for that day. Such was the influence of the divine throne that the cowherd's king was obeyed even by the people, who happend to come near him, and what ever he told came to be true due to the power of the throne and the judgment given by him went never wrong.

By virtue of these things the place as well as the cow-herd king had earned great reputation in the country.

Once a man living in a service of a king in Ujjayini sent a precious gem through the hand of his co-worker to give to his wife in Nepal. The bearer being enticed with the gem did not hand it over to the wife of his friend. The man, on the other hand, arranged a group of witnesses as to give testimony in the court if needed. Ater a long time the man came from Ujjayini to his home in Nepal. By and by, he asked his wife about the gem he had sent, on which she said, "I have not got it yet, I do not know who brought it when, and what is it for ?" The man came to know that he had been cheated. So, he immediately went to the home of his friend and asked him to return the gem soon. Thereupon that fellow said, "It was handed over to your wife the very day I arrived here." "Be careful" he said and went to the court of the king of the country and filed a suit against him. The witnesses having been bribed by the cheat gave false evidence and as a result the king gave the judgement against his fate. Being cheated of the gem and having been deprived of the justice he went sad to his home while he saw the cow-herd king, his ministers and soldiers, on the way, playing their parts. The man noticed it standing near by. The king also heard from him all the matters of his sadness. The cowherd king promised him to give a correct judgement for his case. On being told the matter from the very beginning, the cow-herd king summoned all of them. On comming they were ordered to sit separately and draw a sketch of that gem on the piece of paper without knowledge of the others. None of them except the sender and the bearer of the gem could draw the correct picture of it. The cow-herd king seated on the raised level of a ground under which the throne was hidden giving decison said that the gem was undoubtedly sent and it was brought by that man, but neither it was handed over to the wife of the plaintiff nor was even seen by the witnesses. The judgement given by the cow-herd king, was accepted by the court of Nepal and so the cheat was compelled to give back the gem to the owner of it.

The king being amazed at his genius gave an order to his ministry to excavate the ground sitting whereon the cow-heard boy could give the correct judgement to his subjects. Excavation was done and a throne inscribed with the thirty two figures of the celestial damsels was discovered and then there had been a thorough investigation of it. From this the king and his court came to the conclusion that it was the throne of the world renowned king, Vikramaditya. By order of the king a grand temple was built and an image of Shree Ramchandra was installed on that throne. The temple of Ramaji is called Battishputali. From the story we can draw the conclusion that Vikramaditya was the common king of Ujjayini and Nepal. There is another story of Vikrama told in connection with the Narayanhiti, a water spring near the royal palace of present Maharajadhiraja. (Hiti means natural water spring in Newari language.) Once there happened a drought for twelve years in Nepal. Due to the want of water men and animals began to breathe their last. The king sent for the wise men of his country and asked them the cause and the remedy of this calamity. The wise men advised the king to sacrifice a young man of spotless character as the remedy of the terrible drought. The king sent his men t seek for such a young man of spotless characte for the solution of the drought problem. Th men, in search of such a man, roamed far an wide but in vain. At last they found a hand some young man entirely of spotless characte and body. He was not a son of an ordinar man but of King Vikramaditya himsel Hearing this, the eyes of the prince shone wit delight and he made up his mind to sacrific his life in order to save the country from th calamity of the drought. He liked to go t the Hity every day to enjoy the fresh and pur air of the holy place. One day he heard the wise men say that the Muhurta of sacrifice had come. Having heard this he could hardly go back to his palace. On returning any how to his palace he sent a royal order to the soldiers in charge of the Hiti-temple to slaughter the man who would be found sleeping covered with a white sheet to that night. At night the prince stealthily came out of his palace and went direct to the temple and slept without the knowledge of the soldiers on the duty at that place. At the given time the soldiers slew their prince as they were rdered. As soon as the sacrifice of a spotless young man was done the water began to flow in abundance. The sky was immediataly covered with the dark clouds and rained in torrents. Now the Hiti is known as Narayan Hiti and Vikrama Hiti as well. It is separated by a circuit wall from the main road to the royal palace. It might have been clear that one of the sons of Vikramaditya had given up his life to save the Neples from being doomed.

From these stories a thoughtful man can satisfactorily conclude that Vikrama, a king of Nepal, ruled even over the country of Ujjayani.

Now, it can be asked, what the story has to do with the birth place of Kalidasa. From the above mentioned story it can be infered that the king Vikrama probably took some Brahmin boys from Nepal to Ujjayini; among whom Kalidas and Amar Singh were prominent. Before giving decision of the birth place of our poet, it also is necessary to ponder upon the incidents of his life.

There are hermitages of the sage Kandu and Bharata on the bank of the Malini (Madi) river in the western Nepal. And not in a long distance from Reedi or Rireetirth, is a village known as Alaka (Argha), where our poet was born. On the first day of Ashadha, in the first century B. C. His father lived a holy life. He was not so poor as generally Brahmins are seen. But he was not much happy because he had no son. So he used to go to the bank of the Kali, and prayed her daily to bless him with a son. One night his wife dreamed a dream in which she saw the Kaliganga blessing her with a smile. In the morning she told her dream to her husband. On hearing it his joy knew no bounds. She became pregnant and gave birth to a son, who being given by goddess Ganga Kali, was named Kalidasa. It was that boy who was later on known as poet Kalidasa.

The boy grew up to be a very arrogent young man. He was very strong and staut but in his early age he was quite distitute of intellect. Once, it is said that he went to gather the leaves of the trees for the fodder of the cattle, as the hill people generally do, began to cut down the branches of a tree on which he was seated. At the very time one of the Men of the king Vikramaditya happened to come to the spot. He was surprised to see the wonderful boy cutting down the branches of the tree on which he sat, not minding of the falling down with the branch. The person, who thought to take revenge upon the daughter of the king who had refused to become his wife, promised the thick brained boy to cause the king to marry his beloved daugher to him provided the later would keep silence till the marriage would take place. The boy agreed to do so and then was taken to the palace of the king Vikrama who liked him very much for his beauty, strength and health. He was married to a beautiful princess named Viddyottama a very learned girl. Till the marriage ended Kalidas uttered no word. In the night of honey moon he broke his silence with tears in his eyes.

"What is the matter with you" said she to him. "My 'ka' has become very lean and thin." He said with a stammering voice.

She again enquired of her husband to explain what he meant by becoming lean and thin of the 'ka'. "It was bigger when I read it in the board of my home, now poor 'ka' is reduced very small" he said.

On hearing this, she became very much disappointed and sad. She came to know also the mischief done by the unsatisfied minister. Then she said "oh, fool, you do not recognize even the 'ka' the first letter of Devanagari character ? Be at once out of this palace."

He was turned out of the home by his wife. He asked her where should he go and what should he do. In response of it she said. You illiterate fool, go to the temple of the Kali, near Ruruterth, (now known as Ridi) and do some Upasana to propetiate her. Do not turn your face any more, until you become a learned man".

He thought, a man without the knowledge of the scripture is as useless as an overcoat in Bombay. For days together he kept wondering hither and thither at random as a hotel boy's of Nepal in India. One day he was told that on the bank of the Kali Ganga there is a temple of Goddess Kali within a forest, some distance away from Ridi. He became as happy to know it as a minister in a party. He entered the temple and propiciated Kali within a week. Kali being very pleased with his insisting devotion appered before him and blessed him with a boon. By the grace of mother Kali the veil of obstacle for knowledge had been removed from his mind. His brain became as clear as crystal ready to receive what ever was seen, heard and thought. His sorrow ended. Again one night mother Kali blessed him by putting her hand on his head and ordered him to go to Mithila, the birth place of king Janaka and Yagnyavalkya, Satananda and Seeta.

In those days Janakpur had been the centre of learning. So he went to Mithila, Janakpur. There he came to know of the place where an Upadhya, a great teacher lived. Kalidasa went to the Acharya who accepted him as a disciple. Within a short time he completed the course of study and became a great erudite. There is a temple of Kali in a village named Uchcha in Durbhanga district. It is told that Kalidasa during the time of his study used to go to this temple of Kali. Even now the temple is known to be of the Kali of Kalidasa where the students go every day to be benefited in their study.

After completing his study, he went Ujjayini where his wife with his father were waiting for him with a great impatience. Vidyawati, the wife of the poet Kalidasa was in her room making preparation to receive her husband. As soon as she saw him said अस्ति कश्चिद् वाक् विशेष:, that is "I think, there is a progress in your learning" Then the poet smiled slightly and sat by her side and promised her to present the books beginning with the words spoken by her in the sentence said above. He spoke this in a cultured and lucid sanskrit.

He thought it, his duty to pay homage to his mother country on whose bosom he was brought up. So he wrote at the set up the Kumar Sambhava.

कुमार सम्भव

अस्तुत्तरस्यां दिशि देवतात्मा हिमालयो नाम नगाधिराज:
पूर्वी परौ तोयनिधिर्वगाह्य स्थित: पृथिव्या इव मानदण्ड:

—There is an abode of Gods named Himalaya in the north direction, the king of the mountains, whose two ends are merged into the ocean and is standing as the measuring rod of the Earth. By this stanza every thoughtful person can automatically make an inference for the fact that the lap of the Himalaya was the birth place of Kalidasa. Becouse the patriot, whenever he may have lived, being compelled by the circumstances, recalls his birth place now and then. Our poet felt it his bounden duty to make the world know his birth place Himalaya. So he beginies by the sentence "there is Himalaya," etc. He wants to express his hidden joy before the people of the world. The sublimity of the Himalaya is known to the world. For this reason the kings of Nepal, from the time immemorial are honoured with the epithet of Adhiraja or the kings of kings—supreme rulers. As the king of Nepal today are given the title of Adhiraja. In this stanza, adhering the tradition of Nepal the poet gives the title of supreme ruler to his beloved Himalayas. The Gauri Shanker or the Mount Everest in Nepal is the highest plateau of the mother earth. That is why the poet says that Himalaya is the spinal cord of the Earth.

The world renowened Kirat Pradesh is in Nepal which is given a considerable room in the book of the both poets Kalidasa and Bharavi. Bharavi has a general knowledge of Kiratas while Kalidasa seems to have a particular knowlege of it. The description of the Kiratas in Kumar Shambhava can be told by no means that it was a mere imagination of the poet. It will be clear from the following stanzas that the knowledge that he had of the Kirat Province can not at all be called derived from the books of geography only. No man who had not seen it with his own eyes can describe it so clearly. For example, Kalidasa, expounding the natural beauty of the Kirat region says,

वनेचराणां वनिता सखानां दरीगृहोत्सङ्गनिषक्तभासः ।
भवन्ति यत्रौषधयो रजन्यां अतैलपूराः सुरतप्रदीपाः ॥ १।१० कु०
दिवाकरात् रक्षति यो गुहासु लीनं दिवाभीतमिवान्धकारम् ।
क्षुद्रेऽपि नूनं शरणं प्रपन्ने ममत्वमुच्चैः शिरसां सतीव ॥ १।१२
यत्रां शुकाक्षेपविलज्जितानां यच्चक्षुषा किं पुरुषाङ्गनानाम् ।
दरीगृहद्वारविलम्बिबिम्बस्तिरस्करिण्यो जलदा भवन्ति ॥ १।१४
मनीषिताः सन्ति गृहेषु देवतास्तपः क्ववत्से क्वच तावकं वपुः ।
पदं सहेत भ्रमरस्य पेलवं शिरीषपुष्पं न पुनः पतत्रिणः ॥ ५।४
अथानुरूपाभिनिवेशतोषिणा कृताभ्यनुज्ञा गुरुणा गरीयसी ।
प्रजासु पश्चात् प्रथितं तदाख्यया जगाम गौरीशिखरं शिखण्डिमत् ॥
५।७

For the better understanding of the nature of the hill men the underlined words in the above given stanzas of Kumar Shambhava are worth studying with a great attention अतैलपूराः प्रदीपा— lights without oil etc. Some kinds of woods are seen in the forest of Nepal, shining at night as bright as the day-light, known as Ujeli Kath in Nepali words. People wonder to see the trees emitting bright rays at night enough to read books in it.

The poet seems very proud to be a hill man, so he expresses his noble pride in this stanza.

दिवाकरात् रक्षति यो गुहासुलीनं द्विाभीतमिवान्धकारम् ।
छुद्रेऽपिनूनं शरणं प्रपन्ने ममत्वमुच्चै: शिरसां सतीव ॥

—Our Himalaya is very proud to give shelter to an afraid as he protects the darkness afraid of the rays of the sun giving shelter in his caves. This is the nature of the noble ones to embrace the refugee however mean he may be".

Let us procced a little further and look what an attractive picture is put before us by this "तिरस्करिण्यो जलदा भवन्ति" it means the clouds have served the purpose of a screen for the doors of the Himalayan-caves-dwelling-Kiratas. A man without living in a place for a long time can never describe so actually as the poet in the 1/14 stanza of Kumar-Shambhava. An inexperienced man can, however campare the cloud with the canopy not with the screen, as seen in the folk tales of Nepal "लोला यो हरि को बादल को फरिको न छेक मलाइ" O' sport of lord, door of cloud, do not hide me, (the stanza is sung every where in the hill side of Nepal.) Kalidasa's favowrite flower 'Shirish' and the

Niwari corn which he prefers above all are grown in good deal in the soil of Nepal. Shirish is colled Shiru in Nepali and there is a Hindu tribe named Niwara after the name of Niwar corn and Niwari flowers here. The Himalya is termed as Gauri guru by Kalidas in his 'Shakuntala'. There is Gouri Shanker summit of the Himalaya in Nepal about which very little was and is known to the people of Bengal, Bihar Kashmir and Ujjayini. In the stanza 22 Kumar and 2/65 in Raghubansha, for the break fast of Uma the word parana is used by our poet. In the Magari and Nevari language the same words are used, Uma for the mother and Parana for eating after the religious fast respectively. Ume (उमे) in Magari and 'palan' (पालँ) in Newari is an etymological proof of his beeing a Native of Nepal. Again the day in which Uma took the vow of austerity to attain Shanker as her husband and gained the name Uma is held very auspicious and is observed fast in this day by every woman of Nepal. It was therefore his charecterstic to call up the fasting of the female members of his house of whose he depicted the picture in the name of Uma.

शरीरमाद्यं खलु धर्मसाधनं 5/32
न रत्नमन्विष्यति मृग्यते हि तत् 5/S5

"Body is the first thing amongst the means of Dharma-duty : the gems do not seek man but it is sought by him." This is a great respect towards the Hindu woman. From the first line it is also hinted that the people living in the bossom of Gauri Shanker, which is puryfying by nature, penance is not necessary at all. In this way Kalidasa expresses his deep sentiment towards his mother land Nepal.

In the 6th canto of Kumar Shambhava we come across the word Koshi for the river Koushika flowing in Nepal. But in every books of the Sanskrit literature the word Koushika alon is found except in Nepal Mahatmya. If Kalidas had not been a Nepali by birth he would have used the word Koushika for Koshi as other Sanskrit writers have done.

तत्प्रयातौषधिप्रस्थं सिद्धये हिमवत्पुरम्
महाकोशी प्रपातेऽस्मिं संगमः पुनरेव नः 6/33

The great commentator Mallinath says "महाकोशी नाम तत्रत्या काचिन्नदी" तत्रत्या means नेपालत्या. A river of that place means the river of Nepal, because the river Koshi flows in her eastern part. Again the sages fixed the very place of the bank of Koshi for their further meeting. From this it has been clearly proved that the marriage of Gaurishankar had taken place in Nepal, not in Kashmir and other places.

India is widely regarded and accepted as the instructor of the world विश्वगुरु, but Nepal is held by the poet to be the instructor even of the instructors—"भव विश्वगुरोर्गुरु" ६—८३

See Kumar Sambhava :—

अस्तोतुः स्तूयमानस्य बन्धस्यानन्यवन्दिनः ।
सुतासंबन्ध विधिना भव विश्वगुरोर्गुरुः ॥

In this way we see Kalidasa respecting his mother land estimating the glory of her above all. And in the last stanza of this canto Pashupatinatha, the famous diety of Nepal is remembered by the poet with a great reverence.

The birth-day celebration of Kumara, Kartikeya is observed no where with such preparation as in Nepal. And it also is a thing to be considered that Kartikeya, the son of Pashupatinath is called nowhere Kumara but in Nepal. Kalidasa also is seen habituated to call him by the name of Kumara. If he was born in other countries, he would have given the tittle of Kartikeya janma etc. to his Kabyam—Kumarasambhava. Shithi, the birth-day of Kumaraji is also celebrated by the Newars is Nepal with a great pump and show.

It is called Shithinakha in their dialect From this it can easily be guessed that Kalidasa was mostly influenced by this traditional festival of Nepal; which without being a Nepali was not possible. How the poet like Shree Ramachandra expresses his heartly homage to his mother-land Nepal, will be more clear from the comparative study of the two verse from the Ramayan and Kumarasambhava.

अपि स्वर्णमयी लङ्का न मे लक्ष्मण रोचते ।
जननीजन्मभूमिश्च स्वर्गादपि गरीयसी ॥—राम
दिवं यदि प्रार्थयसे वृथा श्रमः
पितुः प्रदेशो स्तव देवभूमयः ॥ (कालिदास)

"Even the golden Lanka has no charm for me, Mother and mother country are greater than heaven." Rama.

"It is futile to long for heaven for the province of your father itself is the abode of the gods" Kalidasa.

Amples of examples of this sert can be given to prove him Nepali from the Shlokas found in his books.

Evidences from the Raghuwansha also can be given here for firmly establishing the fact that the great poet of the world was a son of Nepal.

It is interesting to note here that our poet at the beginning of the Raghuwansha, which commeness with the second word (वाक्य) of the sentences uttered by his wife, makes an abescience to Parwati the daughter of King Himalaya of Nepal, the mother of the universe. It is not a thing of a little pride for us that the daughter of Nepal is the mother of the world.

The word अधिराज or Supreme ruler, a very much favourite word to the Nepalese, is used two times in one stanza in Raghuwansha.

रघुवंश

इति प्रगल्भं पुरुषाधिराजो मृगाधिराजस्य वचो निशम्य
प्रत्याहतास्त्रो गिरिशप्रभावादात्मन्यवज्ञां शिथिलीचकार

Bhutan, which is geographically a part of Nepal, was called Bhutasthanam in the days of Kalidasa. This place is believed even to day to be the abode of the Bhutas of Pashupatinath. It will be quite clear from the statement given below that our poet was fully acquainted with the Bhutas and their lord.

सम्बन्धमाभाषण पूर्वमाहुर्वृत्तः सनौ संगतयोर्वेनान्ते ।
तद्भूतनाथानुग नार्हसित्वं सम्बन्धिनो मे प्रणयं विहन्तुम् ॥2/58

The hermitage of Bashistha where the King

Raghu with his queen, Sudakshina, stayed to tend the Cow, Kamadhenu, is found on the bank of Koshi in the tarai of Nepal. As to the nature of the language of Kalidasa have been already given an account from his Verse in the explanatory remark of Kumar Sombhava. Hence I try to give some satisfactory examples to make clear the prooves given above. It is the nature of Nepali language that two words are not used separatly to denote eating and drinking. For example they say. भात खानु, फल खानु, दूध खानु, पानी खानु etc, ie—rice, fruits, milk and water are all eaten not drunk. In Sanskrit, as mostly in other languages there are two words used separately for eating and drinking as खादति, पिबति—are separately used for the separate things. But our poet, Kalidasa being habituated with the nature of Nepali language, uses sometimes Bhuncho (भुङ्क्ष्व) verb from a root to eat even for milk as we read in Raghuwansa.

सन्तान कामाय तथेति कामं राज्ञे प्रतिश्रुत्य पयस्विनी सा दुग्ध्वा पयः पत्रपुटे मदीयं पुत्रोपभुङ्क्ष्वेति तमादिदेश 2/85

The cow said to the king "O son, extract my milk in a vessel of leaves (खोची) and drink it up (eat it up) भुंक्ष्व—पिप - मल्लीनाथ Milk is not eaten but drunk. Nowhere except in Nepal water and milk are eaten. Though in Maithili and Bengali language the words जलखई and पनखई too are used but with the जलपन or water-there must be something eadible, such as chyura and bread. Only for water is पिना – PINA to be drunk not खाना–KHANA to be eaten.

From these stanzas we can infere that the nature of Nepali Language was deeply inrooted in the heart of Kalidasa. It is perfectly plain to every intelligent persons who is interested in where about of the great man of the world that the birth-place of Kalidasa must decidedly be Nepal.

The Bengalese claim Kalidasa to be the native of their province, but to quote a stanza from the fourth canto of the Raghuwansha will be enough to disprove the claim of them.

वंगान्नुत्खात्य तरसा नेता नौ साधनोद्यतान् ।
निचखान जयस्तम्भान् गंगास्रोतोन्तरेषु स: ॥ ४—३६

Having uprooted the kings of Bengalese who were ready to face him, by his skill the leader (Raghu) erected the Jaya Stambha, a pillar as the sign of his victory in the islands surrounded by the current of the Ganges. If Bengal was the birth-place of the poet he would never have tolerated the victory of a king from outside over his birth place.

Raghu went up to the back of OXUS or anshu वन्शु, the river in Pamir Platue, which eets the Oral Sea.

विनीताध्वश्रमास्तस्य वंक्षुतीरविचेष्टनै: ।
दुधुवुर्वाजिन: स्कन्ध लग्नान् कुङ्कुमकेसरान् ॥ ४—६७

Then he defeated Hunaj as is seen in four/ sixty-eight 4/68 stanza of Raghuwansha and conquering the princes of Kamboja returned towards the Everest.

On the way to it the gentle breeze charged with the particles of the water of the Gangese served him. The poet showing his forvent affection for his birth-place—says, "In the shade of Nameru trees the army of Raghu took rest." The fact is not in dispute that every-body having returned from his adventured journey takes rest in his own home-land. And at the time of Raghu the western part of Nepal was regarded as a part of Avadha that is why he took rest there. The evidence of his taking rest here endicates clearly that Nepal must had been the home of Kalidasa. To make the fact clear I quote some of the stanzas here from the Raghubansha.

विश्रश्रमुर्नमेरूणां छायास्वध्वास्य सैनिका:
इषदोबासितोत्सङ्गनिषण्णमृगनाभिभि: 4/74.

सरलासक्तमादङ्गमैवेयस्फुरितत्विषः
आसन्नोषधयो नेतुर्नेक्तमस्नेहदीपकाः 4/57.

They rested, the herbs (उजेली काठ) served their purposes of lamps without oils at night.

The shining herbs are found profusely in the region from Kumayoo to Kirat province only.

Talking rest for some days in his home-land, Nepal, he again left his haunting abode and proceeded towards Kirat Pradesh. तत्रजन्यं अभूत् — 4177. There a battle had been fought between the armirs of Raghu and Kirats and defeating whom the King Raghu marched towards Asam where the foot of Raghu had been worshiped by the king Kama Roopa of Asam. Thus conquering the quarters of the earth the victorious Raghu returned back to his birth place back to again. From these accounts there will be left not the slightest ground to distrust the views that Kalidas was one of the worthy sons of Nepal.

It can be said that at the occasion of the स्वयंवर or the choice of a husband by the princess of Vidarva, where suitors assembled for that purpose, no name of a prince of Nepal is mentioned there, therefore it is emproper to say that Kalidasa was a native of Nepal. But an adequate froof can be given here to verify the certainty of his being a Nepali. It was even already said that the time of Ramayan Nepal was included in Utter Koshala. From Gandaki (Saryu) Mahatmya we can get ample of evidence for this.

Muktichetra lies on the foot of Himalaya in western Nepal which even now is visited by the thousands of pilgrimes every year. This is even now said Koshala. Raghu the prince of Koshala or Nepal was elected by the princess of Vidharva. The genious of our poet Kalidas never failed to bring the truth into light. He says :—

इक्ष्वाकु वंश्यः ककुदं नृपाणां ककुत्थ उत्पादित लक्ष्णोऽभूत् ।
काकुत्थ शब्दं यत् उन्नतेच्छाः इलाघ्यंदधुरुत्तरकोशलेन्द्राः ॥ ६।७
असौ कुमारस्तमजोनुजातस्त्रिविष्टपस्येव पतिं जयन्तः ।
गुर्वीं धुरं यो भुवनस्य पित्रा धुर्येणदम्यः सदृशं विभर्ति ॥ ६।७८
सा चूर्णगौरं रघुनन्दनस्य धात्रीकराभ्यां करभोपमोरूः ।
आसजयामास यथा प्रदेशं कण्ठे गुणं मूर्तिमिवानुरागन् ॥ ६।८३

The Prince of Uttara Koshala—Northern Koshala, that is the son of Western Nepal, was garlanded by the princess. The World (चूर्ण गौरं) red-auspicious powder is widely used on the occasion of marriage ceremony particularly in Nepal.

If we impartially judge the thoughts given

in Meghaduta by the poet, it will not be difficult for us to decide vividly the birth place of the poet.

As the month of Ashadha is of very importance in the life of Nepali. Seeing the claud above their head in the month of Ashadha, every Nepali dances with joy. The Nepalies hold it to be an auspecious matter and to be besmeared with the mud of this month.

They eat and as well feed their keeth and kinnes the curd and bitten rice as the sign of rejoycing at its arrival. And in this accession every Nepali wants to be in house and enjoy the cheerful month with his family. Being a Nepali Kalidas naturally remembers his beloved wife living in his house Alaka in Nepal.

तस्मिन्नद्रौ कतिचिदबला विप्रयुक्तः स कामी ।
नीत्वा मासान्कनकवलयभ्रंशरिक्तप्रकोष्ठः ॥
आषाढस्य प्रथमदिवसे मेघमाश्लिष्टसानुं ।
वप्रक्रीडापरिणतगजप्रेक्षणीयं ददर्श ॥ ५ मे० २

The passionate Yachchya in the separation of his wife spent several months in that mountain. His wrist began to seem empty due to the slipping away of the golden bangle from it. At the first day of the Ashadha month he saw the cloud on the top of the hill as an elephant engaged in butting the ground.

It is obvious to all that the cloud on the top of a mountain resembles an elephant and seems very pretty to be looked at. It is an old belief among the Nepali people that the cloud dripping in their auspicion days is a good omen. From this also it can be guessed that the Ist day of Asar was the birth day of our poet which reminded him his darling and home. The newly appeared cloud made him more passionate than usual.

The village of Alaka now known Argha is in western Nepapl near Reedi, as has been said in the begining of this book. Kalidas himself says it to be his Birth place :—

तत्रागारं धनपति गृहादुत्तरेणास्मदीयम्
दुराल्क्ष्यं सुरपतिधनुश्चारुणातोरणेन
यस्योपान्ते कृतकतनयः कान्तयार्धितो मे
हस्तप्राप्यस्तवकनमितो वालमन्दारवृक्षः ॥१२॥

तत्र there धनपतिगृहान्नुत्तरेण to the North from the house of Kubera अस्मदीयं our आगारं home is. In this sloka Kali Das is telling us his house openly.

The expressions chandeshwara in the 33 Stanza Pashupati in 36, and Brahmavarta in

48, are worth to be contemplated upon. In Pauranic period the name of Nepal was Brahmavarta. "ब्रह्मावर्तों हि कालेन पुण्यो नपपाल संज्ञक." as seen in Gandaki Mahatmya.—The holy place of Brahmavarta has been changed [into Nepal with the change of time. The temple of Chandeshwara situated on the bank of Punyamati in Nepal is world renowned. Kailasha, the fovourite term of the poet is near Pashupati Nath. About six miles north from Kathmandu, there is a temple of Gokerneshwara sitting in the cave of which Ravana cut his heads to propetiate Lord Shiva. Having receined the boon from Shiva he tried to test his strength by lifting up the mountain Kailash. The samething is said by the poet in this stanza.

गत्वा चोर्ध्वं दशमुखभुजोच्छासितप्रस्थसंधे:
कैलासस्य त्रिदशवनितादर्पणस्या तिथि: स्या:
श्रृंगोच्छायै: कुमुदविशदैर्यो वितत्य स्थित: स्वं
राशीभूत: प्रतिदिनमिव त्र्यम्बकस्याट्टहास: ५८

To confirm the idea said above a Sloka can be quoted from an inscripton of Lichchavi time, which is written in commemoration of Kailas.

कैलाशकूट भवनाद् भुवनप्रकाशात्
ज्योत्स्नावमृष्ट हिमवच्छिखरोरुदीप्ते: ।

आसागरप्रसृतशुभ्रयशो ध्वजानाम्
राज्ञां कुलांवरशशिर्मुवि लिच्छवीनाम् ॥

If these lines of the poet are studied by with an unbased mind there would be no trace of doubt left to say that Nepal was not his birth place.

Kalidasa, of course, praised Ujjayini from the bottom of his heart. This implies to the best that he was matrimonially connected with her and he spent a long period of his life there and he made her the field of his literary perfermances. The main thihg to be noted is that the poet's method of expressing the feelings towards Nepal and Ujjeyini differs not only in style but also in tenor.

As a psychological proof :—

Describing Alka, his home, his eyes are filled with tears and paying homage to Ujjayini he seems to be fired by passion. The two different sentiments in describing the two countries show how much he is clear about his father land and father in-law's land—

वक्रपन्था यदपि भवतः प्रस्थितस्योत्तराशां
सौधोत्सङ्ग प्रणय विमुखो मास्मभूरूज्जयिन्याः
विद्युदामस्फुरितचकितैस्तत्र पौराङ्गनानाम्
लोलापाङ्गैर्यदि न रमसे लोचनैर्वञ्चितोऽसि

The word प्रणय is used specially for the carnal love as प्रणयी, प्रणयिनी means one who loves for the sake of carnal pleasure. Here the poet becomes a sentimental and suggests his friend Megha to go to Ujjayini and enjoy the side long glances of the ladies of there. This is the psychological truth that even the memory of the house of father-in-low makes a man unconsciously jolly.

He seems quite different when he advises his friend to see his wife in Alaka. (Argha)

तां जानीथाः! परिमितकथा जीवीतं मे द्वितीयम्
दुरीभूते मयिसहचरे चक्रवाकीमिवैकाम्
गाढोत्कण्ठां गुरुषुदिवसेष्वेषु गच्छत्सु वालाम्
जातां मन्ये शिशिर मथितां पद्मिनीं वान्यरूपाम् ।३२०।
सव्यापारामहनि न तथा पीडयेन्मन्वियोगः
शङ्के रात्रौ गुरुतरशुचं निर्विनोदां सखीं ते
मत्सन्देशैः सुखयितुमलं पश्य साध्वीं निशीथे
तामुन्निद्रामवनिशयनां सौधवातायनस्थः ।३२६।

Even the entry to his house is forbidden to his friend. His friend is asked to stay on the window of his house. Here the poet does not advise Megha to enjoy the side-glance of her, on the contrary he orders his friend to narrate his message to his wife from a distance. And he praises the chastity of his wife before him. He believes his wife as chaste as Seetaji—"इत्याख्याने पवन तनयं मैथिली वोन्मुखी सा". It is resounded from the word पवनतन (Mahabeer) that he wants his friend to be as holy as Hanuman ; because every body wants a man sent near his wife to be chaste.

If we take a little effort to think over the words used by the poet above we can easily decide that this miraculous expression is not mere hypothesis on the part of the poet.

Let us cast a glance again at the Shakuntalam, the masterpiece of Kalidasa. The fourth act of this drama is anonymously held as an unique act before the existing dramas of the world. Here we find a compound word (होम-वेला) used for the time of offering the oblations which is exactly Nepali in charactor. The Nepalis never say हवन समय as the people in all the other provinces of India do. The Nepalise use always होम बेला and बेला for time and होम for oblation which are used without any change by Kalidasa.

KanduAshram is stationed on the bank of Madi in the Western Part of Nepal, where Shakuntala was brought up. The word उद्धातिनी भूमि also denotes that the place must had been a hilly one and not a plain. There is a place known as 'Apsarakunda or Shachitirth on the way to Muktichhetra wherefrom Shakuntala was taken off by her mother Menaka (See Himabat Khand). Again in the 6th act of Shakuntala we see a picture of mother Nepal drawn by our poet Kalidasa through the hands of King Dushyanta :—

कार्या सैकतलीनहंसमिथुना स्रोतोवहामालिनी
पादास्तामभितो निषण्ण हरिणा गौरीगुरो: पावना:
शाखालंबितवल्कलस्य च तरोर्निर्मोतुमिच्छाम्यध:
शृङ्गे कृष्णमृगस्य वामनयनं कण्डूयमानां मृगीम् ।

In this very stanza the Ashrama of the Sage Kandu is depicted to be on the bank of Malini or Madi river on the holy foot of Himalaya within the Kingdom of Nepal. See, Kasyapeswar in Himavat Khand.

The hermitage of Kashyapa also has been recently discovered. It is on a mountain near Gosaikund, on the foot of Himalaya, where Shakuntala, with her Son Bharat, was met by king Dushyanta, on returning back from Trivistapa (TIBBET). Here Matali, the charioteer, was asked by the king what mountain was that. In reply he said "एष खलु हेमकूटो नाम किंपुरुष पर्वत:" this is the mountain of Kimpurusha named Hemokuta. According to

Bhagvat Puran, Kimpurush is a part of Tibbet, the Nothern part of it was in Nepal then.

उत्तरेषु च कुरुषु यज्ञपुरुषः—भा ५/१८
किं पुरुषे वर्षे भगवत्तमादि पुरुषं ५/१९

One part of the Kimpurush which is now within the territory of Nepal, was in those days under the reign of Dushyanta, which is seen. क्षणादायुष्मान् स्वाधिकारभूमौ वर्तिष्यते You will be shortly in the region governed by you, O, long lived king" said Matali.

Kalidasa is not so clear in the discription of the other countries as is seen that of Nepal. This is evident from the said statements. It is another fact to proof that Kalidasa was brought up in Nepal, otherwise it was not possible for him to describe so minutely every details of Nepal. So it is not an exgaggeration to think that Kalidas had migreted to Ujjayini from Nepal with King Vikrama.

There are other incidents worth to be mentioned from Ritusamhara, Vikramorvashiya, and Malavikagnimitra and the other works of Kalidasa. Our poet being an inhabitant of a cold country feels much hot in Ujjayini. For that reason he begings his Ritusambara with the discription of the hot region ग्रीष्म and naturally there comes out of his mouth the word प्रचण्ड—terrible as an adjective for the sun of summer and at the same time he sings for the rays of the moon.

ऋतु संहार

प्रचण्ड सूर्य: स्पृहणीय चन्द्रमा
सदावगाहक्षतवारि संचय:
दिनान्त रम्योऽभ्युपशान्तमन्मथो
निदाघकालोऽयमुपागत: प्रिये ॥१॥

There are other proofs in his being a Nepali. We find some miracles in his poems for the discription of Hemanta as well as Shishira, which are saturated with the sap of the Nepalese life.

नवप्रवालोद्गमसस्य रम्य:
प्रफुल्ललोध्र: परिपक्वशालि:
विलीनपद्मप्रतपतुषारो
हेमन्तकाल: समुपागतोऽयम् (१)
बहु गुण रमणीयो योषितां चित्तहारी
परिणत बहुशाली व्याकुलग्रामसीमा
सतत मति मनोज्ञ: क्रौञ्चमाला परीत:
प्रदिशतु हिमयुक्त: काल एष: सुखं व: (१८)

The first month of Hemanta,—marga is the harvest season of Nepal. In this season the peasents of Nepal reap and gather the corn, especially the shali paddy, the crop of this season. Every child of Nepal knows that the rice of shali paddy is the most tasteful grain and therefore held holy. The Lodhrā flower begins to bloom in this season and also a little frost begins to fall. This is seen only in Nepal in this season. Shali, Lodhra, tushara are in vogue in Nepali without any defermation. The Kraunch birds are seen over the sky flying from South to North in rows then In Nepal. We pass hence to some stanzas in which the cold season is described.

निरुद्ध वातायनमन्दिरोदरं
हुताशनो भानुमतो: गभस्तय:
गुरूणि वामांस्यबला: स यौवना:
प्रयान्ति कालेऽत्र जनस्य सेव्यताम् ॥

Here the words निरुद्ध वातायन closed window of the houses of the country is note worthy. Because in the cold season very few windows of the houses of Nepal are left open. The picture of cold season drawn by Kalidasa is also a proof of his being a Nepali.

The fifth act of Vikramorvashiya is also important regarding the birth place Kalidasa. The drama ends with the union of the king Pururava with his bereaved wife Urvashi and her son.

The forest of Ambika, where Sudimna had been Changed into a woman named Illa, is near Bhaktapura. The forest is known a Phulchoki. The son of Illa was Pururaba, who fell in love with Urvashi. The mythological story from Mahabharat is adopted in the drama. The scene and scenaries of the drama are mostly taken from the Natures available in Nepal. The meeting place of Pururaba with his beloved son and wife Urvashi happened to be near the hermitage of sage Chyevana—च्यवनाश्रमात् कुमारं गृहीत्वा तापसी संप्राप्ता The female hermit reached here from Chyevanashram (5th act of B. U.) This Ashrama is now found on the bank of "Betrabati" near west-number 1, Nuvakota, where the celestial saint Narada had come to give the message of Indra to king Pururava. There is also another proof to confirm this idea. It is this that we read here

अद्रिः प्रक्षालितो मणिः कर्मै प्रदीताम्
राजा-किराति, अग्नि शुद्धमेनं कृत्वा पेटकं प्रवेशय ॥

(किराती) Kirats is a tribe of old times living in the eastern part of Nepal on the foot of Himalayas. The tribe even now devotes with the business of Kasturi, Chauri, and Rubbies.

The word Kirti is used no where but in Nepal.

In Nepal the people say Kirati no Kerata to donote both sex of that tribe. Thus the thoughful men are naturally compelled to believe that the poet was a nattive of Nepal. It is surprising that Kalidas has miraculously woven a net of literature with the warp and waft of the materials of Nepal. His love for Ujjayini and geographical knowledge of the world was vast and what he wrote wrote correctly but not so minutely, that is why it has become very difficult with the critics to say difinite by about his native place.

The style of Malavikagni mitra is some what different from those of his other works. This drama is social in charactor. In this drama the poet warns to those who have made their habits of giving their judgements indiscriminately. Here Paribrajica Kaushika, an ascetic woman is given the authority to distinguish a better student from the two students. In those days the fine art, dance, music, and stage crafts had developed a lot in Nepal. Even now Nateswara Shiva and Nateswari Parvati are worshipped in every quarters of Nepal with the hope gaining efficiency in music and dance. Every girl of Newar family attaining the stage of puberty is called Lyashe for their efficiency in the dance of lasyam. This very fact is staged in Malabikagnimitram by the poet. He wants to give a new colour to the old spirit of the dance in this drama. So he say :—

पुराणमित्येव न साधुसर्वं
न चापि काव्यं नवमित्यवद्यम्
सन्तः परीक्ष्यान्यतरद्भजन्ते
मूठः पर-प्रत्यय-नेय बुद्धिः

"Everything can not be said to be good because of its antiquity, niether the new literature can be said all perfect due to its newness. A man of discrimination uses his own wisdon, while the fools blindly follow others". The drama Malavikagnimitram was staged on the occasion of the coming of the spring. Even now the custom of Vasantotsava is in vogue in Nepal. This festival is accompanied by dances and musics. In this way allmost all the plots in the books of Kalidasa are miraculously derived from the book of nature of Nepal. Though this drama is doubted to be a composition of that very Kalidasa because there are few things that can be said of Nepal.

CPSIA information can be obtained
at www.ICGtesting.com
Printed in the USA
BVHW091329241218
536332BV00036B/2457/P

Real Analysis and Algebra

Revised First Edition

By Tamás Forgács and Oscar Vega

California State University, Fresno

Bassim Hamadeh, CEO and Publisher
Christopher Foster, General Vice President
Michael Simpson, Vice President of Acquisitions
Jessica Knott, Managing Editor
Kevin Fahey, Cognella Marketing Manager
Jess Busch, Senior Graphic Designer

Copyright © 2013 by Cognella, Inc. All rights reserved. No part of this publication may be reprinted, reproduced, transmitted, or utilized in any form or by any electronic, mechanical, or other means, now known or hereafter invented, including photocopying, microfilming, and recording, or in any information retrieval system without the written permission of Cognella, Inc.

First published in the United States of America in 2013 by Cognella, Inc.

Trademark Notice: Product or corporate names may be trademarks or registered trademarks, and are used only for identification and explanation without intent to infringe.

Printed in the United States of America

ISBN: 978-1-935551-64-5 (pbk)

Contents

1 Things You Should Know 1
   1.1 Sets and Functions 1
   1.2 Numbers 4
       1.2.1 The Integers 4
       1.2.2 The Rationals 6
       1.2.3 The Reals 6
       1.2.4 The Completeness Axiom and Related Topics 7
       1.2.5 The Complex Numbers 10
   1.3 Linear Algebra 10
       1.3.1 Matrices 12
   1.4 Algebraic Vs. Transcendental Numbers 13

2 Groups 15
   Problems 15
   2.1 A Short Historical Exposition 15
   2.2 The Basics of Group Theory 16
   2.3 $\mathbb{Z}_n$ 20
   2.4 Orders of Groups and Elements 21
   2.5 Subgroups 22
   2.6 Cosets 24
   2.7 Permutations 25
   2.8 Homomorphisms and Normal Subgroups 31
   2.9 Quotient Groups and Products of Groups 35

3 Rings 41
   Problems 41
   3.1 A Short Historical Exposition 41
   3.2 $\mathbb{Z}$ and $\mathbb{Z}_n$ Revisited 42
   3.3 Basic Properties, and Examples, of Rings 43
   3.4 Homomorphisms and Ideals 47
   3.5 Quotient Rings and Direct Products 51
   3.6 Polynomial Rings 54

4 Approximation 57
   Problems 57
   4.1 Metrics and Metric Spaces 57
   4.2 Sequences 59
       4.2.1 On the Convergence of Real Sequences 61

		4.2.2 Recursive Sequences	62
		4.2.3 Sequences of Functions	64
	4.3	Subsequences	66
		4.3.1 Subsequential Limits	66

5 Add Infinitum 69
Problems 69
5.1 A Short Historical Exposition 69
5.2 Infinite Series 70
5.3 Computing Values of Convergent Series 72
    5.3.1 Geometric Series 72
    5.3.2 Telescoping Series 73
    5.3.3 Series With a Given Value 74
    5.3.4 The Harmonic Series 75
5.4 Operations and Rearrangements 76
5.5 Convergence Tests 77

6 Continuity 81
Problems 81
6.1 Continuity - A Pointwise Notion 81
    6.1.1 The Algebra of Continuous Functions 82
    6.1.2 Further Fun Facts About Continuous Functions 83
6.2 Uniform Continuity - The Setwise Notion 84
6.3 Lipschitz Continuity 86

7 Differentiability 89
Problems 89
7.1 A Short Historical Exposition 89
7.2 Differentiability - The Basics 90
7.3 Higher Order Derivatives 91
7.4 Fractional Derivatives 92
    7.4.1 The Fourier Transform 92
    7.4.2 The Complex Perspective 93
    7.4.3 Riemann-Liouville fractional derivatives 93
7.5 Further Properties of Differentiable Functions 94

8 Integrability 99
Problems 99
8.1 Partitions, The Basics 99
8.2 Riemann Integrability 102
8.3 Properties of the Integral 103

9 Polynomials 107
Problems 107
9.1 A Short Historical Exposition 108
9.2 Polynomials 108
    9.2.1 The Zeros of Polynomials 109
    9.2.2 Coefficients and Zeros 115
    9.2.3 Approximating the Zeros of a Polynomial 119
9.3 A More Algebraic Look at Polynomials 119

References 125

Preface

Most mathematics programs in the country have several different courses in their curriculum to teach topics in analysis and algebra. The California State University, Fresno is no exception. The authors have been involved with the teaching of these courses for the past several semesters. It was the students' insistence on trying to keep materials in different courses separate that prompted the idea for this book. The goal of this book is twofold. The authors wanted to present the standard material in real analysis and abstract algebra in a way that encourages the students to bring concepts and ideas together from different areas of mathematics. Many of the examples and exercises serve as foundation for this approach, asking the student to cross 'boundaries' between analysis and algebra to varying extent. The students will also find that theorems which are analytic/algebraic in flavor may have algebraic/analytic proofs thereby further strengthening the feeling that their course is a study of ideas in mathematics rather than in analysis and algebra.

The authors also wanted to present the material in a way which induces a healthy amount of independent learning. For this reason we seldom give a full proof of a theorem. Instead, there are exercises after the statements of the results, in which the students are guided through the development of the proofs. In case we do give a proof of a theorem, we amend the discussion with an exercise in which the student can work out an alternative proof. In addition, the statements of some results are purposefully incomplete focusing mainly on the gist of a particular result. The authors found that this helps students remember the content of the theorems better, and also serves as a great opportunity to ask students to investigate, experiment and make the statements of results as precise as necessary. The authors want to emphasize that this approach seems to work well at Fresno State, while recognizing that other universities and colleges may have student bodies requiring more or less precision in the treatment of the material.

There are several places in the text where the reader is encouraged to explore and do further readings. Should the reader come to a concept s/he is unfamiliar with, s/he is encouraged and expected to find alternative resources and brush up on the gaps in his or her background. It is the hope of the authors that Chapter 1 will help some in this regard. However they make no claim that Chapter 1 is comprehensive in any way when it comes to results and concepts that may be called upon throughout the text.

All chapters have some problems at the very beginning which are intended to get the students' hands dirty. These are problems which are unsolvable at the time the students first encounter them, but should be easily solvable after the completion of the chapter. The authors found that on many occasions these problems trigger novel and excellent ideas from students, and also serve as motivation for what is to come.

Some chapters also have a short historical exposition section. These are intended to make mathematics come alive. Sometimes students get intimidated by the material. Giving them a chance to read about the development of said material - how long it took, how many of the giants of mathematics struggled to get to where we are now, what personal interactions there might have been between mathematicians working on the same ideas - should provide perspective and will hopefully help in engaging the material.

There is more material in the book than can be covered in a given semester. Therefore this book is not intended to be covered in its entirety. Rather, it is up to the instructor's discretion how much and what part of the material to use. There are some dependencies between chapters, but there is a large degree of freedom on the part of the instructor as to how to best present the material.

The authors hope that this exposition will shape the approach students take to learning new mathematics and will make for an enjoyable experience in taking the class/es using this book as a text.

Tamás Forgács and Oscar Vega
Fresno, CA. June, 2012

Chapter 1
Things You Should Know

Abstract This chapter summarizes results, mostly without proof, which we consider necessary background for the material presented in the book. Please review this chapter and use it in the spirit it was intended: as a refresher for the things you know, and as a quick guide to learn the things you haven't yet seen but will need to succeed.

> Mathematical discoveries, small or great are never born of spontaneous generation. They always presuppose a soil seeded with preliminary knowledge and well prepared by labour, both conscious and subconscious.

Jules Henri Poincaré (1854 –1912)

1.1 Sets and Functions

We all have an intuitive idea of what a set is. Still, the concept of a set has given headaches to several mathematicians throughout the history of the subject. This is perhaps because the intuitive definition of a set is too simple, yet easily misunderstood and misused. Consider for instance the following set A:

$$A = \{x;\ x \notin A\}$$

This set gives rise to what is called Russell's Paradox, named after Bertrand Russell (1872-1970), a Welsh mathematician and philosopher. It is easy to see that if an element belongs to A then it cannot belong to A, and if an element does not belong to A, then it must be in A. Quite a conundrum!

Sets like these called for a formal development of set theory. The achievement of this task is mostly attributed to Georg Cantor (1845-1918), and it is said to have driven Cantor certifiably insane. Many of the deep problems and questions of set theory are beyond the scope of this course, although a few of them are disturbingly 'obvious'. Let us say for example that we want to choose an even number larger than the number of atoms in the world. This is something we assume feasible but... how would you find such a number? In order to deal with such questions the ability of choosing one element with a certain characteristic out of a set must be assumed. We use this assumption whenever we say 'let', 'choose', etc. in constructing a mathematical argument.

The assumption mentioned above is called *the axiom of choice*, and is an essential building block of most of modern mathematics. This axiom is equivalent, or closely related, to many other important assumptions we make everyday when doing mathematics. We list here some of the more commonly used ones: Trichotomy (a real number is either positive, negative or zero), every vector space has a basis, the well-ordering principle (Axiom 1 on page 4), the cartesian product (see Definition 1.2) of two non-empty sets is non-empty, etc.

For this course we will not need much beyond what the following simple definition of a set will allow us to do.

Definition 1.1.
(a) A set is a non-ordered collection of items.
(b) The cardinality of a set A, denoted $|A|$, is the number of elements in A (if A has finitely many elements) or ∞ if the set contains infinitely many elements.

Definition 1.2. Let A and B be non-empty sets.
(a) A subset of a set A is a set which only contains elements that also belong to A.
(b) $A \cup B$, the union of A and B, is the set that contains all the elements belonging to A or B.

(c) $A \cap B$, the intersection of A and B, is the set that contains all the elements belonging to A and B simultaneously.
(d) A partition of A is a collection of subsets of A such that their union is A and the intersection of any two of these subsets is empty.
(e) The cartesian product of A and B, denoted $A \times B$ is the set

$$A \times B = \{(a,b); a \in A, b \in B\}$$

Remark 1.3. The union and intersection of A and B can also be presented as

$$A \cup B = \{x; x \in A \text{ or } x \in B\}$$

$$A \cap B = \{x; x \in A \text{ and } x \in B\}$$

Although you might be used to thinking that a function is only something that looks like $f(x) = sin(x)$ or $g(x) = e^x + x^2 - 3$, you need to recall that a function is just a special type of a relation, and that a relation is nothing but a subset of a cartesian product of two sets. We formalize these ideas in the following definition.

Definition 1.4. Let A and B be non-empty sets.
(a) A relation R from A to B is a subset of $A \times B$.
An element $(a,b) \in R$ can also be thought of as $R(a) = b$. In other words, b would be an image of a under R. Yet another way to write $(a,b) \in R$ would be aRb (mostly when $A = B$). In the latter case it is customary to denote a relation by $\sim$.
(b) If F is a relation from A to B such that every element in A has a unique image in B, then F is said to be a function.

In the particular case of a relation from A to A, relations might have special properties.

Definition 1.5. Let $\sim$ be a relation on a set A. We define the following types of relations:
(a) $\sim$ is reflexive if $a \sim a$ for all $a \in A$,
(b) $\sim$ is symmetric if $a \sim b$ necessarily implies that $b \sim a$,
(c) $\sim$ is transitive if whenever $a \sim b$ and $b \sim c$ then it must happen that $a \sim c$.

If a relation $\sim$ satisfies the three properties listed above, then $\sim$ is said to be an equivalence relation on A.

Exercise 1.6. Show that the following relations are equivalence relations on the indicated sets:
(a) $\sim$ defined on $\mathbb{R}$ by $a \sim b$ if and only if $|a| = |b|$.
(b) $\sim$ defined on $\mathbb{Z}$ by $a \sim b$ if and only if both are even or both are odd.
(c) $\sim$ defined on $\mathbb{Z}$ by $a \sim b$ if and only if, for a fixed integer $n > 1$, a and b have the same remainder when dividing by n.
(d) $\sim$ defined on the set of all differentiable functions $\mathbb{R} \to \mathbb{R}$, where $f \sim g$ if and only if $f'(x) = g'(x)$ for all $x \in \mathbb{R}$.
(e) $\sim$ defined on the set of all functions $f : A \to B$ by $f \sim g$ if and only if $f(a) = g(a)$ for some *fixed* $a \in A$.
(f) $\sim$ defined on the set of all triangles by $\triangle ABC \sim \triangle DEF$ if and only if the two triangles are congruent. Would this relation still be an equivalence relation if we replaced 'congruent' by 'similar'?
(g) $\sim$ defined on $\mathbb{R}^3$ by $P \sim Q$ if and only if $P - Q$ is orthogonal to the vector $(1,1,1)$.
(h) $\sim$ defined on the set of all finite sets by $A \sim B$ if and only if there exists a bijective function $f : A \to B$.
(i) $\sim$ defined on $\mathbb{Z}$ by $a \sim b$ if and only if, for a fixed $n \in \mathbb{Z}$, $\gcd(a,n) = \gcd(b,n)$ (see Definition 1.16).

Exercise 1.7. Let $\sim$ be an equivalence relation on a set A.
(a) Assume that $|A| = n < \infty$ and that every equivalence class under $\sim$ contains exactly k elements. Prove that $k \mid n$.
(b) How many equivalence classes are there in part **(a)**?
(c) Assume that $A = \mathbb{Z}$ and that $\sim$ is defined by:

1.1 Sets and Functions

$$a \sim b \text{ if and only if } a \equiv b \pmod{n}$$

Prove that if C_1 and C_2 are (any) two equivalence classes of $\sim$ then there is a bijective function $f : C_1 \to C_2$.

Example 1.8. Let A be a set and $f : A \to A$ a *fixed* bijective function. We set
(a) $f^0 = id$,
(b) $f^n = \underbrace{f \circ f \circ \cdots \circ f}_{n \text{ times}}$, and
(c) $f^{-n} = (f^n)^{-1}$, for all $n \in \mathbb{N}$, where $(f^n)^{-1}$ is the inverse function of f^n.

Consider the relation $\sim$ on A by $a \sim b$ if and only if $f^k(a) = b$, for some $k \in \mathbb{Z}$. We next show that $\sim$ is an equivalence relation.
Let $x, y, z \in A$.

- $\sim$ **is reflexive:** In order to check reflexivity we need to prove that $x \sim x$. That is, we need to find a power k of f such that $f^k(x) = x$, but since f^0 is the identity function then $f^0(x) = x$. It follows that $x \sim x$.
- $\sim$ **is symmetric:** Suppose $x \sim y$. This means that we know that there exists $k \in \mathbb{Z}$ such that $f^k(x) = y$ (definition of the relation). In order to show that $y \sim x$ we need to find a power of f such $f^r(y) = x$. Since f^k sends x to y, then the inverse function of f^k sends y back to x (the inverse function exists because f is bijective, and thus any composition of f's is also bijective, in particular f^k). It follows that $(f^k)^{-1}(y) = x$. But, according to the definition above $(f^k)^{-1} = f^{-k}$. It follows that $f^{-k}(y) = x$. Since $k \in \mathbb{Z}$ then so is $-k$, therefore $y \sim x$.
- $\sim$ **is transitive:** Suppose $x \sim y$ and $y \sim z$. This means that there exist $k, l \in \mathbb{Z}$ such that $f^k(x) = y$ and $f^l(y) = z$. Since

$$f^l f^k = \left(\underbrace{f \circ f \circ \cdots \circ f}_{l \text{ times}} \right) \circ \left(\underbrace{f \circ f \circ \cdots \circ f}_{k \text{ times}} \right) = \underbrace{f \circ f \circ \cdots \circ f}_{k+l \text{ times}}$$

then $f^{k+l}(x) = f^l f^k(x) = f^l(f^k(x)) = f^l(y) = z$. Since $k, l \in \mathbb{Z}$ then $k + l \in \mathbb{Z}$, and thus $x \sim z$.

Equivalence relations are very important in mathematics because, among other things, they create partitions on the set where they are defined. If R is an equivalence relation on A, then for a fixed $a \in A$, we define the (equivalence) class of a as

$$[a] = \{b \in A;\ aRb\}$$

Exercise 1.9. Suppose that R is an equivalence relation on a set A. Show that the set of all distinct equivalence classes is a partition of A.

Exercise 1.10. What are the equivalence classes determined by the relations in Exercise 1.6?

As we mentioned before, a function is just a special type of a relation. We present here a few special classes of functions which will be in used frequently later.
A function $f : A \to B$ is
(a) injective (or one-to-one) if every two distinct elements in the domain have distinct images. This is an intuitive definition, which we make more precise as follows: a function f is injective if and only if

$$\text{whenever } f(x) = f(y), \text{ we must have } x = y.$$

(b) surjective (or onto) if every element in B has a pre-image (in A). More formally, f is onto if and only if

$$\forall b \in B,\ \exists a \in A \text{ such that } f(a) = b$$

(c) bijective if f is both injective and surjective.

Note that if the sets A and B are finite, and $f : A \to B$ is a bijection then A and B must necessarily have the same number of elements. In fact, injectivity, surjectivity and the condition that $|A| = |B|$ are even more closely related when the sets A and B are finite.

Theorem 1.11 (The 2 out of 3 property). *If $f : A \to B$ is a function, and A and B are finite sets, then any two of the following three properties implies the third.*
(a) f is injective.
(b) f is surjective.
(c) $|A| = |B|$

Exercise 1.12. Prove Theorem 1.11.

We close this section on sets and functions setting some notation that will be useful in later chapters.

Definition 1.13. For a function $f : \mathbb{R} \to \mathbb{R}$ we define

$$f^+ = \max\{f, 0\}$$
$$f^- = \max\{-f, 0\}$$

Exercise 1.14. Show that $f = f^+ - f^-$ and that $|f| = f^+ + f^-$.

1.2 Numbers

Mathematics owes most of its modern advances to the study of equations. The evolution of what mathematicians consider to be 'all numbers' has gone from counting to ten, to considering zero as a number, to accepting that negative numbers are not 'false', to rational numbers, to numbers that cannot be written as fractions of whole numbers, to the complex numbers and beyond: to the abstractions of these numbers. The study of equations also helped the development of Calculus. Questions of what type of a number solves a particular equation transformed to questions of how many solutions are there, where are those solutions found in the complex plane, how are the solutions related to solutions to derived equations and the like. Many of the answers to these questions came after the development of calculus, and used 'cutting edge' methods from analysis.

In the following sections we will take a look at sets of numbers, keeping in mind that these sets are always more than just a collection of items. They all satisfy algebraic and analytic properties, without which their study would be much less interesting.

1.2.1 The Integers

We note that $\leq$ defines a relation on $\mathbb{N}$ which is reflexive, transitive, but not symmetric. In fact, it is antisymmetric, as $a \leq b$ and $b \leq a$ force $a = b$. This relation yields the following important axiom (or, corollary of the axiom of choice).

Axiom 1 (Well-ordering principle) *Every non-empty set A of positive integers contains a least element. That is, if A is a non-empty set of positive integers, then there exists $x \in A$ such that $x \leq y$ for all $y \in A$.*

We will see an application of this axiom in the proof of Theorem 1.28, and there are many more applications of this axiom to come.

Besides the obvious properties of the integers, such as $xy = yx$, for all $x, y \in \mathbb{Z}$, there are others that are also well-known but deserve to be remarked.

Remark 1.15. The number 1 is the only integer with the property: $x \cdot 1 = 1 \cdot x = x$ for all $x \in \mathbb{Z}$. Similarly, 0 is the only integer with the property $x \cdot 0 = 0 \cdot x = 0$ for all $x \in \mathbb{Z}$. The second property can also be phrased as

$$(\dagger) \quad \text{If } xy = 0 \text{ then } x = 0 \text{ or } y = 0.$$

1.2 Numbers

This property (as phrased in (†)) will reappear in Chapter 3 when we discuss rings with and without zero divisors.

One very important concept in Number Theory, and in a good part of Abstract Algebra, is that of divisibility.

Definition 1.16.
(a) Let $a, b \in \mathbb{Z}$. We say that a divides b (in $\mathbb{Z}$) if there is an integer c such that $b = ac$. We write $a|b$ when a divides b, and $a \nmid b$ when a does not divide b.
(b) A positive integer is said to be prime if it is divisible by exactly two distinct positive integers, 1 and itself.
(c) An integer d that divides two integers a and b at the same time is called a common divisor of a and b. The largest (positive) common divisor of a and b is called the greatest common divisor of a and b. This is denoted $d = \gcd(a,b)$, or just $d = (a,b)$.
(d) If $\gcd(a,b) = 1$ then a and b are said to be relatively prime.

Exercise 1.17. Let $a, b, c \in \mathbb{Z}$. Prove the following claims.
(a) $1|a$, for all $a \in \mathbb{Z}$.
(b) $a|0$, for all $a \in \mathbb{Z} \setminus \{0\}$.
(c) If $a|b$ and $b|c$, then $a|c$.
(d) If $c|a$ and $c|b$, then $c|(ma+nb)$ for all $m, n \in \mathbb{Z}$.
(e) If $d = \gcd(a,b)$ then $\dfrac{a}{d}$ and $\dfrac{b}{d}$ are relatively prime.
(f) $d = \gcd(a,b)$ is the largest element in the set $C(a,b) = \{c \in \mathbb{Z};\ c|a \text{ and } c|b\}$.
(g) $\gcd(a,b) = \gcd(a,-b)$, for all $a, b \in \mathbb{Z}$.
Hint: Compare the sets $C(a,b)$ and $C(a,-b)$.
(h) $\gcd(ac,bc) = |c|\gcd(a,b)$, for all $a, b, c \in \mathbb{Z}$.
(i) $\gcd(a,b) = \gcd(a, b+ac)$, for all $a, b, c \in \mathbb{Z}$.

Every positive integer can be written as a product of powers of primes. Such a presentation of an integer is called the prime decomposition (or factorization) of that number. For example

$$24 = 2^3 \cdot 3 \qquad\qquad 60 = 2^2 \cdot 3 \cdot 5 \qquad\qquad 375 = 3 \cdot 5^3$$

It is important to remark that the prime factorization of a positive integer is always possible, and that it is unique up to rearrangement of the factors. This important result is called *The Fundamental Theorem of Arithmetic*. The following result generalizes, in some way, the notion of divisibility. A proof for this theorem may be found in most Number Theory texts, see for example [21].

Theorem 1.18 (The division theorem). *If $a, b \in \mathbb{Z}$ and $b \neq 0$, then there exist two unique integers q and r such that*
$$a = bq + r \qquad\qquad \text{with } 0 \leq r < b.$$
The number q is called the quotient, and r is called the remainder (of the division of a by b).

Remark 1.19. Most of the theorems that can be proved for integers using the division theorem (above) can also be proved for polynomials using Theorem 9.22.

Exercise 1.20. Let $a, b, q, r \in \mathbb{Z}$ be as in the division theorem above. Prove the following claims.
(a) $b|a$ if and only if $r = 0$.
(b) $\gcd(a,b) = \gcd(b,r)$.

Now we can use the division theorem repeatedly, together with the previous exercise to find the *gcd* of any pair of integers. The idea is that we can look for the $\gcd(a,b)$ by looking for the $\gcd(b,r)$ where r is the remainder of the division of a by b. What we gain with this is that now the numbers we will work with are smaller than the initial ones. Moreover, we can repeat this idea (dividing repeatedly) until the result is obvious or very easy to find. The process of using remainders to find the greatest common divisor of any two given integers is called *The Euclidean Algorithm*.

Exercise 1.21. Use the Euclidean algorithm to find $gcd(a,b)$ for

(a) $a = 100, \ b = 31$
(b) $a = 12,000,000, \ b = 9,100,000$

Lastly we consider is the following theorem, due to the French mathematician Étienne Bézout (1730–1783).

Theorem 1.22 (Bézout's lemma). *If $d = \gcd(a,b)$, then there are integers m and n such that*
$$d = ma + nb$$

Exercise 1.23. The following exercises will lead you to prove Bézout's lemma.
Let $a, b \in \mathbb{Z}$ and set
$$S(a,b) = \{c \in \mathbb{Z}; \ c = am + bn, \text{ for some } m, n \in \mathbb{Z}\}.$$

Prove the following.
(a) Both a and b are elements of $S(a,b)$.
(b) If c is a common divisor of a and b then c divides every element in $S(a,b)$.
(c) $S(a,b)$ contains a least positive element. Call it D.
(d) D divides every element in $S(a,b)$.
Hint: Use the division theorem.
(e) $D = \gcd(a,b)$.

Exercise 1.24. Let $a, b, c \in \mathbb{Z}$. Prove the following claims.
(a) If $c|(ab)$ and $\gcd(c,a) = 1$ then $c|b$.
(b) If p is a prime and $p|(ab)$ then $p|a$ or $p|b$.
(c) If $1 = am + bn$ then $\gcd(a,b) = 1$.
(d) If $\gcd(a,b) = d$ and $d = am + bn$, then $\gcd(m,n) = 1$.

Exercise 1.25. That is: Let $a,b \in \mathbb{Z}$.
(a) Give an example of $a,b,d,m,n \in \mathbb{Z}$ such that $d = am + bn$ but $\gcd(a,b) \neq d$.
(b) Find $m \in \mathbb{Z}$ such that 13 divides $8m - 1$. In other words: find an integer m such that $8m \equiv 1 \pmod{13}$.

1.2.2 The Rationals

The notation we use for the set of rational numbers is $\mathbb{Q}$. The development of this set from $\mathbb{Z}$ is straightforward. First, we form all quotients of the form p/q, where $p, q \in \mathbb{Z}$, and $q \neq 0$. We put these quotients in a set, and call it Q. Then we define the relation R on Q as follows. If $r_1 = p_1/q_1 \in Q$ and $r_2 = p_2/q_2 \in Q$, then $r_1 R r_2$ if and only if $q_2 p_1 = q_1 p_2$. It is easy to verify that this is an equivalence relation on Q. Denoting by $[r]$ the equivalence class of r in Q, we define
$$\mathbb{Q} := \{[r] \mid r \in Q\}$$

Thus $\mathbb{Q}$ is really a set of equivalence classes. With this understanding, we will ignore the equivalence class notation, and write r instead of $[r]$ for the generic rational number.

Exercise 1.26. Prove that the equivalence relation given above is, in fact, an equivalence relation.

1.2.3 The Reals

The notation we use for the set of real numbers is $\mathbb{R}$. To shed some light on the difference between $\mathbb{Q}$ and $\mathbb{R}$, consider the following set:

1.2 Numbers

$$S = \{s \in \mathbb{Q} \mid s^2 < 2\}$$

It only takes a few moments to realize that this set does not have a largest element. Furthermore, if $r \in \mathbb{Q}$ is such that $s < r$ for all $s \in S$, then there exists $\tilde{r} \in \mathbb{Q}$ such that $\tilde{r} < r$ and $s < \tilde{r}$ for all $s \in S$. In other words, the set

$$U := \{r \in \mathbb{Q} \mid s < r, \forall s \in S\}$$

has no smallest element. Since S and U do not intersect (convince yourself of this), there seems to be a gap between the two sets. One way to think about the set of real numbers, is that it is the smallest set which has no such gaps, and which contains all rational numbers. Subsection 1.2.4 explains in what exact sense we 'plug' these gaps.

One could ask what could be said about the integers and their relation with the real numbers. Although these two sets of numbers are very different from each other, we can still find important properties which relate them.

Theorem 1.27 (Archimedean property). *Given two positive real numbers a, b there is an integer n such that $b < na$.*

Proof. If $b < a$ take $n = 1$. Suppose then that $a \leq b$. Consider the set

$$E = \{k \in \mathbb{N} \mid ka \leq b\} = \left\{k \in \mathbb{N} \mid k \leq \frac{b}{a}\right\}.$$

The set E is non-empty, in particular $1 \in E$. Furthermore E is bounded above by b/a. Thus E has a largest element, say n. Consider the number $n+1$. First, it is an integer since $n \in E$. Second, $n+1 \notin E$ since $n = \max E$. This means that

$$n + 1 > \frac{b}{a} \quad \rightarrow \quad b < (n+1)a.$$

$\square$

It is very clear that the integers sit fairly far apart from each other on the real line (they are at distance at least one from each other). On the other hand the rationals seem to be very close to each other. The question is: How tightly arranged are they?

Theorem 1.28 (Density of $\mathbb{Q}$ in $\mathbb{R}$). *Between any two real numbers there is a rational number.*

Proof. Let $a < b \in \mathbb{R}$ be given. Since $b - a > 0$ there exists an n such that $1 < n(b - a)$.

- **Case 1** Assume $b > 0$. Consider the set $E = \{k \in \mathbb{N} \mid b \leq k/n\}$. By Theorem 1.27 E is non-empty. By the well ordering principle (axiom 1 in page 4) E has a least element, say k_0. Set $m = k_0 - 1$ and $q = m/n$. First notice that $m \notin E$. Thus either $m \leq 0$ or $b > m/n$. In either case we have $q < b$. On the other hand, since $k_0 \in E$ we have

$$a = b - (b-a) < \frac{k_0}{n} - \frac{1}{n} = \frac{k_0 - 1}{n} = q.$$

- **Case 2** $b \leq 0$. Choose $k \in \mathbb{N}$ such that $k + b > 0$. Since $k + a < k + b$ we get by the analysis in case 1 a number $r \in \mathbb{Q}$ such that $k + a < r < k + b$. Since $r - k \in \mathbb{Q}$, the proof is complete. $\square$

1.2.4 The Completeness Axiom and Related Topics

We take a moment to emphasize that sequences can be used to approximate elements of metric spaces. If $X = \mathbb{R}$, and $(f_n)_{n=1}^\infty$ is a sequence of real numbers that converges to π, we can use this sequence to give successively better approximations to π. Furthermore in Example 4.19 we will see that we could take a sequence of rational numbers to approximate an irrational number. This raises a very important question. Suppose that $f : \mathbb{N} \to X$ is convergent with limit L. When does L belong to X? We have seen that this need not be the case when $X = \mathbb{Q}$. We could define the set of real numbers $\mathbb{R}$ to be the smallest set equipped with the Euclidean distance containing $\mathbb{Q}$ which has the property that every convergent sequence in $\mathbb{R}$ has its limit in $\mathbb{R}$. Metric spaces with this property are called *complete*. In general the smallest metric space X containing an incomplete metric space Y is called the *completion* of Y. Thus, $\mathbb{R}$ is simply the completion of $\mathbb{Q}$ and the completion of a complete metric space is simply itself.

Note that there are alternative (and equivalent) ways to define the set of real numbers starting with the ordered field $\mathbb{Q}$, for example one could use Dedekind cuts to do so (these are just subsets of $\mathbb{Q}$ of a particular form). Here we follow the approach of axiomatizing $\mathbb{R}$. We point out that the Completeness axiom can be proved to hold for $\mathbb{R}$ if one develops $\mathbb{R}$ using either of the above mentioned methods. For more details on the development of $\mathbb{R}$ and for further references see for example [22].

Definition 1.29. Let S be a non-empty subset of $\mathbb{R}$. We say that a real number ℓ is a lower bound for S, if $\ell \leq s$ for all $s \in S$. We say that a real number u is an upper bound for S, if $u \geq s$ for all $s \in S$.

Definition 1.30. If a non-empty set S of real numbers has a lower/upper bound, we say that S is bounded from below/above. If a non-empty set S of real numbers is bounded both from above and from below, we simply say that S is bounded.

The Completeness Axiom Let $S \subseteq \mathbb{R}$ be a non-empty set which is bounded from below. Then there exist a number m^* such that **(a)** m^* is a lower bound for S, and **(b)** every lower bound m for S satisfies $m \leq m^*$. The number m^* is called the *greatest lower bound* or *infimum* of S, and is denoted by $\inf S$.

Exercise 1.31. Show using the Completeness Axiom that for every non-empty subset S of $\mathbb{R}$ which is bounded from above there exists a real number M^* such that if M is any upper bound for S then $M^* \leq M$. M^* is called the least upper bound (or supremum) of S and is denoted by $\sup S$.

Exercise 1.32. Suppose that $S \neq \emptyset$, $S \subseteq T$ and T is a bounded subset of $\mathbb{R}$. Say all you can (with proof) about the relations between $\inf T$, $\inf S$, $\sup T$, $\sup S$, $\inf(T \cap S)$, $\sup(T \cap S)$, $\inf(T \cup S)$, $\sup(T \cup S)$.

Exercise 1.33. Suppose that $S \subseteq \mathbb{R}^2$ is non-empty and bounded. That is, there exists some $R \in \mathbb{R}$ such that if $s \in S$, then the distance from s to the origin is less than R. Develop a notion of $\inf S$ and $\sup S$ and compare the properties of these to those of infima/suprema of subsets of the real numbers. Does your notion for subsets of $\mathbb{R}^2$ generalize to higher dimensions?

Exercise 1.34. Suppose that S is any non-empty bounded subset of $\mathbb{Z}$. Use the Well Ordering Principle to show that $\inf S \in S$ and $\sup S \in S$.

Exercise 1.35. Given two subsets S, T of $\mathbb{R}$ we can define a new set

$$S + T := \{s + t \mid s \in S, t \in T\}$$

Order the following if possible: $\inf T, \inf S, \inf T + S, \sup S, \sup T, \sup(S + T), \sup(S \cap T)$ and $\sup(S \cup T)$ for the following sets T and S:
(a) $T = \mathbb{Q} \cap [0, 1]$ and $S = \mathbb{Z} \cap [-10, 10]$.
(b) $T = \{1/n \mid n \in \mathbb{N}\}$ and $S = \{2, \sqrt{2}\}$.
(c) $T = \cap_{n=1}^{\infty} [1 - 1/n, 1 + 1/n]$ and $S = \cup_{n=1}^{\infty} [1 - 1/n, 1 + 1/n]$.
(d) $T = \mathbb{Q} \cap [0, 1]$, $S = \mathbb{Z} \cap (-10, 10)$.

Exercise 1.36. Let S, T be two non-empty subsets of $\mathbb{R}$. Make a claim about how $\sup(T + S)$ and $\inf(T + S)$ are related to $\inf T$, $\inf S$, $\sup T$ and $\sup S$ and prove your claim.

Given a function $f : \mathbb{R} \to \mathbb{R}$, and a non-empty set $T \subseteq \mathbb{R}$, we can define the set

$$f(T) = \{f(x) \mid x \in T\}.$$

$f(T)$ is then a non-empty set of real numbers, so it makes sense to ask whether $\sup f(T)$, and $\inf f(T)$ exist. The following exercise will get you to think a little bit about these quantities.

1.2 Numbers

Exercise 1.37. Give examples of functions $f : \mathbb{R} \to \mathbb{R}$ and non-empty sets T such that
(a) T is bounded, but $f(T)$ is not.
(b) $f(T)$ is bounded, but T is not.
(c) Neither T, nor $f(T)$ is bounded.
(d) Both T and $f(T)$ are bounded, and
    (i) $\inf T < \inf f(T) < \sup f(T) < \sup T$
    (ii) $\inf T < \inf f(T) < \sup T < \sup f(T)$
    (iii) $\inf f(T) < \inf T < \sup f(T) < \sup T$
    (iv) $\inf f(T) < \inf T < \sup T < \sup f(T)$
    (v) $\inf T < \sup T < \inf f(T) < \sup f(T)$

1.2.4.1 Inequalities

Solving equations is something the reader has most likely engaged in the past, and will do so again in greater depth as she studies abstract and/or linear algebra. The study of inequalities is as important to analysis as the study of equations and their solvability is to algebra. In this section we refresh some of the ideas involved in working with inequalities, with a focus on real valued functions.

The reader will recall that if $r, s \in \mathbb{R}$, then $r = s \iff (r \geq s \land s \geq r)$. The following is a similar result using strict inequalities:

Exercise 1.38. Suppose $r, s \in \mathbb{R}$. Prove that if $r < s + \frac{1}{n}$ for all $n \in \mathbb{N}$, then $r \leq s$, and if $s - \frac{1}{n} < r$ for all $n \in \mathbb{N}$, then $r \geq s$.
Hint: Theorem 1.27 may be useful.

When it comes to working with inequalities which involve functions, the first order of business is to really understand what it means for a function to be "greater" than another function. For example, what could the statement $f \geq g$ possibly mean? The most natural interpretation is that $f(x) \geq g(x)$ for $x \in S$, where S is a subset of the domains of both f and g. A moment's consideration reveals that such a statement is *always true* for any two functions f, g and *some* set S (remember the empty set!). Things get somewhat more interesting however if we don't allow S to be empty. Now all of the sudden the statement that

$$f(x) \geq g(x) \quad \forall x \in S$$

may or may not be true for a given $S \subseteq Dom(f) \cap Dom(g)$.

Exercise 1.39. Suppose that $f(x) = e^x$ and $g(x) = e^{-x}$.
(a) Find a non-empty S such that $f(x) \geq g(x)$ for all $x \in S$.
(b) Find the largest set S such that $f(x) \geq g(x)$ for all $x \in S$. (By this we mean that if T is a set such that $f(x) \geq g(x)$ for all $x \in T$, then $T \subseteq S$.)
(c) Repeat parts (a) and (b) but now for $g(x) \geq f(x)$.

As you work with inequalities, the following are important facts to keep in mind. If

$$f(x) \geq g(x) \geq 0 \quad \text{for} \quad x \in S,$$

then
(i) $h(x) \geq 0$ for $x \in S \implies h(x) + f(x) \geq g(x)$ for $x \in S$.
(ii) $h(x) \geq 1$ for $x \in S \implies h(x)f(x) \geq g(x)$ for $x \in S$.
(iii) $h(x) \geq k(x) > 0$ for $x \in S \implies \dfrac{f(x)}{k(x)} \geq \dfrac{g(x)}{h(x)}$ for $x \in S$.
(iv) $h(x)$ monotone increasing on $S \implies h(f(x)) \geq h(g(x))$ for $x \in S$.

Exercise 1.40. Prove the following statements:
(a) $e^x > 1 + \sqrt{x} + \frac{x}{2}$ for $x \geq 1$
(b) $\left|\frac{x}{1+x} - \frac{1}{2}\right| \leq 1$ for $x \in [0, 2]$
(c) $\sin x \geq \frac{2x}{\pi}$ for $x \in [0, \pi/2]$.
(d) $(1+x)^n \geq \left(1 + \left(\frac{n-1}{2}\right)x\right)^2$ for all $n \in \mathbb{N}$ and $x \geq 0$.
(e) $\ln(x) \leq kx$ for any $k > 0$ and $x \gg 1$ (this means for large enough x).
(e') A more difficult question is the following: given $k > 0$ find $S_k \subseteq \mathbb{R}$ such that $\ln(x) \leq kx$ for all $x \in S_k$.

Exercise 1.41. Suppose $f, g : D \to \mathbb{R}$. Set $h(x) = \max\{f(x), g(x)\}$. Show that
(a) $h(x) \geq f(x)$ and $h(x) \geq g(x)$ for all $x \in D$, and
(b) if $t(x) \geq f(x)$ and $t(x) \geq g(x)$ for all $x \in D$, then $t(x) \geq h(x)$ for all $x \in D$.

1.2.5 The Complex Numbers

The notation we use for the set of complex numbers is $\mathbb{C}$. One can describe this set in a number of different ways. Here we choose to define $\mathbb{C}$ as a set of ordered pairs of real numbers equipped with two operations, '+' and '·'. To be more specific, we set

$$\mathbb{C} = \{(a,b) \mid a, b \in \mathbb{R}\}$$

and make the following definitions for the binary operations '+' and '·':

$$(a,b) + (c,d) := (a+c, b+d)$$
$$(a,b) \cdot (c,d) := (ac - bd, bc + ad).$$

One could also define $\mathbb{C}$ as the set of all expressions of the form $a + ib$, where $a, b \in \mathbb{R}$ and $\mathbf{i}^2 = -1$. One can then define '+' and '·' for these expressions the obvious way. A third characterization of $\mathbb{C}$, using 2×2 matrices, is given as Exercise 1.52. It is easy to see that the three definitions lead to a set of the same cardinality. Furthermore, (once you delve into the exciting world of fields), it is also easy to see that there are isomorphisms between the three structures.

1.3 Linear Algebra

Linear Algebra is one of the most important areas of mathematics, as it is used in most of the other areas of math, in particular in Abstract Algebra and Analysis. This topic is vast, and we hope that you have been exposed to this material already. In what follows we only cover the basic ideas, for more please go to your library and get a good book on this subject (you may want to look at the list of references in the back before you make the trip to the library, we especially recommend [2] and [12]).

Definition 1.42. A set V equipped with an operation $+$ and a (scalar) multiplication $\cdot$ is a vector space (over $\mathbb{R}$) if

1. $\mathbf{v} + \mathbf{w} \in V$, for all $\mathbf{v}, \mathbf{w} \in V$.
2. $\mathbf{v} + \mathbf{w} = \mathbf{w} + \mathbf{v}$, for all $\mathbf{v}, \mathbf{w} \in V$.
3. $(\mathbf{v} + \mathbf{w}) + \mathbf{u} = \mathbf{v} + (\mathbf{w} + \mathbf{u})$, for all $\mathbf{u}, \mathbf{v}, \mathbf{w} \in V$.
4. There is a vector, we will call $\mathbf{0}_V$, such that $\mathbf{v} + \mathbf{0}_V = \mathbf{0}_V + \mathbf{v} = \mathbf{v}$, for all $v \in V$. Most of the times we will write $\mathbf{0}$ instead of $\mathbf{0}_v$, as long as the context allows us to do so.

1.3 Linear Algebra

5. $\alpha \cdot \mathbf{v} \in V$, for all $\alpha \in \mathbb{R}$ and $\mathbf{v} \in V$
6. $(\alpha + \beta) \cdot \mathbf{v} = \alpha \cdot \mathbf{v} + \beta \cdot \mathbf{v}$, for all $\alpha, \beta \in \mathbb{R}$ and $\mathbf{v} \in V$.
7. $\alpha \cdot (\mathbf{v} + \mathbf{w}) = \alpha \cdot \mathbf{v} + \alpha \cdot \mathbf{w}$, for all $\alpha \in \mathbb{R}$ and $\mathbf{v}, \mathbf{w} \in V$.
8. $(\alpha \beta) \cdot \mathbf{v} = \alpha \cdot (\beta \cdot \mathbf{v})$, for all $\alpha, \beta \in \mathbb{R}$ and $\mathbf{v} \in V$.
9. $1 \cdot \mathbf{v} = \mathbf{v}$, for all $\mathbf{v} \in V$.

Exercise 1.43. Check that the following sets, with their operations, define vector spaces over $\mathbb{R}$.
(a) The n-dimensional Euclidean space is the set of ordered n-tuples

$$\mathbb{R}^n = \{(x_1, x_2, x_3, \cdots, x_n); x_i \in \mathbb{R}\}$$

with the standard sum

$$(x_1, x_2, \cdots, x_n) + (y_1, y_2, \cdots, y_n) = (x_1 + y_1, x_2 + y_2, \cdots, x_n + y_n)$$

and scalar multiplication

$$\alpha \cdot (x_1, x_2, \cdots, x_n) = (\alpha x_1, \alpha x_2, \cdots, \alpha x_n)$$

for all $\alpha \in \mathbb{R}$.
(b) The set of all polynomials with coefficients in $\mathbb{R}$,

$$\mathbb{R}[x] = \{a_0 + a_1 x + \cdots + a_n x^n; a_i \in \mathbb{R} \text{ for all } i = 0, 1, 2, \cdots, n\}$$

with the standard operations learned in high school algebra.
(c) The set of all $m \times n$ matrices with entries in $\mathbb{R}$,

$$M_{m \times n}(\mathbb{R}) = \{(a_{ij}); a_{ij} \in \mathbb{R} \text{ for all } i = 1, 2, \cdots, m, \ j = 1, 2, \cdots, n\}$$

with the usual addition of matrices and scalar multiplication.
(d) Let V be the set of all functions $f : \mathbb{R} \to \mathbb{R}$ such that $f(\pi) = 0$. The operations are

$$(f + g)(x) = f(x) + g(x) \qquad (\alpha f)(x) = \alpha f(x)$$

for all $\alpha \in \mathbb{R}$.
(e) If $S \subseteq \mathbb{R}$ is a non-empty set, the continuous functions on S form a real vector space. Functions which are differentiable on S also form a real vector space. For both of these spaces the operations of addition and scalar multiplication are defined the same way as in part **(d)** above.

Throughout the rest of the book you will find many other examples of vector spaces.

Definition 1.44. For $\mathbf{x}, \mathbf{y} \in \mathbb{R}^n$ we define the dot product of the two vectors to be the real number

$$\mathbf{x} \cdot \mathbf{y} = \sum_{i=1}^{n} x_i y_i.$$

The notation $\langle \mathbf{x}, \mathbf{y} \rangle$ is also used to denote the dot product of $\mathbf{x}$ and $\mathbf{y}$.

Exercise 1.45. Fix a vector $\mathbf{v} \in \mathbb{R}^n$, define the set

$$S = \{\mathbf{x} \in \mathbb{R}^n; \mathbf{v} \cdot \mathbf{x} = 0\}$$

Show that S is a vector space over $\mathbb{R}$.

Finally, consider the set S of solutions (in $\mathbb{R}^2$) of the equation $ax + by = 0$. This is a vector space as well! Similarly, the set of solutions of a system of linear equations of the form

$$ax + by = 0$$
$$cx + dy = 0$$

is also a vector space. In general, the solution set of a homogeneous system of linear equations is always a vector space, no matter how many equations or variables one considers.

1.3.1 Matrices

Matrices are ubiquitous in mathematics. For instance, the systems of linear equations mentioned in the previous section can be solved using matrices, one can solve differential equations using matrices, the Jacobian, Hessian and the differential of a smooth function is a matrix, etc (a long etc.).
In this book, matrices will be looked at from several different angles, so for now we will just review a few important definitions and results.

Definition 1.46.
(a) An $n \times m$ matrix M with entries in, let us say $\mathbb{C}$, is an array of elements of $\mathbb{C}$ consisting of n rows and m columns.
(b) The element in the intersection of the ith row and j^{th} column of M is denoted by m_{ij}. Also, a matrix can be denoted by $M = (m_{ij})$.
(c) The set of all $n \times n$ matrices with entries in $\mathbb{C}$ is denoted by $M_n(\mathbb{C})$.
(d) The sum of matrices is defined entry-wise:

$$(m_{ij}) + (n_{ij}) = (m_{ij} + n_{ij})$$

(e) The product of matrices is only defined if (m_{ij}) is an $m \times n$ matrix and (n_{ij}) is an $n \times r$ matrix. In such case, the product is defined as 'rows times columns'. That is, the element in the intersection of the ith row and j^{th} column of the product matrix $(m_{ij})(n_{ij})$ is given by

$$\sum_{k=1}^{n} m_{ik} n_{kj}$$

The resulting matrix is an $m \times r$ matrix.
(f) The $n \times n$ matrix with 1 on the diagonal and zeros everywhere else is called the identity matrix, and it is denoted by I. It is easy to see that $IM = MI = M$ for all $M \in M_n(\mathbb{C})$.
(g) A matrix $M \in M_n(\mathbb{C})$ is said to be invertible if and only if there exists a matrix $N \in M_n(\mathbb{C})$ such that $MN = NM = I$.

Remark 1.47. We could replace $\mathbb{C}$ in Definition 1.46 by $\mathbb{Z}$, $\mathbb{Q}$, $\mathbb{R}$, and many other sets.

A linear transformation with the same domain and range, let us say $T : \mathbb{R}^3 \to \mathbb{R}^3$, can be represented by a 3×3 matrix with entries in $\mathbb{R}$ (after choosing a basis of $\mathbb{R}^3$, of course). This matrix somehow summarizes all the properties of T. For example, T will be bijective if and only if the matrix representing T is invertible, which is something that can be easily checked (without finding the inverse matrix explicitly) by computing the determinant of the matrix.

Definition 1.48. Let $\mathbb{K}$ be either $\mathbb{Q}$, $\mathbb{R}$, or $\mathbb{C}$.
(a) The determinant is a function $det : M_n(\mathbb{K}) \to \mathbb{K} \setminus \{0\}$ such that (among many other important properties)
 (i) $det(AB) = det(A)det(B)$, which means that det is multiplicative.
 (ii) $det(I) = 1$.
(b) The trace is the function $tr : M_n(\mathbb{K}) \to \mathbb{K}$ defined by

$$tr(m_{ij}) = \sum_{i=1}^{n} m_{ii}$$

Remark 1.49. The determinant doesn't respect sums very well, and the trace behaves terribly with respect to products.

Exercise 1.50. Let $M, N \in M_n(\mathbb{K})$.
(a) Assume that M is invertible. Prove that $det(M^{-1}) = det(M)^{-1}$.
(b) det is invariant under conjugation, i.e. $det(M) = det(NMN^{-1})$, if N is invertible.
(c) $tr(M+N) = tr(M) + tr(N)$, which means that tr is additive.
(d) $tr(MN) = tr(NM)$.
(e) tr is invariant under conjugation, i.e. $tr(M) = tr(NMN^{-1})$, if N is invertible.

Theorem 1.51. *Let $M \in M_n(\mathbb{K})$. M is invertible if and only if $det(M) \neq 0$.*

There are times when one can write known sets as sets of matrices. Before looking at the following example we point out that the set of matrices have a 'one', which is the identity matrix, and a 'zero', which is the zero matrix. These matrices behave exactly as the *numbers* one and zero behave.
Consider now the matrix
$$\mathbb{I} = \begin{bmatrix} 0 & -1 \\ 1 & 0 \end{bmatrix}.$$
This matrix has the property that $\mathbb{I}^2 = -I$, thus reminding us of the complex number i. These ideas lead us to consider matrices of the form $aI + b\mathbb{I}$ in hopes of describing $\mathbb{C}$ as a set of matrices. It turns out that the set of all 2×2 matrices of the form
$$\begin{bmatrix} a & -b \\ b & a \end{bmatrix}$$
can be put in 1-1 correspondence with the complex numbers as the following exercise will have you do.

Exercise 1.52. Show that we could use the set
$$C = \left\{ \begin{bmatrix} a & -b \\ b & a \end{bmatrix} ; a, b \in \mathbb{R} \right\}$$
with the standard matrix operations to define the complex numbers via the association
$$\begin{bmatrix} a & -b \\ b & a \end{bmatrix} \longleftrightarrow a + bi$$
Verify that using this association one gets the usually defined addition and multiplication of complex numbers. Furthermore, verify that every non-zero element of C is invertible and find the inverse.

1.4 Algebraic Vs. Transcendental Numbers

It is clear that there are many numbers in $\mathbb{R}$ which do not live in $\mathbb{Q}$. You probably have already seen the proof of the fact that $\sqrt{2}$ is not rational in a previous course, and know that π, and e are also not rational. However, $\sqrt{2}$ is 'closer' to being rational than π and e because $\sqrt{2}$ is a solution of the equation $x^2 - 2 = 0$ while π nor e are not solutions of any polynomial equation with rational coefficients (it is not easy to prove this fact, though).

Definition 1.53. Let $\alpha \in \mathbb{C}$. If there is a polynomial $p(x)$ with rational coefficients such that $p(\alpha) = 0$, then α is said to be an algebraic element (or algebraic over $\mathbb{Q}$).
If there is no polynomial with rational coefficients such that $p(\alpha) = 0$ then α is said to be transcendental (over $\mathbb{Q}$).

Remark 1.54. The numbers π and e are transcendental over $\mathbb{Q}$. The proofs for these facts (due to Lindemann and Hermite respectively) are highly nontrivial. They can be found in [25], and you might also want to see [1] and [26] for topics related to this fact.

The definition of algebraic and transcendental is not always restricted to $\mathbb{Q}$. In fact, since i is a solution of $x^2 + 1 = 0$, then i is algebraic over $\mathbb{R}$.

Exercise 1.55. Show that if $k, n \in \mathbb{N}$, then $\sqrt[k]{n}$ is an algebraic element over $\mathbb{Q}$. Show that i is algebraic over $\mathbb{Q}$.

The content of the fundamental theorem of algebra (see Chapter 9) is that every polynomial equation with coefficients in $\mathbb{C}$ has all its solutions in $\mathbb{C}$. On the other hand, there are polynomial equations with coefficients in $\mathbb{Q}$, or $\mathbb{R}$, that do not have solutions in $\mathbb{Q}$, or $\mathbb{R}$. These radically different situations prompt the following definition.

Definition 1.56. Let $\mathbb{F}$ be a field (see Definition 3.23).
(a) $\mathbb{F}$ is said to be algebraically closed if every polynomial equation with coefficients in $\mathbb{F}$ has a solution in $\mathbb{F}$.
(b) In case $\mathbb{F}$ is not algebraically closed, the smallest algebraically closed field containing $\mathbb{F}$ is called the algebraic closure of $\mathbb{F}$. Note that every polynomial equation with coefficients in $\mathbb{F}$ will have solutions in the algebraic closure of $\mathbb{F}$.

Remark 1.57. The fundamental theorem of algebra says that $\mathbb{C}$ is algebraically closed. Furthermore $\mathbb{C}$ is the algebraic closure of both $\mathbb{Q}$ and $\mathbb{R}$.

Chapter 2
Groups

Abstract This chapter is about one of the most basic and ubiquitous algebraic structures: groups. Groups arise in functional analysis, the study of the Rubik's cube, algebraic topology, the study of the structure of crystals, and in the art of M.C. Escher (among other places).
We will study groups not just by looking at the properties of their elements and subsets but also by using special functions (called homomorphisms) defined on groups which preserve most of the algebraic properties of their domains.

> The Theory of Groups is a branch of mathematics in which one does something to something and then compares the result with the result obtained from doing the same thing to something else, or something else to the same thing.

James Roy Newman (1907–1966)

Getting Your Hands Dirty

Exercise 2.1. What properties are needed to solve the equation $ax = b$?

Exercise 2.2. Do you know of any functions which transform sums into products or products into sums?

Exercise 2.3. Is it always true that $a \cdot b = b \cdot a$?

Exercise 2.4. Let $\times$ be the standard cross-product in $\mathbb{R}^3$. Is
$$u \times (v \times w) = (u \times v) \times w \ ?$$
for any $u, v, w \in \mathbb{R}^3$?

2.1 A Short Historical Exposition

The study of (mostly polynomial) equations, and their solutions, has been the focus of mathematics through the centuries. Group Theory is not an exception. In fact, shortly before dying in a duel, the exceptional French mathematician Évariste Galois (1811 - 1832) used the idea of a group to study a specific set of functions that permute the solutions of a given equation.
Galois' ideas were later used by the Norwegian Niels Henrik Abel (1802 - 1829), another mathematician who died young. Abel proved that it was impossible to find a formula to solve any equation of degree 5 (formulas to solve quadratic, cubic, and quartic equations were already known by then).
The fact that a group 'does something to something' (see quote above, which describes what is called a *group action*) is a common occurrence in mathematics. This is how groups may be associated to art, chemistry, quantum mechanics, etc. Sometimes additional structure is given to groups to match the needs of different areas of mathematics, and giving rise to objects such as Lie groups (named after Sophus Lie: 1842 - 1899).

The study of groups is still an active area of research. The most recent monumental result was the classification of finite simple groups (finished in 1983), determining all possible 'building blocks' (simple groups) needed to construct all possible finite groups. This work took about 20 years and involved more than 100 mathematicians.

2.2 The Basics of Group Theory

Definition 2.5. We will say that $(G, *)$ is a group if G is a set with an operation $*$, defined for any pair of elements in G, such that
(a) $a * b \in G$, for all $a, b \in G$, (**Closure of** $*$)
(b) $a * (b * c) = (a * b) * c$, for all $a, b, c \in G$, (**Associativity**)
(c) there is an element $e \in G$ such that $a * e = e * a = a$, for all $a \in G$,
(d) for every $a \in G$ there exists an element b in G such that $a * b = b * a = e$.
The element e introduced in (c) is called the identity of G. Sometimes it is denoted e_G. The element b introduced in (d) is called the inverse of a and it is denoted by a^{-1}.

Remark 2.6. Most of the times we will not write the symbol $*$ for the operation in the group, that is, instead of writing $a * b$ we will write ab. Similarly, when the operation used to define a group is clear by the context we will just say G is a group, not mentioning the operation.

Note that in Definition 2.5 we say "**the** identity of G" and "**the** inverse of g", but we have not yet proved that these elements are unique!

Exercise 2.7. Let G be a group. Prove the following claims:
(a) The identity of G is unique.
(b) If $g \in G$ is such that $g^2 = g$, then $g = e_G$.
(c) For any given $g \in G$, g^{-1} is unique.
(d) $(g^{-1})^{-1} = g$, for all $g \in G$.
(e) Let $g_1, g_2, \ldots, g_n \in G$. Then $(g_1 g_2 \cdots g_n)^{-1} = g_n^{-1} g_{n-1}^{-1} \cdots g_2^{-1} g_1^{-1}$.
(f) $(hgh^{-1})^k = hg^k h^{-1}$, for all $g, h \in G$ and $k \in \mathbb{N}$.

Note that the operation in a group does not need to be a standard multiplication, it could be something defined specifically for the set in consideration. Thus, you should not assume properties of the standard multiplication which have not been given to you, or have not been proved. In particular, try to avoid using $a * b = b * a$ unless you are certain that this equality holds.

Remark 2.8. It is very common to want to prove that a set H contained in a group G is also a group (with the operation of G). Since the associativity law holds in G then it must also hold in H. We will say that the associative law in H is inherited from G.

Exercise 2.9. Are the following pairs $(G, *)$ groups?

(a) $(\{0\}, +)$ (b) $(\{1\}, +)$ (c) $(\{1\}, \cdot)$ (d) $(\{1, -1\}, \cdot)$

Definition 2.10. If $(G, *)$ is a group such that $a * b = b * a$ for all $a, b \in G$, then G is said to be an Abelian group.

Remark 2.11. What is mentioned in Remark 2.8 for the associative law is also valid for the commutative law. That is, the commutative law is also 'inheritable'.

Exercise 2.12. Let $\mathbb{K} = \mathbb{Q}, \mathbb{R}$ or $\mathbb{C}$. Prove what is claimed below.
(a) $\mathbb{K}$ is an Abelian group with the standard addition.
(b) $\mathbb{Z}$ is an Abelian group with the standard addition.
(c) $\mathbb{K}$ with the standard multiplication is not a group.

2.2 The Basics of Group Theory

(d) $\mathbb{Z}$ with the standard multiplication is not a group.
(e) $(\mathbb{K}\setminus\{0\},\cdot)$ is an Abelian group. $(\mathbb{Z}\setminus\{0\},\cdot)$ is not an Abelian group.
(f) $M_n(\mathbb{K})$ (see Definition 1.46) is an Abelian group with the standard matrix addition. Is $M_n(\mathbb{K})$ a group under the standard matrix multiplication?

Exercise 2.13. Let G be a group. Prove that the following three statements are equivalent:
(a) $(ab)^2 = a^2b^2$ for all $a,b \in G$.
(b) $(ab)^{-1} = a^{-1}b^{-1}$ for all $a,b \in G$.
(c) G is Abelian.

Now that the simpler examples of groups are out of the way (and, hopefully, clear), we will look at more complex groups. We will start with a couple of interesting groups of functions.

Exercise 2.14. Solve the following problems on groups of functions:
(a) For a *fixed* set A define $\mathscr{F}_A = \{f : A \to A;\ f \text{ is a bijective function}\}$. Prove $\mathscr{F}_A$ is a group under the operation given by composition of functions. Is this group always Abelian?
(b) Let $G = \mathbb{R} \setminus \{0\}$ and $H = \mathbb{Z}$. Consider the set

$$F(G,H) = \{f : G \to H;\ f \text{ is a function}\}$$

Show that $F(G,H)$ is a group with the operation $*$ defined by

$$(f*g)(x) = f(x) + g(x) \quad \forall x \in G.$$

Is this group Abelian?
(c) Now we will generalize the previous problem. Let $(G,*)$ and $(H,\star)$ be groups. Prove that

$$F(G,H) = \{f : G \to H;\ f \text{ is a function}\}$$

is a group with the operation $*$ defined by

$$(f*g)(x) = f(x) \star g(x) \quad \forall x \in G.$$

(d) Find necessary and sufficient conditions for $F(G,H)$, as in part **(c)**, to be Abelian.

Exercise 2.15. Let G be a group and $k \in \mathbb{N}$ be fixed. Define $G_k = \{g \in G;\ g^k = e\}$. Prove that if G is Abelian then G_k is a group.

Remark 2.16. If G is not Abelian G_k is not necessarily a group. An example of this may be found in Exercise 2.82.

We continue with a group with a very colorful history, the quaternions, denoted by Q_8. This group is the set $\{1,-1,i,-i,j,-j,k,-k\}$ with the multiplication given by

$$i^2 = j^2 = k^2 = -1 \qquad \text{and}$$

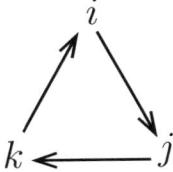

where the diagram above means $ij = k,\ jk = i,\ ki = j,\ ji = -k,\ kj = -i,\ ik = -j$. Note that $ij \neq ji$, which implies that Q_8 is not Abelian.

Exercise 2.17.
(a) Prove that Q_8 is a group.
(b) How many solutions of the equation $x^2 = -1$ are there in Q_8?
(c) Find G_k, for $k = 1,2,\ldots,8$ and $G = Q_8$ (G_k was defined in Exercise 2.15).

In Exercise 2.12 you proved that, for $\mathbb{K} = \mathbb{Q}, \mathbb{R}$ or $\mathbb{C}$, $(M_n(\mathbb{K}), +)$ is a group but $(M_n(\mathbb{K}), \cdot)$ is not. The problem with $(M_n(\mathbb{K}), \cdot)$ is that there are many matrices which are not invertible. This leads us to consider

$$GL_n(\mathbb{K}) = \{M \in M_n(\mathbb{K}); M \text{ is invertible}\} = \{M \in M_n(\mathbb{K}); \det(M) \neq 0\}.$$

We will call $GL_n(\mathbb{K})$ the general linear group (with entries in $\mathbb{K}$).

Exercise 2.18. Prove the following claims. As usual, $\mathbb{K} = \mathbb{C}, \mathbb{R}$, or $\mathbb{Q}$.
(a) $GL_n(\mathbb{K})$ is an infinite multiplicative group.
(b) $GL_n(\mathbb{K})$ is not Abelian unless $n = 1$.
(c) $GL_n(\mathbb{K})$ is not an additive group.
(d) For any $c \in \mathbb{K} \setminus \{0\}$, there is an $M_c \in GL_n(\mathbb{K})$ such that $\det(M_c) = c$.
(e) The set

$$SL_n(\mathbb{K}) = \{M \in GL_n(\mathbb{K}); \det(M) = 1\}$$

is a group under the operation of $GL_n(\mathbb{K})$.
We call this group *the special linear group*.

Exercise 2.19. Label the corners of an equilateral triangle using 1, 2, and 3. Now consider the set of (rigid) motions of the triangle that move corners to corners (and leave the shape, and position, of the triangle intact). This set of rigid motions is denoted D_3.
(a) Find the 6 elements in D_3.
(b) Let m_1 and m_2 be two elements in D_3 and let $m_1 * m_2$ be the rigid motion of the equilateral triangle obtained by performing the motion m_2 and then follow it by the motion m_1.
Prove that D_3 is a non-Abelian group under this operation.
(c) Repeat this process with a square, and find the 8 elements in D_4 (rigid motions of the square now), and prove that D_4 is a non-Abelian group.

This construction can be repeated with any regular polygon to get D_n: the group of rigid motions of a regular n-gon.
(d) Find $2n$ elements in D_n (n rotations and n reflections), assume there are no other elements and convince yourself of the fact that D_n is a non-Abelian group.

Definition 2.20. The group D_n defined in Exercise 2.19 is called the dihedral group of order $2n$.

Exercise 2.21. Let $\mathbb{Q}(\sqrt{2}) = \{a + b\sqrt{2}; a, b \in \mathbb{Q}\}$, and let $+$ and $\cdot$ be the standard addition and multiplication in $\mathbb{R}$.
(a) Show that $\mathbb{Q} \subseteq \mathbb{Q}(\sqrt{2}) \subseteq \mathbb{R}$.
(b) Show that $(\mathbb{Q}(\sqrt{2}), +)$ is an Abelian group.
(c) Show that $(\mathbb{Q}(\sqrt{2}) \setminus \{0\}, \cdot)$ is an Abelian group.
(d) Show that $(\mathbb{Q}(\sqrt{2}) \setminus \{-1\}, *)$, where $a * b = a + b + ab$, is an Abelian group.

The following two exercises are similar, in that both of them construct groups of functions over groups. It is very important to understand these functions, as they are essential tools in the study of groups.

Exercise 2.22. Let G be a group. For any fixed $g \in G$ define $\varphi_g : G \to G$ by $\varphi_g(x) = gx$.
(a) Show that φ_g is bijective for all $g \in G$.
(b) Show that $S = \{\varphi_g; g \in G\}$ is a group under composition.
(c) Find necessary and sufficient conditions for S to be Abelian. Prove your claim.

Exercise 2.23. Let G be a group, and let g be a fixed element in G. Consider the function $\phi_g : G \to G$ defined by $\phi_g(x) = gxg^{-1}$.
(a) Show that ϕ_g is bijective for all $g \in G$.
(b) Show that $S = \{\phi_g; g \in G\}$ is a group under composition.
(c) Find, with proof, sufficient conditions for S to be Abelian.

2.2 The Basics of Group Theory

Definition 2.24. Let $(G, *)$ be a group and let g be a fixed element in G. Consider the set

$$<g> = \{g^i; i \in \mathbb{Z}\}$$

where, for each $i \in \mathbb{N}$, g^i denotes $\underbrace{g * g * \cdots * g}_{i \text{ times}}$, g^{-i} denotes the inverse of g^i, and g^0 denotes e_G.
$<g>$ is called *the group generated by g*.

Exercise 2.25. Let G be a group and let g be any element in G. With the notation as in Definition 2.24, show that:
(a) $<g>$ is a subset of G (not necessarily proper).
(b) $<g>$ is a group with the same operation of G.
(c) The rules of exponents are valid in $<g>$ (and thus in G). That is, show that

$$g^n g^m = g^{n+m} \qquad\qquad (g^n)^m = g^{nm}$$

for all $n, m \in \mathbb{Z}$.
(d) If G is finite then so is $<g>$.

Exercise 2.26. With the notation as in Definition 2.24, find explicitly what $<g>$ is in the group G, where:
(a) $g = e_G$ and G is any group.
(b) $g = 5$ and $G = \mathbb{Z}$.
(c) $g = 1$ and G is the group described in Exercise 2.21 part (d).
(d) $g = i$ and $G = Q_8$.
(e) $g = i$ and $G = \mathbb{C}$.
(f) $g = i$ and $G = \mathbb{C} \setminus \{0\}$.
(g) $g = R$ and $G = D_3$.
(h) g is the function $f(x) = x^3$ and $G = \mathscr{F}_\mathbb{R}$.
(i) g is the function $g(x) = \dfrac{x}{x-1}$ and $G = \mathscr{F}_{\mathbb{R} \setminus \{1\}}$.
(j) $g = \begin{bmatrix} 1 & 1 \\ 0 & 1 \end{bmatrix}$ and $G = SL_n(\mathbb{R})$.
(k) $g = \begin{bmatrix} 0 & 1 \\ 1 & 0 \end{bmatrix}$ and $G = GL_n(\mathbb{R})$.
(l) g is the function $g(x) = \cos(\pi x)$ and $G = F(\mathbb{Z}, \mathbb{R} \setminus \{0\})$ (see Exercise 2.14).

Definition 2.27. If a group G contains an element g such that $G = <g>$, then G is said to be *cyclic generated by g*.

Exercise 2.28. Prove the following claims:
(a) If G is cyclic then it is Abelian.
(b) $\mathbb{Q} \setminus \{0\}$ is Abelian but not cyclic.
(c) $\mathbb{Z}$ is cyclic. It can be generated by either 1 or -1.
(d) $\mathbb{Z}$ cannot be generated by any element different from ± 1. Which group does $n \neq \pm 1$ generate?

The next exercise provides examples of cyclic groups containing any number of elements one might want. In order to be able to work with these groups you should feel comfortable working with complex numbers (Subsection 1.2.5).

Exercise 2.29. Let $G = \mathbb{C} \setminus \{0\}$. Recall that G is a group under the standard multiplication in $\mathbb{C}$.
(a) Show that the unit circle is a group, contained in G, under the operation in G.
(b) Let $n \in \mathbb{N}$. Consider $H_n = <g_n>$, where $g_n = e^{2\pi i/n}$. Prove that H_n is a cyclic group containing exactly n elements.

An important family of cyclic groups we want to understand well is the additive version of the groups H_n in Exercise 2.29. We will now construct these groups and study their properties.

2.3 $\mathbb{Z}_n$

Let $n \in \mathbb{N}$ be given, and consider the relation $\sim_n$ in $\mathbb{Z}$ defined by $a \sim_n b$ if and only if $n \mid (b-a)$.

Exercise 2.30. Fix $n \in \mathbb{Z}, n > 1$.
(a) Prove that $\sim_n$ is an equivalence relation on $\mathbb{Z}$.
(b) Show that there are exactly n equivalence classes.
(c) Let $a \in \mathbb{Z}$. Show that $[a] = [r]$, where r is the remainder of a when dividing by n.

Definition 2.31. Let $n \in \mathbb{Z}, n > 1$. We define the set $\mathbb{Z}_n$ as the classes of the n possible remainders (after division by n). That is,
$$\mathbb{Z}_n = \{[0], [1], [2], \cdots, [n-2], [n-1]\}$$

Now we need to define operations in this set. Since $\mathbb{Z}_n$ is constructed from elements in $\mathbb{Z}$ it makes sense to mimic the operations in the integers when defining the operations in $\mathbb{Z}_n$.

Example 2.32. Let us consider $n = 6$ and take two elements in the class of $[5]$, for instance, 5 and 11. Now take two elements in the class of $[3]$, for instance -3 and 15. Observe the following operations between an element of $[5]$ and an element of $[3]$:

$$5 + (-3) = 2 \qquad 5 + 15 = 20 \qquad 11 + (-3) = 8 \qquad 11 + 15 = 26$$

The results are $2, 20, 8$ and 26, which are all in the class of $[2]$. Let us generalize this by considering the sum of an element $6a + 5$ in $[5]$ and any element $6b + 3$ in $[3]$. The result is $6a + 5 + 6b + 3 = 6(a + b + 1) + 2$, which always lives in $[2]$. Hence, it makes sense to conclude that $[5] + [3] = [2]$.
Similarly, we can see that

$$(6a+5)(6b+3) = 36ab + 18a + 30b + 15 = 6(6ab + 3a + 5b + 2) + 3$$

which is an element of $[3]$. Hence, we conclude that $[5] \cdot [3] = [3]$.
Note that all this is consistent with defining the operations modulo n.

We now generalize the addition and multiplication rules obtained for $\mathbb{Z}_6$ to $\mathbb{Z}_n$ as follows

$$\forall\, [a], [b] \in \mathbb{Z}_n, \qquad [a] + [b] = [a+b] \qquad \text{and} \qquad [a] \cdot [b] = [ab]$$

where the congruence classes are now modulo n.

Exercise 2.33. Show that the operations defined above are well-defined. That is, show that if

$$[a] = [b] \qquad \text{and} \qquad [c] = [d]$$

then

$$[a] + [c] = [b] + [d] \qquad \text{and} \qquad [a] \cdot [c] = [b] \cdot [d].$$

Exercise 2.34. Show that $(\mathbb{Z}_n, +)$ is a cyclic group of order n, but $(\mathbb{Z}_n, \cdot)$ is not a group.

The next theorem helps us understand which group is generated by an element in $\mathbb{Z}_n$.

Theorem 2.35. *Fix $n \in \mathbb{Z}, n > 1$, and let $k \in \mathbb{Z}$. The element $[k] \in \mathbb{Z}_n$ generates a group of order $\dfrac{n}{d}$, where $\gcd(k, n) = d$. Moreover, $<[k]> = <[d]>$.*

Exercise 2.36. In this exercise we will prove Theorem 2.35. Let $d = \gcd(k, n)$ and $H = <[k]>$.
(a) Prove that if $[x] \in H$, then $an + bk = x$, for some $a, b \in \mathbb{Z}$.
(b) Use Bézout's lemma (Theorem 1.22) to show that $[d] \in H$.

(c) Prove that d divides all numbers of the form $an+bk$, where $a,b \in \mathbb{Z}$.
(d) Conclude that $<[d]> = H$.
(e) Use that d divides n to prove $|H| = \dfrac{n}{d}$.

Exercise 2.37. Let $k,n \in \mathbb{Z}$.
(a) Prove that $\gcd(k,n) = 1$ if and only if $<[k]> = \mathbb{Z}_n$.
(b) Let p be an odd prime and $[k] \in \mathbb{Z}_p$. What can $<[k]>$ be?
(c) Find $<[k]>$ for all elements $[k] \in \mathbb{Z}_{12}$.

2.4 Orders of Groups and Elements.

Sometimes knowing information about the number of elements in a group might be useful. As an example, consider the next theorem.

Theorem 2.38. *All groups containing at most 4 elements are Abelian.*

Proof. Let G be a group containing n elements. Note that if $a,b \in G$ and $ab = e$, then $a^{-1}(ab) = a^{-1}$ (by multiplying by a^{-1} on the left), which forces $b = a^{-1}$. Multiplying by a on the right, we get $ba = a^{-1}a$. Hence, if $ab = e$ then $ba = e$.
We continue by considering the cases $n = 1,2,3,$ and 4.

- If $n = 1$, then $G = \{e_G\}$, which is Abelian.
- If $n = 2$ then $G = \{e,a\}$. Since $a^2 \in G$ then either $a^2 = e$ or $a^2 = a$. The latter case is impossible, as Exercise 2.7 would imply $a = e$. Hence, $a^2 = e$ and thus $G = <a>$, which is cyclic, and thus Abelian.
- If $n = 3$ then $G = \{e,a,b\}$. Then we have three possibilities for ab: if $ab = a$ then multiplying both sides (on the left) by a^{-1} we get $b = e$, a contradiction. Similarly, assuming $ab = b$ yields a contradiction. It follows that $ab = e$ and thus $b = a^{-1}$. Hence, $G = <a>$, which is Abelian.
- If $n = 4$. then $G = \{e,a,b,c\}$. By contradiction, suppose that $ab \neq ba$. Since ab (and ba) can be neither a nor b (otherwise we get a contradiction) then we must have $ab = e$ and $ba = c$ (or vice versa), which is a contradiction.

The proof is complete. $\square$

The previous theorem leads us to the following definition.

Definition 2.39. Let G be a group.
(a) The order of G, denoted by $|G|$, is the cardinality of the set G.
(b) The order of $g \in G$, denoted by $|g|$, is the least positive exponent k (if it exists) such that $g^k = e_G$. In case this number does not exist, we say that the order of g is infinite.

Exercise 2.40.
(a) Find the orders of the groups and elements given in previous exercises and examples.
(b) Prove by example that there are cyclic groups of any finite order, and of infinite order as well.
(c) Do all elements in an infinite group have infinite order?

Exercise 2.41. Prove the following claims.
(a) All nonzero elements in $\mathbb{Z}$ have infinite order.
(b) The only two elements in $(\mathbb{Q} \setminus \{0\}, \cdot)$ with finite order are ± 1. In fact, 1 has order one and -1 has order two.
(c) The order of an element is a quantity that depends on the operation of the group containing the element, not solely on the element.
(d) The elements in $(\mathbb{C} \setminus \{0\}, \cdot)$ with finite order live on the unit circle. Describe all these elements.
(e) Does every element in $(\mathbb{C} \setminus \{0\}, \cdot)$ on the unit circle have finite order?

Exercise 2.42. Let G be a group. Prove the following claims.
(a) $|g^{-1}| = |g|$, for all $g \in G$.
(b) $|hgh^{-1}| = |g|$, for all $g, h \in G$.
Hint: Exercise 2.7.
(c) $|g| = |<g>|$, for all $g \in G$.
(d) If $|G| = n < \infty$ and G contains an element of order n then G is cyclic. Is this result true if $n = \infty$?
(e) Prove that every element in a finite group has finite order.

The following theorem is often used to study orders of elements.

Theorem 2.43. *Let g be an element of finite order in a group G and assume $g^x = e$, for some $x \in \mathbb{Z}$. Then, $|g|$ divides x.*

Exercise 2.44. The objective of this exercise is to prove Theorem 2.43. Assume all notation and hypotheses given in Theorem 2.43.
(a) Show that WLOG we can assume $x \in \mathbb{N}$.
(b) Show that $|g| \leq x$.
(c) Using the division theorem (Theorem 1.18) we get $x = |g|q + r$, where $q, r \in \mathbb{Z}$ and $0 \leq r < |g|$. Prove Theorem 2.43 by contradicting the definition of $|g|$ if $r \neq 0$.

Exercise 2.45.
(a) Is the converse of Theorem 2.43 true?
(b) Prove that if $|g|$ is even, then $|g^2| = \dfrac{|g|}{2}$, and that if $|g|$ is odd, then $|g^2| = |g|$.
(c) Generalize part (b) by finding $|g^p|$, where p is any prime.

2.5 Subgroups.

We have seen in several examples/exercises (for instance, Exercises 2.25 and 2.29) that there are groups whose operation is 'inherited' from a larger group. This idea is generalized in the following definition.

Definition 2.46. Let $(G, *)$ be a group. If $H \subseteq G$ and $(H, *)$ is a group then H is said to be a subgroup of G, denoted $H \leq G$.

Theorem 2.47 (Subgroup Test). *Let G be a group and $\emptyset \neq H \subseteq G$. $H \leq G$ if and only if*
(a) $e_G \in H$, and
(b) $gh^{-1} \in H$, for all $g, h \in H$.

Exercise 2.48. Prove Theorem 2.47.
Hint: Remark 2.8.

Remark 2.49. Exercise 2.25 gives us a way to find subgroups in any given group G by just considering $<g>$ for any element $g \in G$.

Exercise 2.50. Prove the following statements.
(a) Let G be any group. $\{e_G\}$ and G are subgroups of G.
We will say that these subgroups of G are *trivial* (many authors use the word trivial only to refer to $\{e_G\}$).
(b) The set of all multiples of a given integer is a subgroup of $\mathbb{Z}$.
(c) If $k, n \in \mathbb{N}$ are such that $k \mid n$, then $H_k \leq H_n$ (defined in Exercise 2.29).
(d) If G is an Abelian group and $H \leq G$ then H is Abelian. Is the converse true?
(e) If H and K are subgroups of G, then $(H \cap K) \leq G$.
(f) $SL_n(\mathbb{K}) \leq GL_n(\mathbb{K})$, where $\mathbb{K} = \mathbb{Q}, \mathbb{R}, \mathbb{C}$.
(g) $H = \{f \in \mathscr{F}_\mathbb{R}; f(0) = 0\}$ is a subgroup of $\mathscr{F}_\mathbb{R}$ (see Exercise 2.14).

2.5 Subgroups.

Theorem 2.51. *If G is a cyclic group and $H \leq G$ then H is also cyclic.*

Exercise 2.52. The objective of this exercise is to prove Theorem 2.51. Let $G = <g>$.
(a) Use Axiom 1 in page 4 to prove that there exist a smallest positive integer x such that $g^x \in H$.
(b) Take any element $h \in H$ and write it as g^y, for some $y \in \mathbb{Z}$. Use the ideas in the proof of Theorem 2.43 to prove that x must divide y.
(c) Conclude that H is cyclic by showing that $H = <g^x>$.

Exercise 2.53. Give a counterexample to the following statement:

$$\text{If } H \leq G \text{ and } H \text{ is cyclic, then } G \text{ must be cyclic.}$$

In Exercises 2.28 and 2.41 you proved that $(\mathbb{Z}, +)$ is a cyclic group. Moreover, Theorem 2.51 characterizes all subgroups of $\mathbb{Z}$ to be of the form $<n>$ for some $n \in \mathbb{N}$. In the next exercises we use these facts to learn a little more about the algebraic structure of $\mathbb{Z}$.

Exercise 2.54.
(a) Prove that $\{x \in \mathbb{Z};\ x = 6n + 8m \text{ for some } m, n \in \mathbb{Z}\} = <2>$.
Hint: Bézout's lemma.
(b) Let $a, b \in \mathbb{Z}$ be fixed. Prove that $S = \{x \in \mathbb{Z};\ x = am + bn \text{ for some } m, n \in \mathbb{Z}\}$ is a subgroup of $\mathbb{Z}$. Moreover, $S = <\gcd(a,b)>$.

Exercise 2.55. Let H and K be two non-trivial subgroups of $\mathbb{Z}$.
(a) Prove that $a | b$ if and only if $<b> \subseteq <a>$.
(b) Prove that $<a> = <b>$ if and only if $a = \pm b$.
(c) Describe in detail what $H \cap K$ is.
Hint: Exercise 2.50.
(d) Give an example of two subgroups of $\mathbb{Z}$ such that their union is not a subgroup of $\mathbb{Z}$.
(e) What is the smallest subgroup of $\mathbb{Z}$ that contains both H and K simultaneously?
(f) Construct an infinite sequence $H_1, H_2, \cdots$ of subgroups of $\mathbb{Z}$ (all distinct) such that

$$\cdots \subsetneq H_3 \subsetneq H_2 \subsetneq H_1 \subsetneq \mathbb{Z}$$

(g) Prove that there is no infinite sequence $H_1, H_2, \cdots$ of subgroups of $\mathbb{Z}$ (all distinct) such that

$$H_1 \subsetneq H_2 \subsetneq H_3 \subsetneq \cdots$$

Now that we understand the subgroups of $\mathbb{Z}$, the obvious next step is to study the subgroups of $\mathbb{Z}_n$. In order to do this we look back at Theorem 2.35 and realize that it gives an explicit description of what the subgroups of $\mathbb{Z}_n$ look like.

Exercise 2.56.
(a) Let p be an odd prime. Find all subgroups of $\mathbb{Z}_p$.
Hint: Exercise 2.37.
(b) Find all subgroups of $\mathbb{Z}_{24}$.
(c) Let $n \in \mathbb{N}$. Prove that $\mathbb{Z}_n$ contains subgroups of order k, for all $k \in \mathbb{N}$ such that $k | n$.
(d) Let $k, n \in \mathbb{N}$ be such that $k | n$. How many subgroups of order k does $\mathbb{Z}_n$ have?
(e) Let $n \in \mathbb{N}$. Does $\mathbb{Z}_n$ have any subgroups of order k if $k \nmid n$?
(f) Find G_k (defined in Exercise 2.15) for several different values of k and $G = \mathbb{Z}_{24}$. Repeat this problem for $G = \mathbb{Z}_n$. Any comments?

2.6 Cosets.

We move on to a concept that is more general than that of subgroup.

Definition 2.57. Let H be a subgroup of G, and let $g \in G$. The set

$$gH = \{gh;\ h \in H\}$$

is called the (left) coset of H with representative g. The (right) coset Hg is similarly defined. Whenever we say coset we mean *left* coset.
Also, for $g, k \in G$ we define

$$gHk = \{ghk;\ h \in H\}.$$

Remark 2.58. For any fixed $g \in G$, define $\phi_g : G \to G$ by $\phi_g(x) = gx$. It follows that $\phi_g(H) = gH$, for all $H \leq G$.

Exercise 2.59. Prove that for any fixed $g \in G$, the function ϕ_g defined above is bijective.

Exercise 2.60. Let G be a group, $H \leq G$, and $g, k \in G$. Prove the following:
(a) $g \in gH$.
(b) $|gH| = |kH| = |H| = |Hg|$.
(c) $(gk)H = g(kH)$.
(d) $(gH)k = g(Hk)$.
(e) $gH = Hk$ if and only if $gHk^{-1} = H$.

Exercise 2.61. Let G be a group and let $H \leq G$. Prove the following claims.
(a) The only coset of H in G which is a group is H itself.
(b) If $h \in H$ then $hH = H$. If $h \notin H$ then hH and H are disjoint.
(c) Let $k = gh$ for some $k, g, h \in G$. If $h \in H$ then $kH = gH$. If $h \notin H$ then kH and gH are disjoint.
(d) $gH = kH$ if and only if $k^{-1}g \in H$.

Remark 2.62. Exercises 2.60 part (a) and 2.61 part (b) justify that gH is sometimes called *the* coset of H in G containing g.
Also, Exercise 2.61 implies that the set of all cosets of a given subgroup of G is a partition of G.

Most of the times, partitions are obtained when working with equivalence classes. Now we have obtained a partition by considering cosets. It makes sense to wonder if there is any relation between these two approaches. The next exercise makes the connection between cosets and equivalence classes clear.

Exercise 2.63. Let G be a group and let $H \leq G$. Define a relation on G by $g \sim h$ if and only if $gH = hH$.
(a) Prove that $\sim$ is an equivalence relation.
(b) Prove that the equivalence classes of $\sim$ are exactly the cosets of H in G.

Exercise 2.64. Find (explicitly, if possible) all cosets of H in G, where:
(a) $H = <[18]>$ and $G = \mathbb{Z}_{24}$.
(b) $H = <6>$ and $G = \mathbb{Z}$.
(c) $H = <-1>$ and $G = Q_8$.
(d) $H = <R>$ and $G = D_3$.
(e) $H = SL_n(\mathbb{R})$ and $G = GL_n(\mathbb{R})$.
Hint: Prove that $det(M) = det(N)$ if and only if $M = NA$, for some $A \in SL_n(K)$.
(f) $H = \{f \in \mathscr{F}_{\mathbb{R}};\ f(0) = 0\}$ and $G = \mathscr{F}_{\mathbb{R}}$.
Hint: Exercise 2.50, and prove that $f(0) = g(0)$ if and only if $g = fh$, for some $h \in H$.
(g) $H = \{e_G\}$ and G is any group.
(h) $H = G$ and G is any group.

Definition 2.65. Let G be a group and let $H \leq G$. The number (possibly infinite) of cosets of H in G is called the index of H in G, and it is denoted by $[G : H]$.

Exercise 2.66.
(a) Find $[G:H]$ for all the pairs $\{H,G\}$ in Exercise 2.64.
(b) Prove that for every k dividing n, $\mathbb{Z}_n$ has a subgroup H_k such that $[\mathbb{Z}_n : H_k] = k$.

Theorem 2.67 (Lagrange's theorem). *If G be a finite group and $H \leq G$, then*

$$[G:H] = \frac{|G|}{|H|}.$$

In particular, $|H|$ divides $|G|$.

Exercise 2.68. Prove Lagrange's theorem.
Hint: Exercise 2.60.

Remark 2.69. It is known that the converse of Lagrange's theorem is true for finite Abelian groups. However, in Section 2.7 we will get a counterexample to such converse as a corollary of Theorem 2.97.
It is also known that if p is prime and $|G| = p^a m$, where $\gcd(p,m) = 1$, then G always contains subgroups of order p^i, for all $i = 1, 2, \cdots, a$ (this is one of the famous Sylow Theorems).

Exercise 2.70. Prove that the converse of Lagrange's theorem is true if $G = \mathbb{Z}_n$. That is, prove that if $k|n$ then there is a subgroup of $\mathbb{Z}_n$ of order k.

Theorem 2.71. *If G is a finite group, and $g \in G$, then $|g|$ divides $|G|$.*

Exercise 2.72. Prove the previous theorem as a corollary of Lagrange's theorem.

Exercise 2.73. Let G be a group.
(a) Let $H, K \leq G$. Prove that if $\gcd(|H|, |K|) = 1$, then $H \cap K = \{e\}$.
(b) Assume that $|G| = 30$ and that $g \in G$. Find $|g|$ if $g \neq e$ and $g^9 = e$.
Hint: Theorem 2.43.

Theorem 2.74. *Every group of prime order is cyclic.*

Exercise 2.75. Prove Theorem 2.74.
Hint: Exercise 2.42.

In Definition 2.57 we introduced sets of the form gHk. In later sections, sets of the form gHg^{-1}, for $g \in G$ and $H \leq G$ will be of most importance. The following exercises will help us prepare for that.

Exercise 2.76. Assume G is a group, $g \in G$ and $H, K \leq G$. Prove the following.
(a) $gHg^{-1} \leq G$.
(b) $|H| = |gHg^{-1}|$.
(c) $gH = Hg$ if and only if $gHg^{-1} = H$.
(d) $|ghg^{-1}| = |h|$, for all $h \in G$. Do this by proving that if $H = <h>$ then $gHg^{-1} = <ghg^{-1}>$.
(e) If $M \in GL_n(\mathbb{K})$, then $MSL_n(\mathbb{K})M^{-1} = SL_n(\mathbb{K})$.

2.7 Permutations

In Exercise 2.14 we learned that for any fixed set A, the set

$$S_A = \{f : A \to A; \ f \text{ is a bijective function}\}$$

is a group under the operation $\circ$ (composition). In this section we will focus on the special case when A is a finite set.

Exercise 2.77. Give an example of a set A such that S_A is not Abelian. Then give an example of a set B such that S_B is Abelian. Prove your claims.

Definition 2.78. If A contains a finite number of elements, say $|A| = n$, we identify A with the set $\{1, 2, 3, \cdots, n\}$. In this case S_A is denoted S_n and it is called *the symmetric group in n objects*. The elements of S_n are called *permutations*.

Exercise 2.79. Prove that the order of S_n is $n!$.
Hint: Use the fundamental counting principle to decide how many choices there are for the images of the elements 1, 2, etc.

There are, pretty much, two ways of representing permutations. For example, the function $\sigma \in S_4$ mapping

$$\sigma : \begin{array}{c} 1 \mapsto 4 \\ 2 \mapsto 1 \\ 3 \mapsto 2 \\ 4 \mapsto 3 \end{array}$$

can be written as

$$\sigma = \begin{pmatrix} 1 & 2 & 3 & 4 \\ 4 & 1 & 2 & 3 \end{pmatrix},$$

where each element in the bottom row is the image under σ of the element that is right above it. The problem with this notation is that it might take too long to write a function in, for example, S_{100}. So, we use instead the following way to represent the same permutation:

$$\sigma = (1432)$$

where each element is mapped to the element directly to their right, and the last element is sent to the first one. So, in this example, 1 is mapped to 4, 4 is mapped to 3, 3 is mapped to 2 and finally 2 gets mapped to 1. One needs to be careful with the second way of writing permutations when the function fixes a number. For example, the function

$$\tau : \begin{array}{c} 1 \mapsto 1 \\ 2 \mapsto 4 \\ 3 \mapsto 3 \\ 4 \mapsto 2 \end{array}$$

is represented by $\tau = (24)$ (implying that 1 and 3 are fixed). The identity function (which fixes everything) will be denoted by the symbol e.

By definition, permutations are functions and the operation in S_n is function composition. On the other hand, the second way to write permutations gives permutations a 'non-function' look. As expected these two approaches will, at the beginning, create some problems. For instance, the product of permutations might, at the beginning, appear to be a little counterintuitive, even backwards. However, the idea is simple: one has to keep track of where all numbers are being mapped by the functions to determine where it is sent by the composition. For example

$$(153)(43)(5312)(15)(24) = (14352)$$

because 1 is fixed by (24), then it is mapped to 5 by (15), 5 is sent to 3 by (5312), 3 is sent to 4 by (43), and finally 4 is fixed by (153), thus 1 is sent to 4. The other entries for the permutation can be obtained similarly.

Exercise 2.80. In this exercise all permutations live in S_n, for some large n.
(a) Compute the following, and write your answer as a product of disjoint cycles.

(i) $(123)^4$ (ii) $(123)^{1000000}$ (iii) $(45)^5$ (iv) $[(123)(45)]^{1000000}$ (v) $[(123)(34)]^{1000000}$

2.7 Permutations

(b) Find the order of the following permutations:

(i) $(123)(45)$ (ii) $(45)(123)(45)$ (iii) $(45)(34)$ (iv) $(123)(34)$

(v) (1234) (vi) $(1234)^2$ (vii) $(1234)^3$ (viii) $(1234)^{10000001}$

Exercise 2.81. Find the groups generated (in S_3) by

(a) (12) (b) (13) (c) (23) (d) (123) (e) (132)

Now use this to find the order, and the inverse, of every element in S_3.

Exercise 2.82.
(a) Prove that S_3 is a non-Abelian group of order 6.
(b) Can you identify elements in S_3 with elements in D_3 to get a perfect matching?
(c) Repeat part **(b)** by presenting the elements in D_4 as elements in S_4.
(d) Prove that G_2 is not a group for $G = S_3$ (G_2 was defined in Exercise 2.15).
(e) Prove that $S_k \leq S_n$ if and only if $k \leq n$.

Definition 2.83. Let $\sigma \in S_n$.
(a) If $\sigma = (a_1 a_2 \cdots a_k)$ then σ is said to be a k-cycle.
(b) A 2-cycle is also called a transposition.
(c) Let $\sigma = (a_1 a_2 \cdots a_k), \tau = (b_1 b_2 \cdots b_m) \in S_n$. If $a_i \neq b_j$ for all $i = 1, \ldots, k$ and $j = 1, \ldots, m$, then σ and τ are said to be disjoint.

Remark 2.84. A k-cycle may be written in many different ways. For instance,

$$(1234) = (2341) = (3412) = (4123)$$

Exercise 2.85. Prove the following claims for cycles in S_n.
(a) The order of a k-cycle is k.
(b) If σ and τ are two disjoint cycles then $|\sigma\tau| = lcm(|\sigma|, |\tau|)$.

Theorem 2.86. *Every permutation is the product of disjoint cycles.*

Proof. Let $\sigma \in S_n$. We define a relation on $A = \{1, 2, \ldots, n\}$ by $x \sim y$ if and only if $\sigma^k(x) = y$ for some $k \in \mathbb{Z}$. In Example 1.8 we proved that $\sim$ is an equivalence relation on A.
Now we consider the equivalence classes created by $\sim$ on A. For each class C choose an element a in C and notice that the elements in C are $a, \sigma(a), \sigma^2(a), \ldots$. It follows that the class C yields the cycle $\sigma_{[a]} = (a \; \sigma(a) \; \sigma^2(a) \cdots)$. Moreover, $\sigma(a) = \sigma_{[a]}(a)$.
Finally, we notice that the cycles constructed above are all disjoint (because their corresponding equivalent classes are disjoint), and thus they commute. It follows that σ is the product of all the cycles associated to the classes of $\sim$. □

Exercise 2.87. Prove the following claims for permutations in S_n.
(a) If σ and τ are two disjoint cycles then $\sigma\tau = \tau\sigma$.
(b) If $\sigma = (a_1 \; a_2 \; \cdots \; a_k), \tau \in S_n$, then $\tau\sigma\tau^{-1} = (\tau(a_1) \; \tau(a_2) \; \cdots \; \tau(a_k))$, where $\tau(a_i)$ is the image of a_i under τ. Note that both σ and $\tau\sigma\tau^{-1}$ are k-cycles.
(c) $(a_1 \; a_2 \; \cdots \; a_k) = (a_1 \; a_k) \cdots (a_1 \; a_3)(a_1 \; a_2) = (a_1 \; a_2)(a_2 \; a_3) \cdots (a_{k-1} \; a_k)$.
Conclude that every permutation can be written as a product of transpositions.
(d) Any permutation can be written as a product of the transpositions $(1 \; 2), (1 \; 3), \ldots, (1 \; n)$ (repetitions are allowed).
(e) Any transposition is a product of (12) and $(1 \; 2 \; 3 \; \cdots \; n)$ (repetitions are allowed).
Hint: Part **(b)** of this exercise.
(f) Any permutation is a product of (12) and $(1 \; 2 \; 3 \; \cdots \; n)$ (repetitions are allowed).

Exercise 2.88. Use Exercise 2.87 to solve the following problems:
(a) Find $\tau \in S_5$ such that $\tau(1234)\tau^{-1} = (1352)$. Can you find more than one such τ?
(b) Find $\tau \in S_6$ such that $\tau(351)(46)\tau^{-1} = (123)(45)$. Can you find more than one such τ?
(c) How would the answers to parts (a) and (b) change if τ is asked to be in S_{10}? Explain.

As mentioned in the Exercise 2.87, every permutation is the product of transpositions. It is easy to see that there are many ways to write the same permutation as a product of distinct sets of transpositions. Moreover, for a fixed permutation σ, the number of transpositions needed to obtain σ will always be either even or odd. We will prove this fact in the following lemmas and exercises.

Exercise 2.89. Check that the following identities hold in S_n ($n \geq 4$) for any $a, b, c,$ and d distinct elements in $\{1, 2, \cdots, n\}$.
(a) $(c\ d)(a\ b) = (a\ b)(c\ d)$.
(b) $(b\ c)(a\ b) = (a\ c)(c\ b)$.
(c) $(a\ c)(a\ b) = (a\ b)(b\ c)$.
(d) $(a\ b)(a\ b) = e$.

Note that in the identities above, the product on the right always has a only in the far left transposition, or a is cancelled out. Note also that the parity of the product does not change. Finally, observe that using these identities we can 'move' a from any product of transpositions all the way to the far left without leaving any a's in the other transpositions.

Lemma 2.90. *e cannot be written as a product of an odd number of transpositions.*

Proof. By contradiction, assume that e can be written as a product of an odd number of transpositions. Let k be the smallest odd number such that e as a product of k transpositions.
Consider any number a in a cycle of this product. We can move a to the left using the techniques proved in Exercise 2.89. However, we cannot have just one appearance of a in the transpositions used to write e, otherwise a would not be fixed. That means that a is cancelled out on its way to the left using an identity of the form $(ab)(ab) = e$. Since this diminishes the number of transpositions by 2 we have found a shorter product of an odd number of transpositions. This is a contradiction to the minimality of k. □

Theorem 2.91. *A product of an even number of transpositions cannot be written as the product of an odd number of transpositions.*

Exercise 2.92.
(a) Give an example of a permutation in S_6 that can be written as a product of k transpositions, for $k = 4, 6, 8, 10$.
(b) Prove Theorem 2.91.
Hint: Lemma 2.90.

Definition 2.93. $\sigma \in S_n$ is said to be an even permutation if it can be written as a product of an even number of transpositions. In case this is not possible, then σ is said to be odd. The identity is considered to be even.

Note that theorem 2.91 says that a permutation is either even or odd.

Definition 2.94. The set $A_n = \{\sigma \in S_n; \sigma \text{ is even}\}$ (which is a subgroup of S_n, see Exercise 2.95) is called the alternating group (in n elements).

Exercise 2.95. Prove the following claims about permutations in S_n.
(a) A k-cycle is even if and only if k is odd.
(b) A_n is a subgroup of S_n.
(c) Let $\sigma \in S_n$, then $\sigma A_n \sigma^{-1} = A_n$.
Hint: Exercise 2.76.
(d) Prove that $V = \{e, (12)(34), (13)(24), (14)(23)\} \leq A_4$, and that V is of order 4, Abelian but not cyclic.

2.7 Permutations

Now that we know that $A_n \leq S_n$, we move on to study the index of A_n in S_n. The following exercise accomplishes this task.

Exercise 2.96. Let $n \in \mathbb{N}$.
(a) Prove that the product of any two odd permutations in S_n is even.
(b) Prove that the product of an odd permutation in S_n with an even permutation in S_n is odd.
(c) Represent the set of all odd permutations in S_4 as σA_4 for a specific $\sigma \in S_4$. What is the set σT, where T is the set of all odd permutations in S_4?
(d) Fix $\tau \in S_n$ an odd permutation. Show that every element in S_n is either in A_n or it has the form $\tau \sigma$ with $\sigma \in A_n$.
(e) Prove that the set of all odd permutations in S_n is a coset of A_n.
(f) Conclude that $|A_n| = \dfrac{n!}{2}$.

The alternating groups are of great importance in group theory. For instance, the simplest example of a group for which the converse of Lagrange's theorem is false is A_4. In fact, the following theorem says that, even though 6 divides the order of A_4, there is no subgroup of A_4 of order 6.

Theorem 2.97. A_4 *does not have a subgroup of order 6.*

Proof (See [5]). By contradiction. Assume $[A_4 : H] = 2$. Then we have $A_4 = H \cup gH$ for all $g \in A_4 \setminus H$. Consider the coset $g^2 H$, which could be equal to either H or gH. If $g^2 H = gH$, then $gH = H$, which is a contradiction. It follows that $g^2 H = H$ for all $g \in A_4 \setminus H$. Now, since $h^2 H = H$ for all $h \in H$ then $g^2 H = H$ for all $g \in A_4$. Thus, H contains all the squares of elements in A_4.
We now look at squares of elements in A_4, we get at least the following 7 distinct elements

$$e = e^2 \qquad (132) = (123)^2 \qquad (123) = (132)^2 \qquad (142) = (124)^2$$

$$(124) = (142)^2 \qquad (134) = (143)^2 \qquad (143) = (134)^2$$

which contradicts $|H| = 6$. □

Exercise 2.98. For all parts of this problem $\sigma = (135)(24) \in S_5$ and $H = \{e, (12)(34), (13)(24), (14)(23)\}$.
(a) Find $|\sigma|$, σ^{-1}, σ^{2011}, and $<\sigma>$.
(b) Prove that $H \leq S_5$.
(c) Are $<\sigma>$ and/or H Abelian? Prove your claim.
(d) Describe explicitly the coset of H containing σ
(e) How many cosets does $<\sigma>$ have in S_5?
(f) Describe five distinct cosets of $<\sigma>$ in S_5. Explain how to find them all.
(g) Find all the subgroups of $<\sigma>$.
(h) Find all the subgroups of H.
(i) Write σ as a product of transpositions in two different ways.
(j) Find $\tau \in S_5$ such that $\tau \sigma \tau^{-1} = (123)(45)$. Can you find more than one such a τ in S_5? What about finding more such τ's in S_{10}?
(k) Describe the set of all elements in S_5 that can be written as $\tau \sigma \tau^{-1}$, for some $\tau \in S_5$.

One of the most important properties of the symmetric groups is that we can use them to represent complicated groups in an easy way. In fact, Cayley proved (Theorem 2.104) that every finite group can be thought of as a subgroup of S_n, for some . The following few results give a good example of this property.

In Exercise 2.19 and Definition 2.20, the dihedral groups are defined as the group of rigid motions of a regular n-gon. The following figures represent such a polygon for n odd and n even:

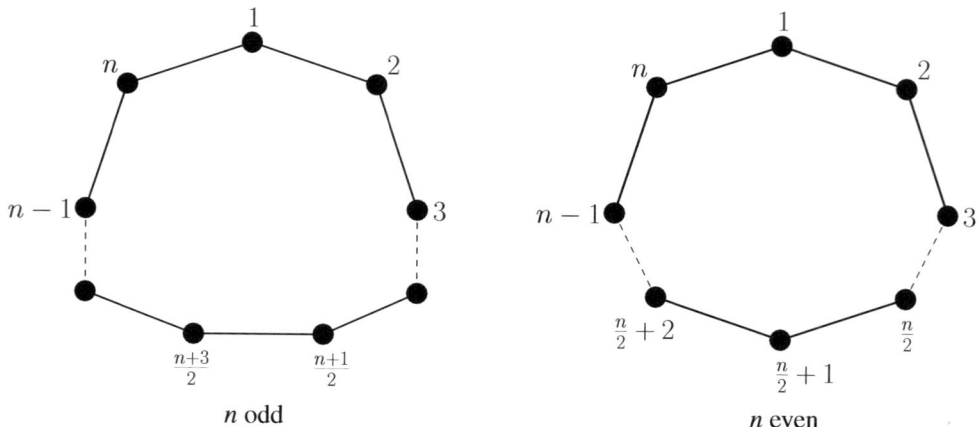

n odd n even

Let us give a different presentation for D_n using permutations. First we label the n vertices of a regular n-gon as in the figure above. Then a rotation by $\frac{2\pi}{n}$ radians can be represented as a function sending $1 \mapsto 2 \mapsto 3 \mapsto \cdots \mapsto n \mapsto 1$. This can be represented by the n-cycle $R = (1\ 2\ 3\ \cdots\ n)$. Hence, all the rotations of the n-gon correspond to the elements in $<R>$.

Since we know that $|D_n| = 2n$, and $|<R>| = n$, all the non-rotation elements (i.e. reflections) of D_n will be given by a rotation followed by a fixed reflection (we will call this fixed reflection F). Now we need to know what F looks like, as a permutation.

If n is odd, F must fix a vertex. WLOG we will assume F fixes 1. In this case F looks like:

$$F = (2\ n)(3\ n-1)\cdots \left(\frac{n+1}{2}\ \frac{n+3}{2}\right)$$

If n is even, then F fixes 1 and $\frac{n}{2}+1$, then

$$F = (2\ n)(3\ n-1)\cdots \left(\frac{n}{2}\ \frac{n}{2}+2\right)$$

So,

$$D_n = \{F^i R^j;\ i = 0, 1 \text{ and } j = 1, 2, \cdots, n\}$$

The previous presentation is very useful, as it allows us to avoid the geometry associated to the first definition of D_n. However, since D_n is a group, then the product of two elements of the form $F^i R^j$ should also be of that form, a fact that is not so obvious. The following exercise will give us the tools we need to perform multiplications in D_n with ease.

Exercise 2.99.
(a) What is the element in S_5 that corresponds to the clockwise rotation in 144° (which is an element of D_5)?
(b) Give an example of an element in S_5 that does not correspond to an element in D_5.

Exercise 2.100. Let R and F be as defined above.
(a) Understand the product FRF geometrically to conclude that FRF is a rotation.
(b) Prove that $FRF = R^{-1}$.
(c) Prove that $R^j F = F R^{-j}$, for all $j = 1, 2, \cdots, n$.

Remark 2.101. The equality $R^j F = F R^{-j}$ allows us to 'commute' powers of R and F. This was the last piece we needed to be able to perform multiplication in D_n. It follows that D_n can be presented as

2.8 Homomorphisms and Normal Subgroups

$$D_n = \{F^i R^j; \; F^2 = R^n = e, \; RF = FR^{-1}\}$$

Exercise 2.102.
(a) Compute the product $(RF)(R^2 F)(R^3 F)(R^4 F)$ in D_6 and write your answer in the form $F^i R^j$.
(b) Describe an algorithm to show that any product of R's and F's in D_6 can be written as $F^i R^j$ for some $i = 0, 1$ and $j = 1, 2, \cdots, 6$.
(c) Compute the following products in D_6:

(i) R^{1003} (ii) $R^{-2011} J^{100}$ (iii) RJR^{-1} (iv) $R^4 J^3 R^3 J$

(v) $R^{-2} JR^{-1} J$ (vi) $(RJR^{-1})^{10000}$ (vii) $RJR^2 JR^3 JR^4 J$

Exercise 2.103. Let $n > 4$. Prove that $FR^3 = R^{-3} F$, where R is the rotation by $\dfrac{2\pi}{n}$ and F is a flip (both in D_n).

The idea of being able to present groups as groups of permutations is justified by the following very important result. We give a proof here that should be re-read after finishing reading Section 2.8.

Theorem 2.104 (Cayley). *If G is a finite group of order n, then G can be considered to be a subgroup of S_n.*

Proof. Exercise 2.59 says that the function $\phi_g : G \to G$, defined by $\phi_g(x) = gx$, is bijective, for all $g \in G$. It follows that ϕ_g is an element of $\mathscr{F}_G$, for all $g \in G$. Since G has order n, then we can identify $\mathscr{F}_G$ with S_n. It follows that, under the identification $\mathscr{F}_G \leftrightarrow S_n$, we can think ϕ_g to be an element of S_n. Hence, the function $\phi : G \to S_n$, mapping g to ϕ_g, is well-defined, and is clearly injective.

(*What follows will make more sense after section 2.8*)

The function described above also preserves the operations in the groups G and S_n, as

$$\phi(gh)(x) = \phi_{gh}(x) = (gh)x = g(hx) = g(\phi_h(x)) = (\phi_g \circ \phi_h)(x)$$

for all $x \in G$. This means that the element gh corresponds to the element $\phi_g \circ \phi_h$. Hence, the group G is 'copied' into S_n, in a one-to-one way, by identifying an element g with the element ϕ_g. □

2.8 Homomorphisms and Normal Subgroups

In previous sections we have stumbled upon the idea of a group having different presentations. For instance, the original definition of D_n does not involve permutations, but later we did develop a way to work with elements in D_n as permutations. Similarly, in the proof of Cayley's theorem (Theorem 2.104) groups are being 'matched' with certain functions, which are also thought of as permutations of the elements in the group.
The following exercise gives an idea on how to connect the structures of two distinct groups by using a function.

Exercise 2.105. Consider the Abelian group $(\mathbb{R}, +)$ and the set of positive real numbers $\mathbb{R}_+$. The function $f : \mathbb{R} \to \mathbb{R}_+$, defined by $f(x) = e^x$ is bijective and has the property $f(x+y) = f(x)f(y)$ for all $x, y \in \mathbb{R}$.
(a) Use the operation in $(\mathbb{R}, +)$ and the exponential function to obtain an operation $*$ in $\mathbb{R}_+$ which makes it a group.
(b) Is there any relation between the identities of $(\mathbb{R}, +)$ and $(\mathbb{R}_+, *)$?
(c) Any relation between inverses in $(\mathbb{R}, +)$ and inverses in $(\mathbb{R}_+, *)$?

Next we will formalize all these ideas.

Definition 2.106. Let $(G, *)$ and $(H, \star)$ be groups.
(a) A function $f : G \to H$ such that $f(x * y) = f(x) \star f(y)$ is said to be a group homomorphism.
(b) A homomorphism of groups that is also bijective is called an isomorphism of groups.
(c) If there is an isomorphism $\phi : G \to H$ then the groups G and H are said to be isomorphic to each other, denoted $G \cong H$.

Exercise 2.107. Check that the following functions are homomorphisms of groups. Are they injective? Are they surjective?
(a) $\phi : \mathbb{Z}_{14} \to \mathbb{Z}_7$, defined by $\phi[x] = [3x]$.
(b) $\phi : \mathbb{Z} \to \mathbb{Z}_{18}$, defined by $\phi(x) = [x]$.
(c) $\phi : S_n \to \{\pm 1\}$, defined by $\phi(\sigma) = \begin{cases} 1 & \text{if } \sigma \text{ is even} \\ -1 & \text{if } \sigma \text{ is odd} \end{cases}$.
(d) The (determinant) function $\det : GL_n(\mathbb{C}) \to \mathbb{C} \setminus \{0\}$.
(e) The trace function $tr : M_n(\mathbb{C}) \to \mathbb{C}$ (see Definition 1.48 in page 12).

Groups are sets together with an operation. Considering functions between groups without regard to their operations would mean that we lose all algebraic properties of the groups. In order to avoid such situations, we study homomorphisms. Two groups that are isomorphic will be 'equal' in all senses, as an isomorphism defines a perfect matching between the elements of the groups which preserves all the operation-related properties of the group.

Exercise 2.108. Let G, H be groups and let $\phi : G \to H$ be a homomorphism. Prove the following:
(a) $\phi(e_G) = e_H$.
(b) $\phi(g^{-1}) = \phi(g)^{-1}$.
(c) $\phi(g^i) = \phi(g)^i$, for all $i \in \mathbb{Z}$.
(d) $|\phi(g)|$ divides $|g|$.
(e) If ϕ is injective then $|\phi(g)| = |g|$.
(f) If ϕ is surjective and $K \leq G$, then $\phi(K) \leq H$.
(g) If ϕ is an isomorphism and $K \leq G$, then $K \cong \phi(K)$.
(h) If ϕ is onto and G is Abelian then H is Abelian.
(i) If ϕ is onto and G is cyclic then H is cyclic.

Not only can properties of a group be 'moved' onto another group, sometimes properties of a group are equivalent to the existence of certain homomorphisms. The following exercise exemplifies this situation.

Exercise 2.109. Let G be a group. Prove that the following are equivalent:
(a) G is Abelian
(b) $\phi : G \to G$ defined by $\phi(x) = x^{-1}$ is a homomorphism.
(c) $\varphi : G \to G$ defined by $\varphi(x) = x^2$ is a homomorphism.

We have seen how well-behaved homomorphisms are respecting group structures. Now we will see that homomorphisms also behave well under composition.

Exercise 2.110. Let G, H, and K be groups, and $f : G \to H$ and $g : H \to K$ be group homomorphisms. Prove the following:
(a) $g \circ f$ is a group homomorphism.
(b) If f and g are group isomorphisms, then so are $g \circ f$, f^{-1} and g^{-1}.

Exercise 2.111. The objective of this exercise is to prove that the function ϕ used in the proof of Cayley's theorem (Theorem 2.104) is an injective homomorphism.
(a) Prove that ϕ is an injective homomorphism, in the particular case when $G = \mathbb{Z}_4$.
(b) Prove that ϕ is an injective homomorphism, in full generality.
(c) Conclude that G is isomorphic to a subgroup of S_n.

2.8 Homomorphisms and Normal Subgroups

Theorem 2.112. *If G is a cyclic group, then G is isomorphic to either $\mathbb{Z}$ or $\mathbb{Z}_n$, for some $n \in \mathbb{N}$.*

Exercise 2.113. The objective of this exercise is to prove Theorem 2.112. Let $G = <g>$.
(a) Assume that G is infinite. Prove that $\phi : G \to \mathbb{Z}$ defined by $\phi(g^i) = i$ is an isomorphism.
(b) Assume that G is finite of order n. Prove that $\phi : G \to \mathbb{Z}_n$ defined by $\phi(g^i) = [i]$ is an isomorphism.

Exercise 2.114. Let G be a group.
(a) Assume G is isomorphic to all its non-trivial cyclic subgroups. Prove that G is cyclic.
(b) Assume that G is isomorphic to all its non-trivial subgroups. Prove that G is isomorphic to either $\mathbb{Z}$ or $\mathbb{Z}_p$, for some prime p.

In order to prove that two groups are isomorphic we need to construct an isomorphism between them. On the other hand, in order to check that two groups are not isomorphic to each other we need to proceed by contradiction, and use properties like the ones proved in Exercise 2.108.

Exercise 2.115. Prove that $G \not\cong H$, for:

(a) $G = S_3$, $H = \mathbb{Z}_6$ (b) $G = D_4$, $H = \mathbb{Z}_8$ (c) $G = Q_8$, $H = \mathbb{Z}_8$

(d) $G = A_4$, $H = \mathbb{Z}_{12}$ (e) $G = Q_8$, $H = D_4$ (f) $G = \mathbb{Z}_{10}$, $H = D_5$

Next we introduce an important set that will help us work better with homomorphisms.

Definition 2.116. Let $\phi : G \to H$ be a homomorphism of groups. The kernel of ϕ is the set

$$Ker(\phi) = \{g \in G;\ \phi(g) = e\}.$$

Exercise 2.117. Find the kernel of every homomorphism in Exercise 2.107.

Exercise 2.118. In this exercise you will give examples of homomorphisms. Be creative!
(a) Construct two distinct homomorphisms $\phi, \psi : \mathbb{Z} \to \mathbb{Z}$ such that $Ker(\phi) \neq \{0\} \neq Ker(\psi)$.
(b) Construct a homomorphism $\phi : \mathbb{Z}_6 \to \mathbb{Z}_6$ such that $Ker(\phi) = <[2]>$. What is $\phi(\mathbb{Z}_6)$?
(c) Construct a non-trivial homomorphism $\phi : Q_8 \to D_4$. Find the kernel of such a homomorphism.

Exercise 2.119. The objective of this exercise is to see what one needs to do to find all homomorphisms between $\mathbb{Z}_n$ and $\mathbb{Z}_m$. We start by considering a homomorphism $\phi : \mathbb{Z}_{10} \to \mathbb{Z}_{12}$.
(a) Prove that ϕ is uniquely determined by what $\phi[1]$ is.
(b) Use Exercise 2.108 and Theorem 2.71 to find what could the order of $\phi[1]$ be.
(c) Use Theorem 2.35 to find what could $\phi[1]$ be.
(d) Generate an algorithm to find all homomorphisms from $\mathbb{Z}_n$ to $\mathbb{Z}_m$.
(e) Find all homomorphisms from $\mathbb{Z}_p$ to $\mathbb{Z}_{2p}$, where p is prime.

Exercise 2.120.
(a) Prove that $H = \{\sigma \in S_n;\ \sigma(n) = n,\ \sigma(n-1) = n-1\}$ is a subgroup of S_n.
(b) Show that H is the image of a homomorphism $\phi : S_m \to S_n$, for some $m \in \mathbb{N}$.
(c) Can you find the kernel of such a homomorphism?
(d) Prove that S_n contains a subgroup (namely $\phi(S_m)$) that is isomorphic to S_m.
(e) Conclude that if $r < t$ then S_r is 'contained' in S_t, in the same way $S_m \subseteq S_n$ in part (d).

Theorem 2.121. *Let $\phi : G \to H$ be a group homomorphism. Then, the following hold:*
(a) $Ker(\phi) \leq G$.
(b) ϕ is injective if and only if $Ker(\phi) = \{e\}$.

Exercise 2.122. Prove Theorem 2.121.

Exercise 2.123. Let G and H be groups, and $\phi : G \to H$ be a group homomorphism. Prove the following:
(a) $\phi(g) = \phi(h)$ if and only if g and h belong to the same coset of $Ker(\phi)$.
(b) $gKer(\phi)g^{-1} = Ker(\phi)$ for all $g \in G$.

In the previous exercise we encountered an expression of the form gHg^{-1}. It is time to study why expressions of this type are so ubiquitous.

Exercise 2.124. Let G be a group and fix $g \in G$. Define $\varphi_g : G \to G$ by $\varphi_g(x) = gxg^{-1}$ (called the *conjugation by g function*).
(a) Prove that φ_g is an isomorphism for $G = D_4$ and g the reflection represented (as an element in S_4) by (13).
(b) Prove that φ_g is an isomorphism, for any group G and element $g \in G$.
Hint: Exercise 2.23.
(c) Prove that if $H \leq G$ then $gHg^{-1} \leq G$.

Just as we did in Exercise 2.23, where we constructed a group of functions over a group, we will now construct a group of homomorphisms over a group.

Exercise 2.125. Let G be a group.
(a) Show that the set
$$Aut(G) = \{\phi : G \to G;\ \phi \text{ is an isomorphism}\}$$
is a group under composition.
(b) Consider the set
$$Inn(G) = \{\varphi_g \in Aut(G);\ g \in G\},$$
where $\varphi_g(x) = gxg^{-1}$, for all $x \in G$. Show that $Inn(G) \leq Aut(G)$.
(c) Is the set
$$Hom(G) = \{\phi : G \to G;\ \phi \text{ is a homomorphism}\}$$
a group under composition?

Definition 2.126. Let G be a group. If $H \leq G$ and $gHg^{-1} = H$ for all $g \in G$, then H is said to be a normal subgroup of G. If H is a normal subgroup of G, we will write $H \trianglelefteq G$.

Remark 2.127. Exercise 2.123 says that if $\phi : G \to H$ is a group homomorphism then $Ker(\phi)$ is normal in G. This fact could be used to prove normality without going through the definition of a normal subgroup.
You are encouraged to look back at previous exercises (such as Exercise 2.117) and obtain your own examples of normal subgroups.

Exercise 2.128. Prove the following claims.
(a) $A_n \trianglelefteq S_n$.
(b) $SL_n(\mathbb{K}) \trianglelefteq GL_n(\mathbb{K})$, where $\mathbb{K} = \mathbb{C}, \mathbb{R}, \mathbb{Q}$, or $\mathbb{Z}$.
(c) All subgroups of Q_8 are normal in Q_8.
(d) Consider $V = \{e, (12)(34), (13)(24), (14)(23)\}$. Show that $V \trianglelefteq A_4$, $V \trianglelefteq S_4$, and $V \trianglelefteq D_4$ (where we represent the elements of D_4 as elements in S_4).
Hint: Exercise 2.95.
(e) $D_3 \trianglelefteq D_6$ (think about these groups as subgroups of S_6).
(f) $S_k \ntrianglelefteq S_n$, for all $1 < k < n$.

Exercise 2.129. Let G, H, K be groups. Prove the following claims.
(a) If H and K are normal subgroups of G, then $H \cap K \trianglelefteq G$.
(b) If $H \leq K \leq G$ and $H \trianglelefteq G$, then $H \trianglelefteq K$.
(c) If $H \leq G$ and $K \trianglelefteq G$, then $H \cap K \trianglelefteq H$.
(d) If $H \leq G$ and G is Abelian, then $H \trianglelefteq G$.

Sometimes checking normality is not easy. Hence, we need results that could help us to do this. Such results follow.

Lemma 2.130. *Let G be a group and $H \leq G$. $H \trianglelefteq G$ if and only if every left coset of H is a right coset of H.*

Proof. Assume $H \trianglelefteq G$. Then, by Exercise 2.76, we get that $gH = Hg$, for all $g \in G$.
Now assume that for every $g \in G$ there is a $k \in G$ such that $gH = Hk$. Since $g \in gH$, then $g \in Hk$. It follows that $Hk = Hg$, implying $gH = Hg$, for all $g \in G$. Exercise 2.76 finishes the proof. $\square$

Theorem 2.131. *Let G be a group and $H \leq G$. If $[G:H] = 2$ then $H \trianglelefteq G$.*

Exercise 2.132. Let G be a group and $H \leq G$.
(a) Prove Theorem 2.131 as a corollary of Lemma 2.130.
(b) Assume that $ghg^{-1} \in H$, for all $g \in G$ and $h \in H$. Prove $H \trianglelefteq G$.
Hint: Fix $g \in G$. Prove $gHg^{-1} \subseteq H$. Next, for every $h \in H$ prove $(g^{-1})h(g^{-1})^{-1} = k \in H$, and thus $h \in gHg^{-1}$.

We finish this section with another important concept which will prove to be a further source of normal subgroups.

Definition 2.133. Let G be a group. We define the center of G by

$$Z(G) = \{g \in G;\ gh = hg \text{ for all } h \in G\}.$$

The center of G is the set of all the elements in G that commute with every element in G.

Exercise 2.134. Prove the following claims.
(a) $Z(G) = G$ if and only if G is Abelian.
(b) $Z(G)$ is an Abelian group.
(c) $Z(Z(G)) = Z(G)$.
(d) $Z(G) \trianglelefteq G$.
(e) $Z(Q_8) = <-1>$.
(f) $Z(D_4) = <R^2>$.
(g) $Z(S_3) = \{e\}$.
(h) $Z(S_n) = \{e\}$ for all $n > 2$.
Hint: Let $\sigma \in Z(S_n)$. If $\sigma(n) = n$ use an induction argument. If $\sigma(n) \neq n$ you will need to find $\tau \in S_n$ that does not commute with σ. You might need to take cases.

2.9 Quotient Groups and Products of Groups

Sometimes it is possible to construct groups from other groups. In this section we will learn about two very distinct techniques, both of which accomplish this.

Definition 2.135. Let G be a group and $H \leq G$. The quotient of G by H is the set of cosets of H in G:

$$G/H = \{gH;\ g \in G\},$$

where $gH = hH$ if and only if $h^{-1}g \in H$ (see Exercise 2.61).
We define a multiplication on G/H by

$$(gH)(hH) = (gh)H$$

In case G/H is a group with this multiplication, we call it a quotient group.

Exercise 2.136. Prove the following claims.
(a) If G/H is a group, then $e_{G/H} = H$.

(b) G/N is a quotient group if and only if $N \trianglelefteq G$.
Hint: Don't forget to check that the operation is well-defined!
(c) Let G be a group and $N, K \trianglelefteq G$, with $K \subseteq N \subseteq G$. Then, $(N/K) \trianglelefteq (G/K)$.

Exercise 2.137.
(a) Let $H = <(26)(35)>$ be a subgroup of D_6. Find D_6/H explicitly. What is $(123456)H$ times $(14)(23)(56)H$?
(b) Prove by example that, since D_4 is not normal in S_4, S_4/D_4 is not a group.
(c) What is $Q_8/Z(Q_8)$ isomorphic to? What is $D_4/Z(D_4)$ isomorphic to?

Exercise 2.138. Prove or give a counterexample for each of the directions of the following three 'if and only if' statements. Let G be a group and N be a proper normal subgroup of G.
(a) G is Abelian if and only if G/N is Abelian.
(b) G is cyclic if and only if G/N is cyclic.
(c) G is non-Abelian if and only if G/N is non-Abelian.

Exercise 2.139. Give two examples per part; one which satisfies the statement, and one which provides a counterexample:
(a) If $H \leq G$ such that $G \cong H$ then $G/H = \{eH\}$.
(b) If $G_1 \cong G_2$ and $H_1 \cong H_2$ and $H_i \trianglelefteq G_i$, for $i = 1, 2$ then $G_1/H_1 \cong G_2/H_2$.
(c) $Z(G/H) \cong Z(G)/Z(H)$.

Exercise 2.140. Let $G = \mathbb{Z}_{24}$, $[18] \in \mathbb{Z}_{24}$, and $H = <[18]>$.
(a) Prove that G/H is cyclic.
(b) Find an isomorphism between G/H and $\mathbb{Z}_k$, where $|G/H| = k$. Prove your claim.

Remark 2.141. We can re-phrase Lagrange's theorem (Theorem 2.67) as,

$$|G/H| = \frac{|G|}{|H|}$$

for every finite group G and $H \leq G$.

Exercise 2.142. Let G be a finite group, $H \trianglelefteq G$, and $g \in G$. Prove that the order of gH (in G/H) divides the order of g (in G).

As you might have already realized, it is fairly hard to prove that two groups are isomorphic. It is even harder to prove that two groups are isomorphic when one of them is a quotient group. One way to avoid working with quotients, even if one wants to prove a quotient group is isomorphic to another group, is by using isomorphism theorems.

Theorem 2.143 (First Isomorphism Theorem). *Let $\phi : G \to H$ be a surjective group homomorphism. Then*

$$G/Ker(\phi) \cong H.$$

Exercise 2.144. The objective of this exercise is to prove the first isomorphism theorem.
(a) Prove that $\overline{\phi} : G/Ker(\phi) \to H$, defined by

$$\overline{\phi}(gKer(\phi)) = \phi(g)$$

is a well-defined function.
Hint: Exercise 2.123.
(b) Prove that $\overline{\phi}$ is bijective.
(c) Show that $\overline{\phi}$ is an isomorphism.

2.9 Quotient Groups and Products of Groups

Exercise 2.145. In this exercise we look at applications of the first isomorphism theorem.
(a) Show that $\mathbb{Z}/<n> \cong \mathbb{Z}_n$, for all $n \in \mathbb{Z}, n > 1$.
(b) Show that $GL_n(\mathbb{C})/SL_n(\mathbb{C}) \cong \mathbb{C} \setminus \{0\}$.
Hint: Exercise 2.107.

Corollary 2.146 (Third Isomorphism Theorem). *Let G be a group and $N, K \trianglelefteq G$, with $K \subseteq N \subseteq G$. Then,*
$$(G/K)/(N/K) \cong G/N.$$

Exercise 2.147. The goal of this exercise is to prove the third isomorphism theorem (Corollary 2.146).
(a) Check that in order for $(G/K)/(N/K) \cong G/N$ to make sense we need the hypothesis given.
(b) Prove the third isomorphism theorem.
Hint: First isomorphism theorem and Exercise 2.136.

The following exercise provides a very useful characterization of the subgroups of G/H.

Exercise 2.148. Let G be a group and $H \trianglelefteq G$. Consider $\varphi : G \to G/H$ defined by $\varphi(g) = gH$.
(a) Prove that φ is a homomorphism.
(b) Prove that φ is onto, and that $Ker(\varphi) = H$.
(c) Let $H \leq K \leq G$, prove that $\varphi(K) \leq G/H$. Moreover, $K \trianglelefteq G$ if and only if $\varphi(K) \trianglelefteq G/H$.
(d) Let $J \leq G/H$. Prove that $J = \varphi(K)$ for some $H \leq K \leq G$.
(e) Show that all subgroups of $\mathbb{Z}_n$ are cyclic by using that $\mathbb{Z}_n \cong \mathbb{Z}/<n>$ and the ideas used in previous parts of this problem.
(f) Prove that every normal subgroup of G is the kernel of some group homomorphism.

Exercise 2.149. Let S^1 be the unit circle (see Exercise 2.41). Consider the function $\phi : \mathbb{R} \to S^1$, defined by $\phi(x) = e^{2\pi i x}$.
(a) Prove that ϕ is a group homomorphism.
(b) Find $Ker(\phi)$.
(c) Use the first isomorphism theorem to prove that $\mathbb{R}/\mathbb{Z} \cong S^1$.
(d) Draw the real line and identify the cosets/elements in $\mathbb{R}/\mathbb{Z}$. Can you explain geometrically the isomorphism in part (c)?

We will now see another way to construct groups from known ones. By doing so we will learn about the 'missing' second isomorphism theorem.

Definition 2.150. Let $H, K \leq G$. The set HK is defined as
$$HK = \{hk;\ h \in H,\ k \in K\}.$$

Exercise 2.151. Let $H, K \leq G$ with $H \trianglelefteq G$. Prove the following claims:
(a) $HK \leq G$.
(b) $H \trianglelefteq (HK)$.
(c) (Second Isomorphism Theorem).
$$K/(H \cap K) \cong (HK)/H.$$

Hint: First isomorphism theorem.

Exercise 2.152.
(a) Consider $H = <(123)>$ and $K = <(134)>$, both as subgroups of S_4. Find HK explicitly, and show that HK is not a group.
(b) Give an example of two distinct groups H and K, both subgroups of a group G, such that neither one is normal in G but $HK \leq G$.

Definition 2.153. Let $(G,*)$ and $(H,\star)$ be groups. The (external) direct product of G and H, denoted $G \times H$, is the group with elements of the form (g,h) where $g \in G$ and $h \in H$, and operation

$$(g_1,h_1)(g_2,h_2) = (g_1 * g_2, h_1 \star h_2).$$

Exercise 2.154. Let G and H be groups.
(a) Prove that $G \times H$ is a group.
(b) Prove that $G \times H \cong H \times G$.
(c) Prove that $G \times H$ is Abelian if and only if G and H are Abelian.
(d) Find $Z(G \times H)$.
(e) Prove that if G and H are finite then so is $G \times H$. Moreover, $|G \times H| = |G||H|$.
(f) Prove that $G \cong G \times \{e_H\}$ and $H \cong \{e_G\} \times H$.
(g) Prove that both $G \times \{e_H\}$ and $\{e_G\} \times H$ are normal subgroups of $G \times H$.
(h) Prove that if $K \leq G$ and $J \leq H$, then $(K \times J) \leq (G \times H)$.
(i) Prove that $\mathbb{Z} \times \mathbb{Z}$ has subgroups that are not of the form given in the previous part.
(j) Show that $\mathbb{Z} \times \mathbb{Z}$ is not cyclic.
(k) Find all subgroups of $\mathbb{Z}_p \times \mathbb{Z}_p$, where p is prime.
(l) If p is prime, show that $\mathbb{Z}_p \times \mathbb{Z}_p$ is not cyclic.
(m) Find all subgroups of $\mathbb{Z}_6 \times \mathbb{Z}_6$.
(n) Find all subgroups of $\mathbb{Z}_{10} \times \mathbb{Z}_6$.
(o) Find the order of every element in $S_3 \times \mathbb{Z}_4$. How big can these orders get?
(p) Find a couple of subgroups of $S_3 \times \mathbb{Z}_4$. Can you find all of them?

Exercise 2.155. Prove the following claims:
(a) $\mathbb{Z}_4$ is not isomorphic to $\mathbb{Z}_2 \times \mathbb{Z}_2$.
(b) $V = \{e, (12)(34), (13)(24), (14)(23)\}$ is isomorphic to $\mathbb{Z}_2 \times \mathbb{Z}_2$.
(c) $G = \mathbb{Z}_4 \times \mathbb{Z}_2$, $H = \mathbb{Z}_2 \times \mathbb{Z}_2 \times \mathbb{Z}_2$ and $\mathbb{Z}_8$ are not isomorphic to each other.
(d) $\mathbb{Z}_9$ is not isomorphic to $\mathbb{Z}_3 \times \mathbb{Z}_3$.
(e) $\mathbb{Z}_6 \cong \mathbb{Z}_2 \times \mathbb{Z}_3$.
(f) The groups $G = \mathbb{Z}_3 \times \mathbb{Z}_4$, $H = \mathbb{Z}_2 \times \mathbb{Z}_6$, and A_4 are not isomorphic to each other.
(g) S_3 is not the direct product of any of its subgroups. Prove the same result for Q_8 and D_4.

Exercise 2.156. Show that $S_3 \times \mathbb{Z}_2 \cong D_6$.
Hint: Find $R \in S_3 \times \mathbb{Z}_2$ of order 6, then check that for every element $F \in (S_3 \times \mathbb{Z}_2) \setminus <R>$ of order 2 we get $FRF = R^{-1}$.

Theorem 2.157. *Let $m, n \in \mathbb{Z}$, $m, n > 1$. Then*

$$\mathbb{Z}_{mn} \cong \mathbb{Z}_m \times \mathbb{Z}_n$$

if and only if $\gcd(m,n) = 1$.

Exercise 2.158. Prove Theorem 2.157 by showing that $(m,n) = 1$ if and only if $([1],[1])$ has order mn.

Remark 2.159. It is known that every finite Abelian group is isomorphic to a direct product of, possibly many different, $\mathbb{Z}_n$'s. That is, if G is finite Abelian, then

$$G \cong \mathbb{Z}_{n_1} \times \mathbb{Z}_{n_2} \times \cdots \times \mathbb{Z}_{n_t}$$

for some $t, n_1, n_2, \cdots, n_t \in \mathbb{N}$.

2.9 Quotient Groups and Products of Groups

We now summarize all that we have learned about groups in the following table. This table contains all groups of small order (up to order 15), which is almost all the groups there are (there is a non-Abelian group of order 12 missing in the table). We consider two isomorphic groups to be the same, thus all groups appearing in the table are non-isomorphic to each other, unless it is clearly stated. We also include a summary on $\mathbb{Z}$ and $\mathbb{Z}_n$ in the spirit of reviewing all that we have learned.

Order	Group	Abelian?	More info about this group
2	$\mathbb{Z}_2$	✓	Theorems 2.38 & 2.74
3	$\mathbb{Z}_3 \cong A_3$	✓	Theorem 2.38. Definition 2.94
4	$\mathbb{Z}_2 \times \mathbb{Z}_2$	✓	Theorem 2.38. Exercise 2.154
4	$\mathbb{Z}_4$	✓	Theorem 2.38. Exercise 2.155
5	$\mathbb{Z}_5$	✓	Theorem 2.74
6	$\mathbb{Z}_2 \times \mathbb{Z}_3 \cong \mathbb{Z}_6$	✓	Theorem 2.157
6	$D_3 \cong S_3$	No	Exercise 2.19. Section 2.7. Remark 2.101
7	$\mathbb{Z}_7$	✓	Theorem 2.74
8	$\mathbb{Z}_2 \times \mathbb{Z}_2 \times \mathbb{Z}_2$	✓	Exercise 2.155
8	$\mathbb{Z}_2 \times \mathbb{Z}_4$	✓	Exercise 2.155
8	$\mathbb{Z}_8$	✓	Exercise 2.155
8	D_4	No	Exercise 2.19. Remark 2.101
8	Q_8	No	Exercises 2.17 & 2.128
9	$\mathbb{Z}_3 \times \mathbb{Z}_3$	✓	Exercise 2.154
9	$\mathbb{Z}_9$	✓	Exercise 2.155
10	$\mathbb{Z}_2 \times \mathbb{Z}_5 \cong \mathbb{Z}_{10}$	✓	Theorem 2.157
10	D_5	No	Exercise 2.19. Remark 2.101
11	$\mathbb{Z}_{11}$	✓	Theorem 2.74
12	$\mathbb{Z}_2 \times \mathbb{Z}_2 \times \mathbb{Z}_3 \cong \mathbb{Z}_2 \times \mathbb{Z}_6$	✓	Theorem 2.157. Exercise 2.155
12	$\mathbb{Z}_3 \times \mathbb{Z}_4 \cong \mathbb{Z}_{12}$	✓	Theorem 2.157. Exercise 2.155
12	$D_6 \cong S_3 \times \mathbb{Z}_2$	No	Exercises 2.19, 2.102 & 2.156. Remark 2.101
12	A_4	No	Thm. 2.97. Ex. 2.115 & 2.155. Def. 2.94
13	$\mathbb{Z}_{13}$	✓	Theorem 2.74.
14	$\mathbb{Z}_2 \times \mathbb{Z}_7 \cong \mathbb{Z}_{14}$	✓	Theorem 2.157
14	D_7	No	Exercise 2.19
15	$\mathbb{Z}_3 \times \mathbb{Z}_5 \cong \mathbb{Z}_{15}$	✓	Theorem 2.157
n	$\mathbb{Z}_n$	✓	Exercises 2.33 – 2.37 Theorems 2.35 & 2.112 Exercises 2.56, 2.70 & 2.145
∞	$\mathbb{Z}$	✓	Exercises 2.28, 2.41 & 2.55 Theorem 2.112 Exercise 2.145

Exercise 2.160. Check all that is (sometimes implicit) in the table. In other words:
(a) Prove that all the groups that are claimed to be isomorphic are, in fact, isomorphic.
(b) Prove that all groups that are not explicitly denoted as isomorphic are, in fact, non-isomorphic to each other.
(c) *Check* that we do not know of any other groups that should be in this table.
Note that *proving* that there are no other groups (up to isomorphism) of order 10 but those in the table is a very hard problem.

Chapter 3
Rings

Abstract This chapter covers the basics of Ring Theory, a vast area of mathematics that applies to a very wide spectrum of algebraic structures which could behave in completely different ways. For instance, a field will be a type of ring in which one can do all the algebraic manipulations one is used to doing with the real and complex numbers. On the other hand, there are rings without identity, rings that are non-commutative and rings in which we don't have the ability to divide. Such a large variety makes rings difficult to study. Fret not! Since most of the algebraic structures you have worked with so far are in fact rings, we expect that you will experience many *aha!* moments as you revisit these structures from a more abstract perspective.

> Three Rings for the Elven-kings under the sky,
> Seven for the Dwarf-lords in their halls of stone,
> Nine for Mortal Men doomed to die,
> One for the Dark Lord on his dark throne
> In the Land of Mordor where the Shadows lie.
> One Ring to rule them all, One Ring to find them,
> One Ring to bring them all and in the darkness bind them
> In the Land of Mordor where the Shadows lie.
>
> J.R.R Tolkien (1892 – 1973)

Getting Your Hands Dirty

Exercise 3.1. What properties are needed to solve the equation
$$ax+c = bx+d?$$

Exercise 3.2. Why is it forbidden to divide by zero?

Exercise 3.3. Is it always true that the only way to get $ab = 0$ is by having $a = 0$ or $b = 0$?

Exercise 3.4. How many solutions does $x^2 - 1 = 0$ have?

Exercise 3.5. Give as many examples as you can of sets having two operations.

3.1 A Short Historical Exposition

Mathematicians have been working with the integers for centuries, and also have been using matrices for quite a while. The first ring one usually encounters with a slightly more 'artificial' construction is the integers modulo n (what we call $\mathbb{Z}_n$), developed by (Johann) Carl Friedrich Gauss (1777 - 1855) in his book Disquisitiones Arithmeticae published in 1801. We note that $\mathbb{Z}_n$ was already well-studied by other mathematicians, such as

Pierre de Fermat (1601 - 1665), Adrien-Marie Legendre (1752 - 1833), but mostly by the Swiss genius Leonhard Euler (1707 - 1783).

Ring theory has evolved to address more abstract sets and properties, many of them too complex to be considered in this book. In an effort to name a few important contributors to the theory of rings (and modules) we mention here Sir William Rowan Hamilton (1805 - 1865), Joseph Henry Maclagan Wedderburn (1882 - 1948), Emil Artin (1898 - 1962), (Amalie) Emmy Noether (1882 - 1935), Nathan Jacobson (1910 - 1999), Ernst Eduard Kummer (1810 - 1893), Julius Wilhelm Richard Dedekind (1831 - 1916), Leopold Kronecker (1823 - 1891), Arthur Cayley (1821 - 1895), William Kingdon Clifford (1845 - 1879), and Leonard Eugene Dickson (1874 - 1954).

Note that many of these mathematicians were alive a mere hundred years ago!

3.2 $\mathbb{Z}$ and $\mathbb{Z}_n$ Revisited

Sometimes it is important to study groups in a framework which transcends group theory, in order to exploit additional properties/structure the groups may have. The integers serve as a great example of this phenomenon. We have seen that $(\mathbb{Z},+)$ is a group but $(\mathbb{Z},\cdot)$ is not a group. Does this mean that we should forget about the multiplication? No!! What we should do is to consider both operations together and see how they complement each other.

We will now look at properties of $+$ and $\cdot$ in $\mathbb{Z}$ which will be generalized later in this chapter. We know that the basic algebraic properties of $\mathbb{Z}$ are the following:

(i) $(\mathbb{Z},+)$ is an Abelian group.

(ii) The multiplication in $\mathbb{Z}$ is closed.

(iii) The number 1 has the property: $x \cdot 1 = 1 \cdot x = x$ for all $x \in \mathbb{Z}$.

In other words, even though $\mathbb{Z}$ is not a multiplicative group it has an identity for its multiplication.

(iv) $(xy)z = x(yz)$ for all $x, y, z \in \mathbb{Z}$. This is the associative law for the multiplication.

(v) For every $x, y, z \in \mathbb{Z}$ we have

$$x(y+z) = xy + xz \qquad (y+z)x = yx + zx$$

This is just the distributive law, describing how $+$ and $\cdot$ can be combined.

Also, the following properties are secondary but important

(vi) $xy = yx$ for all $x, y \in \mathbb{Z}$ (commutativity for the multiplication).

(vii) The only elements in $\mathbb{Z}$ having multiplicative inverses are ± 1.

(viii) If $xy = 0$ then $x = 0$ or $y = 0$.

Exercise 3.6. Prove the eight properties above using the fact that $(\mathbb{Z},+)$ is cyclic generated by 1, and that multiplication is nothing but notation for a repeated sum.

The operations in $\mathbb{Z}_n$ behave pretty much like the operations in $\mathbb{Z}$, which shouldn't be surprising, since we have defined how to add and multiply elements in $\mathbb{Z}_n$ based on how their representatives are added and multiplied in $\mathbb{Z}$.

Theorem 3.7. *The operations $+$ and $\cdot$ in $\mathbb{Z}_n$ satisfy the following:*
(i) $(\mathbb{Z}_n, +)$ is an Abelian group.
(ii) Multiplication is closed.
(iii) Multiplication is associative.
(iv) The distributive law holds.
(v) There is a multiplicative identity.
(vi) Multiplication is commutative.

As mentioned before, the properties listed in Theorem 3.7 are similar to the ones we described for $\mathbb{Z}$. However, $\mathbb{Z}_n$ seems to be missing two of the eight properties we listed for $\mathbb{Z}$. The following proposition takes care of one of these properties. Note that the situation will depend on whether or not n is prime.

Proposition 3.8. *The following properties hold in $\mathbb{Z}_n$.*
(a) If $[x][y] = [0]$ in $\mathbb{Z}_n$, then $[x] = [0]$, $[y] = [0]$ or $n | (xy)$.
(b) If n is prime, $[x][y] = [0]$ implies $[x] = [0]$ or $[y] = [0]$.

Exercise 3.9. Prove Theorem 3.7 and Proposition 3.8.
Hint: Use properties of $\mathbb{Z}$ and Exercise 2.33.

We now investigate which elements are invertible in $\mathbb{Z}_n$. Consider $[k] \in \mathbb{Z}_n$ with $gcd(k,n) = 1$. Using Bézout's lemma we know there are integers α and β such that

$$\alpha k + \beta n = 1$$

which implies $[\alpha k] = [1]$. Since $[\alpha k] = [\alpha][k]$, we have that $[\alpha]$ is the inverse of $[k]$ in $\mathbb{Z}_n$. This argument can be easily reversed giving us the following theorem.

Theorem 3.10. *An element $[k] \in \mathbb{Z}_n$ has a multiplicative inverse if and only if $gcd(k,n) = 1$.*

Corollary 3.11. *If n is prime, then every non-zero element in $\mathbb{Z}_n$ has a multiplicative inverse.*

Exercise 3.12.
(a) Prove Theorem 3.10 and Corollary 3.11.
(b) Show that if $[0] \neq [k] \in \mathbb{Z}_n$ with $gcd(k,n) = d \neq 1$ then there is an integer $[a] \neq [0]$ such that $[a][d] = [0]$, and thus $[a][k] = [0]$.
Hint: Proposition 3.8.
(c) Prove that $[a]$ is invertible in $\mathbb{Z}_n$ if and only if $<[a]> = \mathbb{Z}_n$.
Hint: Exercise 2.37.

3.3 Basic Properties, and Examples, of Rings

A ring is a generalization of the algebraic structure we found in $\mathbb{Z}$ and $\mathbb{Z}_n$. So far, we have found six common properties (the ones listed in Theorem 3.7) but we do not know if these are all the properties we want to consider. Hence, before giving the definition of a ring, we should check whether or not there are other sets with two operations that have those six properties.

Exercise 3.13. Consider the following sets:
(a) $R = <2>$ (the even numbers).
(b) $R = M_2(\mathbb{R})$ (under standard matrix addition and multiplication).
Which of the six properties common to $\mathbb{Z}$ and $\mathbb{Z}_n$ (Theorem 3.7) does the set R also have?

After having seen two more examples of sets with two operations, we now have a better idea of which properties we want a ring to possess.

Definition 3.14. A ring is a non-empty set R endowed with two operations $+$ and $\cdot$ (we write this as $(R, +, \cdot)$) such that
(i) $(R, +)$ is an Abelian group.
(ii) R is closed under multiplication.

(iii) Multiplication is associative.
(iv) Both distributive laws hold. That is,

$$a(b+c) = ab+ac \qquad \text{and} \qquad (b+c)a = ba+ca$$

hold for all $a,b,c \in R$.

Example 3.15. $\mathbb{Z}, \mathbb{Q}, \mathbb{R}$ and $\mathbb{C}$, with the usual addition and multiplication, are rings. Similarly, $\mathbb{Z}_n$, $M_2(\mathbb{R})$, and $<2>$ are also rings under their standard operations.

Definition 3.16. Let R be a ring.
(a) The additive identity in R is denoted by 0. The additive inverse of r is denoted $-r$.
(b) If $ab = ba$ for all $a,b \in R$ then R is called a commutative ring.
(c) If an element x in a ring R is such that $xr = r$, for all $r \in R$ then x is called a left identity. A right identity is defined similarly. If R has an element that is both a left and a right identity then it will be denoted by 1, and R will be said to be a ring with 1.
(d) An element x such that $xr = 1$ is called a left inverse of r. A right inverse of r is defined similarly. If a double-sided inverse of r exists it will be denoted r^{-1}. An element $r \in R$ is said to be invertible (or a unit of R) if it has a (double-sided) multiplicative inverse.

Remark 3.17. Given that the additive structure of a ring R is always an Abelian group, most of the adjectives given to a ring will refer to the ring's multiplicative structure. Specifically:
(a) When we say that a has an inverse in R (or that a is invertible in R) we mean that a has a *multiplicative* inverse in R.
(b) When we mention the identity of R, we mean the *multiplicative* identity (if it exists).
(c) A ring is commutative if the *multiplication* in R satisfies the commutative law.

Example 3.18. $\mathbb{Z}, \mathbb{Q}, \mathbb{R}, \mathbb{C}$ and $\mathbb{Z}_n$ are commutative rings with one. $M_2(\mathbb{R})$ is a non-commutative ring with one. $<2>$ is a commutative ring without one.

Exercise 3.19. Check the claims in the previous examples. What are the units in these rings?

Proposition 3.20. *The following properties hold in any ring R.*
(a) If R has a left identity x, and a right identity y, then $x = y$.
(b) If R has a one, then it is unique.
(c) If R is a ring with 1 and $r \in R$ has a left inverse x, and a right inverse y, then $x = y$.
(d) If $r \in R$ has an inverse in R then it is unique.
(e) $-(-r) = r$ for all $r \in R$.
(f) $(-r)s = -(rs) = r(-s)$, for all $r,s \in R$.
(g) $r \cdot 0 = 0 \cdot r = 0$ for all $r \in R$.
(h) $(r^{-1})^{-1} = r$, for all invertible $r \in R$.
(i) If R has a one and r is a unit in R, then $(r^a)^{-1} = (r^{-1})^a$, where $a \in \mathbb{N}$.
(j) For all $r \in R$, and $a,b \in \mathbb{N}$,

$$r^a \cdot r^b = r^{a+b} \qquad \text{and} \qquad (r^a)^b = r^{ab}$$

Moreover, if R has a one, and under the convention $r^{-a} = (r^a)^{-1}$ and $r^0 = 1$, then the exponents rules are valid for all $a,b \in \mathbb{Z}$.

Exercise 3.21. Prove Proposition 3.20.

We have already seen a few examples of rings. It is time to look at some more complex examples.

3.3 Basic Properties, and Examples, of Rings

Example 3.22. Consider $R = M_2(\mathbb{Q})$. We know R is an additive Abelian group, we only have to check the multiplicative properties in the definition of a ring. Clearly, the multiplicative identity of the ring is the identity matrix
$$I = \begin{bmatrix} 1 & 0 \\ 0 & 1 \end{bmatrix}$$
Since $M_2(\mathbb{R})$ is a ring (see Exercise 3.13) and $M_2(\mathbb{Q}) \subseteq M_2(\mathbb{R})$, $M_2(\mathbb{Q})$ inherits the (multiplicative) associative and distributive laws from $M_2(\mathbb{R})$.

Finally, because of the closure of the rationals under addition and multiplication we have that the product of two matrices in $M_2(\mathbb{Q})$ is again in $M_2(\mathbb{Q})$. Hence, $M_2(\mathbb{Q})$ is a ring with one. Since
$$\begin{bmatrix} 1 & 0 \\ 0 & 2 \end{bmatrix} \begin{bmatrix} 1 & 1 \\ 1 & 1 \end{bmatrix} \neq \begin{bmatrix} 1 & 1 \\ 1 & 1 \end{bmatrix} \begin{bmatrix} 1 & 0 \\ 0 & 2 \end{bmatrix},$$
we see that $M_2(\mathbb{Q})$ is a non-commutative ring.

Definition 3.23. Let R be a commutative ring with one in which $0 \neq 1$. If every nonzero element of R has a multiplicative inverse then R is called a field.

Example 3.24. $\mathbb{Q}, \mathbb{R}$, and $\mathbb{C}$ are fields, while $\mathbb{Z}, <2>$, and $M_2(\mathbb{Q})$ are not fields.

Exercise 3.25. Prove the following claims.
(a) $R = \{0\}$ with the obvious sum and multiplication is a commutative ring with one (which turns out being equal to 0).
(b) If we assume $1 = 0$ in a ring R, then $R = \{0\}$.
(c) An (additive) Abelian group $G \neq \{0\}$ with the product defined by $ab = 0$ for all $a,b \in G$ is a commutative ring without one.
(d) Prove what is claimed in Example 3.24.
(e) $R = \{[0], [3]\} \subseteq \mathbb{Z}_6$ is a ring with one (under the operations of $\mathbb{Z}_6$).
(f) $\mathbb{Z}_n$ is a field if and only if n is prime.
Hint: Section 3.2.
(g) If $n > 1$, $M_n(\mathbb{K})$ for $\mathbb{K} = \mathbb{C}, \mathbb{R}, \mathbb{Q}$ or $\mathbb{Z}$ is a non-commutative ring with one. $GL_n(\mathbb{C})$ is not a ring with the standard addition and multiplication of matrices.
(h) The set of all even integers is a commutative ring without one.
(i) The set of matrices with entries in the set of even numbers is a non-commutative ring without one.
(j) Let F be a ring. F is a field if and only if $(F \setminus \{0\}, \cdot)$ is an Abelian group.
(k) Let R be a ring and let $\mathscr{F}$ be the set of all functions $f : \mathbb{R} \to \mathbb{R}$ with operations given by:
$$(f+g)(x) = f(x) + g(x) \qquad (fg)(x) = f(x)g(x)$$
for all $x \in \mathbb{R}$.
$\mathscr{F}$ is a ring with the operations described above.
(l) The set $\mathbb{Q}(\sqrt{2}) = \{a + b\sqrt{2}; a,b \in \mathbb{Q}\}$ is a field, under the standard operations in $\mathbb{R}$.
Hint: Exploit similarities with $\mathbb{C}$.
(m) The set $\mathbb{Z}(\sqrt{2}) = \{a + b\sqrt{2}; a,b \in \mathbb{Z}\}$ is a commutative ring with one, but not a field (under the standard operations in $\mathbb{R}$).

Theorem 3.26. *The set of units of a ring with one form a multiplicative group.*

Definition 3.27. The group formed by the units of a ring with one is denoted by R^* (some authors use $U(R)$). Most of the time R^* is called the multiplicative group of R.

Exercise 3.28. Prove Theorem 3.26

Exercise 3.29.
(a) Let R be a commutative ring with one. Show that R^* is Abelian.
(b) Give an example of a ring R such that R^* is not Abelian.
(c) Show that the groups $\mathbb{Z}_n^*$ for $n = 2,3,4,5,6,7$ are all cyclic. What is $\mathbb{Z}_8^*$ isomorphic to?
(d) Let $S = \{[x]^2; \ [x] \in \mathbb{Z}_n^*\}$. Show that S is a subgroup of $\mathbb{Z}_n^*$.
(e) Show that $M_2(\mathbb{Z}_2)^* \cong S_3$. Note that you are proving that $GL_2(\mathbb{Z}_2) \cong S_3$.

Definition 3.30. Let R be a ring. A nonzero element $a \in R$ is said to be a zero divisor if there is $0 \neq b \in R$ such that $ab = 0$.

Exercise 3.31. Prove the following claims.
(a) A nonzero element in $\mathbb{Z}_n$ is either a unit or a zero divisor. In fact, $[a] \in \mathbb{Z}_n$ is invertible if and only if $gcd(a,n) = 1$.
Hint: Exercise 3.12.
(b) The only units in $\mathbb{Z}$ form the multiplicative group $\{\pm 1\}$. $\mathbb{Z}$ contains no zero divisors.
(c) A unit cannot be a zero divisor. Conclude that a field contains no zero divisors.
(d) The units of $M_n(\mathbb{C})$ form the multiplicative group $GL_n(\mathbb{C})$. Elements of $M_n(\mathbb{C})$ which are not in $GL_n(\mathbb{C})$ are all zero divisors in $M_n(\mathbb{C})$.
Hint: Use linear algebra.
The statement above would remain true if $\mathbb{C}$ were replaced by any other field.
(e) Any nonzero function $f \in \mathscr{F}$ (see Exercise 3.25) is either invertible or a zero divisor.
Hint: Consider two cases: when there is $x \in \mathbb{R}$ such that $f(x) = 0$ and when such x does not exist.
(f) Generalize the previous exercise by first re-defining $\mathscr{F}$ for functions over any field $\mathbb{F}$ (instead of over $\mathbb{R}$), proving that the re-defined $\mathscr{F}$ is a ring, and then showing that every nonzero function $f \in \mathscr{F}$ is either invertible or a zero divisor.

Theorem 3.32. *Let R be a finite ring with 1, and let r be a nonzero element in R. Then, r is either a zero divisor or a unit.*

Exercise 3.33. The goal of this exercise is to prove the previous theorem. Fix r to be any nonzero element in R and consider the function $f_r : R \to R$ given by $f(x) = rx$.
(a) Prove that f_r is injective.
(b) Prove that f_r is bijective.
(c) Conclude there is an element $x \in R$ such that $rx = 1$.
(d) Prove that x is the inverse of r.

Definition 3.34. Let R be a ring.
(a) If R is commutative, has one ($1 \neq 0$), and no zero divisors, then R is said to be an integral domain.
(b) If R has one ($1 \neq 0$) and every nonzero element of R has a multiplicative inverse, then R is said to be a division ring.
(c) A commutative division ring is called a field.

Exercise 3.35. Prove the following claims.
(a) Definitions of field 3.23 and 3.34 are equivalent.
(b) If R is a field then it is a division ring. If R is a commutative division ring then it is an integral domain.
(c) $\mathbb{Z}$ is an integral domain but is not a division ring.

Remark 3.36. The set
$$\mathscr{H} = \{a + bi + cj + dk; \ a,b,c,d \in \mathbb{R}\}$$
with the operations given by standard (polynomial-like) addition and distribution, and
$$i^2 = j^2 = k^2 = -1 \qquad\qquad ij = k, \ jk = i, \ ki = j, ji = -k, \ kj = -i, \ ik = -j$$

is a division ring which is not a field.

Note the similarity between the definition of the multiplication in $\mathscr{H}$ with how the operation in Q_8 was defined. $\mathscr{H}$ is called 'the real quaternions', or 'the Hamiltonians'.

In the following table we summarize all the information gathered in this section about the most important rings we have studied.

	C-ring?	One?	Zero divisors	Units
$\mathscr{F}$	Yes	Yes	f such that $f(a) = 0$ for some $a \in \mathbb{R}$	f such that $f(x) \neq 0, \forall x \in \mathbb{R}$
$\mathbb{Z}_n$	Yes	Yes	$[x]$ such that $gcd(x,n) \neq 1$	$[x]$ such that $gcd(x,n) = 1$
$\mathbb{Z}$	Yes	Yes	None	± 1
A field $\mathbb{K}$	Yes	Yes	None	$\mathbb{K} \setminus \{0\}$
$M_n(\mathbb{R})$	No	Yes	$M \neq 0$ such that $det(M) = 0$	M such that $det(M) \neq 0$
$<2>$	Yes	No	None	None

3.4 Homomorphisms and Ideals

We will now study operation-preserving functions between rings.

Definition 3.37. Let $(R, +, \times)$ and $(S, \oplus, \otimes)$ be rings. A function $\phi : R \to S$ such that

$$\phi(r+s) = \phi(r) \oplus \phi(s) \qquad \phi(r \times s) = \phi(r) \otimes \phi(s)$$

for all $r, s \in R$ is said to be a homomorphism of rings.

If a homomorphism is bijective, it is called an isomorphism. In case there is an isomorphism between two rings R and S we will say that R and S are isomorphic to each other, and we will write $R \cong S$.

Exercise 3.38. Let $\phi : R \to S$ be a homomorphism of rings. Prove the following claims.
(a) If R has a one, ϕ is onto and $\phi(1) \neq 0$ then S has a one and $\phi(1) = 1$.
(b) Part (a) is false if one does not assume $\phi(1) \neq 0$.
(c) Part (a) is false if one does not assume ϕ is onto.
(d) ϕ is a homomorphism between the Abelian groups $(R, +)$ and $(S, +)$.
(e) $\psi : (\mathbb{Z}_2, +) \to (\mathbb{Z}_4, +)$ defined by $\psi(1) = 2$ is a homomorphism of Abelian groups that is not a homomorphism of rings (with standard operations in $\mathbb{Z}_n$).
(f) Find all ring homomorphisms between $\mathbb{Z}_{10}$ and $\mathbb{Z}_{12}$.
Hint: Exercise 2.119.

Exercise 3.39. Let $\phi, \psi : R \to R$ be two ring homomorphisms. Is $\psi \circ \phi$ a ring homomorphism?

We study homomorphisms of rings (just like we did with groups) because they preserve the operations on the rings. In addition, if two rings, R and S, are isomorphic, S will have all the algebraic properties of R, and vice versa (after the re-labeling given by the isomorphism). The following exercises make these claims explicit.

Exercise 3.40. Let $\phi : R \to S$ be an isomorphism of rings, where R and S are nonzero rings. Prove the following claims.
(a) R is commutative if and only if S is commutative.
(b) R has a one if and only if S has a one.
(c) If r is a unit in R then $\phi(r)$ is a unit in S. Is the converse true?
(d) If r is a zero divisor in R then $\phi(r)$ is a zero divisor in S. Is the converse true?

Definition 3.41. Let $\phi : R \to S$ be a homomorphism of rings. The kernel of ϕ is defined to be the set

$$Ker(\phi) = \{r \in R;\ \phi(r) = 0\}$$

Remark 3.42. The kernel of ϕ is also the kernel of the homomorphism of Abelian groups associated to ϕ.

Exercise 3.43. Let $\phi : R \to S$ be a homomorphism of rings. Prove the following claims.
(a) $\phi(r+s) = \phi(r)$ for all $r \in R$ and $s \in Ker(\phi)$. Moreover, if $\phi(r) = \phi(t)$, then $t = r+s$ for some $s \in Ker(\phi)$.
(b) $Ker(\phi) = \{0\}$ if and only if ϕ is injective.
(c) $Ker(\phi)$ is a ring with the operations of R.
(d) $Ker(\phi)$ has the following 'absorption' property: If $r \in R$ and $s \in Ker(\phi)$ then both rs and sr are in $Ker(\phi)$.

Definition 3.44. Let R be a ring and let $S \subseteq R$. We say that S is a subring of R if S is a ring with the same operations of R.

Note that if S is a subset of a ring R then many of the properties S must satisfy to be a ring are 'inherited' from R. For instance, the associative law holds for *all* elements in R, thus it holds for all elements in S. This consideration leads to the following criterion.

Theorem 3.45. *Let R be a ring and let S be a nonempty subset of R. S is a subring of R if and only if*
(i) $r - s \in S$, for all $r, s \in S$.
(ii) $rs \in S$, for all $r, s \in S$.

Exercise 3.46. Prove Theorem 3.45.

The concept of a subring is to rings what subgroups were to groups. However, the role subrings play in the study of rings is minor, as the concept of subrings is eclipsed by the concept of ideals, which we will define shortly. First, we define the equivalent of the center of a group, which is the center of a ring.

Definition 3.47. Let R be a ring. The center of R is the set

$$C(R) = \{r \in R;\ rs = sr \text{ for all } s \in R\}$$

Exercise 3.48. Let R be a ring. Prove the following claims:
(a) $C(R)$ is a subring of R.
(b) $C(R)$ is a commutative ring.
(c) If S is a subring of R that is not contained in $C(R)$ then S is not commutative.

Definition 3.49. Let R be a ring and let $\emptyset \neq I \subseteq R$ such that $(I, +) \leq (R, +)$.
(a) If $ra \in I$ for all $a \in I$ and $r \in R$ we will say that I is a left ideal of R.
(b) If $ar \in I$ for all $a \in I$ and $r \in R$ we will say that I is a right ideal of R.
(c) If ra and ar are in I for all $a \in I$ and $r \in R$ then we will say that I is a double-sided ideal, or simply an ideal of R.
If I is an ideal of R, we will write $I \leq R$.

3.4 Homomorphisms and Ideals

Exercise 3.50. Let R be a commutative ring.
(a) Prove that all left ideals of R are also right ideals, and thus ideals of R.
(b) Does the same situation hold for right ideals?
(c) Find a left ideal that is not a right ideal in $M_2(\mathbb{Z}_2)$.

Exercise 3.51. Let R be a ring such that $|R| = 200$. Can R have an ideal of cardinality 30? Discuss.

Note that if I is an ideal of R then I is also a subring of R. Thus, by modifying Theorem 3.45, we obtain the following criterion:

Theorem 3.52 (Ideal Test). *Let R be a ring, and let $\emptyset \neq I \subseteq R$. $I \leq R$ if and only if*
(i) $r - s \in I$, for all $r, s \in I$.
(ii) ra and ar are in I, for all $r \in R$ and $a \in I$.

Exercise 3.53. Prove Theorem 3.52.

Exercise 3.54. Let R be a ring. Prove the following claims.
(a) $\{0\}$ and R are ideals of R. We will call these ideals 'trivial'.
(b) $I = \{f \in \mathscr{F};\ f(0) = 0\}$ is an ideal of $\mathscr{F}$, where $\mathscr{F}$ is the ring given in Exercise 3.25.
(c) I is an ideal of $\mathbb{Z}$ if and only if I is a subgroup of $\mathbb{Z}$.
Note that this implies that any ideal of $\mathbb{Z}$ is the set of multiples of some fixed integer.
(d) $I \leq R$ if and only if $M_n(I) \leq M_n(R)$.
(e) Let $a \in R$. $I_{left} = \{r \in R;\ ra = 0\}$ is a left ideal of R and $I_{right} = \{r \in R;\ ar = 0\}$ is a right ideal of R.
(f) If $\phi : R \to S$ be a homomorphism of rings, then $Ker(\phi) \leq R$.
(g) Let R be a finite ring that has no non-trivial ideals and let $\phi : R \to R$ be a ring homomorphism. Prove that either ϕ is an isomorphism or ϕ is the zero function.
(h) $I = \{2a + \sqrt{2}b;\ a, b \in \mathbb{Z}\} \leq \mathbb{Z}(\sqrt{2})$.
(i) $I = \{M \in M_2(\mathbb{R});\ M\mathbf{v} = \mathbf{0}\}$ is a left ideal of $M_2(\mathbb{R})$, where $\mathbf{0}$ represent the zero vector in $\mathbb{R}^2$ and $\mathbf{v}$ is a fixed vector in $\mathbb{R}^2$. Is I an ideal of $M_2(\mathbb{R})$?

Exercise 3.55. Let R be a ring. Prove the following claims.
(a) Let $I \leq R$. If $1 \in I$ then $I = R$.
(b) Let $I \leq R$. If I contains a unit of R, then $I = R$.
(c) A field has no non-trivial ideals.

Definition 3.56. Let R be a ring.
(a) An element $r \in R$ is said to be nilpotent if $r^n = 0$ for some positive integer n.
(b) An element $r \in R$ is said to be idempotent if $r^2 = r$.
(c) If R is commutative, the set of all nilpotent elements in R is called the nilradical of R.
(d) If R is commutative and $a \in R$, the annihilator of a is the set

$$Ann(a) = \{r \in R;\ ra = 0\}$$

Exercise 3.57. Let R and S be rings.
(a) Prove that if R is commutative then the nilradical of R is an ideal of R.
(b) Find a non-commutative ring such that the set of nilpotent elements is not an ideal.
(c) Find the nilradical of a few $\mathbb{Z}_n$'s such as $\mathbb{Z}_4, \mathbb{Z}_{12}$, and $\mathbb{Z}_{18}$. Conjecture and prove what the nilradical of $\mathbb{Z}_n$ is.
(d) Find the nilradical of $\mathbb{Z}$.
(e) Let $\phi : R \to S$ be a homomorphism. Show that if $r \in R$ is nilpotent, then so is $\phi(r)$.
(f) Show that if r is nilpotent, then $1 - r$ is a unit.
(g) Show that e is idempotent if and only if $1 - e$ is idempotent.
(h) Show that the set of idempotents of R contains exactly one unit. The rest of its elements are all zero divisors.
(i) Assume every element in R is an idempotent. Show that $a = -a$, for all $a \in R$.

Hint: Compute $(a+a)^2$ in two different ways.
(j) Assume every element in R is an idempotent. Show that R is commutative.
Hint: Compute $(a+b)^2$ in two different ways.
(k) Prove that if R is commutative then $Ann(a)$ is an ideal of R. What are the conditions on a for $Ann(a)$ to be non-trivial?

Remark 3.58. Let a be a nonzero element in a ring R. From the definition of an ideal we can see that ar and ra must be contained in every ideal containing a, for all $r \in R$. Moreover, if R is commutative with one then the set

$$aR = \{ar; \text{ where } r \in R\}$$

is an ideal of R which contains a.
It follows that *if R is a commutative ring with one* then aR is contained in every ideal which contains a, and hence aR is the smallest ideal of R containing a.

Exercise 3.59. Prove all the claims in Remark 3.58.

Exercise 3.60. Let $R = \mathbb{Z}_{240}$ and $r = [9] \in R$. Let

$$I = \{s \in R; \; sr = 0\}$$

(a) Prove that I is an ideal of R.
(b) Give an example of an ideal of R that contains r.

Definition 3.61. Let R be a ring.
(a) The smallest ideal containing $r \in R$ denoted by (r), it is called the principal ideal generated by r.
(b) Any ideal that can be generated by some element r is called a principal ideal.
(c) If R is an integral domain, and every ideal of R is principal, then R is said to be a principal ideal domain (PID).

Exercise 3.62.
(a) Show that the principal ideal generated by n in $\mathbb{Z}$ is the set of all multiples of n.
(b) Prove that $\mathbb{Z}$ is a PID.
Hint: Exercise 3.54.
(c) Let R be an integral domain, and let $a, b \in R$. Prove that $(a) \subseteq (b)$ if and only if there is a $c \in R$ such that $a = bc$ (in other words, b divides a in R). Moreover, $(a) = (b)$ if and only if c is a unit.
(d) Prove that a field is always a PID.

Remark 3.63. Not all ideals are principal. For an example we need a more sophisticated type of ring, and thus we will need to wait until Exercise 3.90. For now, the reader should keep in mind that an ideal is not necessarily principal unless it is specifically assumed, or proved to be.

In Exercise 3.62 you proved that $\mathbb{Z}$ was a PID. As usual, the properties of $\mathbb{Z}_n$ are related to similar properties of $\mathbb{Z}$. Hence, it makes sense to study whether or not $\mathbb{Z}_n$ is a PID.

Exercise 3.64. Let $n \in \mathbb{N}$. Prove the following claims.
(a) All ideals in $\mathbb{Z}_n$ are principal.
(b) For every $I \leq \mathbb{Z}_n$ there exists an element $[a] \in \mathbb{Z}_n$ such that $I = Ann([a])$.
(c) $\mathbb{Z}_n$ is a PID if and only if n is prime. That is, $\mathbb{Z}_n$ is a PID if and only if $\mathbb{Z}_n$ is a field.
(d) $\mathbb{Z}_n$ has non-trivial ideals if and only if n is composite.

Definition 3.65. Let R be a ring. $M \leq R$ is said to be maximal if there are no non-trivial ideals of R containing M other than M itself.

Exercise 3.66.
(a) Prove that the maximal ideals of $\mathbb{Z}$ are of the form $p\mathbb{Z}$, where p is prime.
(b) Find the intersection of all the maximal ideals of $\mathbb{Z}_n$, for $n = 18, 24$, and 30 (and other smaller numbers if necessary). Conjecture and prove what the intersection of all the maximal ideals of $\mathbb{Z}_n$ is.
(c) Let R be a ring with one, and let I be an ideal of R. Prove that I is maximal if and only if for every $r \in R \setminus I$ there is an element $s \in R$ such that $1 - rs \in I$.
Hint: Use the ideas discussed in Remark 3.58.

3.5 Quotient Rings and Direct Products

Ideals are to rings what normal subgroups are to groups. Thus it is no surprise that we will use ideals to construct quotient rings, the same way we used normal subgroups to construct quotient groups in Chapter 2.

Definition 3.67. Let I be an ideal of a ring R. The set

$$R/I = \{r + I;\ r \in R\}$$

with the operations

$$(a+I) + (b+I) = (a+b) + I \qquad (a+I)(b+I) = (ab) + I$$

is called the quotient ring of R over I.

Exercise 3.68. Show that R/I defined as above is indeed a ring. Don't forget to check that the given operations are well-defined!

Exercise 3.69. Let R be a ring and let $I \leq R$.
(a) What are the quotient rings for the rings and ideals in exercises 3.54, 3.62 (a), 3.60, and 3.66?
(b) Let R be a commutative ring with one. Show that I is maximal if and only if R/I is a field.
(c) Show that the function $\pi : R \to R/I$ defined by $\pi(r) = r + I$, for all $r \in R$, is a surjective homomorphism of rings. Moreover, $Ker(\pi) = I$.
(d) Show that every ideal is the kernel of some homomorphism of rings.
(e) Use the function π defined above to prove that there is a one-to-one correspondence between the ideals of R/I and the ideals of R containing I.
Hint: Exercise 2.148.
(f) Show that every quotient of a PID is a commutative ring with one for which every ideal is a principal ideal.
(g) Is every quotient of a PID also a PID?

Theorem 3.70 (First Isomorphism Theorem). *If $\phi : R \to S$ be a surjective homomorphism of rings, then*

$$R/Ker(\phi) \cong S$$

Proof. Let $I = Ker(\phi)$. The function $\overline{\phi} : R/I \to S$, defined by $\overline{\phi}(r+I) = \phi(r)$ is an isomorphism. □

Exercise 3.71. Complete the proof of the first isomorphism theorem by proving:
(a) $\overline{\phi}$ is well-defined.
(b) $\overline{\phi}$ is a homomorphism.
(c) $\overline{\phi}$ is bijective.

Exercise 3.72.
(a) Prove that $\mathbb{Z}/n\mathbb{Z} \cong \mathbb{Z}_n$ (as rings), for all $n \in \mathbb{N}$.
(b) Consider the ring $\mathscr{F}$ in Exercise 3.25. Prove that the ideal (see Exercise 3.54)

$$I = \{f \in \mathscr{F};\ f(0) = 0\}$$

is a maximal ideal of $\mathscr{F}$.
Hint: What is $\mathscr{F}/I$? Use Exercise 3.69.

Theorem 3.73. *Let R be a ring, let S a subring of R, and let $I, J \leq R$.*
(a) (Second Isomorphism Theorem) If $S+I = \{s+i;\ s \in S, i \in I\}$, then $S+I$ is a subring of R, $S \cap I \leq S$, and

$$(S+I)/I \cong S/(S \cap I)$$

(b) (Third Isomorphism Theorem) Assume $I \subseteq J$. Then, $J/I \leq R/I$, and

$$(R/I)/(J/I) \cong R/J$$

Exercise 3.74. Prove Theorem 3.73.
Hint: Look at how the second and third isomorphism theorems were proved for groups.

Just as we did for groups, given two rings R and S we can construct a third ring containing both R and S. We will now see this construction.

Definition 3.75. Let R and S be non-empty rings. We define the direct product of R and S to be the set $R \times S$ with the operations

$$(r,s) + (a,b) = (r+a, s+b) \qquad (r,s)(a,b) = (ra, sb)$$

Exercise 3.76. Prove the following claims.
(a) $R \times S$ as described above is a ring.
(b) $R \times S \cong S \times R$.
(c) $R \cong R \times \{0\}$ and $S \cong \{0\} \times S$.
(d) $R \times \{0\} \leq R \times S$ and $\{0\} \times S \leq R \times S$.
(e) $(R \times \{0\}) \cap (\{0\} \times S) = \{(0,0)\}$ and $(R \times \{0\}) + (\{0\} \times S) = R \times S$.

The previous problems give an idea of the structure of $R \times S$. The next ones show that $R \times S$, R and S have 'similar' properties.

Exercise 3.77. Let R and S be rings. Prove the following claims
(a) R and S are commutative if and only if $R \times S$ is commutative.
(b) R and S have a one if and only if $R \times S$ has a one.
(c) If R and S are nonzero then $R \times S$ is never an integral domain.
(d) $(r,s) \in R \times S$ is a unit if and only if $r \in R$ and $s \in S$ are units.

In order to obtain a result similar to Theorem 2.157 we need an important result about systems of modular equations.

Theorem 3.78 (The Chinese Remainder Theorem). *Let $\gcd(m,n) = 1$, and $a, b \in \mathbb{Z}$. Then,*

$$x \equiv a \pmod{m}$$
$$x \equiv b \pmod{n}$$

has a unique solution modulo mn.

Proof. It is clear that if there is a solution then it must be unique modulo mn. Now we need to find one solution. Consider

$$x = aAn + bBm$$

3.5 Quotient Rings and Direct Products

where A is an inverse of n modulo m, and B is an inverse of m modulo n. It is easy to see that x is a solution of the system. $\square$

Exercise 3.79.
(a) Fill all the details in the proof of Theorem 3.78. Among other things, be sure to check that A and B exist, and that the solution is unique modulo mn.
(b) Prove that if $\gcd(m,n) = 1$, then $\mathbb{Z}_m \times \mathbb{Z}_n \cong \mathbb{Z}_{mn}$.
Hint: Consider the map $\phi : \mathbb{Z}_{mn} \to \mathbb{Z}_m \times \mathbb{Z}_n$ *defined by* $\phi(x) = (x \mod m, x \mod n)$. *Use Theorem 3.78 to check ϕ is onto.*
(c) Prove that if $\gcd(m,n) = 1$, then $\mathbb{Z}_m^* \times \mathbb{Z}_n^* \cong \mathbb{Z}_{mn}^*$.
Hint: Modify the function used in part (b).

Exercise 3.80. Let R be a commutative ring with one. Assume that e is idempotent (Definition 3.56). Prove the following claims:
(a) e is idempotent if and only if $1-e$ is idempotent.
(b) $R \cong eR \times (1-e)R$.
(c) Both e and $1-e$ are zero divisors in R. However, they are the identities of eR and $(1-e)R$, respectively.

Definition 3.81. Let R be a commutative ring with one. The characteristic of R, denoted $char(R)$ is the least positive integer n such that $\underbrace{1+1+\cdots+1}_{n \text{ times}} = 0$.

In case that such a number does not exist, then we say that the characteristic of the ring is zero.

Remark 3.82. $char(R) \neq 0$ if and only if the order of 1 in $(R,+)$ coincides with $char(R)$. Similarly, $char(R) = 0$ if and only if the order of 1 in $(R,+)$ is ∞.

Exercise 3.83. Let R and S be rings. Prove the following claims.
(a) $char(\mathbb{Z}_n) = n$.
(b) $char(\mathbb{Z}) = char(\mathbb{Q}) = char(\mathbb{R}) = char(\mathbb{C}) = 0$.
(c) $char(\mathbb{Z}_p \times \mathbb{Z}_p) = p$
(d) If F is a field, then its characteristic is either zero (such as $\mathbb{Q}$) or a prime number p (such as $\mathbb{Z}_p$). Is the converse true?
(e) If $char(R) = 0$ then $char(R \times S) = 0$ for any ring S.
(f) If $char(R) = n$ and $char(S) = m$, with $nm \neq 0$, then $char(R \times S) = lcm[n,m]$.
(g) If $\phi : R \to S$ is a ring homomorphism then $char(S)$ divides $char(R)$.

Exercise 3.84. Let p be prime.
(a) Show that p divides $\binom{p}{k}$ for $k = 1, 2, \ldots, p-1$.
(b) Show that in $\mathbb{Z}_p$ the following holds:
$$(a+b)^p = a^p + b^p$$
(c) Show that $\phi : \mathbb{Z}_p \to \mathbb{Z}_p$ defined by $\phi(x) = x^p$ is a homomorphism of rings (fields, in fact).
(d) Show that ϕ is an isomorphism.

We now apply what we have learned so far to study rings of matrices in more depth. An earlier exercise asked you to show that $I \leq R$ if and only if $M_n(I) \leq M_n(R)$ (if you have not proved this result yet, now would be a good time to do so). This result helps us find the ideals of the matrix ring $M_n(R)$.

Exercise 3.85. Let R be a ring and let $n \geq 2$. Prove the following claims.
(a) The center of $M_n(R)$ is the set of all scalar matrices. That is,

$$C(M_n(R)) = \{M \in M_n(R);\ M = rId,\ \text{for some } r \in R\}$$

(b) The set

$$C = \left\{ \begin{bmatrix} a & b \\ -b & a \end{bmatrix};\ a,b \in \mathbb{R} \right\}$$

is a subring of $M_2(\mathbb{R})$ which is isomorphic to $\mathbb{C}$.
Hint: Exercise 1.52.
(c) The set of all upper (or lower) triangular $n \times n$ matrices is a subring of $M_n(R)$.
(d) The set of all matrices of the form

$$\begin{bmatrix} a_1 & a_2 & a_3 & \cdots & a_{n-2} & a_{n-1} & a_n \\ 0 & a_1 & a_2 & a_3 & \cdots & a_{n-2} & a_{n-1} \\ 0 & 0 & a_1 & a_2 & a_3 & \cdots & a_{n-2} \\ \vdots & 0 & 0 & \ddots & \ddots & \ddots & \vdots \\ \vdots & \vdots & & \ddots & \ddots & \ddots & a_3 \\ 0 & 0 & 0 & \cdots & 0 & a_1 & a_2 \\ 0 & 0 & 0 & \cdots & 0 & 0 & a_1 \end{bmatrix}$$

where $a_i \in R$ for all $i = 1, 2, \ldots, n$, is a subring of $M_n(R)$, and of the ring in part **(c)**.
(e) The previous two subrings of $M_n(R)$ are not ideals of $M_n(R)$.
(f) If $I \leq R$, then $M_n(R/I) \cong M_n(R)/M_n(I)$.
(g) The matrix A of $M_2(R)$ given by

$$A = \begin{bmatrix} 0 & 0 \\ 1 & 1 \end{bmatrix}$$

is idempotent. Moreover, any block matrix which has either A or some size of identity or zero matrix for a block and zeros everywhere else, such as

$$diag(0_1, A, A, I_2) = \begin{bmatrix} 0 & & & & & & \\ & 0 & 0 & & & & \\ & 1 & 1 & & & & \\ & & & 1 & 0 & & \\ & & & 0 & 1 & & \\ & & & & & 1 & 0 \\ & & & & & 0 & 1 \end{bmatrix}$$

is also idempotent.
(h) An upper (or lower) diagonal matrix with zeros in its diagonal is nilpotent.
(i) Using that $e^x = \sum_{n=0}^{\infty} \dfrac{x^n}{n!}$ find e^M, where $M = \begin{bmatrix} 0 & 1 & 2 & 0 & 3 \\ 0 & 0 & 1 & 2 & 0 \\ 0 & 0 & 0 & 1 & 2 \\ 0 & 0 & 0 & 0 & 1 \\ 0 & 0 & 0 & 0 & 0 \end{bmatrix}$.

3.6 Polynomial Rings

A very important family of rings is the family of polynomial rings. We will meet these rings often in this book. Chapter 9 is dedicated exclusively to polynomials.

Definition 3.86. The set of all polynomials with coefficients in a ring R is denoted by $R[x]$. The elements in $R[x]$ look like

3.6 Polynomial Rings

$$p(x) = a_n x^n + a_{n-1} x^{n-1} + \cdots + a_1 x + a_0$$

where $a_i \in R$ for all i.
The addition and multiplication of polynomials are the standard ones: combining like terms and FOILing while using the operations in R.
The largest exponent of x in $p(x)$ is called the degree of $p(x)$. The zero polynomial has no degree.

Example 3.87. $\mathbb{Z}_2[x]$ is a commutative ring. The elements here look like

$$a_n x^n + a_{n-1} x^{n-1} + \cdots + a_1 x + a_0$$

where $a_i \in \mathbb{Z}_2$ for all i (in other words $a_i = 0$ or 1, for all i).
The addition and multiplication of polynomials is the standard one, keeping in mind that '$2 = 0$' since the coefficients live in $\mathbb{Z}_2$. It is easy to see that $\mathbb{Z}_2[x]$ is closed under addition and multiplication (due to closure in $\mathbb{Z}_2$). Similarly, commutativity for $+$ is automatic. Distributivity is just a corollary of the same law in $\mathbb{Z}_2$. Finally, 0 and 1 are the two identities, and the additive inverse of a polynomial is itself.

Exercise 3.88. Let R be a ring.
(a) Prove that $R[x]$ is, in fact, a ring under the operations described in Definition 3.86.
(b) Check all the claims made in Example 3.87.
(c) Find all the elements in the set of polynomials of $\mathbb{Z}_3[x]$ of degree at most two (union the zero polynomial). Is this set a ring?
(d) Construct an injective ring homomorphism $\phi : R \to R[x]$. Note that this shows that '$R \subseteq R[x]$'.
(e) Prove that $R[x]$ has a one if and only of R has a one.
(f) Prove that $R[x]$ is commutative if and only if R is commutative.
(g) Is the set of polynomials in $\mathbb{Q}[x]$ of degree at most 10, union the zero polynomial, a ring?
(h) Prove that $\psi : R[x] \to R$, defined by $\psi(p(x)) = p(0)$ is a ring homomorphism.
(i) If we defined ψ in part (g) by $\psi(p(x)) = p(r)$, for a fixed $r \in R$, would ψ still be a ring homomorphism?

Exercise 3.89. Prove the following claims.
(a) If R has no zero divisors then

$$deg(p(x)q(x)) = deg(p(x)) + deg(q(x)) \quad \forall p(x), q(x) \in R[x].$$

(b) If R has zero divisors then there are polynomials $p(x), q(x) \in R[x]$ such that

$$deg(p(x)q(x)) \neq deg(p(x)) + deg(q(x)).$$

(c) If R has no zero divisors then $R[x]$ has no zero divisors either and $R[x]^* = R^*$.

Exercise 3.90.
(a) The ideal (x) of $\mathbb{R}[x]$ is maximal.
(b) Let $p(x) \in \mathbb{R}[x]$ be of degree one. Prove that the ideal generated by $p(x)$ is maximal in $\mathbb{R}[x]$.
Hint: Consider first the case $p(x) = x - a$. Use the Taylor expansion of a polynomial in $\mathbb{R}[x]$ centered at $x = a$.
(c) The set

$$I = \{2p(x) + xq(x) \in \mathbb{Z}[x];\ p(x), q(x) \in \mathbb{Z}[x]\}$$

is not a principal ideal. Conclude that $\mathbb{Z}[x]$ is not a PID.
(d) Show that $\mathbb{Z}[x]/(x) \cong \mathbb{Z}$, and thus is a PID.

Chapter 4
Approximation

Abstract In this chapter we investigate what it means for two objects to be "close". To this end we will think about how to measure distance between various objects. We introduce the notion of a sequence and use it to carry out approximations. Our main focus will be on sequences of real numbers and sequences of real valued functions, but we will venture into some less commonly considered sequences, such as sequences of groups and matrices.

Sometimes it is useful to know how large your zero is.

Anonymous

Getting Your Hands Dirty

Exercise 4.1. Find a number close to 2011. Now find another even closer. Do the same for π.

Exercise 4.2. Find functions that are close to e^x. Be specific in what sense your choice of functions are close to e^x.

Exercise 4.3. Discuss various *different* ways that two points in $\mathbb{R}^2$ can be considered to be close.

Exercise 4.4. Give an approximation for the number $\sqrt[5]{2}$. Now give successive rational approximations for this number.

Exercise 4.5. In your previous courses you have been exposed to the process of exponentiation and learned about exponential functions. Suppose that A is an $n \times n$ matrix. Given what you know about exponentiation, give a definition of what it might mean to exponentiate A, i.e. what could expression e^A possibly mean?

4.1 Metrics and Metric Spaces

Definition 4.6. A metric space (X, d) consists of a set X and a function $d : X \times X \to \mathbb{R}$ satisfying the following conditions:
(a) $d(x, y) = 0$ if and only if $x = y$.
(b) $d(x, y) = d(y, x)$ for all $x, y \in X$
(c) $d(x, z) \leq d(x, y) + d(y, z)$ for all $x, y, z \in X$.

Exercise 4.7. Show that if $d : X \times X \to \mathbb{R}$ is a metric, then $d(x, y) \geq 0$ for all $x, y \in X$.

Exercise 4.8. Suppose d_1 and d_2 are both metrics on a set X. Which of the following (if any) is then also a metric on $X \times X$?
(a) $d_1 \cdot d_2$

(b) $d_1 + d_2$
(c) $d_1 2^{d_2}$

Exercise 4.9. (*Discrete Metric*) Let X be any non-empty set. Define $d : X \times X \to \mathbb{R}$ by

$$d(x,y) = \begin{cases} 0 & \text{if } x = y \\ 1 & \text{otherwise} \end{cases}$$

Show that d is a metric on X.

Exercise 4.10. (*Euclidean space*) Show that $(\mathbb{R}, d)$ is a metric space where $d(x,y) = |x-y|$. If

$$D(\mathbf{u}, \mathbf{v}) = \sqrt{\langle \mathbf{u}-\mathbf{v}, \mathbf{u}-\mathbf{v} \rangle},$$

show that $(\mathbb{R}^n, D)$ is a metric space (see Definition 1.44 for the definition of the dot product of two vectors).

Exercise 4.11. (*The Poincaré disk*) Let $X = \{(x,y) \in \mathbb{R}^2 \mid x^2 + y^2 < 1\}$. Let

$$\delta(\mathbf{u},\mathbf{v}) := 2 \frac{\langle \mathbf{u}-\mathbf{v}, \mathbf{u}-\mathbf{v} \rangle}{(1-\langle \mathbf{v},\mathbf{v}\rangle)(1-\langle \mathbf{u},\mathbf{u}\rangle)}.$$

Show that $(X, \operatorname{arccosh}(1 + \delta(\cdot, \cdot)))$ is a metric space. Is $(X, \delta(\cdot, \cdot))$ a metric space?

Definition 4.12. The *open metric ball* with radius $r \in \mathbb{R}$ centered at a point x in a metric space (X,d) is the set defined as

$$B_{r,d}(x) = \{y \in X \mid d(x,y) < r\}.$$

In this definition we may allow $r = +\infty$, in which case we get the entire space: $B_{+\infty,d}(x) = X$ for any $x \in X$! In the case the metric d is clear from the context we will write $B_r(x)$ instead of $B_{r,d}(x)$.

Definition 4.13. Two metrics d_1 and d_2 on a set X are called *equivalent* if for any $x \in X$, and $r \in \mathbb{R}^+$, there exists $R_1, R_2 \in \mathbb{R}^+$ such that

$$B_{R_1,d_1}(x) \subseteq B_{r,d_2}(x) \quad \text{and}$$
$$B_{R_2,d_2}(x) \subseteq B_{r,d_1}(x)$$

Exercise 4.14. Show that "being equivalent metrics" is an equivalence relation on the set of metrics on a set X.

Exercise 4.15. Let $X = M_2(\mathbb{R})$. Define a function $d : X \times X \to \mathbb{R}$ which makes (X,d) into a metric space.

Exercise 4.16. Let $X = \mathbb{Z}_p$ where p is a prime number. Try to define a metric on this space. Discuss your ideas in detail.

If (X,d) is a metric space, it makes sense to talk about distances between various elements of X. In other words, given two elements $x,y \in X$ we can declare that the *distance* between x and y is $d(x,y)$. It is clear that the numerical value of the distance between two points in a metric space inherently depends on the particular metric we use. However, other metric-dependent notions, like continuity for example, only depend on the equivalence class of the metric.

Exercise 4.17. Determine whether the following statement is true. If true give a proof, if not, give a counterexample.

Two metrics d_1 and d_2 are equivalent if and only if there exist constants c and C such that $cd_1(x,y) \leq d_2(x,y) \leq Cd_1(x,y)$ for every $x,y \in X$.

Exercise 4.18. Suppose (X,d) is a metric space. Show that $d(\cdot,\cdot)$ and

(a) $\dfrac{d(\cdot,\cdot)}{s}$, where $s \in \mathbb{R}^+$

(b) $\dfrac{d(\cdot,\cdot)}{1+d(\cdot,\cdot)}$

are equivalent metrics.

4.2 Sequences

Now that we have an idea of how to think about certain objects being close, we can think about successive approximations to points in various spaces.

Example 4.19. The rational numbers $1.4, 1.41, 1.414, 1.4142$ are successively better approximations to $\sqrt{2}$. We use the word "better" in the following sense:

$$|\sqrt{2} - 1.4142| < |\sqrt{2} - 1.414| < |\sqrt{2} - 1.41| < |\sqrt{2} - 1.4|$$

If you think about counting the successive approximations, there are four in Example 4.19, but there is no reason why we could not have more - we could have (countably) infinitely many. A sequence is in essence no more than a countably infinite succession of objects in a space. For example, if we continued our successive approximations of $\sqrt{2}$, we would have obtained a sequence of real numbers.

Definition 4.20. A *sequence* is a function $f : \mathbb{N} \to X$ where X is any set of objects. Typically, instead of denoting the values of f as $f(n)$, we write f_n for notational ease. In line with this notational convention we will write $(f_n)_{n=1}^\infty$ for a sequence. Note that each f_n is an element of X.

Example 4.21. Let X contain the permutation group S_n on n numbers for each $n = 1, 2, 3, \ldots$. Then $f : \mathbb{N} \to X$ given by $(S_n)_{n=1}^\infty$ is a sequence of permutation groups. A different sequence is given by $f_n = S_5$ for all n.

Exercise 4.22. Give an explicit example of a sequence of elements in $\mathbb{Z}_{11}$.

Definition 4.23. A *sequence of real numbers* is merely a sequence whose range is a subset of $\mathbb{R}$.

Example 4.24. Let X contain all real numbers between 0 and 1. Then $f : \mathbb{N} \to X$ given by $f(n) = \dfrac{1}{n^n}$ is a sequence of real numbers.

Definition 4.25. Let (X, d) be a metric space. A sequence $f : \mathbb{N} \to X$ is *bounded*, if there exists a metric ball $B_r(x) \subseteq X$ such that $f_n \in B_r(x)$ for all $n \in \mathbb{N}$. If no such ball exists, we call the sequence *unbounded*.

Exercise 4.26. Show that a sequence $(f_n)_{n=1}^\infty$ is bounded if and only if there exists a metric ball $B_r(x)$ and a number N, such that $f_n \in B_r(x)$ for all $n \geq N$.

Exercise 4.27. Show that the sequence $(n)_{n=1}^\infty$ is unbounded where the natural numbers are understood as a metric space $(\mathbb{N}, |\cdot|)$.

In case $(X, d) = (\mathbb{R}, |x - y|)$, a metric ball is simply an interval. Consequently, a set of real numbers S is said to be bounded if and only if there is an interval $I = (a, b)$ with $-\infty < a < b < \infty$, such that $S \subseteq I$. In the setting of $(\mathbb{R}, |\cdot|)$ it makes sense to talk about a set being bounded from below and/or above, but this is not the case in general.

As the reader will recall from her introductory calculus course, a real valued function f has a limit L at infinity, if for all $\varepsilon > 0$ there exists an M, such that $x > M$ implies that $|f(x) - L| < \varepsilon$. One can of course extend this idea to functions which map to a metric space (X, d). In particular, we can look at functions $f : \mathbb{N} \to X$ and say that such an f has a limit L at infinity if for all $\varepsilon > 0$ there exists an $N \in \mathbb{N}$, such that $n > N$ implies that $d(f(n), L) < \varepsilon$.

Definition 4.28. Let (X, d) be a metric space. A sequence $f : \mathbb{N} \to X$ is called *convergent* if f has a limit L at infinity. In this case we say that $(f_n)_{n=1}^\infty$ converges to L and will use the notation $f_n \to L$ or $\lim\limits_{n \to \infty} f_n = L$.

Exercise 4.29. Give two different examples of convergent sequences.

Example 4.30. Here we provide some examples to illustrate that not every bounded sequence is convergent.
(a) First consider the space $(\mathbb{R}, |x-y|)$ and the sequence $f(n) = (-1)^n$. The only possible limits for this sequence are 1 and -1. We show that 1 can't be a limit. To this end let $\varepsilon = 1/2$ and consider the ball $B_\varepsilon(1)$. We see that if $n = 2k+1$, then $f(n) \notin B_\varepsilon(1)$. In other words there does not exist an $N \in \mathbb{N}$ such that $n > N$ would imply that $|f(n) - 1| < 1/2$. This means that 1 can't be the limit of the sequence. The argument for -1 goes similarly, and establishes that f is not convergent. However, $f(n) \in B_2(0)$ for all n, hence f is a bounded sequence.
(b) Our next example is the sequence $f : \mathbb{N} \to \mathbb{C}$ given by

$$f(n) = e^{(2\pi i n)/k} \quad \text{for some} \quad k \in \mathbb{N}.$$

It follows that every term of the sequence is a kth root of unity. (The metric we are using is the Euclidean metric on $\mathbb{C}$). Since $|f(n)| = 1$ for all n, we see that $f(n) \subseteq B_2(0)$, as the illustration shows. In other words this is a bounded sequence. The reason why it is not convergent is analogous to that of our first example. The details are left to the reader.

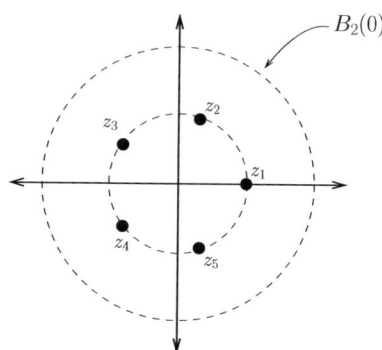

Fig. 4.1 The five 5th roots of unity $(k=5)$

(c) Consider the set $\mathbb{Z}_5$ equipped with the discrete metric. Let $f : \mathbb{N} \to \mathbb{Z}_5$ be defined by $f(n) = [n]$. Since $\mathbb{Z}_5 \subseteq B_{2,d}([4])$, we see that f is a bounded sequence. However it is not convergent for reasons similar to those in 1 and 2. We leave the details as an exercise.

Exercise 4.31. Prove that the sequence in Example 4.30 part 2 is not convergent. Start by reasoning why the only possible limits are the k distinct k^{th} roots of unity.

Exercise 4.32. Prove that the sequence in Example 4.30 part 3 is not convergent. Start by reasoning why the only possible limits are $[0], [1], [2], [3]$ and $[4]$.

Lemma 4.33. *Every convergent sequence of real numbers is bounded.*

Exercise 4.34. Prove Lemma 4.33. Note that there is no metric specified in the statement of the lemma, thus you will have to work with the generic metric space $(\mathbb{R}, d)$. Follow this outline:
(a) Use the definition of convergence to show that a metric ball centered at the limit of the sequence with radius 1 contains all but finitely many terms of the sequence.
(b) Show that one can increase the radius from 1 to a sufficiently large number so as to include all remaining terms.

Definition 4.35. Let X be an ordered set. A sequence $f : \mathbb{N} \to X$ is called *increasing* if $f_n \leq f_{n+1}$ for all n. If the inequality holds strictly for all n, we call the sequence *strictly increasing*. A sequence $f : \mathbb{N} \to X$ is called *decreasing* if $f_n \geq f_{n+1}$ for all n. If the inequality holds strictly for all n, we call the sequence *strictly decreasing*.

4.2 Sequences

Definition 4.36. Let X be an ordered set. A sequence $f : \mathbb{N} \to X$ is called *monotone* if it is either increasing or decreasing.

Exercise 4.37. Give an example of each of the following:
(a) an increasing sequence of real numbers
(b) a strictly decreasing sequence of real numbers
(c) a not monotone bounded sequence of real numbers
(d) a not monotone convergent sequence of real numbers

4.2.1 On the Convergence of Real Sequences

Proposition 4.38. *Let $(s_n)_{n=1}^\infty$ and $(t_n)_{n=1}^\infty$ be convergent sequences of real numbers with limits s and t respectively and let $c \in \mathbb{R}$. The sequences*
(a) $(s_n + t_n)_{n=1}^\infty$
(b) $(s_n t_n)_{n=1}^\infty$
(c) $(cs_n)_{n=1}^\infty$
are all convergent with the obvious limits.

Exercise 4.39. Prove Proposition 4.38.
Hint: you may want to make use of Lemma 4.33 and the following equalities:
(a) $(s_n + t_n) - (s + t) = (s_n - s) + (t_n - t)$
(b) $(s_n t_n - st) = (s_n t_n - st_n) + (st_n - st)$, and
(c) $(cs_n) - (cs) = c(s_n - s)$.

As we have demonstrated in Example 4.30, not every bounded sequence of real numbers is convergent. However, if we impose one extra condition in addition to being bounded, we will get an equivalent characterization of a sequence of real numbers being convergent.

Theorem 4.40 (Monotone Convergence). *Let $(s_n)_{n=1}^\infty$ be a sequence of real numbers. If $(s_n)_{n=1}^\infty$ is bounded and monotone then it is convergent.*

Exercise 4.41. Prove Theorem 4.40 by following this outline:
(a) Argue why we may assume WLOG that $(s_n)_{n=1}^\infty$ is increasing.
(b) Argue why $S = \sup\{s_n \mid n \in \mathbb{N}\}$ is finite.
(c) Show that $\lim_{n \to \infty} s_n = S$.

Exercise 4.42. Prove that the set of convergent sequences of real numbers equipped with the natural operations is a commutative ring with one. Is this ring a field?

Exercise 4.43. Is the set of all nonzero sequences of real numbers a ring? Is it a field?

Definition 4.44. A sequence of real numbers $(s_n)_{n=1}^\infty$ is called a *Cauchy sequence* if for all $\varepsilon > 0$ there exists an $N \in \mathbb{N}$ such that $m, n > N$ implies that
$$|s_n - s_m| < \varepsilon.$$

It turns out that for sequences of real numbers being convergent is equivalent to being Cauchy, which is the content of the following proposition.

Proposition 4.45. *If $(s_n)_{n=1}^\infty$ is a sequence in $(\mathbb{R}, |\cdot|)$ then $(s_n)_{n=1}^\infty$ is convergent if and only if it is Cauchy.*

Exercise 4.46. Prove Proposition 4.45. Convergent $\Longrightarrow$ Cauchy is fairly easy, using the 'trick'

$$|s_n - s_m| \leq |s_n - s| + |s_m - s|.$$

For the reverse suppose that (s_n) is Cauchy.
(a) Prove that (s_n) is bounded.
(b) Use Proposition 4.78 to conclude that (s_n) has a convergent subsequence, say $s_{n_k} \to a$.
(c) Use the Cauchy property to establish that (s_n) itself converges to a.
Does this proposition hold for a sequence $(s_n)_{n=1}^{\infty}$ in *any* metric space (X, d)?

Example 4.47. The equivalence of being convergent and Cauchy can be used to show that a particular sequence $(s_n)_{n=1}^{\infty}$ is convergent. Of course, one could go the other way. For example, to determine that the sequence $(x_n)_{n=1}^{\infty}$ satisfying

$$|x_n| \leq \frac{1+n}{1+2n+n^2}, \quad \forall n$$

is Cauchy, all one would have to observe is that this sequence is convergent, since

$$0 \leq |x_n| < \frac{1+n}{1+2n+n^2} \to 0 \quad \text{as} \quad n \to \infty.$$

Exercise 4.48. Prove that a Cauchy sequence of integers is eventually constant. In other words show that if $(s_n)_{n=1}^{\infty}$ is a Cauchy sequence of integers, then there is an $N \in \mathbb{N}$ such that $n > N$ implies $s_n = s_N$.

Exercise 4.49. Suppose that $(x_n)_{n=1}^{\infty}$ is a convergent sequence of real numbers. Prove that the limit of this sequence is unique.

Exercise 4.50 (Squeeze Theorem). Suppose that $(x_n)_{n=1}^{\infty}$ and $(y_n)_{n=1}^{\infty}$ are two convergent sequences of real numbers, such that

$$(\dagger) \quad \lim_{n \to \infty} x_n = \lim_{n \to \infty} y_n = 0.$$

Prove that if $(z_n)_{n=1}^{\infty}$ is a sequence of real numbers such that $x_n \leq z_n \leq y_n$ for all $n \in \mathbb{N}$, then $\lim_{n \to \infty} z_n = 0$. How badly can this result fail if we remove the condition $(\dagger)$?

4.2.2 Recursive Sequences

It is possible to define a sequence by specifying the nth term as a function of the first $n - 1$ terms. Of course when doing so, we need to give the value of at least the first term. Perhaps one of the most famous recursive sequences is the Fibonacci sequence given by $F_0 = F_1 = 1$ and $F_{n+2} = F_{n+1} + F_n$ for $n \geq 0$. Although the Fibonacci sequence is *not* convergent, the following limit does exist:

$$\lim_{n \to \infty} \frac{F_{n+1}}{F_n} = \phi,$$

and is called the *Golden Ratio*. An excellent source of fun information on the Golden Ratio is [32].

Exercise 4.51 (this exercise is involved so be patient!). *Assuming* that the above limit exists, find the value of ϕ using the recurrence relation $F_{n+2} = F_{n+1} + F_n$.
Now prove that the limit exists. Here are two alternative ways (after [7]) to do this:
(a) Find the *generating function* $f(x)$ for F_n by considering the series

$$(\dagger) \quad f(x) = \sum_{n=0}^{\infty} F_n x^n$$

4.2 Sequences

and using the recurrence relation. Having found the radius of convergence for (†), use the ratio test to find the limit in question.

(a) Derive Binet's Formula for F_n and compute the limit in question directly.

Example 4.52. Let $s_1 = 1$ and let $s_n = \sum_{k=1}^{n-1} s_k$. This sequence has terms

$$1, 1, 2, 4, 8, 16, 32, \ldots$$

so we could instead of the recursive definition give the sequence as $s_1 = 1$ and $s_n = 2^{n-1}$, for $n \geq 2$.

Example 4.53. Let $a_1 = 1$ and $a_n = 3\sqrt{a_{n-1}} - 1$, for $n \geq 2$. Then the first few terms of the sequence are given by

$$1, 2, 3\sqrt{2} - 1, 9\sqrt{2} - 4, \ldots$$

With the help of Theorem 4.40 one can find out whether a given sequence is convergent, without having to know all or any of its terms.

Example 4.54. Let $0 < a < 1$ be any real number. Set $s_1 = a$ and $s_{n+1} = 1 - \sqrt{1 - s_n}$, for $n \geq 2$. We claim that this sequence converges to 0. To establish this fact we show that $(s_n)_{n=1}^{\infty}$ is bounded and monotone decreasing. It is clear that $0 < s_1 < 1$. If $0 < s_n < 1$, we see that $0 < 1 - s_n < 1$, and consequently $1 - s_n < \sqrt{1 - s_n}$. But then $1 > s_n > 1 - \sqrt{1 - s_n} =: s_{n+1} > 0$. This establishes monotonicity and boundedness. A simple calculation shows that the only two possible limits are 0 and 1, and since the sequence is decreasing, we conclude that it must converge to 0.

Exercise 4.55. Let $a_1 = 3$ and $a_{n+1} = \frac{1}{3}\sqrt{a_n + 1}$, for $n \geq 2$. Determine whether $(a_n)_{n=1}^{\infty}$ is convergent.

Exercise 4.56. Let $(s_n)_{n=1}^{\infty}$ be a sequence of real numbers satisfying

$$|s_{n+1} - s_n| < \frac{1}{2011^n} \qquad \text{for all} \quad n \in \mathbb{N}. \tag{4.1}$$

Prove that $(s_n)_{n=1}^{\infty}$ is convergent.
Hint: show that $(s_n)_{n=1}^{\infty}$ is Cauchy and use Proposition 4.45.

Example 4.57. The reader should contemplate the exact role of the number $\frac{1}{2011}$ in Exercise 4.55. One can (and should) easily verify that any number $a > 1$ could take the place of 2011 without changing the conclusion. In more generality, any exponential growth in the denominator suffices.

The point of this example is to show that we *cannot* replace exponential growth by polynomial growth. To this end consider the sequence $s_n = \ln n$. This sequence is not convergent, since $\lim_{n \to \infty} s_n = +\infty$. However, it does satisfy a bound with polynomial growth in the denominator, as we next demonstrate:

$$|s_{n+1} - s_n| = \ln(n+1) - \ln n = \ln\left(\frac{n+1}{n}\right) = \ln\left(1 + \frac{1}{n}\right).$$

Since for any $n = 1, 2, 3, \ldots$

$$e^{\frac{1}{n}} = 1 + \frac{1}{n} + \sum_{k=2}^{\infty} \frac{1}{n^k k!} > 1 + \frac{1}{n},$$

after taking the natural log of both sides we see that

$$\frac{1}{n} > \ln\left(1 + \frac{1}{n}\right) = |s_{n+1} - s_n|.$$

Exercise 4.58. Consider the following family of sequences. $s_1 = a$ and $s_{n+1} = k_1 s_n + k_2 b$, for $n \geq 2$, for some $a, b \in \mathbb{R}$ and $k_1, k_2 \in [0, 1]$. Determine the convergence properties of these sequences as a function of a, b, k_1 and k_2. (Think of the vector (a, b, k_1, k_2) as a point in $\mathbb{R}^4$, and determine the subset of $\mathbb{R}^4$ that gives convergence). Is this subset of $\mathbb{R}^4$ actually a subspace of $\mathbb{R}^4$?

4.2.3 Sequences of Functions

In order for us to be able to talk about sequences of functions, we need to clarify which kinds of functions we mean, i.e. we need to specify the metric space (X,d) where we find our "sequence of functions".

Definition 4.59 (Uniform Metric). Let X be the space of real valued continuous functions on $[0,1]$. We define the following function $d_u : X \times X \to \mathbb{R}$:

$$d_u(f,g) := \max_{x \in [0,1]} |f(x) - g(x)|.$$

Exercise 4.60. Prove that the function d_u in Definition 4.59 is a metric on the set of continuous functions on $[0,1]$.

Exercise 4.61. Show that the word "continuous" in the previous exercise could be replaced by "bounded" and we would still have a metric space.

Note that every constant function $f_n = \dfrac{1}{n}$ is in X, and there are many more. (X,d) is a metric space (Exercise 4.67), thus it makes sense to talk about a sequence $f : \mathbb{N} \to X$. Terms of this sequence are real valued functions! Thus

$$f_1, f_2, f_3, \ldots, f_{2011}, \ldots$$

are all functions. If $e_a : X \to \mathbb{R}$ is the evaluation map at $a \in [0,1]$ (i.e. $e_a(f) = f(a)$), then $e_a \circ f : \mathbb{N} \to \mathbb{R}$ is a sequence of real numbers:

$$f_1(a), f_2(a), f_3(a), \ldots, f_{2011}(a), \ldots$$

We can talk about the convergence of both of these sequences but we need to make sure that we are working in the correct space! In fact, corresponding to the two different sequences we associate two different notions of convergence of a sequence of functions.

Definition 4.62. Let (X,d) be the space as in Example 4.59. We say that a sequence $f : \mathbb{N} \to X$ *converges uniformly* to a function g on $[0,1]$ if and only if

$$\lim_{n \to \infty} d_u(f_n, g) = 0.$$

Definition 4.63. Let (X,d) be as in Example 4.59. We say that a sequence $f : \mathbb{N} \to X$ is *uniformly Cauchy* on $[0,1]$ if and only if for all $\varepsilon > 0$ there exists an $N \in \mathbb{N}$ such that $m, n \geq N$ implies that

$$d_u(f_n, f_m) < \varepsilon.$$

Proposition 4.64. *Let (X,d) be as in Example 4.59. A sequence $f : \mathbb{N} \to (X, d_u)$ converges if and only if it is uniformly Cauchy.*

Exercise 4.65. Prove Proposition 4.64.

Exercise 4.66. Can the space (X,d) in Example 4.59 be generalized so that Definition 4.62, Definition 4.63 and Proposition 4.64 still make sense and hold? If so, give such a generalization, if not explain why.

Exercise 4.67. Show that the space (X,d) in Example 4.59 is a metric space.

Exercise 4.68. Consider the setup of Example 4.59. Find a sequence of functions $(f_n)_{n=1}^{\infty}$ and an evaluation map e_a such that the sequence of real numbers $(f_n(a))$ is convergent, but the sequence $(f_n)_{n=1}^{\infty}$ is not. You may want to take a look at Figure 4.2 at the top of next page).

4.2 Sequences

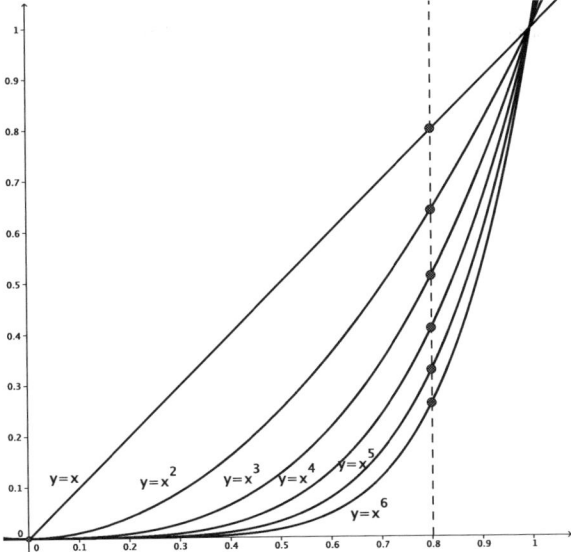

Fig. 4.2 The first few terms of $f_n(x) = x^n$ on the interval $[0,1]$

Example 4.69. Let (X,d) be the space of continuous functions on the interval $[0,1]$. Define

$$d(f,g) := \int_0^1 |f(x) - g(x)| dx.$$

There is no question of the meaning of the integral in this case, since we know from introductory Calculus that all continuous functions are integrable on a closed and bounded interval. The verification of the fact that (X,d) is a metric space is simple, except for perhaps one direction of property **(a)** in Definition 4.6. We defer this part until our discussion of integration. (See Exercise 8.54).

Exercise 4.70. Prove that the space in Example 4.69 is a metric space (except for one direction of **(a)** of Definition 4.6).

Exercise 4.71. Let $f(x) = e^x$, and let X be the set of all piecewise continuous functions on $[0,1]$. Using the function $d(\cdot,\cdot)$ as in Example 4.69, find a sequence of functions f_n converging to f. See Definition 6.51 for the definition of a piecewise continuous function.

We close this section with the discussion of a different type of convergence associated to a sequence of functions. To this end let's suppose that $(f_n)_{n=1}^\infty$ is a sequence of function on the interval $[0,1]$ (the choice of the interval isn't important here - it could be replaced by any subset of the real numbers). Choose and *fix* a point $a \in [0,1]$ and consider the sequence of real numbers $e_a \circ f : \mathbb{N} \to \mathbb{R}$. This sequence may or may not converge, it doesn't even need to be bounded a priori. The convergence of this sequence is what we refer to as *pointwise convergence*.

Definition 4.72. A sequence of real valued functions $(f_n)_{n=1}^\infty$ on an interval I is said to converge pointwise on I if and only if $e_a \circ f_n$ is a convergent sequence of real numbers for every $a \in I$.

Exercise 4.73 (Uniform convergence implies pointwise convergence). Suppose that $(f_n)_{n=1}^\infty$ is a convergent sequence in $([0,1], d_u)$ where d_u is as in Example 4.59. Show that this sequence is pointwise convergent on $[0,1]$.

Exercise 4.74. Show by example that the converse of Exercise 4.73 is false. That is, exhibit a sequence of functions $(f_n)_{n=1}^\infty$ that converges pointwise everywhere on $[0,1]$ but which does not converge in d_u.

4.3 Subsequences

Definition 4.75. Let $f : \mathbb{N} \to X$ be any sequence and let $g : \mathbb{N} \to \mathbb{N}$ be a strictly increasing function. Then $f \circ g$ is also a sequence in X. We call $f \circ g$ a *subsequence* of f. Sometimes we use the notation s_{n_k} for a subsequence of the sequence s_n.

Lemma 4.76. *If $f : \mathbb{N} \to \mathbb{R}$ is a convergent sequence then any subsequence $f \circ g$ of f is also convergent, and*
$$\lim_{n \to \infty} f(n) = \lim_{n \to \infty} (f \circ g)(n).$$

It is clear that every sequence has as many subsequences as there are increasing functions from $\mathbb{N}$ to $\mathbb{N}$. What can we say about the existence of convergent subsequences?

Exercise 4.77. Give an example of a sequence which does not have any convergent subsequences.
Hint: This sequence can't possibly be convergent...

Proposition 4.78 (Bolzano-Weierstrass). *Every bounded sequence of real numbers has a convergent subsequence.*

Definition 4.79. Let $(s_n)_{n=1}^\infty$ be a sequence of real numbers. A term s_k is called *dominant* if $s_k \geq s_\ell$ for all $k \leq \ell$.

Exercise 4.80. Prove Proposition 4.78. As a first step, establish that every sequence of real numbers has a monotone subsequence. To do this, you may want to use the notion of a dominant term. Consider then the case of **(a)** finite and **(b)** infinitely many dominant terms in a sequence.
Once you have established that every sequence of real numbers has a monotone subsequence, invoke Theorem 4.40.

Exercise 4.81. Prove Theorem 4.78 using the following outline:
(a) Show that if $(a_n)_{n=1}^\infty, (b_n)_{n=1}^\infty$ are sequences of real numbers such that $a_n < b_n$ for all n and $[a_{n+1}, b_{n+1}] \subseteq [a_n, b_n]$ then
$$\bigcap_{n=1}^\infty [a_n, b_n] \neq \emptyset.$$
(b) If $(s_n)_{n=1}^\infty$ is bounded, then there is some interval $[a,b]$ such that $s_n \in [a,b]$. Dividing $[a,b]$ into two equal length subintervals yields $[a, \frac{b-a}{2}]$ and $[\frac{b-a}{2}, b]$ (at least) one of which contains infinitely many of the s_ns. Continue subdividing, and pick a subsequence judiciously.

4.3.1 Subsequential Limits

In light of Proposition 4.78, even non-convergent sequences, if they are bounded, have convergent subsequences. Let $SL(s_n)$ denote the set of all limits of all convergent subsequences of a sequence $(s_n)_{n=1}^\infty$:
$$SL(s_n) := \{r \in \mathbb{R} \mid \exists (s_{n_k}) \to r\}.$$

Proposition 4.78 says that if $(s_n)_{n=1}^\infty$ is bounded, then $SL(s_n)$ is non-empty. It is also clear that if $(s_n)_{n=1}^\infty$ is bounded, so is the set $SL(s_n)$. Hence both $\inf SL(s_n)$ and $\sup SL(s_n)$ are finite real numbers.

Exercise 4.82. Is the following statement true or false? Justify your answer.

If $(s_n)_{n=1}^\infty$ is unbounded, then $SL(s_n)$ must be empty.

4.3 Subsequences

Exercise 4.83. Determine whether or not the following is true or false. Suppose that $S \subseteq \mathbb{R}$ is bounded and non-empty. Suppose further that $\inf S \notin S$ and $\sup S \notin S$. Then there are sequences $f : \mathbb{N} \to S$ and $g : \mathbb{N} \to S$ converging to $\inf S$ and $\sup S$, respectively.

For notational ease we make the following definition:

Definition 4.84. Let $(s_n)_{n=1}^\infty$ be a sequence of real numbers. We define the *limit supremum* and *limit infimum* of $(s_n)_{n=1}^\infty$ to be

$$\liminf_{n \to \infty} s_n := \inf SL(s_n)$$

$$\limsup_{n \to \infty} s_n := \sup SL(s_n).$$

Alternatively, given a sequence of real numbers $(s_n)_{n=1}^\infty$, one can define $\liminf s_n$ and $\limsup s_n$ as

$$\liminf_{n \to \infty} s_n := \lim_{n \to \infty} \inf\{s_k \mid k \geq n\}$$

$$\limsup_{n \to \infty} s_n := \lim_{n \to \infty} \sup\{s_k \mid k \geq n\}.$$

Exercise 4.85. Prove that
(a) $\limsup\limits_{n \to \infty} s_n \in SL(s_n)$
(b) $\liminf\limits_{n \to \infty} s_n \in SL(s_n)$
Hint: You may want to make use of the result of Exercise 4.83.

Exercise 4.86. Prove that the two definitions given above are actually equivalent.

Exercise 4.87. Suppose that $(s_n)_{n=1}^\infty$ is a sequence of real numbers such that $|SL(s_n)| = 1$. Must $(s_n)_{n=1}^\infty$ be convergent? If so, prove it. If not, give a counterexample.

Exercise 4.88. Suppose $(s_n)_{n=1}^\infty$ is a discrete sequence of real numbers. Say all you can about what kind of a set $SL(s_n)$ can possibly be.

Exercise 4.89. Prove that if $(s_n)_{n=1}^\infty$ is a sequence of real numbers such that $|SL(s_n)| \geq 1$, then $(s_n)_{n=1}^\infty$ has infinitely many convergent subsequences.

Exercise 4.90. Let $N \in \mathbb{N}$ be given (and fixed). Construct a sequence of real numbers $(s_n)_{n=1}^\infty$ with *exactly N* distinct subsequential limits (i.e. such that $|SL(s_n)| = N$).
Hint: How many subsequential limits does the sequence $s_n = (-1)^n$ have?

Exercise 4.91. Construct a sequence of real numbers $(t_n)_{n=1}^\infty$ such that every $N \in \mathbb{N}$ is a subsequential limit of $(t_n)_{n=1}^\infty$.
Hint: Cantor's diagonalization argument might come in handy here.

Exercise 4.92. Give an example of a sequence $(s_n)_{n=1}^\infty$ such that
(a) $SL(s_n) = \{1/N, N \in \mathbb{N}\}$.
(b) $SL(s_n) = \{1/N, N \in \mathbb{N}\} \cup \{0\}$.

Exercise 4.93. Is it possible to construct a sequence of real numbers $(s_n)_{n=1}^\infty$ so that every real number is a subsequential limit of $(s_n)_{n=1}^\infty$? Substantiate your answer as well as you can.

Exercise 4.94. Suppose that $f : \mathbb{N} \to \mathbb{R}$ is a sequence of real numbers. Define

$$A(f,n) := \left|\left\{k \in \mathbb{N} \,\middle|\, k \leq |f(n)|\right\}\right|.$$

Prove the following claims:

(a) If f is a convergent sequence, then $\lim\limits_{n \to \infty} \dfrac{A(f,n)}{n} = 0$ but the converse is false.

(b) If $\lim\limits_{n\to\infty} \dfrac{A(f,n)}{n} = 1$ then f must be unbounded, but the converse is false.

(c) For any non-negative rational number q there exists a sequence $f_q : \mathbb{N} \to \mathbb{R}$ such that
$$\lim_{n\to\infty} \frac{A(f_q,n)}{n} = q.$$

Exercise 4.95. Let $f : \mathbb{N} \to \mathbb{R}$ be any sequence, and let $g : \mathbb{N} \to \mathbb{N}$ be any increasing function. We will write $f \circ g^k$ for the subsequence $f \circ \underbrace{g \circ \cdots \circ g}_{k-\text{times}}$. Prove that
$$\lim_{k\to\infty} \limsup_{n\to\infty} \frac{A(f \circ g^k, n)}{n} = 0.$$

Exercise 4.96. Let $f : \mathbb{N} \to \mathbb{R}$ be a sequence. Suppose that there exist $k, m \in \mathbb{N}$ such that
$$(f \circ g^k)(n) = (f \circ g^m)(n), \quad \forall n \in \mathbb{N}.$$

Say all you can about the function g. Does your answer change if we require that f be a convergent sequence?

Chapter 5
Add Infinitum

Abstract In this chapter we discuss infinite sums of real numbers. First we will make precise what it means to 'add infinitely many numbers together'. Then we will develop tools which will allow us to decide when such a procedure leads to a finite number, and we will spend some time on the evaluations of such sums. We will also mention some open problems in this area - sums whose analytical evaluation has eluded mathematicians to this date.

> The infinite! No other question has ever moved so profoundly the spirit of man.

David Hilbert (1862–1943)

Getting Your Hands Dirty

Exercise 5.1. What is the sum of all rational numbers in the interval $[0, 1/2]$?

Exercise 5.2. Fix an $n \in \mathbb{N}$. What is the sum of all rational numbers in the interval $[0, \frac{1}{n}]$?

Exercise 5.3. How far can a stack of n books protrude over the edge of a table without the stack falling over?

Exercise 5.4. What is the value of the infinite sum

$$1 - 1 + 1 - 1 + 1 - 1 + 1 - 1 + \cdots?$$

5.1 A Short Historical Exposition

Perhaps one of the most famous of Aesop's fables is the one involving the race between the tortoise and the hare. While the fable is an important one for its original message, it is just as important as its *alter* fable, the one it usually gets confused with. In the original fable the hare can't beat the tortoise because he falls asleep in the middle of the race, while in the confused version the hare can't beat the tortoise because it has to cover the distance to the previous position of the tortoise, during which time the tortoise has already moved ahead. This latter problem is one of Zeno's several paradoxes called "Achilles and the Tortoise". There are several others focusing on the same mathematical and philosophical problems. Some of these include the dichotomy paradox, and the arrow paradox. For a lot more on these the site [29] is an excellent source.

In any case, the central problem of these paradoxes addresses the issue of the connection between infinite and finite quantities. We should point out that there are two problems hidden here: **(a)** can a sum of infinite positive quantities be finite? **(b)** and can an infinite amount of tasks ever be carried out? We address the first (mathematical issue) in our dealings with infinite series. As Burton points out in the *History of Mathematics*:

> Although Zeno's argument confounded his contemporaries, a satisfactory explanation incorporates a now-familiar idea, the notion of a 'convergent infinite series.' [6]

Another bemusing problem that involves infinite series, is the problem of card or book stacking. The problem is certainly not as old as Zeno's paradoxes. The problem

> ...appears in physics and engineering textbooks from as early as the mid 19th century. The problem was apparently first brought to the attention of the mathematical community in 1923 when J.G. Coffin posed it in the "Problems and Solutions" section of the American Mathematical Monthly [C1923]; no solution was presented there. [20]

The question is fairly simple: How far can a stack of n books protrude over the edge of a table without the stack falling over? There is of course physics involved in the solution here, namely, you would have to stack the books so that the center of gravity of the stacked books is precisely at the edge of the table. As it turns out, the "length" of the protrusion is exactly one half of the nth harmonic number, or as we will call it later, the Nth partial sum of the harmonic series. As we shall see, the harmonic series is divergent, and as a result, one can protrude as far over the edge of the table as one wishes, given enough books.

So when did infinite series first occupy the minds of philosophers? According to Giorgio T. Bagni

> A first notion of infinite series may well have a very ancient source: Aristotle of Stagria (384-322 BC) implicitly underlined that the sum of a series of infinitely many addends (potentially considered) can be a finite quantity (*Physics*, III, VI, 206 b, 1-33). In his *Quadrature Parabolae*, Archimedes of Syracuse (287-212 BC) considered implicitly a geometric series. [4]

Thus the ancient greeks have already given infinite series some thought, but as it is the case with a lot of notions in mathematics, the first occurrences of infinite series precluded some of the analytical ideas needed to rigorously treat them. For example, the series

$$1 - 1 + 1 - 1 + 1 - 1 + 1 \cdots$$

(called Grandi's series) has been the source of some heated discussions between mathematicians, with regards to its value. Is it 0? Is it 1? Or perhaps 0.5?

> After the late 17th-century introduction of calculus in Europe, but before the advent of modern rigor, the tension between these answers fueled what has been characterized as an "endless" and "violent" dispute between mathematicians. [30]

Once the modern tools of analysis have been introduced, the debate no longer had fuel and the answer is now known to all students of calculus. We conclude the historical overview by pointing out that many mathematical giants occupied themselves with pondering questions involving infinite series. The brothers Jakob and Johann Bernoulli have both given thought to the divergence of the harmonic series. The younger Johann claimed that it took him the grueling and laborious work of a night to establish the divergence of the harmonic series, a feat his brother could not accomplish with years of work. Even Euler, whose activities were widespread across the various fields of mathematics, can count as one of his major results the evaluation of the series $\sum_{n=1}^{\infty} \frac{1}{n^2}$ (for more details see Chapter 8 of [9]).

5.2 Infinite Series

Definition 5.5. Let $(s_n)_{n=1}^{\infty}$ be a sequence of real numbers. Define

$$S_N := s_1 + s_2 + \cdots + s_N.$$

We define the infinite series $\sum_{n=1}^{\infty} s_n$ to be the limit of the sequence $(S_N)_{N=1}^{\infty}$:

$$\sum_{n=1}^{\infty} s_n := \lim_{N \to \infty} S_N. \tag{5.1}$$

5.2 Infinite Series

We call S_N the Nth partial sum of the series $\sum_{n=1}^{\infty} s_n$. If the limit in (5.1) does not exist, we say that the series diverges. Otherwise we say that the series converges. We shall use the notation $\sum_{n=1}^{\infty} s_n < +\infty$ to indicate that the series is convergent.

Example 5.6. The series $\sum_{n=1}^{\infty} \frac{1}{2^n} = \frac{1}{2} + \frac{1}{4} + \frac{1}{8} + \ldots$ converges to 1. A graphical illustration of this fact is shown in Figure 5.1

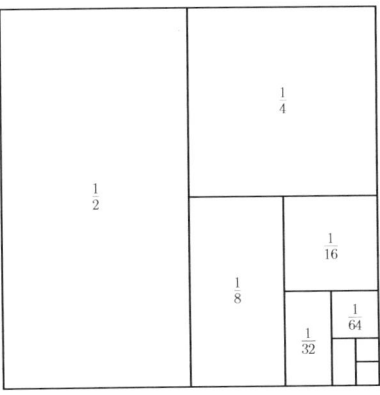

Fig. 5.1 A geometric interpretation of the series $\sum_{n=1}^{\infty} \frac{1}{2^n}$

Exercise 5.7. What effect, if any, does the convergence of the sequence $(s_n)_{n=1}^{\infty}$ have on the convergence of the series $\sum_{n=1}^{\infty} s_n$?

Definition 5.8. A series $\sum_{n=1}^{\infty} s_n$ is called absolutely convergent if and only if

$$\sum_{n=1}^{\infty} |s_n| < +\infty.$$

Exercise 5.9. Is it possible for a series of negative terms to converge, but not absolutely? Is it possible for a series to have infinitely many negative terms and to converge but not absolutely?

Definition 5.10. A series is called conditionally convergent if it is convergent, but not absolutely convergent.

Exercise 5.11. Show that a conditionally convergent series must have infinitely many positive and infinitely many negative terms.

Before we go on to investigate the relationship between the various notions of convergence for series, we make the following observation.

Lemma 5.12. *A series* $\sum_{n=1}^{\infty} s_n$ *is convergent if and only if* $\sum_{n=N}^{\infty} s_n < +\infty$ *for some* $N \in \mathbb{N}$.

Exercise 5.13. Prove Lemma 5.12.

Definition 5.14. [Cauchy Criterion] A series $\sum_{n=1}^{\infty} s_n$ is said to satisfy the Cauchy criterion if and only if for any $\varepsilon > 0$ there exists an N, such that $n \geq m \geq N$ implies that

$$\left| \sum_{k=m}^{n} s_k \right| < \varepsilon.$$

Proposition 5.15. *A series of real numbers $\sum_{n=1}^{\infty} s_n$ satisfies the Cauchy criterion if and only if it is convergent.*

Exercise 5.16. Prove Proposition 5.15.

Proposition 5.17. *If a series of real numbers $\sum_{n=1}^{\infty} s_n$ is absolutely convergent, then it is convergent.*

Exercise 5.18. Prove Proposition 5.17. Also show by example that the converse of the proposition is false.

Lemma 5.19. *If a series of real numbers $\sum_{n=1}^{\infty} s_n$ is convergent, then $\lim_{n \to \infty} s_n = 0$.*

Exercise 5.20. Prove Lemma 5.19 and also show that the converse of the lemma is false.

5.3 Computing Values of Convergent Series

Until now we might have been able to decide whether a particular series converged, but we don't have a wide variety of tools to decide what the value of the series might be. It turns out that in most cases the computation involves reducing a given series to a modification of the *geometric series*.

5.3.1 Geometric Series

Definition 5.21. Let $a \in \mathbb{R}$ be given. The geometric series in a is the series given by

$$\sum_{n=0}^{\infty} a^n.$$

Whether the above series converges depends crucially on the absolute value of a.

Exercise 5.22. Prove that the series $\sum_{n=0}^{\infty} a^n$ is convergent if and only if $|a| < 1$. To this end, you may want to develop a formula for the Nth partial sum of this series. Once you have done that, you will find that for $|a| < 1$ the series in fact sums to

$$\sum_{n=0}^{\infty} a^n = \frac{1}{1-a}.$$

Exercise 5.23. What, if anything, is wrong with the following argument?

5.3 Computing Values of Convergent Series

Let's write the series in question as
$$\sum_{n=0}^{\infty} a^n = S = 1 + a + a^2 + \cdots + a^n + \cdots$$

Then
$$aS = a + a^2 + a^3 + \cdots + a^{n+1} + \cdots$$

Forming the difference of the two we obtain
$$S(1-a) = 1.$$

It follows that
$$\sum_{n=0}^{\infty} a^n = S = \frac{1}{1-a}.$$

Exercise 5.24. Compute the value of the following series:

(a) $\sum_{n=1}^{\infty} \left(\frac{2}{3}\right)^{2n}$; (b) $\sum_{n=0}^{\infty} (\sqrt{2}-1)^{n+2}$; (c) $\sum_{n=1}^{\infty} \frac{1}{2^{n+1}}$

5.3.2 Telescoping Series

Another family of series whose values are easy to compute is the family of telescoping series.

Definition 5.25. Suppose that a series $\sum_{n=1}^{\infty} s_n$ has terms of the form $s_n = t_n - t_{n+1}$. Then the Nth partial sum of the series is given by
$$S_N = t_1 - t_N. \qquad \text{(verify this)}$$

Consequently, the value of the series is equal to $t_1 - \lim_{N \to \infty} t_N$. Such series are called telescoping, since the sum of the first N term collapses (like a telescope) to the difference of the first and N^{th} terms.

Typically, when one is trying to express a series as a telescoping series, partial fraction techniques will inevitably have to be used.

Example 5.26. The expression $\frac{n-1}{2^{n+1}}$ can be written as
$$\frac{n-1}{2^{n+1}} = \frac{n}{2^n} - \frac{n+1}{2^{n+1}}.$$

As a result, the series made up of these terms can be computed as outlined above.

Exercise 5.27. Find the value of the following series:

(a) $\sum_{n=1}^{\infty} \frac{n-1}{2^{n+1}}$; (b) $\sum_{n=1}^{\infty} \frac{n}{2^{n+1}}$; (c) $\sum_{n=1}^{\infty} \frac{1}{n(n+1)}$

Remark 5.28. The series (c) is related to the so called *triangular numbers*
$$1, 3, 6, 10, 15, 21, \ldots$$

It was Leibniz who showed that the infinite sum of the reciprocals of the triangular numbers equals 2.

Exercise 5.29. Compute
$$\sum_{n=0}^{\infty} \frac{1}{(\alpha+n)(\alpha+n+1)}$$

for any $\alpha \neq 0, -1, -2, -3, \ldots$.

For further examples we refer the reader to [17]. We have seen that the value of a telescoping series is fairly simple to compute. Yet even small changes in the generic terms of a telescoping series can make the computation of the value significantly more involved. Consider now the series given below.

$$(\dagger) \qquad \sum_{n=1}^{\infty} \frac{1}{n^2} = \zeta(2).$$

Even though it looks a lot like the last series in Exercise 5.27, computing the value of the series in ($\dagger$) is a bit more difficult. The problem of computing the sum analytically has first been proposed by Pietro Mengoli in 1644 [33]. The correct answer was first furbished by Euler in 1735. Exercise 5.30 outlines one way to compute the sum.

Exercise 5.30. Show that

$$\sum_{n=1}^{\infty} \frac{1}{n^2} = \frac{\pi^2}{6}$$

by considering the Fourier series expansion of the function $f(x) = x^{2n}$. Show that in this expansion the sine coefficients are all zero, and make the substitution $x = \pi$ to arrive to the answer.

Exercise 5.31. Explain how the result in Exercise 5.30 can be interpreted as a rational approximation of a transcendental number over $\mathbb{Q}$.

Exercise 5.32. Show that $\zeta(2) = \dfrac{\pi^2}{6}$ by expanding the function $\sin(x)$ in a **(a)** Taylor series, and **(b)** an infinite product, and equate the coefficients of x^2 in the two different representations.

Remarkably, the technique described in Exercise 5.32 works for other sums. For example,

$$\sum_{n=1}^{\infty} \frac{1}{n^4} = \frac{\pi^4}{90}, \qquad \sum_{n=1}^{\infty} \frac{1}{n^6} = \frac{\pi^6}{945}, \qquad \sum_{n=1}^{\infty} \frac{1}{n^{26}} = \frac{1315862}{11094481976030578125}\pi^{26}.$$

However, sums of the reciprocals of odd powers of the natural numbers have proven much more difficult to evaluate. To date, no one knows how to compute analytically even the simplest of these sums, namely $\sum_{n=1}^{\infty} \dfrac{1}{n^3}$. The long standing conjecture is that the sum should have the form $\dfrac{p}{q}\pi^3$ for some $p, q \in \mathbb{N}$, an assertion that undoubtedly already occurred to the reader.

5.3.3 Series With a Given Value

It is clear that we can, without much difficulty, construct series with a given sum. For example, if the desired sum is 2011, then we can simply declare $a_1 = 2011$ and $a_n = 0$ for $n \geq 2$ to get a series which sums to 2011. If we want to achieve this feat with a series with infinitely many non-zero terms we can do so. To see this select *any* sequence $a_n \to 2011$. We create a telescoping series from the terms of the sequence:

$$a_0 + (a_1 - a_0) + (a_2 - a_1) + (a_3 - a_2) + \cdots (a_n - a_{n-1}) + \cdots$$

Since the nth partial sum of this series is precisely a_n, we see that the series converges to 2011. To get our hands on a suitable sequence a_n we may consider any sequence $s_n \to 0$ and form

$$a_n = 2011 - s_n.$$

For example if we choose $s_n = \dfrac{1}{\sqrt{n+1}}$, then $a_n = 2011 - \dfrac{1}{\sqrt{n+1}}$ and consequently

5.3 Computing Values of Convergent Series

$$a_n - a_{n-1} = \frac{1}{\sqrt{n}} - \frac{1}{\sqrt{n+1}}$$

giving in turn

$$2011 = a_0 + \sum_{n=1}^{\infty}(a_n - a_{n-1}) = a_0 + \sum_{n=1}^{\infty}\left(\frac{1}{\sqrt{n}} - \frac{1}{\sqrt{n+1}}\right) = 2010 + \sum_{n=1}^{\infty}\frac{1}{\sqrt{n(n+1)}(\sqrt{n+1} + \sqrt{n})}.$$

Notice that in our calculations we have identified a series which sums to 1:

$$\sum_{n=1}^{\infty} \frac{1}{\sqrt{n(n+1)}(\sqrt{n+1} + \sqrt{n})} = 1.$$

Exercise 5.33. Find three more *distinct* series of non-zero terms that sum to 1.
Hint: identify three distinct sequences all converging to zero and use 1 in place of 2011.

5.3.4 The Harmonic Series

Although Mengoli failed to find the sum $\zeta(2)$, he did manage to show that the harmonic series diverges and he did so about 40 years before Johann Bernoulli did [9].

Theorem 5.34. *The harmonic series* $H = \sum_{n=1}^{\infty} \frac{1}{n}$ *diverges.*

Exercise 5.35. Prove Theorem 5.34 by using the following steps.
(a) Show that if $a > 1$, then $\frac{1}{a-1} + \frac{1}{a} + \frac{1}{a+1} > \frac{3}{a}$.
(b) Use your result in (a) to give increasing lower bounds for the harmonic series in the following fashion:

$$H = 1 + \left(\frac{1}{2} + \frac{1}{3} + \frac{1}{4}\right) + \left(\frac{1}{5} + \frac{1}{6} + \frac{1}{7}\right) + \frac{1}{8} + \cdots$$
$$> 1 + \frac{3}{3} + \frac{3}{6} + \cdots$$
$$= 1 + 1 + \frac{1}{2} + \frac{1}{3} + \frac{1}{4} + \cdots$$

Bernoulli's proof of Theorem 5.34 boils down to showing that $H = H + 1$, an equation that is absurd should H be any real number (see for example [9]). Thus H must be infinite. A little later we will find yet another way to prove this theorem. Namely we will show that the divergence of the integral

$$\int_1^{\infty} \frac{1}{x} dx$$

implies the divergence of the harmonic series.

5.4 Operations and Rearrangements

Definition 5.36. Let $\sum_{n=1}^{\infty} s_n$ be a series of real numbers, and let $b : \mathbb{N} \to \mathbb{N}$ be a bijection. We call $\sum_{n=1}^{\infty} s_{b(n)}$ a rearrangement of the original series $\sum_{n=1}^{\infty} s_n$.

In our opinion, one of the most astonishing facts about infinite series of real numbers is contained in the following two propositions.

Proposition 5.37. *Suppose that $\sum_{n=1}^{\infty} s_n$ is absolutely convergent with $\sum_{n=1}^{\infty} s_n = v$. Then for any bijection $b : \mathbb{N} \to \mathbb{N}$ we also have*
$$\sum_{n=1}^{\infty} s_{b(n)} = v.$$

In other words, any rearrangement of an absolutely convergent series is convergent, and all rearrangements converge to the same value.

Exercise 5.38. Prove Proposition 5.37 using the following outline:
Suppose that $\sum_{n=1}^{\infty} s_n$ is absolutely convergent with value S, and let $\sigma : \mathbb{N} \to \mathbb{N}$ be a bijection.
(a) Show that for any $N \in \mathbb{N}$, there exists an $M \in \mathbb{N}$ such that $n \geq M$ implies $\sigma(n) \geq N$.
(b) Show that $\{s_1, s_2, \ldots, s_N\} \subseteq \{s_{\sigma(1)}, s_{\sigma(2)}, \ldots, s_{\sigma(M)}\}$
(c) Show that for any $\varepsilon > 0$, there is an $N \in \mathbb{N}$ such that
$$\sum_{n=N+1}^{\infty} |s_n| < \frac{\varepsilon}{2}$$
(d) Let $T_K = \sum_{n=1}^{K} s_{\sigma(n)}$. Show that if $K \gg 1$, then
$$|T_K - S| \leq |T_K - S_N| + |S_N - S| < \varepsilon$$

Proposition 5.39. *Suppose that the series $\sum_{n=1}^{\infty} s_n$ is conditionally convergent. Given any real number $r \in \mathbb{R}$, there exists a rearrangement of this series converging to r.*

Exercise 5.40. Prove Proposition 5.39 using the following outline: Suppose that $\sum_{n=1}^{\infty} s_n$ is conditionally convergent, and let $r \in \mathbb{R}$ be given. We will give an outline of the proof that there is a rearrangement of $\sum_{n=1}^{\infty} s_n$ which converges to r.
(a) Show that if $s_n = s_n^+ - s_n^-$, then we have $\sum_{n=1}^{\infty} s_n^+ = \sum_{n=1}^{\infty} s_n^- = +\infty$
(b) Show that there is a sequence of integers $k_1 < m_1 < k_2 < m_2 < \ldots$ such that
$$|S_{k_j} - r| \leq s^+_{k_1 + \sum_{i=1}^{j-1}(k_{i+1} - m_i)} \qquad \forall j \in \mathbb{N}$$
and
$$|S_{m_j} - r| \leq s^-_{\sum_{i=1}^{j}(m_i - k_i)} \qquad \forall j \in \mathbb{N}$$

5.5 Convergence Tests

(c) Show that as $j \to \infty$, we have
$$s^+_{k_1 + \sum_{i=1}^{j-1}(k_{i+1} - m_i)} \to 0$$
and
$$s^-_{\sum_{i=1}^{j}(m_i - k_i)} \to 0.$$

(d) Show that $S_{k_j} \to r$ and $S_{m_j} \to r$.

(e) Argue that S_{k_j} and S_{m_j} are partial sums of a rearrangement of $\sum_{n=1}^{\infty} s_n$.

(f) Prove that if $n \in \mathbb{N}$, then there exists a $j \in \mathbb{N}$ such that
$$S_{m_j} \leq S_n \leq \max\{S_{k_j}, S_{k_{j+1}}\}$$

(g) Conclude that $\lim_{n \to \infty} S_n = r$

Exercise 5.41. Prove that if $\sum_{n=1}^{\infty} s_n$ is a conditionally convergent series with $\sum_{n=1}^{\infty} s_n = v$, then there are infinitely many rearrangements of this series which all converge to v.

Example 5.42. The alternating harmonic series converges:
$$\sum_{n=1}^{\infty} \frac{(-1)^n}{n} = -\ln 2.$$

Since this is a conditionally convergent series, rearranging the terms could have an effect on the value (and the convergence) of the series. For example, we can rearrange this series as follows:
$$-1 + \left(\frac{1}{2} + \frac{1}{4} + \frac{1}{6} + \frac{1}{8}\right) - \frac{1}{3} + \left(\frac{1}{10} + \frac{1}{12} + \frac{1}{14} + \frac{1}{16} + \frac{1}{18}\right) - \frac{1}{5} + \cdots$$

If we do so, we necessarily obtain either a divergent series, or one whose value is positive.

Lemma 5.43. *Suppose that $\sum_{n=1}^{\infty} s_n$ and $\sum_{n=1}^{\infty} t_n$ are two convergent series. Let $r \in \mathbb{R}$ be any real number. Then*
(a) $\sum_{n=1}^{\infty}(s_n + t_n)$ *is convergent.*
(b) $\sum_{n=1}^{\infty} r s_n$ *is convergent.*

Exercise 5.44. Prove Lemma 5.43.

Exercise 5.45. If $\sum_{n=1}^{\infty} s_n$ and $\sum_{n=1}^{\infty} t_n$ are convergent series, must the series $\sum_{n=1}^{\infty} s_n t_n$ converge? If so, prove it. If not, give a counterexample. (You may want to think about how this problem is related to Theorem 5.57)

5.5 Convergence Tests

We begin with perhaps the simplest of all convergence tests, the comparison test.

Theorem 5.46. *[Comparison Test] Let $\sum_{n=1}^{\infty} a_n$ be a series of non-negative terms and let $b_k \geq a_k$ for large k. Then*

(a) If $\sum_{n=1}^{\infty} b_n < +\infty$ then $\sum_{n=1}^{\infty} a_n < +\infty$

(b) If $\sum_{n=1}^{\infty} a_n = +\infty$ then $\sum_{n=1}^{\infty} b_n = +\infty$.

Exercise 5.47. Prove Theorem 5.46.

Theorem 5.48. *[Root Test] Let $\sum_{n=1}^{\infty} a_n$ be a series of real numbers. Set $r := \limsup_{n \to \infty} |a_n|^{1/n}$.*

(a) If $r < 1$ then $\sum_{n=1}^{\infty} a_n$ converges absolutely.

(b) If $r > 1$, then $\sum_{n=1}^{\infty} a_n$ diverges.

Exercise 5.49. Prove Theorem 5.48 by establishing first the following auxiliary result: for a real sequence $(x_n)_{n=1}^{\infty}$,
(a) if $\limsup_{n \to \infty} x_n < x$, then $x_n < x$ for $k \gg 1$.
(b) if $\limsup_{n \to \infty} x_n > x$, then $x_n > x$ for infinitely many n.
(c) if $x_n \to x$, then $\limsup_{n \to \infty} x_n = x$.

Theorem 5.50. *[Ratio Test] Let $\sum_{n=1}^{\infty} a_n$ be a series with non-zero terms. Then*

(a) The series converges absolutely if $\limsup_{n \to \infty} \frac{|a_{n+1}|}{|a_n|} < 1$.

(b) The series diverges if $\liminf_{n \to \infty} \frac{|a_{n+1}|}{|a_n|} > 1$.

(c) If $\liminf_{n \to \infty} \frac{|a_{n+1}|}{|a_n|} \leq 1 \leq \limsup_{n \to \infty} \frac{|a_{n+1}|}{|a_n|}$ the test is inconclusive.

Exercise 5.51. Show that if $(a_n)_{n=1}^{\infty}$ is any sequence of non-zero numbers, then

$$\liminf_{n \to \infty} \frac{|a_{n+1}|}{|a_n|} \leq \liminf_{n \to \infty} |a_n|^{1/n} \leq \limsup_{n \to \infty} |a_n|^{1/n} \leq \limsup_{n \to \infty} \frac{|a_{n+1}|}{|a_n|}$$

Exercise 5.52. Use the result in Exercise 5.51 to prove Theorem 5.50.

We have spent some time on showing that the harmonic series diverges. However, as we claimed in Example 5.42, the series $\sum_{n=1}^{\infty} \frac{(-1)^n}{n}$ converges to the value $-\ln 2$. The following tests will help us decide whether a series with terms whose signs alternate converges.

Theorem 5.53. *[Dirichlet's Test] For $k \in \mathbb{N}$, let $a_k, b_k \in \mathbb{R}$. If the sequence of partial sums $s_n := \sum_{k=1}^{n} a_k$ is bounded and $b_k \searrow 0$ as $k \to \infty$, then $\sum_{k=1}^{\infty} a_k b_k < +\infty$.*

5.5 Convergence Tests

Exercise 5.54. Prove Theorem 5.53 by showing that under the hypotheses of the test, the sequence of partial sums of $\sum_{k=1}^{\infty} a_k b_k$ form a Cauchy sequence. In your effort you may want to make use of the following (which you also have to prove): if $A_{n,m} = \sum_{k=m}^{n} a_k$, then

$$\sum_{k=m}^{n} a_k b_k = A_{n,m} b_n - \sum_{k=m}^{n-1} A_{k,m}(b_{k+1} - b_k).$$

Theorem 5.55. *[Alternating Series Test] If $a_n \searrow 0$ as $n \to \infty$, then the series $\sum_{n=0}^{\infty} (-1)^n a_n$ converges.*

Exercise 5.56. Prove Theorem 5.55.

In light of Theorem 5.55 it is immediate that $\sum_{n=1}^{\infty} \frac{(-1)^n}{n}$ converges, since $\frac{1}{n} \to 0$ as $n \to \infty$.

Theorem 5.57. *[Abel's Test] Suppose that $\sum_{n=1}^{\infty} a_n$ converges, and that $b_n \searrow 0$ as $n \to \infty$. Then $\sum_{n=1}^{\infty} a_n b_n$ converges.*

Exercise 5.58. Prove Theorem 5.57.

The last test we consider in this chapter is the integral test.

Theorem 5.59. *[Integral Test] Let $f : [1, \infty) \to \mathbb{R}$ be a positive, decreasing function. The series $\sum_{n=1}^{\infty} f(n)$ converges if and only if the integral*

$$\int_{1}^{\infty} f(x) dx$$

is finite.

Exercise 5.60. Prove Theorem 5.59. Pictures will be worth a thousand words here...

We close the chapter by exhibiting that the harmonic series diverges using the integral test. In fact,

$$\int_{1}^{\infty} \frac{1}{x} dx = \lim_{b \to \infty} \ln(b) - \ln(1) = \lim_{b \to \infty} \ln(b) = +\infty$$

and the result follows.

Chapter 6
Continuity

Abstract This chapter focuses on developing a notion of continuity. We discuss three different but related versions of this notion, while changing the focus from local to global: pointwise, uniform and Lipschitz continuity.

> Once upon a time
> Somebody say to me
> (This is the dog talkin' now)
> What is your, conceptual, continuity?

Frank Zappa (1940 – 1993)

Getting Your Hands Dirty

Exercise 6.1. It is likely that at one point in your mathematical career you have been told that a function is continuous if you can draw its graph without lifting your pencil. Do you think the 'converse' is true? In other words, if you can't draw the graph of a function (without lifting your pencil or otherwise) then the function must be discontinuous?

Exercise 6.2. Suppose that you have a sequence of functions f_n on $[0,1]$ whose graphs are getting closer and closer to the graph of another function f. If all f_n's are continuous on $[0,1]$, must f be also continuous on $[0,1]$?

Exercise 6.3. Is continuity preserved under functional composition? How about the converse: if the functional composition $f \circ g$ is continuous, must either/one/both functions f, g be continuous?

6.1 Continuity - A Pointwise Notion

Definition 6.4. Let f be a real valued function defined on an interval I containing a. We say that f is continuous at the point a if and only if $\lim_{x \to a} f(x) = f(a)$.

Definition 6.5. If f is continuous at every point of a set S we say that f is continuous on S.

Definition 6.6. If f is not continuous at a point, we say that it is discontinuous there, or alternatively, we say that f has a discontinuity at that point.

Remark 6.7. Though it is clear from the definition, we emphasize that we can only talk about continuity of a function at a point $a \in \text{Dom}(f)$, even though we can talk about the limit of a function at a point where the function is not defined.

Exercise 6.8. Give three examples of functions f_1, f_2, f_3 and corresponding points a_1, a_2, a_3 such that f_i is not defined at a_i but it has a limit there for $i = 1, 2, 3$.

Exercise 6.9. Recall the definition of the limit of a function: $\lim_{x \to a} f(x) = L$ if and only if for any $\varepsilon > 0$ there exists a $\delta > 0$ such that $|x - a| < \delta$, $x \in \text{Dom}(f)$ implies that $|f(x) - L| < \varepsilon$. The logical statement reads as follows:
$$\forall \varepsilon > 0 \, \exists \delta > 0 \, (|x - a| < \delta \wedge x \in \text{Dom}(f) \implies |f(x) - L| < \varepsilon)$$
Write the logical statement for a function to be discontinuous at a point a in its domain.

Exercise 6.10. Prove that if f is continuous at x_0 and $f(x_0) \neq 0$, then f is nonzero on some open interval containing x_0.

Exercise 6.11. Prove that the function $f(x) = \lfloor x \rfloor$ is discontinuous at every integer. (Recall that this function assigns to every real number its integer part).

Exercise 6.12. Assume f has a jump at $x = a$, $\lim_{x \to a^-} f(x) = b$ and $\lim_{x \to a^+} f(x) = c$. Consider $g(x) = (x - b)(x - c)$. What are the left and right limits of $(g \circ f)$ at a?

The next theorem gives an equivalent characterization of continuity.

Theorem 6.13. *[Sequential Characterization] A function f is continuous at a point a if and only if $\lim_{n \to \infty} f(x_n) = f(a)$, for any sequence of points $(x_n)_{n=1}^{\infty}$ converging to a in the domain of f.*

Exercise 6.14. Prove Theorem 6.13. You will have to think about how to relate growing indices n to small numbers ε.

Exercise 6.15. Consider the Dirichlet function
$$f(x) = \begin{cases} 1 & x \in \mathbb{Q} \\ 0 & x \notin \mathbb{Q} \end{cases}$$
Show that this function is discontinuous at every point $x \in \mathbb{R}$.

Exercise 6.16. A variant of the Dirichlet function is the following slightly more complicated creature:
$$f(x) = \begin{cases} 1 & x = 0 \\ \dfrac{1}{q} & \text{if } x = \dfrac{p}{q} \in \mathbb{Q} \setminus \{0\}, \text{ in reduced form} \\ 0 & \text{if } x \notin \mathbb{Q} \end{cases}$$
Prove that this function is discontinuous at every rational point in $\mathbb{R}$ and that it is continuous at every irrational point.

6.1.1 The Algebra of Continuous Functions

Theorem 6.17. *Let f, g be continuous functions and let $r \in \mathbb{R}$. The following are true:*
(a) $f + g$ is continuous
(b) fg is continuous
(c) f/g is continuous
(d) $f \circ g$ and $g \circ f$ are continuous.
(e) rf is continuous.

Exercise 6.18. Prove Theorem 6.17. Note that you need to make precise at which point these various functions are continuous and add hypotheses if necessary to ensure that the statement you are trying to prove actually makes sense. For the details of the proof you may want to use the results of Proposition 4.38.

6.1 Continuity - A Pointwise Notion

Exercise 6.19. Show by example that we can have $f \circ g$ and/or $g \circ f$ continuous at a point even if one or both of the functions f and g are discontinuous at that point.

6.1.2 Further Fun Facts About Continuous Functions

In what follows, the notion of a closed and bounded interval will be crucial. We already know what it means for a set of real numbers to be bounded. Next we define what it means for a set of real numbers to be closed.

Definition 6.20. A set of real numbers S is *closed*, if and only if every convergent sequence $(s_n)_{n=1}^\infty$ in S converges to a point in S. In other words, if $s_n \in S$ for all $n \in \mathbb{N}$ and $s_n \to s$, then $s \in S$.

Note that a closed set S need not be bounded. For example, the set $[0, +\infty)$ is closed but not bounded. In addition, a closed and bounded set need not be an interval: the set $\{0, 1, 2, \pi, e, 2011\}$ is closed and bounded, but is certainly not an interval. Finally, a closed and bounded set which is not an interval need not be discrete: the set $\left\{\dfrac{1}{n} \;\middle|\; n \in \mathbb{N}\right\} \cup \{0\}$ is closed and bounded, but is not discrete.

Proposition 6.21. *If f is a continuous function on a closed and bounded set S, then f is bounded on S.*

Exercise 6.22. Prove Proposition 6.21. One way to do this is by assuming that f is not bounded, creating a convergent sequence and using the sequential characterization of continuity to arrive at a contradiction.

Exercise 6.23. Show that Proposition 6.21 fails if we omit either 'closed' or 'bounded' in the statement of the proposition.

Exercise 6.24. Give an example of a continuous function f on a set S and a Cauchy sequence in $(x_n)_{n=1}^\infty \in S$ such that $f(x_n)$ is *not* convergent.

Proposition 6.25. *[Intermediate Value Theorem] Suppose that f is continuous on an interval of the form $[a,b]$. Then f takes on every value between $f(a)$ and $f(b)$.*

Exercise 6.26. Prove Proposition 6.25. First treat the case when $f(a) = f(b)$. If $f(a) < f(b)$, select an m so that $f(a) < m < f(b)$ and consider the set

$$S = \{x \in [a,b] \mid f(x) < m\}.$$

Establish that $\sup S$ exists, and find $f(\sup S)$.

Corollary 6.27. *[Brouwer's Fixed Point Theorem] Suppose $f : [0,1] \to [0,1]$ is continuous. Then there exists at least one point $x_0 \in [0,1]$ such that $f(x_0) = x_0$.*

We remark that this theorem holds in much larger generality, in particular, one can replace the interval $[0,1]$ with any compact set in $\mathbb{R}^n$.

Exercise 6.28. Prove Corollary 6.27 by applying the Intermediate Value Theorem to the function $g(x) = f(x) - x$.

Theorem 6.29. *[Extreme Value Theorem] Suppose f is continuous on an interval of the form $[a,b]$. Then f obtains both its maximum and its minimum on the interval $[a,b]$.*

Proof. Since f is continuous on $[a,b]$, by Proposition 6.21 we get that f is bounded on $[a,b]$. This means that the set
$$S = \{f(x) \mid x \in [a,b]\}$$
is bounded. Since it is also trivially non-empty, by the Completeness Axiom, $M = \sup S$ is a finite real number. We claim that there is some $x_0 \in [a,b]$ such that $f(x_0) = M$. For if not, then $f(x) \neq M$ for any $x \in [a,b]$ and hence the function
$$g(x) = \frac{1}{M - f(x)}$$
is defined on all of $[a,b]$ and is continuous there. A repeated application of Proposition 6.21 shows that g is bounded on $[a,b]$. In other words, there exists some $k > 0$, such that
$$\left|\frac{1}{M - f(x)}\right| \leq k \quad \forall \ x \in [a,b],$$
or equivalently
$$\frac{1}{k} \leq M - f(x) \quad \forall \ x \in [a,b].$$
Hence
$$f(x) \leq M - \frac{1}{k} \quad \forall \ x \in [a,b],$$
contradicting the fact that M is the least upper bound of S. The proof is complete.

Exercise 6.30. Does the conclusion of Theorem 6.29 still hold if we replace the interval $[a,b]$ by an arbitrary closed set? What if we replace the interval $[a,b]$ by a closed and bounded set?

Exercise 6.31. Prove Theorem 6.29 in a constructive way using sequential techniques.

6.2 Uniform Continuity - The Setwise Notion

As we pointed out earlier, continuity is a pointwise notion. What this means in essence is that when we examine whether a function is continuous at a point, the 'delta' appearing in the definition is allowed to depend on the point in question.

Example 6.32. Consider the function $f(x) = \frac{1}{x}$ on the interval $(0,1)$. If $x_0 \in (0,1)$, and $\varepsilon > 0$, then
$$\left|\frac{1}{x} - \frac{1}{x_0}\right| = \left|\frac{x_0 - x}{x_0 x}\right| \leq |x - x_0| \frac{2}{x_0^2}$$
as long as we require that $|x - x_0| < \frac{1}{2} x_0$. To get the expression on the right to be less than ε we replace the expression $|x - x_0|$ by δ and solve for δ in the equation
$$\delta \frac{2}{x_0^2} = \varepsilon$$
to obtain $\delta = \frac{x_0^2 \varepsilon}{2}$. The write-up is then as follows: Given $\varepsilon > 0$ set
$$\delta = \min\left\{\frac{x_0}{2}, \frac{x_0^2 \varepsilon}{2}, 1 - x_0\right\}.$$

6.2 Uniform Continuity - The Setwise Notion

Then $|x - x_0| < \delta$ implies that $|f(x) - f(x_0)| < \varepsilon$ and hence f is continuous at x_0. (Note: the $1 - x_0$ expression is included in the candidates for the minimum in order to ensure that the interval $(x_0 - \delta, x_0 + \delta)$ is actually a subset of $(0, 1)$.

The previous example shows explicitly that the choice of δ may depend on x_0 (it will almost always depend on ε except for some special cases of functions).

Exercise 6.33. Give an example of a family of continuous functions where your choice of δ in proving continuity is independent of ε.

Exercise 6.34. Can the dependence of δ on x_0 in Example 6.32 be removed? Describe the difficulty encountered when trying to do so.

We are now in the position to formulate a stronger notion of continuity.

Definition 6.35. A function f is called *uniformly continuous* on a set S if for any $\varepsilon > 0$ there exists a $\delta > 0$ so that for all $x, y \in S$, $|x - y| < \delta$ implies $|f(x) - f(y)| < \varepsilon$.

Exercise 6.36. You should think hard about how this notion is different from the notion of continuity at a point.

Exercise 6.37. Write the logical statement for a function *not* to be uniformly continuous on a set S. (see Exercise 6.9)

Theorem 6.38. *Suppose $f : [a, b] \to \mathbb{R}$ is continuous. Then f is in fact uniformly continuous on $[a, b]$.*

Exercise 6.39. Prove Theorem 6.38 by contradiction using your work in Exercise 6.37.

Exercise 6.40. Suppose that f is uniformly continuous on the closed and bounded intervals S and T. Prove that f is uniformly continuous on $S \cup T$. Does this result extend to arbitrary sets S and T? Discuss.

Example 6.41. The function $f(x) = \dfrac{1}{x}$ is *not* uniformly continuous on its entire domain $\mathbb{R} \setminus \{0\}$. It is however uniformly continuous on $[1, \infty)$ and on $(-\infty, -1]$, and in fact, f is uniformly continuous on $(-\infty, -1] \cup [1, \infty)$.

Exercise 6.42. Suppose that $f : \mathbb{R} \to \mathbb{R}$ is a periodic function with a finite period. Prove that if f is uniformly continuous on any set containing a full period of f, then f is uniformly continuous on all of $\mathbb{R}$. As a quick corollary of this exercise we see that both $\sin x$ and $\cos x$ are uniformly continuous on $\mathbb{R}$.

Proposition 6.43. *Suppose $f : S \to \mathbb{R}$ is uniformly continuous and that $(x_n)_{n=1}^{\infty} \in S$ is a Cauchy sequence. Then $(f(x_n))_{n=1}^{\infty}$ is also a Cauchy sequence.*

Exercise 6.44. Prove Proposition 6.43. What is the candidate for $\lim_{n \to \infty} f(s_n)$?

Before we can state our next theorem, we need to make the following definition.

Definition 6.45. Let $f : S \to \mathbb{R}$ be any function and let D be any set such that $S \subseteq D$. If $F : D \to \mathbb{R}$ satisfies $F(x) = f(x)$ for all $x \in S$, we call F an extension of f to D. We will use the notation $F\big|_S = f$ to indicate such an extension.

Similar to the notion of the extension of a function is the restriction of a function.

Definition 6.46. Let $f : D \to \mathbb{R}$ be any function and let S be any set such that $S \subseteq D$. If $f : S \to \mathbb{R}$ satisfies $F(x) = f(x)$ for all $x \in S$, we call f the restriction of f to S.

Note that the restriction of a function to a subset of its domain is always unique, whereas there are infinitely many extensions of a function to any set containing its domain.

Theorem 6.47. *Let $f : (a,b) \to \mathbb{R}$ be a function. Then f is uniformly continuous on (a,b) if and only if f has a continuous extension to the closed interval $[a,b]$.*

Exercise 6.48. Prove Theorem 6.47. One direction is essentially and application of Theorem 6.38. For the other, use sequences to show that one can uniquely define function values at the endpoints a and b in such a fashion that the resulting extension is continuous on $[a,b]$. You will need to use Proposition 6.43.

Exercise 6.49. Show that if f is unbounded on the interval $(0,1)$, then f is not uniformly continuous on $(0,1)$.

Example 6.50. Consider the function $f(x) = \sin\left(\dfrac{1}{x}\right)$ on the interval $(0,1)$. It is clearly continuous there, but it has no continuous extension to the closed interval $[0,1]$. In fact, for any $\ell \in [-1,1]$ we can find a sequence x_n^ℓ such that $\lim_{n \to \infty} f(x_n^\ell) = \ell$.

We close this section with a definition that uses the notion of uniform continuity. We shall make use of this definition in Chapter 8 when we discuss integration.

Definition 6.51. Let $f : \mathbb{R} \to \mathbb{R}$ be a function. We say that f is piecewise continuous on $\mathbb{R}$ if there exist points $\ldots, x_{-n}, x_{-n+1}, \ldots, x_0, x_1, \ldots x_n, x_{n+1}, \ldots$ such that

$$f\big|_{(x_j, x_{j+1})}$$

is uniformly continuous on (x_j, x_{j+1}) for all $j \in \mathbb{Z}$.

6.3 Lipschitz Continuity

Definition 6.52. Let $f : S \to \mathbb{R}$ be a real valued function. We say that f is Lipschitz continuous of order α on S if there exist positive constants k, α such that

$$|f(x) - f(y)| \leq k|x-y|^\alpha \qquad \forall\, x, y \in S.$$

We will write $f \in Lip(\alpha)$ if f is Lipschitz continuous of order α.

Remark 6.53. You may find alternative sources on this topic which restrict the notion of Lipschitz continuity to the case $\alpha = 1$. In that convention, f is called Hölder continuous of order α if $0 < \alpha < 1$.

Example 6.54. The function $f(x) = \sin x$ is in $Lip(1)$ on the interval $[0, 2\pi]$. In fact, it is in $Lip(1)$ on the entire real line.

Example 6.55. The function $f(x) = |x|$ is in $Lip(1)$ with $k = 1$. To see this, simply note that the triangle inequality for real numbers gives

$$||x| - |y|| \leq |x - y|$$

for all x, y in $\mathbb{R}$, which by definition means that $f(x) = |x|$ is in $Lip(1)$ on all of $\mathbb{R}$.

Example 6.56. The function $f(x) = x^2$ is not in $Lip(\alpha)$ on all of $\mathbb{R}$ for any $\alpha > 0$. To show this we assume, to the contrary, that $f(x) = x^2$ is in $Lip(\alpha)$ for some $\alpha > 0$. Then there exists $k > 0$, such that

$$|x^2 - y^2| \leq k|x - y|^\alpha \qquad \forall\ x, y \in \mathbb{R}.$$

Choosing $x = k$ and $y = k+1$ transforms the above equation into

$$2k + 1 \leq k,$$

6.3 Lipschitz Continuity

which is obviously false for any $k > 0$. Thus we conclude that $f(x) = x^2$ is not in $Lip(\alpha)$ on all of $\mathbb{R}$ for any $\alpha > 0$.

It turns out that being Lipschitz continuous of order $\alpha > 1$ is a very restrictive condition, as the next example shows.

Example 6.57. We show that if $f \in Lip(\alpha)$ for some $\alpha > 1$ on all of $\mathbb{R}$, then f is a constant function, i.e. $f(x) = c$ for all $x \in \mathbb{R}$ and some $c \in \mathbb{R}$.

Proof. Select $x < y \in \mathbb{R}$, and let $d = |x - y|$. Divide $[x, y]$ into n equal subintervals, and let $x_1 < x_2 < \cdots < x_n = y$ be the right endpoints of these subintervals. Then

$$|f(x) - f(y)| \leq |f(x) - f(x_1)| + |f(x_1) - f(x_2)| + \cdots + |f(x_{n-1}) - f(y)|$$
$$\leq kn \left(\frac{d}{n}\right)^\alpha = k \frac{d^\alpha}{n^{\alpha-1}}.$$

Note $\dfrac{d^\alpha}{n^{\alpha-1}} \to 0$ as $n \to \infty$ (recall that $\alpha > 1$), hence $f(x) = f(y)$. Since x and y were arbitrary, we see that f is a constant function.

Exercise 6.58. Prove the above result using the definition of the derivative, and by showing that the derivative of f at every point in $\mathbb{R}$ is zero.

Exercise 6.59. Determine whether the following functions are in $Lip(\alpha)$ for any $\alpha > 0$ on the indicated set:
(a) $f(x) = \sqrt{x}$ on $[0, 1]$
(b)
$$f(x) = \begin{cases} x^{3/2} \sin\left(\dfrac{1}{x}\right) & x \neq 0 \\ 0 & x = 0 \end{cases}$$

on the interval $[-1, 1]$.
(c) $f(x) = \sqrt{x^2 + 5}$ on all of $\mathbb{R}$.

Exercise 6.60. Determine the relationship between Lipschitz continuity, uniform continuity and continuity at a point. Which of these notions imply which others (if any)?

Exercise 6.61. Show that if f is differentiable on an open interval (a, b) with a bounded derivative, then $f \in Lip(1)$ on (a, b). For this exercise you may want to look at Theorem 7.31.

Exercise 6.62. Show that given a set $S \subseteq \mathbb{R}$, the set of functions in $Lip(\alpha)$ is a vector space over $\mathbb{R}$ for every α. If you feel like thinking some more about this question, try to determine the dimension of this vector space. How does the dimension depend on α?

Chapter 7
Differentiability

Abstract In this chapter we introduce a stricter notion of smoothness - that of differentiability. We delve into understanding how functions can fail to be differentiable, and look into one possible way to differentiate functions not once or twice, but fractional times.

> The Mean Value Theorem is the midwife of calculus – not very important or glamorous by itself, but often helping to deliver other theorems that are of major significance.
>
> Edwin Purcell and Dale Varberg

Getting Your Hands Dirty

Exercise 7.1. We have seen in Chapter 6 that the Dirichlet function is discontinuous at every point in $\mathbb{R}$. Is there a function which is continuous everywhere on $\mathbb{R}$ but is not differentiable anywhere?

Exercise 7.2. We know from calculus that it is possible to take the first, second, third, etc. derivatives of a (smooth enough) function. What would it mean to take the $\frac{3^{th}}{2}$ derivative of a function?

Exercise 7.3. Give an example of a sequence of functions $f_n(x)$ so that each f_n is *exactly* n-times differentiable at the origin.

7.1 A Short Historical Exposition

The history of mathematics is full of feuds. Famous are the disputes between Cardano and Tartaglia, Cauchy and Galois, and Klein and Poincaré, just to name a few. Hence, it is not surprising that differentiation started with a quarrel (and a lot of name-calling) between two great mathematicians.
Differentiation was originally developed, independently, by Sir Isaac Newton (1643 - 1727) in England and Gottfried Wilhelm Leibniz (1646 - 1716) in Germany. Newton was the first to find applications of differentiation to physics and engineering, and was responsible for the product rule, chain rule, and many other important results. On the other hand, Leibniz was responsible for the notation we currently use (dy/dx), which is not just 'convenient', but also expresses the deep understanding Leibniz had of infinitesimals and the rigorousity of his work. Now both Newton and Leibniz are credited with the creation of Calculus, but back in the 1600's the accusations of "stolen work" were commonplace between these two great minds.
Another spicy story surrounds L'Hôpital's rule. The French Guillaume François Antoine Marquis de L'Hôpital (1661 - 1704) published 'his' theorem in his book *Analyse des Infiniment Petits pour l'Intelligence des Lignes Courbes* (Analysis of the Infinitely Small for the Understanding of Curved Lines) in 1696, which was the first textbook on differential calculus. However, L'Hôpital had hired Johann Bernoulli (1667 - 1748) to tutor him in mathematics and to keep him up to date with all of Bernoulli's mathematical discoveries. Since the Marquis was very rich and Bernoulli was not, this deal was not something the young Johann could reject. The arrangement

eventually lead to L'Hôpital publishing 'his' book containing several of Bernoulli's results, among them what we call today L'Hôpital's rule.

7.2 Differentiability - The Basics

Definition 7.4. A function f is called differentiable at a point x_0 in its domain if the limit

$$\lim_{h \to 0} \frac{f(x_0 + h) - f(x_0)}{h}$$

exists. A function is differentiable on a set S if it is differentiable at each $s \in S$. If a function is differentiable on a set, we can talk about its derivative function on that set. If the above limit doesn't exist, we say that f is not differentiable at x_0.

Remark 7.5. We will use the following notations interchangeably to denote the derivative of a function f:

$$f'(x), \quad \frac{d}{dx}(f), \quad Df, \quad f_x(x)$$

Example 7.6. All polynomials are differentiable on $\mathbb{R}$. You will prove this in Exercise 7.7.

Exercise 7.7. Prove that any monomial $f(x) = x^n$ is differentiable on $\mathbb{R}$. Then prove that the set of functions which are differentiable at a point x_0 is closed under addition and multiplication by a real number. Conclude that all polynomials are differentiable on $\mathbb{R}$.

Example 7.8. Let $f(x) = \sin x$. Then

$$\lim_{h \to 0} \frac{\sin(x+h) - \sin x}{h} = \lim_{h \to 0} \frac{\sin x \cos h + \cos x \sin h - \sin x}{h}$$
$$= \lim_{h \to 0} \sin x \frac{\cos h - 1}{h} + \cos x \frac{\sin h}{h}$$
$$= \cos x.$$

This shows that $\sin x$ is differentiable on $\mathbb{R}$ and its derivative is $\cos x$.

Example 7.9. Let $f(x) = \frac{1}{x}$. Then

$$\lim_{h \to 0} \frac{\frac{1}{x+h} - \frac{1}{x}}{h} = \lim_{h \to 0} -\frac{1}{x(x+h)} = -\frac{1}{x^2}.$$

Exercise 7.10. Prove that

$$\frac{d}{dx}\left(\frac{1}{x^n}\right) = -\frac{n}{x^{n+1}} \quad n = 1, 2, 3, \ldots.$$

Theorem 7.11. *Suppose f, g are differentiable functions. Then $f \pm g$, $f \cdot g$, f/g and $f \circ g$ are all differentiable.*

Exercise 7.12. Prove Theorem 7.11. You will have to make precise the statement and add restrictions if necessary. Also, you may need to make use of 'clever' additions of zero, such as

$$(fg)(x_0 + h) - (fg)(x_0) = (fg)(x_0 + h) - f(x_0)g(x_0 + h) + f(x_0)g(x_0 + h) - (fg)(x_0).$$

Exercise 7.13. Suppose that $f : S \to \mathbb{R}$ is differentiable at $a \in S$. Prove that f must be continuous at $a \in S$.

Exercise 7.14. Find the fault in the following argument:

Suppose
$$f(x) = \begin{cases} x+1 & x \geq 0 \\ x & x < 0 \end{cases}$$

The slopes at any point are equal (all one), so f should be differentiable but it is not continuous at 0.

Exercise 7.15. Consider the family of functions indexed by $n, k \in \mathbb{N}$ given by

$$f_{n,k}(x) = \begin{cases} x^n \sin\left(\dfrac{1}{x^k}\right) & x \neq 0 \\ 0 & x = 0 \end{cases}$$

For which pairs (n,k) is the above function differentiable on $\mathbb{R}$? For which pairs (n,k) is the above function continuously differentiable on $\mathbb{R}$?

Exercise 7.16. In section 6.3 we introduced the notion of a Lipschitz-continuous function of order α on a set S. For which α does the following statement hold ?

If f is Lipschitz-continuous of order α on S then f is differentiable on S.

7.3 Higher Order Derivatives

Definition 7.17. Suppose that we know what the kth derivative $f^{(k)}$ is of a function f. Then the $(k+1)^{st}$ derivative is simply

$$\lim_{h \to 0} \frac{f^{(k)}(x+h) - f^{(k)}(x)}{h}$$

if the limit exists. We say that f is exactly k-times differentiable at a point a if $f^{(k)}(a)$ exists, but $f^{(k+1)}(a)$ does not.

Consider now a k-times differentiable function. How "badly" is the kth derivative non-differentiable? We may have that the kth derivative is continuous but is no longer differentiable. Functions of the this variety are called k-times continuously differentiable functions and are denoted by $\mathscr{C}^k$. For example the set $\mathscr{C}^{2011}([0,1])$ denotes all 2011 times continuously differentiable functions on the interval $[0,1]$.

Exercise 7.18. Show that the function $f(x) = x|x|$ is in $\mathscr{C}^1(\mathbb{R})$.

With Exercise 7.18 under our belt, we can now give an answer to Problem 7.3. In fact, if we let f_1 to be the function f as in Exercise 7.18, then we may take

$$f_2(x) := \int_0^x f_1(t)dt + g_2(x)$$

where $g_2 \in \mathscr{C}^\ell(\mathbb{R})$, for some $\ell \geq 2$. Having defined $f_{n-1}(x)$, we then set

$$f_n(x) := \int_0^x f_{n-1}(t)dt + g_n(x)$$

where $g_n \in \mathscr{C}^\ell(\mathbb{R})$ for some $\ell \geq n$.

In some cases We may have that the kth derivative exists and is discontinuous! Take for example the function

$$f(x) = \begin{cases} x^2 \sin\left(\dfrac{1}{x}\right) & x \neq 0 \\ 0 & x = 0 \end{cases}$$

This function is differentiable on all of $\mathbb{R}$, but the derivative is discontinuous at $x = 0$.

Exercise 7.19. Prove the assertion above.

Similarly to the solution to Exercise 7.3, we can then create for any $k \in \mathbb{N}$ a function that has derivatives of the $1^{st}, 2^{nd}, \ldots, k^{th}$ order and no derivatives of higher order, with the kth order derivative being discontinuous at a given point.

7.4 Fractional Derivatives

At a first glance it might seem a weird notion to try to take a half a derivative of a function. And yet, Leibniz has already described this notion in 1695. Since then, the theory of fractional calculus has developed tremendously. The fact that there are several different fractional derivatives in the literature (Riemann-Liouville, Hadamard, and Riesz fractional derivatives just to name a few) suggests that there are different possible ways to define this notion. In this light we offer three possibilities and let the reader discover the properties of each. For more on fractional calculus we direct the reader to [16].

7.4.1 The Fourier Transform

Definition 7.20. Let f be continuous on $[a,b]$ and differentiable on (a,b). We define the forward *Fourier transform* of f as

$$\mathscr{F}[f(x)](k) := \frac{1}{\sqrt{2\pi}} \int_{-\infty}^{\infty} f(x) e^{-2\pi i k x} dx.$$

The reader may object (rightly so) to the fact that f wasn't defined on all of $\mathbb{R}$. To circumvent this problem, we extend f to be identically zero outside $[a,b]$. If we were to take the forward Fourier transform of f' and integrate by parts we would get

$$\begin{aligned}
\mathscr{F}[f'(x)](k) &= \frac{1}{\sqrt{2\pi}} \int_{-\infty}^{\infty} f'(x) e^{-2\pi i k x} dx \\
&= \frac{1}{\sqrt{2\pi}} f(x) e^{-2\pi i k x} \Big|_{-\infty}^{\infty} + 2\pi i k \frac{1}{\sqrt{2\pi}} \int_{-\infty}^{\infty} f(x) e^{-2\pi i k x} dx \\
&= 2\pi i k \frac{1}{\sqrt{2\pi}} \int_{-\infty}^{\infty} f(x) e^{-2\pi i k x} dx \\
&= 2\pi i k \mathscr{F}[f(x)](k).
\end{aligned}$$

Exercise 7.21. Show that

$$\mathscr{F}[f^{(n)}(x)](k) = (2\pi i k)^n \mathscr{F}[f(x)](k).$$

With Exercise 7.21 as inspiration, we could make the following implicit definition for a qth derivative of a function f for any $q \in \mathbb{Q}$:

$$\mathscr{F}[f^{(q)}(x)](k) = (2\pi i k)^q \mathscr{F}[f(x)](k).$$

The question then becomes how to recover the qth derivative from here. If we could 'invert' the Fourier transform, we would be in business. It turns out that we can!

Definition 7.22. The inverse Fourier transform of a function $F(k)$ is given by

$$\mathscr{F}^{-1}[F(k)](x) = \frac{1}{\sqrt{2\pi}} \int_{-\infty}^{\infty} F(k) e^{2\pi i x k} dk.$$

Putting all this together, then we can make the following definition for a fractional derivative. To this end, we denote multiplication by the qth power of $2\pi i k$ by M_q.

7.4 Fractional Derivatives

Definition 7.23. Suppose $f : [a,b] \to \mathbb{R}$ is continuous and differentiable on (a,b) and let $q \in \mathbb{Q}$ be given. Then the qth derivative of f is defined as

$$f^{(q)}(x) := \mathscr{F}^{-1} \circ M_q \circ \mathscr{F}[f(x)].$$

Example 7.24. Let $f(x) = 1$. We know that $f'(x) = 0$. In fact we have that for any $0 < q < 1$, $q \in \mathbb{Q}$ we have $f^{(q)}(x) = 0$. But of course, $f^{(0)}(x) = 1$.

7.4.2 The Complex Perspective

Let $f : \mathbb{C} \to \mathbb{C}$ be a function. We say that f is differentiable at a point $z \in \mathbb{C}$ if the limit $\lim_{h \to 0} \frac{f(z+h) - f(z)}{h}$ exists (note that $h \in \mathbb{C}$!). If f has a derivative at every point $z \in D$ for some domain D, we say that f is analytic on D. One of the most useful results in the theory of complex functions is the following.

Theorem 7.25. *[The Cauchy Integral Formula] Suppose that $f : \mathbb{C} \to \mathbb{C}$ is analytic on a domain D, and that γ is a circle contained in D centered at z_0. Then for all $k \geq 0$ we have*

$$f^{(k)}(z_0) = \frac{k!}{2\pi i} \int_\gamma \frac{f(\zeta)}{(\zeta - z_0)^{k+1}} d\zeta.$$

In light of Theorem 7.25 we could define the qth derivative for an analytic function for all $q \geq 0$ to be

$$f^{(q)}(z) = \frac{\Gamma(q+1)}{2\pi i} \int_\gamma \frac{f(\zeta)}{(\zeta - z)^{q+1}} d\zeta.$$

For example, the function $f(z) = z^2$ is analytic on $\mathbb{C}$. We compute the qth derivative of f at $z = 0$. Let γ be the unit circle oriented counterclockwise. If $0 \leq q < 2$ then

$$f^{(q)}(0) = \frac{\Gamma(q+1)}{2\pi i} \int_\gamma \frac{\zeta^2}{\zeta^{1+q}} d\zeta$$

$$= \frac{\Gamma(q+1)}{2\pi} \int_0^{2\pi} e^{(2-q)i\theta} d\theta$$

$$= \frac{\Gamma(q+1)}{2\pi} \left(\frac{\sin((2-q)2\pi)}{2-q} - i \frac{\cos((2-q)2\pi)}{2-q} + \frac{i}{2-q} \right).$$

Notice that when $q = 1$, we get $f'(0) = 0$, as expected. If $q = 2$, we get

$$f''(0) = \frac{\Gamma(3)}{2\pi i} \int_\gamma \frac{1}{\zeta} d\zeta = \frac{2}{2\pi i} \int_0^{2\pi} \frac{1}{e^{i\theta}} i e^{i\theta} d\theta = 2,$$

and if $q > 2$ we simply have

$$f^{(q)}(0) = \frac{\Gamma(q+1)}{2\pi i} \int_0^{2\pi} e^{-(q-2)i\theta} d\theta = 0.$$

7.4.3 Riemann-Liouville fractional derivatives

Lastly, we consider the Riemann-Liouville fractional derivatives. To this end, let $f : \mathbb{R} \to \mathbb{R}$ be a function, let $a \in \mathbb{R}$, and let $q > 0$. Define the right and left sided fractional integrals of f as

$$(I_{a+}^q f)(x) := \frac{1}{\Gamma(q)} \int_a^x (x-\tau)^{q-1} f(\tau) d\tau, \qquad x > a$$

and

$$(I_{a-}^q f)(x) := \frac{1}{\Gamma(q)} \int_x^a (\tau-x)^{q-1} f(\tau) d\tau, \qquad x < a.$$

With these integrals at hand, we define the right and left Riemann-Liouville fractional derivative of f of order q as

$$(D_{a+}^q f)(x) := \left(\frac{d}{dx}\right)^n \left(I_{a+}^{n-q} f)(x)\right), \qquad x > a$$

and

$$(D_{a-}^q f)(x) := \left(-\frac{d}{dx}\right)^n \left(I_{a-}^{n-q} f)(x)\right), \qquad x < a,$$

where $n = \lceil q \rceil$.

Example 7.26. In this example we compute the first derivative of $f(x) = x^2$ using the Riemann-Liouville derivative with $a = 0$. Thus, $q = 1$, $\lceil q \rceil = 2$, and

$$(D_{0+}^1 f)(x) = \left(\frac{d}{dx}\right)^2 (I_{0+}^1 f)(x) = \left(\frac{d}{dx}\right)^2 \int_0^x \tau^2 d\tau = \frac{d}{dx}(x^2) = 2x.$$

A similar calculation shows that $(D_{0-}^1 f)(x) = 2x$. Using a different point for a, we conclude that the derivative of x^2 is equal to $2x$ on all of $\mathbb{R}$.

Exercise 7.27. Compute the Riemann-Liouville derivative of order $q = 1/2$ of the function $f(x) = x^2$.

Exercise 7.28. In the definition of the Riemann-Liouville fractional derivatives we have been deliberately vague about the properties of f (continuity, differentiability, etc). What properties does f require in order for its fractional derivatives to be well defined?

7.5 Further Properties of Differentiable Functions

Proposition 7.29. *Suppose that f is differentiable at a point x_0 and that f has a local extremum at x_0. Then $f'(x_0) = 0$.*

Exercise 7.30. Prove Proposition 7.29. It might be fruitful to examine quotients of the type

$$\frac{f(x) - f(x_0)}{x - x_0}$$

for **(a)** $x > x_0$, and **(b)** $x < x_0$.

Theorem 7.31. *[The Mean Value Theorem] Let f be continuous on $[a,b]$ and differentiable on (a,b). Then there exists at least one $x \in (a,b)$ such that*

$$f'(x) = \frac{f(b) - f(a)}{b - a}.$$

Exercise 7.32. Prove Theorem 7.31.
Hint: You might find useful the function $L(x)$ whose graph is a line through the points $(a, f(a))$ and $(b, f(b))$. What can you say about $g(x) = f(x) - L(x)$?

7.5 Further Properties of Differentiable Functions

Exercise 7.33. Show that if f is differentiable on (a,b) and $f'(x) = 0$ for all $x \in (a,b)$, then f is a constant function.

Exercise 7.34. Show that if f, g are differentiable on (a,b) and $f'(x) = g'(x)$ for all $x \in (a,b)$, then $f(x) = g(x) + k$ for some constant $k \in \mathbb{R}$.

You should take a look at Theorem 9.42 and complete the two exercises following the theorem. You should then prove that Theorems 7.31 & 9.42 are equivalent: given either of the two, you can prove the other.

Theorem 7.35. *[The Generalized Mean Value Theorem] Let f, g be continuous functions on $[a,b]$ and differentiable on (a,b). Then there exists at least one $x \in (a,b)$ such that*

$$f'(x)[g(b) - g(a)] = g'(x)[f(b) - f(a)].$$

Exercise 7.36. Prove Theorem 7.35 by considering the function

$$h(x) = f(x)[g(b) - g(a)] - g(x)[f(b) - f(a)].$$

In 1875 the French mathematician Jean-Gaston Darboux (1842 – 1917) proved the intermediate value theorem for derivatives. For the original source see [8]. Here we present an elegant proof by Nadler appearing in [18].

Theorem 7.37. *[The Intermediate Value Theorem for the derivative] Let I be an interval and let f be a differentiable real valued function on I. Then the derivative f' satisfies the conclusion of Proposition 6.25, namely every number between two values of f' is also a value of f'.*

Remark 7.38. The proof we gave for Proposition 6.25 does not carry over, since it would require the continuity of f' restricting the validity of the proof to those functions that are $\mathscr{C}^k$ for some k.

Proof. Let V denote the set of all values of f'. The proof's goal is to show that V is in fact an interval. Let C denote the set of all slopes of chords joining distinct points on the graph of f. By Theorem 7.31 we see that $C \subseteq V$. Since every point in V is a limit of a sequence of points in C, we conclude that $V \subseteq \bar{C}$. Thus it suffices to show that C is an interval, as this would imply immediately that V is also an interval. To show that C is an interval, we show that any two points of C can be connected by an interval in C. To this end, let p, q be two points in C given by

$$p = \frac{f(b) - f(a)}{b - a}, \qquad q = \frac{f(d) - f(c)}{d - c} \qquad (a < b, \ c < d)$$

Define $g : [0,1] \to C$ as follows:

$$g(t) = \frac{f((1-t)b + td) - f((1-t)a + tc)}{((1-t)b + td) - ((1-t)a + tc)}$$

Then $g(0) = p$ (slope of the line segment $\overline{t_0 s_0}$ in Figure 7.1) and $g(1) = q$ (slope of the line segment $\overline{t_5 s_5}$ in Figure 7.1). Moreover, g is continuous, hence by Theorem 6.25, g takes on every value between p and q, which in turn means that there is an interval connecting p and q in C. The proof is complete.

Exercise 7.39. Give a different proof of Theorem 7.37, using the following line of argument: suppose $p = f'(x)$ and $q = f'(y)$ are values of the derivative, and let w be a number between p and q. Show that the function $F(x) = f(x) - wx$ has an extreme value in the interior of I somewhere. Conclude that f' must equal w at that point.

Proposition 7.40. *Suppose that $f : (a,b) \to \mathbb{R}$ is differentiable with a bounded derivative, i.e. there exists an $M \in \mathbb{R}$ such that $|f'(x)| \leq M$ for all $x \in (a,b)$. Then f is uniformly continuous on (a,b) and hence on $[a,b]$.*

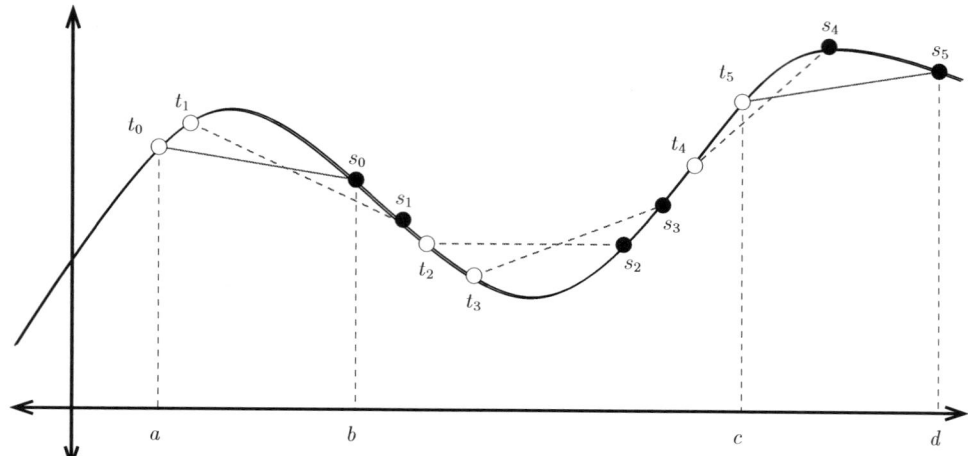

Fig. 7.1 Chords as in the proof of Theorem 7.37

Exercise 7.41. Prove Proposition 7.40 using the Mean Value Theorem.

Proposition 7.42. *Suppose that f is invertible on an interval I. If f is differentiable at $x_0 \in I$ and $f'(x_0) \neq 0$, then f^{-1} is differentiable at $f(x_0)$ and*

$$(f^{-1})'(f(x_0)) = \frac{1}{f'(x_0)}.$$

Exercise 7.43. Prove Proposition 7.42 by filling in the details of the following argument:
(a) Show that there is a $\delta > 0$, such that for $0 < |h| < \delta$, $f^{-1}(f(x_0) + h)$ is defined.
(b) For such an h, set $x = f^{-1}(f(x_0) + h)$, and compute

$$\lim_{h \to 0} \frac{f^{-1}(f(x_0) + h) - f^{-1}(f(x_0))}{h}.$$

Theorem 7.44. *[L'Hôpital Rule] Suppose f, g are differentiable functions for which*

$$\lim_{x \to s} \frac{f'(x)}{g'(x)} = L$$

exists (the limit can be one sided, or a limit at infinity). If either
(a) $\lim_{x \to s} f(x) = \lim_{x \to s} g(x) = 0$, *or*
(b) $\lim_{x \to s} |g(x)| = \infty$,
then

$$\lim_{x \to s} \frac{f(x)}{g(x)} = L.$$

Exercise 7.45. Show that in case (b) of Theorem 7.44 we must have that $\lim_{x \to s} |f(x)| = \infty$.

We stated L'Hôpital's rule in its greatest generality, but in line with the philosophy of the book, the next exercise will ask the student to prove a version of the theorem, which is a bit more easily handled.

7.5 Further Properties of Differentiable Functions

Exercise 7.46. Prove the following version of L'Hôpital's rule: Suppose f,g are continuous on $[a,b]$ and differentiable on (a,b). If $x_0 \in (a,b)$ is such that
(a) $g'(x) \neq 0$ for all $x \in (a,b)$,
(b) $f(x_0) = g(x_0) = 0$, and
(c) $\lim\limits_{x \to x_0} \dfrac{f'(x)}{g'(x)} = L$ exists,

then
$$\lim_{x \to x_0} \frac{f(x)}{g(x)} = L.$$

To start, take x_0, select $(x_n)_{n=1}^{\infty}$ converging to x_0, and choose c_n between x and x_n as in Theorem 7.35. Rearrange the conclusion of Theorem 7.35 and let $n \to \infty$.

Chapter 8
Integrability

Abstract In this chapter we investigate the notion of integrability. We will base the development of the theory on Darboux-sums, and will show that integrability in the Darboux and Riemann sense are equivalent notions. We will exhibit the existence of non Darboux-integrable functions and give various conditions on real valued functions that are equivalent to being integrable.

> But just as much as it is easy to find the differential of a given quantity, so it is difficult to find the integral of a given differential. Moreover, sometimes we cannot say with certainty whether the integral of a given quantity can be found or not.

Johann Bernoulli (1667 – 1748)

Getting Your Hands Dirty

Exercise 8.1. Consider the Dirichlet function on the interval $[0,1]$. Using what you know about integration, determine what the integral
$$\int_0^1 \chi_{\mathbb{Q}} dx$$
should be.

Exercise 8.2. Consider the functions $f(x) = \lfloor x \rfloor$ and $g(x) = x - \lfloor x \rfloor$. Find the integrals
$$\int_0^3 f(x)dx, \quad \int_{-2}^3 g(x)dx$$
by giving a geometric interpretation for the integrals.

8.1 Partitions, The Basics

Definition 8.3. Let $[a,b]$ be an interval in $\mathbb{R}$ with $a \neq b$. Let $P = \{x_0, x_1, x_2, \ldots, x_n\}$ be a finite collection of real numbers. If
(a) $x_0 = a$ and $x_n = b$, and
(b) $x_0 < x_1 < x_2 < x_3 < \ldots < x_n$
then we call P a partition of the interval $[a,b]$.

Remark 8.4. It is obvious from the definition, yet is worth reiterating, that a partition P of an interval $[a,b]$ is a *set* of points in $[a,b]$.

Definition 8.5. If P and Q are partitions of the same interval $[a,b]$, we say that Q is a refinement of P if $P \subseteq Q$.

Exercise 8.6. Let P and Q be partitions of $[a,b]$. Show that $P \cap Q$ and $P \cup Q$ are also partitions of $[a,b]$.

Exercise 8.7. Give an example of two partitions P and Q of the interval $[0,2]$ so that P and Q are distinct but neither is a refinement of the other. Then construct a partition that refines them both.

Definition 8.8. Suppose $P = \{x_0, x_1, x_2, \ldots, x_n\}$ is a partition of the interval $[a,b]$ and let $f : [a,b] \to \mathbb{R}$ be a bounded real valued function. We define the upper and lower Darboux sums of f associated to P as

$$U(f,P) := \sum_{i=1}^{n} \left(\sup_{[x_{i-1}, x_i]} f \right) \cdot (x_i - x_{i-1})$$

$$L(f,P) := \sum_{i=1}^{n} \left(\inf_{[x_{i-1}, x_i]} f \right) \cdot (x_i - x_{i-1})$$

Exercise 8.9. Let f be a bounded real valued function and let $S \subseteq T \subseteq \text{Dom}(f)$. Show that
(a) $\inf_S f(x) \geq \inf_T f(x)$
(b) $\sup_S f(x) \leq \sup_T f(x)$.

Exercise 8.10. Prove that if f is continuous on $[a,b]$ and there exists a partition P of $[a,b]$ such that $U(f,P) = L(f,P)$, then f is a constant function. Is the assumption that f is continuous necessary?

Exercise 8.11. Give an example of a function $f(x)$ and a partition P of $[0,1]$ so that $L(f,p) < U(f,P)$.

Exercise 8.12. Let $[a,b]$ be an interval in $\mathbb{R}$. Prove that $L(f,P) \leq U(f,P)$ for all f and P.

Exercise 8.13. Suppose that P and Q are partitions of $[a,b]$ and that Q is a refinement of P. Prove that
(a) $L(f,P) \leq L(f,Q)$
(b) $U(f,Q) \leq U(f,P)$.
Hint: Argue first that we may assume without loss of generality that Q has exactly one more point than P.

Proposition 8.14. *Suppose $f : [a,b] \to \mathbb{R}$ is a real valued function and that P and Q are partitions of $[a,b]$. Then*

$$L(f,P) \leq U(f,Q).$$

Exercise 8.15. Prove Proposition 8.14 by putting together the results of Exercise 8.12 and Exercise 8.13.

Definition 8.16. Let $f : [a,b] \to \mathbb{R}$ be a real valued function. We define the upper and lower Darboux integral of f over the interval $[a,b]$ to be

$$U(f) = \inf_{P \text{ a partition of } [a,b]} U(f,P)$$

$$L(f) = \sup_{P \text{ a partition of } [a,b]} L(f,P).$$

Exercise 8.17. Prove that if $f : [a,b] \to \mathbb{R}$ is bounded, then $L(f) \leq U(f)$.

Definition 8.18. A bounded function $f : [a,b] \to \mathbb{R}$ is called integrable in the Darboux sense on $[a,b]$ if and only if $U(f) = L(f)$. In this case we will write

$$\int_a^b f(x)\,dx$$

for the common value $U(f) = L(f)$.

Exercise 8.19. Suppose $f : [a,b] \to \mathbb{R}$ is bounded. Prove that if there exists a partition P such that $L(f,P) = U(f,P)$, then f is integrable on $[a,b]$.

8.1 Partitions, The Basics

Exercise 8.20. Let $f(x) \equiv 1$ on $[0,1]$ and let

$$g(x) = \begin{cases} 0 & x = 0 \\ 1 & 0 < x \leq 1 \end{cases}$$

Prove that f and g are integrable in the Darboux sense, and show that $\int_0^1 f(x)dx = \int_0^1 g(x)dx$.

Exercise 8.21. We have seen the function

$$f(x) = \begin{cases} 0 & x = 0 \\ x \sin\left(\frac{1}{x}\right) & x \neq 0 \end{cases}$$

before in connection with differentiability. Is this function integrable, in the Darboux sense, on $[0,1]$? Justify your answer.

Exercise 8.22. Suppose f is integrable on $[a,b]$. Show that f^2 is also integrable. Is the converse true?
Hint: You may want to prove that

$$U(f^2, P) - L(f^2, P) \leq 2B(U(f,P) - L(f,P))$$

for all partitions P of $[a,b]$, where $|f(x)| \leq B$ for all $x \in [a,b]$.

Exercise 8.23. Prove that the function $f(x) = x$ is integrable in the Darboux sense on the interval $[0,2]$ and find its integral. You may want to consider the partitions

$$P_n = \left\{0, \frac{1}{n}, \frac{2}{n}, \ldots, \frac{2n-1}{n}, 2\right\}.$$

Exercise 8.24. Using the same partitions as in the previous exercise, show that $f(x) = x^2$ is integrable in the Darboux sense on the interval $[0,2]$.

Example 8.25. Consider the function

$$f(x) = \begin{cases} x & \text{if } x \in \mathbb{Q} \\ 0 & \text{if } x \notin \mathbb{Q} \end{cases}$$

This function is not integrable on $[0,1]$. To see this, note that by the density of the irrationals in $\mathbb{R}$, $L(f,P) = 0$ for *any* partition P of $[0,1]$, and hence $L(f) = 0$. Next, we consider any partition P of $[0,1]$ containing $x = 1/2$. Since $\sup_{[1/2,1]} f \geq 1/2$, we see that $U(f,P) \geq 1/4$ and hence $U(f) \geq 1/4$, which establishes that f is not integrable on $[0,1]$ in the Darboux sense. (Fill in the details of this argument so as to completely convince yourself).

Exercise 8.26. Consider $f(x) = \chi_\mathbb{Q}$ on the interval $[0,1]$. Prove that f is integrable in the Darboux sense on $[0,1]$ if and only if $1 - f(x)$ is integrable in the Darboux sense on $[0,1]$, and that in this case, f and $1 - f$ have the same integral. Use Theorem 8.4 to conclude that the common integral must be $1/2$. Compute any lower Darboux sum for f on $[0,1]$, and use the result to conclude that f (and hence $1 - f$) are not integrable in the Darboux sense on $[0,1]$.

Exercise 8.27. Show that if f is a periodic function with period N and it is integrable in the Darboux sense on any interval of length at least N, then it is integrable on any closed and bounded interval.

Exercise 8.28. Prove or disprove the following statement:

If f, g are integrable on $[a,b]$ and $f(x) \geq h(x) \geq g(x)$ for all $x \in [a,b]$, then h is integrable on $[a,b]$ and

$$\int_a^b f(x)dx \geq \int_a^b h(x)dx \geq \int_a^b g(x)dx.$$

If you think that the statement is false, give sufficient conditions on h so that the statement becomes true.

Theorem 8.29. *Let f be bounded on $[a,b]$. Then f is integrable on $[a,b]$ in the Darboux sense if and only if for all $\varepsilon > 0$ there exists a partition P of $[a,b]$ such that*

$$U(f,P) - L(f,P) < \varepsilon.$$

Exercise 8.30. Prove Theorem 8.29 using the following outline:

[$\Leftarrow$] Argue that for each $n \in \mathbb{N}$ there are partitions P_n such that $U(f,P_n) - L(f,P_n) < 1/n$. Use this to conclude that $U(f) \leq L(f)$.

[$\Rightarrow$] For $\varepsilon > 0$, use the definition of the infimum to conclude that there is a partition P such that $U(f,P) - U(f) < \varepsilon/2$. Prove a similar statement about lower sums, and put the two together.

8.2 Riemann Integrability

Definition 8.31. Let $P = \{x_0, x_1, \ldots, x_n\}$ be a partition of $[a,b]$. We define the mesh of the partition P as

$$\text{mesh}(P) = \max_i |x_i - x_{i-1}|.$$

Theorem 8.32. *A bounded function f is integrable in the Darboux sense on $[a,b]$ if and only if for all $\varepsilon > 0$ there exists a $\delta > 0$ such that*

$$\text{mesh}(P) < \delta \quad \Longrightarrow \quad U(f,P) - L(f,P) < \varepsilon$$

for all partitions P of $[a,b]$.

Exercise 8.33. Prove Theorem 8.32. One direction is a quick consequence of Theorem 8.29. For the other direction, you will need to think about how many different components one has to control in a sum like

$$\sum_{i=1}^{n} \left(\sup_{[x_{i-1}, x_i]} f - \inf_{[x_{i-1}, x_i]} f \right)(x_i - x_{i-1})$$

and how to go about controlling them?

Definition 8.34. Let f be a bounded function on $[a,b]$ and let $P = \{x_0, x_1, \ldots, x_n\}$ be a partition of $[a,b]$. A Riemann sum of f associated to the partition P is a sum of the form

$$S = \sum_{i=1}^{n} f(t_i)(x_i - x_{i-1})$$

where $t_i \in [x_{i-1}, x_i]$ for all $i = 0, 1, \ldots n$.

Definition 8.35. We say that f is Riemann integrable on $[a,b]$ if there exists a real number r such that given any $\varepsilon > 0$ there is a $\delta > 0$, so that $\text{mesh}(P) < \delta$ implies

$$|S - r| < \varepsilon$$

for every Riemann sum S of f associated to the partition P. If such r exists, we call it the Riemann integral of f over $[a,b]$ and write

$$r = R\int_a^b f(x)dx.$$

Theorem 8.36. *A bounded function f is Darboux integrable on $[a,b]$ if and only if it is Riemann integrable there, in which case the two integrals coincide.*

Exercise 8.37. Prove Theorem 8.36. For one direction, Theorem 8.32 will come in handy, along with the choice $r = U(f) = L(f)$. For the other direction, assume f is Riemann integrable with integral r. Use properties of the infimum to show that given $\varepsilon > 0$, $L(f) > r - O(\varepsilon)$, and prove a similar inequality for $U(f)$.

We close the section by pointing out that there are other ways to define the notion of integrability for real valued functions. One such is the notion of Lebesgue integrability. With this notion we can integrate functions that are not Riemann integrable, such as the Dirichlet function and many others. For a reference of Lebesgue integration see [23].

8.3 Properties of the Integral

Theorem 8.38. *Suppose $f, g : [a,b] \to \mathbb{R}$ are integrable, and let $r \in \mathbb{R}$ be any real number. Then*
(a) $f + g$ is integrable on $[a,b]$.
(b) $f - g$ is integrable on $[a,b]$.
(c) $f \cdot g$ is integrable on $[a,b]$.
(d) $r \cdot f$ is integrable on $[a,b]$.

Exercise 8.39. Prove Theorem 8.38 by thinking about the following key relations:

(a) $\inf_{[x_{i-1}, x_i]} (f+g)(x) \overset{?}{\sim} \inf_{[x_{i-1}, x_i]} f(x) + \inf_{[x_{i-1}, x_i]} g(x)$

$\sup_{[x_{i-1}, x_i]} (f+g)(x) \overset{?}{\sim} \sup_{[x_{i-1}, x_i]} f(x) + \sup_{[x_{i-1}, x_i]} g(x)$

(b) $\sup_{[x_{i-1}, x_i]} (-f)(x) - \inf_{[x_{i-1}, x_i]} (-f)(x) \overset{?}{\sim} \sup_{[x_{i-1}, x_i]} f(x) - \inf_{[x_{i-1}, x_i]} f(x)$

(c) Exercise 8.22 together with $(f+g)^2 = f^2 + 2f \cdot g + g^2$

Exercise 8.40. Show that functions which are integrable on $[a,b]$ form a real vector space (see Definition 1.42). Show that integration is a linear transformation on the vector space of integrable functions on the interval $[a,b]$.

Corollary 8.41. *Every polynomial is integrable on any interval of the form $[a,b]$.*

This corollary gives us a large collection of functions which are integrable on closed and bounded intervals, but in fact, we can do much much better.

Theorem 8.42. *If $f : [a,b] \to \mathbb{R}$ is continuous, then f is integrable on $[a,b]$.*

Exercise 8.43. Prove Theorem 8.42 using Theorem 8.29 and the fact that continuous functions on closed and bounded intervals are uniformly continuous!

Another way to get new integrable functions from old ones is to modify them only slightly.

Theorem 8.44. *Suppose f is integrable on $[a,b]$. Let $d \in [a,b]$ and let $\tilde{f}$ be a function on $[a,b]$ so that $f(x) = \tilde{f}(x)$ for all $x \in [a,b] \setminus \{d\}$. Then $\tilde{f}$ is integrable on $[a,b]$ and*

$$\int_a^b f(x)dx = \int_a^b \tilde{f}(x)dx.$$

Exercise 8.45. Prove Theorem 8.44. You can do this directly, or you can prove the following first, and then use it: if $g : [a,b] \to \mathbb{R}$ is given by
$$g(x) = \begin{cases} 1 & x = x_0 \\ 0 & \text{else} \end{cases}$$
for some $x_0 \in [a,b]$, then g is integrable on $[a,b]$ and $\int_a^b g(x)dx = 0$. Of course, if $\tilde{f} = f$ except at $x_0 \in [a,b]$, then we can write $\tilde{f} = f + g$ for a function g as above and use Theorem 8.38 to obtain the required result.

Corollary 8.46. *Suppose f is integrable on $[a,b]$. If g is any function on $[a,b]$ which differs from f at at most finitely many points, then g is integrable on $[a,b]$ and*
$$\int_a^b f(x)dx = \int_a^b g(x)dx.$$

Exercise 8.47. Prove Corollary 8.46.

In the same spirit we have the following Theorem.

Theorem 8.48. *Suppose that $f : [a,b] \to \mathbb{R}$ is piecewise continuous. Then f is integrable on $[a,b]$.*

Exercise 8.49. Prove Theorem 8.48. You may want to revisit Definition 6.51, and use Theorem 8.42 repeatedly. Note that this result is an immediate consequence of Theorem 8.57, but you are not allowed to use that theorem at this point.

Exercise 8.50. Suppose that $f : [a,b] \to \mathbb{R}$ is integrable. Then for any $c \in [a,b]$ we have
$$\int_a^b f(x)dx = \int_a^c f(x)dx + \int_c^b f(x)dx.$$

Theorem 8.51. *If f is integrable on $[a,b]$ then $|f|$ is integrable on $[a,b]$. Furthermore,*
$$\left| \int_a^b f(x)dx \right| \leq \int_a^b |f(x)|dx.$$

Exercise 8.52. Prove Theorem 8.51. It might be useful for this purpose to consider the functions f^+ and f^- (See Definition 1.13 on page 4). Another way to go about this particular problem, is to show that
$$\sup_{[x_{i-1}, x_i]} |f| - \inf_{[x_{i-1}, x_i]} |f| \leq \sup_{[x_{i-1}, x_i]} f - \inf_{[x_{i-1}, x_i]} f$$
by noting that
$$|f(x)| - |f(y)| \leq |f(x) - f(y)| \leq \sup_{[x_{i-1}, x_i]} f - \inf_{[x_{i-1}, x_i]} f$$
for any $x, y \in [x_{i-1}, x_i]$.

Proposition 8.53. *Suppose $g \geq 0$ on $[a,b]$ and that g is integrable. Then $\int_a^b g(x)dx \geq 0$.*

Exercise 8.54. Use Theorem 8.51 and Proposition 8.53 to complete Exercise 4.70.

8.3 Properties of the Integral

Exercise 8.55. Suppose $g \geq 0$ on $[a,b]$ and that g is continuous. Show that if there exists $x_0 \in [a,b]$ such that $g(x_0) > 0$, then $\int_a^b g(x)dx > 0$. Conclude that if $g \geq 0$ on $[a,b]$ and g is continuous with $\int_a^b g(x)dx = 0$, then $g(x) = 0$ for all $x \in [a,b]$.

Exercise 8.56. Suppose $f, g : [a,b] \to \mathbb{R}$ are integrable. Show that if $0 \leq f \leq g$ then
$$\int_a^b f(x)dx \leq \int_a^b g(x)dx.$$

Theorem 8.57. *Suppose that $f : [a,b] \to \mathbb{R}$ is bounded, and that there exists $a < c < b$ such that f is integrable on $[a,c]$ and on $[c,b]$. Then f is integrable on $[a,b]$ and*
$$\int_a^c f(x)dx + \int_c^b f(x)dx = \int_a^b f(x)dx.$$

Exercise 8.58. Prove Theorem 8.57. One way to start is by trying to establish the conclusion of Theorem 8.29 for f on $[a,b]$. You can accomplish this by using the same theorem for f on $[a,c]$ and $[c,b]$.

The following proposition gives a 'converse' of Theorem 8.57 in the sense that integrability on a set S implies integrability on any smaller set contained in S.

Proposition 8.59. *Suppose $f : [a,b] \to \mathbb{R}$ is integrable. Then f is integrable on any subinterval $[c,d] \subseteq [a,b]$.*

Exercise 8.60. Prove Proposition 8.59. In the context of the proposition, must $\int_c^d f(x)dx$ be less than $\int_a^b f(x)dx$? If so, give a proof, if not give a counterexample.

Theorem 8.61 (The Fundamental Theorem of Calculus I). *Suppose that $f : [a,b] \to \mathbb{R}$ is a continuous function which is differentiable on (a,b). If f' is integrable on $[a,b]$, then*
$$\int_a^b f'(x)dx = f(b) - f(a).$$

Exercise 8.62. Prove Theorem 8.61. You may want to make use of the Mean Value Theorem as you try to relate values of the derivative to values of the function itself.

Exercise 8.63. Give an example of an interval $[a,b]$ and a continuous $f : [a,b] \to \mathbb{R}$, which is differentiable on (a,b), but such that f' is NOT integrable on $[a,b]$.

Theorem 8.64 (The Fundamental Theorem of Calculus II). *Suppose that $f : [a,b] \to \mathbb{R}$ is integrable. For $x \in [a,b]$ set*
$$F(x) = \int_0^x f(t)dt.$$
Then F is continuous on $[a,b]$. In addition, if f is continuous on $[a,b]$, then F is differentiable on $[a,b]$ and $F'(x) = f(x)$.

Exercise 8.65. Prove Theorem 8.64 using the following hints:
(a) Compute $|F(x) - F(y)|$ and use simple bounds for the integral.
(b) For the differentiability part, use the identity

$$f(x_0) = \frac{1}{x-x_0} \int_{x_0}^{x} f(x_0)dt$$

to compute

$$\left| \frac{F(x) - F(x_0)}{x - x_0} - f(x_0) \right|.$$

Remark 8.66. Take a moment to appreciate the content of Theorem 8.64. It says that integration 'moves functions up the smoothness chain': from integrable to continuous, and from continuous to differentiable. It is in this sense that various authors may refer to integration as a smoothing operator. You can find much more on this in any functional analysis book.

Exercise 8.67. The formulas for integration by parts and u-substitution can be worked out using Theorems 8.61 & 8.64. Derive these formulas.

Chapter 9
Polynomials

Abstract This chapter focuses on various issues related to the roots of polynomials. We highlight the interplay between analysis and algebra with two theorems: the Fundamental Theorem of Algebra, which has an analytic proof, and an interpolation theorem about polynomials, which has an algebraic proof.

> The shortest path between two truths in the real domain passes through the complex domain.
>
> Jacques Hadamard (1865 – 1963)

Getting Your Hands Dirty

Exercise 9.1. How many **(a)** integer; **(b)** rational; **(c)** real; **(d)** complex solutions are there for the equation $x^2 - ax + 1 = 0$, where a is a real number?

Exercise 9.2. Let D denote differentiation, and let $p(x)$ be a polynomial. By considering several specific examples for p, form a conjecture about the relative location of the zeros of $p(x)$ and $D[p(x)]$.

Exercise 9.3. How many pairs of integers (x,y) satisfy the equation $y = \dfrac{x+12}{2x-1}$? What does this have to do with roots of polynomials?

Exercise 9.4. Construct a polynomial p so that $p(1) = p(2) = p(3) = 0$. Now construct a polynomial q so that $q(1) = q(2) = q(3) = q'(1) = q'(2) = q'(3) = 0$. Try to give a general formula for a polynomial p so that p and p' both vanish at the points $r_1, r_2, \ldots, r_m$.

Exercise 9.5. Consider the set $P_{10} := \{p(x) \in \mathbb{R}[x] \mid \deg p(x) \leq 10\}$. Let $F : \mathbb{R}^{11} \times P_{10} \to P_{10}$ be defined as follows: if $p(x) = \sum_{i=0}^{10} a_i x^i$ and $\mathbf{v} = (v_0, v_1, v_2, \ldots, v_{10})$, then

$$F(\mathbf{v}, p(x)) = \sum_{i=0}^{10} a_i v_i x^i.$$

Find a vector $\mathbf{s}$ such that $F(\mathbf{s}, p)$ has only positive real zeros if p has only positive real zeros. Can you find *all* such $\mathbf{s} \in \mathbb{R}^{11}$?

Exercise 9.6. Let $q(x) = (x-1)(x-2)(x-3)$. Find all real numbers r such that $r^3 q\left(\dfrac{1}{r}\right) = 0$. Do the same for some other polynomials of your choice. Make a conjecture about the general case: if $\deg p = m$, think about how the zeros of $p(x)$ and $x^m p\left(\dfrac{1}{x}\right)$ are related.

Exercise 9.7. Find a cubic polynomial that goes through the points $(-1,-1)$, $(0,0)$, $(1,0)$ and $(2,0)$. Now try to find a *different* cubic polynomial going through these same points.

Exercise 9.8. Let $q(x) = (x-2010)(x-43)(x+7)$. Let $g(p,q) = \sum |r_i - t_i|$ where p,q are polynomials and r_i, t_i are their respective roots. Is g a metric? Make this question exact. Now write q in the form

$$q(x) = a_3 x^3 + a_2 x^2 + a_1 x + a_0,$$

and consider sequences $s_{n,i} \to a_i$. What can you say about the zeros of

$$q_n(x) = s_{n,3} x^3 + s_{n,2} x^2 + s_{n,1} x + s_{n,0}?$$

9.1 A Short Historical Exposition

As we give a short overview of the history of solving polynomial equations, we rely heavily on the source [28]. Solving equations occupied the minds of many ancient mathematicians. The Babylonians solved quadratic equations, and Euclid, author of the *Elements* demonstrated a geometrical solution for solving a quadratic equation ca. 300 bc. In 1539 Girolamo Cardano (1501-1576) gave the complete solution of cubics in his book, *The Great Art, or the Rules of Algebra*. Before Cardano, complex numbers had been rejected as solutions for quadratics as absurd, but they are needed in Cardano's formula to express real solutions. *The Great Art* also included the solution of the quartic equation by Ludovico Ferrari (1522-1565), but it was played down because it was believed to be absurd to take a quantity to the fourth power, given that there are only three dimensions. In 1637 Descartes gave his rule of signs determining the possible number of positive and negative roots of a polynomial with real coefficients. In 1666 Sir Isaac Newton introduced what we now know as Newton sums - a recursive way of expressing the sum of the roots to a given power in terms of the coefficients. Three years later he introduced his iterative method for the numerical approximation of roots - an idea you know from your calculus course as Newton's Method. In 1796 Jean Baptiste Joseph Fourier (1768-1830) determined the maximum number of roots in an interval, and in 1799 Carl Friedrich Gauss (1777-1855) proved the fundamental theorem of algebra (see Theorem 9.30).

While there are formulas for solving the general quadratic and cubic polynomial equations, only special quintics can be solved algebraically. In 1821 Niels Henrik Abel

> showed that there is no general algebraic solution for the roots of a quintic equation, or any general polynomial equation of degree greater than four, in terms of explicit algebraic operations. ...[He] sent a paper on the unsolvability of the quintic equation to Gauss, who proceeded to discard without a glance what he believed to be the worthless work of a crank. [31]

In 1831 Augustin-Louis Cauchy (1789-1857) determined how many roots of a polynomial lie inside a given contour in the complex plane. A year later Évariste Galois (1811-1832) wrote down the main ideas of his theory in a letter to Auguste Chevalier the day before he died in a duel. There are many others who have contributed greatly to the development of solving polynomial equations, we mention here lastly Emil Artin (1898-1962) who used field theory to develop the modern theory of algebraic equations.

9.2 Polynomials

Definition 9.9. A *polynomial in x with coefficients a_i from a ring R* (i.e. $p(x) \in R[x]$) is a formal sum of the form

$$p(x) = \sum_{i=0}^{n} a_i x^i,$$

9.2 Polynomials

where $a_n \neq 0$, unless $p(x) \equiv 0$. The coefficients a_0 and a_n are called the constant term and the leading coefficient of p, respectively. Also, if $n \neq 0$, we say that p is of *degree* n. A polynomial is said to be *monic* if its leading coefficient is equal to 1.

Remark 9.10. The degree of a nonzero constant polynomial is zero, but the degree of the zero polynomial is undefined. Two polynomials $p(x) = \sum_{i=0}^{n} a_i x^i$ and $q(x) = \sum_{i=0}^{k} a_i x^i$ are the same if and only if $k = n$ and $a_i = b_i$ for all i.

Polynomials can also be thought of as functions, in which case distinct polynomials could yield the same polynomial function (see Exercise 9.15 below). This situation is impossible if polynomials are understood as functions over $\mathbb{C}$.

From an algebraic point of view, x is an indeterminate. The structure of $R[x]$ is mostly inherited from the structure of the ring R and the rules that guide how to add and multiply indeterminates. (i.e. there has to be an *addition* operation and a *multiplication* operation defined for the indeterminates). Past that however, x could be considered, by evaluation, to be anything from points in a vector space to functions to matrices.

Exercise 9.11. Write out a degree 3 polynomial
(a) in 2×2 matrices with real entries with coefficients from $\mathbb{Z}_2$.
(b) in vectors in $\mathbb{R}^n$ with coefficients from $\mathbb{Q}$.
(c) in polynomials with coefficients from $M_2(\mathbb{C})$.

When it comes to analysis, most of the time we consider polynomial *functions* but we usually just call them polynomials. We give a fairly general such definition here:

Definition 9.12. A *complex polynomial* is a function $p : \mathbb{C} \to \mathbb{C}$ of the form

$$p(x) = \sum_{i=0}^{n} a_i x^i,$$

where $a_i \in \mathbb{C}$ for all i.

Exercise 9.13. Say all you can about the structure that real valued polynomials with real coefficients have as a substructure of complex polynomials.

9.2.1 The Zeros of Polynomials

Both in Analysis and Algebra, the zeros of a polynomial play an important role in determining the properties of the polynomial itself. Hence, in this section we will focus our attention on them.

Definition 9.14. Let R be a ring, and let $p(x) \in R[x]$. If $d \in R$ is such that $p(d) = 0$ then d is called a zero (or a root) of $p(x)$ in R.

Exercise 9.15.
(a) Find all roots of $p(x) = x^2 - 2$ and of $q(x) = x^2 + 1$ in $\mathbb{Q}$, $\mathbb{R}$, $\mathbb{C}$, and $\mathbb{Z}_3$.
(b) Find all the roots in $\mathbb{Z}_5$ of $p(x) = x^5 - x$.

The following results all concern various aspects and properties of the zeros of polynomials. As Definition 9.9 indicates, one can consider polynomials with coefficients from any ring R. The next result concerns polynomials belonging to $\mathbb{Z}[x]$.

Theorem 9.16 (The Rational Zeros Theorem). *Let $p(x) \in \mathbb{Z}[x]$ be of the form*

$$p(x) = a_n x^n + a_{n-1} x^{n-1} + \cdots + a_1 x + a_0.$$

If $r = \dfrac{t}{q}$ is in reduced form, $t, q \in \mathbb{Z}$, and r is a zero of p, then $t | a_0$ and $q | a_n$.

Proof. We have that

$$0 = p(r) = a_n \left(\frac{t}{q}\right)^n + a_{n-1} \left(\frac{t}{q}\right)^{n-1} + \cdots + a_1 \left(\frac{t}{q}\right) + a_0.$$

Multiplying by q^n and rearranging in two different ways we obtain the equations

$$-a_0 q^n = a_n t^n + a_{n-1} t^{n-1} q + \cdots + a_1 q^{n-1} t \qquad (9.1)$$
$$-a_n t^n = a_{n-1} t^{n-1} q + \cdots + a_1 q^{n-1} t + a_0 q^n \qquad (9.2)$$

Observe that t is a factor in every term on the right hand side of (9.1), and as a result, t divides the right hand side of (9.1). Consequently t must divide $-a_0 q^n$. Since t and q are relatively prime, we obtain that $t | a_0$. A similar argument shows using (9.2) that $q | a_n$. The proof is complete. □

Theorem 9.16 helps one decide whether a given polynomial has rational zeros. For example, in the case of the polynomial $p(x) = 2x^3 - 3x + 1$, if t/q were a zero of p, we would have to have $q | 2$ and $t | 1$. As a result, the only possible rational roots are $r = \pm 1$ and $r = \pm 2$. Substituting these values into p we see that $p(1) = 0$, $p(-1) = 2$, $p(2) = 13$ and $p(-2) = -11$, thus the only rational root is $r_1 = 1$. Factoring $p(x)$ as

$$p(x) = (x - 1)(2x^2 + 2x - 1)$$

and using the quadratic formula we obtain the other roots: $r_{2,3} = \dfrac{-1 \pm \sqrt{3}}{2}$. Since these numbers can't be rational, they must be irrational! (This fact quickly implies that $\sqrt{3}$ is irrational)

Exercise 9.17. Use the polynomial $p(x) = x^2 - 2$ and Theorem 9.16 to show that $\sqrt{2}$ is irrational.

Suppose that we have established that a polynomial p has no rational roots. Is there anything further one can say about **(a)** what kind of roots p has and **(b)** where the roots may lie? The answer to these questions is yes. The following are some results partially addressing these questions.

In *La Géométrie*, Descartes makes the following statement without any proof:

> We can determine also the number of true and false roots that any equation can have, as follows: An equation can have as many true roots as it contains changes of sign, from + to - or from - to +; and as many false roots as the number of times two + signs or two - signs are found in succession. [27]

True roots here are a reference to positive roots, while false roots refer to negative roots. It is probably this statement that lead to the naming of the following theorem.

Theorem 9.18 (Descartes' Rule of Signs). *The number of positive roots of a polynomial $p(x)$ with real coefficients is equal to the number of "changes of sign" in the list of coefficients, or is less than this number by a multiple of 2. The number of negative roots of $f(x)$ is equal to the number of sign changes in the list of coefficients of $f(-x)$, or is less than this number by a multiple of 2.*

Example 9.19. Consider the polynomial $p(x) = 3x^3 - \pi x^2 + 4$. The coefficients are $3, -\pi, 0, 4$ with sign sequence $+, -, 0, +$. Since we don't count 0 as a change, there is a total of 2 sign changes here. We conclude that p has 2 or zero positive real roots. (By Theorem 9.21 complex roots come in pairs so this comes as no surprise).

Next we look at $p(-x) = -3x^3 - \pi x^3 + 4$, with the sign sequence $-,-,0,+$. There is only 1 sign change, thus we conclude that p must have exactly 1 negative real zero. And now for the truth: the zeros of p are

$$r_1 = -0.840445, \qquad r_{2,3} = 0.943821 \pm i \cdot 0.834063.$$

Exercise 9.20. Prove Theorem 9.18 for linear and quadratic polynomials.

Theorem 9.21 (Complex Roots come in conjugate pairs). *Suppose that $p(x) \in \mathbb{R}[x]$. If $p(r) = 0$, then $p(\bar{r}) = 0$ as well, where $\bar{r}$ is the complex conjugate of r.*

Proof. Write $p(x) = a_n x^n + a_{n-1} x^{n-1} + \cdots + a_1 x + a_0$ and suppose that $p(r) = 0$. Then the following string of equalities holds:

$$\begin{aligned}
0 &= p(r) \\
&= \overline{p(r)} \\
&= \overline{a_n} \, \overline{r^n} + \overline{a_{n-1}} \, \overline{r^{n-1}} + \cdots + \overline{a_1} \, \overline{r} + \overline{a_0} \\
&= a_n \bar{r}^n + a_{n-1} \bar{r}^{n-1} + \cdots + a_1 \bar{r} + a_0 \\
&= p(\bar{r}).
\end{aligned}$$

The proof is complete. □

We then see that every polynomial with real coefficients must have an even number of complex roots. You may wonder whether this statement is still true for $p(x) \in \mathbb{C}[x]$, i.e. for polynomials with complex coefficients. The answer is no, which one can easily see by constructing for example a quadratic polynomial with one real and one complex root:

$$q(x) = (x-i)(x-1) = x^2 - (1+i)x + i.$$

This theorem together with the next few results will yield a characterization of polynomials in $\mathbb{R}[x]$ (see Exercise 9.29).

Theorem 9.22 (The Division Theorem). *Let F be a field, and let $p(x), d(x) \in F[x]$ with $d(x) \neq 0$. Then there exist $q(x), r(x) \in F[x]$ such that*

$$p(x) = q(x)d(x) + r(x)$$

where $r(x) = 0$ or $\deg(r) < \deg(d)$. Moreover, $r(x)$ is unique.

This result should remind you of the division theorem in $\mathbb{Z}$, discussed in Chapter 1. In fact, most of the results proved in $\mathbb{Z}$ using that theorem can be re-written as theorems for polynomials (with coefficients on a field). Go ahead, try it out!

Exercise 9.23. Find an example that shows that the division algorithm does not hold in $\mathbb{Z}[x]$.

Exercise 9.24. Prove that $F[x]$, where F is a field is a PID.
Hint: Use the same ideas that were used to prove that $\mathbb{Z}$ is a PID. You will need to 'translate' that proof from its context in the integers to this new context, in $F[x]$. In order to do this, compare Theorems 1.18 and 9.22.

Theorem 9.25 (The Factor Theorem). *Let $p(x)$ be a polynomial in $F[x]$, where F is a field. Then $(x-d)$, for $d \in F$, is a factor of $p(x)$ if and only if d is a zero of $p(x)$.*

Exercise 9.26.
(a) Prove the factor theorem using the division theorem.
(b) Prove that $I = \{p(x) \in \mathbb{C}[x];\ p(0) = 0\}$ is a principal ideal of $\mathbb{C}[x]$ by finding explicitly a generator of I.
(c) Show that $I = \{p(x) \in \mathbb{R}[x];\ p(0) = p(1) = 0\}$ is a principal ideal of $\mathbb{R}[x]$ by finding explicitly a generator of I.

Theorem 9.27 (The Remainder Theorem). *Let $p(x)$ be a polynomial function with coefficients in a field F. Then $p(d)$ is equal to the remainder of dividing $p(x)$ by $x - d$.*

Exercise 9.28. Show that the factor theorem and the remainder theorem are equivalent. In other words show that you can prove either one using the other.

Now let us put all this machinery to work in order to understand the structure of $\mathbb{R}[x]$.

Exercise 9.29. Let $p(x) \in \mathbb{R}[x]$ be of degree at least one.
(a) Assume that all the roots of $p(x)$ are real. Prove that $p(x)$ factors into linear polynomials in $\mathbb{R}[x]$.
(b) Suppose $p(x)$ has a non-real root say $z = a + bi$. Show that $p(x) = (x^2 - 2ax + a^2 + b^2)q(x)$, where $q(x) \in \mathbb{R}[x]$. Note that $x^2 - 2ax + a^2 + b^2 \in \mathbb{R}[x]$ as well.
(c) Show that $p(x)$ can always be written as a product of polynomials in $\mathbb{R}[x]$ all of which have degree either one or two.

Theorem 9.30 (The Fundamental Theorem of Algebra). *For any non-constant $p(x) \in \mathbb{C}[x]$ there exists at least one point $r \in \mathbb{C}$ such that $p(r) = 0$. In other words, every complex polynomial one has a complex root.*

Remark 9.31. Before we embark on the proof of this theorem, we should mention that as A. R. Schep points out in [24], there are at least 28 articles written in just the American Mathematical Monthly on the proof of the Fundamental Theorem of Algebra. The most commonly cited one using complex analysis is based on Liouville's theorem. Here we adopt Schep's proof which we find quite appealing.

Proof. We proceed by contradiction. To this end let $p : \mathbb{C} \to \mathbb{C}$ be given by

$$p(z) = z^n + a_{n-1} z^{n-1} + \cdots a_1 z + a_0,$$

and suppose that $p(z) \neq 0$ for all $z \in \mathbb{C}$. Since

$$|p(z)| \geq |z|^n \left(\left(1 - \frac{|a_{n-1}|}{|z|} - \cdots - \frac{|a_1|}{|z^{n-1}|} - \frac{|a_0|}{|z^n|}\right) \right),$$

we see that

$$\lim_{z \to \infty} |p(z)| = \infty.$$

Since $p(z) \neq 0$ for all $z \in \mathbb{C}$, the function $1/p(z)$ is entire. Applying the Cauchy integral formula to this function we obtain

$$\frac{1}{p(0)} = \frac{1}{2\pi i} \int_{|z|=r} \frac{dz}{z p(z)}.$$

Rearranging the above equation we get

$$\int_{|z|=r} \frac{dz}{z p(z)} = \frac{2\pi i}{p(0)} \neq 0.$$

On the other hand,

9.2 Polynomials

$$\left| \int_{|z|=r} \frac{dz}{zp(z)} \right| \leq 2\pi r \max_{|z|=r} \frac{1}{|z||p(z)|} = \frac{2\pi}{\min_{|z|=r} |p(z)|} \to 0 \quad \text{as} \quad r \to \infty,$$

clearly a contradiction. The proof is complete. $\square$

Exercise 9.32. In the proof of the Fundamental Theorem of Algebra we only consider polynomials whose leading coefficient is 1. Explain why there is no loss of generality doing so.

Exercise 9.33. Let $p(x) \in \mathbb{C}[x]$ be a polynomial of degree $n \geq 1$. Show that $p(x)$ has all its roots in $\mathbb{C}$.

For additional reading we suggest Theo de Jong's paper [14] which gives a proof of the Fundamental Theorem of Algebra using Lagrange multipliers.
The Fundamental Theorem of Algebra actually implies that a degree n polynomial has exactly n roots. It is also immediate that, given n distinct points $a_1, a_2, \ldots, a_n$ in $\mathbb{C}$, the polynomials having these points as zeros are all constant multiples of each other. Thus, n data points are not enough to uniquely determine a degree n polynomial. $n+1$ points will however suffice, as the following theorem shows.

Theorem 9.34. *Given $n+1$ points $(x_0, y_0), (x_1, y_1), \ldots, (x_n, y_n)$ in $\mathbb{C} \times \mathbb{C}$ with all the x_i's distinct, there is a unique polynomial p of degree at most n such that $p(x_i) = y_i$ for $i = 1, 2, \ldots, n$.*

Proof. Consider the generic n degree polynomial $p(x) = a_n x^n + a_{n-1} x^{n-1} + \cdots + a_1 x + a_0$. The requirement that $p(x_i) = y_i$ for $i = 1, 2, \ldots, n$ can be rewritten in the matrix equation

$$\begin{bmatrix} x_0^n & x_0^{n-1} & \cdots & 1 \\ x_1^n & x_1^{n-1} & \cdots & 1 \\ \vdots & \vdots & \ddots & \vdots \\ x_n^n & x_n^{n-1} & \cdots & 1 \end{bmatrix} \begin{bmatrix} a_n \\ a_{n-1} \\ \vdots \\ a_0 \end{bmatrix} = \begin{bmatrix} y_0 \\ y_1 \\ \vdots \\ y_n \end{bmatrix}. \tag{9.3}$$

Observe that

$$\det \begin{bmatrix} x_0^n & x_0^{n-1} & \cdots & 1 \\ x_1^n & x_1^{n-1} & \cdots & 1 \\ \vdots & \vdots & \ddots & \vdots \\ x_n^n & x_n^{n-1} & \cdots & 1 \end{bmatrix} = \prod_{\substack{i,j=0 \\ i<j}}^{n} (x_i - x_j) \neq 0 \tag{9.4}$$

since $x_i \neq x_j$ if $i \neq j$. It follows that the system in (9.3) has a unique solution and the proof is complete. $\square$

Exercise 9.35. Prove the first equality presented in equation 9.4.

Corollary 9.36. *If two n degree polynomials agree at $n+1$ distinct points, then they must be the same polynomial.*

Exercise 9.37. Prove Corollary 9.36 by assuming that there are two polynomials say p and q of degree n that agree at $n+1$ points. Then consider the polynomial $h = p - q$.

This is quite a striking result: it says that all we need is $n+1$ values of an n-degree polynomial in order to uniquely identify it!
Although Theorem 9.34 assures us of the existence of a unique interpolating polynomial, it does not actually tell us what the polynomial looks like. We now fill in this gap by constructing the interpolating polynomial called the *Lagrange interpolating polynomial*.
Let $(x_0, y_0), (x_1, y_1), \ldots, (x_n, y_n)$ be $n+1$ points with distinct first coordinates. Define

$$L(x) = \sum_{k=0}^{n} y_j \prod_{\substack{i=0 \\ j \neq i}}^{n} \frac{x - x_i}{x_j - x_i}. \tag{9.5}$$

One can then check that $L(x_i) = y_i$ for $i = 0, 1, \ldots, n$.

Exercise 9.38. Show that the function defined in equation (9.5) satisfies $L(x_i) = y_i$ for $i = 0, 1, \ldots, n$.

Exercise 9.39. Let ξ_1, ξ_2, ξ_3 and ξ_4 denote the 4 distinct fourth roots of unity. Identify the unique cubic polynomial $q \in \mathbb{C}[x]$ that satisfies $q(\xi_i) = 1$ for $i = 1, 2, 3, 4$.

Sometimes, when the space considered is not the complex numbers, it is possible to take the idea of interpolation even further. For instance, any function $f : \mathbb{Z}_p \to \mathbb{Z}_p$, where p is prime, can be represented as a polynomial. This is in stark contrast with some facts you learned in Calculus courses. Namely, some functions, such as $f(x) = e^x$ could not be represented as polynomials, although they could be approximated by polynomials arbitrarily well.

Exercise 9.40. Think about what role the finiteness of $\mathbb{Z}_p$ plays in the difference of representability of functions as polynomials. Think! Discuss it with your peers!

Exercise 9.41. Let p be a prime number.
(a) Have you ever heard of Fermat's Last Theorem? Well, there is another theorem, actually proven by Fermat, that has the same acronym (FLT). Find out/research/look for what Fermat's Little Theorem says.
(b) Use FLT (the little one) to prove that if $f : \mathbb{Z}_p \to \mathbb{Z}_p$ is a function then

$$f(x) = \sum_{k=0}^{p-1} f(k)(1 - (k-x)^{p-1})$$

for all $x \in \mathbb{Z}_p$.
(c) Conclude that any function $f : \mathbb{Z}_p \to \mathbb{Z}_p$ can be represented as a polynomial of degree at most $p - 1$.

Theorem 9.42 (Rolle's Theorem). *Let $f : [a,b] \to \mathbb{R}$ be continuous on $[a,b]$ and differentiable on (a,b). If $f(a) = f(b)$, then there must be a point $c \in (a,b)$ such that $f'(c) = 0$.*

Proof. There are two essential components of this proof: **(a)** a continuous function attains its maximum and its minimum on a closed interval; **(b)** if x_0 is a local extremum for a differentiable function f, then $f'(x_0) = 0$.

(a) This is the content of the Extreme Value Theorem. If an extreme value lies in (a,b), then by **(b)** we have what we need. Otherwise the extreme values occur at the endpoints. Since $f(a) = f(b)$, we see that in fact we must have $f(x) = f(a)$ for all $x \in [a,b]$. Thus $f'(x) = 0$ for every $x \in (a,b)$.

(b) Let x_0 be a local maximum (the case of the local minimum is assigned in Exercise 9.44). Then there exists an $\varepsilon > 0$ such that $f(x) \leq f(x_0)$ if $|x - x_0| < \varepsilon$. It follows that

$$\frac{f(x_0) - f(x)}{x_0 - x} \geq 0 \quad \text{if} \quad 0 < x_0 - x < \varepsilon, \quad \text{and}$$

$$\frac{f(x_0) - f(x)}{x_0 - x} \leq 0 \quad \text{if} \quad 0 < x - x_0 < \varepsilon.$$

Consequently

$$\lim_{x \to x_0^-} \frac{f(x_0) - f(x)}{x_0 - x} \geq 0 \geq \lim_{x \to x_0^+} \frac{f(x_0) - f(x)}{x_0 - x}. \tag{9.6}$$

Since f is differentiable at x_0, both inequalities in (9.6) hold with equality, implying that $f'(x_0) = 0$. $\square$

9.2 Polynomials

Exercise 9.43. Complete **(a)** in the proof of Theorem 9.42 by addressing the case of the minimum.

Exercise 9.44. Complete **(b)** in the proof of Theorem 9.42 by addressing the case of a local minimum.

Corollary 9.45. *Between any two zeros of a differentiable function lies a zero of its derivative.*

Exercise 9.46. Suppose f satisfies the conditions of Rolle's theorem, with $f(a) = f(b)$. Can there be *more than one* root of the derivative between a and b?

Corollary 9.47. *If a degree n polynomial $p(x) \in \mathbb{R}[x]$ has only real zeros, then $p^{(k)}(x)$ has exactly $n-k$ real zeros for any $0 \leq k \leq n$, where $p^{(k)}(x)$ denotes the kth derivative of p. (By convention we say that any degree zero polynomial has only real zeros, encompassing both the case when such a polynomial has no zeros, and the case when it has infinitely many).*

9.2.2 Coefficients and Zeros

Naturally, given any polynomial $p(x) = a_n x^n + a_{n-1} x^{n-1} + \cdots + a_1 x + a_0$, the roots have to be related to the coefficients somehow. In other words, in the case of a degree n polynomial there should be n functions $f_1, \ldots, f_n$ such that

$$f_i(a_0, a_1, a_2, \ldots, a_n) = r_i \quad \text{for} \quad i = 1, 2, \ldots, n.$$

In the case of the quadratic polynomial $q(x) = ax^2 + bx + c$, these functions are

$$f_1(a,b,c) = \frac{-b + \sqrt{b^2 - 4ac}}{2a}$$

$$f_2(a,b,c) = \frac{-b - \sqrt{b^2 - 4ac}}{2a}$$

as you well know. Although such functions do not exists in general for polynomials of degree 5 or higher, some relations between the coefficients and the roots can be obtained. Efforts of Newton to find such relations resulted in (among others) what we now call Newton sums. Recall that the zeros of

$$p(x) = a_n x^n + a_{n-1} x^{n-1} + \cdots + a_1 x + a_0$$

and those of

$$\tilde{p}(x) = x^n + \frac{a_{n-1}}{a_n} x^{n-1} + \cdots + \frac{a_1}{a_n} x + \frac{a_0}{a_n}$$

are the same. Thus we will only consider monic polynomials in what follows.

Theorem 9.48 (Newton's Sums). Let $p(x) = x^n + a_{n-1} x^{n-1} + \cdots + a_1 x + a_0$, and suppose that p has roots $t_1, t_2, \ldots, t_n$. For $k = 1, 2, \ldots$ define the Newton sum of order k as

$$s_k := \sum_{i=1}^{n} t_i^k.$$

Then

$$s_k + a_{n-1} s_{k-1} + \cdots + a_0 s_{k-n} = 0 \quad (k > n) \tag{9.7}$$

$$s_k + a_{n-1} s_{k-1} + \cdots + a_{n-k+1} s_1 = -k a_{n-k} \quad (1 \leq k \leq n) \tag{9.8}$$

Example 9.49. Let $q(x) = x^4 - 3x^3 + x - 4$. Then $a_3 = -3, a_2 = 0, a_1 = 1$ and $a_1 = -4$. It follows that

$$\begin{aligned} s_1 &= -(-3) = 3, \\ s_2 &= -(-3s_1) - 2(0) = 9, \\ s_3 &= -(-3s_2) - 0(s_1) - 3(1) = 24, \\ s_4 &= -(-3s_3) - 0(s_2) - 1(s_1) - 4(-4) = 72 - 3 + 16 = 85. \end{aligned}$$

The fact that all of these numbers are real is a consequence of Theorem 9.21 and the following simple exercise.

Exercise 9.50. Let $z \in \mathbb{C}$ be any complex number. Show that for any $n \in \mathbb{N}$ we have that $z^n + \bar{z}^n \in \mathbb{R}$.

Proof. (Of Newton's Sums) Let C be the companion matrix of p. Then the roots of p are the eigenvalues of C and in fact the kth powers of the roots are the eigenvalues of C^k. Since the sum of the eigenvalues equals the trace of a matrix, we see that

$$s_k = \text{tr}(C^k)$$

Thus equation (9.7) can be reformulated as

$$\text{tr}(C^k) + a_{n-1}\text{tr}(C^{k-1}) + \cdots + a_0\text{tr}(C^{k-n}) = 0.$$

To prove this equality notice that

$$\begin{aligned} \text{tr}(C^k) + a_{n-1}\text{tr}(C^{k-1}) + \cdots + a_0\text{tr}(C^{k-n}) &= \text{tr}(C^k + a_{n-1}C^{k-1} + \cdots + a_0 C^{k-n}) \\ &= \text{tr}(C^{k-n}p(C)). \end{aligned}$$

By the Cayley-Hamilton theorem (see for example page 86 of [13]) we have $p(C) = 0$ and the proof of the case $k > n$ is complete.

We now turn our attention to the case $1 \leq k \leq n$. The trace version of equation (9.8) is given by

$$\text{tr}(C^k + a_{n-1}C^{k-1} + \cdots + a_{n-k+1}C) = -ka_{n-k}$$

which, after adding na_{n-k} to both sides gives

$$\text{tr}(C^k + a_{n-1}C^{k-1} + \cdots + a_{n-k+1}C + a_{n-k}I) = (n-k)a_{n-k}. \tag{9.9}$$

In order to derive equation (9.9), which involves the coefficients of p, it is natural that we would start by working with p. Recall that $p(C) = 0$, in other words C is a zero of p. Writing $X = xI$ we may factor $p(X)$ as

$$\begin{aligned} p(X) = (X - C)\big[&X^{n-1} + (C + a_{n-1}I)X^{n-2} + (C^2 + a_{n-1}C + a_{n-2}I)X^{n-3} \\ &+ \cdots + (C^{n-1} + a_{n-1}C^{n-2} + \cdots + a_1 I)I\big]. \end{aligned} \tag{9.10}$$

We would like to take the trace of both sides, but the trace operation does not behave well with respect to multiplication. Thus we first multiply both sides by $(X - C)^{-1}$ to obtain

$$\begin{aligned} (X - C)^{-1}p(X) = &X^{n-1} + (C + a_{n-1}I)X^{n-2} + (C^2 + a_{n-1}C + a_{n-2}I)X^{n-3} \\ &+ \cdots + (C^{n-1} + a_{n-1}C^{n-2} + \cdots + a_1 I)I. \end{aligned} \tag{9.11}$$

If we now take the trace of both sides, we arrive at the equation

$$\begin{aligned} \text{tr}\big((X - C)^{-1}p(X)\big) = &nx^{n-1} + \text{tr}(C + a_{n-1}I)x^{n-2} \\ &+ \cdots + \text{tr}(C^{n-1} + a_{n-1}C^{n-2} + \cdots + a_1 I). \end{aligned} \tag{9.12}$$

Notice that expressions of the same type as the one on the left hand side of equation (9.9) appear on the right hand side of equation (9.12) *as coefficients* of various powers of x. If we could show that the left hand side of

9.2 Polynomials

(9.12) is equal to $p'(x)$, our proof will be complete by simply equating coefficients. To this end note that

$$(X-C)^{-1}p(X) = p(x)(xI-C)^{-1}$$

and consequently

$$\operatorname{tr}((X-C)^{-1}p(X)) = p(x)\operatorname{tr}((xI-C)^{-1}).$$

Now the eigenvalues of $(xI-C)^{-1}$ are just the fractions

$$\frac{1}{x-t_1}, \frac{1}{x-t_2}, \ldots, \frac{1}{x-t_n}$$

and the trace of a matrix is equal to the sum of its eigenvalues. Putting these facts together we see that

$$\begin{aligned}\operatorname{tr}((X-C)^{-1}p(X)) &= p(x)\operatorname{tr}((xI-C)^{-1}) \\ &= p(x)\left(\frac{1}{x-t_1}+\frac{1}{x-t_2}+\cdots+\frac{1}{x-t_n}\right) \\ &= p'(x).\end{aligned}$$

The proof is thus complete. □

Exercise 9.51. Discuss what makes equation 9.10 work. Provide an example of such a factorization.

Exercise 9.52. Verify the step that leads from equation (9.11) to equation (9.12).

Exercise 9.53. Let p be a degree n polynomial with roots $t_1, t_2, \ldots, t_n$. Let C be the companion matrix of p (for the definition of a companion matrix see for example [13], page 147). Prove that the eigenvalues of $(xI-C)^{-1}$ are the fractions

$$\frac{1}{x-t_1}, \frac{1}{x-t_2}, \ldots, \frac{1}{x-t_n}.$$

Exercise 9.54. Suppose $p(x) = x^4 + a_3 x^3 + a_2 x^2 + a_1 x + a_0$. Write down the explicit formulas for s_1, s_2 and s_3 in terms of a_0, a_1, a_2, a_3.

Exercise 9.55. Using anything you want from this chapter say all you can about the zeros of the polynomial $p(x) = 4x^3 - 2x + 1$.

Newton sums provide relations between the coefficients and the zeros of a given polynomial. These relations can in some special cases be used to say something about the location of the zeros. As little as that information may help finding the zeros, it is in nature similar to the following two results, both of which give information on the location of the zeros of a complex polynomial.

Theorem 9.56 (The Argument Principle). *Let $p(x) \in \mathbb{C}[x]$ be given. Let γ be any positively oriented piecewise smooth simple closed curve in $\mathbb{C}$ which does not go through any of the zeros of p. Then*

$$\frac{1}{2\pi}\left\{\begin{array}{c}\text{change in }\arg p(z) \\ \text{as }z\text{ traverses }\gamma\end{array}\right\} = \left\{\begin{array}{c}\text{number of zeros} \\ \text{of }p\text{ inside }\gamma\end{array}\right\}$$

counting multiplicities.

The proof of this theorem would take us too far afield from our investigations, so omit it here. For a complete proof we refer the interested reader to [10] or [19].

Example 9.57. We determine the number of zeros of the function $f(x) = x^4 + 3ix^2 + x - 2 + i$ in the upper half plane. To this end, consider a contour γ as shown in Figure 9.1.

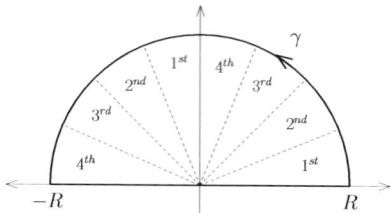

Fig. 9.1 The path γ

We need to find out the net change in the argument of $f(x)$ as x traverses the contour. By letting $R \to \infty$, we will find the number of zeros of f in the upper half plane. Consider now $f(x)$ for $x \in [-R, R]$. Since $x \in \mathbb{R}$ and

$$\Re(f(x)) = x^4 + x - 2, \qquad \Im(f(x)) = 3x^2 + 1.$$

we get that the imaginary part of $f(x)$ remains positive as x goes from $-R$ to R, while the real part of $f(x)$ starts out positive, turns negative and then turns positive again. So $f(x)$ goes from the first quadrant to the second quadrant and back to the first quadrant as x goes from $-R$ to R. Next we consider the arc piece of γ. Writing $x = Re^{i\theta}$, we see that for x on the arc,

$$f(Re^{i\theta}) = \left(R^4 \cos(4\theta) - 3R^2 \sin(2\theta) + R\cos(\theta) - 2\right) \\ + i\left(R^4 \sin(4\theta) + 3R^2 \cos(2\theta) + R\sin(\theta) + 1\right).$$

It follows that for large R, the sign of $\Re(f(x))$ is determined by that of $\cos(4\theta)$, and the sign of $\Im(f(x))$ is determined by that of $\sin(4\theta)$. It is now easy to see that, as θ goes from 0 to π, $f(Re^{i\theta})$ starts in the first quadrant, goes through the $2^{nd}, 3^{rd}$ and 4th quadrants, returns to the first quadrant and repeats this course. Thus the argument of $f(x)$ changes by 4π as θ goes from 0 to π. Putting this all together and applying the argument principle we see that $f(x)$ has 2 zeros in the upper half plane. It is perhaps reassuring to know that the zeros of f are

$$r_1 = -1.38257 + i0.797123, \quad r_2 = -0.237604 + i1.02998 \\ r_3 = 0.477885 - i0.653031, \quad r_4 = 1.14229 - i1.17407.$$

Exercise 9.58. What is wrong with the following argument?

Well, $f(x) = x^4 + 3ix^2 + x - 2 + i$ is a degree four polynomial. Since $3x^2 + 1$ is never zero if $x \in \mathbb{R}$, we see that f has no real roots. Since complex roots come in conjugate pairs, it is automatic that f has two roots in the upper half plane.

Theorem 9.59 (Rouché's Theorem). *Suppose that f and g are analytic on an open set containing a piecewise smooth simple closed curve γ and its interior. If*

$$|f(z) + g(z)| \leq |f(z)| \qquad \text{for all} \qquad z \in \gamma,$$

then f and g have an equal number of zeros inside γ, counting multiplicities.

We remark that one can use Rouché's theorem to prove the Fundamental Theorem of Algebra. Fisher gives such a proof in [10]. Since we have already proven the FTA, we elect not to present another proof here.

Example 9.60. We show here that $g(x) = x^3 + x^2 + x + 3$ has no zeros inside the unit disk. To see this set $f(x) = -3$. Then $f(x) + g(x) = x^3 + x^2 + x$. On the circle $|x| = 1$ we have

$$|x^3 + x^2 + x| \leq |x|^3 + |x|^2 + |x| = 3 = |f(x)|$$

9.3 A More Algebraic Look at Polynomials

Thus, by Rouché's theorem, g and f have the same number of zeros in the unit disk, namely none. The zeros of g are $r_1 = -1.574$ and $r_{2,3} = 0.2873 \pm 1.35i$.

Exercise 9.61. Determine how many zeros $p(x) = x^3 - 3x + 1$ has in the annulus $1 < |x| < 2$ (x here is treated as a complex variable).

9.2.3 Approximating the Zeros of a Polynomial

So far we have been focusing mostly on the zeros of polynomials. One of the first approximation ideas you have seen is called Newton's method. This method helps you approximate a zero of a polynomial in a certain way. It is a very useful method, but here we want shift our perspective and think about how changing the coefficients of a polynomial may affect its zeros. To this end consider a function that maps from $\mathbb{C}^n$ to $\mathbb{C}^n$ assigning to an n-tuple of coefficients, an n-tuple of zeros. If we could show that this function is continuous at some $\mathbf{v} \in \mathbb{C}^n$, it would mean that the roots of the polynomial whose coefficients are the coordinate entries of $\mathbf{v}$ can be approximated nicely by zeros of polynomials whose coefficients are "close" to the coordinate entries of $\mathbf{v}$.

It turns out that the situation is even simpler than one might think at first. Consider first linear polynomials of the form $p(x) = ax + b$, $a \neq 0$. If we have $(a_n)_{n=1}^\infty$ and $(b_n)_{n=1}^\infty$ converging to a and b respectively, then the zeros of $p_n(x) = a_n x + b_n$ (given by $r_n = -\dfrac{b_n}{a_n}$) converge to the zero of p, namely b/a.

Proposition 9.62. *Suppose $p_n(x), q_n(x), p(x)$ and $q(x)$ are complex polynomials such that $\deg p_n(x) = \deg p(x)$ and $\deg q_n(x) = \deg q(x)$, for all $n \in \mathbb{N}$. Suppose that that the zeros of $p_n(x)$ converge to those of $p(x)$, and the zeros of $q_n(x)$ converge to those of $q(x)$. Then the zeros of $p_n(x)q_n(x)$ converge to those of $p(x)q(x)$.*

Proof. The proof uses the limit theorems for convergent sequences and Theorem 9.48. □

Exercise 9.63. Give the proof of Proposition 9.62.

Example 9.64. Consider the function $f(x) = \sin x$. This function has a Taylor series expansion with infinite radius of convergence. In fact,
$$f(x) = \sum_{n=0}^\infty (-1)^n \frac{x^{2n+1}}{(2n+1)!} \qquad \forall x \in \mathbb{R}.$$

This of course means that
$$\lim_{n \to \infty} S_n(x) = \lim_{n \to \infty} \sum_{k=0}^n (-1)^k \frac{x^{2k+1}}{(2k+1)!} = f(x) \qquad \forall x \in \mathbb{R}.$$

Notice that $(i) f(x) = 0$ if and only if $x = m\pi$ for some $m \in \mathbb{Z}$, and $(ii) S_k(x) \in \mathbb{Q}[x]$ for all $k \geq 0$. Hence no zero of f is a zero of *any* of the approximating polynomials, since π is transcendental over $\mathbb{Q}$. As a side note we remark that this example shows that one can approximate transcendental numbers with algebraic ones, though this fact can easily be shown with much less machinery.

Similar situations can easily arise for real analytic functions with only real zeros, insofar as it is quite possible to have approximating polynomials whose zeros are all complex.

9.3 A More Algebraic Look at Polynomials

The algebraic study of polynomials focuses mostly on the structure of the set $R[x]$, where R is a ring. It turns out that if R is a field then the algebraic structure of $R[x]$ is extremely similar to that of $\mathbb{Z}$. For instance, Theorems

1.18 and 9.22 give us division theorems for $\mathbb{Z}$ and $R[x]$, when R is a field, respectively. Similarly, Exercise 9.24 says that when R is a field, $R[x]$ is a PID. The proof in this exercise is almost the same as the one we used to obtain that $\mathbb{Z}$ is a PID.

There are many similarities between $\mathbb{Z}$ and a ring of polynomials $R[x]$. In both rings one can divide (and thus factor), a fact made transparent by the comparison between Theorem 1.18 and Theorem 9.22. When we factor in $\mathbb{Z}$, we use prime numbers as basic building blocks. Until now these building blocks have not been matched to any particular elements in $R[x]$. In this section we study the 'prime numbers' in $R[x]$: irreducible polynomials.

ioned before, $R[x]$ has two operations that are mainly inherited from the structure of R. The following exercises show this relation.

Exercise 9.65. Let R be a ring. Show that $R[x]$ is a ring. Moreover,
(a) Show that if R has a 1, then so does $R[x]$.
(b) Show that R is commutative if and only if $R[x]$ is commutative.
(c) Consider $p(x) = 2x^7 + 1$ and $q(x) = 3x^{12} + 5$ in $\mathbb{Z}_6[x]$. Compute $p(x)q(x) \in \mathbb{Z}_6[x]$.
(d) Prove that $R[x]$ is an integral domain if and only if R is an integral domain.
(e) Let $p(x), q(x) \in R[x]$. Prove that
  (i) $deg(pq) \leq deg(p) + deg(q)$. When does equality hold?
  (ii) $deg(p+q) \leq max\{deg(p), deg(q)\}$. When does equality hold?
(f) Show that $R[x]$ is not a field.
Hint: Try to find an inverse for $p(x) = x$.
(g) Show that $p(x) = 2x + 1$ is a unit of $\mathbb{Z}_4[x]$.
(h) Assume R is an integral domain. Find all polynomials in $R[x]$ that have inverses in $R[x]$. Note that this question asks you to find $R[x]^*$.

Exercise 9.66. Let R be a ring.
(a) Does the set of polynomials of even degree in x (union zero) form a ring?
(b) Does the set of all polynomials of the form

$$p(x) = a_0 + a_1 x^2 + a_2 (x^2)^2 + \cdots + a_n (x^2)^n$$

where $a_i \in R$ for all i, form a ring?
(c) What is the principal ideal in $R[x]$ generated by x^2?
(d) Does the set of all polynomials of degree at most 10 (union zero) form a ring?

We have already mentioned factors of polynomials. We now give the precise algebraic definition of a factor, which we will call a divisor. Recall that 'divisors' were already defined for integers, further underlining the parallels between integers and polynomials.

Definition 9.67. Let R be a ring and $p(x), q(x) \in R[x]$.
(a) We will say that $p(x)$ divides $q(x)$ in $R[x]$ if there is a polynomial $f(x) \in R[x]$ such that $q(x) = p(x)f(x)$. We will write this down as $p(x)|q(x)$.
(b) A polynomial $f(x)$ that divides two polynomials $p(x)$ and $q(x)$ is said to be a common divisor of $p(x)$ and $q(x)$.
(c) A common divisor $d(x)$ of $p(x)$ and $q(x)$ such that any other common divisor of $p(x)$ and $q(x)$ also divides $d(x)$ is called **a** greatest common divisor of $p(x)$ and $q(x)$, or $d(x) = \gcd(p(x), q(x))$.
(d) Two polynomials $p(x)$ and $q(x)$ will be said to be associates if $p(x)|q(x)$, and $q(x)|p(x)$.

Exercise 9.68. Give an example of two polynomials that are associates in $\mathbb{R}[x]$ but are not associates in $\mathbb{Z}[x]$.

Definition 9.69. A proper (or non-trivial) factorization of a polynomial $p(x) \in R[x]$ is $p(x) = q(x)r(x)$, where $q(x)$ and $r(x)$ are non-invertible elements in $R[x]$.
A polynomial $p(x)$ of degree $n > 0$ is irreducible in $R[x]$ if it has no proper factorization in $R[x]$.

The same polynomial may be irreducible over a ring and reducible over another. Let us see some examples,

9.3 A More Algebraic Look at Polynomials

Example 9.70.
(a) The polynomial
$$p(x) = x^2 - 2 \in \mathbb{Z}[x]$$
is reducible over $\mathbb{R}[x]$ but not over $\mathbb{Q}[x]$.
(b) The polynomial
$$q(x) = x^2 + 1 \in \mathbb{Z}[x]$$
is reducible over $\mathbb{C}[x]$ but not over $\mathbb{R}[x]$.
(c) A more bizarre example: Consider
$$r(x) = x^2 + x + 1 \in \mathbb{Z}_2[x]$$

In this case, we cannot use the quadratic formula to find the zeros of $r(x)$, as we can't divide by 2 in the field $\mathbb{Z}_2$. So, we use that $\mathbb{Z}_2$ has only two elements to check if there is any zero of $r(x)$ in $\mathbb{Z}_2$. As $r(0) = r(1) = 1$ in $\mathbb{Z}_2$, then there are no roots of $r(x)$ in $\mathbb{Z}_2$, thus $r(x)$ is irreducible over $\mathbb{Z}_2[x]$.

Note that in the examples above we used that a reducible polynomial of degree two must have a root. This fact does not extend to polynomials having degree larger than 3.

Example 9.71.
$$s(x) = x^4 + 2x^2 + 1 = (x^2 + 1)^2 \in \mathbb{R}[x]$$
is reducible in $\mathbb{R}[x]$, however it does not have any roots in $\mathbb{R}$.

Exercise 9.72. Show that there are no irreducible polynomials of degree larger than 2 in $\mathbb{R}[x]$.
Hint: Invoke Theorem 9.21 and show that for any $z \in \mathbb{C}$, $(x-z)(x-\bar{z})$ is a polynomial in $\mathbb{R}[x]$.

We move on to a couple of classical results for polynomials in $\mathbb{Z}[x]$. These seem fairly technical but give essentially the only tool to check irreducibility in $\mathbb{Q}[x]$. We start with a lemma by Gauss, for which we will not provide a proof, as it would go beyond the scope of this book. A nice proof can be found in [3].

Theorem 9.73 (Lemma of Gauss). *Let $p(x) = a_0 + a_1 x + \cdots + a_n x^n \in \mathbb{Z}[x]$. Let the only common divisors of $a_0, a_1, \cdots, a_n$ be ± 1 (that is $\gcd(a_0, a_1, \cdots, a_n) = 1$). If $p(x)$ can be factored in $\mathbb{Q}[x]$, then $p(x)$ can be factored in $\mathbb{Z}[x]$.*

One of the most important properties of irreducible polynomials is their similarity with prime numbers in $\mathbb{Z}$. We know that every positive integer can be written in an essentially unique way (up to re-arrangement of the factors) as a product of powers of primes. The equivalent result for polynomials (with coefficients restricted to fields and $\mathbb{Z}$) is contained in the following theorem.

Theorem 9.74. *Let R be a ring that is a field or $\mathbb{Z}$. Then every polynomial $p(x) \in R[x]$ can be factored as a product of powers of irreducible polynomials. This product is unique up to re-arrangement of the factors and use of associates (or, if you prefer, of invertible elements).*

The idea of the first part of the proof is recursive. First we assume that $p(x)$ is not irreducible, then we can factor it, then we look at one of these factors and see if it is irreducible or not, and the process repeats until we obtain only irreducible polynomials. The main idea is that the process of "reducible then factor" applied repeatedly leads to, in a **finite** number of steps, a product of irreducible polynomials. This process stops because the degree of $p(x)$ is a finite number, and because F is a field.

The second part uses 'prime-like' properties of irreducible polynomials. You probably want to see the proof of the Fundamental Theorem of Arithmetic and 'translate' it into the language of polynomials for a nice and simple proof.

The big question is, how can we determine whether or not a polynomial is irreducible? Well, for irreducibility in $\mathbb{Q}[x]$ we have the following criterion.

Theorem 9.75 (Eisenstein's Criterion). *Let $f(x) = a_0 + a_1 x + \cdots + a_n x^n \in \mathbb{Z}[x]$, whose coefficients a_i for $0 \leq i \leq n-1$ are all divisible by a prime number p. If p does NOT divide a_n and p^2 does NOT divide a_0 then $f(x)$ is irreducible in $\mathbb{Q}[x]$*

Proof. Using the previous theorem, we just need to show that $f(x)$ is irreducible in $\mathbb{Z}[x]$. We assume that $f(x)$ factors in $\mathbb{Z}[x]$ into two polynomials, $q(x)$ and $d(x)$, of degree larger than zero.

$$q(x) = b_0 + b_1 x + \cdots + b_i x^i \qquad d(x) = c_0 + c_1 x + \cdots + c_j x^j$$

Note that after multiplying $q(x)$ and $d(x)$, the coefficient a_n equals $b_i c_j$, thus $p \nmid a_n$ implies $p \nmid b_i$ and $p \nmid c_j$. Similarly, $a_0 = b_0 c_0$, and $p | a_0$ thus $p | b_0$ or $p | c_0$, but as $p^2 \nmid a_0$, then p cannot divide both b_0 and c_0. Let us say that $p | b_0$ and $p \nmid c_0$.

Now assume that $p | b_0, p | b_1, \cdots, p | b_{s-1}$ and $p \nmid b_s$. Note that

$$a_s = b_s c_0 + b_{s-1} a_1 + \cdots + b_0 c_s.$$

Using that p divides a_s and all terms of the right hand side, except (possibly) for $b_s c_0$, we obtain that $p | b_s c_0$. Using that $p \nmid b_s$ we get that $p | c_0$, a contradiction. It follows that p divides all coefficients of $q(x)$, and consequently all coefficients of $f(x)$ would be divisible by p, contradicting $p \nmid a_n$. □

Remark 9.76. If p is not prime, then the result does not necessarily hold. Consider $q(x) = x^2 + 4x + 4$, and note that $p = 4$ satisfies the hypothesis of the criterion (except being prime). But $q(x)$ is reducible, as $q(x) = (x+2)^2$.

Corollary 9.77. *Let p be a prime number. The polynomial*

$$\Phi_p(x) = x^{p-1} + x^{p-2} + \cdots + x^2 + x + 1$$

is irreducible in $\mathbb{Q}[x]$.

The polynomial Φ_p is called the pth cyclotomic polynomial.

Proof. The result is trivial for $p = 2$, so suppose $p > 2$. Note that

$$q(x) = x^{p-1} + x^{p-2} + \cdots + x^2 + x + 1 = \frac{x^p - 1}{x - 1}$$

Consider $r(x) = q(x+1)$. With this choice

$$\begin{aligned} r(x) = q(x+1) &= \frac{(x+1)^p - 1}{(x+1) - 1} \\ &= \frac{(x+1)^p - 1}{x} \\ &= \frac{\left(\sum_{k=0}^{p} \binom{p}{k} x^k\right) - 1}{x} \\ &= \frac{\sum_{k=1}^{p} \binom{p}{k} x^k}{x} \\ &= \sum_{k=1}^{p} \binom{p}{k} x^{k-1}. \end{aligned}$$

Now we use Exercise 3.84 to see that p divides all coefficients of $r(x)$ except for the coefficient of x^{p-1}, and that p^2 does not divide the coefficient free of x. Hence, by Eisenstein's criterion, $r(x)$ is irreducible in $\mathbb{Q}[x]$. It follows that $q(x)$ is irreducible in $\mathbb{Q}[x]$. □

9.3 A More Algebraic Look at Polynomials

Remark 9.78. Note that the previous result does not necessarily hold for p not a prime number. For example, for $p = 4$,
$$x^3 + x^2 + x + 1 = (x^2 + 1)(x + 1)$$

We next generalize the conjugate root theorem for real polynomials (Theorem 9.21) by replacing the conjugation of complex numbers by a function that has 'all the right properties'.

Let $\mathbb{F}$ be a field contained in another field $\mathbb{E}$, e.g. $\mathbb{Q} \subseteq \mathbb{R}$ or $\mathbb{R} \subseteq \mathbb{C}$, and let $\phi : \mathbb{E} \to \mathbb{E}$ be a function such that
(a) $\phi(z+w) = \phi(z) + \phi(w)$ for all $z, w \in \mathbb{E}$,
(b) $\phi(zw) = \phi(z)\phi(w)$ for all $z, w \in \mathbb{E}$, and
(c) $\phi(x) = x$ for all $x \in \mathbb{F}$.
Note that conjugation of complex numbers does exactly this in the case $\mathbb{F} = \mathbb{R}$ and $\mathbb{E} = \mathbb{C}$.

Let us now consider $p(x) = a_0 + a_1 x + \cdots a_n x^n \in \mathbb{F}[x]$. Suppose z is a zero of $p(x)$, then
$$0 = p(z) = a_0 + a_1 z + \cdots a_n z^n.$$

Applying ϕ both sides yields

$$\begin{aligned}
\phi(0) &= \phi(a_0 + a_1 z + a_2 z^2 + \cdots a_n z^n) \\
0 &= \phi(a_0) + \phi(a_1 z) + \phi(a_2 z^2) + \cdots \phi(a_n z^n) && \text{using \textbf{(a)} above} \\
&= \phi(a_0) + \phi(a_1)\phi(z) + \phi(a_2)\phi(z^2) + \cdots \phi(a_n)\phi(z^n) && \text{using \textbf{(b)} above} \\
&= \phi(a_0) + \phi(a_1)\phi(z) + \phi(a_2)\phi(z)^2 + \cdots \phi(a_n)\phi(z)^n && \text{using \textbf{(b)} again} \\
&= a_0 + a_1 \phi(z) + a_2 \phi(z)^2 + \cdots a_n \phi(z)^n && \text{using \textbf{(c)} above, and that } a_i \in \mathbb{F} \\
&= p(\phi(z))
\end{aligned}$$

Hence, $0 = p(\phi(z))$, which means that if z is a zero of $p(x)$ then so is $\phi(z)$.

Remark 9.79. The previous idea can be used to show that if $p(x) \in \mathbb{Q}[x]$ and $a + b\sqrt{5}$, where $a, b \in \mathbb{Q}$, is a root of $p(x)$ then so is $a - b\sqrt{5}$.

References

1. Aigner, M. & Ziegler, G., *Proofs from THE BOOK*. Springer, 2nd ed. 2000.
2. Axler, S., *Linear Algebra Done Right*. Springer, 2nd ed. 1997.
3. Beachy, J. & Blair, W. *Abstract Algebra*, Waveland Press, Inc. 3rd ed. 2006.
4. Bagni, G.T. (2005). "Mathematics education and historical references: Guido Grandis infinite series". *Normat Nordisk Matematisk Tidsskrift*, 53, 4, 173 - 185.
5. Brennan, M. & Machale, D., "Variations on a Theme: A_4 Definitely Has no Subgroup of Order Six!", *Mathematics Magazine*, Vol. 73, No. 1. (2000), pp. 36-40.
6. Burton, David, *A History of Mathematics: An Introduction*, McGraw Hill, 2010.
7. D'Angelo, J. P., *An Introduction to Complex Analysis and Geometry*, AMS Pure and Applied Undergraduate Texts, no. 12., 2010.
8. Darboux, G., (1875) "Mémoire sur les fonctions discontinues", *Ann. Sci. École Norm. Sup.*, **4**:57-122
9. Dunham, William *Journey through Genius*, Penguin Books, 1991.
10. Fisher, S. D., *Complex Variables*, Dover Publications, 2nd ed. 1990.
11. Green, B., "Approximate groups and their applications: Work of Bourgain, Gamburd, Helfgott and Sarnak", *AMS Current Events Bulletin*, 2010, 1-25.
12. Hoffman & Kunze, *Linear Algebra*, Prentice Hall, 2nd ed. 1971.
13. Horn, R. A., and Johnson, C. C., *Matrix Analysis*, Cambridge University Press, 1985.
14. de Jong, T., "Lagrange Multipliers and the Fundamental Theorem of Algebra", *American Mathematical Monthly*, **116** (9): 828-830.
15. Kalman, D., "A Matrix Proof of Newton's Identities", *Mathematics Magazine* **73**, (4): 313-315.
16. Katugampola, U. N., "Mellin Transforms of the Generalized Fractional Integrals and Derivatives", arXiv: 1112.6031v1
17. Knopp, K., *Theory and application of infinite series*, Dover Publications Inc., New York., 1990
18. Nadler, S., B. Jr., "A Proof of Darboux's Theorem", *American Mathematical Monthly*, **117**(2): 174-175.
19. Palka, B. P., *An Introduction to Complex Function Theory*, Springer. 1995.
20. Paterson, M., Peres, Y., Thorup, M., Winkler, P. & Zwick, U., "Maximum Overhang" *American Mathematical Monthly* **116** (9); 763 - 787.
21. Rosen, K., *Elementary Number Theory and Its Applications*, Addison Wesley; 5 ed, 2004.
22. Ross, K. A., *Elementary Analysis: The Theory of Calculus*, Springer, 1980.
23. Royden, *Real Analysis*. Prentice Hall; 3 ed, 1988.
24. Schep, A. R., "A Simple Complex Analysis and an Advanced Calculus Proof of the Fundamental Theorem of Algebra" *American Mathematical Monthly*, **116** (1): 67-68.
25. Stewart, I., *Galois Theory*. Chapman and Hall/CRC, 3 ed. 2003.
26. Young, J.W.A. (Editor). Monographs on Topics of Modern Mathematics. Dover, 1955.
27. http://www.cut-the-knot.org/fta/ROS2.shtml
28. http://www.vimagic.de/hope/
29. http://en.wikipedia.org/wiki/Achilles_and_the_Tortoise
30. http://en.wikipedia.org/wiki/Grandi%27s_series
31. http://scienceworld.wolfram.com/biography/Abel.html
32. http://mathworld.wolfram.com/GoldenRatio.html
33. http://mathworld.wolfram.com/RiemannZetaFunctionZeta2.html

The Law Commission
(LAW COM No 248)

Legislating the Criminal Code
CORRUPTION

Item 11 of the Sixth Programme of Law Reform:
Criminal Law

Laid before Parliament by the Lord High Chancellor pursuant to section 3(2) of the Law Commissions Act 1965

Ordered by The House of Commons *to be printed*
2 March 1998

LONDON: The Stationery Office
£16.35

HC 524

The Law Commission was set up by section 1 of the Law Commissions Act 1965 for the purpose of promoting the reform of the law.

The Commissioners are:
> The Honourable Mrs Justice Arden DBE, *Chairman*
> Professor Andrew Burrows
> Miss Diana Faber
> Mr Charles Harpum
> Mr Stephen Silber QC

The Secretary of the Law Commission is Mr Michael Sayers and its offices are at Conquest House, 37-38 John Street, Theobalds Road, London, WC1N 2BQ.

The terms of this report were agreed on 15 January 1998.

The text of this report is available on the Internet at:
> http://www.open.gov.uk/lawcomm/

EXECUTIVE SUMMARY

1. In this report we make recommendations for the reform of the criminal law of corruption, which is currently to be found in a number of common law offences and statutes, of which the most important are the Prevention of Corruption Acts 1889 to 1916.[1] This legislation suffers from a number of defects: it is obscure, complex, inconsistent and insufficiently comprehensive. In this report we recommend[2] that the common law of bribery and the Prevention of Corruption Acts should be replaced by a modern statute, and we attach a draft Bill for this purpose.

2. Although one of the uncertainties of the present law lies in its application to Members of Parliament, this report does not deal with that issue. This is because the issue is currently under consideration by the Home Office and the Joint Committee on Parliamentary Privilege. In these circumstances we have decided that it would be a wasteful duplication of effort for us to examine it as well.

3. The present law draws a distinction between "public bodies" and others. This distinction is relevant for several different purposes. The most important is that in the case of public bodies there is a "presumption" of corruption.[3] This means that a benefit conferred on an employee of a public body, by someone who holds or is seeking a contract with any such body, is *deemed* to be corrupt unless it is proved to have been innocent.

4. This state of affairs gives rise to two distinct but connected difficulties. First, it is no longer sufficiently clear which bodies are "public".[4] For example, many organisations which used to be publicly owned have now been privatised. They still provide a public service, but now do so for the profit of their shareholders (who may include the Government). Are they still public bodies?

5. Secondly, it is debatable whether (even in the case of public bodies) the presumption of corruption is still necessary or justifiable.[5] It constitutes a radical exception to the general rule that it is for the prosecution to prove guilt, not for the defendant to prove innocence. For this reason it might be held to contravene Article 6(2) of the European Convention on Human Rights, which provides that a person charged with a criminal offence must be presumed innocent until proved guilty.[6] Moreover, it may no longer be necessary, since the Criminal Justice and Public Order Act 1994 now allows the court or the jury to draw adverse inferences if the defendant fails to answer questions or to give evidence at trial.[7] If

---

[1] The collective name given to the Public Bodies Corrupt Practices Act 1889, the Prevention of Corruption Act 1906 and the Prevention of Corruption Act 1916 by s 4(1) of the 1916 Act.

[2] Para 2.33 (Recommendation 1).

[3] Paras 3.8 – 3.9.

[4] Paras 3.19 – 3.25.

[5] See Part IV.

[6] Paras 4.21 – 4.29.

[7] Paras 4.31 – 4.48.

a defendant has participated in an apparently corrupt transaction but declines to offer an explanation, appropriate conclusions can now be drawn.

6. We were impressed by the fact that respondents with great experience of criminal trials supported our provisional view that corruption is no harder to prove than some other offences.[8] This was the view of the General Council of the Bar, the Criminal Bar Association and the SFO.[9] Furthermore, we bore in mind that those prosecuting corruption cases will often rely on conspiracy to defraud or common law bribery, neither of which have the benefit of the presumption, and we are not aware that the prosecution of those offences presents any particular difficulties.

7. We were also impressed by the views of the Australian Model Criminal Code Officers Committee ("the MCCOC") who said in their report that it was difficult to accept that corruption was any more difficult to prove than theft or fraud.[10]

8. For these reasons we conclude that there is no longer any need for a presumption of corruption, and that the prosecution should be required to prove corruption (like any other offence) beyond reasonable doubt. And since it is only in relation to the presumption that the distinction between public and non-public bodies has much practical significance, we believe that that distinction can also be discarded – thus solving the problem of how to define a public body.[11]

9. We begin our analysis of corruption in terms of an "agent" being tempted by bribery to betray the trust owed to his or her principal, using the concept of "agent" in a broad sense of someone who has agreed to perform functions for another person – the agent's "principal". We then extend the analysis to include those who have been entrusted to perform functions not for an identifiable principal but for the *public*, whether the public of the United Kingdom or elsewhere.[12]

10. Our recommendations would involve replacing the existing law of corruption with a modern statute creating four offences, namely:
    ◆ corruptly conferring, or offering or agreeing to confer, an advantage;[13]
    ◆ corruptly obtaining, soliciting or agreeing to obtain an advantage;[14]
    ◆ corrupt performance by an agent of his or her functions as an agent;[15] and
    ◆ receipt by an agent of a benefit which consists of, or is derived from, an advantage which the agent knows or believes to have been corruptly obtained.[16]

---

[8] Para 4.71.

[9] Para 4.72.

[10] Para 4.76.

[11] Para 4.78 (Recommendation 2).

[12] Paras 5.15 – 5.36 (Recommendation 3).

[13] Para 5.50 (Recommendation 6); for the definition of "advantage" see para 5.43 (Recommendation 5).

[14] Para 5.53 (Recommendation 7).

[15] Para 5.58 (Recommendation 8).

[16] Para 5.61 (Recommendation 9).

11. The third and fourth of these offences represent a strengthening of the law. At present an agent commits an offence by accepting a corrupt bribe or reward, but not by acting in return for the bribe or attempting to earn a reward. This seems illogical. What makes bribery wrong is that it tempts an agent to betray his or her principal's trust; yet, while the acceptance of the bribe is an offence, the betrayal itself is not. At present, prosecutions sometimes fail because there is evidence that an agent acted in breach of his or her duty, but not that a bribe or reward was paid or even agreed. Under our recommendations it would be sufficient to prove that the agent's conduct was motivated by the *hope* of a corrupt reward, whether or not there was any agreement to that effect.[17]

12. Again, it is not clear at present that an agent commits an offence by accepting part of a bribe paid to a third party in return for favour to be shown by the agent. Under our recommendations this would be an offence.

13. Central to all these offences is the concept of doing something *corruptly*. This word is used in the existing legislation but is not defined, and its precise meaning is unclear. We believe that it should be defined, and have therefore attempted to analyse what it means.[18] Our conclusion is that the essence of corruption lies in the influencing of an "agent" (that is, a person who has agreed to perform functions for another person – the agent's "principal" – or for the public) to perform those functions in a certain way, and to do so in return (or at least *primarily* in return) for the conferring of an advantage on the agent or a third party. Thus we recommend that a person who confers an advantage should be regarded as doing so *corruptly* if he or she intends a person, in performing his or her functions as an agent, to do an act or make an omission, *and* he or she believes that, if that person did so, it would probably be *primarily in return for the conferring of the advantage.*[19]

14. The concept of corruption in the other offences that we recommend would build upon the concept of corruptly conferring an advantage. For example, we recommend that a person who obtains an advantage should be regarded as obtaining it corruptly if he or she knows or believes that the person conferring it *confers* it corruptly, and he or she either requests it or at least consents to obtaining it.[20]

15. Read literally, these definitions would include an agent's remuneration by his or her principal (or on behalf of the public), and we therefore recommend an express exception for such remuneration.[21] We also recommend a further exception for the case where the agent's principal knows all the material circumstances and consents to what is done.[22]

---

[17] Paras 5.54 – 5.58 (Recommendation 8).

[18] Paras 5.62 – 5.98.

[19] Para 5.99 (Recommendation 10).

[20] Para 5.118 (Recommendation 12).

[21] Paras 5.86 – 5.87; para 5.99 (Recommendation 10).

[22] Paras 5.88 – 5.98; para 5.99 (Recommendation 10).

16. The investigation of corruption is an integral part in successfully proving corruption. We considered whether the powers of the police should be extended, either to meet those of the Serious Fraud Office or in some other way, in cases of corruption.[23] Whilst we accept that an anomaly arises in circumstances where a case falls within the remit of the SFO but, for whatever reason, is investigated by the police, we conclude that extending the powers of the police would create a greater anomaly between cases of corruption and other cases of fraud. Further, we consider that an extension of police powers might be vulnerable to challenge under Article 6 of the European Convention on Human Rights. We therefore do not recommend that the investigative powers of the police should be extended in the case of corruption.[24]

17. We have carefully considered the issue of territorial jurisdiction and the new offences of corruption since, as we point out, corruption has an increasing international element.[25] We recommend that the new offences of corruption should be included in the list of Group A offences for the purposes of Part I of the Criminal Justice Act 1993,[26] which extends the jurisdiction of the English courts over offences of fraud and dishonesty committed abroad.

18. We recommend that prosecutions for the new offences should not require the consent of either the Law Officers or the Director of Public Prosecutions.[27]

19. Although we take the view that corruption is a serious offence, we do not think that it is right to say that *all* instances of corrupt behaviour are sufficiently serious to require trial on indictment. We recommend, therefore, that the new offences should be triable either way.[28]

20. We recommend that the new offences should not have retrospective effect.[29]

21. In the consultation paper, we asked for views on whether the procurement of a breach of duty by deception or threats should be criminal.[30] We conclude that, although the present law may not deal adequately with such circumstances, they should not be caught by the law of *corruption*.[31]

---

[23] See Part VI.

[24] Para 6.29.

[25] Para 7.8.

[26] Para 7.15 (Recommendation 15); Part I of the CJA 1993 is not in force.

[27] Para 7.26 (Recommendation 16).

[28] Para 7.28 (Recommendation 17).

[29] Para 7.33 (Recommendation 18).

[30] See part VIII.

[31] Para 8.20.

THE LAW COMMISSION

LEGISLATING THE CRIMINAL CODE: CORRUPTION

CONTENTS

	Paragraph	Page
PART I: INTRODUCTION		**1**
The background to this report	1.1	1
The significance of this project to the objective of codification of the law	1.8	3
Previous reform proposals	1.11	4
The Home Office consultation document	1.14	6
International initiatives	1.16	6
Scope of the project	1.21	8
Members of Parliament	1.21	8
Scotland	1.23	8
False documents	1.24	9
The structure of the report	1.26	9
PART II: THE PRESENT LAW AND THE NEED FOR CHANGE		**11**
The present law	2.2	11
The common law offence of bribery	2.2	11
The statutory offences	2.6	12
The 1889 Act	2.7	13
The 1906 Act	2.11	14
The 1916 Act	2.15	15
Problems with the present law	2.17	16
"Public bodies"	2.17	16
The presumption	2.18	16
Agents	2.20	16
The inapplicability of the presumption	2.22	17
Persons connected with agents	2.23	17
Persons who have been, or are to become, an agent	2.24	18
Purported agents	2.25	18
Applicability of corruption laws to police officers, judges and local councillors	2.26	18
The meaning of "corruptly"	2.29	19

	Paragraph	Page
Our provisional conclusion, the response on consultation and our recommendation	2.31	20

PART III: THE DISTINCTION BETWEEN PUBLIC BODIES AND OTHERS — 21

	Paragraph	Page
The significance of the distinction in the present law	3.3	21
The 1889 Act	3.4	21
The scope of the offences	3.5	21
Third parties	3.5	21
People who have been, or are to become, agents	3.6	21
Penalties	3.7	22
The presumption	3.8	22
The distinction between a person serving under a public body and a person serving under a non-public body.	3.10	22
The definition of a public body	3.11	23
The Local Government and Housing Act 1989 amendment	3.15	24
Public bodies outside the United Kingdom	3.17	24
Should the distinction be retained?	3.19	25
The diminishing importance of the distinction	3.19	25
Possible arguments in favour of the distinction	3.27	26
Is public sector corruption more serious than private sector corruption ?	3.28	26
Is the public more in need of protection than the private sector?	3.29	27
Is there a need for higher standards of conduct in the public sector?	3.31	27
Our provisional proposals and the response on consultation	3.34	28

PART IV: THE PRESUMPTION OF CORRUPTION — 30

	Paragraph	Page
The presumption	4.5	31
Historical reasons for the creation of the presumption	4.10	32
Justifications for the presumption	4.15	33
The Redcliffe-Maud Committee	4.16	33
The Salmon Commission	4.17	34
The presumption and the European Convention on Human Rights	4.20	34
Does the presumption infringe the ECHR?	4.21	35
The presumption of innocence	4.22	35
Possible implications of the CJPOA	4.31	38
The effects of the presumption and the CJPOA compared	4.37	40
Establishing a case to answer	4.38	41
The existing law	4.38	41
If section 2 were repealed	4.39	41

	Paragraph	Page
Where the defendant adduces no evidence	4.43	42
The existing law	4.43	42
If section 2 were repealed	4.44	42
Where the defendant does not testify but adduces other evidence	4.45	42
The existing law	4.45	42
If section 2 were repealed	4.46	43
Where the defendant testifies	4.47	43
The existing law	4.47	43
If section 2 were repealed	4.48	43
Options for reform, our provisional proposal and the response on consultation	4.49	43
Option 1: extend the presumption	4.52	44
Option 2: the Hong Kong option	4.55	46
Option 3: reduce the weight of the burden imposed by the presumption	4.59	46
Option 4: abolish the presumption	4.61	47
Arguments against the presumption	4.62	47
Arguments from principle	4.63	47
Practical considerations	4.67	48
The shortcomings of the presumption in its present form	4.67	48
The CJPOA	4.68	48
Arguments in favour of the presumption	4.71	49
Corruption is more difficult to prove than other offences	4.71	49
Standards of probity should be higher in the public sector	4.73	50
Consistency with Scotland	4.75	51
Our recommendation	4.76	51

PART V: FORMULATING A MODERN LAW OF CORRUPTION

52

	Paragraph	Page
An outline of the scheme we recommend	5.2	52
The essential character of corruption	5.4	53
Breach of duty	5.5	53
The functions of an agent *as an agent*	5.11	55
The agency relationship	5.15	56
Terminology	5.15	56
Defining the agency relationship	5.17	57
The general definition	5.20	58
Private agency relationships	5.21	58
Quasi-fiduciary relationships	5.25	59
Mixed relationships	5.28	60

	Paragraph	Page
The list of examples	5.29	61
Transnational agency relationships	5.31	61
Agents with private foreign principals	5.31	61
Agents acting for the public interest of another country	5.32	62
Agents acting on behalf of international intergovernmental organisations	5.35	63
Our recommendation	5.36	63
The concept of advantage	5.38	64
The offences	5.44	65
Corruptly conferring, or offering or agreeing to confer, an advantage	5.45	66
Advantage conferred on a person other than the agent	5.47	66
Intermediaries	5.48	66
Our recommendation	5.50	67
Corruptly obtaining, soliciting or agreeing to obtain an advantage	5.51	67
Corruptly performing functions as an agent	5.54	68
Agent receiving benefit from corruption	5.59	69
"Corruptly"	5.62	70
Is a definition necessary?	5.63	70
Corruptly conferring an advantage as an inducement	5.67	71
Corporate hospitality	5.75	73
"Sweeteners"	5.83	75
Bribee unaware of corrupt purpose	5.85	76
Proper remuneration or reimbursement	5.86	76
Things done with the consent of the principal	5.88	76
Our recommendation	5.99	79
Corruptly conferring an advantage as a reward	5.100	80
Former and future agents	5.110	83
Corruptly obtaining or agreeing to obtain an advantage	5.112	84
Corruptly soliciting an advantage	5.119	85
Corruptly performing functions as an agent	5.121	86
Dishonesty	5.123	86
Is corruption an offence of dishonesty?	5.123	86
Should the definition of corruption include a requirement of dishonesty?	5.131	89
Possible defences	5.135	90
Disclosure	5.136	91
No obligation to account	5.138	91
Normal practice	5.139	91
Small value	5.142	92
Entrapment	5.144	93
Public interest	5.151	95

Paragraph   Page

PART VI: THE INVESTIGATION OF CORRUPTION   97

The SFO and section 2 of the Criminal Justice Act 1987	6.3	97
Powers of the SFO	6.3	97
Safeguards for the person investigated	6.8	98
Comparison of the powers of the SFO and the CPS	6.10	99
Comparison of the powers of the SFO and the police	6.11	100
The powers of the DTI	6.12	100
Section 2 and the ECHR	6.18	101
The options	6.23	103
Responses on consultation	6.24	103
Conclusions	6.28	104

PART VII: TERRITORIAL JURISDICTION, THE ATTORNEY-GENERAL'S CONSENT TO PROSECUTION AND OTHER ANCILLARY MATTERS   105

Territorial jurisdiction	7.2	105
The present law	7.3	105
The Criminal Justice Act 1993	7.9	106
The effect of extending the CJA 1993 to corruption	7.11	107
Our provisional proposal and the views of respondents	7.14	107
Our recommendation	7.15	108
Ancillary matters	7.16	108
Requirement of the Attorney-General's consent	7.16	108
The need for consent	7.17	108
Views of respondents	7.22	110
Mode of trial	7.27	111
Sentence	7.29	112
Retrospectivity	7.31	112

PART VIII: CORRUPTION AND BREACH OF DUTY   113

Two basic law reform questions	8.1	113
The first question	8.2	113
Acting contrary to duty: the radical approach	8.2	113
The second question	8.5	114
Corruption by means other than bribery	8.5	114
The present law and its deficiencies	8.9	115
Blackmail and breaches of duty procured by threats	8.9	115
Breaches of duty procured by deception	8.14	116

	Paragraph	Page
Whether offences of procuring a breach of duty by threats or deception are offences of corruption	8.15	116
Our conclusion	8.20	117

PART IX: OUR RECOMMENDATIONS — 118

	Page
The present law and the need for change	118
The presumption of corruption	118
The agency relationship	118
The concept of advantage	119
The offences	119
"Corruptly"	120
Territorial jurisdiction	122
Ancillary matters	122

APPENDIX A: DRAFT CORRUPTION BILL — 125

APPENDIX B: EXTRACTS FROM RELEVANT LEGISLATION — 134

	Page
Public Bodies Corrupt Practices Act 1889	134
Prevention of Corruption Act 1906	136
Prevention of Corruption Act 1916	137
Criminal Justice and Public Order Act 1994	137

APPENDIX C: LIST OF PERSONS AND ORGANISATIONS WHO COMMENTED ON THE CONSULTATION PAPER — 141

ABBREVIATIONS

IN THIS REPORT WE USE THE FOLLOWING ABBREVIATIONS:

ACPO: the Association of Chief Police Officers

the 1889 Act: the Public Bodies Corrupt Practices Act 1889

the 1906 Act: the Prevention of Corruption Act 1906

the 1916 Act: the Prevention of Corruption Act 1916

Archbold: *Archbold – Criminal Pleading, Evidence and Practice* (1997 ed, ed P J Richardson)

the CBI: the Confederation of British Industry

the CJA 1987: the Criminal Justice Act 1987

the CJA 1993: the Criminal Justice Act 1993

the CJPOA: the Criminal Justice and Public Order Act 1994

CIPFA: the Chartered Institute of Public Finance and Accountancy

the CPS: the Crown Prosecution Service

the Convention: the European Convention on Human Rights

the DPP: the Director of Public Prosecutions

the DTI: the Department of Trade and Industry

the ECHR: the European Convention on Human Rights

the FLP: the Financial Law Panel

the Home Office consultation paper: *The Prevention of Corruption – Consolidation and Amendment of the Prevention of Corruption Acts 1889–1916*: A Government Statement (June 1997)

ILEX: the Institute of Legal Executives

the MCCOC: the Model Criminal Code Officers Committee established by the Standing Committee of Attorneys-General of Australia

the MCCOC Report: the final report of the MCCOC, Chapter 3: *Theft, Fraud, Bribery and Related Offences* (1995)

the Neill Committee: the Committee on Standards in Public Life, formerly the Nolan Committee (Chairman: Lord Neill of Bladen QC)

the Nolan Committee: the Committee on Standards in Public Life, now the Neill Committee (Chairman: the Rt Hon the Lord Nolan)

the Nolan Report: *Standards in Public Life*, the first report of the Nolan Committee (1995) Cm 2850

the OECD: the Organisation for Economic Co-operation and Development

PACE: the Police and Criminal Evidence Act 1984

the Redcliffe-Maud Committee: the Prime Minister's Committee on Local Government Rules of Conduct (Chairman: the Rt Hon the Lord Redcliffe-Maud GCB CBE)

the Redcliffe-Maud Report: *Conduct in Local Government*, the report of the Redcliffe-Maud Committee (1974) Cmnd 5636

the SFO: the Serious Fraud Office

the SIB: the Securities and Investments Board

the Salmon Commission: the Royal Commission on Standards of Conduct in Public Life (Chairman: the Rt Hon the Lord Salmon)

the Salmon Report: the report of the Salmon Commission (1976) Cmnd 6524

the Strasbourg Commission: the European Commission of Human Rights

the Strasbourg Court: the European Court of Human Rights

TI (UK): Transparency International (UK)

THE LAW COMMISSION
Item 11 of the Sixth Programme of Law Reform: Criminal Law

LEGISLATING THE CRIMINAL CODE: CORRUPTION

To the Right Honourable the Lord Irvine of Lairg, Lord High Chancellor of Great Britain

PART I
INTRODUCTION

THE BACKGROUND TO THIS REPORT

1.1 In this report we make recommendations for the reform of the criminal law of corruption. We carried out this work in response to calls from two prestigious bodies for a review of this area of the law.[1] We were also very conscious of the consequences of corruption, which strikes at the root of our commercial life and of democracy itself. If public servants require to be bribed, or make decisions in response to bribes, the government is not democratic, and the citizen cannot rely on a principle which we take for granted, namely that those in positions of power will take their decisions in accordance with the law. As a recently published Home Office consultation document says, "corruption is regarded as a more serious crime than simple dishonesty and the international concern about this kind of crime and the difficulty of investigating this offence may justify exceptional measures."[2]

1.2 In the consultation paper on which the recommendations in this report are based,[3] we drew attention to four major defects of the present law. First, we explained that the present law is drawn from a multiplicity of sources, including many overlapping common law offences[4] and at least 11 statutes.[5] We pointed out that much of the legislation was a hasty response to a contemporary problem and, in consequence, was neither comprehensive, clear nor consistent. Therefore, it was

---

[1] The Salmon Commission and the Nolan Committee.

[2] *The Prevention of Corruption: Consolidation and Amendment of the Prevention of Corruption Acts 1889–1916*: A Government Statement (June 1997), para 3.15. See para 1.14 below.

[3] Legislating the Criminal Code: Corruption (1997) Consultation Paper No 145.

[4] These include misconduct in public office (*Llewellyn-Jones* [1968] 1 QB 429) and specific bribery offences (see para 2.2 below) such as embracery (bribing of jurors) (*Pomfriet v Brownsal* (1600) Cro Eliz 736; 78 ER 968); attempts to bribe a privy councillor (*Vaughan* (1769) 4 Burr 2495; 98 ER 308); attempts to bribe a police constable (*Richardson* 111 Cent Crim Ct Sess Pap 612); and the taking of a bribe by a coroner not to hold an inquest (*Harrison* (1800) 1 East PC 383).

[5] Sale of Offices Act 1551; Sale of Offices Act 1809; Public Bodies Corrupt Practices Act 1889; Prevention of Corruption Act 1906; Prevention of Corruption Act 1916; Honours (Prevention of Abuses) Act 1925; Licensing Act 1964, s 178; Criminal Law Act 1967, s 5; Local Government Act 1972, s 117(2); Customs and Excise Management Act 1979, s 15; Representation of the People Act 1983, ss 107, 109 and 111–115.

not surprising that the Salmon Commission recommended the rationalisation of the statute law on bribery,[6] while the Nolan Committee pointed out that since the Government had accepted, but not implemented, that recommendation, it might be a task which this Commission could take forward.[7] In this project we have done so.

1.3 Secondly, we pointed out that another problem with the present law is that it is dependent on a distinction between *public* and *non-public* bodies.[8] The Public Bodies Corrupt Practices Act 1889 ("the 1889 Act") is concerned only with corruption in *public* bodies, while the more narrowly drafted Prevention of Corruption Act 1906 ("the 1906 Act") extends the law of corruption to *all* agents. Significantly, the presumption of corruption under section 2 of the Prevention of Corruption Act 1916 ("the 1916 Act") is similarly limited, applying only to persons "in the employment of [Her] Majesty or any Government Department or a public body".[9] The distinction between public and non-public bodies causes difficulty because of uncertainty as to what constitutes a public body. Many former public bodies have now been privatised, and it is uncertain which of them, if any, can still be regarded as public bodies. An important question which we address in this report is whether this distinction is still justified.

1.4 Thirdly, we drew attention to the difficulty in ascertaining to whom the present legislation applies. For example, it is uncertain whether certain categories of individuals, such as judges, fall within the definition of an agent. It appears that the 1906 Act, unlike the 1889 Act, does not extend to those who accept bribes before or after the currency of their agency, or to third party recipients.[10]

1.5 Fourthly, we discussed whether the rebuttable presumption of corruption under section 2 of the 1916 Act[11] is still justified in the light of sections 34 and 35 of the Criminal Justice and Public Order Act 1994 ("the CJPOA"),[12] which allow adverse inferences to be drawn from a defendant's silence in the course of an investigation or trial. Conversely, we also considered whether the presumption should be not only retained but *extended*, as suggested in the Salmon Report.[13] Furthermore, there is a very significant question whether, in the light of the

---

[6] Salmon Report, para 87.

[7] Nolan Report, para 2.104.

[8] See Part VI of Consultation Paper No 145, and Part III below.

[9] 1916 Act, s 2. See also Part V below. Less significantly, the 1906 Act distinguishes between a person serving under a public body and a person serving under a non-public body: whereas the former is an "agent" for the purpose of the 1906 Act, the latter is not unless he or she falls into the definition of "agent" for some other reason, namely that he or she is "employed by" or "acting for" the non-public body (1906 Act, s 1).

[10] See paras 5.110 and 5.47 respectively.

[11] This rebuttable presumption arises where payment is made to an employee of the Crown, a Government department or a public body by a person holding or seeking to obtain a contract with the Crown, a Government department or a public body (see para 4.6 below).

[12] Paras 4.31 – 4.36 below.

[13] Para 4.15 below.

CJPOA, the presumption is compatible with the European Convention on Human Rights.

1.6 All these matters appear to us to call for a review of the law of corruption. We were fortified in this conclusion not only by the wishes of the Salmon Commission and the Nolan Committee[14] but also by encouragement received from many other sources. In answer to a Parliamentary question asking what action he proposed to take to reform and consolidate the law of corruption (with particular reference to bribery of Members of Parliament), the Home Secretary stated on 9 June 1997 that the Government was committed to tackling corruption in all areas of public and private life, including the bribery of Members of Parliament.[15] He announced that he was publishing on that day a statement on reform of the corruption statutes and that he would consider carefully the results not only of that consultation exercise but also of that conducted by this Commission in Consultation Paper No 145, together with any further recommendations which the Nolan Committee (now the Neill Committee) might make in relation to the criminal law. He expected to make a further statement on the reform of the law early in 1998.[16] These statements confirm our view of the importance of this project.

1.7 We are also mindful of our statutory duty to keep the whole of the law under review "with a view to its systematic development and reform, including ... generally the simplification and modernisation of the law".[17] During the course of this project we have kept in mind the long-held aim of this Commission to make the criminal law more accessible, comprehensible, consistent and certain,[18] an aim which we believe would be furthered by the implementation of our recommendations.

THE SIGNIFICANCE OF THIS PROJECT TO THE OBJECTIVE OF CODIFICATION OF THE LAW

1.8 The Law Commission is charged with the duty to keep the law under review "with a view to its systematic development and reform, including in particular the codification of [the] law".[19] In 1989 we produced a draft Criminal Code,[20] but it was in many respects a statement of the existing law or of fairly recent proposals for reform which were open to criticism. Accordingly, we subsequently adopted a policy of reviewing areas of criminal law so that one by one they would be modernised (where appropriate) before being assembled into a code.[21] In the

---

[14] See nn 6 and 7 above.

[15] Written Answers, *Hansard* (HC) 9 June 1997, vol 295, col 346.

[16] On 27 October 1997, the Minister of State for the Home Office said, in answer to a written Parliamentary question, that the Home Secretary was expected to make a further statement on reform early in 1998: Written Answers, *Hansard* (HC) 27 October 1997, vol 299, col 739.

[17] Law Commissions Act 1965, s 3(1).

[18] Codification of the Criminal Law (1985) Law Com No 143, paras 1.3 – 1.9.

[19] Law Commissions Act 1965, s 3(1).

[20] Criminal Law: A Criminal Code for England and Wales (1989) Law Com No 177.

[21] Law Commission Twenty-Seventh Annual Report (1992) Law Com No 210, p 10.

course of our code project we have completed reports on offences against the person and general principles,[22] involuntary manslaughter,[23] the year and a day rule in homicide,[24] money transfers,[25] rape within marriage,[26] computer misuse,[27] intoxication,[28] and hearsay.[29]

1.9 It is important to stress the main reasons for having a modern criminal code. First, such a code would lead to a substantial saving of both time and money. Secondly, it would ensure compliance with the European Convention on Human Rights. Article 7 of that Convention, as applied by the European Court of Human Rights, requires criminal offences to be defined with reasonable precision. The European Commission of Human Rights has pointed out that such offences must be "adequately accessible and formulated with sufficient precision to enable the citizen to regulate his conduct".[30]

1.10 Thirdly, since criminal law is arguably the most direct expression of the relationship between a State and a citizen and is a matter of constitutional principle, that relationship should be clearly stated in a code laid down by Parliament. Fourthly, the adoption of a policy of codification provides a framework for fundamental and continuing review aimed at modernisation of the criminal law. The existence of a code shows clearly the seriousness with which society regards the control and punishment of crime: a nation whose basic rules of criminal law are muddled is sending the wrong messages to criminals and potential criminals. Finally, a modern criminal code would ensure that the users of the criminal justice system are provided with accessible and comprehensible criminal law.

PREVIOUS REFORM PROPOSALS

1.11 The Redcliffe-Maud Committee reported[31] in 1974 in response to widespread public disquiet about conduct in local government following several prosecutions for offences of corruption. Its terms of reference were to examine local government law and practice and how it might affect the conduct of members and officers in situations involving a conflict of interest between their public functions

---

[22] Legislating the Criminal Code: Offences against the Person and General Principles (1993) Law Com No 218.

[23] Legislating the Criminal Code: Involuntary Manslaughter (1996) Law Com No 237.

[24] Legislating the Criminal Code: The Year and a Day Rule in Homicide (1995) Law Com No 230: implemented by the Law Reform (Year and a Day Rule) Act 1996.

[25] Offences of Dishonesty: Money Transfers (1996) Law Com No 243: implemented by the Theft (Amendment) Act 1996.

[26] Criminal Law: Rape within Marriage (1992) Law Com No 205: implemented by s 142 of the Criminal Justice and Public Order Act 1994.

[27] Criminal Law: Computer Misuse (1989) Law Com No 186: implemented by the Computer Misuse Act 1990.

[28] Legislating the Criminal Code: Intoxication and Criminal Liability (1995) Law Com No 229.

[29] Evidence in Criminal Proceedings: Hearsay and Related Topics (1997) Law Com No 245.

[30] *G v Federal Republic of Germany*, 6 March 1989 (1989) 60 DR 256, 262.

[31] Under the appointment in 1973 of the then Prime Minister, the Rt Hon Edward Heath MP.

and private interests.[32] The Committee made several recommendations for law reform. These included the extension of the presumption of corruption under the 1916 Act. The Government of the time welcomed the report[33] but took little action with regard to it.[34]

1.12 Following the Poulson[35] affair, the Salmon Commission was established in 1974. It reported two years later. Its terms of reference were to examine standards of conduct in central and local government in relation to the problems of conflict of interest and the risk of corruption involving favourable treatment from a public body, and to recommend further safeguards to ensure the highest standard of probity in public life. The Commission examined the law of corruption and recommended that the Prevention of Corruption Acts 1889 to 1916, insofar as they applied to the public sector, ought to be consolidated and amended. However, no further action was taken by the Government.[36]

1.13 More recent public corruption scandals led to the First Report of the Committee on Standards in Public Life in May 1995.[37] Its terms of reference were to examine current concerns about standards of conduct of all holders of public office and to make recommendations to ensure the highest standards of propriety in public life.[38] The Report concentrated on three main areas: issues relating to Members of Parliament; ministers and civil servants; and quangos. In making its recommendations, the Committee thought it important that the general principles of public life be restated; namely, selflessness, integrity, objectivity, accountability, openness, honesty and leadership.[39] The main recommendation that the Committee made in regard to the law of corruption was to request that steps be taken to clarify the law on bribery in relation to the receipt of a bribe by an MP, as recommended by the Salmon Commission, combined with consolidation of the statute law on bribery.[40] These recommendations also led to the Select Committee on Standards in Public Life, charged with considering the Nolan Report, recommending a review of the law on bribery in similar terms.

---

[32] Redcliffe-Maud Report, para 2.

[33] See, eg, Written Answers, *Hansard* (HC) 23 May 1974, vol 874, cols 236–237.

[34] This may have been because by the time the Redcliffe-Maud Report was being debated, publication of the Salmon Report was already anticipated: see *Hansard* (HC) 27 June 1974, vol 875, cols 1719–1720.

[35] Poulson, an architect, had used corrupt methods to obtain work, by bribing councillors and local authorities.

[36] The report was not discussed in the House of Commons, but was fully debated in the House of Lords where it received a mixed response: see *Hansard* (HL) 8 December 1976, vol 378, cols 585–604 and 611–674.

[37] The Committee was appointed by the then Prime Minister, the Rt Hon John Major MP and reported under the charge of the (then) Chairman, the Rt Hon the Lord Nolan. Since 10 November 1997, the Chairman of the Committee has been Lord Neill of Bladen QC.

[38] First Report of the Committee on Standards in Public Life ("the Nolan Report").

[39] Nolan Report, p 14.

[40] Nolan Report, para 2.104.

THE HOME OFFICE CONSULTATION DOCUMENT

1.14 The Home Office produced a consultation document on the prevention of corruption in June 1997.[41] It contained a number of interesting and useful statements concerning the Government's approach. It stated that there may be some justification in having a single offence of corruption,[42] and that there was a case for extending the existing statutes to cover trustees and all situations where a person has a duty, whether express or implied, to use his or her impartial judgment on an issue.[43] It suggested that it was right to consider carefully an extension to the presumption of corruption.[44]

1.15 In the consultation document it was also suggested that a corruption offence should be a serious arrestable offence.[45] It was pointed out that offences under the 1889 and 1906 Act are arrestable under the provisions of section 24 of Police and Criminal Evidence Act 1984 ("PACE"); and they fall within the definition of a "serious arrestable offence" only if it can be shown that the offence in question would lead to substantial financial gain or serious financial loss to any person. That this consequence might follow is not clear at the beginning of an investigation and the purpose of the recommendation would be to ensure that corruption offences were always serious arrestable offences. The Home Office consultation document raised the question of the jurisdiction of the courts to deal with acts of corruption which arise outside the jurisdiction.[46] Finally, it drew attention to the issues that will have to be addressed with regard to the mental element of a corruption offence.[47]

INTERNATIONAL INITIATIVES

1.16 As we have seen, there are a number of initiatives taking place at home aimed at countering corruption, in particular corruption in the public sector. We are aware also of efforts to counter corruption undertaken in a range of international fora such as, for example, the European Union, the Council of Europe, the Organisation for Economic Co-operation and Development ("the OECD"), the Commonwealth Law Ministers, the G7, the United Nations, the World Trade Organisation, the International Monetary Fund[48] and the Organisation of American States.

1.17 In May 1994, the OECD Ministerial Council adopted a recommendation on bribery in international business transactions: "that Member countries take

---

[41] *The Prevention of Corruption: Consolidation and Amendment of the Prevention of Corruption Acts 1889–1916*: A Government Statement (June 1997).

[42] *Ibid*, para 3.6.

[43] *Ibid*, para 3.10.

[44] *Ibid*, para 3.12.

[45] *Ibid*, para 3.14.

[46] *Ibid*, paras 3.16 – 3.19.

[47] *Ibid*, paras 4.1 – 4.9.

[48] See, eg, a publication by the International Monetary Fund, published in February 1997, entitled *Why Worry About Corruption?*

effective measures to deter, prevent and combat the bribery of foreign public officials in connection with international business transactions". On 17 December 1997, an OECD convention was signed,[49] the purpose of which was "to assure a functional equivalence among the measures taken by the Parties to sanction bribery of foreign public officials".[50] Paragraph 1 of Article 1 states:

> Each Party shall take such measures as may be necessary to establish that it is a criminal offence under its law for any person intentionally to offer, promise or give any undue pecuniary or other advantage, whether directly or through intermediaries, to a foreign public official, for that official or for a third party, in order that the official act or refrain from acting in relation to the performance of official duties, in order to obtain or retain business or other improper advantage in the conduct of international business.

1.18 In September 1994, the Committee of Ministers of the Council of Europe set up the Multidisciplinary Group on Corruption ("GMC") with the purpose of examining "what measures might be suitable to be included in a programme of action at international level against corruption".[51] A working group of the GMC dealing with penal law matters has drawn up a draft convention on corruption which is currently before the relevant committees. As presently drafted, the convention would require parties to criminalise amongst other things corruption of domestic public officials, of foreign officials and of officials working in international organisations and corruption in the business sector.

1.19 In June 1995, a convention was signed in the European Union requiring Member States to criminalise acts of fraud committed by their national officials where they affect the Communities' budget. The First Protocol of the convention was adopted in September 1996. It was directed at criminalising acts of *corruption* involving national officials which damage or are likely to damage the Communities' financial interests.[52] Another convention, signed last year, reflected the European Union's concern to combat corruption of public officials whether or not that corruption damaged the Communities' financial interests.[53]

1.20 In addition to these intergovernmental initiatives, there has been, for example, the Lima Declaration in 1997, a result of an international conference against corruption representing the citizens of 93 countries, which called for a concerted

---

[49] Convention on Combating Bribery of Foreign Public Officials in International Business Transactions.

[50] Commentaries on the Convention on Combating Bribery of Foreign Public Officials in International Business Transactions, para 1.

[51] Progress Report on the work of the Multidisciplinary Group on Corruption (GMC) for the attention of the 21st Conference of European Ministers of Justice, Prague, 10–11 June 1997.

[52] The Convention on the Protection of the European Communities' Financial Interests. It has yet to be ratified. The target date for ratification of the convention and its protocols is mid-1998.

[53] The Convention on the Fight Against Corruption Involving Officials of the European Communities or Officials of Member States of the European Union. It has yet to be ratified.

effort by all international organisations, national governments and others to combat corruption.[54]

SCOPE OF THE PROJECT

Members of Parliament

1.21 We explained in the consultation paper that the Government was taking steps to clarify the law relating to the bribery of, or the receipt of a bribe by, a Member of Parliament. In December 1996 the Home Office published a document to achieve this aim. The document, entitled *Clarification of the Law Relating to the Bribery of Members of Parliament*, was addressed to the Select Committee on Standards and Privileges in the House of Commons and the Committee for Privileges in the House of Lords. The Committees were invited to consider four broad options[55] but had not reported by the end of the last Parliament. The matter has now been taken up by the Joint Committee on Parliamentary Privilege.[56]

1.22 We therefore took the view that it would be inappropriate and a waste of our resources for us to look into the position of members of both Houses of Parliament at that stage. We also considered that similar considerations apply to Ministers of the Crown, who are members of either the Commons or the Lords. We believed that the position of Ministers might well be affected by the Home Office review, and therefore did not deal with them in the consultation paper. We have assumed, and continue to assume, that the position of members of either House and of Ministers will be clarified by the Home Office review, and appropriate steps taken. For this reason, the draft Bill annexed to this report does not expressly refer to MPs or Ministers: if our recommendations are accepted, we would expect the draft Bill to be amended in accordance with the conclusions of the current review, and express provision made.

Scotland

1.23 The remit of the Law Commission of course only covers the law of England and Wales, and it is that law that we have had under consideration for possible reform, not that of Scotland as well. The Prevention of Corruption Acts apply in Scotland as in England and Wales, with only very minor differences. However, there are major differences between the relevant rules of evidence and procedure in the two jurisdictions, and the common law is also rather different.

---

[54] It took place during 7–11 September 1997.

[55] Para 9 sets out the four broad options:
  (1) to rely solely on Parliamentary privilege to deal with accusations of the bribery of Members of Parliament;
  (2) subject Members of Parliament to the present corruption statutes in full;
  (3) distinguish between conduct which should be dealt with by the criminal law and that which should be left to Parliament itself;
  (4) make criminal proceedings subject to the approval of the relevant House of Parliament.

[56] Under the chairmanship of the Rt Hon the Lord Nicholls of Birkenhead.

False documents

1.24 Section 1(1) of the 1906 Act contains three paragraphs. The first two paragraphs set out the offences of accepting (or obtaining, or agreeing to accept or attempting to obtain) a bribe and giving (or agreeing to give or offering) a bribe. The third paragraph of section 1 of the 1906 Act provides:

> If any person knowingly gives to any agent, or if any agent knowingly uses with intent to deceive his principal, any receipt, account or other document in respect of which the principal is interested, and which contains any statement which is false or erroneous or defective in any material particular, and which to his knowledge is intended to mislead the principal;
>
> he shall be guilty [of an offence].

1.25 This offence differs from the offences of bribery contained in the first two paragraphs in that, despite its presence in a statute concerned with corrupt practices, the offence created by this paragraph is not in fact one of corruption at all. This point was established by the Divisional Court in *Sage v Eicholz*[57] where the provision was literally construed. We therefore took the view in the consultation paper[58] that this offence, although it appears in the 1906 Act, is not in fact an offence of corruption at all, but one of fraud; and, having concluded that fraud, although a common feature of corruption, was not an essential element of corruption offences, we further concluded that the false documents offence lay outside the scope of this project. We remain of this view.

THE STRUCTURE OF THE REPORT

1.26 In Part II we consider whether there is a need to reform the present law. We proceed in Part III to consider the desirability of distinguishing, as the present law does, between public bodies and others in the law of corruption. We then consider in Part IV whether the reformed law should contain any form of presumption. In Part V we deal with the formulation of modern offences of corruption. In Part VI we look at the investigation of corruption, and in Part VII at the territorial jurisdiction of the new offences and whether there is a need for consent to prosecutions by the Attorney-General or any other person. Also included in this part is discussion of various ancillary matters. Part VIII looks at the possibility of introducing new offences of corruption by means of threats or deception. In Part

---

[57] [1919] 2 KB 171. The defendant made a fraudulent representation in a claim handed to the agent of a water board, so that the defendant could obtain a rebate on his water rates. The agent was not aware that the claim was false, so there was no corruption of the agent. However, the defendant's actions fell squarely within the wording of the third paragraph. He knowingly gave the agent a document in which the agent's principal was interested. That document contained a false statement and was intended to mislead the principal. All the necessary ingredients for liability were therefore present. The Divisional Court considered that the word "corruptly" was deliberately absent from the third paragraph and that the word "knowingly", with which it is replaced, imports no element of *corruption*. In short, the approach in *Sage* was that if Parliament had wanted to confine the offence to the corruption of agents or those who corrupt them, the word "corruptly" would have appeared therein as it does in the earlier paragraphs.

[58] See paras 10.20 – 10.22.

IX we set out our recommendations in full. A copy of a draft Bill embodying our recommendations is set out in Appendix A, with an index. Appendix B contains various statutory provisions and Appendix C is a list of all those who commented on the consultation paper.

1.27 We received a large number of very helpful responses to the consultation paper and we are grateful to those who took the time and trouble to respond. We must also record our gratitude for the help that we have received from the Committee on Standards in Public Life which has been of great value to us; and we are pleased that Lord Nolan, former chairman of that committee, felt able recently to make the following comment to the House of Lords Select Committee on the Public Service:

> We have called for the corruption statutes to be brought up to date and clarified, and the Law Commission, with our support, has produced an excellent consultation paper on this.[59]

1.28 We must also express our gratitude to Professor Sir John Smith CBE, QC, LLD, FBA, Emeritus Professor of Law at the University of Nottingham, who gave us much assistance with this project.

---

[59] These comments were made by Lord Nolan, in his capacity as Chairman of the Committee on Standards in Public Life, to the Select Committee on 28 October 1997.

PART II
THE PRESENT LAW AND THE NEED FOR CHANGE

2.1 In the consultation paper, we considered whether there was a need for a change in the present law.[1] As we have already said, many of the statutory offences were introduced as impulsive reactions to particular problems and scandals[2] and, therefore, not surprisingly, this has led to an uncertain and inconsistent law. In this part we will briefly describe the present law. We shall then summarise the difficulties with the present law which we highlighted in the consultation paper, difficulties which we believe justify the view we took provisionally that the law of corruption is in an unsatisfactory condition and that the common law and statutory offences of corruption should be re-stated in a modern statute.[3] We will then consider the responses that we received on consultation before stating our conclusion.

THE PRESENT LAW

The common law offence of bribery

2.2 Bribery at common law has evolved over time, and opinions differ as to whether it is to be regarded as a general offence (applying to a range of different offices or functions) or whether the common law is comprised of a number of offences of bribery (distinguished by the office or function to which a particular offence applies).[4] *Russell on Crime*, however, provides the following general statement of the offence:

> Bribery is the receiving or offering [of] any undue reward by or to any person whatsoever, in a public office, in order to influence his behaviour in office, and incline him to act contrary to the known rules of honesty and integrity.[5]

2.3 The most widely cited definition of who is to be regarded as a public officer for the purposes of common law bribery is taken from the early twentieth century case of *Whitaker*,[6] in which "a public officer" was defined as

---

[1] Part IV of the consultation paper.

[2] See para 1.2 above.

[3] Consultation Paper No 145, para 4.18.

[4] "[T]he offence underwent a development over the centuries and is often described in terms of a number of individual offences rather than a single offence": D Lanham, "Bribery and Corruption" in *Criminal Law: Essays in Honour of J C Smith* (1987) 92, 92–93. Examples of specific offences, or specific instances of the offence, are bribery of a privy councillor (*Vaughan* (1769) 4 Burr 2494; 98 ER 308) and bribery of a coroner (*Harrison* (1800) 1 East PC 382). See also *Archbold*, para 31–129.

[5] *Russell on Crime* (12th ed 1964) p 381.

[6] [1914] 3 KB 1283, pp 1296–7. The defendant, a colonel, had accepted money from a firm of caterers in return for giving the firm the tenancy of the regiment's canteen. It was argued

an officer who discharges any duty in the discharge of which the public are interested, more clearly so if he is paid out of a fund provided by the public.[7]

2.4 One commentator[8] has argued that the common law is not confined to public officers holding some sort of *permanent* public office but extends also to those who discharge an *ad hoc* public duty. In *Pitt and Mead*,[9] for example, bribing electors at a parliamentary election was held to be a common law offence. In *Lancaster and Worrall*[10] the same decision was reached in respect of the bribery of local government electors. Embracery (the bribery of jurors) is a common law offence, though it is now considered obsolete.[11]

2.5 As for the nature of the bribe, Russell defines a bribe as "any undue reward".[12] However, the benefit conferred in a particular case may be so small that it cannot be considered a reward at all. For example, in the *Bodmin Case*[13] Willes J mentioned that he had been required to swear that he would not take any gift from a man who had a plea pending, unless it was "meat or drink, and that of small value". Russell describes the mental element of common law bribery in terms of a briber intending to influence the behaviour of a public officer with a view to that officer "act[ing] contrary to the known rules of honesty and integrity";[14] it appears, however, from the case law that it is not a necessary feature of the offence that the briber should intend the bribee to commit a *breach* of duty.[15]

The statutory offences

2.6 As we noted in Part I above,[16] offences of corruption can be found in a variety of legislative sources. We have confined our attention, however, to the principal corruption statutes, namely the Prevention of Corruption Acts 1889 to 1916.

---

that the law of bribery applied to "judicial and ministerial officers" and that the colonel belonged to neither category. The Court of Appeal disagreed, holding that every public officer who was not a judicial officer was a ministerial officer.

[7] *Ibid*, at p 1296, *per* Lawrence J.

[8] D Lanham, *op cit*, pp 93–94.

[9] (1762) 3 Burr 1335; 97 ER 861.

[10] (1890) 16 Cox CC 737.

[11] *Owen* [1976] 1 WLR 840.

[12] See para 2.2 above.

[13] (1869) 1 O'M & H 121.

[14] See para 2.2 above.

[15] See, eg, *Gurney* (1867) 10 Cox CC 550. The defendant was charged with attempting to bribe a justice of the peace. The jury was told that if the defendant had an intention to produce *any effect at all* on the justice's decision, that was an attempt to corrupt.

[16] See para 1.2 above.

The 1889 Act

2.7 The 1889 Act was introduced following revelations of malpractice made before a Royal Commission appointed to inquire into the affairs of the Metropolitan Board of Works, the body exercising the powers of local government in London at that time.[17] Section 1 of the Act, closely following the Commission's recommendation,[18] provides:

> (1) Every person who shall by himself or by or in conjunction with any other person, corruptly solicit or receive, or agree to receive, for himself, or for any other person, any gift, loan, fee, reward, or advantage whatever as an inducement to, or reward for, or otherwise on account of any member, officer, or servant of a public body as in this Act defined, doing or forbearing to do anything in respect of any matter or transaction whatsoever, actual or proposed, in which the said public body is concerned, shall be guilty of an offence.
>
> (2) Every person who shall by himself or by or in conjunction with any other person corruptly give, promise, or offer any gift, loan, fee, reward, or advantage whatsoever to any person, whether for the benefit of that person or of another person, as an inducement to or reward for or otherwise on account of any member, officer, or servant of any public body as in this Act defined, doing or forbearing to do anything in respect of any matter or transaction whatsoever, actual or proposed, in which such public body as aforesaid is concerned, shall be guilty of an offence.

2.8 As originally enacted, the 1889 Act was concerned only with *local* public bodies in the United Kingdom. Section 7 provides, in part, that the expression "public body" means

> any council of a county or county of a city or town, any council of a municipal borough, also any board, commissioners, select vestry, or other body which has power to act under and for the purposes of any Act relating to local government, or the public health, or to poor law or otherwise to administer money raised by rates in pursuance of any public general Act, but does not include any public body as above defined existing elsewhere than in the United Kingdom.

Section 4(2) of the 1916 Act extended the definition of "public body" to include "local and public authorities of all descriptions". A further extension is provided for under the Local Government and Housing Act 1989, so that it would include companies which, in accordance with Part V of that Act, are "under the control of one or more local authorities".[19] This provision is not, however, in force.

2.9 The bribe must take the form of "a gift, loan, fee, reward, or advantage". The expression "advantage" includes

---

[17] P Fennell and P A Thomas, "Corruption in England and Wales: An Historical Analysis" (1983) 11 Int J Soc L 167, 172.

[18] Salmon Report, para 44.

[19] Local Government and Housing Act 1989, Sched 11, para 3.

any office or dignity, and any forbearance to demand any money or money's worth or valuable thing, and includes any aid, vote, consent, or influence, or pretended aid, vote, consent, or influence, and also includes any promise or procurement of or agreement or endeavour to procure, or the holding out of any expectation of any gift, loan, fee, reward, or advantage, as before defined.[20]

2.10 The advantage must be given or received as an *inducement* to, or a *reward* for, or *otherwise on account of*, any member, officer or servant of a public body doing or forbearing to do something in respect of any matter or transaction in which the body is concerned.

The 1906 Act

2.11 In 1898 a report was published by the Secret Commissions Committee of the London Chamber of Commerce calling for the criminal law of corruption to be extended into the private sector.[21] Following a number of unsuccessful attempts at legislation,[22] the 1906 Act was passed. The 1906 Act applies to all "agents", whether in the public or the private sector. Section 1(1) provides in part:

> If any agent corruptly accepts or obtains, or agrees to accept or attempts to obtain, from any person, for himself or for any other person, any gift or consideration as an inducement or reward for doing or forbearing to do, or for having after the passing of this Act done or forborne to do, any act in relation to his principal's affairs or business, or for showing or forbearing to show favour or disfavour to any person in relation to his principal's affairs or business; or
>
> If any person corruptly gives or agrees to give or offers any gift or consideration to any agent as an inducement or reward for doing or forbearing to do, or for having after the passing of this Act done or forborne to do, any act in relation to his principal's affairs or business, or for showing or forbearing to show favour or disfavour to any person in relation to his principal's affairs or business
>
> ... he shall be guilty [of an offence].

2.12 "Agent" is defined as including "any person employed by or acting for another"[23] and a person serving under the Crown[24] or any local or public authority.[25] The term "serving under the Crown" does not require *employment* by the Crown. In *Barrett*[26] the Court of Appeal held that, for the purposes of the Act, a

---

[20] 1889 Act, s 7.
[21] P Fennell and P A Thomas, *op cit*, p 174.
[22] Salmon Report, para 45.
[23] 1906 Act, s 1(2).
[24] 1906 Act, s 1(3).
[25] 1906 Act, s 1(3), as amended by 1916 Act, ss 4(2), (3).
[26] [1976] 1 WLR 946.

2.13 superintendent registrar of births, deaths and marriages was serving under the Crown, although he was not appointed, paid or liable to dismissal by it.

2.13 Whereas the 1889 Act describes a bribe as a "gift, loan, fee, reward or advantage", the 1906 Act uses the expression "gift or consideration";[27] and "consideration" is defined as including "valuable consideration of any kind".[28] As under the 1889 Act, the putative bribe must be given or received as a *reward* or *inducement*; but, whereas under the 1889 Act it must be connected to a particular "matter or transaction", under the 1906 Act it may be a general "sweetener" designed to secure *generally* more favourable treatment.

2.14 An ingredient common to the offences under both the 1889 and the 1906 Acts is that the putative bribe should be given or received *corruptly*.[29] The term is not defined in the legislation and has, as a result, been construed through case law.[30]

The 1916 Act

2.15 The 1916 Act was prompted by wartime scandals involving contracts with the War Office,[31] and was passed rapidly through Parliament as an emergency measure. Aside from increasing the maximum sentence for bribery in relation to contracts with the Government or public bodies,[32] and broadening the definition of "public body",[33] it also introduced the presumption of corruption. Section 2 of the 1916 Act provides:

> Where in any proceedings against a person for an offence under the Prevention of Corruption Act 1906, or the Public Bodies Corrupt Practices Act 1889, it is proved that any money, gift, or other consideration has been paid or given to or received by a person in the employment of [Her] Majesty or any Government Department or a public body by or from a person, or agent of a person, holding or seeking to obtain a contract from [Her] Majesty or any Government Department or public body, the money, gift, or consideration shall be deemed to have been paid or given and received corruptly as such inducement or reward as is mentioned in such Act unless the contrary is proved.

---

[27] 1906 Act, s 1(1).

[28] 1906 Act, s 1(2).

[29] A third offence created by the 1906 Act, s 1(1), is the offence of using a false document with the intent to mislead a principal. *This* offence does not require the agent to have acted corruptly (*Sage v Eicholz* [1919] 2 KB 171). See paras 1.24 – 1.25 above.

[30] See paras 2.29 – 2.30 below.

[31] P Fennell and P A Thomas, *op cit*, pp 174 and 185–186. Particularly influential were the comments made by Low J, who presided over two of the War Office cases (*Asseling*, *The Times* 10 September 1916, and *Montague*, *The Times* 18 September 1916). See also the Salmon Report, para 46.

[32] The maximum penalty at the time was 2 years' hard labour, and the effect of the 1916 Act was to increase it to 7 years in cases to which the 1916 Act applied. The disparity in sentencing between the 1889 and 1906 Acts was removed by s 47 of the Criminal Justice Act 1988.

[33] See para 2.17 below.

2.16 The presumption shifts the burden of proof, so that it is for the defence to prove (on the balance of probabilities)[34] that a given payment was not corrupt. It applies only to payments made to employees of public bodies, and only to cases involving contracts.

PROBLEMS WITH THE PRESENT LAW

"Public bodies"

2.17 As we have seen, the present law distinguishes between "public bodies" and others. Initially the law applied only to public bodies;[35] it was then extended to the private sector[36] (although that extension was not reflected in the application of the presumption of corruption[37] which was again confined to public bodies). The rationale for this distinction appears to be the assumption that higher standards of conduct are required in the public sector than in the private sector. We commented in the consultation paper that this justification would be more convincing if the distinction were on the basis of rules requiring higher standards in the public sector,[38] but it is not. It has no direct bearing on the question of which conduct is corrupt but only affects the ease of proving corruption.

The presumption

2.18 The presumption of corruption applies only to employees of public bodies, and only where they receive a benefit from a person holding or seeking a contract with such a body. There are substantial difficulties in practice in defining "public body" which have been compounded by the recent trend towards privatising public bodies and contracting out of work by public bodies to the private sector.

2.19 Furthermore, two important developments have occurred since the enactment of the 1916 Act which together call into question the need for and legality of the presumption: first, the enactment of the CJPOA and, in particular, sections 34 and 35 which allow inferences to be drawn from the silence of a defendant in interview, on being charged or at trial; second, the fact that the United Kingdom is a signatory to the European Convention on Human Rights ("the ECHR") and intends to incorporate the convention into domestic law.

Agents

2.20 As we have seen,[39] the 1906 Act is concerned with the corruption of "agents", defined in the Act as including any person "employed by or acting for another"[40]

---

[34] *Carr-Briant* [1943] KB 607.

[35] Under the 1889 Act.

[36] Under the 1906 Act.

[37] Under the 1916 Act.

[38] Consultation Paper No 145, para 4.3.

[39] See paras 2.11 and 2.12 above.

[40] 1906 Act, s 1(2).

or any person serving under the Crown or any public body.[41] There has been criticism about the uncertainty of the meaning of the word "agent".

2.21 We drew attention to three main areas in which problems arise:

(1) the inapplicability of the presumption to agents (of public bodies) who are not classifiable as employees (of public bodies);[42]

(2) the inconsistency of the law as regards

(a) persons connected with agents,[43]

(b) persons who have been, or are to become, agents,[44] and

(c) purported agents;[45] and

(3) the uncertainty of the law as regards

(a) police officers,[46]

(b) judges,[47] and

(c) local councillors.[48]

The inapplicability of the presumption

2.22 We were concerned that the presumption of corruption under section 2 of the 1916 Act affects only *employees* of the Crown, Government department or public body. The presumption, therefore, will not arise where the individual concerned is an agent who cannot be classified as an employee. This category might include employees of firms engaged in contracted-out work and private sector secondees to Government departments. Our provisional view was that whatever view was taken of the presumption it was clear that if it was to be retained, it should be applied consistently.[49]

Persons connected with agents

2.23 We also noted that whereas the 1889 Act does not require that the person being tempted to act in breach of duty should also be the recipient of the bribe, the 1906 Act requires those roles to be played by the same person. So, under the 1889 Act, in a transaction in which A pays a third party, D, who is distinct from B, the member, officer or servant of a public body whose conduct is sought to be influenced, D may be guilty of an offence if the receipt can be shown to have been

---

[41] 1906 Act, s 1(3) and 1916 Act, ss 4(2), (3).

[42] See para 2.22 below.

[43] See para 2.23 below.

[44] See para 2.24 below.

[45] See para 2.25 below.

[46] See para 2.26 below.

[47] See para 2.27 below.

[48] See para 2.28 below.

[49] Consultation Paper No 145, para 4.8.

an "inducement to" or "reward for" or "otherwise on account of" B's acting (or refraining from acting) in his or her capacity as member, officer or servant of a public body. Under the 1906 Act, on the other hand, in the circumstance where A pays a third party, D, with a view to influencing or rewarding an agent, B, none of the parties would be guilty.[50] In this respect the 1906 Act is more limited than the 1889 Act, and, as a result, although third-party transactions involving those associated with public bodies fall foul of the 1889 Act, the 1906 Act will not bite on similar transactions in the private sector.

Persons who have been, or are to become, an agent

2.24 We drew attention to another example of the inconsistency between the 1889 and 1906 Acts: whereas the 1906 Act appears to require that the bribe be received by the agent during the *currency* of the agency (or that the agreement to receive the bribe should be during the currency of the agency), the 1889 Act applies to circumstances in which the public officer is *no longer* in office at the time of receipt of the bribe (or at the time of the agreement to receive the bribe) or has *yet* to assume office.[51]

Purported agents

2.25 Where a person *purports* to be, but is not in fact, an agent it is doubtful whether that person can be guilty if he or she obtains a corrupt payment.[52] Professor A T H Smith has commented on the balance of the argument as to whether the law *ought* to extend to "purported" agents:

> It is unclear whether a person who merely purports to hold a public office – as opposed to an actual office holder – might be guilty of the offence by receiving a "bribe". One view is that he does, but the conclusion must be open to argument. It is outside the mischief at which the offence, even at its broadest, seems to be aimed, since the recipient of the bribe is in no sense in any position to exploit an office that he does not actually hold. On the other hand, the person handing over the bribe believes that recipient will be influenced by it and is in a position to affect his affairs favourably, and the recipient knows that this is his belief, and it might be said that this should suffice.[53]

Applicability of corruption laws to police officers, judges and local councillors

2.26 It is uncertain whether a police officer is an agent for the purposes of the 1906 Act: an English case suggests that a police officer is a servant of the State[54] but a

---

[50] If, however, D passed A's "bribe" on to B, both B and D would be guilty under the 1906 Act, since D would then be an offeror and B the recipient agent.

[51] Consultation Paper No 145, para 4.10.

[52] Consultation Paper No 145, para 4.11.

[53] A T H Smith, *Property Offences* (1994) pp 792–793, para 25-04 (footnotes omitted).

[54] *Fisher v Oldam Corporation* [1930] 2 KB 364 (a civil case).

Scottish case decided that the police officer was held to be an agent of the Chief Constable.[55]

2.27 Although it is an offence at common law to give a judge, magistrate or other judicial officer any gift or reward intended to influence his or her behaviour,[56] it appears that the statutory offence of corruption might not apply to judicial officers.[57] Any attempt to bribe a judicial officer may, of course, be dealt with either as a contempt of court or an attempt to pervert the course of justice.

2.28 Local councillors are unlikely to be covered by the 1906 Act although they are likely to fall within the common law of corruption and the 1889 Act.[58]

The meaning of "corruptly"

2.29 Each of the principal corruption Acts uses the term "corruptly" but its meaning is by no means clear. In the consultation paper,[59] we set out the two competing strands of judicial interpretation of "corruptly": on the one hand, it describes an act which the law forbids as tending to corrupt,[60] and on the other, a dishonest intention to weaken the loyalty of an agent to his or her principal.[61] We noted Lanham's conclusion that the authorities on this point are in "impressive disarray".[62]

2.30 The reason why the cases are in this state is principally because of the approach of the House of Lords in the leading case of *Cooper v Slade*[63] in which the majority view was that "corruptly" meant "purposely doing an act which the law forbids as intending to corrupt".[64] Professor A T H Smith suggests that such an interpretation would leave the word "devoid of any functional significance".[65] In other cases it has been suggested that the word "corruptly" was akin to an evil mind,[66] while it has also been suggested that the defendant must have *dishonestly* intended to weaken the loyalty of servants to their master and to transfer that

---

[55] *Graham v Hart* 1908 SC (J) 26.

[56] See *Gurney* (1867) 10 Cox CC 550 and n 15 above.

[57] Consultation Paper No 145, para 4.13.

[58] Consultation Paper No 145, para 4.14.

[59] Consultation Paper No 145, paras 2.24 – 2.28, and para 5.65 below.

[60] *Cooper v Slade* (1857) 6 HL Cas 746; 10 ER 1488, and *Wellburn* (1979) 69 Cr App R 254.

[61] *Lindley* [1957] Crim LR 321 and *Callard* [1967] Crim LR 236.

[62] Consultation Paper No 145, para 4.15.

[63] (1857) 6 HL Cas 746: 10 ER 1488.

[64] (1857) 6 HL Cas 746, 773; 10 ER 1488, at 1499 per Wills J.

[65] A T H Smith, *op cit*, para 25-26.

[66] *Per* Martin B in *Bradford Election Case (No 2)* (1869) 19 LT 723, 728.

loyalty to himself or herself.[67] In the consultation paper we provisionally agreed with Lanham that "the position in English law is hardly satisfactory".[68]

OUR PROVISIONAL CONCLUSION, THE RESPONSE ON CONSULTATION AND OUR RECOMMENDATION

2.31 For all these reasons, we provisionally concluded in our consultation paper that the law of corruption was in an unsatisfactory state and that the common law and statutory offences of bribery should be re-stated in a modern statute.[69] The great majority of those respondents who commented on this topic agreed with our conclusions.

2.32 Of those who disagreed, the Financial Law Panel was concerned that the new offence would have no practical effect if applied to the financial sector, which was already subject to a substantial degree of regulation. Significantly, however, it did not disagree with our view about the unsatisfactory state of the present law and the difficulties that arise outside the financial sector. The Securities and Investments Board ("the SIB") did not think that the present law caused many difficulties but conceded that there was an attraction in restating the common law and statutory offences in a modern statute. We have reconsidered our provisional view. We have received no arguments which persuade us that it is wrong.

2.33 We conclude that the present law is in an unsatisfactory state and **we recommend that the common law offence of bribery and statutory offences of corruption should be replaced by a modern statute.**

(Recommendation 1)

---

[67] *Lindley* [1957] Crim LR 321, followed in *Callard* [1967] Crim LR 236.
[68] D Lanham, *op cit*, p 106.
[69] Consultation Paper No 145, para 4.18.

PART III
THE DISTINCTION BETWEEN PUBLIC BODIES AND OTHERS

3.1 An important feature of the present law is the distinction between corruption involving public and non-public bodies. As we explained in the consultation paper, the shape of a reformed law of corruption will depend on whether this distinction is to be retained or altered in any way.[1]

3.2 In this part we will explain the significance of the present law and the reasons for our provisional view that the distinction should be abolished. We will then consider the responses and our final view.

THE SIGNIFICANCE OF THE DISTINCTION IN THE PRESENT LAW

3.3 The distinction is important in two main respects. First, the 1889 Act applies only to corruption in public bodies. Second, under section 2 of the 1916 Act there is a presumption of corruption in certain cases where a public body is concerned. A third, less important, consequence is that there is a distinction between a person *serving under* a public body and a person serving under a non-public body.

The 1889 Act

3.4 The 1889 Act only covers corruption in public bodies: the bribe must relate to the conduct of a member, officer or servant of the public body. Although there is a substantial overlap between the 1889 and 1906 Acts, the earlier Act has certain advantages (for the prosecution) in relation to both the scope of the offences and the penalties available on conviction.

The scope of the offences

THIRD PARTIES

3.5 The first advantage is that, whereas the 1906 Act affects only agents and does not extend to third parties, the 1889 Act, in contrast, sets no limits on the category of persons who may be charged with soliciting or receiving a bribe – it embraces not only a member, officer, or servant of a public body but also *any third party* who solicits or receives a bribe in respect of the conduct of a member, officer, or servant of a public body. Therefore the 1889 Act, but not the 1906 Act, would apply, for example, to a spouse of an employee of a public body who accepts a bribe in return for persuading that employee to commit a breach of duty.

PEOPLE WHO HAVE BEEN, OR ARE TO BECOME, AGENTS

3.6 Secondly, the 1906 Act only applies to acts done by agents or people having dealings with *existing* agents. In contrast, under the 1889 Act, it is possible to prosecute although the agreement to receive the reward or receipt of the reward

---

[1] Consultation Paper No 145, para 6.1.

itself was not until after the relevant agency had terminated.[2] So if the person at the time of the receipt of the reward or at the time he or she agreed to receive the reward is not an agent – either because he or she has not taken office or has ceased to take office – an offence may have been committed under the 1889 Act but not under the 1906 Act.

Penalties

3.7 The powers of the court following conviction under the 1889 Act are far more extensive than the penalties available under the 1906 Act. In addition to imprisonment or a fine (or both), the court may, for example, order a defendant to pay to the public body concerned the amount, or value of the gift, loan, fee or reward received. The court may also disqualify the defendant from public office or order his or her forfeiture of office.[3]

The presumption

3.8 The presumption under section 2 of the 1916 Act, which has the effect that any money, gift or consideration is *deemed* to have been paid or given and received corruptly as an inducement or a reward *unless the contrary is proved*, applies only where the employer of the putative bribee or the body with which the putative briber holds or seeks to hold a contract is the Crown, a Government department or a public body. The presumption applies not only to the 1889 Act, which is itself confined to public bodies, but also to the 1906 Act, which is not.

3.9 Clearly, it will often be of crucial importance to determine whether or not a particular body is a public body as this will decide who bears the burden of proof concerning what will often be the only live issue at the trial.

The distinction between a person serving under a public body and a person serving under a non-public body

3.10 A further, somewhat theoretical, point is that a person serving under a public body counts as an "agent" for the purpose of the 1916 Act[4] whereas a person serving under any other body is an agent only *if employed by* or *acting for* the body in question. As we pointed out in the consultation paper[5] we found it hard to think of any examples of a person who would be "serving under" a public body although neither "employed by" it nor "acting for" it – in which case the person would be an agent within the definition of the 1906 Act, even *without* the extension in the 1916 Act. If such a case exists, the person in question is an "agent" within the meaning of the Acts *solely* because the body under which he or she serves happens to be a "public body". In other words, whether or not a body is a public body is to this extent relevant to the distinction between persons who are and are not "agents" within the meaning of the Acts.

---

[2] *Andrews Weatherfoil Ltd* [1972] 1 WLR 118.

[3] 1889 Act, s 2. See para 7.30 below.

[4] 1916 Act, s 4(3).

[5] Consultation Paper No 145, para 6.9.

THE DEFINITION OF A PUBLIC BODY

3.11 Under the 1889 Act, the phrase "public body" is defined as follows:

> The expression "public body" means any council of a county or county of a city or town, any council of a municipal borough, also any board, commissioners, select vestry, or other body which has power to act under and for the purposes of any Act relating to local government, or the public health, or to poor law, or otherwise to administer money raised by rates in pursuance of any public general Act, but does not include any public body as above defined existing elsewhere than in the United Kingdom.[6]

3.12 It has been said that the definition is "clearly ... confined to local authorities".[7] It consists of a list of various types of local authorities which existed when the 1889 Act was passed, but this list proved too restrictive since some public bodies were not included.[8] Section 4(2) of the 1916 Act extended the term "public body" to include "local and public authorities of all descriptions". Regrettably, there have been serious doubts as to what this means.

3.13 In *Newbould*[9] Winn J cast doubt on the effectiveness of the 1916 amendment by holding that the National Coal Board was not a public body within the meaning of section 4(2). However, this view was expressly overruled by the House of Lords in *DPP v Holly and Manners*,[10] where it was held that the North Thames Gas Board *was* a public body. That expression included

> bodies which have public or statutory duties to perform and which perform those duties and carry out their transactions for the benefit of the public and not for private profit.[11]

Lord Edmund-Davies said it was difficult to imagine wording that could have been wider than that of section 4(2).[12]

3.14 This decision appeared for a time to clarify matters. It became clear that a public body need not be one established by statute. Thus the Criminal Injuries Compensation Board, established by royal prerogative, was a public body within the meaning of the Act[13] because it had public duties to perform.[14] The *Holly* test

---

[6] 1889 Act, s 7.

[7] *Joy and Emmony* (1974) 60 Cr App R 132, 133, *per* Judge Rigg.

[8] Eg, the Port of London Authority and the Water Board: see Lord Buckmaster LC in the debate on what became s 4(2) of the 1916 Act (*Hansard* (HL) 19 December 1916, vol 23, col 987).

[9] [1962] 2 QB 102.

[10] [1978] AC 43.

[11] [1978] AC 43, 53, *per* Lord Diplock.

[12] [1978] AC 43, 54, citing with approval the words of Judge Rigg in *Joy and Emmony* (see n 7 above).

[13] See generally J F Garner, "Public Bodies" (1977) 121 SJ 785. Other examples of public bodies cited in this article were the British Gas Council, the British Airways Board, the National Coal Board and the Post Office.

was simple: did the body carry out its business for the benefit of the public, or for private profit?

The Local Government and Housing Act 1989 amendment

3.15 The concept of a public body will be further widened if paragraph 3 of Schedule 11 to the Local Government and Housing Act 1989 comes into force.[15] That provision adds the following words to section 4(2) of the 1916 Act:

> ... and companies which, in accordance with Part V of the Local Government and Housing Act 1989, are under the control of one or more local authorities.

3.16 Section 68 of the same Act defines a company "under the control of" a local authority. Limited companies which are subsidiaries of local authorities, or the majority of whose shares are owned or controlled by a local authority, will themselves be "public bodies" within the meaning of the Prevention of Corruption Acts.

Public bodies outside the United Kingdom

3.17 It is not entirely clear whether a public body existing outside the United Kingdom is a "public body" within the meaning of the Acts. Section 7 of the 1889 Act expressly excludes such bodies from the definition. Section 4(2) of the 1916 Act extends the definition so as to include "local and public authorities of all descriptions"; but this appears to have been intended only to include certain British bodies which fell outside the original definition. Had Parliament intended to remove the exception for bodies existing outside the United Kingdom, it would surely have done so expressly. We therefore provisionally concluded that the bribery of an employee of a foreign public body would not be an offence under the 1889 Act, and that the presumption under section 2 of the 1916 Act would not apply.

3.18 It has been suggested[16] that, if our interpretation is correct,[17] it creates a "singular anomaly" as regards those who qualify as agents: "neither private agents nor Crown Servants enjoy any similar exemption". However, the anomaly arises only where the official in question, though *serving under* a public body existing outside the United Kingdom, is neither *employed by* nor *acting for* that body. If the official were employed by it or acting for it, he or she would be an agent within the

---

[14] See *R v Criminal Injuries Compensation Board, ex p Lain* [1967] 2 QB 864.

[15] Although Commencement Order No 8 (SI 1990 No 1274) was drafted so that this provision would come into force on 1 July 1990, that order was amended by an order made on 28 June 1990 (SI 1990 No 1335) which omitted reference to Paragraph 3 of Sched 11.

[16] D Lanham, "Bribery and Corruption", *Criminal Law: Essays in Honour of J C Smith* (1987) 92, 101.

[17] Lanham thinks it is arguable that s 4(2) of the 1916 Act removes the exception for foreign bodies in s 7 of the 1889 Act, though "[a]s the exception works in favour of the liberty of the subject" it would probably be held that the exception survives: *ibid*. In our view it was clearly intended to survive.

original definition in section 1(2) of the 1906 Act, and the question of whether the body qualified as a public body would for this purpose be immaterial.

SHOULD THE DISTINCTION BE RETAINED?

The diminishing importance of the distinction

3.19 As we have explained,[18] in *Holly* it was said that a body is a public body only if it performs public or statutory duties for the benefit of the public and not for private profit. This rule has become increasingly arbitrary as a result of developments since that case was decided.

3.20 For example, more and more functions previously performed by organisations of national or local government are being sub-contracted to private companies. In addition, there are an increasing number of joint ventures between local and central government and private companies. Private Finance Initiatives are used to finance projects which had previously been undertaken at public expense. Another major development is that formerly nationalised utilities – such as the water boards, British Gas and electricity generators – have been privatised. These bodies have continued to provide the same service, which had previously been "for the benefit of the public", but now do so for their shareholders' profit.

3.21 The matter is further confused because in some cases the state has retained a controlling influence in those companies through a "golden share". In other cases, they have statutory duties to perform and are regulated by "quangos" answerable to Parliament.[19]

3.22 In the consultation paper we drew attention to the effect of the distinction drawn in *Holly* that an agent of a nationalised industry who accepts a bribe is committing an offence under the 1889 Act and, if prosecuted, must prove that the transaction was not corrupt; but an agent who did the same thing after the industry had been privatised would be committing an offence under the 1906 Act only and the burden of proof would rest entirely on the prosecution.[20] We found it difficult to justify the determining factor as being whether the organisation was acting for profit or not. We are not alone in taking this view. The MCCOC Report concluded:

> Given that the distinction between the functions to be privatised are based primarily on economic criteria, linking the offence of bribery to functions which *happen* to be performed in the public sector for the time being is arbitrary.[21]

3.23 The Home Office in its consultation document took an identical view and thought that the distinction was no longer valid.[22]

---

[18] See para 3.13 above.

[19] Eg, OFGAS and the Office of the Rail Regulator.

[20] Consultation Paper No 145, para 6.28.

[21] MCCOC Report, p 271.

[22] Home Office consultation paper, paras 3.4 and 3.5.

3.24 In defence of the present position it may be argued that the *Holly* test was not in fact formulated against the background of the industrial scene in 1978, which was perhaps the high-water mark of state ownership. Not only were nearly all "public" enterprises then owned by the state, but so were some of a wholly *private* character.[23] The conditions that prevailed between 1889 and 1916 are, in this respect, more like the present than those of 1978. It would have been quite wrong to construe the legislation with 1978 spectacles, and the House of Lords did not do so: it merely confirmed the correctness of a test laid down in 1907,[24] which Parliament must be taken to have had in mind in 1916. Arguably it follows that that test may have become inappropriate by 1978, but has become *less* inappropriate since then.

3.25 There is force in this argument. But the real question, we believe, is not whether developments since 1978 have undermined the reasoning in *Holly*. It is whether, in the light of the experience of the last fifty years, it still seems sensible to apply different rules of criminal law to an enterprise which is currently in state hands and to one which currently is not. With the benefit of that experience, the difference between the two forms of enterprise now seems far less significant than it must have seemed to the legislators of 1889 and 1916. In our view the distinction between public and private can no longer be regarded as a simple dichotomy: rather, it is a spectrum. At one extreme there are quintessentially public bodies, the privatisation of which is scarcely conceivable; at the other, firms which exist to make profits for their owners and for no other purpose. In between, there is a grey area. The bodies in this area may or may not earn profits for shareholders, but their functions are such that it is a matter of public concern that they should be properly run.

3.26 Bearing in mind the difficulty in distinguishing between public bodies and others, we now consider the possible arguments in favour of the retention of the distinction.

Possible arguments in favour of the distinction

3.27 We considered a number of issues relevant to the important question of whether the distinction between public bodies and others is one which should continue to play a crucial part in the law of corruption.

Is public sector corruption more serious than private sector corruption?

3.28 Our provisional view was that, in general, corruption on the part of a public servant was likely to be more damaging to the public interest, and therefore a more serious offence, than corruption in the private sector. However, we did not accept that this was an adequate justification for a *rigid* distinction between the two.[25] We explained that some private sector corruption is very serious whilst some public sector corruption might be trivial, and we thought that the seriousness of

---

[23] Eg Rover and Rolls Royce.

[24] In *The Johannesburg* [1907] P 65.

[25] Consultation Paper No 145, para 6.20.

the offence was a matter more appropriately to be considered at the sentencing stage.[26]

Is the public more in need of protection than the private sector?

3.29 It is arguable that public bodies are more vulnerable to corruption than private bodies because private bodies are more able to look after themselves. Lord Randolph Churchill, the sponsor of the 1889 Bill, said, in response to criticism that it should extend to private bodies:

> There is an essential difference between a private body and a public one. A private body has a direct interest in looking after its servants, but in the case of a public body what is everybody's business is nobody's business.[27]

3.30 However, it must be remembered that this was said at a time when private sector corruption was not a statutory offence at all. Parliament decided in 1906 that the availability of civil remedies was insufficient reason for allowing such corruption to remain outside the criminal law. Nowadays, it is universally accepted that private sector corruption should be criminal and it seems that the point made by Lord Randolph Churchill is not relevant to the issue of whether private sector corruption should continue to be governed by *different rules*.

Is there a need for higher standards of conduct in the public sector?

3.31 There is a long-standing argument that different and higher standards are required and expected of those who work in the public sector. The Redcliffe-Maud Committee, for example, explained that "public life requires a standard of its own and those entering public office for the first time must be made aware of this from the outset".[28]

3.32 We provisionally concluded that this point is irrelevant to the question under consideration.[29] We believed that the distinction between public bodies and others was not an attempt to reflect the view that public life requires higher standards of conduct than private activities.

---

[26] The MCCOC took a similar view. In their report, pp 271–273, they wrote:

> Confining bribery to the public sector assumes that public sector corruption does more harm to the community than private sector corruption. That assumption is questionable. The secret commissions paid to Johns in the Tricontinental Bank case amounted to $2 million. The corrupting effect of a secret commission of that amount on confidence in the *general* commerce and finances of the community were very serious and more harmful than many instances of bribery in the public sector ... The public needs to be able to have confidence in the integrity of both the public and the private sector. It should not be statutorily presumed that corrupt payments in the public sector do more harm than corrupt payments in the private sector. The amount of damage in a particular case should be a question for sentencing rather than the subject of a separate offence.

[27] *Hansard* (HL) 25 March 1889, vol 334, col 810.

[28] Redcliffe-Maud Report, para 76.

[29] Consultation Paper No 145, para 6.25.

3.33 As we have noted,[30] the distinction made by the Prevention of Corruption Acts between public bodies and others is significant for three reasons. First, the 1889 Act applies only to corruption in public bodies. Second, under section 2 of the 1916 Act there is a presumption of corruption in certain cases where a public body is concerned. A third, less important, consequence is that there is a distinction between a person serving under a public body and a person serving under a non-public body. None of these reasons relates to the standard of conduct required of agents in the public and private sectors.

OUR PROVISIONAL PROPOSALS AND THE RESPONSE ON CONSULTATION

3.34 Against that background, our provisional views were:

(1) that the existence of the distinction between public bodies and others, and the various different effects of that distinction, render the law complex and confusing;

(2) that there would therefore be substantial advantages in abandoning it;

(3) that the arguments for retaining the distinction *in principle* are outweighed by the advantages of abandoning it;

(4) that it should therefore be retained *only* if it is necessary to do so for any of the three purposes we have identified;[31] and

(5) that it is not necessary to retain it for the purpose of determining the scope of

(a) certain of the individual offences (in the sense that they can be committed only where a public body is involved) or

(b) the concept of an "agent".[32]

3.35 Of those who responded on this point, about two-thirds were in favour of abandoning the distinction between public and private bodies.

3.36 Those in favour of abandoning the distinction supported our reasoning. They included the CPS who took the view that the law is rendered complex and confusing by the distinction and no persuasive reason in favour of its retention could be found.

3.37 Those who took a different view were generally concerned that a particularly high standard should be expected of those in public life. We have much sympathy for this concern but believe that the appropriate stage at which to mark it is in sentencing. We have not found anything in the minority view which persuades us to abandon or amend our provisional view that the distinction between public and private bodies is unnecessary *unless* it is still needed for the purpose of the presumption of corruption created by section 2 of the 1916 Act. In order to decide whether we can recommend that it should be discarded, we must therefore

---

[30] At paragraph 3.3 above.

[31] Consultation Paper No 145, paras 6.2 – 6.9.

[32] Consultation Paper No 145, para 6.32.

turn to the question of whether the presumption should be retained in its present form. We consider this question in Part IV.

PART IV
THE PRESUMPTION OF CORRUPTION

4.1 Corruption is a difficult crime to prove: it tends by its very nature to be carried out in secret, and its "victims" may never be aware of it. As we have said,[1] in certain circumstances section 2 of the 1916 Act creates a *presumption* of corruption which can simplify the prosecution's task. However, it applies only to cases involving public bodies. In the previous part we concluded that the distinction between public and private bodies is not worth retaining *unless* it is necessary to keep it for the purpose of the presumption. We therefore consider next whether the presumption should be retained in its present form, extended to private bodies or abolished altogether. Only in the first case would it require the preservation of the distinction between public and private bodies.

4.2 Since the presumption was introduced there have been two major developments: first, the United Kingdom has become a signatory to the ECHR; and second, there has been significant change in the law relating to the "right of silence".[2] These developments together raise the question whether the presumption is still justified, or whether it is in breach of the ECHR.

4.3 A third development which will highlight this latter point is the decision of the present Government to incorporate into domestic law the rights and freedoms guaranteed under the ECHR. Under the recently published Human Rights Bill, a Minister of the Crown taking a piece of legislation through Parliament must, before Second Reading, make a statement to the effect that in his or her view the provisions of the new Bill are compatible with the Convention rights set out in that Bill,[3] or make a statement to the effect that, although he or she is unable to make a statement of compatibility, the Government nevertheless wishes the House to proceed with the Bill.[4] We are therefore particularly aware of the need to ensure that our recommendations are compatible with the ECHR. In the light of its significance, we shall deal with this matter in some detail.

4.4 Our first step must be to see whether the presumption is consistent with the ECHR. This entails considering the *combined* effect of the presumption and of those provisions of the Criminal Justice and Public Order Act 1994 ("CJPOA") which enable inferences to be drawn from a defendant's silence. We shall then consider whether the presumption remains necessary in the light of the CJPOA, and revisit the options for reform which we set out in the consultation paper.[5]

---

[1] See paras 1.3 and 3.3 above.

[2] See para 4.31 below.

[3] Under cl 1(1) those rights are defined as being those set out in Articles 2 to 12 and 14 of the ECHR and Articles 1 to 3 of the First Protocol of the ECHR agreed in Paris on 20 March 1952 as read with Articles 16 to 18 of the ECHR. The statement has to be in writing and published in such a manner as the Minister making it considers appropriate (cl 19(2)).

[4] Clause 19(1)(b).

[5] Consultation Paper No 145, paras 11.48 – 11.66.

THE PRESUMPTION

4.5 Section 2 of the 1916 Act provides:

> Where in any proceedings against a person for an offence under the Prevention of Corruption Act 1906, or the Public Bodies Corrupt Practices Act 1889, it is proved that any money, gift, or other consideration[6] has been paid or given to or received by a person in the employment of [Her] Majesty or any other Government Department or a public body by or from a person, or agent of a person, holding or seeking to obtain a contract from [Her] Majesty or any Government Department or public body, the money, gift, or consideration shall be deemed to have been paid or given and received corruptly as such inducement or reward as is mentioned in such Act unless the contrary is proved.[7]

4.6 The prosecution must therefore establish to the usual criminal standard of proof (that is, beyond reasonable doubt)

   (1) that some "money, gift, or other consideration" was paid or given to, or received by, an employee of Her Majesty, a Government department or a public body;[8] and

   (2) that the person providing it (or the person whose agent provided it) was holding, or seeking to obtain, a contract from Her Majesty, a Government department or a public body.

Only when these two requirements are satisfied will the presumption come into effect.

4.7 Section 2 applies even if the charge is brought under the 1906 Act and therefore does not itself require that the payment should have been made to an employee of a public body. But, significantly, it applies only to employees, not members (such as local councillors), and only where a contract is involved rather than the exercise of a discretion (such as a grant of planning permission).[9]

4.8 The effect of the presumption is that

> when the matters in that section have been fulfilled, the burden of proof is lifted from the shoulders of the prosecution and descends on the shoulders of the defence. It then becomes necessary for the defendant to show, on the balance of probabilities, that what was going on was not reception corruptly as inducement or reward. In an appropriate case it is the judge's duty to direct the jury first of all that they must decide whether they are satisfied so they are sure that the

---

[6] The word "consideration" has its legal meaning and connotes the existence of some sort of contract or bargain: see *Braithwaite* [1983] 1 WLR 385, and also *Beaton v H M Advocate* [1993] SCCR 48.

[7] On the balance of probabilities; see *Carr-Briant* [1943] KB 607; *Dunbar* [1958] 1 QB 1; *Hudson* [1966] 1 QB 448.

[8] As defined by the 1889 Act, s 7, and the 1916 Act, s 4(2).

[9] *Dickinson* (1948) 33 Cr App R 5.

defendant received money or a gift or consideration, and then to go on to direct them that if they are so satisfied, then under section 2 of the 1916 Act the burden of proof shifts.[10]

4.9 The presumption can be rebutted only by evidence of an innocent explanation, established on the balance of probabilities, and not merely by the unsupported assertion of the defendant that such an explanation exists.[11]

HISTORICAL REASONS FOR THE CREATION OF THE PRESUMPTION

4.10 Because of the general rule that the burden of proof in a criminal trial is on the prosecution, Parliament and the courts have always been rightly wary about placing a burden of proof on the defence: such burdens can often make the difference between a conviction and an acquittal. We must consider why Parliament thought such a burden to be appropriate in the case of corruption, albeit only in the circumstances described in section 2.

4.11 The 1916 Act was passed in the wake of scandals regarding the Clothing Department of the War Office, which involved the taking of bribes by viewers and inspectors of merchandise. It was presented to Parliament as an emergency wartime measure to deal with the burgeoning number of large government contracts and the resulting opportunities for corruption. Corruption in relation to these wartime contracts was viewed at the time as being particularly serious.

4.12 The immediate cause for the enactment of these provisions was the criticism made by a judge[12] who had presided over two cases[13] of corruption in quick succession. In the first case, the judge had considered it impossible to prosecute a civil servant found to be in possession of banknotes which had been traced to a contractor with whom he had official dealings. This was because the prosecution was unable to prove *why* the money was paid. It was as a result of the particular circumstances of this case that section 2 addressed only the issue of transactions involving contracts, and extended *only* to employees. As the Salmon Commission remarked, "These anomalies doubtless reflect the haste with which the legislation was prepared."[14]

4.13 The reasons given in Parliament for the introduction of the presumption were that it was necessary "to remedy an obvious defect in the law"[15] and would cause no injustice to an accused, as it should be easy for the accused to discharge the burden of proof if he or she was indeed innocent.[16] The Lord Chancellor, Lord Buckmaster, said in the House of Lords:

---

[10] *Braithwaite* [1983] 1 WLR 385, 389, *per* Lord Lane CJ.

[11] *Mills* (1978) 68 Cr App R 154.

[12] Low J.

[13] *Asseling, The Times* 10 September 1916; *Montague, The Times* 18 September 1916.

[14] Salmon Report, para 59.

[15] Hansard (HC) 31 October 1916, vol 86, col 1635 (the Home Secretary, Mr H Samuel).

[16] Hansard (HL) 23 November 1916, vol 23, col 654 (the Lord Chancellor, Lord Buckmaster). Similarly the Home Secretary said:

> I feel satisfied that you will agree with me in thinking that, short of high treason, it is almost impossible to imagine an offence more grave than to corrupt one of these public servants and cause the neglect of his duty.[17]

4.14 Attempts were made to widen the ambit of the Bill, but these were thwarted by a Government preoccupied by war and bent on filling a perceived loophole, rather than reconsidering the law of corruption as a whole. This was admitted by the Government:

> If the times were opportune for a meticulous reconsideration of the whole law relating to corruption, I think the long series of amendments which stand [presented] would deserve the attention of the House. But this is by general agreement a war measure which is rendered necessary by the very large number of Government contracts which have come into existence and by the scandals which were exposed in a recent criminal trial.[18]

JUSTIFICATIONS FOR THE PRESUMPTION

4.15 Both the Redcliffe-Maud Committee and the Salmon Commission thought that the presumption was not only *justified* but should be *extended*.[19]

The Redcliffe-Maud Committee

4.16 The task of the Redcliffe-Maud Committee, appointed in 1973, was to examine local government law and practice following widespread disquiet about the standards of conduct in local government.[20] The Committee concluded that although the standards of conduct were "generally high",[21] it was right to be concerned that corruption exists, bearing in mind the fact that "unless corruption is stopped, it spreads".[22] It made a number of recommendations aimed at safeguarding standards in local government, some of which involved reform of the law. Taking the view that "the need to deal firmly with corruption ... justifie[d] strengthening the provisions of the Prevention of Corruption Acts",[23] the Committee recommended that section 2 of the 1916 Act should be extended to include not only contracts but "other exercises of discretion such as the grant of

---

> I am sure this House will agree that it is both reasonable and equitable to put the burden of proof on the person charged. If the payment was innocently made ... it would be easy to prove in Court ... and there would be no risk of innocent men being unjustly convicted.

Hansard (HC) 31 October 1916, vol 86, col 1636.

[17] *Hansard* (HL) 23 November 1916, vol 23, col 653.

[18] *Hansard* (HC) 9 November 1916, vol 87, col 467 (the Attorney-General, Sir F E Smith QC, MP).

[19] For the possibility of extending it, see paras 4.52 – 4.54 below.

[20] See para 1.11 above.

[21] Redcliffe-Maud Report, para 15.

[22] Redcliffe-Maud Report, para 15.

[23] Redcliffe-Maud Report, para 161.

planning permission or the allocation of a council house; and ... elected members as well as employees of local authorities".[24]

The Salmon Commission

4.17 The Salmon Commission, appointed only a few months after the publication of the Redcliffe-Maud Report, had wider terms of reference than the Redcliffe-Maud Committee. It was charged with enquiring into

> standards of conduct in central and local government and other public bodies in the United Kingdom in relation to the problems of conflict of interest and the risk of corruption involving favourable treatment from a public body.[25]

4.18 Whilst acknowledging that placing the burden of proof on the defence could be justified only for "compelling reasons",[26] the Commission concluded that, as regards corruption, there were such reasons. It was argued that, without the presumption, corruption would be very difficult to prove because bribes were seldom paid in the presence of witnesses;[27] there was often little documentary evidence because those involved in a corrupt transaction would inevitably tend to act secretly;[28] and if there were an innocent explanation for the receipt of a gift by a public servant, it would be easy for the recipient to furnish it.[29]

4.19 The Salmon Commission was "satisfied that a burden of proof on the defence is in the public interest and causes no injustice".[30] Agreeing with the recommendations of the Redcliffe-Maud Committee, it recommended that, for consistency, the presumption should be extended to include members of public bodies as well as employees, and to all official transactions, whether or not they involved contracts.[31]

THE PRESUMPTION AND THE EUROPEAN CONVENTION ON HUMAN RIGHTS

4.20 Both the Redcliffe-Maud Committee and the Salmon Commission recommended an extension of the scope of the presumption. Since the time that those recommendations were made, however, the ECHR has gained in prominence, with legislation incorporating the ECHR into domestic law presently before

---

[24] Redcliffe-Maud Report, para 162.

[25] Salmon Report, para 6.

[26] Salmon Report, para 60.

[27] Any witnesses that are available are likely to be accomplices and unwilling to co-operate.

[28] "Corruption in the public service is a grave social event which is difficult to detect, for those who take part in it will be at pains to cover their tracks": *Public Prosecutor v Yuvaraj* [1970] AC 913, 922E-F, *per* Diplock LJ. This was an appeal from the Federal Court of Malaysia to the Privy Council, and Diplock LJ's comments were made in respect of s 4 of the Prevention of Corruption Act 1961, the equivalent of s 2 of the 1916 Act.

[29] Salmon Report, para 60.

[30] Salmon Report, para 61.

[31] Salmon Report, para 62.

Parliament.[32] We therefore turn now to the relationship between the presumption and the right to a fair trial (including the presumption of innocence) guaranteed under the ECHR, taking into account the implications of the CJPOA.

Does the presumption infringe the ECHR?

4.21 As a State party to the ECHR, the United Kingdom has an obligation in international law to conform its domestic law to the Convention's requirements.[33] United Kingdom citizens have an individual right of petition[34] to the Strasbourg Commission and from there, if their petition is declared admissible, to the Strasbourg Court.[35] Any law reform proposals which we consider must be assessed in the light of the ECHR.

The presumption of innocence

4.22 Our starting point is Article 6 of the ECHR, which guarantees the right to a fair trial. Article 6(2) provides:

> Everyone charged with a criminal offence shall be presumed innocent until proved guilty according to law.[36]

4.23 The presumption of innocence is, like the rights of the defence contained in Article 6(3),[37] a component of the *general* right to a fair trial.[38] It applies only to

---

[32] See para 4.3 above.

[33] "The contracting Parties have undertaken ... to ensure that their domestic legislation is compatible with the Convention and, if need be, to make any necessary adjustments to this end.": European Commission of Human Rights, *Yearbook*, vol 2, 234. In *Ireland v UK* (1978) 2 EHRR 25 the Court said, at p 103 (para 239): "By substituting the words 'shall secure' for the words 'undertake to secure' in the text of Article 1, the drafters of the Convention also intended to make it clear that the rights and freedoms set out in Section 1 would be directly secured to anyone within the jurisdiction of a contracting State."

[34] Under Article 25 of the ECHR. Article 26 of the ECHR states that the Commission will only deal with a matter after all domestic remedies have been exhausted. These rights will remain intact following the changes on 1 November 1998, when the formal structure of the ECHR institutions will change. An applicant will have an individual right of petition to a Court Committee which will consider the admissibility of the case. If this is successful, the case will go before a Chamber of the Court for a decision on both admissibility and merits.

[35] Under Article 45 of the ECHR, the Strasbourg Court has jurisdiction over all cases concerning "the interpretation and application" of the ECHR. A similar procedure will be adopted after 1 November 1998 in relation to cases going before the full Grand Chamber of the Court.

[36] The presumption of innocence can also be found in Article 11 of the Universal Declaration of Human Rights.

[37] Article 6(3) provides:

Everyone charged with a criminal offence has the following minimum rights:

    (a)   to be informed promptly, in a language which he understands and in detail, of the nature and cause of the accusation against him;

    (b)   to have adequate time and facilities for the preparation of his defence;

    (c)   to defend himself in person or through legal assistance of his ownchoosing or, if he has not sufficient means to pay for legal assistance, to be given it free when the interests of justice so require;

criminal proceedings. The Strasbourg Court has recently reaffirmed the principle that Article 6(2), like the other elements of the ECHR, must be interpreted in such a way as to guarantee rights which are practical and effective, rather than theoretical and illusory.[39]

4.24 In *X v United Kingdom*[40] the Strasbourg Commission had to decide whether reverse onus clauses of the kind found in section 2 of the 1916 Act violate Article 6(2). The case involved a United Kingdom applicant who had been convicted of the offence of living on the immoral earnings of prostitutes. The legislation creating this offence contains a reverse onus provision to the effect that a person who is proved to be living with a prostitute, or who is, at the material time, proved to be in the company of a prostitute, or who is proved to have exercised control over the movements of a prostitute, so as to assist or encourage the prostitution, is presumed to have committed the offence of living on the earnings of prostitution *unless the contrary is proved.*[41]

4.25 The Commission held[42] that a reverse onus clause of this kind will not violate Article 6(2) if it creates only a *rebuttable* presumption of fact, which the defence may disprove,[43] and is not unreasonable. However, the Commission went on to recognise that

> this form of provision could, if widely or unreasonably worded, have the same effect as a presumption of guilt. It is not, therefore, sufficient to examine only the form in which the presumption is drafted. It is necessary to examine its substance and its effect.[44]

(d) to examine or have examined witnesses against him and to obtain the attendance and examination of witnesses on his behalf under the same conditions as witnesses against him;

(e) to have the free assistance of an interpreter if he cannot understand or speak the language used in court.

[38] See, inter alia, *De Weer v Belgium* (1979-80) 2 EHRR 439; *Allen de Ribemont v France* (1995) 20 EHRR 557, 574 (para 35).

[39] *Allen de Ribemont v France* (1995) 20 EHRR 557, 574 (para 35).

[40] App 5124/71, (1972) 42 Collection of Decisions 135.

[41] Sexual Offences Act 1956, s 30(2).

[42] After this finding the case did not proceed to the Strasbourg Court. A case cannot be heard by the Court without having been forwarded by the Commission, which had refused the application in this case.

[43] In *X v UK* App 5124/71, (1972) 42 Collection of Decisions 135, the Commission stated:

> The Commission has, *ex officio* examined this complaint under Art 6(2) of the Convention ... The Commission has also studied the statutory provision under which the applicant was convicted. This statutory provision states that, when certain facts are proved by the prosecution, certain other facts shall be presumed. This creates a rebuttable presumption of fact which the defence may, in turn, disprove. The provision in question is not, therefore, as such, a presumption of guilt.

[44] App 5124/71, (1972) 42 Collection of Decisions 135.

4.26 In *Salabiaku v France*[45] the Strasbourg Court[46] affirmed the approach taken by the Commission in *X v United Kingdom*. It emphasised the importance of confining reverse onus clauses within "reasonable limits which take into account ... what is at stake and maintain the rights of the defence".[47]

4.27 The Commission in *X v United Kingdom* also stated, as an alternative ground for its opinion, that it would be extremely difficult for the prosecution to obtain evidence capable of satisfying the criminal standard of proof on the question of whether the applicant was living on immoral earnings.[48] Thus a reverse onus clause may be justified if it relates to matters which are difficult for the prosecution to prove because they are peculiarly within the defendant's own knowledge,[49] provided that it creates only a rebuttable presumption of fact and is restrictively worded.[50]

4.28 One commentator has summarised the issue as follows:

> The question seems to be one of whether there is an alternative route of investigation open to the government. If the answer is negative, the presumption would be reasonable. ... [C]orruption can be such an elusive crime that it is next to impossible for the Crown to adduce evidence with respect to the corrupt source of the moneys involved. ...

---

[45] (1988) 13 EHRR 379.

[46] The Court is a superior tribunal to the Commission, which is responsible for vetting applications and making references. If the Commission, having accepted an application, fails to reach a friendly settlement, it prepares a report stating its opinion as to whether there has been a breach of the Convention. This report goes to the Committee of Ministers and the case may then be brought before the Court within three months. See J G Merrills, *The Development of International Law by the European Court of Human Rights* (2nd ed 1993) pp 2–5.

[47] (1988) 13 EHRR 379, 388 (para 28). See also *Hoang v France* (1992) 16 EHRR 53 and *Bullock v United Kingdom* (1996) 21 EHRR CD 85. In the latter case, a statement of the law included the following passage:

> The Commission also recalls that the Court, in the *Salabiaku* judgement and more recently in the *Pham Hoang* judgment, stated that the Convention does not prohibit presumptions of fact or law in principle, but does require Contracting States to remain within certain reasonable limits as regards criminal law which limits take into account the importance of what is at stake and maintain the rights of the defence.

[48] App 5124/71, (1972) 42 Collection of Decisions 135.

[49] See the Salmon Report, para 60: "It is difficult enough to prove the passing of a gift to a public servant from an interested party but, when it occurs, it is normally strong *prima facie* evidence of corruption. If there is an innocent explanation it should be easy for the giver and the recipient of the gift to furnish it; the facts relating to the gift are peculiarly within their own personal knowledge."

[50] The Court's assessment of the reasonableness of a clause may, it is suggested, be affected by the standard of proof which is borne by a defendant. The fact that, in this jurisdiction, the defendant who bears a legal burden will have only to satisfy the court on the balance of probabilities may suggest that a reverse onus clause is reasonable. It should be noted, however, that neither the Commission (in *X v UK*), nor the Court (in *Salabiaku*) considered the standard of proof issue.

Thus, it seems that the non-existence of a viable alternative would make that presumption of fact not unreasonable.[51]

4.29 It seems that by itself section 2 of the 1916 Act would satisfy this criterion in the same way that the statutory provision considered in *X v United Kingdom* did. Support for this view may be taken from the fact that (to the best of our knowledge) no reference has been made to the Strasbourg Court in relation to section 2. We also note the comments of Lord Woolf in the recent case of *Attorney-General of Hong Kong v Lee Kwong-Kut*,[52] where the Privy Council was asked to consider the compatibility of a reverse onus clause with article 11(1) of the Hong Kong Bill of Rights, which guarantees the right to be presumed innocent until proved guilty. Lord Woolf, giving judgment, considered *Salabiaku v France* and then said:

> There are situations where it is clearly sensible and reasonable that deviations should be allowed from the strict applications of the principle that the prosecution must prove the defendant's guilt beyond reasonable doubt. ...
>
> Some exceptions will be justifiable, others will not. Whether they are justifiable will in the end depend upon whether it remains primarily the responsibility of the prosecution to prove the guilt of an accused to the required standard and whether the exception is reasonably imposed, notwithstanding the importance of maintaining the principle which article 11(1) enshrines. ... If the exception requires certain matters to be presumed until the contrary is shown, then it will be difficult to justify that presumption unless, as was pointed out by the United States Supreme Court in *Leary v United States* (1969) 23 L Ed 2d 57, 82, "it can at least be said with substantial assurance that the presumed fact is more likely than not to flow from the proved fact on which it is made to depend".[53]

4.30 The position may, however, be affected by a subsequent significant development to be found in the provisions of the CJPOA, which we now consider.

Possible implications of the CJPOA

4.31 A person suspected or accused of a criminal offence has traditionally been accorded a "right of silence", which means that there is no obligation to answer questions, either before or during the trial. At common law this right has been supplemented by a further rule that the fact-finders may not be invited to draw any adverse inference from the failure of a defendant to assist the police or to give evidence at trial. Under the CJPOA the accused remains free to maintain silence under interrogation and at trial;[54] but the supplementary rule against the drawing

---

[51] Lyy Ma, "Corruption Offences in Hong Kong: Reverse Onus Clauses and the Bill of Rights" (1991) 21 HKLJ 289, 328–329.

[52] [1993] AC 951.

[53] [1993] AC 953, 969–970.

[54] See *Cowan* [1996] QB 373, 378, *per* Lord Taylor of Gosforth CJ: "It should be made clear that the right of silence remains."

of adverse *inferences* has been abolished, and the Act sets out the circumstances in which "proper" inferences may be drawn.

4.32 Under section 34 of the CJPOA,[55] where the defendant relies on a fact in his or her defence at trial which was not put forward when he or she was questioned or charged, or informed that he or she might be prosecuted, the court or jury "may draw such inferences … as appear proper" in determining whether there is a case to answer or whether the defendant is guilty of the offence charged.

4.33 Potentially more important in relation to the presumption is section 35, which allows the court or the jury to "draw such inferences as appear proper" from a defendant's failure to testify or to answer any particular question.[56] In *Cowan*[57] the Court of Appeal rejected an argument that section 35 should be permitted to operate in exceptional cases only, but stressed that silence cannot be the *only* factor on which a conviction is based,[58] and that the jury must be satisfied that the prosecution has established a case to answer before they can draw inferences from silence. The court took this to mean not only that the case should be fit to be left to the jury, but also that the judge should make clear to the jury that they must be convinced of the existence of a prima facie case before drawing an *adverse* inference from silence. It would require "some evidential basis … or some exceptional factors in the case"[59] for a judge to advise a jury *against* drawing an inference from silence.

4.34 These developments raise the question whether the effect of sections 34 and 35 is to render unnecessary the presumption created by section 2 of the 1916 Act. If so, the Strasbourg Court[60] might take the view that the burden imposed by the presumption cannot be justified under Article 6 of the ECHR. Predicting the decisions of the Strasbourg Court, however, is difficult, given "its disparate composition and the fluid nature of some of its jurisprudence".[61] It is also important to bear in mind that "the court does not see its function as being to judge of the constitutionality of a signatory state's legislation",[62] but rather to

---

[55] Section 34 is set out in Appendix B below.

[56] Section 35 is set out in Appendix B below.

[57] [1996] QB 373.

[58] CJPOA, s 38(3).

[59] See *Cowan* [1996] QB 373, 380.

[60] In Evidence in Criminal Proceedings: Hearsay and Related Topics (1997) Law Com No 245, at para 5.2, we noted the difficulty in attempting to judge the compatibility of domestic legislation with the Convention: "because the Strasbourg Court aims to interpret the Convention as a 'living, developing document', the doctrine of precedent weighs less heavily with the Strasbourg Court than it does in English law".

[61] R Munday, "Inferences from Silence and the European Human Rights Law" [1996] Crim LR 370.

[62] *Ibid*, p 371, where the author cites *Klass v Germany* (A/28) (1978) 2 EHRR 214, para 33. In that case, the Strasbourg Court said:

> Article 25 does not institute for individuals a kind of *actio popularis* for the interpretation of the Convention; it does not permit individuals to complain against a law *in abstracto* simply because they feel that it contravenes the

consider the effect of the legislation being challenged in the particular circumstances of the case before it.

4.35 In relation to sections 34 and 35, some indication of the view likely to be taken was made apparent in *Murray v United Kingdom*[63] (a decision concerning equivalent provisions contained in the Criminal Evidence (Northern Ireland) Order 1988), where the Court stated:

> Whether the drawing of adverse inferences from an accused's silence infringes Article 6 is a matter to be determined in light of all the circumstances of the case, having particular regard to the situations where inferences may be drawn, the weight attached to them by national courts in their assessment of the evidence and the degree of compulsion inherent in the situation.[64]

4.36 It is difficult to be sure whether, in the light of the CJPOA, the section 2 presumption is likely to be regarded as a breach of Article 6. The most important consideration, it would seem, is how much more difficult it would be to prove corruption if the presumption did not exist. Clearly the increased difficulty that prosecutors would face is now less than it would have been before the enactment of the CJPOA, since there is now greater pressure on the defence to offer an explanation where prima facie evidence of an offence is adduced. The question is now whether the presumption is still necessary in spite of these new provisions.

The effects of the presumption and the CJPOA compared

4.37 In the consultation paper, we distinguished two issues. First, how much harder would it be to establish *a case to answer* on a corruption charge if the presumption were abolished? And second, once it has been established that there is a case to answer, how much harder would it be to secure a *conviction*? In the second case, we have to consider separately the defendant who adduces no evidence at all, the defendant who does not testify in person but adduces other evidence, and the defendant who does testify. In each case we shall assume that the prosecution has proved the facts necessary to trigger the presumption, namely that some "money, gift, or other consideration" was paid or given to, or received by, an employee of

---

> Convention ... [I]t is necessary that the law should have been applied to (their) detriment.

See also *McCann v UK* (A/324) (1996) 21 EHRR 97, para 153.

[63] (1996) 22 EHRR 29, para 47.

[64] Although in *Murray* the CJPOA silence provisions survived a challenge in the Strasbourg Court, there is no certainty about the outcome of any future challenges. Despite the Court recognising that in this particular case the prosecution case was very strong indeed, nonetheless five of the 19 judges dissented. One commentator concluded:

> The truth is that no-one can really tell whither the ECHR's peculiar collective mind will move. The only certainty is that the court has a sense of its own destiny that impels it to expand its repertoire of human rights concepts with bewildering rapidity. To the extent that it can be teased out of the judgments in *Murray v United Kingdom*, the ECHR's response to the Northern Ireland Order is one of only qualified approval. (R Munday, *op cit*, p 385. Footnotes omitted.)

Her Majesty, a Government department or a public body,[65] and that the person providing it (or the person whose agent provided it) was holding, or seeking to obtain, a contract from Her Majesty, a Government department or a public body. If these "basic facts" are not proved, the presumption has no application in any event. We believe that it is worthwhile to repeat the exercise.

ESTABLISHING A CASE TO ANSWER

The existing law

4.38 At present, the benefit received by the public servant is deemed to have been a corrupt inducement or reward unless the contrary is proved on the balance of probabilities. It follows that under the existing law there would inevitably be a case to answer once the basic facts are proved, because in the event of the defence tendering no evidence the presumption would not have been rebutted.

If section 2 were repealed

4.39 In deciding whether the prosecution have established a case to answer, the court will follow the principles laid down in *Galbraith*:[66]

> (1) If there is no evidence that the crime alleged has been committed by the defendant, there is no difficulty. The judge will of course stop the case. (2) The difficulty arises where there is some evidence but it is of a tenuous character, for example because of inherent weakness or vagueness or because it is inconsistent with other evidence. (a) Where the judge comes to the conclusion that the prosecution evidence, taken at its highest, is such that a jury properly directed could not properly convict upon it, it is his duty, upon a submission being made, to stop the case. (b) Where however the prosecution evidence is such that its strength or weakness depends on the view to be taken of a witness's reliability, or other matters which are generally speaking within the province of the jury and where on one possible view of the facts there *is* evidence upon which a jury could properly come to the conclusion that the defendant is guilty, then the judge should allow the matter to be tried by the jury.

4.40 Whether a case to answer can be established in the absence of the presumption depends on whether the basic facts amount to a prima facie case. Proof of the basic facts is not *in itself* proof that the transaction involved a corrupt inducement or reward. However, it seems likely that a prima facie case would be held to exist where the basic facts are proved, given that corporate largesse is very much less prevalent in the public sector. This factor, coupled with other evidence which the prosecution can ordinarily be expected to adduce (such as the fact that the agent acted without his or her employer's consent), will often be sufficient to enable the case to be left to the jury.

4.41 Moreover, by virtue of section 34 of the CJPOA, if a defendant exercises the right to silence in the face of police questioning and fails to mention any fact relied

---

[65] As defined by the 1889 Act, s 7 and the 1916 Act, s 4(2).

[66] [1981] 1 WLR 1039, 1042, *per* Lord Lane CJ.

upon in his or her defence, the court may draw such inferences as appear proper in determining whether there is a case to answer.[67]

4.42 We believe that the repeal of section 2 would probably not, in practice, make it significantly more difficult for the prosecution to establish a case to answer.

WHERE THE DEFENDANT ADDUCES NO EVIDENCE

The existing law

4.43 If the basic facts are proved and the defendant adduces no evidence, a conviction should in theory be inevitable: the presumption will not have been rebutted.

If section 2 were repealed

4.44 If the presumption were abolished, it would be for the jury to determine whether the basic facts amounted to proof of corruption beyond reasonable doubt. Under section 35 of the CJPOA, however, the jury could take into account any inference that they thought it proper to draw from the defendant's failure to testify. This might amount to an inference of guilt. In *Murray v DPP*,[68] a House of Lords decision dealing with the analogous Northern Ireland provisions, Lord Slynn explained:

> [I]f aspects of the evidence taken alone or in combination with other facts clearly call for an explanation which the accused ought to be in a position to give, if an explanation exists, then a failure to give any explanation may as a matter of common sense allow the drawing of an inference that there is no explanation and that the accused is guilty.[69]

In practice we think it likely that a conviction would result.

WHERE THE DEFENDANT DOES NOT TESTIFY BUT ADDUCES OTHER EVIDENCE

The existing law

4.45 If the basic facts are proved, and the defendant does not testify but adduces other evidence, the position at present is that a conviction should result unless the evidence adduced shows, on the balance of probabilities, that the transaction was not corrupt. Moreover, in deciding whether the defence has proved this, the jury would be entitled under section 35 of the CJPOA to draw appropriate inferences from the defendant's failure to testify.

---

[67] Although specific provision is made for this circumstance in the CJPOA, s 34(2)(c), it is suggested in *Archbold* at para 15–404 that it is a provision which will be rarely used: "it is difficult to see how reliance upon a fact in the accused's defence can arise before he testifies or calls evidence".

[68] [1994] 1 WLR 1. This is the same case that was eventually taken to the Strasbourg Court under the name *Murray v UK*, see para 4.35 above.

[69] [1994] 1 WLR 1, 11G.

If section 2 were repealed

4.46 If the presumption did not exist, an acquittal should result not only (as at present) if the evidence adduced were such as to show, on the balance of probabilities, that the transaction was not corrupt, but also if it appeared that the transaction probably was corrupt but the evidence was sufficient to raise a reasonable doubt. The jury's verdict would depend on the cogency of the evidence adduced, plus any inferences that the jury thought it proper to draw from the defendant's failure to testify. Obviously, in some cases the adverse inference drawn would be sufficiently strong (and the evidence adduced by the defendant sufficiently weak) as to lead to a conviction. In other cases, the evidence adduced by the defence might be such as to persuade the jury that the Crown had not made out its case beyond reasonable doubt – notwithstanding any inferences that they might choose to draw from the defendant's failure to testify.

WHERE THE DEFENDANT TESTIFIES

The existing law

4.47 Under the present law, a conviction should result unless the evidence adduced by the defence (including the defendant's own evidence) shows, on the balance of probabilities, that the transaction was not corrupt. Section 35 of the CJPOA has no application in this case unless the defendant refuses, without good cause, to answer a question; it will normally therefore be somewhat easier to discharge this burden than if the defendant did not testify. However, appropriate inferences might be drawn under section 34 of the CJPOA if the defendant raises matters in evidence which had not previously been mentioned and which the defendant might reasonably have been expected to mention.

If section 2 were repealed

4.48 In the absence of the presumption, it would be the jury's duty to acquit unless the prosecution had proved corruption beyond reasonable doubt. The jury would have to evaluate the cogency and relevance of the evidence adduced by the defendant and consider whether it was sufficient to raise a reasonable doubt. For this purpose, as at present, they could take account of the defendant's refusal, without good cause, to answer any question,[70] or of the raising in evidence of matters that the defendant might reasonably have been expected to mention previously.[71]

OPTIONS FOR REFORM, OUR PROVISIONAL PROPOSAL AND THE RESPONSE ON CONSULTATION

4.49 In the consultation paper, we pointed out that a consequence of the provisions of sections 34 and 35 of the CJPOA was greatly to reduce the need for a presumption. Before those provisions were enacted, neither the prosecution nor the judge could invite the jury to draw any inferences from the failure of a defendant to give evidence, to answer a particular question, or to mention a

---

[70] CJPOA, s 35(3).

[71] CJPOA, s 34(2).

relevant fact when questioned or charged. We explained in the consultation paper[72] that, although it is difficult to predict the approach of the Strasbourg Court, we thought there were grounds for believing, especially in the light of sections 34 and 35 of the CJPOA, that it might regard the presumption as going further than was necessary. The consequence of that view, we pointed out, was that the Court might well find that Article 6 of the ECHR had been contravened.

4.50 We took the view that, given the seriousness of reverse onus provisions and the need to reconcile them with Article 6(2) of the ECHR, it was for those who advocated the retention of the presumption to justify it. We bore in mind the limited scope of section 2 – a reflection of the wartime conditions in which it was enacted – and challenged the view that the evidential strains of proving corruption were so much greater than those involved in proving other offences (such as conspiracy) as to justify reversing the onus of proof. We concluded that, even if proving corruption were so difficult as to require some sort of assistance to the prosecution, such assistance is now provided by the CJPOA. We provisionally concluded, therefore, that the presumption should be abolished.

4.51 However, we also put forward three other options for comment – namely, the retention and extension of the presumption, the "Hong Kong option", and a reduction in the weight of the burden placed on the defendant. We now revisit these options in the light of the views expressed on consultation, before returning to our provisional proposal that the presumption should be abolished altogether.

Option 1: extend the presumption

4.52 It seems to be largely historical accident that the presumption covers some situations and not others.[73] In the Salmon Report it is suggested that "if a burden on the defence is to be retained … it needs to be extended to ensure that it applies consistently".[74] Whereas the Salmon Commission's recommendation was intended to ensure consistency *in the public sector*, our concern ranges over *all* corrupt transactions, in the public and private sectors alike. We suggested a number of different ways in which the presumption might be extended.[75]

(a) Include *all* transactions in which the putative briber seeks a benefit, not only those involving contracts. Both the Salmon Commission[76] and the Redcliffe-Maud Committee[77] recommended that the presumption should

---

[72] Consultation Paper No 145, para 11.44.

[73] See paras 4.10 – 4.14 above.

[74] Salmon Report, para 62. Because the scope of the Salmon Commission's inquiry was limited to the public sector, this call for consistency was perhaps less radical than at first appears. Whereas justifying a distinction between contracts and discretions might prove difficult, recommending the extension of the presumption to the private sector would inevitably be controversial.

[75] Consultation Paper No 145, para 11.49.

[76] Salmon Report, para 62.

[77] Redcliffe-Maud Report, para 162.

(b) Include any *member or agent* (as well as employees) of the Crown, a Government department or a public body who accepts any gift or consideration from a person seeking any benefit from such a body. This extension was recommended by the Redcliffe-Maud Committee[78] and supported by the Salmon Commission.[79]

(c) Include any agent, whether acting for a public body *or on behalf of a private principal*. This option would have the advantage that it would remove the need to distinguish between public bodies and others. It would also recognise that, as a result of the growing number of "private" bodies performing "public" functions, the number of bodies to which the presumption applies is decreasing for reasons which cannot be justified in principle.

(d) Include any gift or consideration received in connection with the employee's public functions, whether or not made during the period of his or her employment. At present, the presumption applies only to a person who is an employee of a public body at the time when a gift or consideration is received. It is possible to envisage a situation in which a reward for action taken during the currency of a person's employment is given after that employment has terminated.[80]

(e) Apply the presumption to *conspiracy* to commit offences of corruption, as well as the substantive offences. It appears that a charge of conspiracy under section 1 of the Criminal Law Act 1977 to commit offences under the Prevention of Corruption Acts will not attract the presumption, since section 2 applies only to proceedings against a person for an offence under the 1889 Act or the 1906 Act.[81]

4.53 In view of our provisional conclusion that there were no adequate reasons to retain the presumption at all, we did not consider the relative merits of the various ways of extending it.

4.54 This option received a small amount of support. ACPO, in its collective response,[82] argued that, within the range of offences which deal with non-violent

---

[78] Redcliffe-Maud Report, vol 1, recommendation 26(ii)(b).

[79] Salmon Report, para 62.

[80] As to whether the present law covers former agents, see paras 5.10 – 5.11 below.

[81] Support for this argument may be found in the reasoning in *McGowan* [1990] Crim LR 399. Section 28 of the Misuse of Drugs Act 1971 normally has the effect of reversing the usual burden of proof so that it falls upon defendants to prove their lack of knowledge or suspicion that articles found under their control were controlled drugs. However, since s 28(1) listed the offences to which the section applied and conspiracy was not one of them, this burden would not apply to counts of conspiracy. Accordingly, the burden remained upon the prosecution throughout.

[82] Individual contributors to the ACPO response held a variety of views, some agreeing that the presumption should be retained and extended and others arguing that it should either be retained in its present form or abolished.

behaviour, there can be no more serious offence than corruption as "it erodes and finally destroys public confidence in individuals and institutions", and concluded that "corruption is an exceptional offence and requires exceptional measures to deal with it". A similar approach was adopted by the Metropolitan Police Company Fraud Department.

Option 2: the Hong Kong option

4.55 Under section 10(1) of the Hong Kong Prevention of Bribery Ordinance 1970, a person who is or has been a Crown servant is guilty of an offence if he or she maintains a standard of living above that which is commensurate with his or her present or past official emoluments, or is in control of pecuniary resources or property disproportionate to those emoluments, and is unable (in either case) to give a satisfactory explanation to the court. The Hong Kong Court of Appeal has held that the value of section 10 in the fight against serious corruption is well established, and that the reversal of the normal burden of proof is justified because the primary facts on which the defendant's explanation would be based would be matters peculiarly within his or her own knowledge.[83]

4.56 In the consultation paper we doubted whether an English equivalent to the Hong Kong provision would survive a challenge before the Strasbourg Court on the ground that it was unreasonable or out of proportion to the mischief prevented. We therefore rejected this option.

4.57 On consultation, very few respondents disagreed with us. Of those who did, Mr R White, the Assistant Chief Constable of the Royal Ulster Constabulary, favoured this option on the grounds that "experience suggests that there are cases in which a public servant is living beyond his apparent means and suspected of corruption but in the absence of evidence of corruption nothing can be done." The Institute of Legal Executives thought it justified because "corruption strikes at the very heart of society itself".

4.58 Among those who agreed that the Hong Kong option should not be adopted was the SFO, which was not convinced that there is presently a need for such a provision. We attach great importance to the SFO's opinion in view of its experience of corruption cases. No arguments were put forward to undermine our view that the Hong Kong approach would fall foul of the ECHR. We therefore reject this option.

Option 3: reduce the weight of the burden imposed by the presumption

4.59 Under this option, the burden on the defendant would be less onerous than the present requirement to prove the absence of corruption, but more so than that imposed by the prospect of comment on a failure to testify and of the drawing of appropriate inferences. The presumption might take a form similar to that raised by the possession of recently stolen goods:[84] on proof or admission of the basic facts, the jury would be directed that those facts call for an explanation, and if

---

[83] *A-G v Hui Kin Hong* [1995] 1 HKCLR 227.

[84] See *Cash* [1985] QB 801.

none is given (or one is given which they are convinced is untrue) they would, in appropriate circumstances, be entitled to convict. Our provisional view was that this compromise offered little advantage, since under the CJPOA the defendant will in practice be obliged to offer an explanation in any event.

4.60 On consultation, only one respondent, Lord Davidson, supported this option. He thought we had underestimated the importance of the presumption, and argued that it was justified "because in broad terms corruption is concerned with agency or something closely analogous to agency [and] a stage should come when the 'agent' has to account for his action".

Option 4: abolish the presumption

4.61 Approximately two-thirds of those who considered the issue of the presumption supported our provisional proposal that it should be abolished. Many different reasons were put forward for this view. We shall now examine in greater detail the arguments for and against it.

Arguments against the presumption

4.62 There are essentially two kinds of argument that can be levelled against the presumption: arguments from principle, and practical considerations.

ARGUMENTS FROM PRINCIPLE

4.63 The effect of the presumption is profound. When it applies, the defendant is required to prove, on the balance of probabilities, that the transaction was not corrupt. It follows that the jury may properly be directed to convict if they are left in doubt.[85] It is possible, therefore, for a defendant to be convicted although the jury think that his or her explanation may well be true, and indeed is just as likely to be true as the prosecution's.

4.64 Professor Sir John Smith expressed the same concern:

> It is wrong in principle that a person should be found guilty of a serious offence and labelled "corrupt" when a jury has found that, as likely as not, he is completely innocent.

4.65 Sheriff Iain McPhail QC agreed.

> I do not rely on sections 34 and 35 of the CJPOA, since there are no equivalent provisions in Scotland. In my view the presumption is objectionable because it requires the jury to convict even if they entertain a reasonable doubt as to whether the money (etc) was given (etc) corruptly. A juror who believes that the chances are 55:45 that the money *was* given corruptly may be satisfied on a balance of probabilities, but not beyond reasonable doubt: yet in that situation, and even in a situation where he regards the probabilities as only 50:50, the law requires him to vote for conviction. If the jury went back to the court and told the judge they were agreed that the

---

[85] *Evans-Jones* (1923) 17 Cr App R 121.

probabilities were only 50:50 and asked for further directions, the judge would be obliged to direct them that, if they could not decide on the evidence whether the defence case that the money was not given corruptly was more probable than not, they should convict. In other words, if the defence evidence leaves them in a state of mind in which they entertain a reasonable doubt as to the existence of a fact which is essential to guilt, namely whether the money was given corruptly, they must nevertheless convict. I would suggest that such a situation, which may arise wherever a legal or persuasive burden of proof is placed upon the accused, can only be justified by weighty considerations. It is possible, for example, to justify the Scottish rule that the defence has the burden of proving insanity or diminished responsibility. My own view is that the reasons advanced in favour of the presumption created by section 2 are not strong enough to justify its retention.

4.66 It is arguable that an agent (especially a public servant)[86] who accepts a personal benefit should be prepared to give an explanation, and that for that reason the presumption is justified. But the presumption cannot be rebutted merely by *offering* an innocent explanation: there must be *evidence* of it,[87] sufficient to prove that it is more likely to be true than the prosecution's explanation.

PRACTICAL CONSIDERATIONS

The shortcomings of the presumption in its present form

4.67 The present presumption applies only to public bodies, and only to benefits received from persons holding or seeking contracts with such bodies. If the presumption were to be retained, it would be necessary to decide whether it should continue to be confined to public bodies or should be extended to the private sector as well. In the former case there would be substantial difficulties in defining a "public body", difficulties compounded by the recent trend towards privatising public bodies and contracting out of work to the private sector. The latter alternative would mean that any employee attending a function held by a company seeking business from his or her employer would, if charged, have to prove that his or her attendance was not corrupt. We do not think this would be acceptable to the business community.

The CJPOA

4.68 Even if the view were taken that the prosecution needs some degree of assistance in proving corruption, there was a measure of support for the argument advanced in the consultation paper that the CJPOA was an adequate safeguard. In their joint response, the Bar Council and the Criminal Bar Association said that

> the establishment of basic facts by the prosecution and the subsequent failure of a defendant to give evidence, in the absence of a cogent explanation at the police interview stage, would almost invariably lead a properly directed, reasonable jury to convict.

---

[86] Bearing in mind the Nolan Committee's Seven Principles of Public Life which include the principles of integrity, accountability, openness and honesty.

[87] *Mills* (1978) 68 Cr App R 154.

4.69 Some respondents, such as the CPS and the Metropolitan Police Company Fraud Department, disputed the importance of the CJPOA. However, while we suspect that it is premature to come to a definite and final view on the effect of the Act, we believe that it is likely to make a realistic difference in easing the task of the prosecution.

4.70 The London Criminal Courts Solicitors' Association also made the point that, under Part I of the Criminal Procedure and Investigations Act 1996, a defendant's failure to make proper disclosure may justify further inferences, similar to those that may arise under the CJPOA. Under section 5, once primary prosecution disclosure has taken place, the accused must give a defence statement to the prosecutor. This will be a written statement which will set out the general nature of the defence, the matters on which the accused takes issue with the prosecution and the reasons for this. If the defence fails to make disclosure, not only will the prosecution not be obliged to make secondary disclosure, but the Act provides for additional sanctions. Under section 11, deficiencies in the defence's disclosure may be commented upon by the court (or, with the court's leave, by a party) and such inferences may be drawn "as appear proper in deciding on the accused's guilt."[88] This clearly strengthens our argument that the presumption is no longer necessary.

Arguments in favour of the presumption

CORRUPTION IS MORE DIFFICULT TO PROVE THAN OTHER OFFENCES

4.71 The Salmon Commission favoured the retention of the presumption because of the peculiar difficulty of proving corruption: it is a "clandestine offence".[89] This perception of corruption, as an offence which because of its intrinsic secrecy is particularly difficult to prove, is widespread. However, some respondents with great experience of criminal trials supported our provisional view that corruption is no harder to prove than some other offences. This was the view of the General Council of the Bar, the Criminal Bar Association and the SFO, which wrote:

> We agree that the case for the retention of the presumption falls, once a case can no longer be made for regarding corruption in the public sector as inherently more difficult to investigate. Corruption of itself is no more or less difficult to investigate than other forms of financial crime.

---

[88] Section 11(3). For further comment on the provisions of this Act, see an article (in three parts) by David Corker, "Maximising Disclosure (1997) 147 NLJ 885, 961, 1063; see also an article by John Sprack, "The Criminal Procedure and Investigations Act 1996: (1) the Duty of Disclosure" [1997] Crim LR 308. Sprack comments at p 312:

> Given that the court "may" comment and/or "may" draw inferences, the question arises when it is right to do so. The CPIA 1996 contains no general guidance on this point, although there is some assistance given as to the approach to be adopted where a defence different from that set out in the defence statement is put forward at trial (section 11(4)). In such a case, the court is told to have regard "to the extent of the difference in the defences and ... to whether there is any justification for it".

[89] Salmon Report, para 61.

4.72 The CPS, though taking the view that it would be preferable for the presumption to be retained and extended, conceded that "private sector corruption is regularly prosecuted without the benefit of the presumption". Furthermore, those prosecuting corruption cases will often rely on conspiracy to defraud or common law bribery, neither of which have the benefit of the presumption. One respondent pointed out that, almost by definition, "fraudsters are plausible liars, and those who engage in corrupt practices are much the same". We would infer that prosecutions for (for example) theft, deception or conspiracy to defraud are likely to pose much the same evidential problems as those for corruption. After considering with care all the responses, we remain unconvinced that corruption is more in need of a reversed burden of proof than other offences are.

STANDARDS OF PROBITY SHOULD BE HIGHER IN THE PUBLIC SECTOR

4.73 It has been argued that the presumption should be retained in relation to the public sector, because the standards of probity that can reasonably be expected of public servants are particularly high. The Redcliffe-Maud Committee emphasised this.

> It is common practice in the commercial world for the transaction of business to be accompanied by the giving of personal gifts or benefits, ranging from the Christmas bottle of whisky to much more elaborate and lavish provision. Public life requires a standard of its own; and those entering public office for the first time must be made aware of this from the outset.[90]

4.74 In the consultation paper we agreed that "other things being equal, corruption on the part of a public servant is likely to be more damaging to the public interest, and therefore a more serious offence, than corruption in the private sector".[91] But, even if we could devise a satisfactory formulation of the distinction between public and private, we do not believe that the presumption is the proper way to reflect this view. Public sector corruption can sometimes be trivial, while private sector corruption can be very serious indeed. More importantly, high standards in the public sector cannot be achieved by adjusting the rules of evidence. The presumption does not require a *higher standard of conduct* in the public sector: the conduct prohibited is the same whether the agent involved is a public servant or not. Rather, the effect of the presumption is that it makes it *easier to prosecute* those accused of corruption if they are working in the public sector.

CONSISTENCY WITH SCOTLAND

4.75 It is arguable that the abolition of the presumption in England and Wales would create an undesirable inconsistency between the law of this jurisdiction and that of Scotland. We agree that in principle the prosecution should, if possible, be in the same position (that is, either with the benefit of the presumption or without) in both jurisdictions. However, the course of a corruption trial is in any event different in England and Wales from that in Scotland. The rules of evidence are

---

[90] Redcliffe-Maud Report, para 76.
[91] Consultation Paper No 145, para 6.20.

quite different, and, crucially, sections 34 and 35 of the CJPOA do not apply in Scotland. We therefore believe that the abolition of the presumption in England and Wales would not cause the glaring discrepancy that might be feared.

OUR RECOMMENDATION

4.76 In the consultation paper, we said that our provisional approach was very similar to that adopted by the MCCOC, who stated:

> The argument for retention [of the presumption] is the difficulty of proving relevant matters which almost invariably will take place in private. On the other hand, it is difficult to accept that it is any more difficult to prove dishonesty in these cases than in theft or fraud, especially if it is established that the principal did not know about and consent to the payment.... MCCOC concludes that there should not be a reverse onus of proof in the secret commissions offences.[92]

4.77 Having considered all the helpful points made to us, we conclude that there is no evidence that it is harder to obtain convictions for those corruption offences to which the presumption does not apply than for those to which it does. We do not feel able to state with certainty that the use of the presumption, together with the provisions of the CJPOA, would contravene the ECHR. However, for the reasons we have given, we take the view that no special reasons exist to justify the continued existence of the presumption with regard to corruption. And, given our conclusion in Part III that the distinction between public bodies and others is unnecessary for any purpose other than the presumption, it follows that the existence of that distinction is also no longer justified.

4.78 **We recommend that the new law of corruption**

(1) **should not include a presumption comparable to that created by section 2 of the 1916 Act (under which a transaction is in certain circumstances presumed to be corrupt unless the contrary is proved), and**

(2) **should not draw a distinction between public bodies and others.**

(Recommendation 2)

---

[92] MCCOC Report, p 307.

PART V
FORMULATING A MODERN LAW OF CORRUPTION

5.1 In this part we consider how a modern law of corruption should be formulated. For the convenience of the reader we begin with an outline of the scheme of offences and definitions that we recommend.[1] We then attempt to analyse the essential character of corruption, and conclude that it involves things done in connection with the performance of an agent's functions as agent.[2] We then discuss who should qualify as an "agent" for this purpose,[3] and how to define the central concept of an "advantage",[4] before setting out the four offences that we recommend.[5] Next we turn to the crucial and difficult question of how (if at all) the concept of doing something "corruptly" should be defined.[6] Our provisional proposals on this matter were strongly contested and, as a result, our recommendations reflect a substantial change in our approach. We then discuss whether the new offences should be treated as, or defined as, offences of dishonesty.[7] Finally we consider, but reject, a number of possible defences.[8]

AN OUTLINE OF THE SCHEME WE RECOMMEND

5.2 Our recommended scheme involves replacing the Prevention of Corruption Acts 1889 to 1916 and the common law of bribery with a modern statute creating four offences, namely:

(1) corruptly conferring, or offering or agreeing to confer, an advantage;[9]

(2) corruptly obtaining, soliciting or agreeing to obtain an advantage;[10]

(3) corrupt performance by an agent of his or her functions as an agent;[11] and

(4) receipt by an agent of a benefit which consists of, or is derived from, an advantage which the agent knows or believes to have been corruptly obtained.[12]

---

[1] Paras 5.2 – 5.3 below.
[2] Paras 5.4 – 5.14 below.
[3] Paras 5.15 – 5.37 below.
[4] Paras 5.38 – 5.43 below.
[5] Paras 5.44 – 5.61 below.
[6] Paras 5.62 – 5.122 below.
[7] Paras 5.123 – 5.134 below.
[8] Paras 5.135 – 5.152 below.
[9] Paras 5.45 – 5.50 below.
[10] Paras 5.51 – 5.53 below.
[11] Paras 5.54 – 5.58 below.
[12] Paras 5.59 – 5.61 below.

5.3 Concepts crucial to these offences are as follows:

(1) the relationship of "agency" (which is given a specially extended meaning for this purpose)[13] and the functions that an agent performs *as* an agent;[14]

(2) "conferring an advantage",[15] and its mirror-image "obtaining an advantage";[16] and

(3) conferring an advantage *corruptly*.[17] Where a person confers an advantage, the question whether he or she does so *corruptly* would, under our recommendations, depend on his or her intention in respect of the conduct of the agent in question, and his or her assessment of the likelihood of the agent acting in the desired way (if at all) *primarily in return for* the advantage conferred.

The concepts of *obtaining* and *soliciting* an advantage corruptly, and corrupt performance of an agent's functions, are defined by reference to the central concept of corruptly *conferring* an advantage.[18] The offence of receiving a benefit derived from an advantage corruptly obtained is in turn defined by reference to the offence of corruptly obtaining an advantage.[19]

THE ESSENTIAL CHARACTER OF CORRUPTION

5.4 In the consultation paper we tried to analyse the essential character of corruption. We took as the paradigm of corrupt inducement a situation involving three parties – A, the briber; B, the recipient of the bribe (the "bribee"), and C, B's principal. The purpose of A's bribing B is, we suggested, to cause B to act in A's interest, which in turn is likely to involve B acting against the interests of C, to whom B, as C's agent, owes a duty of loyalty.[20] We described the mischief with which the law of corruption is concerned in terms of "the fundamental mischief" (B's breach of duty), and "the mischief of temptation" (A's temptation of B, by bribery, to breach B's duty).

Breach of duty

5.5 At the heart of our conception of corruption, therefore, was the tendency of corrupt conduct to encourage *breach of duty* by agents. A number of respondents[21] were critical of this approach. The SIB, for example, wrote:

---

[13] Paras 5.15 – 5.37 below.

[14] Paras 5.11 – 5.14 below.

[15] Paras 5.45 – 5.50 below.

[16] Paras 5.51 – 5.53 below.

[17] Paras 5.62 – 5.122 below.

[18] Paras 5.45 – 5.50 below.

[19] Paras 5.51 – 5.53 below.

[20] Paras 1.12 – 1.15 of the consultation paper.

[21] Including the Financial Law Panel, the Bar Council and the Criminal Bar Association, the SFO, SIB, Professor Sullivan and the Metropolitan Police Company Fraud Department.

Many commercial and public duties are unspecified and the introduction of these words is liable, in certain cases, to lead to protracted legal argument concerning whether a duty is owed and, if so, what constitutes a breach – matters which, as far as commercial relationships are concerned, are primarily the province of the civil law.

5.6 The Bar Council and the Criminal Bar Association suggested that our proposed approach would create "a major problem":

> In discharging their duty an agent will often have to consider a vast amount of material which may be of a highly technical nature, after which there may be a band of discretion within which two honest agents might arrive at different decisions. If the Crown are obliged to prove a breach of duty … , a jury would have to re-visit all the considered material and second guess what may be highly complicated decisions.

5.7 We have considered trying to meet these objections by making the test subjective rather than objective. This would involve focusing not on whether the conduct in question would in fact be a breach of duty on the part of the agent, but on whether the agent would be doing something which he or she *believes* to be contrary to the interests of his or her principal.

5.8 However, this approach does nothing to meet a different, and more fundamental, objection: that an agent can act corruptly by doing something which is not, and which the agent knows is not, contrary to the principal's interests – for example, by demanding a bribe for doing what the agent's duty to the principal already requires the agent to do. In the consultation paper we argued that such conduct should fall within the offence.

> It would be naive to suppose that there is no harm in A paying B to comply with B's duty to C: if B is free to accept such payments, there is an obvious incentive to insist on them as a precondition for the performance of the duty – in other words, an incentive to act in breach of the duty if payment is not forthcoming. Indeed, it may be that the only reason why A is prepared to pay for the performance of B's duty is that that is the only way to secure it.[22]

Our proposal that this kind of conduct should be covered was generally accepted on consultation; but it does not sit happily with the emphasis placed by our other proposals on the need for a tendency to encourage breach of duty.

5.9 Another example of a bribe which is corrupt although the interests of the agent's principal are not prejudiced is the case where, of a range of choices open to the agent, two or more appear equally advantageous to the principal, and the agent is bribed to choose one of these acceptable options rather than another. For example, where from the principal's point of view there is nothing to choose between two tenders for a contract, the agent may be bribed to select one of them. The agent's decision, though not clearly furthering the principal's interests *better* than the alternatives, is not *contrary* to those interests either. But in our view it is

---

[22] Consultation Paper No 145, para 8.13.

still corrupt, and should be criminal. For these reasons we have concluded that a definition couched in terms of breach of duty, or even in terms of the principal's best interests as the agent perceives them, would be too narrow.

5.10 Professor Sir John Smith proposed an alternative approach: that corrupt conduct should be defined in terms of an intention to influence the agent's conduct as agent. We can see the appeal of this approach on practical grounds: it uses straightforward concepts with which practitioners, magistrates and jurors will be familiar, and it would obviate the need to prove the existence of a duty and an intention to induce a breach of that duty. We have therefore decided to use it as the basis for our recommended definition, though we believe that it needs some qualification.[23]

The functions of an agent *as an agent*

5.11 This approach differs somewhat from that of the existing legislation, in that it focuses on what an agent is intended to do *in his or her capacity as agent*. The 1889 Act refers to a public servant "doing … anything in respect of any matter or transaction whatsoever … in which the … public body is concerned", and the 1906 Act to "any act in relation to [the agent's] principal's affairs or business". These expressions would appear to include the case where A pays B to do something in relation to C's affairs which, though it is B's position as C's agent that enables B to do it (or makes it easier for B to do it than for others), is not itself part of B's functions as C's agent. For example, A might pay B to steal documents belonging to C from the office of another of C's employees. Theft from colleagues' offices is no part of B's functions as C's agent: it is merely something that B is enabled to do by virtue of having access to C's premises in order to perform those functions. The opportunity to steal in this way is incidental to B's functions as agent, not directly conferred by them.

5.12 It is arguable that this situation *should* be caught by the law of corruption. In one sense, like the paradigm case where A bribes B to exercise in A's favour a discretion which B is paid to exercise impartially, it involves a breach of the trust placed in B by C. But in another way, and in our view a more significant one, it resembles the case where the thief paid by A is not C's agent but is for some other reason well placed to commit the theft – for example, a cleaner employed by C's landlord. This latter case seems not to be caught by the existing legislation, since theft from a tenant can hardly be described as an act in relation to the landlord's affairs or business; and it is hard to see why it should make any difference in principle that the thief is employed by the victim of the theft rather than by a third party, if the theft has nothing to do with the job that the thief is employed to do. The fact that B is betraying C's trust is a factor which can be taken into account at the sentencing stage if B is convicted of theft,[24] but we do not believe that it turns B's offence from one of theft into one of corruption.

---

[23] See para 5.74 below.

[24] *Barrick* (1985) 81 Cr App R 78; see also *Clark, The Times* 4 December 1997.

5.13 It is also possible that B may be bribed to do an act in relation to C's business which is not itself an offence. For example, it is not an offence to take a *photograph* of a confidential document. This is not theft, because there is no intention permanently to deprive the document's owner of it, and the information contained in the document is not "property" within the meaning of the Theft Act;[25] and it makes no difference that it is done by an agent of the document's owner. But if the agent is *bribed* to do it, there appears to be an offence of corruption under the 1906 Act; and this is arguably right.[26] In our view it is strongly arguable that such conduct should be an offence in itself, whether or not the person who does it is an agent of the document's owner, and whether that person does it on his or her own account or is paid by a third party.[27] In that case, the law of corruption partially fills a gap in the criminal law. But if the law is to catch such conduct at all, in our view it should do so directly and not by a side wind. Whether it should do so directly, and if so how, is a matter to which we intend to return, particularly in the context of our review of the law of dishonesty. The proper concern of the law of *corruption*, we believe, is with agents who are tempted to do (or not to do) their jobs in particular ways, not with agents who are tempted to harm their principals in ways that have nothing to do with the proper performance of their jobs.

5.14 We have therefore concluded that the new offences should be defined in terms of things done in connection with an agent's performance of his or her functions *as an agent*. This does not imply that the prohibited conduct must relate to any *particular* person, or that, if it does, that person must in fact *be* an agent, let alone that he or she must in fact *perform* any particular functions as an agent: it refers solely to the intentions and expectations of the defendant. If A pays a bribe to B, believing that B knows one of the employees of a particular company and that that employee can be persuaded to show favour to A, A's conduct is no less corrupt merely because A has no particular employee in mind and B does not in fact know any of that company's employees.

THE AGENCY RELATIONSHIP

Terminology

5.15 In the consultation paper our starting point for our analysis of bribery was that it was essentially conduct which threatens the relationship of trust which exists between an agent and principal.[28] We suggested that "the paradigm situation with which we are concerned is one in which A acts in relation to B in such a way as to, and in order to, tempt B to act in breach of an obligation of loyalty which B owes

---

[25] *Oxford v Moss* (1979) 68 Cr App R 183.

[26] The Institute of Chartered Accountants, for example, was concerned at our provisional proposal (which we have now abandoned: see paras 5.48 – 5.49 below) to exclude intermediaries from liability for the full offence, on the ground that this would exclude the "information broker", who bribes company employees to provide information which can be sold to companies hoping to win a tender or negotiate an improved deal.

[27] See our Consultation Paper No 150 (1997) Legislating the Criminal Code: Misuse of Trade Secrets, Part VII.

[28] Parts I and V.

to C",[29] but we recognised that a person could owe an obligation of loyalty not only to an identifiable principal but also (either additionally or solely) to the public interest.[30] And we took the view that offences of bribery should be concerned with both sorts of obligation. We noted that the term "agent" is used in the 1906 and 1916 Acts to describe the category of persons capable, in law, of being a bribee, and provisionally proposed that, although the term was not used in its strict sense,[31] it was the "obvious choice"[32] to describe those affected by bribery offences. Most of those respondents who expressed a view on this provisional proposal agreed with us.

5.16 We acknowledge that in referring to "the agency relationship" in the context of bribery offences we are relying on a meaning of that phrase which is extended in two ways. First, we intend that it should include private relationships of trust other than (but as well as) agency relationships in the strict sense normally understood by lawyers. Second, we intend that it should include relationships involving public duties as well as private ones. Although the phrase "agency relationship" may not be a natural description of a relationship where the "agent" owes a duty to an abstraction (namely the public) rather than an identifiable principal,[33] we still regard "agent" as a convenient term for a person in a position of trust, whether that position involves private or public duties.

Defining the agency relationship

5.17 The defining characteristic of the *private* relationships we sought to cover was, we suggested, that they were broadly fiduciary in nature: "agent", in the extended sense we are using, shares the "distinguishing obligation of a fiduciary" identified by Millett LJ in *Bristol & West Building Society v Mothew (t/a Stapley & Co)*,[34] namely "the obligation of loyalty". And drawing on the test for a fiduciary relationship set out in our consultation paper on fiduciary duties and regulatory rules,[35] we attempted to identify the circumstances in which an agency relationship would arise. We suggested that B should be regarded as an agent of C in circumstances where B has undertaken (expressly or impliedly) to act on behalf of C and that undertaking involves one or more of the following features: B exercising a discretion on C's behalf, B having access to C's assets (irrespective of whether B has been given a discretion by C to act in regard to those assets), or B having influence over C's decisions (as regards C's assets or any other interest of

---

[29] Consultation Paper No 145, para 7.2.

[30] Consultation Paper No 145, para 7.7.

[31] That is, agency as "the relation which exists where one person has an authority or capacity to create legal relations between a person occupying the position of principal and third parties": *Halsbury's Laws of England* (4th ed 1990) vol 1(2), p 4, para 1 (footnote omitted).

[32] MCCOC Report, p 297.

[33] As we shall see in para 5.91 below, the different character and capacities of a private principal compared with the public interest as principal – primarily the fact that a private principal has the capacity to consent – has a direct bearing on our recommendation that a defence of principal's consent should be available.

[34] [1996] 4 All ER 698, 711j–712a.

[35] Fiduciary Duties and Regulatory Rules (1995) Law Com No 236.

C's). This description was not, however, appropriate to describe those persons whose duty of loyalty was not to an identifiable principal but to the *public*. We therefore proposed that B should also be regarded as an agent in circumstances where B has undertaken to discharge a public duty. We called such individuals "*quasi*-fiduciaries".

5.18 Having identified the broad circumstances in which an agency relationship would arise, we suggested that it would be unnecessarily burdensome for the prosecution, on every occasion, to be required to prove that a defendant charged with corruption falls within the general definition of an agent. We noted that there were a number of categories of persons which are "invariably characterised as fiduciary":[36] trustee and beneficiary; agent and principal (in the strict sense understood by lawyers); partner and co-partner; director and company; employee and employer; legal practitioner and client. We therefore made a provisional list of uncontentious categories based on the "classic" examples of fiduciaries. As for quasi-fiduciaries, we suggested that judges, local councillors and police officers should be included.[37]

5.19 A number of respondents were against our approach to describing the agency relationship. The SFO, for example, took the view that there was no need for further elaboration of the terms "agent" and "principal" beyond that provided by section 1(2) of the 1906 Act. Other respondents, such as the CPS, were critical of our proposal that there should be a list of examples, whilst others queried the contents of the list.

The general definition

5.20 Our provisional definition of the agency relationship was divided into two parts, corresponding to those who have a relationship with an identifiable principal and those charged with public duties. Each part consisted of a list of examples followed by a general description of circumstances in which an agency relationship would arise for the purposes of the new law of corruption.

PRIVATE AGENCY RELATIONSHIPS

5.21 Of those respondents who commented on the agency relationship generally, none directly commented on the terms of the definition. Nonetheless, we now take the view that our provisional description of the *private* agency relationship was too complex, and that a simpler approach can be applied. The approach adopted in the consultation paper was to identify a comprehensive set of circumstances which would give rise to the sort of relationship that we intended the new law of

---

[36] R Flannigan, "The Fiduciary Obligation" (1989) 9 OJLS 285, 291.

[37] In the consultation paper, paras 7.42 – 7.49, we set out the reasons why we had decided that it was not appropriate for us to consider the position of Members of Parliament or Ministers of the Crown. A number of respondents expressed regret that we had limited our remit in this way. We remain of the view, however, that, given the establishment of a joint committee of both Houses (under the chairmanship of the Rt Hon the Lord Nicholls of Birkenhead) which will investigate, amongst other things, bribery and Members of Parliament, we were right not to consider Members of Parliament. We shall, therefore, make no recommendations in respect of Members of Parliament.

corruption to cover. This approach, although providing useful guidance by illustrating the meaning of the agency relationship, carries with it the risk that the list might be incomplete, or insufficiently flexible to accommodate changing circumstances.

5.22 An alternative approach is to describe the *essential character* of the agency relationship. Bearing in mind our original analysis of corruption in terms of tempting an agent to breach an obligation of loyalty to his or her principal, we took the view that central to the agency relationship is a relationship of *trust*, in which one person has *entrusted* another to undertake functions on his or her behalf.

5.23 We have considered whether that relationship of trust, for the purposes of a corruption offence, should necessarily involve an obligation that the person entrusted should act in the best interests of the person who has entrusted him or her. If this were the case, an agent could be defined as a person who has undertaken to perform functions on behalf of a principal, and to do so in the best interests of the principal. We took the view that, although it is likely that the bribed agent will, in return for the bribe, do some act which is contrary to the principal's best interests (or omit to do some act which, had it been done, would have been in the principal's best interests), there are circumstances in which an agent undertakes to perform functions on behalf of another but without a collateral undertaking that that function will be performed in the best interests of that other. Where, for example, a philanthropist employs an agent to distribute funds to a range of charities which the agent has a discretion to select, the agent will have a duty to act properly, but not necessarily to act in the philanthropist's own interests. Yet it should clearly be an offence of corruption for the agent to accept a bribe to show favour to a particular charity.

5.24 We have also considered whether the law of corruption should catch not only the agent who performs functions on behalf of another as a result of some sort of agreement or understanding between the parties that the agent should do so, but also a person who unilaterally performs functions for another without the other's agreement that he or she should do so. Bearing in mind our analysis of corruption in terms of the potential breach of a relationship of trust, it seems to us artificial to suggest that a relationship of any sort (let alone a relationship of trust) can exist in the absence of any mutual understanding to that effect.

QUASI-FIDUCIARY RELATIONSHIPS

5.25 In the consultation paper we provisionally proposed a general definition of the quasi-fiduciary along the lines that a person was an agent if he or she "has undertaken to discharge a public duty (whether appointed as public office-holder or to perform a specified public function)". Although very few respondents expressed concern about the absence of any definition of "public duty" or "public function", we have considered whether it would be helpful (or indeed possible) to define those terms further. In the consultation paper we suggested that the law on judicial review would provide the basis for identifying those authorities charged with duties involving a public element.[38] But although it is possible to give guidance as to the sorts of characteristics which are likely to signify a public duty or public function, this is a far cry from identifying a set of necessary and sufficient characteristics. Lord Woolf has argued that the question of whether an issue involves public law depends not on whether the *body* concerned can be described as a public body but on whether the *function* being performed is a public function, and describes the boundary between public and private functions as "indistinct and evolving".[39]

5.26 The fluidity of the boundaries between public and private duties and functions raises the question whether it is, in fact, *appropriate* to define those terms (or terms like them) in more detail. Arguably, it is preferable not to define them further, on the ground that they may then have greater flexibility in dealing with a changing political and economic environment. Furthermore, as we have seen,[40] the United Kingdom is involved in a number of international initiatives on corruption. For example, in December 1997, an OECD convention[41] was signed requiring each party to the convention to criminalise the bribery of public officials. The definition of "public official", for the purposes of the convention, includes "any person exercising a public function for a foreign country". Given that the signatories to the convention may well have different concepts of "public function", it would seem preferable to retain a general definition of the term so as to enable the offences to accommodate international differences.

5.27 We are therefore inclined to the view that the concept of "public function" should not be defined, other than in the general terms used in our recommendation. We note that this approach was adopted in the Human Rights Bill introduced in the House of Lords late last year – giving effect to the European Convention on Human Rights by requiring public authorities to exercise their powers in a manner compatible with the Convention – where "public authority" is not exhaustively defined and is said to include "… any person certain of whose functions are functions of a public nature".[42]

MIXED RELATIONSHIPS

5.28 As we noted in the consultation paper, we recognise that a person may have both private and public functions, or a single function which has both private and public elements.[43] We take account of this in our recommendation.

---

[38] Consultation Paper No 145, para 7.24.

[39] The Rt Hon Lord Woolf of Barnes, "Droit Public – English Style" (1995) PL 57, 64.

[40] See paras 1.16 – 1.20 above.

[41] Convention on Combating Bribery of Foreign Public Officials in International Business Transactions.

[42] Clause 6(3)(c) and see also the White Paper on the Bill entitled Rights Brought Home: The Human Rights Bill (1997) Cm 3782, para 2.2, p 8.

[43] See the Rt Hon Lord Woolf of Barnes, *op cit*, p 62: "[E]ven governments … are recognised by English law, if not the media, as having the capacity to perform private activities". In *R v Lord Chancellor, ex p Hibbit and Saunders (a firm)* [1993] COD 326 it was decided that although the Lord Chancellor's Department had acted unfairly in awarding a contract for

The list of examples

5.29 We provisionally proposed that, in order to circumvent legal argument, the definition of "agent" should include a list of those categories of relationship which would invariably involve a relationship of agency for the purposes of the new offence. The list was intended to provide a short cut in ascertaining whether a particular person fell within the terms of the offence. Some respondents expressed concern about the inclusion of a list, warning that it would cause confusion because of any implication that might attach to a relationship being omitted from the list; that legal argument would not be circumvented but would simply shift from the meaning of the general definition to the meaning of the relationships described in the list; and that the list was unnecessary in any event, because of the presence of a general definition.

5.30 Although the list is intended as an illustration of the obvious categories of relationships falling within the general definition, and the general definition should therefore suffice as a description of private agency relationship for our purposes, nonetheless we remain of the view that the list provides a useful indication of what we mean by "agency relationship" in this context. This is particularly helpful given that we are using the labels "agent" and "principal" which have different, more restrictive meanings in other areas of law. In contrast, however, we would dispense with the list of categories in respect of those discharging public functions. Given that there is no risk of the restricted meaning of agency being applied in this case, we take the view that a list would be of little assistance. On the contrary, it might be read as restricting the meaning of the general definition.

Transnational agency relationships

Agents with private foreign principals

5.31 In the consultation paper, we identified two types of what we called "transnational agency relationship": those where the relationship involves a private foreign principal, and those where the agent acts on behalf of the public interest of another country. We provisionally concluded that an agent of an identifiable foreign principal should be regarded as an "agent" for the purposes of a modern law of corruption.[44] This proved uncontroversial amongst those who responded on the issue, and we remain of the view that agents of foreign principals should not be exempted. Given the internationalisation of trade, to exclude them would encourage a climate of corruption in international business transactions. Furthermore, it would go against a developing resolve within the international community that international corruption, in both the public and private sectors, needs to be addressed. We are aware, for example, that efforts have been made in the European Union to secure agreement about the criminalisation of private

---

court reporting services, its conduct was not amenable to judicial review because it lacked a sufficient public element: "[a] governmental body was free to negotiate contracts, and it would need something additional to the simple fact that the governmental body was negotiating the contract to impose on that authority any public law obligation in addition to any private law obligations or duties there might be."

[44] Consultation Paper No 145, para 7.56.

sector corruption. In a document published in May 1997,[45] the Commission of the European Communities gave strong support for a comprehensive attack on corruption, including private sector corruption. Similarly, the Council of Europe is also actively considering private sector corruption.

Agents acting for the public interest of another country

5.32 In the consultation paper, we had provisionally concluded that agents acting solely on behalf of the public interest of other countries should be exempted from the scope of the new offence. Not only has this conclusion met with significant criticism, but we have also been impressed by the international anti-corruption initiatives undertaken in a range of fora such as, for example, the European Union, the Council of Europe, the OECD, the Commonwealth Law Ministers, the G7, the United Nations, the World Trade Organisation, the International Monetary Fund and the Organisation of American States.[46]

5.33 It is widely recognised that corruption can threaten economic development and the integrity of democratic institutions.[47] In the preamble to a draft[48] of the Council of Europe convention on corruption it is emphasised that "corruption threatens the rule of law, democracy and human rights, undermines good governance, fairness and social justice, distorts competition, hinders economic development, and endangers the stability of democratic institutions and the moral foundations of society". In the preamble of the OECD convention on combating bribery of foreign public officials,[49] it is said that bribery in international business transactions "raises serious moral and political concerns, undermines good governance and economic development, and distorts international competitive conditions". And in the recent Lima Declaration,[50] the view was expressed that corruption "erodes the moral fabric of every society; violates the social and economic rights of the poor and vulnerable; undermines democracy; subverts the rule of law which is the basis of every civilised society; retards development; and

---

[45] Communication from the Commission to the Council and the European Parliament, *A Union Policy Against Corruption*, COM(97) 192 final, Brussels, 21.05.1997:

> One of the essential elements in combating corruption is to ensure that it is criminalised. In addition to its deterrent and repressive effect, criminalising corruption represents a clear and unequivocal statement that those practices are not acceptable and are against the public interest. Ideally the criminal law within the Union should address the bribery of EC officials, the bribery of officials of other member States, the bribery from states outside the Union and private sector corruption. (Para 9).

[46] See Part I, paras 1.16 - 1.20.

[47] In his response, Professor Sullivan commented on "the phenomenon of international commercial bribery, with its malign impact on the politics and economies of developing countries" (See G R Sullivan, "Reformulating the Corruption Laws – the Law Commission Proposals" [1997] Crim LR 730, 739).

[48] As approved by the Working Group on Criminal Law after a second reading of the draft at its 10th meeting (4–6 November 1997).

[49] Convention on Combating Bribery of Foreign Public Officials in International Business Transactions, signed on 17 December 1997.

[50] A declaration of the 8th International Conference Against Corruption held in Lima, Peru from 7–11 September 1997. The conference represented the citizens of 93 countries.

denies societies, and particularly the poor, the benefits of free and open competition."

5.34 We are aware of the global nature of corruption and the importance of international efforts to combat corruption in both the public and private sectors, and we agree that exempting those who are required to act in the public interest of another country would run counter to those efforts. We shall be recommending, therefore, that "agent" should include a person acting on behalf of the public, whether the public of the United Kingdom (or any part of it) or of another country.

Agents acting on behalf of international intergovernmental organisations

5.35 So far, as regards those agents who exercise *public* functions, we have concerned ourselves with those who act on behalf of the public, whether the public of this country or another. There is, however, a further category of public agent which we need to consider, namely the agent who acts not on behalf of a national public but on behalf of the interests of an international intergovernmental organisation.[51] We have in mind an organisation such as the European Union, where it has an interest which cannot be described as the collectivity of the various national interests represented by Member States[52] – even though, it is supposed, the purpose of the organisation will be to promote those national interests. We take the view that the extension of the law of corruption to the officials of particular intergovernmental organisations is not a matter best dealt with by the Law Commission. However, we shall be recommending that consideration should be given to the inclusion of such officials in the definition of "agent" in any future legislation governing offences of corruption.

Our recommendation

5.36 **We recommend that**

   (1) **a person should be regarded as an agent, and another as a principal for whom he or she performs functions, if**

   (a) **the first is a trustee and the second is a beneficiary under the same trust;**

   (b) **the first is a director of a company and the second is the company;**

   (c) **each is a partner in the same partnership;**

   (d) **the first is a professional person (such as lawyer or accountant) and the second is his or her client;**

---

[51] The Council of Europe draft Convention makes specific reference to the criminalisation of bribery of officials and contracted employees of "any international or supranational intergovernmental organisation or body".

[52] Hence, for example, the First Protocol to the Convention on the Protection of the European Communities' Financial Interests is directed at the damaging effect of corruption involving national and Community officials on the European Communities' Financial Interests.

> (e) the first is an agent and the second is his or her principal (taking agent and principal in the sense normally understood by lawyers); or
>
> (f) the first is the employee of the second;
>
> (2) a person who does not fall within any of the categories above should be regarded as an agent, and another as a principal for whom he or she performs functions, if there is an agreement or understanding between them (express or implied) that the first is to perform the functions for the second; and
>
> (3) a person should be regarded as an agent performing functions for the public if the functions he or she performs are of a public nature (whether or not in relation to the United Kingdom).

(Recommendation 3)

5.37 **We recommend that consideration should be given to extending the scope of the corruption offences to include officials of international intergovernmental organisations.**

(Recommendation 4)

THE CONCEPT OF ADVANTAGE

5.38 A concept central to the existing law of corruption, and also to the new offences we recommend, is that of the benefit corruptly conferred, obtained, offered or solicited. Under the present legislation this benefit is variously described as a "gift, loan, fee, reward or advantage" (the 1889 Act) or a "gift or consideration" (the 1906 and 1916 Acts). We suggested that the former expression might be slightly wider than the latter, for example in relation to services as against property.

5.39 Our provisional view was that the new legislation should apply to anything done by one person (A) which another (B) wants A to do, or which is of benefit to B,[53] and also to an omission[54] by A to exercise a right to act to B's disadvantage.[55] We described both kinds of benefit as the conferring of an "advantage" by A on B, and this approach won general approval.[56]

5.40 However, we believe that the formulation we provisionally proposed can be improved upon in certain respects. In the first place, the reference to something done by A which *is of benefit* to B might be understood as excluding the case where A instructs or requests a third party, C, to confer a benefit on B: A's instruction or request is not in itself of any benefit to B (because C might decline to act upon it), and the act done by C which benefits B is arguably done *only* by

---

[53] Consultation Paper No 145, para 8.62.

[54] The consultation paper followed the 1889 and 1906 Acts by referring to a "forbearance", but the plainer word "omission" seems preferable.

[55] Consultation Paper No 145, para 8.64.

[56] But the SFO thought it unnecessary to define the word "advantage".

C, not by A. Our recommendation therefore makes it clear that it is sufficient if A's act or omission *results* (directly or indirectly) in a benefit to B.

5.41 Secondly, our provisional proposal referred to A doing something that B wants A to do *or* which is *otherwise* of benefit to B. This might be thought to imply either that anything B wants A to do is inevitably of benefit to B (which arguably is not the case)[57] or that something which is not of benefit to B is not an advantage even if B wants A to do it (which was not our intention). Similarly, it would follow from our proposal that an omission by A to exercise a legal right, even at B's request, would suffice *only* if it were shown that the exercise of the right would in fact have been disadvantageous to B. Our recommendation makes it clear that, if A's act or omission is in consequence of B's request, it is immaterial whether it *benefits* B.

5.42 Thirdly, the requirement of a *request* is slightly narrower than our original test of whether B *wants* A to act or omit to act: given that there is no additional requirement of benefit to B, we do not think it should be sufficient that B wants A to act or omit to act unless B has communicated that desire to A. But an *implied* request should clearly suffice. We think it should also be made clear that it is sufficient if B requests A to act or omit to act to B's advantage without specifying precisely how or when A is to do so, and A does in due course act or omit to act in accordance with that request.

5.43 **We recommend that**

(1) a person should be regarded as conferring an advantage if

(a) he or she does something or omits to do something which he or she has a right to do, and

(b) the act or omission is done or made in consequence of another's request (express or implied) or with the result (direct or indirect) that another benefits; and

(2) a person should be regarded as obtaining an advantage if

(a) another does something or omits to do something which he or she has a right to do, and

(b) the act or omission is done or made in consequence of the first person's request (express or implied) or with the result (direct or indirect) that the first person benefits.

(Recommendation 5)

THE OFFENCES

5.44 We now consider what kinds of conduct in relation to an "advantage" should be prohibited.

---

[57] But, as Judge Rhys Davies QC pointed out, "even 'gratification' is an advantage."

Corruptly conferring, or offering or agreeing to confer, an advantage

5.45 Under the 1889 Act a person may commit an offence by corruptly giving, offering or promising an advantage; under the 1906 Act, by giving, offering or agreeing to give it. In the consultation paper we provisionally preferred "promise" to "agree", on the ground that we saw no difference between the two terms except that "promise" seemed simpler.[58] The Employment Law Bar Association preferred "agree", which would be consistent with our proposal that it should be an offence to agree to *accept* a corrupt advantage.[59] On reflection we now think that "promise" is the wider term, because it does not imply a need for agreement on the part of the prospective *bribee*.

5.46 On the other hand there seems to be no need for a further expression to cover the situation where A promises a bribe which B does not agree to accept, because that would amount to an "offer". Indeed, strictly speaking there seems no need to cover *any* conduct on A's part other than conferring an advantage and offering to confer one: if B solicits an advantage and A agrees to confer it, both will be guilty of a conspiracy to confer and obtain it.[60] However, given that it is a substantive offence even to *propose* such an agreement (by offering or soliciting a corrupt advantage) it would be odd if the agreement itself were not a substantive offence but only a conspiracy to commit the offences involved in carrying it out. We have therefore concluded that it should be an offence to *agree* to confer a corrupt advantage.

Advantage conferred on a person other than the agent

5.47 The 1889 Act does not require that the agent it is sought to influence should be the person who receives the corrupt advantage; but the 1906 Act does not allow for such a separation of roles. Therefore, the existing law is limited in that it will only bite against third-party transactions involving those associated with *public* bodies.[61] In the consultation paper we noted that the Salmon Report considered the wider approach of the 1889 Act to be the right one, and we provisionally agreed, considering that otherwise there is a gap in the law which could be exploited in circumstances where an agent has an interest in a third party (such as the agent's spouse) receiving a benefit.[62] This view was accepted on consultation, and our recommendation does not require that the person obtaining the advantage should be the agent whose conduct it is sought to influence or reward.

Intermediaries

5.48 The Salmon Report suggested that intermediaries were caught by the broad drafting of the 1889 Act, but by the 1906 Act only "to the extent that [the intermediary] actually gives or offers a corrupt gift or consideration to an agent."[63]

---

[58] Consultation Paper No 145, para 8.68.

[59] Consultation Paper No 145, para 8.67.

[60] Criminal Law Act 1977, s 1.

[61] Consultation Paper No 145, para 8.71.

[62] Consultation Paper No 145, para 8.73.

[63] Salmon Report, para 57.

An intermediary may however be guilty of aiding and abetting the substantive offence.[64]

5.49 Unlike the Salmon Committee, we think that the criminality of an intermediary's conduct is less than that of the other parties to the corrupt transaction, and is adequately reflected by liability for aiding and abetting. We therefore provisionally proposed that an intermediary should not commit the offence *as a principal offender*.[65] A number of respondents agreed with this proposal. However, several others argued that, since the law makes no distinction between an accessory and a principal in the formulation of a charge, the distinction we proposed was artificial and unnecessary, if not positively undesirable. We are persuaded by this argument. If an intermediary actually confers a corrupt advantage, or offers or agrees to do so, there is no compelling reason why he or she should not be guilty as a principal offender rather than merely as an aider and abettor. The fact that he or she is only an intermediary may be taken into account in mitigation. We therefore make no recommendation for any specific provision regarding intermediaries.

Our recommendation

5.50 **We recommend that a person should commit an offence if he or she**

(1) **corruptly confers an advantage, or**

(2) **corruptly offers or agrees to confer an advantage.**

(Recommendation 6)

Corruptly obtaining, soliciting or agreeing to obtain an advantage

5.51 Under the 1889 Act a person may commit an offence by soliciting, receiving or agreeing to receive an advantage; under the 1906 Act, by accepting it or agreeing to do so, obtaining it or attempting to do so. In the consultation paper we suggested that "receive", "accept" and "obtain" were interchangeable, and provisionally proposed that "accept" should be used. On further reflection we believe that "obtain" may be slightly more apt for willing acquiescence in the corrupt conferment of an advantage (for example, the payment of funds into a bank account, or of cash to a third party) as distinct from active co-operation;[66] and we believe that such acquiescence should suffice.

5.52 We also provisionally suggested that it was unnecessary to provide for an attempt to obtain an advantage, because it was hard to see how a person might attempt to obtain an advantage without soliciting one. No-one suggested that we were wrong.

5.53 **We recommend that a person should commit an offence if he or she**

(1) **corruptly obtains an advantage, or**

---

[64] Consultation Paper No 145, para 8.78.

[65] Consultation Paper No 145, para 8.79.

[66] The House of Lords has held in the context of insider dealing that a person "obtains" information merely by coming into possession of it, without taking any active steps to acquire it: *A-G's Reference (No 1 of 1988)* [1989] AC 971.

(2) corruptly solicits or agrees to obtain an advantage.

(Recommendation 7)

Corruptly performing functions as an agent

5.54 The existing legislation catches not only corrupt inducements to show favour in the future but also corrupt rewards for favour shown in the past,[67] and our recommendations would do the same. An agent who performs his or her functions as an agent with a view to obtaining a corrupt reward will be guilty of obtaining the reward corruptly, if he or she obtains it at all. But it would be illogical if the agent's liability rested *solely* on the obtaining of the reward. The real criminality of the agent's conduct lies in the fact that the agent has allowed his or her judgment to be influenced by the *hope* of obtaining a reward, not in the fact that he or she has actually obtained one. Morally, there is little to choose between an agent who acts in the hope of a reward and would gladly accept one if it were offered, but is disappointed, and an agent who acts in the hope of a reward and duly gets one.

5.55 Moreover, even where an agent *has* received a reward, it may be harder to prove this than to prove that the agent acted in the hope of one. There may be evidence of a breach of trust on the part of the agent, but evidence of the reward may be lacking. We see no reason in principle why this should preclude a conviction, since the agent's effort to *earn* a reward is corrupt in itself. Such a case can sometimes be successfully prosecuted as a conspiracy to defraud the agent's principal (or the public), and we have previously identified this possibility as one of the advantages for the prosecution of the continuing existence of conspiracy to defraud.[68] But if the allegation is essentially one of corruption, we think it wrong that the prosecution should be forced to lay the charge as a common law conspiracy, rather than as a substantive offence under the corruption legislation. The creation of such an offence would also eliminate one of the gaps in the substantive criminal law which are currently filled only by conspiracy to defraud. It would thus be a step towards our long-term objective of making it possible to recommend the abolition of that offence in its present form.[69]

5.56 We therefore believe that it should be an offence in itself for an agent to perform his or her functions as an agent in the hope of obtaining a corrupt reward for doing so. If the agent does later obtain a reward, this would amount to a *further* offence; but this would be immaterial to the agent's liability for acting corruptly in the first place.

5.57 This conclusion has implications for the position of the agent whose conduct is influenced not by the hope of reward but by a previous inducement. It would be anomalous if the agent were guilty of an offence in the former case but not the latter. Admittedly there is less *need* for an offence in the latter case, because in that case there is no need to cater for the possibility that the agent may fail to get the

---

[67] See paras 2.10 and 2.13 above.

[68] Criminal Law: Conspiracy to Defraud (1994) Law Com No 228, para 4.56.

[69] Criminal Law: Report on Conspiracy and Criminal Law Reform (1976) Law Com No 76, para 1.16.

advantage desired. But, if the corrupt performance of a duty were an offence *only* when done in the hope of later reward, there would be a practical difficulty: an agent charged with that offence would be entitled to an acquittal unless the prosecution could prove that the agent had acted in the hope of future reward, *as distinct from acting in return for a previous inducement*. This would be so even if it were clear that the agent had done one or the other – in other words, that he or she had acted corruptly.

5.58 We have therefore concluded that the corrupt performance of an agent's functions should be an offence in itself, whether it is done in the hope of later reward or in return for a previous inducement, or both. **We recommend that a person should commit an offence if he or she performs his or her functions as an agent corruptly.**

(Recommendation 8)

Agent receiving benefit from corruption

5.59 Although the existing legislation is directed ultimately at misconduct on the part of agents in the performance of their duties, an agent commits an offence only if he or she solicits, receives or agrees to receive an advantage, aids and abets an offence by another party or is party to a conspiracy.[70] In the consultation paper we suggested that, where the advantage is initially conferred on someone other than the agent whose conduct it is sought to influence, and the agent later corruptly receives some of the proceeds or some other consequential benefit (for example, the remission of a debt), it may be debatable whether the agent commits an offence.[71] We therefore provisionally proposed that there should be express provision for the liability of the agent in this situation, and there was substantial support for this proposal.

5.60 However, it could not be an offence for an agent *merely* to receive a benefit in the knowledge that it is derived from an advantage corruptly obtained, because the receipt of a benefit (like the obtaining of an advantage)[72] need not involve the co-operation or even the consent of the receiver. The offence we recommend would therefore be committed only if the agent gives *consent* (express or implied) to the receipt of the benefit.

5.61 **We recommend that if**

   (1)   **an advantage is obtained corruptly by a person other than the agent concerned,**

   (2)   **the agent receives a benefit (in any form) which consists of or is derived (directly or indirectly) from all or part of the advantage,**

---

[70] Consultation Paper No 145, para 8.74. Recommendation 14 would impose liability on an agent who performs his or her functions corruptly, whether or not he or she has personally obtained an advantage or hopes to do so: see paras 5.121 – 5.122 below.

[71] Consultation Paper No 145, para 8.75.

[72] See para 5.113 below.

> (3) the agent knows or believes that the advantage was obtained corruptly and that the benefit consists of or is derived from all or part of the advantage, and
>
> (4) the agent give his or her express or implied consent to receiving the benefit,
>
> **the agent should commit an offence.**
>
> **(Recommendation 9)**

"CORRUPTLY"

5.62 We have so far been following the example of the existing legislation by referring to persons acting "corruptly" without explaining what we mean by that term. We now turn to the question of how corrupt transactions might be differentiated from those that are not corrupt.

Is a definition necessary?

5.63 At present, the prosecution is required only to prove that the defendant acted "corruptly" – a word which is not defined. We invited views on whether the meaning of the word is clear enough to make a definition unnecessary. Respondents were about equally divided on this question: some thought a definition was necessary, while others thought that "corruptly" is a word which fact-finders can readily recognise and apply.

5.64 Our task would be significantly easier if we were confident that the term "corruptly" would be readily understood in the way that we would want it understood; unfortunately, we do not feel able to adopt this line. For a term which is not statutorily defined to be included in the definition of an offence, we must be confident that its generally understood meaning is unequivocal and that that unequivocal meaning is the meaning we would like imported into the offence. We are in the fortunate position of being able to make an informed decision as to whether "corruptly" passes this test. We have two sources of empirical evidence: the interpretative history of the meaning of "corruptly" in existing corruption legislation, and the comments we received on the definition we provisionally put forward.

5.65 In the consultation paper we set out the two strands of judicial interpretation of "corruptly": on the one hand, it describes an act which the law forbids as tending to corrupt,[73] and on the other, a dishonest intention to weaken the loyalty of an agent to his or her principal.[74] We noted Lanham's conclusion that the authorities are in "impressive disarray".[75] As for the responses, they ranged far and wide, and there was little consensus. The clear implication is that we were right to identify in the consultation paper the uncertainty of the meaning of the word "corruptly" as

---

[73] *Cooper v Slade* (1857) 6 HL Cas 746; 10 ER 1488, and *Wellburn* (1979) 69 Cr App R 254.

[74] *Lindley* [1957] Crim LR 321 and *Callard* [1967] Crim LR 236.

[75] Consultation Paper No 145, para 4.15. D Lanham, "Bribery and Corruption", *Essays in Honour of J C Smith* (1987) 92, 104.

one of the most important defects in the present law. We take the view that this defect can only be remedied by a more precise description of the kind of conduct that falls within the offence, and we conclude that a definition is required.

5.66 We therefore turn to the question of how such a definition should be formulated. For convenience, we consider first how the approach we have now adopted might be developed into a workable test of a corrupt *inducement* to an agent to act (or omit to act) in a particular way in the future, and then how that test might be adapted for corrupt *rewards* to agents for having acted (or omitted to act) in a particular way in the past.

Corruptly conferring an advantage as an inducement

5.67 In the consultation paper, we provisionally concluded that an advantage is a corrupt inducement if it is an inducement to an agent to act or refrain from acting

(1) in breach of duty ("the first limb"), or

(2) in any way, *provided* that the transaction has a substantial tendency to encourage that agent, or others in comparable positions, to act in breach of duty ("the second limb").[76]

5.68 As we have explained,[77] we have been persuaded that our provisional definition was too narrow in that it referred to agents being induced to act in *breach of duty*, rather than being influenced in the performance of their functions as agents.

5.69 We have also reconsidered the second limb of our provisional definition. We took the view in the consultation paper that the second limb was necessary on the ground that gifts not directly linked to a breach of duty can be potentially corruptive in that they "contribute to a climate in which ... inducements or rewards are expected or hoped for".[78] Without the second limb, any circumstance in which an advantage is conferred by A on B, otherwise than as an inducement to B to breach a duty to C (or to refrain from so doing), would not be covered, even if it had a potentially corruptive effect.

5.70 Of those who responded on this issue, the majority criticised the second limb.[79] Some, taking a pragmatic standpoint, suggested that it would be very difficult to prove the "substantial tendency"; others were concerned that the criminality of a transaction would be determined by the presence or absence of circumstances "not intended by the actor and of which he may be unaware";[80] and others were concerned that "substantial tendency" was too vague an expression.[81]

---

[76] Consultation Paper No 145, para 8.19.

[77] Paras 5.5 – 5.10 above.

[78] Consultation Paper No 145, para 8.15.

[79] For example, it was Buxton LJ's only concern.

[80] Buxton LJ.

[81] Professor Sir John Smith, for example, commented: "It seems too uncertain – a question which might be the subject of elaborate sociological research but which will be decided by a jury or magistrate on a hunch."

5.71 We are persuaded that the second limb of our provisional definition should be abandoned. We remain of the view that the conferring of an advantage on an agent by a person other than the agent's principal can contribute to a climate of corruption even if *in the case of that particular agent and principal* no harm is done.[82] On the other hand, we recognise the practical difficulties in criminalising conduct which is not *intrinsically* the sort of conduct we would wish to criminalise but is criminal because of an incidental effect, namely a "substantial tendency" to encourage certain kinds of conduct on the part of persons not directly involved.[83]

5.72 Moreover, in the light of our decision to extend the *first* limb of our provisional definition so as to include the influencing of an agent in the performance of his or her functions as an agent, *whether or not the agent is intended to act in breach of duty*,[84] the need for the second limb is greatly diminished. It would be needed only for the case where an advantage is conferred on an agent and, though not conferred in connection with *that* agent's performance of his or her functions as an agent, it is conferred in such circumstances that it might influence *other* agents in the performance of *their* functions. It is hard to imagine how such a case might arise.

5.73 We have therefore concluded that our definition of corrupt conduct should not include an element corresponding to the second limb of our provisional definition. We are left with the first limb. As modified along the lines suggested by Professor Sir John Smith,[85] and in accordance with our conclusion that the fundamental mischief is not breach of duty by agents but the influencing of an agent's performance of his or her functions as an agent, the first limb may be restated as follows:

> *An advantage is a corrupt inducement if it is intended to influence an agent in the performance of his or her functions as an agent.*

5.74 However, we believe that a definition in terms of an intention to influence an agent, without more, would be too wide. The difficulty is that most commercial enterprises are constantly trying to bring in business in a variety of ways, all of

---

[82] See para 1.25 of Consultation Paper No 145:

> It is in the public interest that people should refrain from conduct which might encourage agents to act in breach of their duty – *whether or not anyone would be defrauded by such conduct.*

[83] Buxton LJ wrote:

> While it would be naif to suppose that there is *no* harm in A paying B to comply with B's duty to C, the harm ... lies in what the parties may do in the future or in what others may infer from the payment. If this "innocent" payment creates a temptation to corruption, is it not enough for the law to hold its hand until that temptation ripens into action?

Lord Davidson took a similar line:

> I am not persuaded that an action which itself is not criminal can credibly be made the subject of prosecution on the ground that it tends to encourage others to commit criminal acts.

[84] See paras 5.5 – 5.10 above.

[85] See para 5.10 above.

which involve "influencing" those agents of other firms whose job it is to decide where their firms' business should go. Some of these ways (such as advertising) are irrelevant for our purposes because they do not involve conferring any "advantage" on the agents of the target firms. Others, however, do carry such advantages.

Corporate hospitality

5.75 An obvious example is "corporate hospitality", which is often expensive to provide and (designedly) enjoyable for those who attend. The provision of such hospitality undeniably constitutes the conferring of an advantage on the agents for whom it is provided, and its purpose is normally to influence their conduct as agents of their employers;[86] but it is generally regarded as an acceptable business activity – or, at least, not so unacceptable as to be criminal. We believe that a definition wide enough to catch ordinary corporate hospitality would be too wide.

5.76 On the other hand we cannot simply exclude *all* corporate hospitality from the offence, because that would open the door to abuse: bribery would simply be dressed up as hospitality. It would be absurd if it were an offence to pay an agent a bribe, but not to provide the agent with an expensive holiday incorporating a few token business presentations.

5.77 We have therefore tried to exclude *acceptable* hospitality without also excluding activities which are truly corrupt. For this purpose we have asked ourselves what it is about "acceptable" hospitality that makes it acceptable. For example, suppose A Ltd, hoping to get business from B Ltd, arranges for B Ltd's managing director to attend an important football match as the guest of A Ltd's directors. The match is preceded by drinks and lunch, during which (among other things) A Ltd's current and future activities are discussed. It is intended that the managing director shall enjoy the event – it is an "advantage" – and be influenced by it; yet most people would hesitate to conclude that the whole exercise is corrupt. Why is this?

5.78 The answer, we believe, lies in the *way* in which it is hoped to influence the agent. It is possible to identify various different ways in which an agent's attendance at a corporate function may influence the agent to make decisions favouring the host company. For example,

---

[86] See *The Financial Times* 2 June 1997:

> My diary this Season is one long blank. I had hoped for a corporate invitation to the Chelsea Flower Show, but no luck. Neither have there been any invitations to Wimbledon, Ascot or anywhere else. This enforced austerity has one advantage: it allows me to criticise those who are knocking back champagne at someone else's expense.
>
> For those of us outside the gates, the system of lavish corporate entertaining looks dodgy. The reason companies invite you to this sort of thing is not because they like you – if that is the only reason they should be fired for wasting their shareholder's money. *They ask you because they want something in return. They are offering you an institutionalised, perfectly legal bribe.* (Italics added)

(1) the agent may acquire information about the host company which militates in that company's favour when the agent comes to compare it with its rivals; or

(2) the agent's existing relationships with employees of the host company may be cemented, and new ones formed, and the agent may have a natural preference for doing business with people he or she knows; or

(3) the agent may simply decide to favour the host company in return for an enjoyable occasion.

5.79 An intention to influence an agent in the first or second of these ways (or both) is clearly legitimate, and something done *solely* with such an intention is not corrupt. An intention to influence an agent in the third way is not legitimate, and something done solely with that intention *is* (we believe) corrupt. The host company's purpose is of course to induce the agent to give it business, and, if the agent does so, it is unlikely to care much about his or her reasons. But if it knows that the agent would not give it business for any reason *except* the third, corrupt reason, then it intends the agent to be corruptly influenced.[87]

5.80 In practice, however, it will rarely be clear that, if the agent did give business to the host company, it could *only* be for the third reason. It may be that, as far as the host company can predict, the agent might do so for any one of these three reasons, or for some other reason, or for a combination of any two or more reasons. In such a case, we think the right approach is to ask what the host company thinks the agent's *main* reason would be, if he or she did give the host company the business it wants. Where the host company thinks that the agent's main reason for giving it business (if the agent did so) would be a legitimate reason, we do not believe that the exercise becomes corrupt merely because the host company also hopes for gratitude for the hospitality it provides. If, on the other hand, the host company thinks that a favourable decision on the part of the agent would be motivated primarily by the third, illegitimate reason, the fact that it also expects other, legitimate reasons to play an incidental part should not, in our view, be a defence.

5.81 The host company may of course realise that a favourable decision, if it came about, *might* be made primarily in return for the hospitality, without being *sure* that it would be. We do not think that liability should be negatived merely because there is a remote possibility of the agent's main reason being a legitimate one. On the other hand, we think it would be going too far to impose liability where there is only a remote possibility of the agent's main reason *not* being legitimate. We have concluded that the crucial question should be whether the provider of the alleged inducement thinks it *probable* (that is, more likely than not) that, if the

---

[87] Cf the definition of "intentionally" in cl 1 of the draft Bill annexed to our report Legislating the Criminal Code: Offences Against the Person and General Principles (1993) Law Com No 218, where it states that a person acts "intentionally" with respect to a result when –

(i) it is his purpose to cause it, or (ii) although it is not his purpose to cause it, he knows that it would occur in the ordinary course of events if he were to succeed in his purpose of causing some other result.

agent were to act in the manner desired, it would be primarily in return for the advantage conferred.

5.82 At first sight this test might seem difficult to apply; but, if so, we believe that any difficulty would arise *only* in the case of corporate hospitality and similar advantages, because in practice an advantage must inevitably be intended to create influence of an "illegitimate" kind unless it involves some sort of interaction between the agent and the donor. Thus, in the case of the football match described above,[88] it would have to be proved that A Ltd thought that, if B Ltd's managing director gave it the contract, it would *probably* be *primarily* in return for a good lunch and a good seat for the match, rather than as a result of a constructive discussion in congenial surroundings. But if the directors of A Ltd simply sent B Ltd's managing director a ticket for the match with their compliments, it would be hard to resist the inference that this was primarily (if not exclusively) a bribe. The practical effect, we believe, would be that corporate hospitality would be the subject of prosecution only where it was blatantly corrupt on any view.

"Sweeteners"

5.83 One of the reasons we gave for including the second limb of our provisional definition was that the first limb was not apt to cover the situation where A confers an advantage on B not in order to cause any *specific* breach of duty by B but as a general "sweetener".[89] Buxton LJ queried whether there is any need to intervene before the inducement has any effect. But this approach would be inconsistent with our principle that a bribe need only be *intended* to influence the agent, and need not actually succeed in doing so. We have concluded that it should be sufficient if the briber's intention is to influence the agent's conduct at some indeterminate future time, even if neither party can yet foresee the exact circumstances in which the agent's conduct may be influenced.

5.84 For this reason we describe the conduct desired of the agent not as the performance of his or her functions as agent *in a particular way* (which might imply, contrary to our intention, that the briber must have in mind the *details* of the desired conduct) but as *doing an act or making an omission* in performing his or her functions as agent; and the draft Bill makes it clear that the nature of the

---

[88] See para 5.77 above.

[89] See para 8.16 of the consultation paper:

> Suppose, for example, that Ms B's job involves awarding contracts on behalf of her employer Mr C. In good faith, and in accordance with the appropriate criteria, she awards a contract to Mr A. A shows his gratitude by offering B a substantial sum of money, which she accepts. Our provisional view is that such a payment would be potentially corruptive, since the recollection of it is likely to influence B in any future dealings with A; indeed, that may well be one of A's purposes in making it. This consideration might perhaps be accommodated by regarding the reward for B's first act, where the question of a breach of duty did not arise, as an inducement to act (or not act) in breach of duty in the future. *But if A has no particular objective in mind, and intends only to cultivate his relationship with B in the hope that it will eventually bear fruit, such an approach would be artificial at best.* (Italics added).

intended act or omission, and the time it is intended to be done or made, need not be known when the advantage is conferred or the offer or agreement is made.[90]

Bribee unaware of corrupt purpose

5.85 In our view the briber is no less culpable merely because, unknown to the briber, the intended bribee is unaware of the briber's corrupt purpose. We expressed this view in the consultation paper,[91] and nearly all the respondents who commented on it agreed.

Proper remuneration or reimbursement

5.86 If a corrupt intention could consist *solely* in an intention to induce an agent to perform his or her functions as agent in return for the promise of an advantage, every agreement to employ an agent would be corrupt: the agent's salary or other remuneration is an advantage, and the prospect of it induces the agent to perform his or her functions. It is therefore necessary to ensure that proper remuneration of an agent is excluded from our definition of "corruptly" conferring an advantage. The same applies to the reimbursement of expenditure incurred by an agent in the performance of his or her functions as an agent.

5.87 The formulation of this exception has to take account of the different kinds of function which an agent may perform – those performed for a principal and those of a public nature. In the case of the former, it is remuneration or reimbursement by or on behalf of the principal that is not corrupt; in the case of the latter, remuneration or reimbursement on behalf of the public. In the case of functions performed for a principal *and* the public, remuneration or reimbursement is not corrupt if it is made on behalf of either.

Things done with the consent of the principal

5.88 A number of respondents[92] took the view that the principal's consent should be a complete defence. Professor Sir John Smith gave his reasons as follows:

> I agree with the Commission's rejection of the specific defences examined except that, where the conferment of the advantage has been properly authorised, there is no breach of duty by the agent and so there should be no offence. A person who confers an advantage on an agent knowing or believing that the conferment of that advantage has been authorised by a person entitled to authorise it, does not do so corruptly. There might, as the Commission suggests, be difficulties of proof: but that does not seem to me to be an adequate reason for allowing the offence to apply in circumstances where in principle, it should not do so.

5.89 In the consultation paper we recognised that it was arguable that, if the agent's principal knows all the relevant facts and consents to what is done by the agent,

---

[90] Clause 6(3).
[91] Consultation Paper No 145, para 8.93.
[92] Longmore J, Professor Sir John Smith, the SFO and SIB.

the briber or both, that conduct ought not in principle to be caught by an offence of corruption; but we provisionally rejected this view. We gave three main reasons for doing so.

5.90 First, we argued that the principal's consent should not be a complete defence because corruption is not an offence of fraud or dishonesty (in the sense in which we use that expression).[93] We pointed out that, although the acceptance of a bribe with the principal's consent cannot amount to a fraud on the principal, it can contribute to a climate in which bribery is common.[94] But this argument seems unconvincing now that we have abandoned the second limb of our provisional definition. We now believe that the offence should be confined to conduct intended to induce a *particular* agent to act primarily in return for an advantage, thus creating a potential conflict with *that* agent's duty to further the interests of his or her principal. It is the function of the law of corruption to avoid such conflicts. But we cannot see how any conflict can exist if the agent's principal knows all the circumstances and consents to what is done.[95]

5.91 Secondly, we pointed out that such a defence could have no application in relation to a duty owed to the *public*, rather than any particular person. This is true, but we also suggested a solution:

> It would be possible for the law to say that it is a defence if *any person to whom the agent may owe a duty* consents to what is done: this defence would by its nature be unavailable to those agents whose duties are not owed to persons at all, but to the public or the state.[96]

If an agent has a public function, something done in order to influence the agent's performance of that function is not excusable merely because the agent also performs functions for one or more *persons* who consent to what is done. If an advocate, for example, is bribed to act in breach of his or her professional duty to the court,[97] the *client's* consent should be immaterial.

---

[93] See paras 5.123 – 5.130 below.

[94] Consultation Paper No 145, para 8.43.

[95] In Fiduciary Duties and Regulatory Rules (1995) Law Com No 236, para 2.13, we noted that fiduciary duties could be modified or displaced if the beneficiary's consent was obtained after disclosure of all material facts.

[96] Consultation Paper No 145, para 8.45.

[97] See Meredith Blake and Andrew Ashworth, "Some Ethical Issues in Prosecuting and Defending Criminal Cases" [1998] Crim LR 16, 16–17.

> [T]he lawyer-client relationship is not simply a private contractual matter, with the lawyer being hired as technical expert. For someone to be allowed to practise as a lawyer, she or he must not merely have passed the required examinations and have served whatever form of apprenticeship is prescribed, but also have made an undertaking or pledge to abide by certain professional standards. In most legal systems, including England and Wales, these standards are not enshrined in legislation. They are drawn up by the professions themselves and enforced by way of self-regulation, even though there is a strong argument that the integrity of the profession is fundamental to the administration of justice, which in turn is a basic constitutional function. The "public" element in the lawyer's duties may be said to reside in the recognition of barristers and solicitors as "officers of the court". That

5.92 Even an agent who has no public duties may have more than one principal, and may be bribed by or with the consent of one of them to act against the interests of another. Clearly it is only the consent of the latter principal that could be a defence. Conversely, it seems immaterial that a third principal does *not* consent, if the only one who is adversely affected does. In principle it should be both necessary and sufficient, for this defence to be successfully invoked, that what is done is done with the consent of the principal (or all those principals) for whom the agent performs *the functions in question* – that is, those functions in whose performance the agent is intended to be influenced by the advantage conferred.

5.93 In the consultation paper we added that such a rule might lead to protracted arguments about precisely who or what the agent's duty is owed to, thus distracting attention from the central issue of whether what was done was corrupt. This is indeed a danger; but it may be inevitable. Given that the offence we recommend is essentially concerned with potential breaches of duty by the particular agent involved, it would be anomalous to impose liability where there is no question of any breach of duty because the principal consents to what is done.

5.94 Finally we pointed out that difficulties may arise where the agent clearly owes a duty to a corporate body, but it is not clear which of the individuals within that body has the authority to consent on its behalf. This too may be an inevitable consequence of focusing on potential breaches of duty by the particular agent involved. Where an agent alleges that what was done was done with the consent of someone who had authority to give such consent, it must in principle be right to enquire whether that person did consent and, if so, whether he or she did have authority to do so.

5.95 However, in this situation there may be a further difficulty. Where an individual (such as the agent's manager) has authority to consent to what the agent does, it is possible that the giving of that consent may itself be corrupt: the manager may also have been bribed, or may hope to be. Alternatively, the *agent* may have authority to consent on the principal's behalf, and may corruptly purport to consent to his or her own corrupt conduct. It may be arguable that such a "consent" would not be valid, but this is a difficult issue which would hardly be suitable for determination in a criminal trial. Clearly the defence of consent should not be available in such circumstances, and we think that express provision should be made for this purpose. We recommend below[98] that an offence should be committed by an agent who *performs* his or her functions as agent corruptly – that is, primarily in return for, or in order to obtain, a corrupt advantage. This enables us to recommend, in respect of the principal's consent, that consent may be given on a principal's behalf by another agent of the principal, but that it does not count if in giving it the other agent performs his or her functions as an agent corruptly.

gives rise to certain duties, most prominently the obligation to assist the court in the fair administration of justice, and not knowingly to deceive or mislead the court. (Footnote omitted).

[98] Recommendation 14, para 5.122 below.

5.96 In accordance with the general principle of criminal law that a defendant who is mistaken about the facts should be treated as if the facts were as he or she believes them to be,[99] a person's conduct should clearly not be regarded as corrupt if he or she mistakenly believes that the relevant principal knows all the material circumstances and consents to what is done.

5.97 A more difficult case is that of the defendant who, while knowing that the principal does not in fact consent, or does consent but does not know all the material circumstances, believes (rightly or wrongly) that the principal *would* consent *if* he or she knew all the material circumstances. In the law of theft it is expressly provided that

> A person's appropriation of property belonging to another is not to be regarded as dishonest … if he appropriates the property in the belief that he would have the other's consent if the other knew of the appropriation and the circumstances of it.[100]

5.98 We do not regard corruption as an offence of dishonesty,[101] and it is arguable that an agent (who is by definition a person in a position of trust) should not be able to escape liability on this basis. If the agent does not have the principal's consent, there will usually be an opportunity to get it; and if the agent chooses to proceed without doing so, it should arguably be at his or her own risk. On the other hand, theft may also be committed by persons in positions of trust; yet such a person has a complete defence if he or she believes that the owner of the property would have consented – even if he or she has an opportunity to find out what the owner's reaction would be, and fails to take it. We do not condone the acceptance of benefits by agents who assume, without reasonable enquiry, that their principals would consent to their actions. But we believe that to hold such agents guilty of *corruption* would be unduly harsh. We have therefore concluded that the corruption legislation should follow the example of the Theft Act, and should provide that it is not corrupt to do an act or make an omission in the belief that the relevant principal would consent to it if he or she knew all the material circumstances.

Our recommendation

5.99 **We recommend**

(1) **that a person who confers an advantage, or offers or agrees to confer an advantage, should be regarded as doing so corruptly if he or she**

   (a) **intends a person, in performing his or her functions as an agent, to do an act or make an omission, and**

---

[99] *Williams (Gladstone)* [1987] 3 All ER 411, *Beckford* [1988] AC 130.

[100] Theft Act 1968, s 2(1)(b).

[101] See paras 5.123 – 5.130 below.

(b) believes that, if the person did so, it would probably be primarily in return for the conferring of the advantage (or the advantage when conferred), whoever obtains it,[102]

unless

(i) the advantage is conferred (or to be conferred) by or on behalf of the agent's principal (or, in the case of functions performed for the public, on behalf of the public) as remuneration or reimbursement in respect of the performance of the functions,[103] *or*

(ii) the functions concerned are performed only for one or more principals (not the public), and each principal is aware of all the material circumstances and consents to the conferring of the advantage or the making of the offer or agreement;

(2) a person should be treated as if he or she were aware of all the material circumstances, and consented to the conferring of the advantage or the making of the offer or agreement, if the defendant believes that that person

(a) is aware of those circumstances and so consents, or

(b) would so consent if aware of those circumstances; and

(3) consent should count for these purposes if given on a principal's behalf by any agent of the principal, unless in giving it the agent performs his or her functions as an agent corruptly.[104]

(Recommendation 10)

Corruptly conferring an advantage as a reward

5.100 In the consultation paper[105] we noted that, as the present legislation recognises, the mischief that we seek to prevent can arise in three main ways:

(1) A confers an advantage on B on the understanding (express or implied) that B will in return act in a particular way.

(2) A promises to confer an advantage on B later, on condition that B *first* acts in a particular way. (The "promise" may of course be implied: that is, A may act in such a way as to lead B to believe that a reward may be forthcoming, without actually saying so.)

(3) A confers an advantage on B as a reward for B's having *already* acted in a particular way.

In the first two cases A's conduct acts as an inducement; in the third it is not an inducement but only a reward.

---

[102] See cl 6(1) of the draft Bill.

[103] See cl 10 of the draft Bill.

[104] See cl 11 of the draft Bill.

[105] Consultation Paper No 145, para 8.5.

5.101 Although we appreciated the force of the view that inducements should be treated more strictly than rewards "on the ground that it is not as self-evidently corruptive to reward people for acting in a particular way in the past as it is to induce them to do so in the future", we provisionally proposed[106] that the re-formulated offence should apply equally to inducements and rewards. We gave the following reasons for our conclusion.

> ... it is not just rewards that may or may not have a tendency to corrupt, depending on the circumstances: the same may apply to inducements. It would be artificial to permit the gift to an agent of a bottle of whisky at Christmas, in recognition of the agent's past assistance, but to prohibit such a gift if made in the hope of a mutually profitable relationship in the following year. This reasoning suggests that the crucial distinction is not that between rewards and inducements, but that between conduct which does and does not tend to encourage breaches of duty.[107]

5.102 About a quarter of respondents expressed a view on this aspect of our provisional definition, and all of them agreed with it. We have no doubt that the new offences should, *in some circumstances*, extend to rewards as well as inducements.

5.103 The more difficult question is how to define the circumstances in which a reward is *corrupt*. Our provisional definition of a corrupt transaction applied to rewards as well as inducements, but, for reasons explained above,[108] we have had to abandon that definition. Our new definition of a corrupt *inducement* requires a belief that, if the agent were to act in the manner desired, it would probably be primarily in return for the advantage conferred. This approach obviously needs adaptation before it can be applied to rewards for favours *already* shown. It is not enough simply to require that the reward be given in return for an act or omission on the part of the agent, because that would include legitimate rewards (such as tips, and other tokens of genuine appreciation) as well as corrupt ones. But how does a corrupt reward differ from a legitimate one?

5.104 Developing the analysis referred to above,[109] we may distinguish four situations.

(1) A leads B to believe that, if B acts in a particular way, A will reward B for doing so. B therefore acts in that way, and A does reward B.

(2) Without any encouragement from A, B nevertheless believes that, if B acts in a particular way, A will reward B for doing so – for example, because B believes that A has rewarded other agents for acting in that way. B therefore acts in that way, and A does reward B.

(3) B acts in a particular way, not as a result of a corrupt inducement and not (or not primarily) with a view to reward. A rewards B for acting in that

---

[106] Consultation Paper No 145, para 8.6.

[107] Consultation Paper No 145, para 8.11. Cf *Richards* 6 October 1994, CA No 94/0534/W5, where it was held that the acceptance of inducements and the acceptance of rewards are not separate offences, merely different ways of acting corruptly.

[108] Paras 5.69 – 5.74 above.

[109] Para 5.100 above.

way, hoping that the reward will influence B to act in a similar way in the future.

(4) B acts in a particular way, not as a result of a corrupt inducement and not (or not primarily) with a view to reward. A rewards B for acting in that way, with no thought of influencing B to act in a similar way in the future.

5.105 Cases (1) and (3) are really forms of inducement, and therefore fall within the principles set out in Recommendation 10 above.[110] In case (1), A is offering to confer an advantage with the intention that B should act in return for that advantage. It is immaterial that that advantage will not yet have been conferred when B acts in return for it, or that it may never be conferred at all. And in case (3) the advantage is an inducement to act corruptly in the future, as well as a reward for doing so in the past. The position is the same if the advantage is conferred (or to be conferred) on a third party connected with B, in the hope of influencing B.

5.106 Case (4) is not, in our view, corrupt at all. The act rewarded is not a corrupt act, because it is not illegitimately influenced by inducements or the hope of reward. The reward for it is therefore not a corrupt reward: there is no mischief in rewarding an agent for the way he or she performs his or her functions, provided that the agent does not do so corruptly and that the reward is not also an inducement for the future. It may be that the agent is under an obligation to account to the principal for the reward, and may even hold it on trust for the principal; and a failure to account for it should arguably be treated as an offence of dishonesty against the principal.[111] This is an issue to which we intend to return in the course of our review of the law of dishonesty; but it is not, we believe, an aspect of the law of *corruption*.

5.107 It is perhaps arguable that, even if case (4) is not in principle corrupt, it cannot safely be exempted because it is too hard to distinguish from cases (1) and (2). If B has given A a valuable contract, and A has rewarded B handsomely for doing so, the defence may assert that B did not expect to be rewarded and that A was motivated by unalloyed gratitude; and, it may be said, such a defence would be impossible to rebut. We believe that this reasoning overstates the difficulty. If the reward is more substantial than a genuine token of gratitude would normally be, and no explanation is offered for that fact, the fact-finders will draw such inferences as appear proper – for example, that B had been promised a reward if A got the contract.[112] And a similar inference is likely to be drawn if an innocent explanation is offered but not believed.

5.108 Case (2), on the other hand, is in our view corrupt. By analogy with our recommendation on the definition of a corrupt inducement, we believe that B's conduct, though not procured by corrupt inducement, is nevertheless corrupt if the hope of reward is B's *primary purpose* in acting in that way. If this is the case,

---

[110] Para 5.99 above.

[111] Cf para 5.138 below.

[112] CJPOA, s 35.

and A rewards B in the knowledge that it is the case, A is implicated in B's corrupt purpose, and the reward is therefore corrupt.

5.109 **We recommend that a person who confers an advantage, or offers or agrees to confer an advantage, should be regarded as doing so corruptly if he or she**

    (1)   **knows or believes that, in performing his or her functions as an agent, a person has done an act or made an omission,**

    (2)   **knows or believes that that person has done the act or made the omission primarily in order to secure that a person obtains an advantage (whoever obtains it), and**

    (3)   **intends the person known or believed to have done the act or made the omission to regard the advantage (or the advantage when conferred) as conferred primarily in return for the act or omission.**[113]

                                                                                  (Recommendation 11)

Former and future agents

5.110 The 1906 Act appears to require that the bribe be received by the agent during the currency of the agency, but the 1889 Act extends to circumstances in which the public officer is no longer in office at the time of receipt of the bribe or has yet to assume office.[114] In the consultation paper we took the view that limiting criminality on the basis of *when* a bribe is offered or received is unnecessarily restrictive and artificial, and noted that the MCCOC had reached a similar conclusion. We therefore provisionally proposed that, where a person corruptly accepts, solicits or agrees to accept, or confers or offers or promises to confer, an advantage in connection with the performance by an agent of his or her duty, that person should be guilty of the offence even if the agent in question is no longer an agent, or is not yet an agent, at the material time.[115] All the respondents who commented on this proposal agreed with it.

5.111 Since our analysis of what is "corrupt" hinges on the intentions and beliefs of the person whose conduct is in issue, we would add that it may be corrupt to confer an advantage

    (1)   as an inducement to, or a reward for, an act or omission on the part of a person *mistakenly believed* to be an agent;

    (2)   as a reward for an act or omission on the part of a person mistakenly believed to have been an agent *at that time*; or

---

[113] See cl 6(2) of the draft Bill.
[114] Consultation Paper No 145, para 8.80.
[115] Consultation Paper No 145, para 8.81.

(3) as an inducement to a person (who is not now an agent) to do an act or make an omission in performance of his or her functions as agent *if and when he or she becomes* an agent.

Corruptly obtaining or agreeing to obtain an advantage

5.112 In the great majority of corruption cases the advantage will be corruptly conferred *and* obtained: each party knows the other's motives, and both are corrupt. For most practical purposes, therefore, it would be sufficient if our definition of corruptly *conferring* an advantage were supplemented by a rule that a person who obtains (or agrees to obtain) an advantage does so corruptly if he or she knows that the person conferring it (or agreeing to confer it) does so corruptly.

5.113 However, our definition of "obtaining an advantage" is very wide: a person obtains an advantage if another person does an act or makes an omission which *either* is done or made in consequence of the first person's request *or* results in a benefit to the first person. In the latter case, the person obtaining the advantage may do so without doing anything at all, or even indicating his or her agreement to the receipt of the benefit. It follows that a person who obtains an advantage cannot be regarded as doing so corruptly *merely* because he or she knows that it is conferred corruptly. Otherwise an offence of corruption would be committed by an agent of unimpeachable honesty who knows that another person is paying funds into the agent's bank account (thus conferring a benefit on the agent) and that those funds are intended as a bribe, even if the agent has no intention of acting upon the bribe, promising to act upon it or keeping it. Our recommendation therefore ensures that, where a person *obtains* an advantage (as against *agreeing* to obtain it) but does not *request* it, he or she will not thereby commit an offence unless he or she *consents* to obtaining it.

5.114 It is possible, however, that the parties may be at cross-purposes, or that the bribee may intend to renege on the agreement. In the consultation paper we provisionally proposed that the bribee should be guilty of the full offence, and not merely an attempt,[116] if he or she mistakenly *believes* that the advantage is conferred corruptly: the bribee's conduct is still corrupt.[117] Among those who commented on the point, there was almost unanimous agreement with our view.

5.115 We provisionally proposed that a person who accepts an advantage, knowing it to be intended as a corrupt inducement, should not escape liability merely because he or she has no intention of acting in the manner desired.[118] This proposal was generally accepted.

5.116 On the other hand we also proposed what would in effect be an exception to this rule – namely that a person should not commit an offence of *corruption* (as distinct from deception) by accepting or soliciting an advantage, ostensibly in connection with a person's performance of his or her duty as an agent, if that person is not in

---

[116] See Criminal Attempts Act 1981, s 1.
[117] Consultation Paper No 145, para 8.89.
[118] Consultation Paper No 145, para 8.94.

fact an agent (or was not, or will not be, an agent at the time when the performance of his or her duty is in question).[119] Although the majority of those who commented on this proposal agreed with it, further reflection has persuaded us that this situation is not distinguishable in principle from that discussed in the previous paragraph, where the person in question *is* an agent but has no intention of acting on the inducement. In neither case is there any influence on an agent's performance of his or her functions as agent; but in both cases the advantage is obtained by a person who knows it to be corruptly conferred, and in principle we now believe that that person can properly be treated as being implicated in the other party's corrupt purpose.

5.117 For these reasons our recommendation refers only to the bribee's knowledge or belief or intentions *with regard to the intentions of the briber* – not to whether the *bribee* intends anyone to be influenced, nor to whether the person the briber intends to influence is in fact, or ever has been or is ever likely to be, an agent at all.

5.118 **We recommend that**

   (1) **a person who obtains an advantage should be regarded as obtaining it corruptly if he or she**

   (a) **knows or believes that the person conferring it confers it corruptly, and**

   (b) **gives his or her express or implied consent to obtaining it (in a case where he or she does not request it);[120] and**

   (2) **a person who agrees to obtain an advantage should be regarded as agreeing to obtain it corruptly if he or she knows or believes that the person agreeing to confer it agrees corruptly.[121]**

   (Recommendation 12)

Corruptly soliciting an advantage

5.119 *Soliciting* an advantage is somewhat analogous to conferring an advantage as a corrupt inducement in the absence of any prior agreement that it will be acted upon.[122] Just as the person providing a corrupt inducement hopes that it will procure the desired act or omission on the part of the agent in question, so the person soliciting an advantage hopes that it will be conferred (or at any rate that the person addressed will *agree* to confer it).[123] But the person providing an inducement probably does not care *why* the agent does the act or makes the

---

[119] Consultation Paper No 145, para 8.99.

[120] See cl 7(1) of the draft Bill.

[121] See cl 7(2) of the draft Bill.

[122] See paras 5.100 – 5.109 above.

[123] It is possible to solicit an advantage with the intention that a person should *agree* to confer it but should not in fact confer it, eg for the purpose of demonstrating that that person is *willing* to confer it. This situation falls within the scope of the offence because we do not recommend a defence of entrapment: see paras 5.144 – 5.150 below.

omission, as long as he or she does so; and we have therefore recommended that such a person should be regarded as acting corruptly if he or she believes that the act or omission would *probably* be primarily done or made in return for the inducement. Similarly, the person soliciting an advantage probably does not care whether it is conferred corruptly or otherwise, as long as it is conferred; and in our view it should therefore be sufficient if he or she believes that, if the advantage were conferred at all, it would probably be conferred corruptly.

5.120 **We recommend that a person who solicits an advantage should be regarded as soliciting it corruptly if he or she**

  (1) **intends a person to confer it or agree to confer it, and**

  (2) **believes that, if the person did so, he or she would probably do so corruptly.**[124]

(Recommendation 13)

Corruptly performing functions as an agent

5.121 As we have explained, we believe that an agent should be guilty of an offence not only if he or she corruptly obtains an advantage, or solicits one or agrees to obtain one, but also if he or she corruptly performs his or her functions as an agent, whether in the hope of reward or in return for a previous inducement.[125]

5.122 **We recommend that a person who performs his or her functions as an agent should be regarded as performing them corruptly if he or she**

  (1) **does an act or makes an omission primarily in order to secure that a person confers an advantage (whoever obtains it), and believes that if the person did so he or she would probably do so corruptly, or**

  (2) **does an act or makes an omission when he or she knows or believes that a person has corruptly conferred an advantage (whoever obtained it), and regards the act or omission as done or made primarily in return for the conferring of the advantage.**[126]

(Recommendation 14)

DISHONESTY

Is corruption an offence of dishonesty?

5.123 In the consultation paper we provisionally concluded that corruption was not an offence of fraud or dishonesty, and should not be treated as such.[127] The implications of this provisional conclusion were, we said, twofold. First, it meant that it would be inappropriate to consider the role of the law of fraud and

---

[124] See cl 7(3) of the draft Bill.

[125] Paras 5.54 – 5.58 above.

[126] See cl 8 of the draft Bill.

[127] Consultation Paper No 145, para 1.28.

dishonesty in controlling corruption. Secondly, it defined our task as one of formulating a *free-standing* offence of corruption, independent of the structure and conventions of the law of fraud and dishonesty.[128]

5.124 Among the respondents who commented on our provisional conclusion that corruption is not, and should not be treated as, an offence of dishonesty or fraud, roughly half agreed and half disagreed. Of those who agreed, many did so without further comment. The tenor of such comments as were made was that corruption should be an autonomous offence, independent of any other branch of the criminal law.[129]

5.125 Among those who disagreed with our provisional conclusion (and gave reasons for doing so) there was a division of opinion. Some argued that dishonesty should be *an element of the offence*, which the prosecution must prove in addition to the other elements. Of these, some[130] thought that the "dishonesty" required should be of the kind defined in the theft case of *Ghosh*,[131] while some thought it should be defined differently,[132] or not at all.[133]

5.126 Others agreed with us that dishonesty should not be an *element* of the offence, but disagreed with our conclusion that it should not be treated as an *offence* of dishonesty. However, these respondents seem to have taken issue with the sense in which we used the term "dishonesty", rather than with our analysis of the nature of corruption. We did not deny that corruption is usually (perhaps always) dishonest in a broad sense: an ordinary person would probably say that it is a dishonest thing to do. But, we suggested, this broad sense is not necessarily helpful for the purpose of classifying and defining criminal offences, or law reform projects. For that purpose, we have found it convenient to bracket together those offences that consist in *damaging or endangering a person's financial interests*, by infringing that person's rights in private law. Offences of this kind have important

---

[128] Consultation Paper No 145, paras 1.29 and 1.30.

[129] Professor Sir John Smith took the view that corruption was "distinct from offences of dishonesty and fraud"; the Common Serjeant of London, similarly, referred to corruption being "an offence in its own right". The Association of Chief Police Officers ("ACPO") wrote that it was "an individual offence, in many ways quite unlike any other offence, which should be capable of being dealt with in isolation, unrestricted by constraints imposed upon it by other legislation".

[130] Eg the Employment Law Bar Association. Liberty suggested that the fault element should be "corruptly" but that the definition of that element should include a reference to "dishonesty", meaning that the putative briber should not be guilty unless he or she was aware that the transaction was improper – which closely resembles the *Ghosh* definition (see n 137 below).

[131] [1982] QB 1053. According to this decision, a person's conduct is dishonest if ordinary people would think it dishonest *and* he or she knows that they would.

[132] Professor Mark Freedland suggested defining it as "a component which is presumed from the basic factual elements of corruption, unless the defendant can show on the balance of probabilities that he or she acted in pursuit of a goal or an interest which he or she reasonably regarded as one which it was legitimate in all the circumstances to pursue."

[133] Phillips LJ proposed "qualifying the offence by the adverb 'dishonestly', a word which a jury can understand without further direction". But the use of the word "dishonestly" without express definition would presumably attract the *Ghosh* definition.

similarities with one another, and important differences from other kinds of offence. Corruption, we argued, is not an offence of this kind. For convenience again, we refer to offences of this kind as "offences of dishonesty"; and it was in this sense that we suggested that corruption is not an offence of dishonesty.

5.127 Those respondents who rejected this conclusion appear to accept that corruption is not an offence of dishonesty *in the sense in which we used that phrase*;[134] but they seem to be using that expression in a wider sense, as including *all* offences, *whether or not against a particular victim*, which an ordinary person would think it dishonest to commit. Liberty and the Employment Law Bar Association both argued that "dishonesty" did not require a victim, but only a state of mind. The CPS thought that "corruption is at least morally if not legally dishonest and is most likely to be regarded as such by the general public", and the Employment Law Committee of the Law Society argued that most people "instinctively" regard bribery as dishonest.[135] Similarly the Society's Criminal Law Committee argued that a company affected by corruption "would have little doubt that a fraud has been perpetrated upon it".

5.128 We are not convinced by these arguments, because they seek to demonstrate something which we did not deny – namely that corruption is usually, perhaps always, dishonest in the ordinary sense of that word. We still believe that, for the purposes of law reform, it is helpful to distinguish offences which typically involve dishonesty in a *general* sense – and may indeed be *defined* in such a way as to require it – from offences of *unlawfully causing loss to specific victims*. For the latter category we have so far been unable to devise a better term than "offences of dishonesty" (though we recognise that this usage risks confusion with offences of the former kind).[136]

5.129 Our provisional view was that corruption is not an offence of dishonesty *in this sense*. We still take that view, for two reasons. First, corruption does not necessarily prejudice the interests of the agent's principal. The agent (B) may be bribed to do something which he or she is in any event bound to do, but which he or she might not do unless bribed; or, where two or more options are equally beneficial to B's principal (C), B may be bribed to choose one option rather than another. In these

---

[134] Similarly, there was general agreement that it is not an offence of *fraud* – though, as the CPS pointed out, corrupt conduct will often in practice amount to a conspiracy to defraud.

[135] The Committee also took the view that it would be confusing for employers if corruption were not regarded as an offence of dishonesty:

> In most workplaces, a serious dishonest act will justify summary dismissal. Employers faced with good evidence of someone wrongfully taking money (or some other inducement) to act against the employer's interests, and told that the act was not dishonest may be confused and uncertain about their position and how to proceed.

But we did not suggest that a corrupt act *cannot* be a dishonest act, only that it is *not necessarily* dishonest (in the sense in which we use that term).

[136] An alternative is "fraud offences", because "fraud" technically includes any dishonest conduct causing loss or the risk of loss. Unfortunately this term too would invite confusion, since its non-legal meaning applies only to certain kinds of such conduct – chiefly deception and white-collar crime.

cases C is no worse off than if B had not been bribed. It is true that, if B keeps the bribe, C is worse off than if B had handed it over, as B is legally bound to do; but it is debatable whether this consideration justifies treating B's conduct as an offence of dishonesty vis-à-vis C. B is not depriving C of what is C's, but failing to confer on C a benefit which C did not expect in the first place.

5.130 A further reason for our view that corruption is not in essence a dishonesty offence is the fact that it is not (and under our recommendations would not be) confined to the agents of identifiable principals. It extends also to what we have called "quasi-fiduciaries" – that is, persons whose duty is to subordinate their own interests to those of the *public* rather than those of any particular person. More importantly, it extends to persons such as police officers and civil servants, who are technically the agents of particular persons (namely their employers) but owe a duty to the public as well. We do not believe that it would be practicable to have one form of criminal liability (based on dishonesty) for agents who owe a duty only to particular persons, and another (not based on dishonesty) for agents whose duty is owed wholly or partly to the public. It follows that the offences we recommend would not be offences of dishonesty (in our sense).

Should the definition of corruption include a requirement of dishonesty?

5.131 It does *not* follow that the fault element of the new corruption offences should not include a requirement of dishonesty *in the ordinary sense*, as in the law of theft. It would then be necessary for the prosecution to show that the transaction in question was dishonest according to the standards of ordinary people, and that the defendant knew it was.[137] The prosecution might be able to do this even where no identifiable person was at risk of financial loss. In other words, it would not be illogical for an offence to include a requirement of dishonesty (in the ordinary sense) although it is not an offence of dishonesty (in our sense).

5.132 In the consultation paper we considered whether or not a moral element should be included in the definition of the new offences, and whether or not this moral element should be defined by way of a requirement of dishonesty.[138] As we pointed out in the consultation paper, "The answer may of course depend on how the other requirements of the offence are defined".[139] Given that we had proposed, as part of the new offences, a definition of the fault element "corruptly" which, we believed, would mean that only truly corruptive conduct would be criminalised, we concluded that a moral element was unnecessary. Even if our proposed definition of "corruptly" were not accepted, however, we still did not favour the inclusion of a dishonesty element since, in our view, it did not accurately identify the "prohibited evil" of corruption offences.

---

[137] In *Ghosh* [1982] QB 1053 the court explained that, to establish that a defendant was dishonest, the prosecution had to prove: (1) that what the defendant did was dishonest according to the ordinary standards of reasonable and honest people, and (2) that the defendant must have realised that what he or she was doing was by those standards dishonest.

[138] Consultation Paper No 145, paras 8.30 – 8.39.

[139] Consultation Paper No 145, para 8.25.

5.133 A requirement of dishonesty would mean that, even if a transaction has all the other characteristics of corruption, a party to it could invite the fact-finders to acquit on the ground that, in all the circumstances, an ordinary person would not regard that transaction as dishonest, or that the defendant did not realise that an ordinary person might so regard it. A person charged with theft or deception already has a right to seek an acquittal on this basis, though it is debatable whether this should continue to be so.[140] A person charged with corruption has no such right, since the word "corruptly" does not imply a requirement of dishonesty.[141] The question is whether such a right should now be conferred, thus making convictions harder to obtain.

5.134 We believe that to include a requirement of *Ghosh* dishonesty[142] would significantly reduce the effectiveness of the law of corruption. As we noted in the consultation paper,

> Corruption, unlike most criminal offences, is highly "culture specific".[143] Whether any given conduct is categorised as corrupt will depend, in part, on both the perspective from which it is viewed[144] and the environment in which it occurs.[145]

There would therefore be a real risk that an offence which included a requirement of *Ghosh* dishonesty, or something resembling it, would fail to catch those who engage in corruption but claim that it is (for example) normal practice. We therefore do not recommend that the definitions of the new offences should include any requirement that the defendant must have acted "dishonestly".

POSSIBLE DEFENCES

5.135 In the consultation paper,[146] we considered five possible defences: that the allegedly corrupt transaction was performed openly, that the agent's principal consented to what was done, that the agent had no obligation to account to the principal for the benefit received, that accepting benefits from third parties was normal practice, and that the benefit received was of small value. We provisionally concluded that none of these facts should be a complete defence, though each of them might well be relevant to the issue of whether what was done was corrupt. A number of respondents disagreed with this conclusion and thought that at least one of these facts should be a defence. As we have already explained, we are now persuaded that something done with the informed consent of the agent's principal

---

[140] We intend to discuss this issue in a forthcoming consultation paper on the law of fraud, theft and deception.

[141] *Cooper v Slade* (1858) 6 HL Cas 746; *Smith* [1960] 2 QB 423; *Wellburn* (1979) 69 Cr App R 254.

[142] See n 137 above.

[143] See D J Lowenstein, "For God, For Country or for Me?" (1986) 74 Calif LR 1479.

[144] See A T H Smith, *Property Offences* (1994) para 25–01.

[145] Consultation Paper No 145, para 8.24.

[146] Consultation Paper No 145, paras 8.40 – 8.58.

cannot properly be regarded as corrupt.[147] We now consider the other defences discussed in the consultation paper, as well as possible defences of entrapment and public interest.

Disclosure

5.136   Morison J, President of the Employment Appeal Tribunal, thought that non-disclosure was a necessary part of corruption.

> The mere fact that a secret payment has been made has corrupted the agent so as to put him under the influence of the donor. The element of non-disclosure is, as a first thought, a necessary; and a donee is not corrupted if he makes disclosure, although the gift may have been made with a corrupt intent so that the donor is guilty. It is the "separation" of the agent from his principal that corrupts.

5.137   It is true that an advantage which is disclosed to the agent's principal will rarely be corrupt; but in our view this is not *because* it is disclosed. If, believing that the principal will be reluctant to take legal action, the agent admits taking a bribe, we see no reason why the admission should preclude a prosecution. Nor, we believe, should it make any difference that the agent tells the principal what is happening *before* the bribe is paid or the agreement made. Where disclosure does render the agent's conduct innocent, this will usually be because it indicates that the principal *consents* to what is done, or at any rate that the agent *thinks* the principal will consent. Where either of these is the case, under our recommendations the agent's conduct would not be corrupt.[148] Where neither is the case, we do not believe that the mere fact of disclosure should be a defence, and we make no such recommendation.

No obligation to account

5.138   Under our recommendations there would be a substantial overlap between the scope of the criminal law of corruption and the circumstances in which an agent is under an obligation to account for the benefit received. However, we remain of the view that the unavailability of a civil remedy should not in itself be a defence to a charge of corruption. As we said in the consultation paper, there are circumstances in which the reason why a civil remedy is unavailable has no bearing on whether the agent has acted corruptly. We therefore do not recommend that the absence of an obligation to account should be a defence.

Normal practice

5.139   A number of respondents[149] favoured a defence of "normal practice". Some expressed concern that, without such a defence, British businesses might suffer a competitive disadvantage, given what the Institute of Directors described as the

---

[147]   See paras 5.88 – 5.98 above.

[148]   See para 5.99 above.

[149]   The Institute of Directors, Kingfisher plc, the Institute of Chartered Accountants, the CBI and The Newspaper Society.

"different moralities of export markets". Other respondents expressed concern that "ordinary business hospitality" might be criminalised.

5.140 However, we remain of the view we expressed in the consultation paper:

> [I]f the normality of the conduct were a complete defence it would follow that, once corrupt practices have taken root in a given environment, they could no longer be regarded as corrupt. We believe that if a transaction is essentially corrupt then it should be criminal, however common such transactions may be.[150]

5.141 A defence of "normal practice" would also be far too vague to be workable, and would largely nullify our attempts to spell out what conduct is and is not corrupt. We accept, of course, that it is common for certain kinds of advantage to be conferred on agents with a view to influencing their conduct: "ordinary business hospitality" is perhaps the best example. But we have argued that the lawfulness of such hospitality rests not on its normality but on the absence of the characteristic feature of corrupt conduct – namely the intention that the agent should favour the provider of the advantage *in return*, as distinct from doing so for a legitimate reason which the provision of the advantage may encourage the agent to take into account.[151] We are confident that ordinary business hospitality would not qualify as corrupt conduct under the definition we recommend; but, if it did, we believe that it should make no difference that it is common practice.

Small value

5.142 The Institute of Chartered Accountants supported a specific defence of "small value":

> Benefits of small value are a regular feature of commerce in most jurisdictions. We recognise that in most cases these will be precluded from definition as an offence because they will not be considered corruptive. However, we do wonder whether the law will be easier to interpret and implement, and court cases shorter, if a specific defence of small value is retained.

5.143 We are not persuaded that this is either a workable or a necessary defence. The practical difficulty lies in deciding how valuable a gift has to be before it becomes a bribe. Perceptions of generosity will vary according to the recipient's standard of living. Equally importantly, we are not convinced that a "small value" defence is necessary in any event. If we define "corruptly" in terms of an intention to induce an agent to act in a certain way, the fact that a gift was of small value will be evidence that it could not have been intended to act as an inducement.

---

[150] Consultation Paper No 145, para 8.52.

[151] See paras 5.75 – 5.82 above.

Entrapment

5.144 In the consultation paper we asked for views on whether an intention to expose corruption should be a defence.[152] We were broadly concerned with two situations:

(1) A gives B a bribe in order to expose B as a bribe-taker; and

(2) B accepts from A a bribe in order to expose A as a bribe-giver.

5.145 We were not troubled by the liability of B in case (1) or of A in case (2): there is clear authority that a defence of entrapment is not known to English law,[153] and, in any event, we do not believe that an agent should escape liability merely because he or she is wrong about the intention with which the advantage is conferred.[154] The question is whether the *entrapper* should be liable.

> In each situation the person who is trying to entrap is aware of the corrupt intent of the other and so, on the principles outlined above, should be guilty. His or her intention is to create the temptation to act in breach of duty, and it is this temptation that we seek to prevent.
>
> The question, therefore, is what role *motive* should play in the law of corruption. Although the entrapper is in the short term encouraging corrupt behaviour, his or her long term goal is to prevent it; it might therefore seem unjust to impose liability. On the other hand there is a danger that a spurious claim to have acted with such a motive might be hard to disprove. Moreover, the existence of such a defence might encourage undesirable activities by investigative journalists and others. If one has reason to believe that a person is corrupt, the appropriate course is to bring the matter to the attention of the police or some other suitable authority, not to take the law into one's own hands.[155]

5.146 As regards the case where B's acceptance is intended to expose A, we have recommended that it should be an offence to obtain an advantage in the knowledge that it is corruptly conferred;[156] it makes no difference that the agent does not intend to do anything in return. The reason we gave in support of this view in the consultation paper was that "by accepting the gift knowing that it is a bribe, B is knowingly encouraging A to act in the same way in the future".[157] Arguably, this reason does not apply in cases where B's intention is not merely to refrain from acting on the bribe but to take positive steps to expose A, since exposure is more likely to *dis*courage A from offering bribes in future.

---

[152] Consultation Paper No 145, para 8.101.

[153] See *Mealey and Sheridan* (1974) 60 Cr App R 59, 62:

> In fact, if one looks at the authorities, it is in our judgment quite clearly established that the so-called defence of entrapment, which finds some place in the law of the United States of America, finds no place in our law here.

[154] See para 5.122 above.

[155] Consultation Paper No 145, paras 8.100 and 8.101.

[156] See paras 5.112 – 5.118.

[157] Consultation Paper No 145, para 8.94.

5.147 Where A bribes B in order to expose *B*, under our recommendations A's liability would depend on A's intention. A would be guilty if A confers an advantage on B with the intention of inducing B, in performing B's functions as an agent, to do an act or make an omission in return for the advantage. Therefore, if A *intends* B actually to do what A is purporting to bribe B to do (albeit with the higher motive of exposing B), A would fall within the offence we recommend, in the absence of a specific defence of entrapment.

5.148 However, A may[158] have no interest in seeing any return for the bribe. A's purpose of exposing B may be satisfied simply by B's *acceptance* of the bribe. If so, and if this means that A does not *intend* B to act in return for the bribe, a specific defence of entrapment would be unnecessary: A's conduct would not be corrupt because the requisite intention would not exist.[159] However, those actively involved in exposing corruption (such as investigative journalists) may be loath to rely on this line of defence, since basic principles of criminal law allow fact-finders to impute an intention even where the result allegedly intended is not what a defendant really wanted to happen. On the other hand, they may take comfort from the fact that it is not enough that that result was a foreseeable by-product of the defendant's actions: it must have been a "virtual certainty", and the defendant must have appreciated that this was the case.[160]

5.149 Of the respondents who expressed a view on this issue, only a small minority were in favour of an entrapment defence. Their principal argument was that the purpose of exposing corruption is a laudable one.[161] The main arguments advanced *against* such a defence were that it would "encourage unauthorised persons to take the law into their own hands";[162] that there is a danger of such a

---

[158] A might want to show B up not only as someone prepared to take bribes but, worse still, as someone who will act on them.

[159] Professor Sullivan comments:

> … it is left open whether an intention to expose corruption should afford a defence. Surely it should. If a defendant sought not to corrupt but to expose the presence of corruption, one cannot say that he or she acted corruptly whatever other criticisms the conduct may, justifiably, attract. Further in this vein, it should be a defence that an inducement or reward was offered to obtain information in order to make that information public should it be in the public interest that it be revealed. It should not be enough that the defendant thought that the information should be publicised. Criteria of what the public interest requires would need to be specified and it should be for the jury to decide in the light of these criteria whether the public interest may be successfully invoked by way of defence on the facts of a particular case.

[160] *Nedrick* [1986] 1 WLR 1025, 1027–8, *per* Lord Lane CJ. See *Blackstone's Criminal Practice* (1997 ed) paras A2.2 and B1.11.

[161] The Employment Law Bar Association wrote, for example:

> We believe that entrapment should be a defence, but that the burden of proof should lie upon the defence. That ought to discourage spurious claims to have acted with the motive of uncovering corrupt activities. Furthermore, we do not regard this sort of activity, providing it is carried out by responsible journalists, as "undesirable" – indeed, it is a valuable independent check on corruption.

The Justices' Clerks' Society took a similar line.

[162] Lord Davidson.

defence being exploited by "unscrupulous people";[163] that corrupt activities are a matter for investigative agencies and not investigative journalists;[164] and that such a defence would positively *encourage* temptation and corruption.[165]

5.150 Whilst it may sometimes be in the public interest for corrupt activity to be exposed by entrapment, we agree with the majority of respondents on this issue that there should be no specific defence. In coming to this view, we were persuaded both by the views of our respondents and also by the further point that, were we to introduce such a defence with respect to corruption offences, this would give rise to a glaring inconsistency between, for example, corruption and conspiracy to defraud.

Public interest

5.151 The Newspaper Society expressed concern not only about entrapment – exposing the *corruption* of another – but also on the more general issue of a public interest defence, for example where an investigative journalist pays an agent to expose wrongdoing of any kind by the agent's principal:

> We take the view ... that a moral element *is* inherent in corruption and that a person's reasons for acting in a particular way should be taken into account. ... [W]e can envisage situations in which journalists might encourage or facilitate a breach of duty and thereby act "corruptly" as defined by the Consultation Paper. This would not necessarily be "entrapment" as envisaged in para 8.101... It could be more in the nature of whistleblowing: a desire to uncover information about the employer's questionable activities, an aim to which no moral obloquy can attach. ... Our concern is that the "criminalising" of this kind of newsgathering could inhibit investigative journalism. This could be avoided either through the exclusion of such conduct from the definition of "corrupt" behaviour or perhaps (less attractively) via the introduction of a public interest defence.

5.152 In our recent consultation paper on the misuse of trade secrets[166] we provisionally proposed that a new offence should not extend to those who use or disclose trade secrets in the public interest (such as for the purpose of preventing crime).[167] We have considered whether it would be appropriate to have a public interest defence in the law of corruption so that a defendant could argue, for example, that the purpose of the alleged bribe was to cause an agent to expose his or her principal's criminal activities.[168] We have decided against recommending such a defence. It

---

[163] R C White, Assistant Chief Constable of the RUC.

[164] Office of the Solicitor at the DSS, and the Police Federation of England and Wales.

[165] ACPO.

[166] Consultation Paper No 150, para 6.54.

[167] We note with interest the Lord Chancellor's Department's recent consultation paper on Payments to Witnesses, para 46, in which a public interest defence was considered.

[168] Although Article 10(1) of the ECHR gives everyone the right "to receive and impart information and ideas without interference by public authority", we doubt that this provision would afford the "bribing" investigative journalist with a defence to a charge of corruption. Not only has the right to receive information been interpreted as "basically prohibit[ing] a

may indeed be the case that an agent should be protected if he or she discloses information for the benefit of the public and not for personal gain. Where an advantage is conferred by one party on another, however, we fear that a public interest defence would positively encourage a climate of corruption.

Government from restricting a person from receiving information *that others wish or may be willing to impart to him*" (*Leander v Sweden* (1987) 9 EHRR 433, 456 (italics added)), but Article 10(2) provides a broad list of public interest purposes which will justify public interference with freedom of expression: it may be circumscribed "in the interests of national security, territorial integrity or public safety, for the prevention of disorder or crime, for the protection of health or morals, for the protection of the reputation or rights of others, for preventing the disclosure of information received in confidence, or for maintaining the authority and impartiality of the judiciary".

PART VI
THE INVESTIGATION OF CORRUPTION

6.1 We are very conscious that corruption presents particular difficulties for prosecutors for two main reasons. First, unlike many other offences, such as offences of violence, the victim may often be unaware that the offence has been committed. Secondly, the information necessary for a successful prosecution for corruption is often deliberately shrouded in secrecy. Thus investigating officers are confronted with especial difficulties. In the case of serious fraud, these have been resolved by granting extensive powers to the Serious Fraud Office ("SFO").

6.2 In the consultation paper we provisionally concluded that in the absence of suitable justification as to why corruption offences should be treated exceptionally, the powers of the police should not be extended.[1] Nevertheless, we asked for views on four options,[2] and we will reconsider these options after we have explained the powers given to the SFO and Department of Trade and Industry ("DTI") and the justification for them.

THE SFO AND SECTION 2 OF THE CRIMINAL JUSTICE ACT 1987

Powers of the SFO

6.3 To understand the SFO's powers it is necessary to bear in mind the problems of investigating fraud. The Roskill Committee described the problems of the investigation of serious fraud in this way:

> Fraud ... must be concealed from its victim if it is to succeed, and indeed may not be identified until long after the event. Even when the fraud is detected, it is to be expected that in serious cases the criminals will have taken steps to conceal the way in which the fraud was perpetrated, so as to make the process of investigation and prosecution more difficult. To this end, documents may be falsified or destroyed, and arrangements may be made for some transactions to take place in other jurisdictions, and for the proceeds of the offence to be removed there later perhaps to be followed by the fraudsters themselves. Thus, in large-scale or complex fraud cases, the task facing investigators is formidable.[3]

6.4 On the recommendations of the Fraud Trials Committee under the chairmanship of Lord Roskill, the SFO was established by the Criminal Justice Act 1987 ("CJA 1987"). One of the functions of the office is to "investigate any suspected offence which appears to [the Director of the SFO] on reasonable grounds to involve serious or complex fraud".[4]

---

[1] Para 12.22.

[2] Paras 12.22 – 12.26 (they are considered in paras 6.23 – 6.27 below).

[3] Fraud Trials Committee Report (1986) (Chairman: Lord Roskill) para 2.2.

[4] CJA 1987, s 1(3).

6.5 Section 2 of the CJA 1987 confers on the Director of the SFO ("the Director") extensive powers of investigation. Section 2(2) provides:

> The Director may by notice in writing require the person whose affairs are to be investigated ... or any other person whom he has reason to believe has relevant information to answer questions or otherwise furnish information with respect to any matter relevant to the investigation ... .

Other provisions under section 2 include the power to require the production of documents,[5] and the power to apply to a justice of the peace for a warrant authorising a constable to search for and seize documents.[6]

6.6 Section 2(13) provides that it is an offence[7] to fail, without reasonable excuse, to comply with any requirement imposed under section 2, and section 2(14) further provides that it is an offence[8] to make, either knowingly or recklessly, a statement that is false or misleading in a material particular in purported compliance with any such requirement.

6.7 When a person is required to attend an interview under section 2(2), the Director is not obliged to provide the interviewee with advance information as to the subject-matter of the interview (although the Director may do so if the view is taken that it would be helpful and not likely to prejudice the investigation).[9] The fact that a statement may incriminate its maker is not a reasonable excuse for failing to comply with the Director's order;[10] nor is the fact that the person required to answer questions is the spouse of a person charged with fraud.[11] The powers conferred by section 2 do not cease on a suspect being charged,[12] and section 2 interviews may be conducted even after the delivery of a case statement by the defence.[13]

Safeguards for the person investigated

6.8 The "reasonable excuse" defence in section 2(13) provides one safeguard against the section 2 powers. Another can be found in section 2(8), which provides that a

---

[5] CJA 1987, s 2(3).

[6] CJA 1987, s 2(4)–(7).

[7] Punishable by up to six months' imprisonment, a fine not exceeding level 5 on the standard scale, or both.

[8] Punishable on indictment with up to two years' imprisonment or a fine or both; or on summary conviction, with up to six months' imprisonment or a fine not exceeding the statutory maximum, or both: s 2(15).

[9] *R v SFO, ex p Maxwell, The Times* 9 October 1992.

[10] *R v Director of SFO, ex p Smith* [1993] AC 1.

[11] *R v Director of SFO, ex p Johnson* [1993] COD 58.

[12] *R v Director of SFO, ex p Smith* [1993] AC 1.

[13] *Nadir* [1993] 1 WLR 1322. It would, of course, be open to the defence to try to exclude a late s 2 interview under s 78 or the common law discretion preserved by s 82(3) of the Police and Criminal Evidence Act 1984.

statement given by a person in compliance with a requirement imposed under section 2 may only be used in evidence against him or her

(1) on a prosecution for an offence under section 2(14) (knowingly or recklessly making a statement which is false or misleading); or

(2) on a prosecution for some other offence where, in giving evidence, he or she makes a statement inconsistent with the statement made under section 2.

6.9 Furthermore, section 2 cannot bite in circumstances where a person would be entitled to refuse to disclose information or to produce a document on the grounds of legal professional privilege.[14] Similarly, a person bound by banking confidentiality will not be required to disclose information or produce a document under the section[15] unless either the person to whom the obligation of confidence is owed consents[16] or the Director has authorised the making of the requirement.[17]

Comparison of the powers of the SFO and the CPS

6.10 The SFO[18] has some resemblance to the CPS but differs in three ways:

(1) The role of the SFO is to investigate possible serious fraud and then, if the evidence justifies it, to initiate proceedings. In order to achieve this, the Director of the SFO is given very wide investigatory powers. The CPS, on the other hand, does not itself investigate crime but takes over the conduct of proceedings after the evidence has been gathered by the police or other investigatory body.[19]

(2) The SFO, unlike the CPS, has an *investigatory* function. It works in conjunction with other interested parties, directing the investigation and assessing the evidence as it emerges.

(3) The SFO is required by statute to concern itself only with "serious or complex fraud".[20]

---

[14] CJA 1987, s 2(9).

[15] CJA 1987, s 2(10).

[16] CJA 1987, s 2(10)(a).

[17] CJA 1987, s 2(10)(b). If it is not practicable for the Director to act personally, a member of the SFO designated by the Director may act instead.

[18] For details of the work of the SFO see John Wood, "The Serious Fraud Office" [1989] Crim LR 175; George Staple, "Serious and Complex Fraud: A New Perspective" (1993) 56 MLR 127. John Wood was the SFO's first Director and George Staple was the Director from April 1992 to April 1997.

[19] However, the SFO is not involved in the *detection* of offences of fraud. It relies on detection by the police, the Department of Trade and Industry, the regulatory bodies and members of the public – although, in the investigation of suspected offences, it may uncover cases of fraud not foreseen when the case was accepted. See George Staple, *op cit*, at p 129.

[20] The "key criterion" for deciding whether the SFO should accept a case is that "the suspected fraud is such that the direction of the investigation should be in the hands of those responsible for the prosecution". In determining whether the key criterion is met several factors are taken into account. These are set out in the SFO publication, *Procedure for*

Comparison of the powers of the SFO and the police

6.11 The powers of the police to interview suspects and gain access to financial records are set out in the Police and Criminal Evidence Act 1984 ("PACE"), and are clearly a great deal more restrictive than the powers of the SFO under section 2 of the CJA 1987. The intention of these provisions was to give police access to financial and other documents, but subject to certain stringent safeguards. In particular, confidential financial records fall into a class of material called "special procedure material" to which special provisions as to access apply.[21] An application must be made to a circuit judge at an inter partes hearing, and for access to be granted, the judge must be satisfied, that, among other things:

(1) there are reasonable grounds for believing that a serious arrestable offence has been committed;[22]

(2) there are reasonable grounds for believing that the requested material is likely to be relevant evidence, and is also likely to be of substantial value to the investigation;

(3) other methods of obtaining the material have been tried without success or have not been tried because it appeared that they were bound to fail; and

(4) it is in the public interest that the material should be produced.

The powers of the DTI

6.12 Under section 431 of the Companies Act 1985, the Secretary of State may appoint inspectors to investigate the affairs of a company and to report on them. Section 432 *requires* the Secretary of State to appoint inspectors if the court orders an investigation.

6.13 Section 434(1) of the 1985 Act provides:

> When inspectors are appointed under section 431 or 432, it is the duty of all officers and agents of the company ...
>
> > (a) to produce to the inspectors all books and documents of or relating to the company ... which are in their custody or power,
> >
> > (b) to attend before the inspectors when required to do so, and
> >
> > (c) otherwise to give the inspectors all assistance in connection with the investigation which they are reasonably able to give.

Section 434(5) provides:

Reference of Cases to the Serious Fraud Office, para 2. See also *Arlidge & Parry on Fraud* (2nd ed 1996) paras 13-009 – 13-011.

[21] PACE, s 9 and Sched 1.

[22] The Home Office consultation paper, paras 3.14 and 3.15, invited views on whether corruption should be a serious arrestable offence.

> An answer given by a person to a question put to him in exercise of powers conferred by this section ... may be used in evidence against him.

6.14 If a person fails to comply with a request made by inspectors acting under either section 431 or section 432, the inspectors may certify his or her refusal to the court. The court will then hear evidence and, if appropriate, may punish the offender as if he or she had been guilty of contempt of court.[23]

6.15 Under section 447 of the 1985 Act the Secretary of State may, if he or she thinks there is good reason to do so, require a company to produce its books or papers for examination by departmental officers. Failure to comply with a requirement under this section is an offence.[24] Information obtained is confidential, and may only be used for purposes specified in section 449.

6.16 Sachs LJ in *Re Pergamon Press Ltd*[25] described the function of the DTI inspectors as follows:

> [Their] function is in essence to conduct an investigation designed to discover whether there are facts which may result in others taking action: it is no part of their function to take a decision as to whether action be taken and a fortiori it is not for them finally to determine such issues as may emerge if some action eventuates.[26]

6.17 In the judgment of the Court of Appeal in *Saunders (No 2)*,[27] Lord Taylor of Gosforth CJ set out the Crown's explanation of the differences between the powers of the police, the SFO and DTI inspectors:

> [T]he explanation lies in the very different regime of interviews by DTI Inspectors compared with that of interviews either by the police or the SFO. DTI Inspectors are investigators; unlike the police or SFO they are not prosecutors or potential prosecutors. Here, typically, the two Inspectors were a Queen's Counsel and a senior accountant. They are bound to act fairly, and to give anyone they propose to condemn or criticise a fair opportunity to answer what is alleged against them (*Re Pergamon Press Ltd* [1971] Ch 388).

SECTION 2 AND THE ECHR

6.18 Article 6(1) of the ECHR provides:

> In the determination of his civil rights or obligations or of any criminal charge against him, everyone is entitled to a fair public hearing within a reasonable time by an independent and impartial tribunal established by law.

---

[23] Companies Act 1985, s 436.
[24] Companies Act 1985, s 447(6).
[25] [1971] Ch 388, 401.
[26] [1971] Ch 388, 401.
[27] [1996] Cr App R 463.

6.19 In *Funke v France*,[28] customs officers, accompanied by a police officer, went to the home of the applicant and his wife to obtain "particulars of their assets abroad". The applicant admitted that he had several bank accounts abroad but said that he did not have any bank statements at his home. Following a search, however, officers discovered bank statements and cheque books relating to accounts abroad. The applicant was then asked to produce three years' statements. He refused and was, as a result, prosecuted and fined (with a daily fine until he produced the documents requested) for his refusal.

6.20 The Strasbourg Court took the view that a provision which compelled a person to produce documents under threat of prosecution if he or she did not do so contravened Article 6(1), which conferred on anyone charged with a criminal offence a right to remain silent and not to self-incriminate.[29] Similar decisions were reached in *Miailhe v France*[30] and *Cremieux v France*.[31]

6.21 On these authorities,[32] it seems that there is a risk that the SFO's section 2 powers are vulnerable to challenge before the Strasbourg Court. This risk is arguably confirmed by the recent decision of the Court in *Saunders v United Kingdom*,[33] in which it was held that the admission, in criminal proceedings, of transcripts of interviews[34] obtained by DTI inspectors under the Companies Act 1985 was an infringement of the right not to self-incriminate[35] and violated Article 6(1) of the Convention.[36] The Court specifically rejected the Government's arguments that

---

[28] (1993) 16 EHRR 297.

[29] The Strasbourg Court also held there to have been a contravention of Article 8:

1. Everyone has the right to respect for his private and family life, his home and his correspondence.

2. There shall be no interference by a public authority with the exercise of this right except such as is in accordance with the law and is necessary in a democratic society in the interests of national security, public safety or the economic well-being of the country, for the prevention of disorder or crime, for the protection of health or morals, or for the protection of the rights and freedoms of others.

[30] (1993) 16 EHRR 332.

[31] (1993) 16 EHRR 332, 357.

[32] These cases have been criticised from a *theoretical* perspective (see, eg, Naismith, "Self-Incrimination – Fairness or Freedom" [1997] EHRLR 229) but there has been no indication that the Court will depart from this line of authority in the future.

[33] (1997) 23 EHRR 313.

[34] The Court took the view that the right not to incriminate oneself did not "extend to the use in criminal proceedings of material which may be obtained from the accused through the use of compulsory powers but which has an existence independent of the will of the suspect such as … documents … , breath, blood and urine samples and bodily tissue for the purpose of DNA testing": *Ibid,* para 69.

[35] *Ibid,* para 76.

[36] *Ibid,* para 81. The Court rejected the Government's argument that the complexity of corporate fraud and the public interest in the investigation of such fraud justified "such a marked departure … from one of the basic principles of a fair procedure"; in the Court's view "the general requirements of fairness contained in Article 6 … apply to criminal proceedings in respect of all types of criminal offences without distinction from the most simple to the most complex." *Ibid,* para 74.

the public interest in investigating complex corporate fraud justified a departure from the Article 6 right. On the other hand, the court did not suggest that the investigatory powers of the DTI inspectors *themselves* contravened Article 6(1).

6.22 It is significant that when referring to the case of *Fayed v United Kingdom*,[37] the court commented that

> a requirement that such a preparatory investigation should be subject to the guarantees of a judicial procedure as set forth in Article 6(1) would in practice unduly hamper the effective regulation in the public interest of complex fraud and commercial activities.[38]

THE OPTIONS

6.23 We now reconsider the four options set out in the consultation paper:[39]

(1) no change;

(2) give the police similar powers to those of the SFO in all corruption cases;

(3) give the police powers similar to those held by the SFO in those corruption cases which, despite falling within the SFO's terms of reference, are, for whatever reason, investigated by the police; or

(4) create further investigative powers for corruption offences, not identical to those in the SFO.

Responses on consultation

6.24 The majority of those respondents who considered the issue were in favour of the option of no change.

6.25 There was some support for the second option, namely giving the police similar powers to those of the SFO in *all* corruption cases. The Bar Council and the Criminal Bar Association favoured it on the grounds that the same difficulties encountered in the investigation of serious fraud might well arise in any corruption case. ACPO took the view that some prosecutions had failed because of the inability of investigators to "penetrate the veil".[40]

6.26 Very few of the respondents favoured the third option of giving the police powers similar to those of the SFO in those corruption cases which, despite falling within the terms of reference of the SFO, are, for whatever reason, investigated not by the SFO but by the police. ACPO pointed out that many cases of corruption did not

---

[37] Judgment of 21 September 1994, Series A no 294-B.

[38] Para 67.

[39] Paras 12.20 – 12.26.

[40] Detective Sergeant Moore of the Merseyside Police, author of *Power and Corruption: the Rotten Core of Government and Big Business* (1997), supported this view, explaining that corruption is "a consensual crime, carried out in secret", which makes it very difficult to prove before a jury, and that the award of greater powers would help to combat these difficulties.

fall within the remit of the SFO;[41] and a consequence of this option would be that some cases of corruption would be investigated with the benefit of additional powers whilst others would not.

6.27 There was very limited support for the fourth option, of creating further investigative powers for corruption offences, but ones which were not identical to those of the SFO. Detective Sergeant Moore of the Merseyside Police, for example, suggested that corruption offences should be serious arrestable offences, that a duty be imposed on the Inland Revenue to provide all necessary assistance to investigations and that there should be a requirement to report corruption within organisations providing a public service. The Inland Revenue made clear, however, in their response the importance of maintaining taxpayer confidentiality.

CONCLUSIONS

6.28 We accept that there is an apparent anomaly in the present law: the substantial investigative powers of the SFO will only be available if a case falls within the statutory criteria *and* it is taken up by the SFO. However, were police powers to match SFO powers in respect of those corruption offences *not* investigated by the SFO, there would be the greater anomaly that corruption would attract exceptional powers denied to the police in *other* cases of fraud.

6.29 Our provisional approach was that in the absence of sufficient justification as to why corruption offences should be treated exceptionally, the powers of the police should not be extended. None of the very helpful responses that we received have persuaded us that we were wrong in this view. We consider that an extension of police powers would be vulnerable to challenge under Article 6 of the ECHR, particularly in the light of the recent decisions in *Saunders* and *Funke* which in the words of Professor Sullivan have had "a considerable chilling effect on the use the SFO makes of its special powers".[42] We agree with the majority of our respondents that existing police powers, together with the new powers introduced by the Police Act 1997,[43] are not so demonstrably inadequate that further extension of police investigatory powers can be justified. We therefore do not recommend that the investigative powers of the police should be extended in the case of corruption.

---

[41] This may be either because there is corruption but not fraud (cf paras 5.123 – 5.130 above) or, more likely, because the fraud is neither serious nor complex.

[42] G R Sullivan, "Reformulating the Corruption Laws – the Law Commission Proposals" [1997] Crim LR 730, 737.

[43] Part III of the Police Act 1997 is not yet in force.

PART VII
TERRITORIAL JURISDICTION, THE ATTORNEY-GENERAL'S CONSENT TO PROSECUTION AND OTHER ANCILLARY MATTERS

7.1 In this part we consider the extent of the territorial jurisdiction of the new offences of corruption, whether prosecutions for corruption should require consent (whether that of the Law Officers or the DPP), mode of trial and whether the new offences should have retrospective effect. We also comment on sentence.

TERRITORIAL JURISDICTION

7.2 In the consultation paper we examined the extent to which corruption offences committed abroad are indictable in England and Wales, and we considered whether any steps should be taken to extend the territorial jurisdiction of the courts in this area.[1]

The present law

7.3 The general rule is that the exercise of criminal jurisdiction does not extend to cover acts committed on land abroad.[2] Jurisdiction over a crime belongs to the country in which it was committed.[3] So, in general, British subjects who commit acts abroad are not amenable to the jurisdiction of the courts in England and Wales.

7.4 Under the present law, the English courts do not have jurisdiction to try a criminal offence unless the last act or event necessary for its completion occurs within the jurisdiction.[4] This general principle applies to both common law and statutory offences, but as Parliament is supreme it may extend the territorial limit of a particular offence.[5]

7.5 In applying the common law jurisdiction rules to any criminal offence, it is first necessary to determine the ingredients of the offence, in order to ascertain the

---

[1] Consultation Paper No 145, Part IX.

[2] See the dissenting speech of Lord Morris of Borth-y-Gest in *Treacy* [1971] AC 537, 552.

[3] *Macleod v A-G of New South Wales* [1891] AC 455, 458.

[4] See Criminal Law: Jurisdiction over Offences of Fraud and Dishonesty with a Foreign Element (1989) Law Com No 180, para 2.1.

[5] See, eg, Offences Against the Person Act 1861, s 9; Customs and Excise Management Act 1979, s 148. The territory of England and Wales is extended so as to include British-controlled aircraft if the act taking place on board would be an offence in this country and is not an authorised act under the law which applies to the territory over which the plane is flying: Civil Aviation Act 1982, s 92. Where any act in relation to property or person done by any master or seaman employed on a UK ship would be an offence if done in the UK, it is treated for the purposes of jurisdiction and trial as if it had been done within the jurisdiction of the Admiralty of England: Merchant Shipping Act 1995, s 282.

acts that must be done before the offence can be said to have been committed. This raises the need to draw a distinction between *"conduct crimes"* and *"result crimes"*: only thus can we identify the last act required by an offence, and so discover the jurisdiction in which that last act occurred.

7.6 A *conduct* crime is one whose ingredients consist solely in the acts that must be done for the offence to be made out, and not the results of those acts. A conduct crime, in general, will only be indictable where the conduct required by the offence occurs within the jurisdiction. This is to be contrasted with *result crimes*, whose ingredients include the *consequences* of what is done. The present position is that no offence occurs in English law unless the offence takes place within the jurisdiction. Therefore, because no offence will have been committed until the result of the act occurs, the offence is committed in the place where the result occurs. This means that it is immaterial that any *conduct* required by the offence happens in this jurisdiction; the required result must occur here as well.[6] Indeed, so long as the result occurs here, it is immaterial that the conduct causing it takes place abroad.[7]

7.7 Corruption is a *conduct* crime in that it consists of the act done without any requirement that particular consequences follow from the act. Parliament may extend the territorial limit for particular offences, and has done so on occasion, but the offences under the Prevention of Corruption Acts have no extended territorial limit. Therefore, provided that the act (of, for example, offering or accepting a bribe or agreeing to accept it) takes place within England and Wales, the court will have jurisdiction over the offence.

7.8 We argued in the consultation paper that the issue of jurisdiction may be particularly relevant to corruption as there is an increasing international element to this crime.[8] We criticised the existing law as unable to deal with some circumstances which arguably should be subject to the criminal law; for example, where a public official takes a bribe while abroad and acts on the bribe after returning to England.

The Criminal Justice Act 1993

7.9 Part I of the Criminal Justice Act 1993 ("CJA 1993"), when it is brought into force, will greatly extend the territorial jurisdiction of the English courts over a number of offences of dishonesty (referred to in the Act as "Group A offences"). Under section 2(3) of the Act,

> A person may be guilty of a Group A offence if any of the events which are relevant events in relation to the offence occurred in England and Wales.

Section 2(1) defines a "relevant event" as

---

[6] *Harden* [1963] 1 QB 8; *R v Governor of Pentonville Prison, ex p Khubchandani* (1980) 71 Cr App R 241.

[7] *Bevan* (1987) 84 Cr App R 143.

[8] Consultation Paper No 145, para 9.8.

any act or omission or other event (including any result of one or more acts or omissions) proof of which is required for conviction of the offence.

7.10 The Group A offences do not include bribery or the offences under the Prevention of Corruption Acts.

The effect of extending the CJA 1993 to corruption

7.11 Including our proposed corruption offences in the list of Group A offences would side-step the limitations imposed by the general jurisdictional rules. If A, who is abroad, telephones B in England and offers B a bribe, it is doubtful whether A would at present be committing an offence under English law. It is clear, however, that the matter would be justiciable here if corruption were brought within the list of Group A offences. Section 4(b) of the 1993 Act provides that, in relation to a Group A offence,

> there is a communication in England and Wales of any information, instruction, request, demand or other matter if it is sent by any means
>
> > (i) from a place in England and Wales to a place elsewhere; or
> >
> > (ii) from a place elsewhere to a place in England and Wales.

7.12 The offer of the bribe is a relevant event for the purposes of our recommended offence of corruptly offering to confer an advantage. Section 4(b) would treat the offer, though it was made from abroad, as having taken place within the jurisdiction. A relevant event has therefore occurred within the jurisdiction, and A could be indicted here.

7.13 The 1993 Act also extends the jurisdiction of the English courts with regard to the inchoate offences of conspiracy,[9] attempt[10] and incitement.[11] If these rules were extended to corruption, it would, for example, be an offence indictable in England and Wales to conspire, outside the jurisdiction, to make a corrupt payment within it. There is authority that this is already the position, irrespective of the 1993 Act;[12] but extending the Act to cases of corruption would put the matter beyond doubt.

Our provisional proposal and the views of respondents

7.14 In the consultation paper,[13] we took the view that the extension of the new rules to corruption would bring some cases within the jurisdiction of the English courts which may at present lie outside it. We provisionally proposed that a modern bribery offence should be a Group A offence for the purposes of the CJA 1993.

---

[9] CJA 1993, s 3(2).

[10] *Ibid*, s 3(3).

[11] *Ibid*, s 3(1)(b).

[12] *Sansom* [1991] 2 QB 130.

[13] Paras 9.15 – 9.18.

Of those respondents who commented on the issue, the substantial majority agreed with us.

Our recommendation

7.15 We have been given no reason to suggest that our provisional proposal was wrong. **We, therefore, recommend that the new offences should be Group A offences for the purposes of the Criminal Justice Act 1993.**[14]

(Recommendation 15)

ANCILLARY MATTERS

Requirement of the Attorney-General's consent

7.16 Prosecutions for corruption under the 1889 Act[15] and the 1906 Act[16] require the consent of the Attorney-General or the Solicitor-General. In the consultation paper we examined the justification for this in order to decide whether our proposed new offences should contain any similar provision.[17] There are two stages in the argument, namely:

(1) whether the consent requirement should be retained at all, and

(2) if it should, whether the required consent should be that of the Law Officers (the Attorney-General and the Solicitor-General) or the Director of Public Prosecutions (the "DPP").

The need for consent

7.17 During the passage of the 1889 Act, the Lord Chancellor justified the provision requiring the consent of the Attorney-General on the ground that there was a risk that blackmailers would threaten prosecutions.[18] Similar points were raised when what became the 1906 Act was passed.[19] Thus "the original reasons for introducing the requirement of the Attorney-General's consent were to avoid bribery, collusion, blackmail and other improper practices which frequently surrounded the private prosecution".[20]

---

[14] Given that Part I of the 1993 Act is not in force, an alternative approach would be to make specific provision within the new legislation so that it had the same effect, as regards corruption offences, as CJA 1993 would have if Part 1 of the CJA were in force and corruption were a Group A offence under that Act.

[15] Section 4.

[16] Section 2(1).

[17] Consultation Paper No 145, paras 10.4 – 10.5.

[18] *Hansard* (HC) 20 April 1889, vol 70, col 25; and see the views of Lord Russell of Killowen at col 22.

[19] See, eg, the comments of the Lord Chancellor, the Earl of Halsbury, *Hansard* (HL) 26 April 1904, vol 133, cols 1168–1170.

[20] P Fennell and P A Thomas, "Corruption in England and Wales: An Historical Analysis" (1983) 11 Int Jo Soc Law 167, 173.

7.18 In order to decide whether there should be a consents provision, we considered reasons that have been put forward in justification of such provisions – in particular, a Home Office memorandum[21] to the Departmental Committee on section 2 of the Official Secrets Act 1911[22] (the Franks Committee). According to the Home Office memorandum, "the basic reason for including in a statute a restriction on the bringing of prosecutions is that otherwise there would be a risk of prosecutions being brought in inappropriate circumstances". Five overlapping reasons were given for considering the inclusion of a consent requirement:

(a) to secure consistency of practice in bringing prosecutions, for example, where it is not possible to define the offence very precisely, so that the law goes wider than the mischief aimed at or is open to a variety of interpretations;

(b) to prevent abuse, or the bringing of the law into disrepute, for example, with the kind of offence which might otherwise result in vexatious private prosecutions or the institution of proceedings in trivial cases;

(c) to enable account to be taken of mitigating factors, which may vary so widely from case to case that they are not susceptible to statutory definition;

(d) to provide some central control over the use of the criminal law when it has to intrude into areas which are particularly sensitive or controversial, such as race relations or censorship;

(e) to ensure that decisions on prosecutions take account of important considerations of public policy or of a political or international nature, such as may arise, for instance, in relation to official secrets or hijacking.

7.19 Since we published our consultation paper on corruption, we have subsequently examined as a separate project[23] the issue of the circumstances in which offences should require the consent of the Law Officers or the DPP as a condition precedent to criminal proceedings. We did not consider whether private prosecutions should be abolished. We bore in mind the constitutional importance of consent provisions – not only do they fetter the right of private prosecutions but also, by their nature, they impose an administrative burden on senior officials and cause additional delay within the criminal justice system.

7.20 Our provisional view in the consultation paper on consent provisions was that a requirement of consent should only be used to control prosecutions for those offences

---

[21] Further memorandum by the Home Office on the control of prosecutions by the Attorney-General and the Director of Public Prosecutions (April 1972).

[22] Report of the Departmental Committee on s 2 of the Official Secrets Act 1911 (1972) Cmnd 5104. A prosecution under s 2 of the 1911 Act also requires the consent of the Law Officers.

[23] Consents to Prosecution (1997) Consultation Paper No 149.

(1) which directly affect freedom of expression;

(2) which may involve the national security or have some international element; or

(3) in respect of which it is particularly likely, given the availability of civil proceedings in respect of the same conduct, that the public interest would not require a prosecution.[24]

7.21 The new corruption offences would not fall within any of these exceptions. Therefore, on the basis of the provisional views put forward in Consultation Paper No 149, no consent should be needed for the new offences.

Views of respondents

7.22 Of those who responded on this issue, the majority did not want a consent provision.

7.23 The principal argument in favour of requiring consent was the sensitivity of certain prosecutions for corruption. For example, the Magistrates' Association believed that

> [i]n view of the potential significance and sensitivity of a corruption case it seems sensible to require the consent of the DPP, to ensure that the matter is considered at a sufficiently high level, and away from the local environment.

7.24 We considered this issue carefully. A matter of substantial importance to us was that for many years it has been possible to prosecute for the common law offence of bribery without any need to obtain the consent of the Law Officers or the DPP. We are not aware of, and nobody has drawn to our attention, any problems that this has caused. In particular, the fear mentioned at the time when the 1889 and 1906 Acts were passed,[25] of blackmailers using corruption offences has not been justified. Similarly, there has been no evidence put forward in relation to the use of the common law offence to support the concern that corruption cases are so sensitive as to require the detached view that the consent provisions give. We therefore conclude that there should be no requirement for consent for the new corruption offences.

7.25 To those who are concerned about indiscriminate and unjustified private prosecutions for the new offences, we would point out that there are existing procedural restraints on private prosecutions, namely:

(1) A magistrate may refuse to issue a summons.

(2) The Attorney-General may terminate proceedings by entering a *nolle prosequi*.

---

[24] See Consultation Paper No 149, para 7.5.

[25] See para 7.17 above.

(3) The Attorney-General may prevent criminal proceedings by vexatious litigants by applying to the High Court for a criminal proceedings order.

(4) The DPP may take over private prosecutions and terminate them, whether by discontinuance, withdrawal or offering no evidence.

(5) Legal aid is not available to a private prosecutor save for the limited purpose of resisting an appeal to the Crown Court.[26] A claim for costs may be made either against central funds or against the accused but, even if granted, an award may not meet the full cost of a prosecution, or if it is against the accused, it may not be recoverable.[27]

(6) A private prosecutor might be sued in the civil courts for malicious prosecution.

7.26 **We recommend that prosecutions for the new offences should not require the consent of either the Law Officers or the Director of Public Prosecutions.**

(**Recommendation 16**)

Mode of trial

7.27 We provisionally proposed that the new offence should be triable either way, as are the existing offences under the Prevention of Corruption Acts.[28] The majority of respondents who replied to this issue agreed without further comment.

7.28 Two respondents disagreed on the ground that corruption cases are too serious to be tried summarily. However, we feel that, although it is true that corruption is *generally* a serious offence, it would be impossible to say that *all* instances of corrupt behaviour are sufficiently serious to be triable only on indictment. **We recommend that the new offences should be triable either way.**

(**Recommendation 17**)

Sentence

7.29 This Commission does not regard itself as qualified to specify the maximum sentence for the new offences on conviction on indictment. We note, however, that the courts have passed heavy sentences for the existing corruption offences: for example, the Court of Appeal (Criminal Division) dismissed an appeal against a total of 11 years' imprisonment imposed on a detective constable who had pleaded guilty to four counts of corruption. The appellant had accepted £18,500

---

[26] Legal Aid Act 1988, s 21(1).

[27] See Consultation Paper No 149, paras 5.13 – 5.15.

[28] Consultation Paper No 145, para 10.16.

from a man who was a subject of criminal proceedings to disclose confidential information about the inquiry and to destroy surveillance logs.[29]

7.30 We believe, however, that consideration should be given to allowing the courts the power to impose penalties other than, or in addition to, imprisonment or a fine: for example, an order that a person convicted of bribery pay the amount or value of any gift, loan, fee or reward received by him or her;[30] disqualification from public office; or forfeiture of any office held by him or her at the time of conviction. Such powers exist under the 1889 Act.[31]

Retrospectivity

7.31 We provisionally proposed that the new law should not have retrospective effect.[32] It is obviously exceptional for any new criminal offence to have such an effect. In any event, an offence with retrospective effect would be likely to be held in breach of Article 7 of the ECHR which provides that no-one shall be guilty of a criminal offence on account of any act or omission which did not constitute a criminal offence at the time when it was committed.

7.32 Only a small number of respondents commented on this provisional proposal. It was entirely uncontroversial. **We recommend that the new offence should not have retrospective effect**.

(**Recommendation 18**)

---

[29] *Donald* [1997] 2 Cr App R(S) 272. In *Hopwood* (1985) 7 Cr App R (S) 402 the Court of Appeal regarded as "exceedingly lucky" an appellant who was sent to prison for a total of three and a half years for acting corruptly in the private sphere. He received sums totalling more than £200,000 and the re-wiring of his house in return for acting against the interests of his employer.

[30] We note that the laundering of the proceeds of corruption is an offence by virtue of ss 29 to 33, 35 and 77 of the CJA 1993, which inserted seven new sections, numbered 93A to 93G, into the CJA 1988. Section 93C makes it an offence to conceal or disguise any property which is, or represents, the proceeds of criminal conduct, or to convert or transfer such property or remove it from the jurisdiction. Under s 93A it is an offence to assist another to retain the benefit of criminal conduct, and under s 93B it is an offence to acquire or possess the proceeds of criminal conduct.

[31] See s 2 of the 1889 Act.

[32] Consultation Paper No 145, para 10.19.

PART VIII
CORRUPTION AND BREACH OF DUTY

TWO BASIC LAW REFORM QUESTIONS

8.1 In the consultation paper, we analysed the functions of the law of corruption in terms of the *fundamental mischief* (the agent's breach of duty) and the *mischief of temptation* (the conduct which is likely to result in such a breach).[1] This, in turn, raised two basic law reform questions: first, should the law prohibit the fundamental mischief *directly*, or should it instead prohibit conduct which is *likely to result* in the fundamental mischief? And second, if we take the latter view, should the law prohibit *all* such conduct or only certain kinds; and, if only certain kinds, which kinds?[2]

THE FIRST QUESTION

Acting contrary to duty: the radical approach

8.2 In the consultation paper, we provisionally rejected the "radical approach" of introducing one or more offences of an agent acting contrary to his or her duty.[3] We argued that if *all* such breaches were penalised, this would be unduly draconian, while if only *some* breaches were penalised this would mean devising criteria for determining which breaches should be criminal. Such criteria, we believed, would inevitably be elusive and controversial. Further, we took the view that criminalising the breach of duty directly would fail to reflect the widely understood meaning of corruption as something which destroys or perverts another's integrity or fidelity.[4] About a third of respondents commented on this conclusion and, of those, the majority agreed with our rejection of the radical approach.

8.3 For the reasons set out above, we remain of the view that it would be wrong to criminalise breaches of duty directly, and we therefore do not recommend the adoption of the radical approach.

8.4 However, as we have explained,[5] our analysis of the nature of corrupt transactions has led us to the conclusion that it should be an offence for an agent to perform his or her functions *corruptly* – that is, either in return for a corrupt inducement or in the hope of a corrupt reward. In most cases the commission of this offence would consist in a breach of the agent's duty to his or her principal; but our recommendation is less radical than the criminalisation of breaches of duty per se, because it requires the element of *corruption* (whether by inducement or by the

---

[1] Consultation Paper No 145, paras 1.12 – 1.16.
[2] Consultation Paper No 145, para 5.1.
[3] Consultation Paper No 145, paras 5.2 – 5.4.
[4] Consultation Paper No 145, para 5.3.
[5] Paras 5.54 – 5.58 above.

hope of reward). On the other hand, for reasons we have explained,[6] it does not require a *breach of duty* at all, but only the corrupt performance of the agent's functions as an agent. Given the element of corruption, it is immaterial whether the agent is also in breach of his or her duty.

THE SECOND QUESTION

Corruption by means other than bribery

8.5 We identified three crude ways in which A can cause B to act in breach of duty – by inducement, by threats or by deception.[7] In each instance B will have breached his or her duty – but with different degrees of *voluntariness*.

8.6 In the consultation paper,[8] we considered whether breaches of duty procured by threats or deception are at present criminal. We concluded that they are to some extent, but that there are anomalies. We thought it was at least arguable that there would be advantages in creating specific new offences: first, it would put the fundamental mischief of breach of duty (as we then considered it to be) at the centre of the criminal law of corruption; and, second, it would provide an opportunity to rectify the anomalies that we had identified. On the other hand, we acknowledged that the view might be taken that the present law, though not perfect, was adequate. We therefore asked for views on whether, in addition to a modern offence of bribery, there should be new offences of procuring a breach of duty by deception and procuring breach of duty by threats.

8.7 About a third of the respondents commented on this issue. The clear majority favoured new offences. Most considered that the law relating to both threats *and* deception was inadequate and therefore wished to see the introduction of two new offences, while the remaining few wished only for a new offence to deal *either* with breach of duty induced by threats *or* with breach of duty procured by deception.[9] Those respondents who did not favour new offences did so primarily for the reason that their subject would be well beyond the traditional boundaries of the offence of corruption: if B does not act voluntarily, it was argued, it is difficult to see how he or she can have been *corrupted*.

8.8 In order to reach a conclusion on this issue, we now consider in more detail the present law, its deficiencies and possible remedies. We begin by asking, first, in what circumstances will the present law fail to cover the procuring of breaches of duty by either threats or deception? And then, should any proposed new offences form part of a modern law of *corruption*?

---

[6] Paras 5.5 – 5.10 above.

[7] Consultation Paper No 145, para 5.5.

[8] Consultation Paper No 145, paras 5.10 – 5.19.

[9] Judge Walsh was content with the current law on blackmail but considered that the current law regarding deception was inadequate, whilst Judge Mark Dyer and Liberty were happy with the law relating to deception but wished to see a new offence to deal with breaches of duty induced by threats.

The present law and its deficiencies

Blackmail and breaches of duty procured by threats

8.9 Section 21 of the Theft Act 1968 provides that it is an offence to make an unwarranted demand with menaces[10] with a view to gain for oneself or another or with intent to cause another loss. A demand is unwarranted unless the maker believes that he or she has reasonable grounds for making the demand *and* that the use of the menaces is a proper means of reinforcing that demand.[11] "Gain" and "loss" are defined[12] as "extending only to gain or loss in money or other property". This provides an important limit to the extent of liability for blackmail, although the courts have interpreted the provision widely.[13]

8.10 Professor Sir John Smith notes that, although blackmail is limited to the protection of economic interests, "[i]n most cases, the blackmailer is trying to obtain money to which he knows he has no right and there will be no doubt about his view to gain".[14] Similarly, although our present concern is with the threatening of any agent into breaching his or her duty, the threat will usually be made with "a view to gain" or "intent to cause loss to another", in terms of money or other property, and will therefore amount to blackmail.

8.11 On the other hand, it is possible to envisage circumstances in which the threat will not be made with a view to gain or intent to cause loss. For example, if A threatens B, an agent, into disclosing personal information about B's principal, C, with a view to damaging C's reputation, then there is no offence of blackmail unless the disclosure of the information is intended to result in *financial* loss to C. Alternatively, the gain or loss might be too remote. Professor Sir John Smith suggests that a person who menaces the headmaster of a public school, with a view to gaining admission for his newly-born son, is literally acting with a view to gain if he is motivated by a belief that his son will one day have greater earning power through having been to that school.

---

[10] The definition of "menaces" has received much judicial attention. In *Clear* [1968] 1 QB 670 Sellers LJ held that threats which would influence the mind of a person of "normal stability and courage" would suffice to be a menace. *Garwood* [1987] 1 WLR 319 added that where the threat in fact affected the victim, but would not have affected an ordinary person, the threat is still a menace provided that the defendant knew the victim would react in the way he or she did. In *Lawrence and Pomroy* [1971] Crim LR 645, the Court of Appeal stated that "menace" was an ordinary English word which any jury could be expected to understand, and only in exceptional cases shall a direction be given.

[11] The test of whether the defendant believes that he or she has reasonable grounds for making the demand is purely subjective. In *Harvey* (1981) 72 Cr App R 139, Bingham J held that the beliefs of the reasonable person were irrelevant, save as may help throw light on the defendant's actual beliefs. However, he added that an act known by the defendant to be unlawful can never be "proper" within the meaning of s 21(1)(b).

[12] Theft Act 1968, s 34(2)(a).

[13] In *Bevans* (1988) 87 Cr App R 64 the court held that there was no material difference between a doctor giving a patient some ampoules of morphine and the doctor injecting them himself. The court held that "property" was still gained by the patient albeit through an unusual medium.

[14] J C Smith, The Law of Theft (8th ed 1997), para 10–13.

Yet the gain is so distant in time and subject to so many contingencies that its connection with the demand with menaces may be thought too remote.[15]

8.12 Broadly, respondents favoured a new offence of procuring a breach of duty by threats because most took the view that, in the words of the Employment Law Bar Association, "although some threats are already criminalised, the law is not sufficiently comprehensive to deal with even the majority of threats."

8.13 We also canvassed opinion as to whether there should be an offence of threatening to breach duty. Most of the respondents who gave their views were against the introduction of such offence on the grounds that this circumstance could be adequately dealt with under the law of blackmail. In our view, whilst this may well be the case on most occasions, the law of blackmail may prove inadequate for the reasons given above.[16]

Breaches of duty procured by deception

8.14 Under the present law it is not an offence for A to deceive B into breaching his or her duty to C unless, incidentally, the circumstances of A's conduct are covered by one of the various offences of deception which are contained in the Theft Acts.[17] An agreement between A and another to practise such a deception may, on the other hand, amount to a conspiracy to defraud.[18] Although it seems unlikely that the procurement of a breach of duty by deception will not fall within one of the existing offences, nonetheless there is no reason to suppose that the present offences cover *all* such cases. In any event, it is arguably preferable not to have to rely on a multiplicity of offences of deception. The majority of the respondents favoured a new offence of procuring a breach of duty by deception.

Whether offences of procuring a breach of duty by threats or deception are offences of corruption

8.15 We think that it is likely that the present law may well not be adequate to deal with circumstances where a breach of duty is procured by threats or deception, and that there is a strong argument for extending it so as to catch such circumstances. However, we question whether it is appropriate to treat such conduct as amounting to offences *of corruption*. We can see arguments on both sides.

8.16 The argument in favour is an argument which we put forward in the consultation paper:

---

[15] *Ibid*, para 10–22.

[16] Para 8.11 above.

[17] For example, obtaining property by deception (Theft Act 1968, s 15), obtaining a money transfer by deception (Theft Act 1968 s 15A, inserted by Theft (Amendment) Act 1996, s 1), obtaining a pecuniary advantage by deception (Theft Act 1968, s 16), or procuring the execution of a valuable security by deception (Theft Act 1968, s 20(2)).

[18] *Welham v DPP* [1961] AC 103 and *Wai Yu-Tsang* [1992] 1 AC 269. See Consultation Paper No 145, para 5.15.

> The benefit of creating new offences of procuring a breach of duty by deception and procuring a breach of duty by threats, in addition to a new offence of bribery, is that this would properly put the *fundamental mischief* at the centre of the criminal law of corruption. It would also provide an opportunity to regularise anomalies in the present law (such as the requirement of blackmail that there should be a view to gain or an intent to cause loss, and the uncertainty as to the scope of conspiracy to defraud in this context).[19]

8.17 The argument against is that, although the essential purpose of the criminal law of corruption is to prevent the "fundamental mischief", it does not follow that all instances of the fundamental mischief, whatever the cause, must be proscribed by the law of corruption. In Part V above, we have defined "corruptly" in terms of an agent being influenced, in the performance of his or functions as an agent, by inducements or by the hope of reward.[20] In our view, the essence of corruption involves not only the fundamental mischief but also the mischief of temptation.

8.18 We note the comments of Professor Sir John Smith, who wrote in his response:

> There would be logic in extending the law to include the procurement of breaches by threats or deception. The principal is equally damaged, whether the breach is caused by bribes, threats or deception. The deceived agent is, however, in no sense "corrupted": nor is that word particularly apt for the agent who gives in to threats. Is not an element of rationale of the offence the inexcusable betrayal by the agent which it seeks to bring about? There is no betrayal by the deceived agent and a possibly excusable one where the agent acts only because of threats ... [M]y instinctive reaction is in favour of confining the offence of corruption to its traditional limits.

8.19 Phillips LJ did not wish to include the new offences "in the interests of simplicity". The SIB thought that the existing law was adequate and that the new offences contemplated were a "long way from the common conception of corruption".

OUR CONCLUSION

8.20 Although we believe that it is likely that the present law does not deal adequately with circumstances in which the fundamental mischief is brought about by threats or deception, we do not recommend that these cases should be caught by the law of corruption. Our reasons stem from our analysis of the nature of corruption. We take the view that these issues would be better considered in the context of our review of the law of dishonesty.[21]

---

[19] Consultation Paper No 145, para 5.19.

[20] Para 5.73 above.

[21] See Criminal Law: Conspiracy to Defraud (1994) Law Com No 228, paras 1.16 – 1.19.

PART IX
OUR RECOMMENDATIONS

In this part we set out our recommendations in full.

THE PRESENT LAW AND THE NEED FOR CHANGE

1. We recommend that the common law offence of bribery and statutory offences of corruption should be replaced by a modern statute.[1]

THE PRESUMPTION OF CORRUPTION AND THE DISTINCTION BETWEEN PUBLIC BODIES AND OTHERS

2. We recommend that the new law of corruption

    (1) should not include a presumption comparable to that created by section 2 of the 1916 Act (under which a transaction is in certain circumstances presumed to be corrupt unless the contrary is proved), and

    (2) should not draw a distinction between public bodies and others.[2]

THE AGENCY RELATIONSHIP

3. We recommend that

    (1) a person should be regarded as an agent, and another as a principal for whom he or she performs functions, if

        (a) the first is a trustee and the second is a beneficiary under the same trust;

        (b) the first is a director of a company and the second is the company;

        (c) each is a partner in the same partnership;

        (d) the first is a professional person (such as lawyer or accountant) and the second is his or her client;

        (e) the first is an agent and the second is his or her principal (taking agent and principal in the sense normally understood by lawyers); or

        (f) the first is the employee of the second;

    (2) a person who does not fall within any of the categories above should be regarded as an agent, and another as a principal for whom he or she performs functions, if there is an agreement or understanding between them (express or implied) that the first is to perform the functions for the second; and

---

[1] Para 2.33 above.

[2] Para 4.78 above.

(3) a person should be regarded as an agent performing functions for the public if the functions he or she performs are of a public nature (whether or not in relation to the United Kingdom).[3]

4. We recommend that consideration should be given to extending the scope of the corruption offences to include officials of international intergovernmental organisations.[4]

THE CONCEPT OF ADVANTAGE

5. We recommend that

   (1) a person should be regarded as conferring an advantage if

   (a) he or she does something or omits to do something which he or she has a right to do, and

   (b) the act or omission is done or made in consequence of another's request (express or implied) or with the result (direct or indirect) that another benefits; and

   (2) a person should be regarded as obtaining an advantage if

   (a) another does something or omits to do something which he or she has a right to do, and

   (b) the act or omission is done or made in consequence of the first person's request (express or implied) or with the result (direct or indirect) that the first person benefits.[5]

THE OFFENCES

Corruptly conferring, or offering or agreeing to confer, an advantage

6. We recommend that a person should commit an offence if he or she

   (1) corruptly confers an advantage, or

   (2) corruptly offers or agrees to confer an advantage.[6]

Corruptly obtaining, soliciting or agreeing to obtain an advantage

7. We recommend that a person should commit an offence if he or she

---

[3] Para 5.36 above.
[4] Para 5.37 above.
[5] Para 5.43 above.
[6] Para 5.50 above.

(1) corruptly obtains an advantage, or

(2) corruptly solicits or agrees to obtain an advantage.[7]

Corruptly performing functions as an agent

8. We recommend that a person should commit an offence if he or she performs his or her functions as an agent corruptly.[8]

Agents receiving benefit from corruption

9. We recommend that if

(1) an advantage is obtained corruptly by a person other than the agent concerned,

(2) the agent receives a benefit (in any form) which consists of or is derived (directly or indirectly) from all or part of the advantage,

(3) the agent knows or believes that the advantage was obtained corruptly and that the benefit consists of or is derived from all or part of the advantage, and

(4) the agent give his or her express or implied consent to receiving the benefit,

the agent should commit an offence.[9]

"CORRUPTLY"

Corruptly conferring an advantage as an inducement

10. We recommend

(1) that a person who confers an advantage, or offers or agrees to confer an advantage, should be regarded as doing so corruptly if he or she

(a) intends a person, in performing his or her functions as an agent, to do an act or make an omission, and

(b) believes that, if the person did so, it would probably be primarily in return for the conferring of the advantage (or the advantage when conferred), whoever obtains it,

unless

(i) the advantage is conferred (or to be conferred) by or on behalf of the agent's principal (or, in the case of functions performed for the public, on behalf of the public) as remuneration or reimbursement in respect of the performance of the functions, *or*

---

[7] Para 5.53 above.

[8] Para 5.58 above.

[9] Para 5.61 above.

(ii) the functions concerned are performed only for one or more principals (not the public), and each principal is aware of all the material circumstances and consents to the conferring of the advantage or the making of the offer or agreement;

(2) a person should be treated as if he or she were aware of all the material circumstances, and consented to the conferring of the advantage or the making of the offer or agreement, if the defendant believes that that person

   (a) is aware of those circumstances and so consents, or

   (b) would so consent if aware of those circumstances; and

(3) consent should count for these purposes if given on a principal's behalf by any agent of the principal, unless in giving it the agent performs his or her functions as an agent corruptly.[10]

Corruptly conferring an advantage as a reward

11. We recommend that a person who confers an advantage, or offers or agrees to confer an advantage, should be regarded as doing so corruptly if he or she

   (1) knows or believes that, in performing his or her functions as an agent, a person has done an act or made an omission,

   (2) knows or believes that that person has done the act or made the omission primarily in order to secure that a person obtains an advantage (whoever obtains it), and

   (3) intends the person known or believed to have done the act or made the omission to regard the advantage (or the advantage when conferred) as conferred primarily in return for the act or omission.[11]

Corruptly obtaining or agreeing to obtain an advantage

12. We recommend that

   (1) a person who obtains an advantage should be regarded as obtaining it corruptly if he or she

      (a) knows or believes that the person conferring it confers it corruptly, and

      (b) gives his or her express or implied consent to obtaining it (in a case where he or she does not request it); and

---

[10] Para 5.99 above.

[11] Para 5.109 above.

(2) a person who agrees to obtain an advantage should be regarded as agreeing to obtain it corruptly if he or she knows or believes that the person agreeing to confer it agrees corruptly.[12]

Corruptly soliciting an advantage

13. We recommend that a person who solicits an advantage should be regarded as soliciting it corruptly if he or she

   (1) intends a person to confer it or agree to confer it, and

   (2) believes that, if the person did so, he or she would probably do so corruptly.[13]

Corruptly performing functions as an agent

14. We recommend that a person who performs his or her functions as an agent should be regarded as performing them corruptly if he or she

   (1) does an act or makes an omission primarily in order to secure that a person confers an advantage (whoever obtains it), and believes that if the person did so he or she would probably do so corruptly, or

   (2) does an act or makes an omission when he or she knows or believes that a person has corruptly conferred an advantage (whoever obtained it), and regards the act or omission as done or made primarily in return for the conferring of the advantage.[14]

TERRITORIAL JURISDICTION

15. We recommend that the new offences should be Group A offences for the purposes of the Criminal Justice Act 1993.[15]

ANCILLARY MATTERS

Requirement of the Attorney-General's consent

16. We recommend that prosecutions for the new offences should not require the consent of either the Law Officers or the Director of Public Prosecutions.[16]

Mode of trial

17. We recommend that the new offences should be triable either way.[17]

---

[12] Para 5.118 above.
[13] Para 5.120 above.
[14] Para 5.122 above.
[15] Para 7.15 above.
[16] Para 7.26 above.
[17] Para 7.28 above.

Retrospectivity

18. We recommend that the new offences should not have retrospective effect.[18]

(*Signed*) MARY ARDEN, *Chairman*
ANDREW BURROWS
DIANA FABER
CHARLES HARPUM
STEPHEN SILBER

MICHAEL SAYERS, *Secretary*
15 January 1998

---

[18] Para 7.32 above.

APPENDIX A
DRAFT CORRUPTION BILL

INDEX

This index shows where in the report each substantive provision of the draft Bill is explained.

Clause of draft Bill	Paragraphs of report
1	5.45 – 5.50
2	5.51 – 5.53
3	5.54 – 5.58
4, 5	5.38 – 5.43
6(1)	5.67 – 5.85, 5.99
6(2)	5.100 – 5.109
6(3)	5.83 – 5.84
7(1), (2)	5.112 – 5.118
7(3)	5.119 – 5.120
8	5.121 – 5.122
9	5.15 – 5.37
10	5.86 – 5.87, 5.99
11	5.88 – 5.99
12	5.59 – 5.61
13	7.29 - 7.30
14	7.2 – 7.15
15	2.17 – 2.33

NOTE: For the reasons set out in paragraphs 1.21 and 1.22 of the report, the draft Bill does not expressly refer to MPs or Ministers.

Corruption Bill

ARRANGEMENT OF CLAUSES

Main offences

Clause
1. Corruptly conferring an advantage.
2. Corruptly obtaining an advantage.
3. Performing functions corruptly.

Conferring and obtaining

4. Meaning of conferring an advantage.
5. Meaning of obtaining an advantage.

Acting corruptly

6. Conferring an advantage: meaning of corruptly.
7. Obtaining an advantage: meaning of corruptly.
8. Performing functions: meaning of corruptly.

Agents and principals

9. Meaning of agent and principal.

Exceptions from offences

10. Remuneration or reimbursement: no corruption.
11. Principal's consent: no corruption.

Miscellaneous

12. Receiving benefit from corruption.
13. Penalties.
14. Jurisdiction.
15. Abolition of existing offences.

General

16. Repeals.
17. Commencement.
18. Extent.
19. Citation.

SCHEDULE:—

Repeals.

DRAFT

OF A

BILL

INTITULED

An Act to make provision about corruption. A.D. 1998.

BE IT ENACTED by the Queen's most Excellent Majesty, by and with the advice and consent of the Lords Spiritual and Temporal, and Commons, in this present Parliament assembled, and by the authority of the same, as follows:—

Main offences

1. A person commits an offence if—

(a) he corruptly confers an advantage, or

(b) he corruptly offers or agrees to confer an advantage.

<small>Corruptly conferring an advantage.</small>

2. A person commits an offence if—

(a) he corruptly obtains an advantage, or

(b) he corruptly solicits or agrees to obtain an advantage.

<small>Corruptly obtaining an advantage.</small>

3. A person commits an offence if he performs his functions as an agent corruptly.

<small>Performing functions corruptly.</small>

Conferring and obtaining

4.—(1) A person confers an advantage if—

(a) he does something or he omits to do something which he has a right to do, and

(b) the act or omission is done or made in consequence of another's request (express or implied) or with the result (direct or indirect) that another benefits.

<small>Meaning of conferring an advantage.</small>

(2) An act or omission may be done or made in consequence of a person's request even if the nature of the act or omission, and the time it is intended to be done or made, are not known at the time of the request.

Meaning of obtaining an advantage.

5.—(1) A person obtains an advantage if—

(a) another does something or he omits to do something which he has a right to do, and

(b) the act or omission is done or made in consequence of the first person's request (express or implied) or with the result (direct or indirect) that the first person benefits.

(2) An act or omission may be done or made in consequence of a person's request even if the nature of the act or omission, and the time it is intended to be done or made, are not known at the time of the request.

Acting corruptly

Conferring an advantage: meaning of corruptly.

6.—(1) A person who confers an advantage, or offers or agrees to confer an advantage, does so corruptly if—

(a) he intends a person in performing his functions as an agent to do an act or make an omission, and

(b) he believes that if the person did so it would probably be primarily in return for the conferring of the advantage (or the advantage when conferred), whoever obtains it.

(2) A person who confers an advantage, or offers or agrees to confer an advantage, does so corruptly if—

(a) he knows or believes that in performing his functions as an agent a person has done an act or made an omission,

(b) he knows or believes that the person has done the act or made the omission primarily in order to secure that a person confers an advantage (whoever obtains it), and

(c) he intends the person known or believed to have done the act or made the omission to regard the advantage (or the advantage when conferred) as conferred primarily in return for the act or omission.

(3) For the purposes of subsection (1) the nature of the intended act or omission, and the time it is intended to be done or made, need not be known when the advantage is conferred or the offer or agreement is made.

(4) This section has effect subject to sections 10 and 11.

Obtaining an advantage: meaning of corruptly.

7.—(1) A person who obtains an advantage obtains it corruptly if—

(a) he knows or believes that the person conferring it confers it corruptly, and

(b) he gives his express or implied consent to obtaining it (in a case where he does not request it).

(2) A person who agrees to obtain an advantage agrees to obtain it corruptly if he knows or believes that the person agreeing to confer it agrees corruptly.

(3) A person who solicits an advantage solicits it corruptly if—

(a) he intends a person to confer it or agree to confer it, and

(b) he believes that if the person did so he would probably do so corruptly.

8.—(1) A person who performs his functions as an agent performs them corruptly if—

    (a) he does an act or makes an omission primarily in order to secure that a person confers an advantage (whoever obtains it), and

    (b) he believes that if the person did so he would probably do so corruptly.

(2) A person who performs his functions as an agent performs them corruptly if—

    (a) he does an act or makes an omission when he knows or believes that a person has corruptly conferred an advantage (whoever obtained it), and

    (b) he regards the act or omission as done or made primarily in return for the conferring of the advantage.

Performing functions: meaning of corruptly.

Agents and principals

9.—(1) A person is an agent, and another is his principal for whom he performs functions, if—

    (a) the first is a trustee and the second is a beneficiary under the same trust;

    (b) the first is a director of a company and the second is the company;

    (c) each is a partner in the same partnership;

    (d) the first is a professional person (such as a lawyer or accountant) and the second is his client;

    (e) the first is an agent and the second is his principal (taking agent and principal in the sense normally understood by lawyers);

    (f) the first is the employee of the second.

(2) If subsection (1) does not apply a person is an agent, and another is his principal for whom he performs functions, if there is an agreement or understanding between them (express or implied) that the first is to perform the functions for the second.

(3) A person is an agent performing functions for the public if the functions he performs are of a public nature.

(4) Subsection (3) has effect even if the person has no connection with the United Kingdom, and "public" is not confined to the public of the United Kingdom or of any part of it.

(5) A person may be an agent performing some functions for a principal and others for the public.

(6) As regards a given function, a person may be an agent performing it for a principal and the public.

Meaning of agent and principal.

Exceptions from offences

10.—(1) If—

    (a) an advantage is conferred or an offer or agreement to confer an advantage is made, and

    (b) any of the following three conditions is satisfied,

the advantage is not conferred corruptly or (as the case may be) the offer or agreement is not made corruptly.

Remuneration or reimbursement: no corruption.

(2) The first condition is that—

(a) the functions concerned are performed only for a principal (and not the public), and

(b) the advantage is conferred (or to be conferred) by or on behalf of the principal as remuneration or reimbursement in respect of the performance of the functions.

(3) The second condition is that—

(a) the functions concerned are performed only for the public (and not a principal), and

(b) the advantage is conferred (or to be conferred) on behalf of the public as remuneration or reimbursement in respect of the performance of the functions.

(4) The third condition is that—

(a) some of the functions concerned are performed for a principal and others are performed for the public, or a given function is performed for a principal and the public, and

(b) each element of the advantage is conferred (or to be conferred) by or on behalf of the principal, or on behalf of the public, as remuneration or reimbursement in respect of the performance of the functions.

(5) The functions concerned are the functions relating to the act or omission which is intended to be done or made, or is known or believed to have been done or made.

(6) References to the public are not confined to the public of the United Kingdom or of any part of it.

Principal's consent: no corruption.

11.—(1) If—

(a) an advantage is conferred or an offer or agreement to confer an advantage is made, and

(b) the following condition is satisfied,

the advantage is not conferred corruptly or (as the case may be) the offer or agreement is not made corruptly.

(2) The condition is that—

(a) the functions concerned are performed only for a principal (and not the public), and

(b) the principal, or each of them if more than one, is aware of all the material circumstances and consents to the conferring of the advantage or the making of the offer or agreement.

(3) A person is to be treated as if he were aware of all the material circumstances, and consented to the conferring of the advantage or the making of the offer or agreement, if the defendant believes that—

(a) he is aware of those circumstances and so consents, or

(b) he would so consent if aware of those circumstances.

(4) For the purposes of subsections (2) and (3) consent may be given on a principal's behalf by any agent of the principal, but it does not count if in giving it the agent performs his functions as an agent corruptly.

(5) The functions concerned are the functions relating to the act or omission which is intended to be done or made, or is known or believed to have been done or made.

Miscellaneous

12.—(1) If—   Receiving benefit from corruption.

    (a) an advantage is obtained corruptly by a person other than the agent concerned,

    (b) the agent receives a benefit which consists of or is derived from all or part of the advantage,

    (c) the agent knows or believes that the advantage was obtained corruptly and that the benefit consists of or is derived from all or part of the advantage, and

    (d) the agent gives his express or implied consent to receiving the benefit,

the agent commits an offence.

(2) The benefit may take any form and may be directly or indirectly derived.

13. A person guilty of an offence under this Act is liable—   Penalties.

    (a) on conviction on indictment, to imprisonment for a term not exceeding [ ] years;

    (b) on summary conviction, to imprisonment for a term not exceeding 6 months or a fine not exceeding the statutory maximum or both.

14. In section 1(2) of the Criminal Justice Act 1993 (group A offences for purposes of provisions about jurisdiction) the following paragraph is inserted after paragraph (c)—   Jurisdiction. 1993 c. 36.

    "(cc) an offence under the Corruption Act 1998;".

15.—(1) The common law offence of bribery is abolished.   Abolition of existing offences.

(2) The Public Bodies Corrupt Practices Act 1889 shall cease to have effect.   1889 c. 69.

(3) In section 1(1) of the Prevention of Corruption Act 1906 (offences relating to corrupt transactions with agents) the words from "If" to "business; or" (in the second place where the latter words occur) are omitted.   1906 c. 34.

(4) Section 2 of the Prevention of Corruption Act 1916 (presumption of corruption for certain offences) is omitted.   1916 c. 64.

General

16. The enactments mentioned in the Schedule are repealed to the extent specified.   Repeals.

17.—(1) Sections 1 to 16 and the Schedule apply in relation to acts or omissions done or made on or after the appointed day.   Commencement.

(2) If an act or omission is alleged to have been done or made over a period of two or more days, or at some time in a period of two or more days, it must be taken for the purposes of this section to have been done or made on the last of those days.

(3) The appointed day is such day as the Secretary of State appoints by order made by statutory instrument.

Extent.

18. This Act extends to England and Wales only.

Citation.

19. This Act may be cited as the Corruption Act 1998.

SCHEDULE

Section 16.

Repeals

Chapter	Short title	Extent of repeal
1889 c. 69.	Public Bodies Corrupt Practices Act 1889.	The whole Act.
1906 c. 34.	Prevention of Corruption Act 1906.	In section 1(1) the words from "If" to "business; or" (in the second place where the latter words occur). In section 1(2) the words "expression "consideration" includes valuable consideration of any kind; the".
1916 c. 64.	Prevention of Corruption Act 1916.	Section 2. In section 4(3) the words from "and the" to the end.
1948 c. 65.	Representation of the People Act 1948.	Section 52(7).
1988 c. 33.	Criminal Justice Act 1988.	Section 47(1).
1995 c. x.	London Local Authorities Act 1995.	In Part I of the Schedule, the entry relating to the Public Bodies Corrupt Practices Act 1889.

APPENDIX B
EXTRACTS FROM RELEVANT LEGISLATION

PUBLIC BODIES CORRUPT PRACTICES ACT 1889

1 **Corruption in office a misdemeanor**

(1) Every person who shall by himself or by or in conjunction with any other person, corruptly solicit or receive, or agree to receive, for himself, or for any other person, any gift, loan, fee, reward, or advantage whatever as an inducement to, or reward for, or otherwise on account of any member, officer, or servant of a public body as in this Act defined, doing or forbearing to do anything in respect of any matter or transaction whatsoever, actual or proposed, in which the said public body is concerned, shall be guilty of a misdemeanor.

(2) Every person who shall by himself or by or in conjunction with any other person corruptly give, promise, or offer any gift, loan, fee, reward, or advantage whatsoever to any person, whether for the benefit of that person or of another person, as an inducement to or reward for or otherwise on account of any member, officer, or servant of any public body as in this Act defined, doing or forbearing to do anything in respect of any matter or transaction whatsoever, actual or proposed, in which such public body as aforesaid is concerned, shall be guilty of a misdemeanour.

2 **Penalty for offences**

(1) Any person on conviction for offending as aforesaid shall, at the discretion of the court before which he is convicted, –

    (a) be liable

        (i) on summary conviction, to imprisonment for a term not exceeding 6 months or to a fine not exceeding the statutory maximum, or to both; and

        (ii) on conviction on indictment, to imprisonment for a term not exceeding 7 years or to a fine, or to both; and

    (b) in addition be liable to be ordered to pay to such body, and in such manner as the court directs, the amount or value of any gift, loan, fee, or reward received by him or any part thereof; and

    (c) be liable to be adjudged incapable of being elected or appointed to any public office for five years from the date of his conviction, and to forfeit any such office held by him at the time of his conviction; and

    (d) in the event of a second conviction for a like offence he shall, in addition for the foregoing penalties, be liable to be adjudged to be for ever incapable of holding any public office, and to be incapable for five years of being registered as an elector, or voting at an election either of members to serve in Parliament or of members of any public body, and the enactments for preventing the voting and registration of persons declared by reason of corrupt practices to be incapable of voting shall apply to a person adjudged in pursuance of this section to be incapable of voting; and

(e) if such person is an officer or servant in the employ of any public body upon such conviction he shall, at the discretion of the court, be liable to forfeit his right and claim to any compensation or pension to which he would otherwise have been entitled.

3 Savings

(1) ...

(2) A person shall not be exempt from punishment under this Act by reason of the invalidity of the appointment or election of a person to a public office.

4 Restriction on prosecution

(1) A prosecution for an offence under this Act shall not be instituted except by or with the consent of the Attorney-General.

(2) In this section the expression "Attorney General" means the Attorney or Solicitor General for England, and as respects Scotland means the Lord Advocate ...

7 Interpretation

In this Act –

The expression "public body" means any council of a county or county of a city or town, any council of a municipal borough, also any board, commissioners, select vestry, or other body which has power to act under and for the purposes of any Act relating to local government, or the public health, or to poor law or otherwise to administer money raised by rates in pursuance of any public general Act, but does not include any public body as above defined existing elsewhere than in the United Kingdom:

The expression "public office" means any office or employment of a person as a member, officer, or servant of such public body:

The expression "person" includes a body of persons, corporate or unincorporate:

The expression "advantage" includes any office or dignity, and any forbearance to demand any money or money's worth or valuable thing, and includes any aid, vote, consent, or influence, or pretended aid, vote, consent, or influence, and also includes any promise or procurement of or agreement or endeavour to procure, or the holding out of any expectation of any gift, loan, fee, reward, or advantage, as before defined.

PREVENTION OF CORRUPTION ACT 1906

1   Punishment of corrupt transactions with agents

(1)   If any agent corruptly accepts or obtains, or agrees to accept or attempts to obtain, from any person, for himself or for any other person, any gift or consideration as an inducement or reward for doing or forbearing to do, or for having after the passing of this Act done or forborne to do, any act in relation to his principal's affairs or business, or for showing or forbearing to show favour or disfavour to any person in relation to his principal's affairs or business; or

If any person corruptly gives or agrees to give or offers any gift or consideration to any agent as an inducement or reward for doing or forbearing to do, or for having after the passing of this Act done or forborne to do, any act in relation to his principal's affairs or business, or for showing or forbearing to show favour or disfavour to any person in relation to his principal's affairs or business; or

If any person knowingly gives to any agent, or if any agent knowingly uses with intent to deceive his principal, any receipt, account, or other document in respect of which the principal is interested, and which contains any statement which is false or erroneous or defective in any material particular, and which to his knowledge is intended to mislead the principal;

he shall be guilty of a misdemeanour, and shall be liable –

(a)   on summary conviction, to imprisonment for a term not exceeding 6 months or to a fine not exceeding the statutory maximum, or to both; and

(b)   on conviction on indictment, to imprisonment for a term not exceeding 7 years or to a fine, or to both.

(2)   For the purposes of this Act the expression "consideration" includes valuable consideration of any kind; the expression "agent" includes any person employed by or acting for another; and the expression "principal" includes an employer.

(3)   A person serving under the Crown or under any corporation or any ... borough, county, or district council, or any board of guardians, is an agent within the meaning of this Act.

2   Prosecution of offences

(1)   A prosecution for an offence under this Act shall not be instituted without the consent, in England of the Attorney-General or Solicitor-General, and in Ireland of the Attorney-General or Solicitor-General for Ireland.

PREVENTION OF CORRUPTION ACT 1916

2   Presumption of corruption in certain cases

Where in any proceedings against a person for an offence under the Prevention of Corruption Act 1906, or the Public Bodies Corrupt Practices Act 1889, it is proved that any money, gift, or other consideration has been paid or given to or received by a person in the employment of His Majesty or any Government Department or a public body by or from a person, or agent of a person, holding or seeking to obtain a contract from [Her] Majesty or any Government Department or public body, the money, gift, or consideration shall be deemed to have been paid or given and received corruptly as such inducement or reward as is mentioned in such Act unless the contrary is proved.

4   Short title and interpretation

(1)   This Act may be cited as the Prevention of Corruption Act 1916, and the Public Bodies Corrupt Practices Act 1889, the Prevention of Corruption Act 1906, and this Act may be cited together as the Prevention of Corruption Acts 1889 to 1916.

(2)   In this Act and in the Public Bodies Corrupt Practices Act 1889, the expression "public body" includes, in addition to the bodies mentioned in the last-mentioned Act, local and public authorities of all descriptions.

(3)   A person serving under any such public body is an agent within the meaning of the Prevention of Corruption Act 1906, and the expressions "agent" and "consideration" in this Act have the same meaning as in the Prevention of Corruption Act 1906, as amended by this Act.

CRIMINAL JUSTICE AND PUBLIC ORDER ACT 1994

34   Effect of accused's failure to mention facts when questioned or charged

(1)   Where, in any proceedings against a person for an offence, evidence is given that the accused –

   (a)   at any time before he was charged with the offence, on being questioned under caution by a constable trying to discover whether or by whom the offence had been committed, failed to mention any fact relied on in his defence in those proceedings; or

   (b)   on being charged with the offence or officially informed that he might be prosecuted for it, failed to mention any such fact,

being a fact which in the circumstances existing at the time the accused could reasonably have been expected to mention when so questioned, charged or informed, as the case may be, subsection (2) below applies.

(2)   Where this subsection applies –

(a) a magistrates' court inquiring into the offence as examining justices;

(b) a judge, in deciding whether to grant an application made by the accused under –

(i) section 6 of the Criminal Justice Act 1987 (application for dismissal of charge of serious fraud in respect of which notice of transfer has been given under section 4 of that Act); or

(ii) paragraph 5 of Schedule 6 to the Criminal Justice Act 1991 (application for dismissal of charge of violent or sexual offence involving child in respect of which notice of transfer has been given under section 53 of that Act);

(c) the court, in determining whether there is a case to answer; and

(d) the court or jury, in determining whether the accused is guilty of the offence charged,

may draw such inferences from the failure as appear proper.

(3) Subject to any directions by the court, evidence tending to establish the failure may be given before or after evidence tending to establish the fact which the accused is alleged to have failed to mention.

(4) This section applies in relation to questioning by persons (other than constables) charged with the duty of investigating offences or charging offenders as it applies in relation to questioning by constables; and in subsection (1) above "officially informed" means informed by a constable or any such person.

(5) This section does not –

(a) prejudice the admissibility in evidence of the silence or other reaction of the accused in the face of anything said in his presence relating to the conduct in respect of which he is charged, in so far as evidence thereof would be admissible apart from this section; or

(b) preclude the drawing of any inference from any such silence or other reaction of the accused which could properly be drawn apart from this section.

(6) This section does not apply in relation to a failure to mention a fact if the failure occurred before the commencement of this section.

35 Effect of accused's silence at trial

(1) At the trial of any person who has attained the age of fourteen years for an offence, subsections (2) and (3) below apply unless –

(a) the accused's guilt is not in issue; or

(b) it appears to the court that the physical or mental condition of the accused makes it undesirable for him to give evidence;

but subsection (2) below does not apply if, at the conclusion of the evidence for the prosecution, his legal representative informs the court that the accused will give evidence or, where he is unrepresented, the court ascertains from him that he will give evidence.

(2) Where this subsection applies, the court shall, at the conclusion of the evidence for the prosecution, satisfy itself (in the case of proceedings on indictment, in the presence of the jury) that the accused is aware that the stage has been reached at which evidence can be given for the defence and that he can, if he wishes, give evidence and that, if he chooses not to give evidence, or having been sworn, without good cause refuses to answer any question, it will be permissible for the court or jury to draw such inferences as appear proper from his failure to give evidence or his refusal, without good cause, to answer any question.

(3) Where this subsection applies, the court or jury, in determining whether the accused is guilty of the offence charged, may draw such inferences as appear proper from the failure of the accused to give evidence or his refusal, without good cause, to answer any question.

(4) This section does not render the accused compellable to give evidence on his own behalf, and he shall accordingly not be guilty of contempt of court by reason of a failure to do so.

(5) For the purposes of this section a person who, having been sworn, refuses to answer any question shall be taken to do so without good cause unless –

(a) he is entitled to refuse to answer the question by virtue of any enactment, whenever passed or made, or on the ground of privilege; or

(b) the court in the exercise of its general discretion excuses him from answering it.

(6) Where the age of any person is material for the purpose of subsection (1) above, his age shall for those purposes be taken to be that which appears to the court to be his age.

(7) This section applies –

(a) in relation to proceedings on indictment for an offence, only if the person charged with the offence is arraigned on or after the commencement of this section;

(b) in relation to proceedings in a magistrates' court, only if the time when the court begins to receive evidence in the proceedings falls after the commencement of this section.

38 Interpretation and savings for sections 34, 35, 36 and 37

(1) In sections 34, 35, 36 and 37 of this Act –

"legal representative" means an authorised advocate or authorised litigator, as defined by section 119(1) or the Courts and Legal Services Act 1990; and

"place" includes any building or part of a building, any vehicle, vessel, aircraft or hovercraft and any other place whatsoever.

(2) In sections 34(2), 35(3), 36(2) and 37(2), references to an offence charged include references to any other offence of which the accused could lawfully be convicted on that charge.

(3) A person shall not have the proceedings against him transferred to the Crown Court for trial, have a case to answer or be convicted of an offence solely on an inference drawn from such a failure or refusal as is mentioned in section 34(2), 35(3), 36(2) or 37(2).

(4) A judge shall not refuse to grant such an application as is mentioned in section 34(2)(b), 36(2)(b) and 37(2)(b) solely on an inference drawn from such a failure as is mentioned in section 34(2), 36(2) or 37(2).

(5) Nothing in sections 34, 35, 36 or 37 prejudices the operation of a provision of any enactment which provides (in whatever words) that any answer or evidence given by a person in specified circumstances shall not be admissible in evidence against him or some other person in any proceedings or class of proceedings (however described, and whether civil or criminal).

In this subsection, the reference to giving evidence is a reference to giving evidence in any manner, whether by furnishing information, making discovery, producing documents or otherwise.

(6) Nothing in sections 34, 35, 36 or 37 prejudices any power of a court, in any proceedings, to exclude evidence (whether by preventing questions being put or otherwise) at its discretion.

APPENDIX C
LIST OF PERSONS AND ORGANISATIONS WHO COMMENTED ON THE CONSULTATION PAPER

Individuals

Mr Anthony Arlidge QC

Mrs Wendy Armitstead

Mr R V Ashman

Mr Cyril Brenner

Mr Justice Buxton (now Lord Justice Buxton)

Mrs Pat Bygramed

Judge David Clarke QC, Recorder of Liverpool

Judge Bruce Coles QC

Mr John Colthorp

Dr Stephen Cretney

Lord Davidson

Judge Rhys Davies QC, Recorder of Manchester

Judge Neil Denison QC, Common Serjeant of London

Judge Mark Dyer, Recorder of Bristol

Mr A C Farran

Mr J Flynn

Mr A S Foster

Professor Mark Freedland

Mr Justice Garland

Mrs Elizabeth Gaskell Syms

Mr John H Jeffrey

Mr Roger Jones

Mr Brian Leveson QC

Mr Justice Longmore

Mr E A Marsh

Mr Justice McCullough

Sheriff Iain McPhail QC

Mrs D Mitchell

Mr Frank Mitchell

Mr Justice Morison

Mr Colin Peters

Lord Justice Phillips

G R Poulton

Mr P Prankerd

Mrs Joan Mary Prescott

Mr D Rippen

Mr Norman Scarth

Mr Geoffrey Scriven

Mr Justice Sedley

Mr David Sheldon

Professor Sir John Smith CBE QC LLD FBA

Mr R S Smith

R E Spalton

Professor G R Sullivan

Mrs Janet Talbot

Mr James Tonelli

Mr Neville D Vandyk

Judge Walsh QC, Recorder of Leeds

Mrs M A Watts

Mr John Williams

An anonymous respondent

Organisations

Association of Chief Police Officers

Chartered Institute of Public Finance and Accountancy

Confederation of British Industry

Criminal Bar Association

Crown Office

Crown Prosecution Service

Department of Social Security, Solicitor's Office

Department of Trade and Industry, Business Law Unit

Employment Law Bar Association

Employment Appeal Tribunal

Financial Law Panel

General Council of the Bar

Greater Manchester Police

Hampshire Constabulary

HM Customs and Excise, Criminal Policy Unit

Inland Revenue

Institute of Chartered Accountants of England and Wales

Institute of Directors

Institute of Legal Executives

Joint Council of the Metropolitan and Provincial Stipendiary Magistrates

Justices' Clerks' Society

Kingfisher plc

Labour Group of Councillors, Borough of Reigate and Banstead

Law Society, Criminal Law Committee

Law Society, Employment Law Committee

Liberty

Lloyd's

London Criminal Courts Solicitors' Association

Magistrates' Association

Merseyside Police

Metropolitan Police, Company Fraud Department

Ministry of Defence Police

Newspaper Society

Office of the Judge Advocate General

Police Federation of England and Wales

Royal Town Planning Institute

Royal Ulster Constabulary

Securities and Investments Board

Serious Fraud Office

Transparency International (UK)

Trinidad and Tobago Law Commission